2nd International Conference on Sustainable Infrastructural Development (ICSID 2020)

IOP Conference Series: Materials Science and Engineering
Volume 1036

Ota, Nigeria
27 - 28 July 2020

Editors:

David O. Olukanni
Anthony N. Ede
Isaac I. Akinwumi
Paul O. Awoyera
Chukwueloka A. Okeke
Oluwarotimi M. Olofinnade

ISBN: 978-1-7138-4561-4
ISSN: 1757-8981

Printed from e-media with permission by:

Curran Associates, Inc.
57 Morehouse Lane
Red Hook, NY 12571

Some format issues inherent in the e-media version may also appear in this print version.

Institute of Physics
Dirac House, Temple Back
Bristol BS1 6BE UK

Phone: 44 1 17 929 7481
Fax: 44 1 17 920 0979

techtracking@iop.org

Additional copies of this publication are available from:

Curran Associates, Inc.
57 Morehouse Lane
Red Hook, NY 12571 USA
Phone: 845-758-0400
Fax: 845-758-2633
Email: curran@proceedings.com
Web: www.proceedings.com

TABLE OF CONTENTS

IOP Conf. Series: Materials Science and Engineering 1036 (2021) 011001 doi:10.1088/1757-899X/1036/1/011001

Preface for ICSID 2020 Proceedings

The 2nd edition of the International Conference on Sustainable Infrastructural Development (ICSID 2020) which brought together notable scholars, experts, governmental agencies, policy makers, industrialists, professional bodies and students for the exchange and sharing of experience and ideas on issues pertinent to sustainable infrastructural development, held on July 27th and 28th, 2020. The Conference was organised by the Department of Civil Engineering, Covenant University, Ota, Nigeria.

Due to the Covid-19 Pandemic and the consequent world-wide safety measures such as social distancing and ban on traveling, this year's conference held virtually and provided opportunities for participants to interact and present papers on:

Sustainable Water Resources and Environment,
Sustainable Housing and Construction Management,
Sustainable Construction Materials,
Sustainable Infrastructure, Project Management and Computational Modelling,
Geotechnics and Transportation Infrastructure,
Social Development Issues.

The papers presented in the ICSID 2020 are contained within in this proceeding.

Two plenary sessions addressed different aspects of Sustainable Infrastructural Development on each of the days of the conference. On day 1, international speakers made their presentations. Prof Innocent Musonda from the University of Johannesburg, South Africa spoke on "Building Sustainable Infrastructure in Post Covid-19 Era". while Dr. Nadine Ibrahim of the University of Waterloo, Canada, spoke on "Sustainability of Urban Infrastructure in Cities, Megacities, and Megaregions". On day 2, Dr. Thomas B. Eyo from the Federal Roads Maintenance Agency, FERMA, Abuja, Nigeria, presented on "Research and Development: Adoption of Innovations and Technologies for Sustainable Road Infrastructure".

Furthermore, of the over 160 abstracts and full papers received for the conference, 152 were accepted for the conference and packaged in the conference book of abstracts. Six breakout sessions were created on the basis of the conference themes and 76 presentations were scheduled for each day of the conference, while 82 contributed papers were packaged in this proceeding.

It is our earnest vision that the presentations documented in this proceeding will serve as a catalyst to collective engagements by researcher, policymakers, industrialists and other stakeholders towards achieving the Sustainable Infrastructural Development that will enrich the quality of life in developing nations.

Dr Anthony Nkem Ede
Chair, ICSID 2020 Conference Organizing Committee

IOP Conf. Series: Materials Science and Engineering 1036 (2021) 011001 doi:10.1088/1757-899X/1036/1/011001

EDITORS

Prof. David O. Olukanni

Dr. Anthony N. Ede

Dr. Isaac I. Akinwumi

Dr. Paul O. Awoyera

Dr. Chukwueloka A. Okeke

Dr. Oluwarotimi M. Olofinnade

IOP Conf. Series: Materials Science and Engineering 1036 (2021) 011001 doi:10.1088/1757-899X/1036/1/011001

ICSID 2020 CONFERENCE ORGANIZING COMMITTEE

CONFERENCE CHAIR
Prof. David O. Olukanni, Covenant University, Ota, Nigeria

CHAIR, ORGANIZING COMMITTEE
Dr. Anthony N. Ede, Covenant University, Ota, Nigeria

CONFERENCE SECRETARY
Dr. (Mrs) Olatokunbo M. Ofuyatan, Covenant University, Ota, Nigeria

TECHNICAL CHAIR
Dr. Isaac I. Akinwumi, Covenant University, Ota, Nigeria

TECHNICAL CO-CHAIR
Dr. Paul O. Awoyera, Covenant University, Ota, Nigeria

TECHNICAL SECRETARIES
Dr. Oluwarotimi M. Olofinnade, Covenant University, Ota, Nigeria
Dr. Chukwueloka A. Okeke, Covenant University, Ota, Nigeria

BUDGET AND FUNDING COMMITTEE CHAIR
Dr. Gideon O. Bamigboye, Covenant University, Ota, Nigeria

EVENT AND LOGISTICS CHAIR
Dr. PraiseGod C. Emenike, Covenant University, Ota, Nigeria

PUBLICITY SECRETARY
Mr. Irewole Oniemayin, Covenant University, Ota, Nigeria

FINANCIAL SECRETARY
Engr. (Mrs.) Dunmininu D. Adegoke, Covenant University, Ota, Nigeria

WELFARE
Mrs. Kofoworola Daramola, Covenant University, Ota, Nigeria

IOP Conf. Series: Materials Science and Engineering 1036 (2021) 011001 doi:10.1088/1757-899X/1036/1/011001

Remarks of the Vice-Chancellor, Covenant University, Professor AAA. Atayero at the 2nd International Conference on Sustainable Infrastructural Development (ICSID) on July 27, 2020

Protocols:

It is my pleasure to welcome you to Hebron – The birthplace of Kings and Queens, venue of the 2nd international conference on sustainable infrastructural development with the theme, "Sustainable Infrastructural Development".

Development experts and stakeholders well recognise the vital role of infrastructure in enhancing socio-economic development of communities and nations. Infrastructure systems provide societies with essential services such as energy, transport, telecommunication, and waste management, which enhance people's wellbeing. According to the United Nations Development Programme (UNDP), rising populations and increasing migration have led to the rapid growth of cities and boom in mega-cities, especially in the developing world. Ensuing slums are becoming a more significant feature of urban life. It is estimated that by 2050, two-thirds of all humanity 6.5 billion people will be in urban cities. This development calls for intentional actions to build resilient societies and economies to reverse the low quality of life arising from this growth. King cities sustainable

The transformation of the economies of African countries to achieve the creation of resilient communities requires extensive access to contemporary and sustainable infrastructure. Despite the recognition of the accruing benefits of sustainable infrastructure, Africa as a continent still lags in infrastructural development in terms of quality and quantity. Africans as a rule lack access to affordable energy and housing, efficient transport and reliable communication systems. The infrastructure gap in Africa is estimated to be over USD 100 billion annually. Our quest to create smart cities and connected communities require investment in sustainable infrastructure to improve competitiveness, enable domestic and international trade, while at the same time boosting the Continent's integration into the global economy.

IOP Conf. Series: Materials Science and Engineering 1036 (2021) 011001 doi:10.1088/1757-899X/1036/1/011001

The potential for building sustainable infrastructure exists in Africa. She is abundantly blessed with clean energy resources such as solar, hydropower, biomass, geothermal, compressed natural gas and wind, among others. And most importantly, Africa is blessed with the required intellectual capacity and manpower as visibly demonstrated in the participants of this conference.

It is for this reason, we expect that at the end of this conference, viable strategies on how to harness the green resources of the Continent for sustainable infrastructure development, would be put forward, to enable efficient use of these resources for sound economic development, urban planning and management as well as create business and job opportunities, safe and affordable housing and public transportation systems.

Covenant is actively involved in research efforts aimed at evolving innovative solutions in facilitating the achievement of African Union agenda 2063 and the Sustainable Development Goals. The resent ranking of the University in the 401-600 bracket of the 2020 Impact Ranking of Times Higher Education (THE), attests to our contributions in this direction.

By this conference, we seek to encourage multidisciplinary partnerships in research efforts aimed at promoting sustainable infrastructural development for our beloved Continent. Our doors are open to opportunities for research collaboration with universities, research institutes, and other stakeholders.

I wish you a most impactful conference that will become a point of reference for sustainable infrastructural development in Africa.

Thank you.

Peer review declaration

- Type of peer review: All manuscripts received were sent to at least two (2) persons for peer-review. Some manuscripts received two peer-review reports while others got a single report. Overall, a minimum of single-blind peer review system was used.

- Conference submission management system: Conference Email

- Number of submissions received: 163

- Number of submissions sent for review: 163

- Number of submissions accepted: 81

- Acceptance Rate (Number of Submissions Accepted / Number of Submissions Received X 100): 49.7%

- Average number of reviews per paper: >1

- Total number of reviewers involved: 72

- Any additional info on review process: None

- Contact person for queries: Dr Isaac Akinwumi

Water, Sanitation and Hygiene Practices in Ogun State: Impacts and Implications for Post-COVID-19 Era

D O Olukanni[1]*, D O Iyiola[2] and C O Esu[3]
[123]Department of Civil Engineering, Covenant University, PMB 1023, Canaan land, Ota, Nigeria
*Corresponding Author: david.olukanni@covenantuniversity.edu.ng

Abstract. Globally, access to safe Water, Sanitation, and Hygiene (WaSH) services play an essential role in safeguarding the health and well-being of individuals and communities. This study focuses on the assessment of WaSH Practices in Ogun State, Nigeria. A survey was carried out with a total of 500 questionnaires administered evenly in 5 selected Local Governments Areas (LGAs) based on population, landmass, spatial location, and distribution. The convenience sampling technique was used to select locations within the study areas. The evaluation of WaSH practices was based on the availability of treated water, methods of waste disposal, sanitation and hygiene-related facilities, and awareness program in the communities, respectively. The result shows that 61.2 % of the respondents have access to treated water. Thirty-six percent of the respondents agreed to the use of open dumping system for their waste disposal, 24.6% use collection trucks while the rest bury or burn their wastes. Forty-five percent of the respondents are unaware and unexposed to hygienic conditions. The WaSH standard attained by these communities implies an average healthy living standard for the people, which complies with the data obtained from health centers showing the rate of compliance associated with WaSH related ailments. The results of this study provide useful information on the consequences of poor hygiene practices and sanitation facilities for both residents and authorities. The data can serve as a blueprint to government or private organizations working towards upgrading standard of WaSH practices, which would assist in providing appropriate facilities towards enforcing safe hygiene practices at all levels against the spread of the current global pandemic-COVID-19.

Keywords: Water Supply, Sanitation Practices, Hygiene study, Environmental Sustainability, Community Health, Post COVID-19

1 Introduction

The development and occurrence of water, sanitation, and hygiene problems in many countries in West Africa, and Nigeria in particular, has become more pronounced in recent years. This challenge focuses heavily on low income and affects disadvantaged groups, such as the poor and disabled in developing countries, who make a significant contribution to the poverty cycle [1, 2]. Global access to water, sanitation, and hygiene (WaSH) services plays a vital role in protecting the health and well-being of individuals and society [1, 2]. Water is one of the precious natural resources and an integral part of our lives. It is the most immediate and critical limiting factor for the well-being of humans and the environment [3]. Water availability and quality have always played an important role, not only in determining where people can live but also in their quality of life. Sanitation refers to the hygienic means of protecting human contact from the dangers of waste to promote good health [4, 5]. Sanitation includes the provision of facilities and services for the safe disposal of waste. Hygiene refers to good practices that prevent disease and lead to good health, especially in terms of cleanliness, proper disposal of wastewater, and supply of clean drinking water.

Access to water and sanitation is not just a basic need; it is a human right. Safe, affordable, and sufficient drinking water for all by 2030 with adequate sanitation requires massive investment in infrastructure, provision of sanitation facilities, and encouragement of hygiene practice, which has implications on the Sustainable Development Goals [6]. WaSH practices include a supply of clean water for drinking, washing and cleaning, safe disposal of waste (toilets and garbage disposal), and health promotion activities promoting healthy practices among affected populations [6]. According to [7], potable water supply and sanitation is fundamental for living a healthy life. Sufficient and quality water supply is not only a vital environmental health necessity to live but plays significant roles in the socio-economic and political advancement of the human population. The WHO also notes that water and sanitation for communities and households do not only contribute to physical growth and human development but also people's socio-economic development. [6] reports that over 70 percent of the world population (about 5.2 billion people) who had safely managed drinking water in 2015, still lacked essential drinking water while 39 percent of the world population that had safe sanitation still lacked basic sanitation with worrisome statistics of about 892 million people still practicing open defecation.

Improved access to clean water and improved hygiene practices results in good health, increased socio-cultural development, and promoting economic balance. Access to water and sanitation in developing countries is lacking with millions of people in rural and urban areas most vunerable, causing WaSH-related infectious diseases [7]. The development and occurrence of water, sanitation, and hygiene problems in many countries in West Africa, and Nigeria in particular, has become more pronounced in recent years. This challenge focuses heavily on low income affects disadvantaged groups, such as the poor and disabled in developing countries, who make a significant contribution to the poverty cycle [2]. Over the past decade, Nigeria has made mixed progress in providing safe water and sanitation in rural areas. At the household and school levels, there are still significant concerns about the quality and use of the WaSH facilities [8]. According to report by [9] on water, sanitation, and hygiene, almost one third (29%) of hospitals and clinics in Nigeria do not have toilets. The report shows that one in six (16%) has no room to wash hands with soap.
Shortage of water has made women and school-age children, particularly girls, spend hours in search of water. They walk long distances during the dry season to obtain water from rivers, streams, and wells. These sources are often contaminated with human and animal wastes, giving rise to the occurrence of water and sanitation-related diseases, with some resulting in death [10]. Report from WHO additionally pointed to the fact that potable water supply and sanitation is a pre-requisite for the realization of other human rights; for example, protecting the well-being of women and children in terms of reduction in water and sanitation-related aliments which represent most sickness and demise among individuals in developing countries including Nigeria. Water serves other essential

functions in agriculture, such as farming, fishing, animal husbandry, and food processing [11]. The people who are sixty years and above but have access to potable water supply and sanitation facilities live longer [11]. According to [12, 13, 14], water is utilized for irrigation, dams, hydroelectric power supply, household, industrial and commercial purposes, and satisfies thirst. [14] listed the benefits of potable water in terms of psychological satisfaction and happiness. By 2025, it is estimated that 3.5 billion people (about 6.5 times the figure of 2000) will live in water-stressed countries. By the middle of the century, at worst, 7 billion people in 60 countries and at best 2 billion in 48 countries will be faced with the scourge of water shortage [6, 15].

While the World Health Organization (WHO) recommends frequent hand-washing with soap and water for 20 seconds, amongst other preventive measures, to protect against COVID-19 [16], the pace at which various governments of developing countries are working hard on public awareness is impressive. Most studies have shared different perspectives to WaSH issues. Still, the underlining factor that borders on its impact and implication for the residents of Ogun State in southwest Nigeria has not been dealt with. Therefore, this study aims to assess the effects and impact of Water, Sanitation, and hygiene practices in Ogun State for the post-COVID-19 era. It also assesses the hygiene knowledge and practices, waste disposal operations, and maintenance systems in the state through the following objectives:

i. Assessing the status of Water, Sanitation, and Hygiene practices in selected areas of Ogun State.
ii. Identifying gaps and challenges in the provision of Water, Sanitation and Hygiene practices and offer recommendations, and
iii. Evaluating the impacts and implications of the assessed WaSH practices in Ogun State.

2 Methodology

2.1 *Study Area*

Ogun State is located on longitudes 2° 45'E; and latitudes 6° 15'N and 7° 60'N in South-West of Nigeria, with a land area of 16,980.55 km^2 and a population of 3,751,140, with a density of 220/km^2, according to the last official census in 2006. The state shares borders with the Republic of Benin to the West, Lagos state to the south, and a 20km stretch of the Atlantic Ocean to the South and Ondo and Osun states to East. Data was obtained primarily via questionnaires administered in places such as households, schools, marketplaces, business centers, workshops, and public places within the study area where the survey could be easily assessed as it is easier to meet people and interact. The convenience (availability) sampling technique was used to select locations within the study areas. This type of sampling method relies on data collection from population members that are available to participate in the survey. A total number of 500 questionnaires were administered evenly among the five selected local government areas, and all were ensured to be completed in the course of the survey. Data collected through the questionnaire survey were on the following variables: Socio-Demographics, Household Statistics, Water-Related Practices, Sanitation Related Practices, Hygiene Related Practices, and Impacts and Implications of Wash Practices. Facilities were also checked for their presence, functionality, and maintenance. The facilities accessed were: water source, drinking water facilities, urinals, hand-washing facilities, dust bins, septic tanks. The data from the study were analyzed using descriptive statistics. Points 1, 2, 3, 4, 5 on the map are used to represent the selected 5 LGAs; Ado-Odo/Ota, Obafemi-Owode, Odeda, Ijebu East, and Yewa South, respectively (Figure

IOP Conf. Series: Materials Science and Engineering 1036 (2021) 012001 doi:10.1088/1757-899X/1036/1/012001

1). The population under study comprise of males and females from age 15 and above involved in different socio-economic activities in these locations. Assistance was given to those unable to read and write. Oral discussions in the form of interviews were engaged with residents, market women, and adults in general concerning the subject.

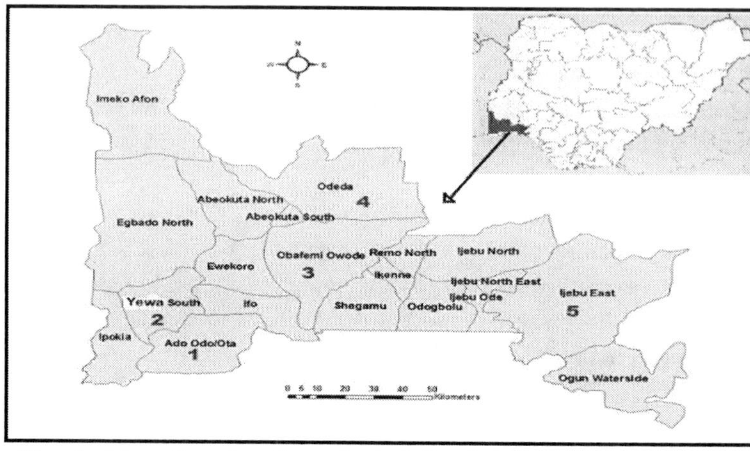

Fig. 1: Map showing the selected five local government areas

At the time of developing this paper, deductions were made based on Water, Sanitation, and Hygiene practices, respectively. That of water-related practices was based on the availability of treated water; sanitation was based on methods of waste disposal, and hygiene-related practices were evaluated based on the existence of sanitation and hygiene-related facilities and awareness program in the community [5, 17].

2a

2b

2c

Fig. 2 a, b and c: Commuters fetching water at a public borehole at Obafemi-Owode-; Open -dump site at Obafemi-Owode-, and Private dug-well at a household at Yewa-South-Local government area, respectively.

2.2 Questionnaire Structure, Data Collection, and Analysis

The questionnaire contained both open and closed-ended questions that are easy to understand, allowing the respondents to express their views without complications. The survey was classified into five significant sections, which were:

Section A- Inquired about the general information such as age, sex, marital status, education level, occupation, the estimate of the annual income level of the respondents

Section B- Inquired about the household statistics like the type of residence, total number of people in the home.

Section C- Examined the source of water supply in the household. It also established the cost and treatment of the water source.

Section D- Examined the sanitation practices and if they are being adhered to.

Section E- Examined the various essential hygiene practice and impacts of WaSH

A significant number of respondents interviewed had little or no fluency in the English Language. Communication for such was, therefore, by interpretation in native Yoruba dialect or Pidgin English, with their responses ticked for them. Some individuals were unwilling to respond as they felt the researchers were government officials, having ulterior motives like tax collector officials, housing and urban planning officials, etc. In the group discussions held among traders, a good number of them admitted to not observing some essential hygiene practices such as washing their hands. Most of the respondents admitted to observing the monthly Environmental Sanitation exercise. The data obtained were analyzed using Descriptive Statistics. The International Business Machine IBM Statistical Package for the Social Sciences (IBM SPSS 20) was used to do the analysis and to generate relevant illustrations, tables, and charts to facilitate a better understanding of the study.

3 Result and Discussion

This section presents the results of the Water, Sanitation, and Hygiene practices survey that was conducted in the selected LGAs of Ogun State. Tabulation of data was adopted for easy understanding, and comparison between the five (5) chosen LGAs. The results were presented through the use of Tables and Charts.

3.1 Presentation of Data from the Age Distribution of Respondents

Table 1: The age distribution of respondents

Age of Respondents	Frequency	Percentage (%)
0-17 years	100	20.0
18-24 years	110	22.0
25-34 years	121	24.2
35-44 years	124	24.8
45-54 years	29	5.8
55 years and above	16	3.2
Total	**500**	**100.0**

Table 1 presents the data on respondents' age. 20% of the respondents are between 0-17 years, 22% between 18-24 years, 24.2% between 25-34 years, and 24.8% between respondents aged 35-44 years, 5.8% represented 45-54 years and over 55-year-olds represents 2.8%, respectively. Table 1 shows that the respondent's age composition is majorly mature persons who are willing to provide information based on their individual experiences. Figure 3 shows the breakdown of respondents' statistics within the respective LGAs.

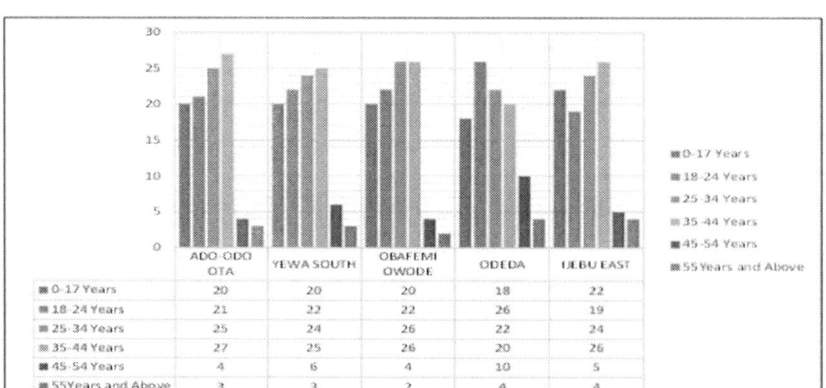

Fig. 3 Age distribution of respondents

IOP Conf. Series: Materials Science and Engineering 1036 (2021) 012001 doi:10.1088/1757-899X/1036/1/012001

Figure 4 shows the breakdown of respondents' statistics within the respective LGAs with 9.2% constituting those with no formal educational background, 19.6% represents those with primary school leaving certificate, 38% have their A Level/O' Level Certificate, and 28.2% of the respondent have their First degree. In comparison, others with probably informal education were 5%.

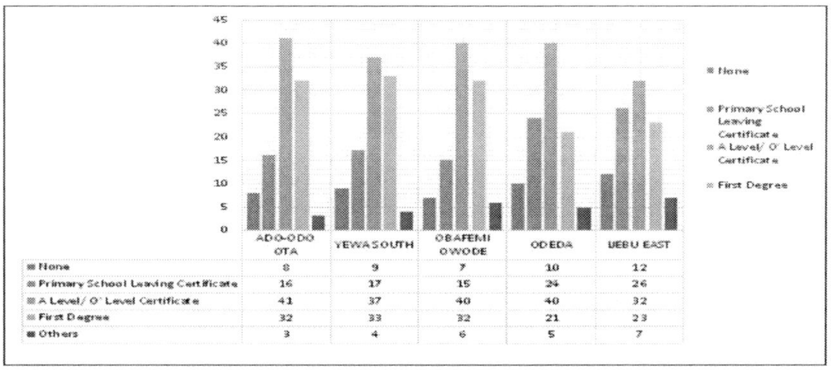

	ADO-ODO OTA	YEWASOUTH	OBAFEMI OWODE	ODEDA	IJEBU EAST
None	8	9	7	10	12
Primary School Leaving Certificate	16	17	15	24	26
A Level/ O' Level Certificate	41	37	40	40	32
First Degree	32	33	32	21	23
Others	3	4	6	5	7

Fig. 4 Response to the Educational Level of Respondents

The number of people living in a household determines the usage of the water supply facilities, which affects hygienic behaviors and also determines the quantity of waste generated from such households [2]. Figure 5 shows the representative household statistics for each of the LGAs. Those respondents who stay alone represents 7.2%, 58.0% represents 2-5 persons living together, 25.4% represents 6-9 persons living together, and 9.4% represents ten persons and above living together.

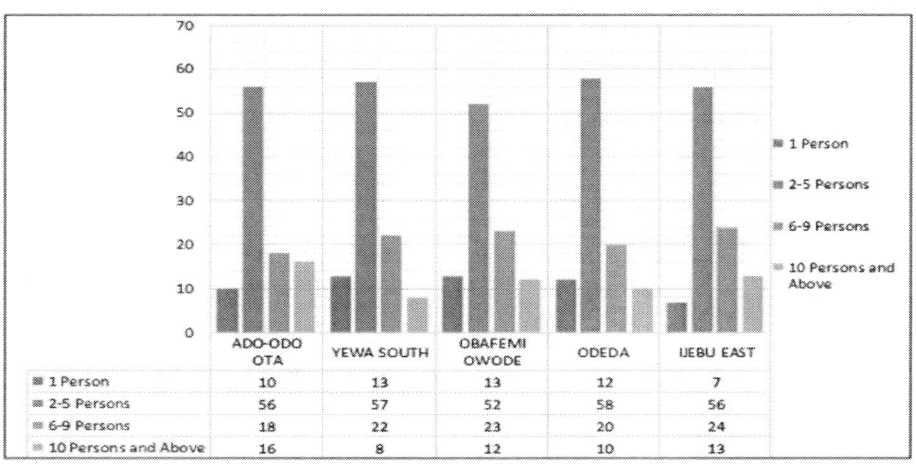

	ADO-ODO OTA	YEWA SOUTH	OBAFEMI OWODE	ODEDA	IJEBU EAST
1 Person	10	13	13	12	7
2-5 Persons	56	57	52	58	56
6-9 Persons	18	22	23	20	24
10 Persons and Above	16	8	12	10	13

Fig. 5 House Statistics of the respondents in each LGAs

Figure 6 shows the graphical representation of the primary sources of water in the state. It shows that private and commercial boreholes are the most used source of water. There are a few boreholes that supply free water. The people complained of a lack of power supply to pump water into the overhead tanks, therefore, rendering the source sometimes useless when there is

IOP Conf. Series: Materials Science and Engineering 1036 (2021) 012001 doi:10.1088/1757-899X/1036/1/012001

no power supply. The primary source of water for households determines the quality of water used by the household [2]. This, in turn, affects the amount of water intake available to members of the household. As shown in Figure 6, the number of responses for private borehole is the highest and also with the highest percentage showing that the primary source of water for most residents is the borehole water system amongst the other responses chosen by the respondents.

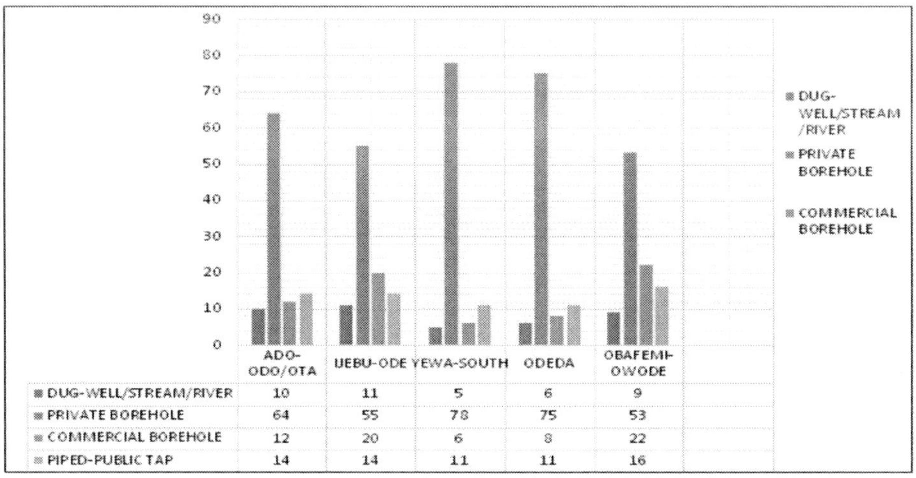

Fig. 6 Response to the primary sources of water available in the neighborhood.

Figure 7 shows that 9.8% of the population sees the cost of getting water as expensive, while 20.2% of the respondents find it moderately steep. A large percent (46.4%) of the respondents get their water supply at no cost. This is because private borehole is the primary source of water supply to the respondents, which they arrive at no charge. However, water is meant to be cheap for all. The government still needs to provide free water for all.

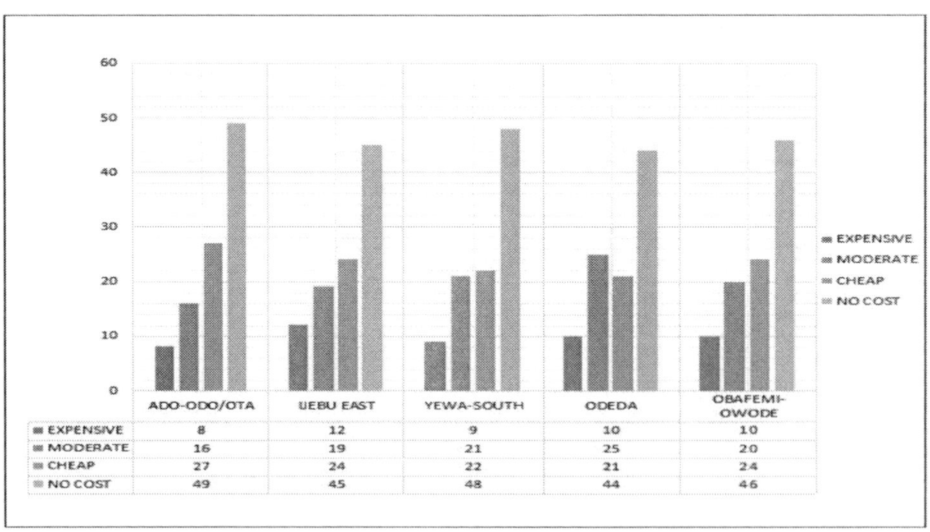

Fig. 7. Response on the cost of getting access to the water source

Figure 8 presents the responses and comparison of the proximity to the water source. Relatively the distance from the source of water is very close. Since the majority of the respondents get their water supply from private boreholes, which most times is available within the respondent's residence, this reduces the distance of getting water supply to the barest minimum.

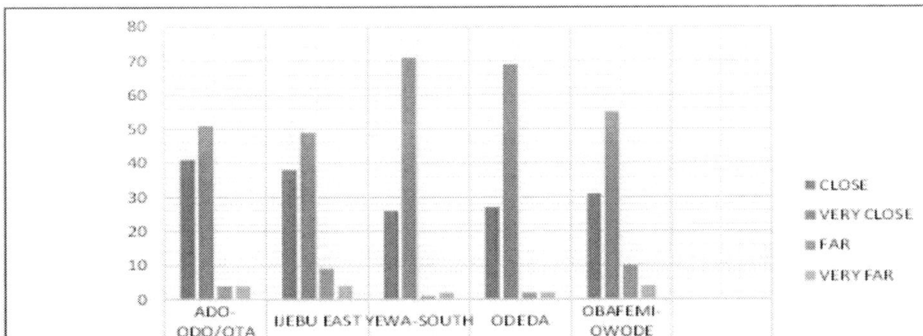

	ADO-ODO/OTA	IJEBU EAST	YEWA-SOUTH	ODEDA	OBAFEMI-OWODE
CLOSE	41	38	26	27	31
VERY CLOSE	51	49	71	69	55
FAR	4	9	1	2	10
VERY FAR	4	4	2	2	4

Fig. 8 Response on proximity to the water source

Figure 9 shows responses to water treatment by the representative respondents. 61.2 percent of the population sometimes treat their water before use. The significant percentage of the people that drink the borehole water sometimes do not address the water because they assume that the borehole water is already clean for usage. While those that always treat the water do so through boiling. Other persons who were unsure stated that they buy sachet water rather than drink the borehole water because they are not sure of the pure nature of the borehole water source.

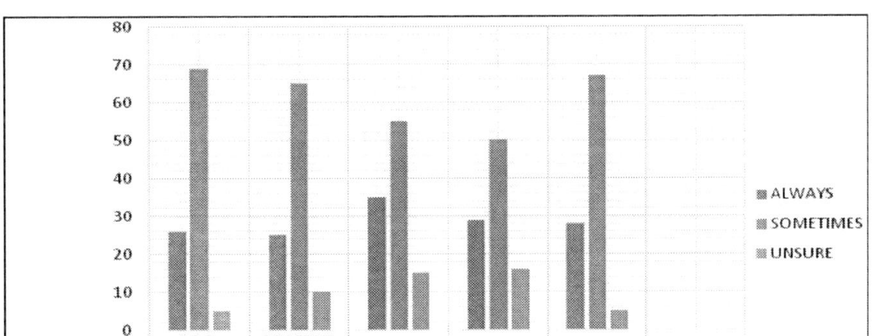

	ADO-ODO/OTA	OBAFEMI-OWODE	YEWA-SOUTH	ODEDA	IJEBU EAST
ALWAYS	26	25	35	29	28
SOMETIMES	69	65	55	50	67
UNSURE	5	10	15	16	5

Fig. 9 Response on water treatment

The results in Figure 10 shows that the vast majority of the respondents do not store their water. The contributing factor is the fact there is a constant supply of water from private and commercial boreholes. The continuous water supply is because they make use of their generators to pump water when there is power supply.

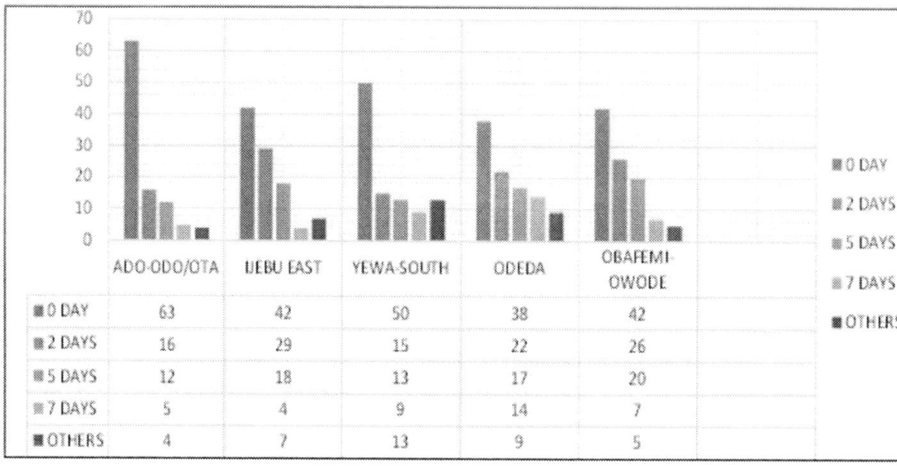

	ADO-ODO/OTA	IJEBU EAST	YEWA-SOUTH	ODEDA	OBAFEMI-OWODE
0 DAY	63	42	50	38	42
2 DAYS	16	29	15	22	26
5 DAYS	12	18	13	17	20
7 DAYS	5	4	9	14	7
OTHERS	4	7	13	9	5

Fig. 10 Response on the duration of water storage

Figure 11 shows a comparison of the method of waste disposal among the five selected LGAs. 36.6% agreed to the use of the open dumping system for heir waste disposal. This system of waste disposal implies that it does not only destroy the aesthetics of the environment, but it also attracts vectors of various diseases, which can pose a severe health risk to the residents of such municipality [18]. 29.8% of the respondents burn their wastes, and this causes air pollution in the environment. 24.6% of the respondent says their waste is being collected by waste collection trucks that come at a scheduled period, which could be daily, weekly, or monthly.

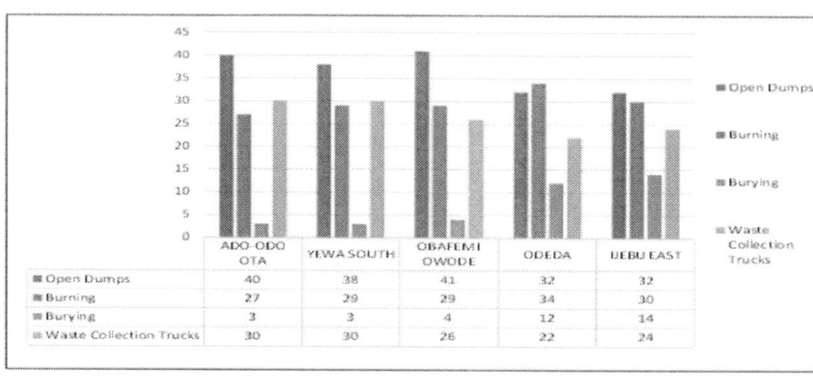

	ADO-ODO OTA	YEWA SOUTH	OBAFEMI OWODE	ODEDA	IJEBU EAST
Open Dumps	40	38	41	32	32
Burning	27	29	29	34	30
Burying	3	3	4	12	14
Waste Collection Trucks	30	30	26	22	24

Fig. 11. Response on Method of waste disposal

Figure 12 displays the results of the frequency of waste disposal with a large number of the respondents (40.4%) disposing of their waste always, and 35.4% disposing of their waste frequently.

Waste disposal should be a daily routine to prevent accumulation, which is harmful to the environment at large.

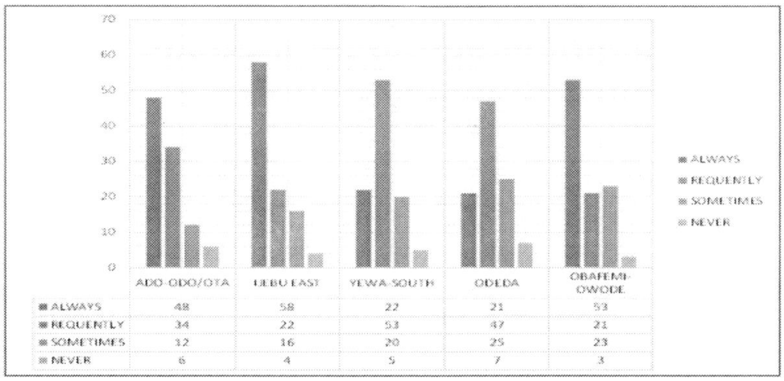

Fig. 12. Response on the frequency of waste disposal.

Figure 13 shows the comparison of how people in the five LGAs observe the sanitation day. While researchers are studying people's motivation and behavior for improving sanitation, its impact on health brings about both social and economic benefits are well understood [2, 19]. It shows that 85.6% of the population observes the Environmental Sanitation program as instructed by the state government. This is probably because there is a task force that enforces the implementation of the law for people to observe the sanitation day. The 14.4 percent of the respondents that do not follow this sanitation day are those that indicated they have jobs that require their attention, such as doctors, bankers, etc.

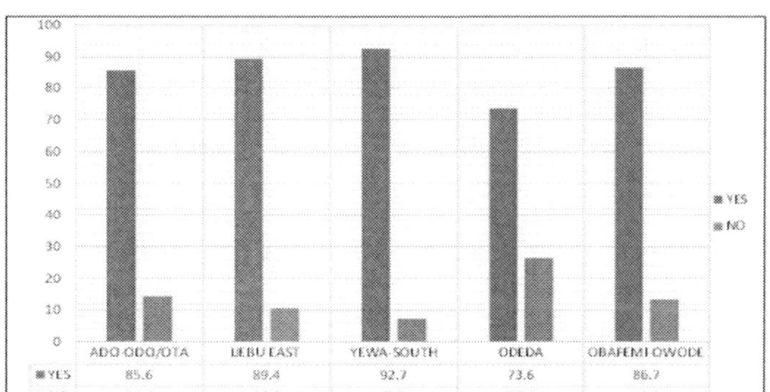

Fig. 13 Response on observation of environmental sanitation day

Living in a clean environment is not only important for physical health but also overall well-being. A higher percentage of the respondents interviewed, clean their environment and surroundings daily. Generally, the respondents clean their environment, which results in safer and healthier surroundings free from related diseases (Figure 14). Stagnant water causes many health hazards and is a breeding ground for many types of bacterial fungi (molds) and parasites and mosquitoes that bring these diseases to humans (WHO).

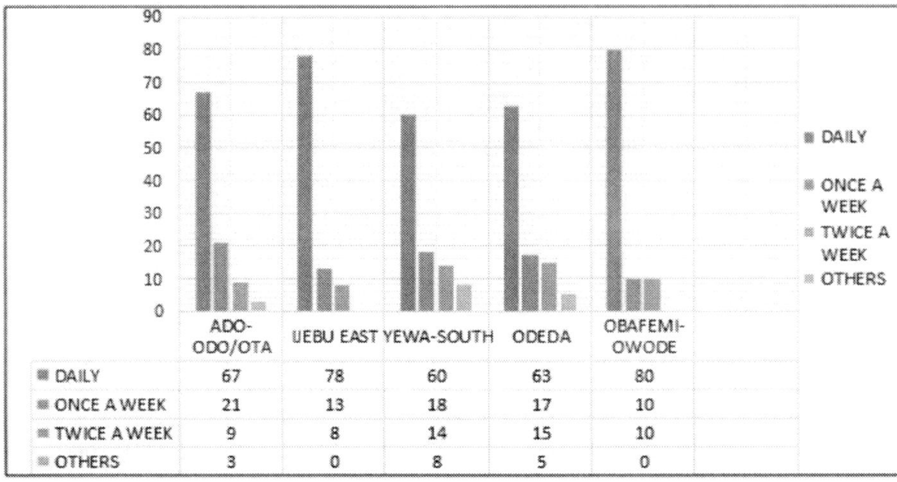

Fig. 14 Response on the frequency of cleaning environment, toilet, and surroundings

Data representation of the respondents, as shown in Figure 15, shows that the majority of the population make use of the Modern Flush toilet system, which is an excellent WaSH facility. The facility is available because the majority of the population lives in apartments that have been constructed with modernized toilet facilities. While it is still uncertain how the role of public and shared toilets would play in the spread of COVID-19, but the study by [19] highlights the need to assess the threats of communal areas, especially as areas begin to lift stay-at-home orders. Figure 15 shows Ado-Odo/Ota having the highest number of respondents using the sit down with a water flush toilet.

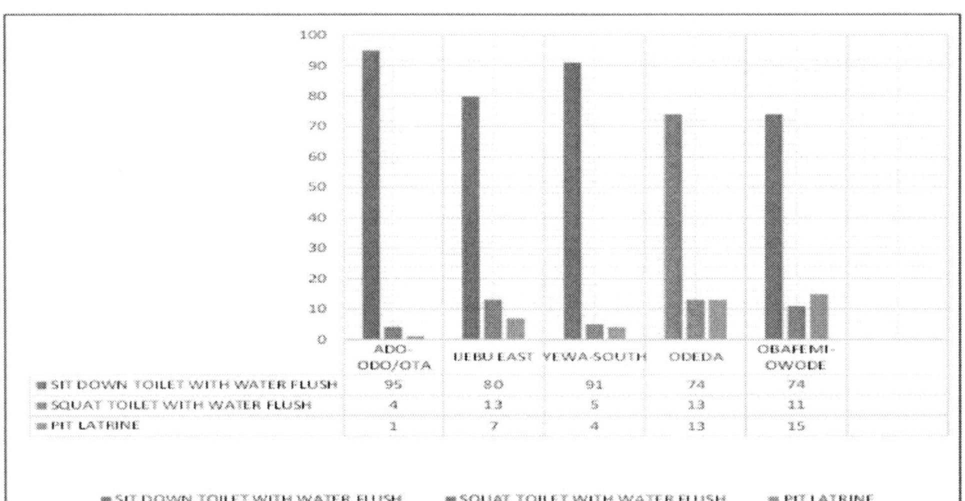

Fig. 15 Response on type of toilet facility in residence

Improved hand hygiene has been recognized as an essential public health measure. Proper hand-washing remains the most effective way of removing germs and harmful bacteria from our hands. This, in turn, helps to prevent the spread of diseases and keeps our environment safe, fresh, and clean. Figure 16 shows the frequency of how the respondents wash their hands after visiting public places. On average, 57.4 percent claimed always to wash their hands, 12.2 percent wash their hands frequently, 25.6 wash their hands sometimes, and a small 4.8 percent admitted to not washing their hands at all. Improved hand hygiene has been recognized as an essential public health measure.

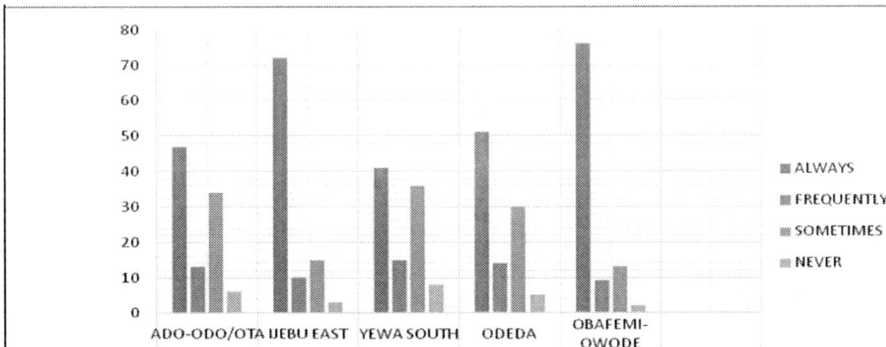

	ADO-ODO/OTA	IJEBU EAST	YEWA SOUTH	ODEDA	OBAFEMI-OWODE
ALWAYS	47	72	41	51	76
FREQUENTLY	13	10	15	14	9
SOMETIMES	34	15	36	30	13
NEVER	6	3	8	5	2

Fig. 16 Response on the frequency of washing hands after visiting public places

One of the main drivers for adequate sanitation and hygiene is proper knowledge about it. Campaigns in public places can increase the chance of people practicing sound sanitation and increase their willingness to follow appropriate hygiene practices [5, 18]. It also improves their knowledge of the potential risks and dangers of poor hygiene and sanitation. Figure 17 shows that the majority of the respondents do not get sick or infected. This can be attributed to the fact that they have good and treated water supply, good hand-washing practices, and they effectively dispose of their waste.

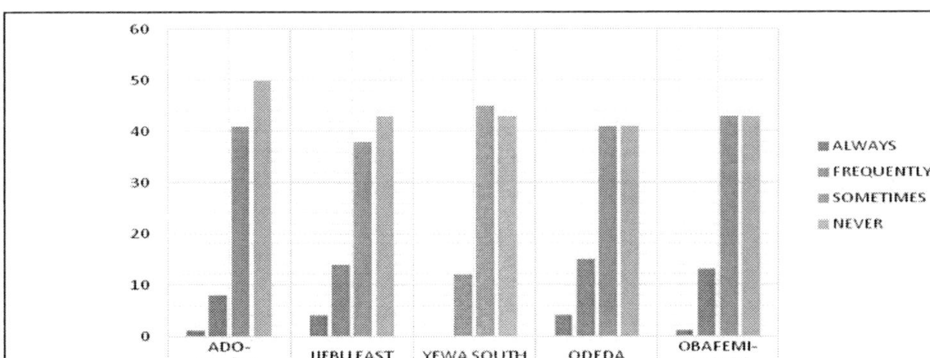

	ADO-ODO/OTA	IJEBU EAST	YEWA SOUTH	ODEDA	OBAFEMI-OWODE
ALWAYS	1	4	0	4	1
FREQUENTLY	8	14	12	15	13
SOMETIMES	41	38	45	41	43
NEVER	50	43	43	41	43

Fig. 17 Response on the frequency of getting sickness/infections/diseases within the last

4 Conclusion and Recommendation

The data gives an overview of the state of water, sanitation, and hygiene practices in the state with the following conclusions drawn: water supply in Ogun state is adequate majorly because a large number of the people get their water supply from boreholes. However, the absence of a constant power supply for pumping water from the boreholes to the overhead tanks is still a significant issue. The system of waste disposal/management in Ogun state is not adequate, and also from observations, indiscriminate dumping of waste is rampant in the communities. This indiscriminate dumping of wastes causes a high health risk to the public with the emission of a foul odor to the environment. It has the potential to contaminate the groundwater. Provision should be made for waste bins in the environment to allow people to dispose of their waste appropriately. The government should introduce the use of sanitary landfills to prevent the indiscriminate burning of waste, which causes air pollution. Enforcement of the practice of safe hygiene at all levels for all is vital to reducing the spread of current global pandemic-COVID-19. This study provides documented information about WaSH knowledge and practices of people in Ogun State and, as a result, stimulate measures to improve the hygiene situations of people in the state. The results of this study have well provided useful information on the consequences of poor hygiene practices and sanitation facilities for both residents and authorities. The data can serve as a blueprint to the government or private organizations working towards upgrading the standard of WaSH practices in the state. It will also help in identifying gaps and challenges in the provision of Water, Sanitation, and Hygiene practices and offer recommendations.

Acknowledgments
The authors wish to thank the management of Covenant University for the platform made available for this research.

References

[1] Olukanni DO 2013 Assessment of WASH Program in Public Secondary Schools in South-Western Nigeria. *ARPN Journal of Engineering and Applied Sciences* **8** (3)
[2] Olukanni D and Okorie U 2015 Empirical Assessment of Water Sanitation, and Hygiene Practices in a Semi-Urban Setting: A Socio-Economic and Cultural Mirror. *Journal of Scientific Research & Reports* **8**(7) 1-11
[3] Paolo FR, Richard CR, Goldstein R and Smith W 2001 Global Water Quality, Supply, and Demand: Implications for Megacities International Seminar on Nuclear War and Planetary Emergencies 25th Session.
[4] Mara D, Lane J, Scott B and Trouba D 2010 Sanitation and health. PLoS Medicine **7**(11) Available at: http://www.pubmedcentral.nih.gov (Accessed 20 February 2020)
[5] Olukanni DO, Pius-Imue BF and Joseph OS 2020 Public Perception of Solid Waste Management Practices in Nigeria: Ogun State Experience *Recycling* **5**(8)
[6] United Nations Development Programme 2020 Goal 6: Clean water and sanitation https://www.undp.org/content/undp/en/home/sustainable-development-goals/goal-6-clean-water-and-sanitation.html (Accessed 20 June, 2020).
[7] WHO 2017. Guidelines for drinking-water quality, 4th edition, incorporating the 1st addendum. Available at

https://www.who.int/water_sanitation_health/publications/drinkingwater-quality-guidelines-4-including-1st-addendum/en/ (Assessed 20 June 2020).
[8] Sofoluwe GO, Schram R and Ogunmekan DA 1996 Principles and Practice of Public Health in Africa (1), Ibadan: University Press plc.
[9] UNICEF and WHO (2020) Progress on Sanitation and Drinking Water: 2015 Update and MDG Assessment. Available at https://www.unicef.org/publications/index_82419.html (Assessed 20 June 2020)
[10] Emon MV 2015 The importance of water quality in livestock production. Available at https://www.drovers.com/article/importance-water-quality-livestock-production (Accessed 20 June 2020)
[11] WHO 2009 WHO guidelines on hand hygiene in health care: first global patient safety challenge Clean care is safer care World Health Organization: Geneva http://www.who.int/gpsc/5may/ tools/9789241597906/en/
[12] Olukanni DO, Adejumo TA, Adedeji AA and Salami AW 2018 Optimization-Based Reliability of a Multipurpose Reservoir by Genetic Algorithms for Jebba Hydropower Dam Nigeria. *Cogent Engineering* **5**(1)
[13] Olukanni DO, Adejumo TA and Adedeji AA and Salami AW 2016 Assessment of Jebba Hydropower Dam Operation for Improved Energy Production and Flood Management. *ARPN Journal of Engineering and Applied Sciences,* **11** (3) 8450-8467
[14] Ogbuji CN 2003 Indigenous water sourcing technology in Ovoko Rural Community, Enugu State: Implication for Health Education: *Nigerian Journal of Health Education* **11(**1**)** 119-133.
[15] World Health Organization (2020) Coronavirus disease (COVID-19) advice for the public https://www.who.int/emergencies/diseases/novel-coronavirus-2019/advice-for-public (Accessed 20 June 2020).
[16] Li Y, Wang J and Chen X 2020 Can a toilet promote virus transmission? From a fluid dynamics perspective. *Phys. Fluids* **32**, 065107 DOI: 1063/5.0013318
[17] Olukanni DO and Nwafor CO 2019 Public-Private Sector Involvement in Providing Efficient Solid Waste Management Services in Nigeria. *Recycling*, **4 (19)** 1-9
[18] Olukanni DO and Oresanya OO 2018 Progression in Waste Management Processes in Lagos State, Nigeria. *Journal of Engineering Research in Africa (JERA)*. **35** 11-23
[19] Waitea RC, Vellemana Y, Woodsb G, Chittyc A and Freemand MC 2016 Integration of Water, Sanitation, and Hygiene for the Control of Neglected Tropical Diseases: A Review of Progress and the Way Forward. *Int. Health* **8** **(**1**)** 22–27

Rain water harvesting: a sustainable alternative for domestic water supply in South –Western Nigeria: in the case study of Ado-Odo/Ota LGA Ogun State

AS Ogbiye[1]*, E Fagbenle[1], AA Busari[1], OO Onakunle[1,2] and DO Omole[1]

[1]Department of Civil engineering, Covenant University, Ogun State, Nigeria
[2]Department of Civil Engineering, University of Lagos, Lagos State, Nigeria
*corresponding author email: ade.ogbiye@covenantuniversity.edu.ng

Abstracts. Nigeria is currently facing various environmental challenges such as water scarcity and water quality; which is one of the largest environmental problems. However, the available supply of renewable water is short of the demand in Nigeria. Researchers have studied many alternative to better solve the emerging water supply problem. This study considers rainwater harvesting as an alternative source of water to ameliorate the possible scarcity. To better understand common practises in the rain water harvesting community and motivation for collecting harvested rainwater a socio-demographic survey was conducted in four communities in Ado-Odo/Ota Local Government Ogun State Nigeria to determine the rate of water consumption and supply from current water sources. A total of 400 questionnaires were administered to various household and the data collected through the survey were analysed using SPSS and selected statistical methods. Rainfall data for Ota was obtained and it was to calculate the cumulative runoff in a year was determined for roof sizes are $\leq$ 75m^2.This study demonstrated among other things that rainwater harvesting has the potential of contributing significantly to solving the problems of water shortage suffered not only by the rural-poor communities but also in the urban areas where conventional modes of water supply have failed to provide the desired level of service.
Keywords: Rainwater, supply

1. Introduction

Beside air, water is undoubtedly man's greatest need, one which will ever remain indispensable as long as there is life. Throughout antiquity and even in the modern times, the goal of providing good quality water in quantities sufficient for the sustenance of man and animals. This invariably constitutes one of the greatest challenges of human civilization. Rain water is one of the basic sources of portable water. The process of collecting water for valuable use or water infiltration into the ground increasing groundwater capacity in the aquifer is known as rain water harvesting [1].

Rain water harvesting ends up being a feasible option for local water supply in south-western Nigeria because of the fact that Nigeria is a developing nation and public water supply to the populace obviously gives a ruin sought after. At the point when there is an incredible imbalance between the regular extraction of water and common release over some undefined time frame, the drop of the water table results in lessening the yield. The main option for the present age is to enhance the recharge of water back into the ground or surface water bodies well beyond the congenital procedures. Rain water harvesting and recharge shows very high prospects and also have artificial recharge methods [2-4].

In south western Nigeria, over the years, surface water via: rivers, streams etc. and ground water have been exploited in various forms to meet water needs [5]. Dams and reservoirs have been built to provide water for different societies. In central urban areas, there are piped borne water facilities which supply water to households, while rural dwellers obtain water from streams and rivers within their environs [6]. However, recent developments in these areas have shown that water scarcity will remain a grave problem, a situation which shows that conventional water supply initiatives do not possess the magic for automatically solving all water problems. Besides the problem of water scarcity, available water bodies within this part of Nigeria are being indiscriminately polluted at an alarming rate due to unhealthy industrial practices.

Rainwater harvesting research has been an overwhelming examination region at the worldwide science. [7] in 1982 audited 170 articles on rainwater harvesting in the vicinity of 1970 and 1980. Likewise, there have been an expansive number of concentrates that have examined rainwater harvesting in all the distinctive parts of the world. Endeavours have been made in Nigeria [8], Brazil, South Africa, and Kenya and some other developing countries. Rainwater harvesting (RWH) is an expanding basic practice everywhere in the world. In Nigeria, populace development, contamination of groundwater caused by poor waste administration and low yield of wells amid the dry season [9] make collected rainwater an appealing option for consumable and non-consumable (e.g. water system, clothing) employments. There is an expansive use of rainwater in numerous parts of the world including, US, Australia, Bangladesh and Korea [9]. People keep on modifying their abodes and devise frameworks to use harvested rainwater because of population increase worries over environmental change and weight on water demand.

There is an issue of absence of supply of pipe-borne water in Nigeria. Thus, numerous homes have wells sited around the house some distance away from their septic tank. The shortage of pipe-borne water has influenced people of the community to discover other water sources; groundwater sources being a reliable source. Wells are a typical groundwater source promptly investigated to meet the society's water necessities or address deficiencies [10]. The difficulties in expanding access to enhanced drinking water is additionally convoluted by disparities in arrangement, which might be topographical (amongst urban and rural settlement); financially (between poor people and all the more monetarily hindered) or identified with the unbalanced spotlight on water in correlation with sanitation. The essential explanation behind this examination is to have a goal at embracing rainwater harvesting as an option for local water supply and by so counteracting water shortage now and in the closest future.

2. Methodology

The study area is Ogun State; is located in the south western region. The study area of this research includes Ilogbo, Balogun, Idiagbon and Olorunda Agonan which are villages located in Ado-Odo/Ota local government in south-western part of Ogun State as shown in Figures 1 & 2. The local government has a population of 526,565 residents (2006, National population Bureau, of statistics). The catchment occupied an area of about 87.8km^2. Below are the characteristics of the villages visited within the local government. The mean annual rainfall for the study area is about 1900mm. Rainfall usually begins in the month of February or March and declines in the months of October to November of each year.

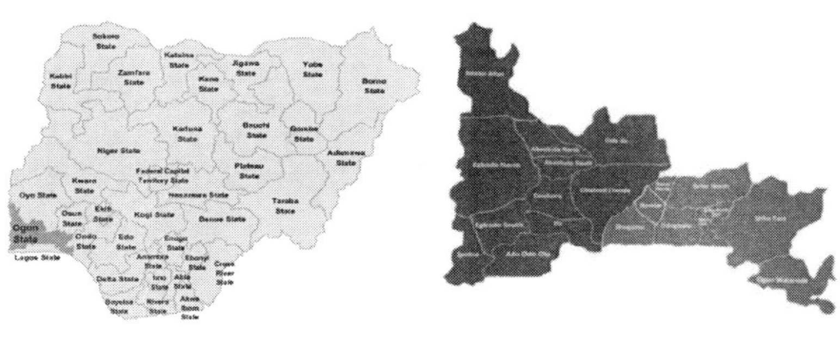

Figure 1. Map of Nigeria and Ogun State Respectively

The study is aimed at detailed description of the socio economic effect of rain water harvesting and common water supply practices. The survey was conducted in villages within Ado-Odo local government with the use of a well-defined questionnaire to obtain information for analysis on the sources of water in the study area, its usage and how to develop a culture of harvesting rain water in every household.

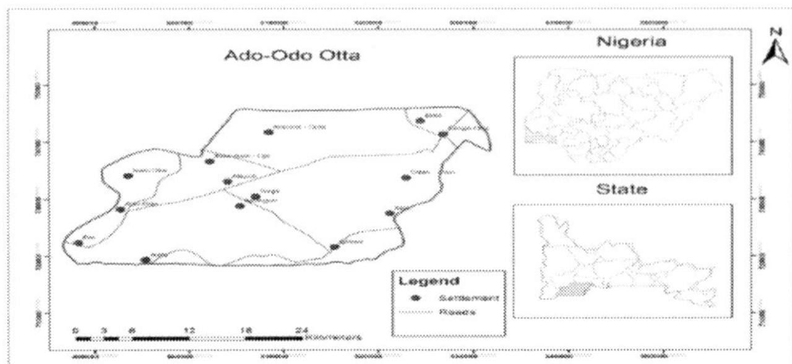

Figure 2. Maps Showing the Study Area, of Ado-Odo/Ota LGA

2.1. *Data Collection Method*
In data collection, two sets of data were obtained in the study, they include primary and secondary data. Primary data/information on rainwater harvesting as an alternative source of water in the Ogun state region which was acquired by meeting inside and out with the utilization of questionnaires as a guide. This technique unfurled respondent perception and information by speaking and taking down the subtle elements and in circumstances where respondents can actually round out the data, they are permitted to do as such. Restrictions experienced incorporate the necessity of respondents' collaboration and powerlessness to give solid answers. A great deal of time was likewise associated with examination of the information as significant exertion is put into its collation.
Secondary data used which is readily available from other archives. These data were obtained from the following sources: Journals, Articles, Reports, Website pages, Conference Proceedings Newspapers.

2.2. *Method of Data Analysis*
The data gotten from the questionnaires was analysed with the use of computer applications such as Microsoft excel and Statistical Package for Social Science (SPSS). The data was then neatly entered into tables in order to permit meaningful statistical analysis and proper interpretation. Statistical tools like charts (bar chart and pie chart), graphs, and tables will also be used for the data analysis. Laboratory test will also be carried out on water samples gotten from rooftop rainwater harvesting technique.

3. **Results and Discussion**
The first section will summarise the general findings of the research with respect to water supply/use scenario within the study area of Ado-Odo Ota local government in Ota town, Ogun state as well as the attitude of the generality of people to domestic rainwater harvesting with the help of the questionnaires shared. In the second section, the cumulative harvested water for each month is calculated using data from the types of roof obtained from the questionnaire.

3.1 *Characteristics of the Respondents*
A total of 400 people were interviewed. 51% of the individuals interviewed are males and 49% are females. The individuals interviewed were within the age group of 17 to 55 and above. Most of the respondents were within the age group of 35-44 which accounts for 27% of the total people interviewed. A very large percentage of the respondents are married and a few were single or divorced. Most of the respondents had one occupation or the other leaving only 4.7% of the respondents unemployed. Most

of the respondent's highest level of education is primary school certificate which explains why most of them selected unskilled labour for their occupation.

Table 1. Locations where the questionnaire were distributed.

Location	Number of questionnaires shared
ILOGBO	50
BALOGUN	50
IDIAGON	50
OLORUNDA	50
LUISADA	50
ISHAGA	50
OGUNTED	50
AKINWUNMI	50

Table 2. Sex distribution of respondents.

Variable N = 400 Sex	Frequency (n)	Percentage (%)
Male	204	51
Female	196	49

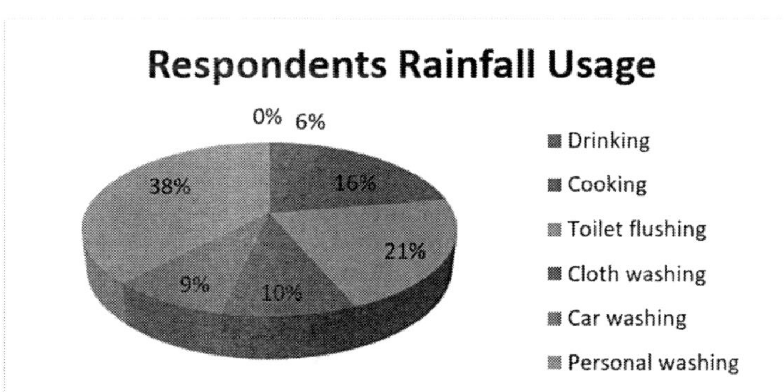

Figure 3. Sex Distribution of Respondent

Table 3. Age distribution.

Variable N = 400 Age (in years)	Frequency (n)	Percentage (%)
0 – 17 years	6	1
18 – 24 years	26	7
25 - 34 years	80	20
35 – 44 years	108	27
45 – 55 years	92	23
55+ years	88	22

3.2. Respondents Roofing Material

From the results obtained, it was noticed that a larger percentage uses corrugated iron sheet for roofing their houses because it is more affordable and economical. This roof is prone to corrosion, so the harvested water needs to be treated with chlorine or other substance if it's to be used for portable use. A result of the research done in the villages visited in Ado-Odo/Ota local government was that the average area roof area is 75 m^2. Now working with volumes in terms of runoff, a 75m^2 roof can harvest 77,657L/yr. This signifies that 212L/day assuming water is gotten constantly.

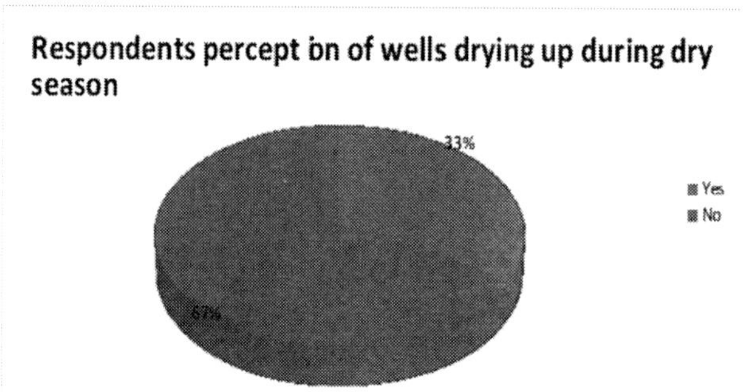

Figure 4. Roofing Material Used by Respondents

3.3. Respondents Rainfall Usage

The fig below shows the usage of rainwater harvested by the respondents (Starting from highest order of usage) personal washing, toilet flushing, cooking, cloth washing, car washing, and general outdoor use respectively. These signifies that majority of the water harvested is used for personal washing and washing of toilets. These two activities consume a lot of water as 55L/day is used for personal washing and 35L/day per day for flushing of toilet.

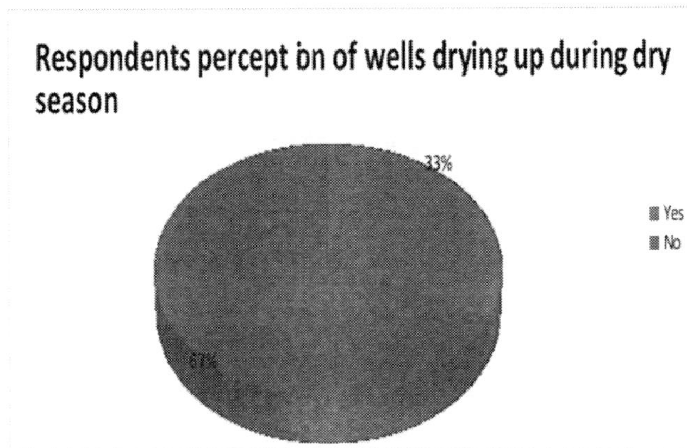

Figure 5. Respondents Rainfall Usage

3.4. *Respondent Awareness to Rainwater Harvesting Technologies*

The chart below shows if the respondents are aware about rainwater harvesting. A very large percentage of 82% are not aware about this and just 18% was aware of this. The respondents were also asked if they would like to buy or rent rainwater technologies if available. 71% of the respondents were interested in buying or renting rainwater technologies. This shows there is still a huge potential for rainwater harvesting.

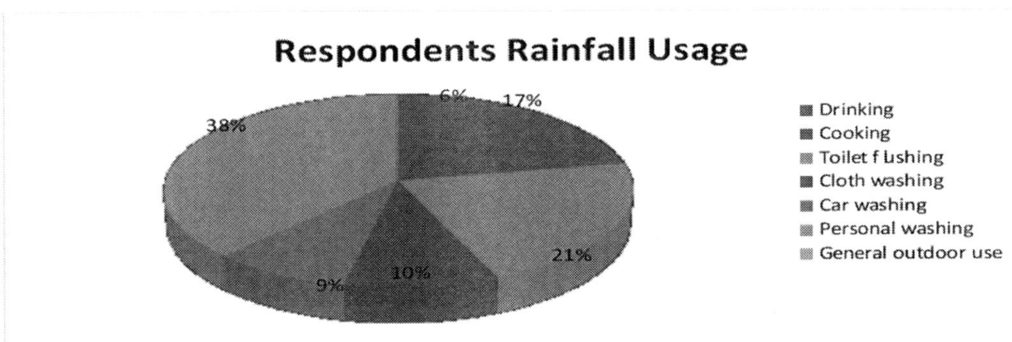

Figure 6. Respondent Awareness to Rainwater Harvesting Technologies

3.5. *Major Health Hazard Associated with Drinking Contaminated Water*

From fig 7, a large percentage (73%) of the respondents suffer from typhoid disease which is quite alarming. 6% of the respondents also suffer from cholera and just 1% suffer from diarrhoea. This means that 80% of the respondents suffer from one disease or the other from consuming water from their various water sources. The result explains that a new portable water source should be found as an alternative water source for the respondents.

These chart shows the percentage of wells that runs dry during dry season. Since a larger percentage of the respondents selected dug wells as their major source of water supply, that means a larger percentage suffers the drying up of well. When also asked about their level of satisfaction with hand-dug wells in terms of its ability to meet their daily water needs as well as quantity, the respondents expressed a high level of frustration with it because of quality problems taste, colour and hardness were the main problems identified with it. Respondents that selected borehole also claimed that water level reduces during dry season when asked. However, a good number of the respondents were not so enthusiastic about its quality. Many openly registered their displeasure with its taste, colour and saltiness.

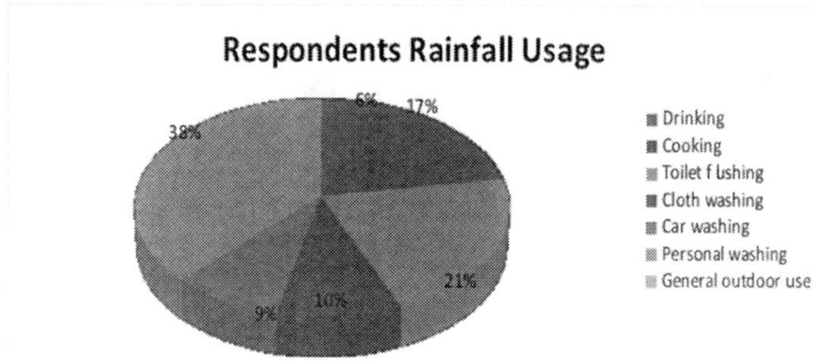

Figure 7. Major Health Hazard Associated with Drinking Contaminated Water.

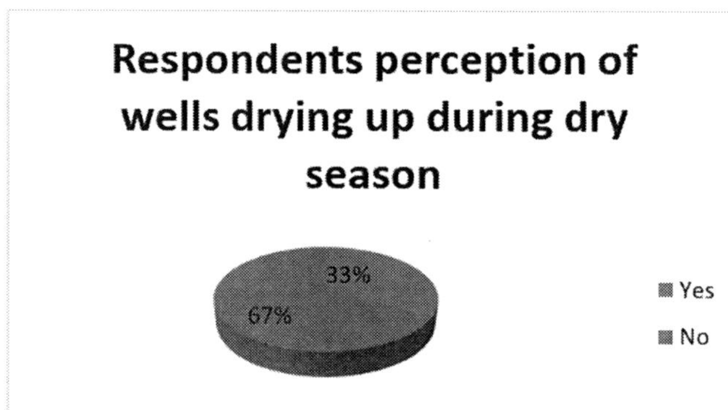

Figure 8. Respondents perception of wells drying up during dry season

3.6. *Rainwater harvesting*

The following are the typical representations for the rainwater harvesting system. Table 4 and Table 5 are shown below for the coefficient for various roofs and the cumulative harvested water.

Table 4. Typical runoff coefficients for various roofs.

Surface	Type	Coefficient
Roof	Pitch roof tiles	0.75-0.90
	Flat roof with smooth surface	0.5
	Flat roof with gravel layer or thin turf (<150mm)	0.4-0.5

IOP Conf. Series: Materials Science and Engineering 1036 (2021) 012002 doi:10.1088/1757-899X/1036/1/012002

Table 5. Cumulative Harvested Water.

Month	Average rainfall [I] (mm)	Runoff coefficient [C]	Roof area [A] (m^2)	Monthly runoff harvested [C*I*A] (m^3)	Cumulative runoff harvested (m^3)
January	3.1	0.825	75	0.19	0.19
February	34.6	0.825	75	2.14	2.33
March	80	0.825	75	5.0	7.33
April	115.5	0.825	75	7.2	14.53
May	150.5	0.825	75	9.3	23,83
June	190.6	0.825	75	11.8	35.63
July	180.4	0.825	75	11.2	46.83
August	146.2	0.825	75	9.01	55.84
September	192.6	0.825	75	12.0	67.84
October	180.9	0.825	75	11.2	79.04
November	20.4	0.825	75	1.3	80.34
December	8.2	0.825	75	0.5	80.84

From the table, a total of 80.84m^3 of water can be harvested from rooftops assuming a constant withdrawal.

4. Conclusion and Recommendations

This study has demonstrated among other things that rainwater harvesting has the potential of contributing significantly to solving the problems of water shortage suffered not by the rural-poor communities but also in the urban areas where conventional modes of water supply have failed to provide the desired level of service. The developing nations of the third world where inadequate water supply continues to be a recurrent feature stand the chance of benefitting enormously from the often neglected "wisdom" inherent in this initiative. The awareness survey carried out revealed a high degree of ignorance about formal rainwater harvesting on the part of the respondents and by implication the generality of the people. Not many are aware of the process involved. However, when asked about their willingness to adopt such systems, a good number of them sounded rather enthusiastic about the idea which is an indication of the initiative holds huge prospects if only there is proper enlightenment. Rain water harvesting could be practised in south western Nigeria from the results gotten from the case study Ado-Odo Ota due to the sustainable level of rainfall. About 88% of the rooftops are still in a fair state and made of appropriate materials. Results gotten the research in the study area shows that the average roof is within 75m^2 range and will collect 77,657L/yr (42L/person/day) for a household of 5 people. This value is close to the water demand for cooking and drinking. Hence, the storage capacity for storing rain water harvested for multipurpose use must be large enough to store a large quantity of rainfall. Due to the nature and type roofs used by people, rainwater must be properly treated before it's used for portable use. The research also shows that Ogun state in general has huge potential of rain water harvesting.

Acknowledgement
The Authors acknowledge Covenant University for the conference support payment

References

[1] Alegre N, Jeffrey P, Mcintosh B, Thomas J S, Hardwick I and Riley S 2004 Strategic options for sustainable water management at new developments: the application of a simulation model to explore potential water savings. *Water Science and Technology*, ***50*** (2), 9–15
[2] Abuubaiddha S and Liyanage Lanka RS 2003 Low-cost tanks and filters in Sri-Lanka, *29th WEDC international conference Abuja, Nigeria*
[3] Ariyabandu R 1999 Rainwater harvesting for domestic use, *South Asia Green File*, **4**(8) pp. 30-31
[4] AriyanandaT 2005 Recommendations for Improving Rain Water Quality, (A Study Conducted in 2 Districts in Sri Lanka), *proceedings of 12th International Rainwater Catchments Systems (IRCSA) Conference at New Delhi, India.*
[5] Akanni A, Ogbiye A and Onakunle O 2019 The Impact assessment of abattoir waste facility discharge on water in Osogbo, Nigeria, *Cogent Engineering* **6**(1), DOI: 10.1080/23311916.2019.1614317
[6] Ogbiye AS, Omole DO, Ade-Balogun KD, Olumuyiwa Onakunle O and Elemile OO 2018 Treatment of brewery wastewater using electro-Fenton and granulated activated carbon, *Cogent Engineering* **5**(1) DOI: 10.1080/23311916.2018.1447224
[7] Boers TM and Ben-Asher J 1982 Review of Rainwater Harvesting Agricultural *Water Management* **5**: 145-158
[8] Efe SI 2006 Quality of rainwater harvesting for rural communities of Delta State, Nigeria *Environmentalist* ***26*** 175–181
[9] Lade OO, Coker AO and Scridhar MKC 2012 Sustainable water supply for domestic use: Application of roof- harvested water for groundwater recharge. *Journal of Environmental Science and Engineering A,* **1** (5) *581-588*
[10] Adelekan BA 2010 Water quality of domestic wells in typical African communities: Case studies from Nigeria. *International Journal of Water Resources and Environmental Engineering* **2** (6) 137-147

Modelling the relationship between climatic variables and land use/land cover classes in Yewa South Local Government Area of Ogun State, Nigeria

S. A. Ojo[1], J. O. Olusina[2], B. U. Ngene[1] and D. O. Olukanni[1]
[1]Department of Civil Engineering, Covenant University, Ota, Ogun State. Nigeria
[2]Department of Surveying & Geoinformatics, University of Lagos. Nigeria

*Corresponding author email: aderemi.ojo@covenantuniversity.edu.ng

Abstract. Climate change and land use/land cover (LULC) change are among the global environmental issues that have recently dominated international debates in order to monitor earth resources. The burning of fossils fuels, clearing of land and burning of bush has given rise to the highest levels of greenhouse pollution in the atmosphere. It is on this note that this study was carried out to determine the effect of climate change on land use/land cover change of Yewa South Local Government Area of Ogun State in Nigeria. The supervised parallelepiped classification scheme was used to classify land use land cover into five classes namely agriculture land, built-up, vegetation, wetland and water body. A multivariate linear regression model was developed and program in MATLAB to model the relationship between climatic variables namely, rainfall (mm), average temperature (°C), evaporation (mm), relative humidity (%), average soil temperature (°C) and some particular land use/land cover variables. After statistical analysis between climatic variables and land use/land cover classification, results indicated a significant relationship between climatic variables and land use/land cover change. The statistical analysis between the variables shows that the climatic parameters has an accuracy of 84.12% prediction model for land use/land cover classification.

Keywords: Land Use/Land Cover, Climate Change, Pollution Abatement, Environmental Sustainability, Multivariate regression model.

1. Introduction

Climate change and land use/land cover are two environmental concerns. The two are global problems that require monitoring of earth resources. The adverse effect of climate change such as: unequal weather pattern, temperature rise, increased atmospheric menaces and rise in sea level are likely to increase the event and strength of exciting event that threaten human health and welfare [1]. The effects of the climate change vary greatly on land use and land cover changes. For example, deforestation may lead to temperature increase in the tropic zone, while it may course regional temperature decrease in the Frigid Zone [2].

Drivers of climate change include natural processes (Biogeographical) and human activities (Anthropogenic). [3,4]. The natural processes are the astronomical and the extraterrestrial factors which change the eccentricity of the earth's orbit and solar radiation quantity. The anthropogenic driver of the climate change are the activities of human beings which produce large amount of greenhouse gases into the atmosphere [3,4]. In other words, the land-cover changes rapidly and abruptly due to anthropogenic activities. Increase in population and activities of human being also increases the pressure on the limited land and soil resources for human needs [5]. Due to increasing population, agricultural land is being converted to build up areas and forest areas which are converted to agricultural land and vice versa. Change in land use and land cover, and the associated habitat loss and fragmentation are major causes of biodiversity loss [6].

Introduction of Remote Sensing (RS) and Geographical Information Systems (GIS) performances, mapping of land use/Land cover are designed to classify an area to be of use for agricultural, residential and for urban and/or industrial areas [7]. The availability of Landsat (Satellite) images data served as an important tool in the classification of land use and land cover, and the application of remote sensing made it possible to study the changes in land use and land cover in a short time with greater accuracy for future planning of the study area. GIS technique offers appropriate platform for data analysis, update and retrieval [8,9]. Predicting future climate trends relies on a large network of meteorological stations and modelling outputs from satellite data [10,11]. In fact, throughout history, the prosperity of nations has always been known to correlate very closely with the well-being of future generations depending largely on its wise management. Multivariate regression model techniques were used to understand the spatial relationship between climatic variables and land use /land cover [12].

To improve social economics activities of a study area without further disintegrating the environment, it is necessary to obtain qualitative data to model the relationship between climate change and land use/land cover change for effective future planning of land use/land cover. To understand this relationship, information about how the changes occurred, the rate and the extent at which it occurs is needed [13]. This information would be in the form of map and statistical data that is very vital for spatial planning, controlling and exploitation of land. As a result, mapping LULC and its change detection as well as updating it overtime has been recognized by various researchers; for example, mapping of land use and land detection analysis in Yewa South Local Government by [14] and monitoring of land cover dynamic by [15].

To achieve the relationship, a program in MATLAB was developed to take a set of input data (Climatic variables) namely, rainfall (mm), average temperature (°C), evaporation (mm), relative humidity (%) and average soil temperature (°C) to model for particular land use class namely water body, wetland, vegetation, built up area and agricultural land. Therefore, the study is an attempt to apply the use of GIS technology, remote sensing and multivariate regression model to assess and model the temporal changes in land use and land cover as a result of environmental dynamic induced by climate change in the past and relate it to the future planning of land use/land cover of the study area.

2. Methodology

2.1. The Study Area

Yewa South Local Government cover with a total land area of 629.38 square kilometers, with population of 150,850 (NPC, 2006) in Ogun State. It is one of the regional Local Government in Ogun State of Nigeria. It is located between 2°47′24″E and 3°6′48″E and 6°37′46″N and 6°55′42″N of the Equator. The area is surrounded in the East by Ifo and Ado – Odo/ Ota Local Government and in the West by Ipokia Local Government, north by Yewa North and in the South by the ocean. The Local government area is populated primarily by the Yoruba people of Nigeria [15]. Figure 1 shows the map of Yewa South LGA of Ogun State. The Local Government is divided into ten (10) wards (Ilaro I, Ilaro II, Ilaro III, Iwoye, Idogo, Owode I, Owode II, Ilobi/Erinja, Oke – Odan and Ajilete). Furthermore, the faming

IOP Conf. Series: Materials Science and Engineering 1036 (2021) 012003 doi:10.1088/1757-899X/1036/1/012003

system in this area is mixed crop faming and livestock. The area is characterized to be moisture sufficient, even with some flooding events in some part of the town [16].

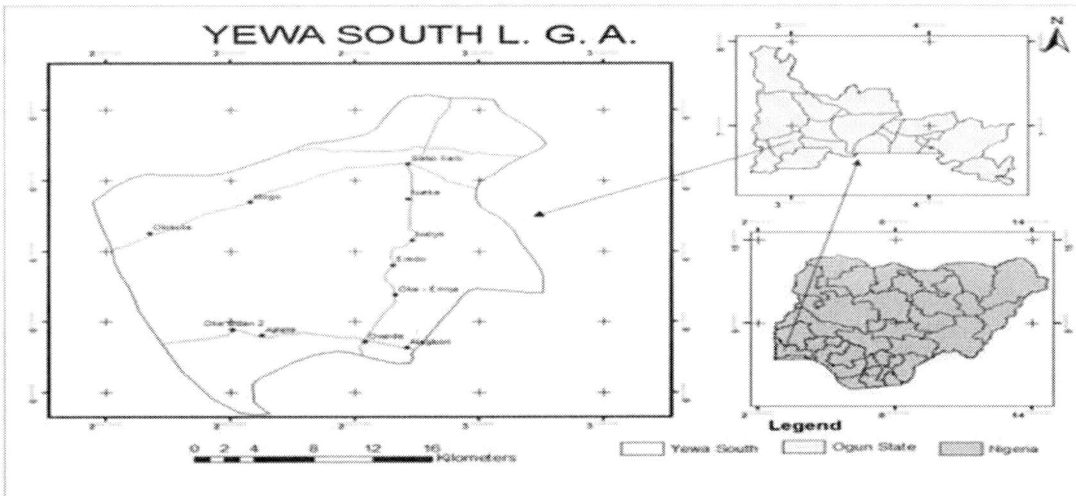

Figure 1. Yewa South Local Government, Ogun State.

Figures 2 and 3 show the flow diagram to illustrate the framework of the methodology process used in modelling the effect of climate variables on land use/land cover change of the study area, and also provide understanding of the method used in data processing. The first part of the work was on image processing, image classification, LULC change detection and preparation of thematic map. This section employed the techniques of satellite Remote Sensing with Geographic Information Systems in the generation and analysis of geospatial data on LULC dynamics. The second part was the use of multivariate linear regression model for the relationship between climatic parameters (independent variables) and patterns of land use land cover change (dependent variables) to predict for the future, with reasonable accuracy.

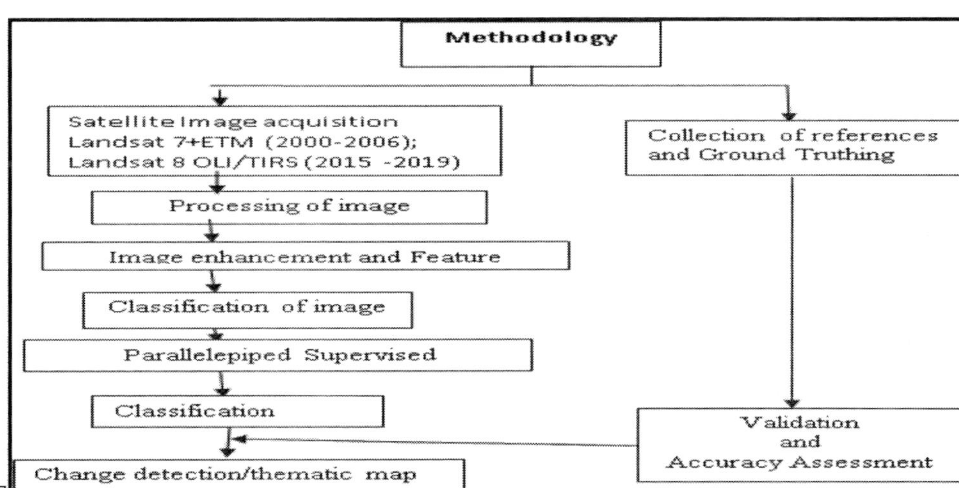

E

Figure 2. Flow chart of change Detection from imageries

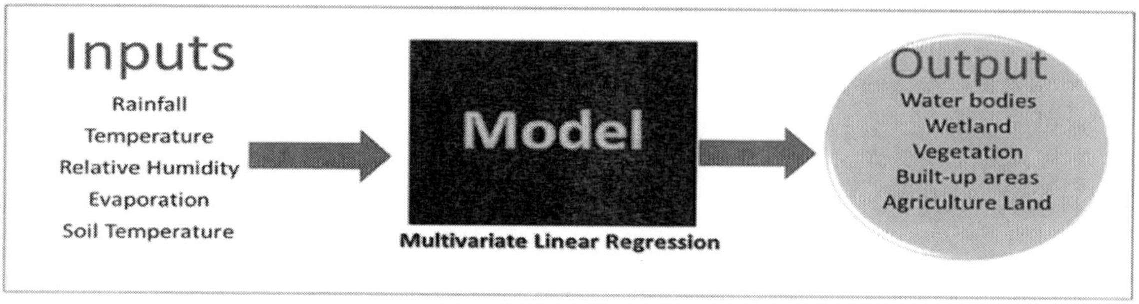

Figure 3. Flow Chart for the Multivariate Modelling

2.2. Data Source

The following data of the study area were obtained from various departments of relevant organizations. These are:

(a) Acquisition of Satellite Images downloaded from the USGS website www.earthexplorer.usgs.gov (Table 1)
(b) Climatic data such as rainfall, temperature, and relative humidity were collected from Ogun – Osun River Basin Development Authority) (Table 2)
(c) The Nigerian Administrative map downloaded in form of shape file from DIVA-GIS. The shape file is a geospatial vector format and exported into the Arc GIS environment.

2.2.1. Acquisition of Satellite Images for LULC Analysis

The properties of the Landsat images are presented in Table 1. These images have similar spatial and spectral resolution of 8 and 11 bands with 30m x 30m resolution.

Table 1: Attribute of the Landsat Data Set

Sensor	Date	Acquisition source	Data format	Spatial Resolution	No. of Bands	Datum
LANDSAT 7ETM+	06/02/2000	USGS	GeoTIFF	MS: 30m Pan: 15m	8	WGS1984
LANDSAT 7ETM+	28/12/2002	USGS	GeoTIFF	MS: 30m Pan: 15m	8	WGS1984
LANDSAT 7ETM+	07/12/2006	USGS	GeoTIFF	MS: 30m Pan: 15m	8	WGS1984
LANDSAT 7ETM+	16/01/2011	USGS	GeoTIFF	MS: 30m Pan: 15m	8	WGS1984
LANDSAT 8 OLI/TIRS	23/12/2013	USGS	GeoTIFF	MS: 30m Pan: 15m	11	WGS1984
LANDSAT 8 OLI/TIRS	26/12/2015	USGS	GeoTIFF	MS: 30m Pan: 15m	11	WGS1984
LANDSAT 8 OLI/TIRS	22/01/2019	USGS	GeoTIFF	MS: 30m Pan: 15m	11	WGS1984

2.2.2. Climatic Data

The climatic data (rainfall, temperature, evaporation, and relative humidity and soil temperature) was obtained from weather stations across the study area (Yewa local government) between 2000 and 2019 from Ogun - Osun River Basin Development Authority and presented in Table 2.

IOP Conf. Series: Materials Science and Engineering 1036 (2021) 012003 doi:10.1088/1757-899X/1036/1/012003

Table 2: Average agro meteorological observational data

Year	Rainfall(mm)	Temperature (^{0}C)	Evaporation (ml)	Relative Humidity (%)	Average soil temperature (^{0}C)
2000	79.03	24.90	6.90	87.30	32.48
2001	79.21	24.10	5.99	74.90	32.12
2002	144.71	21.60	5.89	77.8	31.75
2003	100.56	23.50	6.27	78.20	32.13
2004	92.73	23.20	4.06	81.40	32.51
2005	91.14	29.78	4.875	71.30	32.89
2006	112.6	25.78	5.69	72.05	33.27
2007	162.16	27.4	5.02	72.8	32.49
2008	106.11	26.31	5.45	66.66	31.71
2009	118.98	26.70	5.01	68.78	30.94
2010	102.83	34.35	6.20	90.26	30.16
2011	251.69	30.80	3.96	87.91	29.38
2012	270.25	32.00	4.92	87.01	29.39
2013	77.44	28.52	4.21	97.71	29.40
2014	103.45	32.33	4.09	78.57	29.36
2015	97.35	33.68	4.20	78.22	29.32
2016	91.59	33.74	4.10	79.52	29.38
2017	85.83	33.80	4.08	80.02	29.40
2018	90.11	33.40	4.15	79.98	29.50
2019	74.31	33.93	4.30	83.42	29.55

2.3. Image Enhancement and land use/ land cover Classification

The supervised Parallelepiped classification scheme was used to develop the classified land use land cover classification. The satellite images were converted into digital format and enhanced image. To improve the classification accuracy, the spatial resolution of the multispectral images (layer-stacked images) of all the epochs of the study area was improved from 30 meters to 15 meters by executing a panchromatic sharpening process. The Panchromatic sharpening algorithm used in the ENVI environment was the Gram-Schmidt Spectral sharpening which employs the insertion of the low-resolution image (Multispectral image) and the high-resolution image (Panchromatic band). The image enhancement in the form of pan-sharpening was done for all the epochs .The subset of the multispectral images of the study area for this research was overlaid on the image scenes of the Landsat products (Landsat 7ETM+ and Landsat 8 OLI/TIRS) and the subset of the study area was created, The Nigeria administrative map as downloaded in form of shapefile from DIVA-GIS was brought into the ArcMap 10.3.1 environment where the clipping tool was used to extract the study area (Yewa-South LGA). The resulting raster from image classification was used to create thematic maps as showing in Figures 9 (a-g). Training sites were created to identify homogeneous groups of pixels, which represent various land cover classes of interest in the study area. The combine process of visual image interpretation of tones/colours, patterns, shape, size, and texture of the imageries and digital image processing was used for the identification [17]. The supervised Parallelepiped classification scheme was used to develop the classified land use land cover classification. The classes are vegetation, wetland, waterbody, built-up and agricultural land as described in Table 3.

After classification, the feature classes were transferred to ArcGIS 10.3.1 for editing of classes and removal misclassified sections due to the imperfection of the classification algorithms, elimination of spurious clusters and refinement of the output. Having done the necessary editing and ground-trothing, specific colours were used for the classes as selected from the ramp of colours in ArcGIS. The colours include Quetzel green for wetland, Tzavorite green for agricultural land, Fir green for vegetation, Cretan blue for waterbody and Mars red for built-up as shown in Figure 4.

The classification shape files for the five classes for all the epochs were presented in ArcGIS and the computation of the area for each class using the "calculate geometry" was obtained. The classification accuracies for all the epochs were computed on ArcGIS 10.3.1 using the Frequency tool and the Pivot table tool. The Classification data was successfully classified with overall accuracy of the classification's ranges from 70% overall accuracy to 86% accuracy on average.

Table 3. Land use/Land cover classification scheme

S/N	Class	Description
1	Vegetation	Cropland and pasture fields, grassland, greenhouses, and fallow land
2	Wetland	Marsh or swamp
3	Waterbody	Sea, rivers, ponds and a small lake
4	Built-up	Residential, commercial and industrial areas,
5	Agricultural land	Farmlands

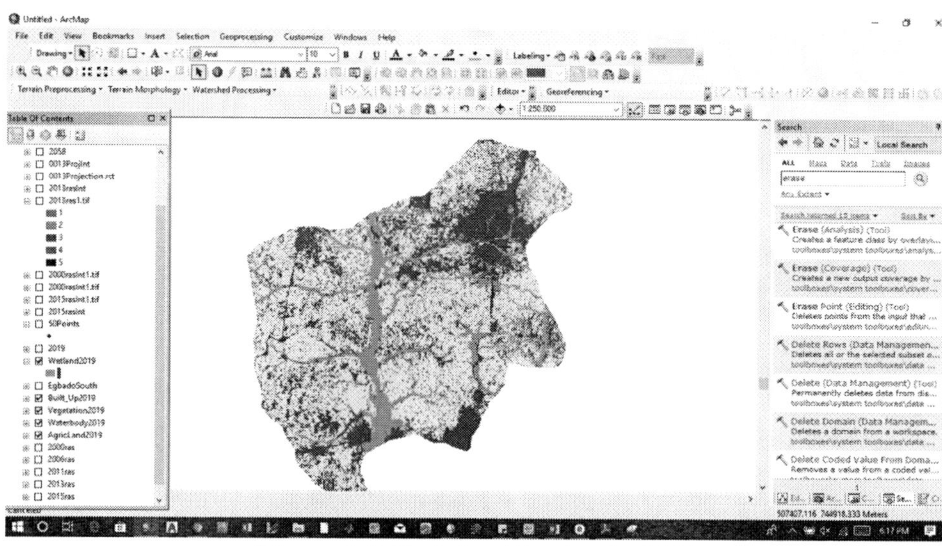

Figure 4: Edited classification for 2019

2.4. *Land use/Land Cover Change Detection Analysis*

The classification shape files for the five classes of all the epochs were presented in ArcGIS and an area field for the computation of the area was added to each class. The area was calculated using the "calculate geometry" function to populate the field with area values. The statistics showing the sum of the area was obtained. The area values for all the classes were obtained and displayed in Microsoft excel worksheet as showing in Table 4 for the calculation of the changes that have occurred over the years.

2.4.1 *Linear Interpolation*

This study embarked on linear interpolation to compensate for the missing data in the land use land cover classification. According to [18] linear interpolation is the process of using points with known values to estimate values at (unknown points). The formula is applied in Equation 1. Two data points (x_0, y_0) and (x_1, y_1) are known and value of y is to be estimated for given value of x somewhat in-

IOP Conf. Series: Materials Science and Engineering 1036 (2021) 012003 doi:10.1088/1757-899X/1036/1/012003

between these two data points. Figure 5 shows the interpolated value of x, which is shown in blue color, within the range of known data points, which are shown in red color. To determine unknown value of y, first the function (x) that passes through these known points are to be calculated with the known techniques and once it is known, the function value for any value of x can be determined [19].

$$Y = \frac{(X-X_0)(Y_1-Y_0)}{(X_1-X_0)} + Y_0 \quad (1).$$

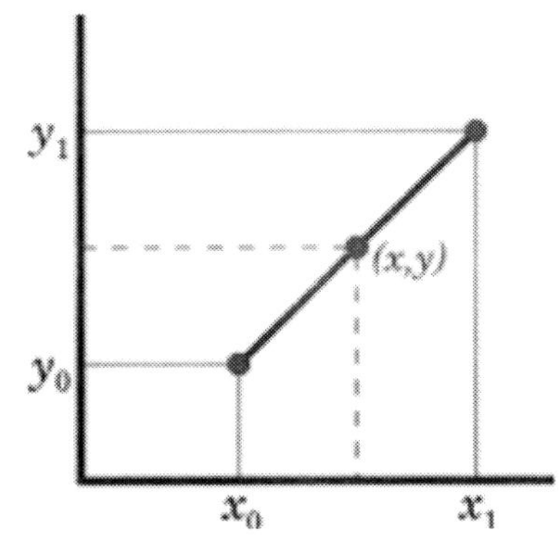

Figure 5: Interpolation.

where X_0 and Y_0 define the points to perform the interpolation. But Y is unknown and the interpolated value. X, X_1, X_0, Y_0, and Y_1 are other points of known value. From the linear interpolation, the missing data in Table 4 was developed for the land use classification per km^2 in different land use/ land cover change in the Yewa South LGA between years 2000 and 2019.

Table 4 Area of change in different land use/ land cover categories in the Yewa South LGA between years 2000 to 2019.

Point	Years	Water body (km^2)	Wetland (km^2)	Vegetation (km^2)	Build up Area (km^2)	Agric land (km^2)
1	2000	0.32	147.10	262.01	25.28	222.17
2	2001	0.38	121.05	224.41	25.50	285.41
3	2002	0.44	95.00	186.81	25.72	348.65
4	2003	0.47	90.76	175.45	26.39	363.60
5	2004	0.48	86.52	164.03	27.05	378.55
6	2005	0.49	82.28	152.69	27.72	393.50
7	2006	0.50	78.04	141.35	28.38	408.45
8	2007	0.50	76.26	136.53	29.23	414.13
9	2008	0.50	74.48	131.72	30.08	419.81
10	2009	0.50	72.70	126.90	30.93	425.49
11	2010	0.50	70.93	122.09	31.79	431.17
12	2011	0.50	69.15	117.27	32.64	436.84
13	2012	o.65	68.26	141.46	37.53	408 62
14	2013	0.79	67.36	165.65	42.41	380.39
15	2014	0.80	80.50	202.85	44.58	327.98
16	2015	0.80	93.64	240.04	46.74	275.57
17	2016	0.75	90.36	211.52	52.01	302.12
18	2017	0.69	87.08	183.00	57.28	328.67
19	2018	0.64	83.81	154.49	62.56	355.22
20	2019	0.58	80.53	125.97	67.83	381.76

Multivariate Linear Regression

In order to determine the effect of climate change on land use/land cover (LULC) change and to create a regression model to predict for land use/land cover change, a multivariate linear regression model was developed. This is based on the statistical principle of multivariate statistics, which involves observation and analysis of more than one statistical outcome variable at a time. Consider the following in multivariate linear regression;

$$y = \begin{bmatrix} y_1 \\ y_2 \\ \vdots \\ y_n \end{bmatrix} = f(x_1, x_2 \dots x_n) \qquad (2)$$

where $f(x_1, x_2 \dots x_n)$ are the causal variables or Independent variables or inputs and y_1 = the response variable or dependent variable or output. Note: y has been changed to a column vector which contains all output variables.

Key points from Equation 2 is that:

i. There is a many to many relationships between the vector y and function $(x_1, x_2 \dots x_n)$
ii. Dimensionality should agree in this case.

The general form for the multivariate linear regression is given as:

$$Y_1 = \beta_0 + \beta_1 X_1 + \beta_2 X_2 \cdots\cdots\cdots + \beta_k X_k + \varepsilon \qquad (3)$$

where

$\beta_{0,}$ β_1,β_2, β_k, and ε are unknown and are to be estimated from the data. These parameters are usually called regression coefficients

β_0 is the constant term or intercept in the model and term Least square estimate and ε is the random error or error estimate and are normally distributed. The error component ε represents the difference between the true (predicted data) and observed data of y.

2.5. *Working Variables*

To design the graph theory for the process of modelling, Table 5 which is the working variables for the model equation was used.

Table 5 Definition of Working Variables

VARIABLES	SYMBOL	TYPE
Rainfall	x_1	INPUT DATA
Temperature	x_2	
Evaporation	x_3	
Relative Humidity	x_4	
Average Soil Temperature	x_5	
Waterbody	y_1	OUTPUT DATA
Wetland	y_2	
Vegetation	y_3	
Built-up	y_4	
Agric Land	y_5	

2.6. *Concept of Graph Theory*

At the beginning of the modelling process, the concepts of graph theory was employed as shown in Figure 6

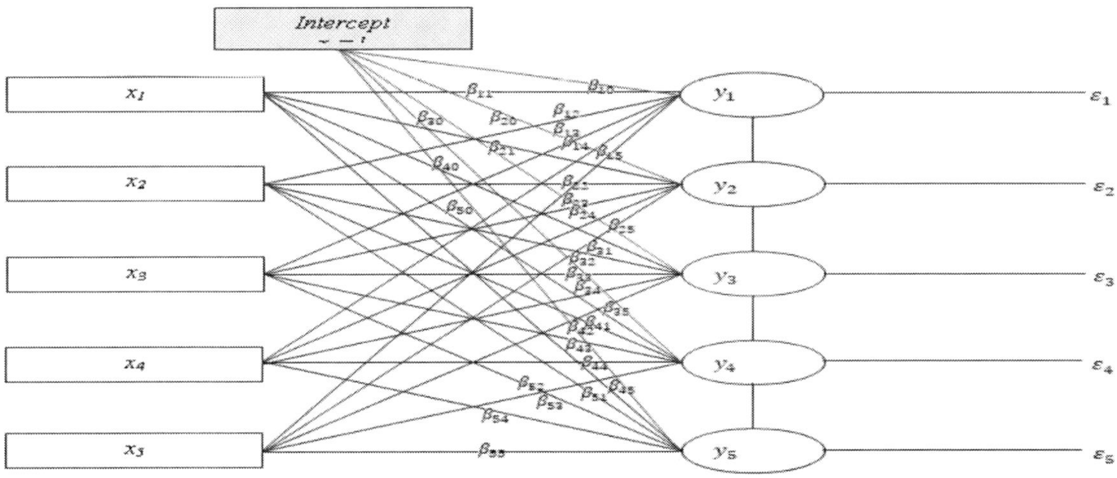

Figure 6 Synapsis of Multivariate Linear Regression

From 6, the following relationships are deduced

$$y_1 = \beta_{10} + \beta_{11}x_1 + \beta_{12}x_2 + \beta_{13}x_3 + \beta_{14}x_4 + \beta_{15}x_5 + \varepsilon_1$$
$$y_2 = \beta_{20} + \beta_{21}x_1 + \beta_{22}x_2 + \beta_{23}x_3 + \beta_{24}x_4 + \beta_{25}x_5 + \varepsilon_2$$
$$y_3 = \beta_{30} + \beta_{31}x_1 + \beta_{32}x_2 + \beta_{33}x_3 + \beta_{34}x_4 + \beta_{35}x_5 + \varepsilon_3 \quad (4)$$
$$y_4 = \beta_{40} + \beta_{41}x_1 + \beta_{42}x_2 + \beta_{43}x_3 + \beta_{44}x_4 + \beta_{45}x_5 + \varepsilon_4 y_5$$
$$= \beta_{50} + \beta_{51}x_1 + \beta_{52}x_2 + \beta_{53}x_3 + \beta_{54}x_4 + \beta_{55}x_5 + \varepsilon_5$$

where y = y_1………………… y_5

y = Prediction for each dependent variable.

X_i = Independent variable or design matrix

β = Least Square Estimate

ϵ = Error Estimate. It is also an additive constant to the model

where i = 1, 2, 3…n

The Least square Estimates ($\boldsymbol{\beta}$)and Error of Estimation (Residual) ($\hat{\boldsymbol{\epsilon}}$) can now be computed using the following least squared residual approach

$$\hat{\beta} = (X'X)^{-1}X'Y \quad (5)$$
$$\hat{\epsilon} = Y - \hat{Y} \quad (6)$$

2.7. Model development

2.7.1 *Assumptions in Multivariate linear regression Analysis*

Before continuing with the Multivariate Linear Regression, the data should be satisfied with the following assumptions:

i. The errors $\epsilon_{n \times q}$ are multivariate normal. This assumption indicated that the expected value of the error term should be zero, that is, $\boldsymbol{\varepsilon} = 0$.
ii. Error variances are equal (homogenous) across observation.

Errors have common covariance structures across observation.

Using multivariate linear regression to examine the effect of climatic variables on land use /land cover change and to create a regression model to predict for land use/land cover change from climatic parameters, a program in MATLAB was develop and a set of input data were taken (causal variables namely, rainfall (mm), temperature (°C), evaporation (mm), relative humidity (%) and average soil temperature (°C) to model for particular land use class namely water body, wetland, vegetation, built up area and agricultural. We may sometimes find it more reasonable to refer climatic variables as independent variables or "predictors", and land use land cover as dependent variables or "response (outcome) or Sample output A least square Estimate $\boldsymbol{\beta}$ and Error Estimate $\boldsymbol{\epsilon}$ as well as the predicted model to determine the effect of climatic variables on land use/land cover change as stated in Equation 7, 8 and 9

$$\beta = \begin{bmatrix} 0.006014 & -0.06665 & -0.000745 & -0.001364 & -0.05505 \\ -1.288 & 9.685 & -0.09898 & 0.38 & -0.07351 \\ -2.843 & 9.729 & -0.3835 & 0.4064 & -11.78 \\ 0.4234 & -2.143 & -0.03448 & -0.07503 & -3.237 \\ 3.703 & -17.19 & 0.5166 & -0.7102 & 15.16 \end{bmatrix} \quad (7)$$

$$\epsilon = \begin{bmatrix} 2.596 \\ 75.8 \\ 588.7 \\ 143.2 \\ -120.5 \end{bmatrix} \quad (8)$$

$$\begin{aligned} \boldsymbol{Wb} &= 0.006014\boldsymbol{Te} - 0.06665\boldsymbol{Ev} - 0.000745\boldsymbol{Ra} - 0.001364\boldsymbol{Rh} - 0.05505\boldsymbol{A} + 2.596 \\ \boldsymbol{Wl} &= -1.288\boldsymbol{Te} + 9.685\boldsymbol{Ev} - 0.09898\boldsymbol{Ra} + 0.38\boldsymbol{Rh} - 0.07351\boldsymbol{As} + 75.8 \\ \boldsymbol{Vg} &= -2.843\boldsymbol{Te} + 9.729\boldsymbol{Ev} - 0.3835\boldsymbol{Ra} + 0.4064\boldsymbol{Rh} - 11.78\boldsymbol{As} + 588.7 \\ \boldsymbol{Bu} &= 0.4234\boldsymbol{Te} - 2.143\boldsymbol{Ev} - 0.03448\boldsymbol{Ra} - 0.07503\boldsymbol{Rh} - 3.237\boldsymbol{As} + 143.2 \\ \boldsymbol{Al} &= 3.703\boldsymbol{Te} - 17.19\boldsymbol{Ev} + 0.5166\boldsymbol{Ra} - 0.7102\boldsymbol{Rh} + 15.16\boldsymbol{As} - 120.5 \end{aligned} \quad (9)$$

where:

Wb = Water body, (ii) Wl = Wetland, (iii) Vg = Vegetation, (iv) Bu= Built-up and (v) Al = Agriculture land while Te= Temperature, Ev = Evaporation, a= Rainfall, Rh= Relative humidity, and As = average soil temperature

2.8. Model Verification

Model Verification is the process of determining whether the model meets its specification. When using regression to create a predictive model. Several criteria are used to evaluate model performance and the difference between simulated and observed data is determined. The criteria used for evaluating model performance are: model validation, correlation analysis, the root means squared error (RMSE) and error estimate called residual.

2.8.1. Model Validation

This is the process of representing the model specification. The degree of agreement between the result of the model and real life data was established. Table 6 shows the data usage.

Table 6: Data Usage

Item	Year	Duration	Percentage
Model Development	2000 – 2015	15 years	80%
Testing and Validation	2016 – 2019	4 years	20%

To ascertain the degree of accuracy of each parameter in the developed model, the set of equation in 10 was used to judge the accuracy of the regression prediction model:

$$\delta_i = \frac{|O_i - P_i|}{O_i} \times 100\% \qquad (10)$$

$$\bar{\delta} = \frac{\sum_{i=1}^{n} \delta_i}{n}$$

$$\text{Accuracy} = (100 - \delta)\%$$

where $\boldsymbol{\delta_i}$ = percentage deviation of single sample data, $\boldsymbol{O_i}$ = the observed data, $\boldsymbol{P_i}$ = the predicted data, n = sample size and $\bar{\boldsymbol{\delta}}$ = total deviation.

3. Results and Discussion

3.1. Result for Model validation

The model performance was assessed by comparing the (sample output) observed data of 2016 – 2019 to the predicted output data 2016 – 2019 using the validated module in MATLAB. The result of the validation is presented in Tables 7a and 7b, and the residual is presented in Table 7c indicating the extent to which the observed data and the predicted data agreed.

Table 7a: Result for Model validation (Output)

Years	Wb (km^2)	Wl (km^2)	Vg (km^2)	Bu (km^2)	Al (km^2)
2016	0.49270	71.6610	220.668	49.6971	330.341
2017	0.57941	127.466	233.601	55.932	379.881
2018	0.72512	94.6925	186.951	61.586	399.051
2019	0.66075	82.5436	175.594	65.3955	285.915

Table 7b Model Validation Observed data (Sample output)

Years	Wb ($_1$) (km^2)	Wl(y_2) (km^2)	Vg (y_3) (km^2)	Bu(y_4) (km^2)	A l(y_5) (km^2)
2016	0.75	90.36	211.52	52.01	302.12
2017	0.69	87.08	183.00	57.28	328.67
2018	0.64	83.81	154.49	62.56	355.22
2019	0.58	80.53	125.97	67.83	381.76

Table 7c: Residual

Year	Water body	Wetland	Vegetation	Built-up area	Agric Land
2016	0.257299	18.69897	9.147965	2.312948	28.22118
2017	0.110585	40.38626	50.60144	1.347786	51.21059
2018	0.085122	10.88253	32.46133	0.973632	43.83146
2019	0.080748	2.013666	49.62433	2.434509	95.84509

The model validation performance values ranging from 74% to 95% accuracy

3.2. Result of Correlation Analysis

This aspect was used to study the strength of a relationship between two successive land use classes. The Pearson's product-moment correlation coefficient was used to determine the correlation coefficients between the model-predicted and observed for each climatic variable in the region.

$$\rho_{PQ} = \frac{cov(P, O)}{\sigma_P \cdot \sigma_O} \quad (11)$$

where $\boldsymbol{\rho_{PO}}$ is the Correlation Coefficient, $\boldsymbol{cov(P, O)}$ is the covariance between P (Predicted) and O (Observed) and $\boldsymbol{\sigma_P \cdot \sigma_O}$ are the standard deviations of *P* and *O* respectively. The Correlation Coefficient is a dimensionless quantity which satisfies the inequality $-\mathbf{1} \leq \boldsymbol{\rho_{PO}} \leq \mathbf{1}$ and standardizes the measure of interdependence between two variables and, consequently, tells you how closely the two variables move. The result of the Correlation Analysis is in Table 8 and Figure 7.

Table 8 Correlation Analysis between land use classes

	Waterbody	Wetland	Vegetation	Built-up	Agric Land
Waterbody	1.00	-0.66	0.09	0.98	0.08
Wetland	-0.66	1.00	0.68	-0.66	-0.80
Vegetation	0.09	0.68	1.00	0.05	-0.98
Built-Up	0.98	-0.66	0.05	1.00	0.11
Agric. Land	0.08	-0.80	-0.98	0.11	1.00

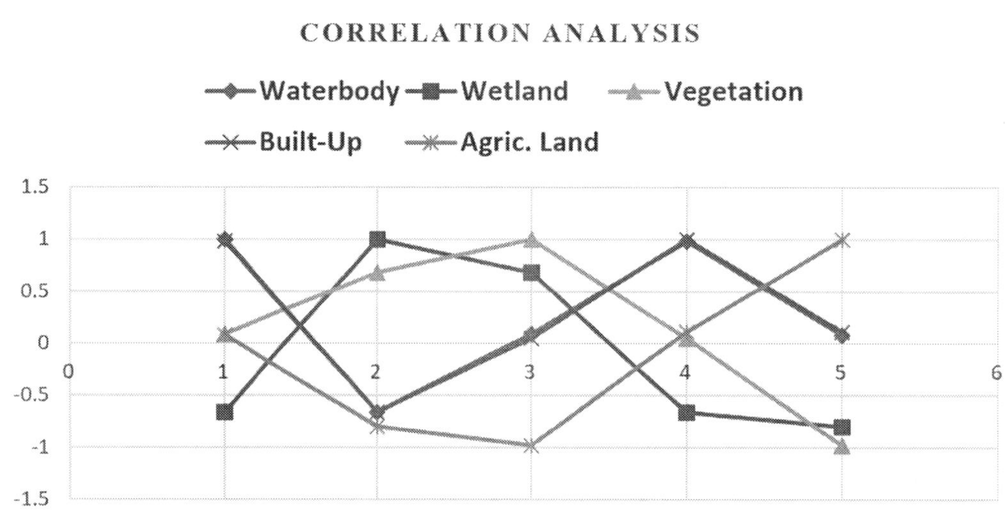

Figure 7: Correlation Analysis between land use classes

From the Table 7 and Figure 7, there is a negative correlation between waterbody and wetland and a positive correlation between Waterbody and Vegetation, Waterbody and Built-up and Waterbody and Agriculture Land. It should be noted that this analysis is limited to the data used in the study area.

3.3. Result of Root Mean Square Error (RMSE)

This was used to determine the magnitude of the error in the model output as the correlation technique is however limited when it comes to this. The Root Mean Square Error represents the square root of the differences between predicted values and observed values which are termed as residuals. It was used to determinite the magnitude of the error in the model. Equation 12 was used to compute the RMSE:

$$RMSE = \left[\frac{1}{n}\sum_{i=1}^{n}(P_i - O_i)^2\right]^{\frac{1}{2}} \tag{12}$$

where Pi = predicted, O_i = Observed, n = size of the data.

3.4. The Root Mean Square Error (RMSE)

The Root Mean Square Error (RMSE) was used to determinite the magnitude of the error in the the model output as the correlation technique is however limited when it comes to this. It measures the difference between fitted and observed values and this was used to evaluate the systematic bias of the model [20]. The result of the Root Mean Square Error (RMSE) is in Figure 8. It represents the differences between the square root of the predicted values and observed values that are termed as residuals.

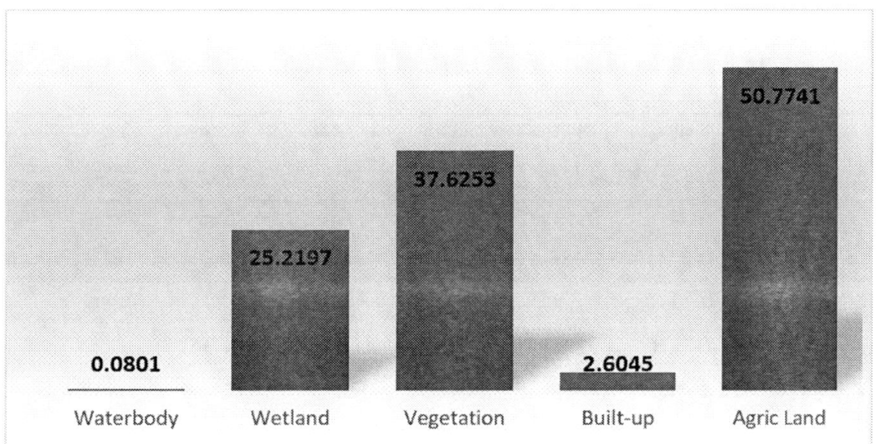

Figure 8: The Root Mean Square Error.

Table 9 shows that agriculture land has the highest root mean square error of 50.7741 followed by vegetation (37.6253). This is due to the random nature of the data constituting agriculture land whereas Waterbody has the least (0.0801).

3.5. Classification of Land Use Land Cover

Land use/land cover was classified successfully with overall accuracy ranging from 70% to 86% accuracy. Five different classes of land use/land cover types and land cover map were developed. The land cover maps are shown in Figures 9a to 9g. The magnitude of change at different land use/land cover classes in the study area between years 2000 – 2019 is showing in Tables 9a and 9b. The resulting land cover types were Water body/watershed, built-up, Vegetation, wetland, and agricultural land in the study area.

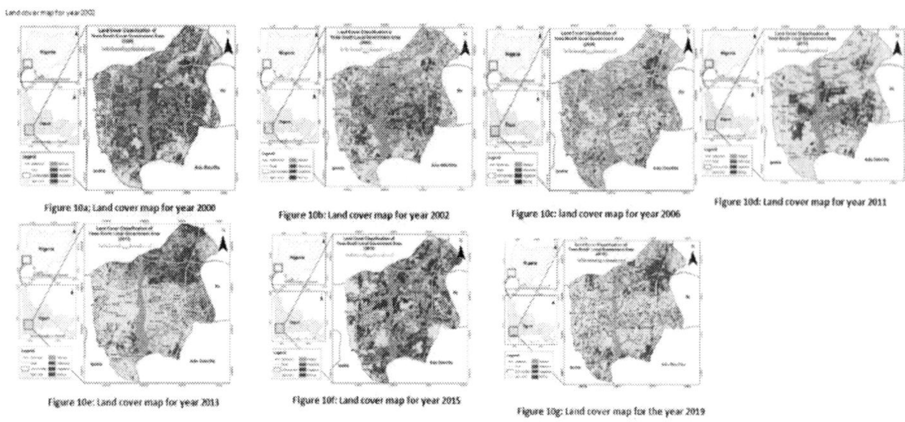

Figure 9 a-g: The land cover maps

Table 9a: The area of the different land use/land cover categories in the study area between years 2000 – 2019

	Year 2000	Year 2002	Year 2006	Year 2011	Year 2013	Year 2015	Year 2019
Land cover	Area (km^2)	Area (km^2)	Area (km^2)	Area (km^2)	Area (km^2)	Area (km^2)	Area (km^2)
Waterbody	0.32	0.44	0.50	0.46	0.79	0.80	0.58
Wetland	147.10	95.02	78.04	69.15	67.36	93.64	80.53
Vegetation	262.01	186.81	141.35	117.27	165.65	240.04	125.97
Built-up	25.28	25.72	28.38	32.64	42.41	46.74	67.83
Agric land	222.17	348.652	408.45	436.84	380.39	275.57	381.76
Total	656.88	656.64	656.72	656.36	656.59	656.80	656.68

Table 9b: The percentage of the different land use/land cover categories in the study area between years 2000 – 2019.

	Year 2000	Year 2002	Year 2006	Year 2011	Year 2013	Year 2015	Year 2019
Land cover	%	%	%	%	%	%	%
Waterbody	0.05	0.06	0.08	0.07	0.12	0.12	0.09
Wetland	22.39	14.47	11.88	10.53	10.26	14.26	12.26
Vegetation	39.89	28.45	21.52	17.87	25.23	36.54	19.18
Built-up	3.85	3.92	4.32	4.97	6.46	7.12	10.33
Agric land	33.82	53.10	62.10	66.56	57.93	41.96	58.14
Total	100.00	100.00	100.00	100.00	100.00	100.00	100.00

The result from the model indicated the relationship between the predicted land use/land cover change and the classified land use/land cover change (Observed) is displayed in Table 10a &10b

Table 10a: Relationship between Predicted value (output) and Observed (sample)Data

	Predicted value (output)					Observed (sample)Data				
Years	Wb (km^2)	Wl (km^2)	Vg (km^2)	Bu (km^2)	Al (km^2)	Wb_1 (km^2)	Wl_2 (km^2)	Vg_3 (km^2)	Bu_4 (km^2)	Al_5 (km^2)
2000	0.31793	134.577	206.146	24.5314	324.414	0.32	147.10	262.01	25.28	222.17
2001	0.41094	121.834	202.065	28.2496	340.501	0.38	121.05	224.41	25.50	285.41
2002	0.36401	118.633	186.757	26.1091	359.247	0.44	95.00	186.81	25.72	348.65
2003	0.36514	124.512	198.055	26.3626	342.383	0.47	90.76	175.45	26.39	363.60
2004	0.49808	104.739	177.982	29.7718	378.687	0.48	86.52	164.03	27.05	378.55
2005	0.46903	100.701	160.523	30.3989	401.152	0.49	82.28	152.69	27.72	393.50
2006	0.35180	112.033	165.094	24.9325	388.658	0.50	78.04	141.35	28.38	408.45
2007	0.40451	98.7136	144.934	27.8060	419.484	0.50	76.26	136.53	29.23	414.13
2008	0.46781	107.842	183.599	31.3327	371.672	0.50	74.48	131.72	30.08	419.81
2009	0.52867	102.603	184.183	34.3538	374.044	0.50	72.70	126.90	30.93	425.49
2010	0.51729	114.642	195.919	36.5017	346.454	0.50	70.93	122.09	31.79	431.17
2011	0.56821	81.1017	134.502	37.3595	438.696	0.50	69.15	117.27	32.64	436.84
2012	0.80807	85.8947	127.913	34.8568	428.762	o.65	68.26	141.46	37.53	408 62
2013	0.66545	105.783	211.431	40.9621	341.966	0.79	67.36	165.65	42.41	380.39
2014	0.70838	92.1225	171.003	43.5664	287.572	0.80	80.50	202.85	44.58	327.98
2015	0.61524	91.2993	199.135	45.3885	366.027	0.80	93.64	240.04	46.74	275.57

Note that; For predicted value, Wb= Waterbody, Wl= Wetland, Vg= Vegetable, Bu= Built-up Al = Agric land, and for sample value, Wb_1= Waterbody, Wl_2= Wetland, Vg_3= Vegetable, Bu_4= Built-up and Al_5 = Agric land

Table 10b: Residual (Error Estimates)

years	wb	wl	Vg	Bu	Al
2000	0.0021	12.5228	55.86451	0.749633	102.239903
2001	0.031941	0.7840	22.3447	2.749594	55.091261
2002	0.075986	23.63267	0.053289	0.38850	10.596381
2003	0.10486	33.74789	22.60497	0.027369	21.226664

2004	0.01773	18.2189	13.95206	2.721795	0.136903
2005	0.021468	18.42112	7.832549	2.678906	7.652334
2006	0.148203	33.98851	23.74411	3.444708	19.791675
2007	0.095488	22.44636	8.403876	1.424007	5.353456
2008	0.032192	33.36202	51.87867	1.252678	48.138341
2009	0.02867	29.90305	57.27262	3.423776	51.445808
2010	0.017587	43.71147	73.82878	4.711713	84.715789
2011	0.067505	11.95172	17.23206	4.716536	1.860242
2012	0.158078	17.63477	13.54695	2.673148	20.14155
2013	0.124565	38.42316	45.78142	1.447986	38.42367
2014	0.091162	11.62725	31.84734	1.013532	40.40754
2015	0.184751	2.340645	40.90532	1.351406	90.45691

Note that; Wb= Waterbody, Wl= Wetland, Vg= Vegetable, Bu= Built-up

3.6. Land use/land cover variations

The study clarified five periods of land use land cover based on the supervised classification known as Parallelepiped classification method. Table 4.1a showed the temporal changes in waterbodies, wetland, vegetation, built up land and agricultural land. These were obtained from seven periods Landsat images. The Land cover pattern has shown much and steady increase for some classes. The Classification data was successfully classified with overall accuracy of the classification's ranges from 70% overall accuracy to 86% accuracy on average.

3.7. *Statistical Analysis of climate and land use/land cover change*

The assessment of LULC classification in the study area was supplemented by the multivariate regression analysis to determine the effect of climate change on land use land cover classification. The land use and land cover classification and the climatic variables were subjected to statistical analysis and tested for, among others to ensure that the data meets the necessary accuracy and utility standards. The following criteria were used to evaluate model performance and the difference between simulated and observed data was determined [20]. Such criteria are model validation, correlation analysis, the root means squared error (RMSE) and error estimate called residual. After addressing and satisfied that the assumptions for multivariate regression model were met, Multivariate Linear Regression model was utilized.

From the model performance, the climatic data has the following accuracy for the land use land cover prediction model:

i. Parameter Rainfall has accuracy of 87.35% Prediction Model for water body
ii. Parameter Temperature has accuracy of 71.83% Prediction Model for Wetland
iii. Parameter Evaporation has accuracy of 80.66% Prediction Model for Vegetation
iv. Parameter Relative humidity has accuracy of 92.91% Prediction Model for Built up area
v. Parameter Average soil temperature has accuracy of 87.84% Prediction Model for Agric land

The above analysis confirmed the relationship between Climatic variables and land use/land cover change. With this we can say that climate changes are closely related to changes in land use/land cover.

In particular, it is widely acknowledged that the climatic variables are the significant factors that influencing the land use/land cover change in any region.

4. Conclusion

The research work analyzed the impact of climate change on land use/land cover using LANDSAT 7ETM+ & LANDSAT 8 OLI/TIRS images and climatic variables and further demonstrated a relationship between them. The supervised Parallelepiped classification helped in this research to determine the changes in LULC cover for a period of 19 years from 2000 to 2019 during which significant environmental, vegetation and human changes have taken place in the study area. A multivariate model was developed to determine the relationship between climatic variables and land use/land cover classification. It is known that there is a relationship between land use land cover (LULC) with climatic variables as an input data. After statistical analysis of the relationship between climatic variables and land use /land cover classification, it was found that 97.9359% of the original uncertainty has been explained by the model. The result supports the conclusion that the multivariate model represents an excellent fit as indicated in the graphs. The multivariate model was successfully tested with the help of MATLAB 2017 on the Windows 10 operating system supports the conclusion that the model represents "an excellent" fit as indicated in the assumption and the graphs. The findings in this research shows that there is a relationship between climatic variables with land use land cover (LULC). From the analysis, the climatic parameter has an average accuracy of 84.12% prediction for land use/land cover. The result supports the conclusion that the multivariate model was successfully tested and represents an excellent fit as indicated in the graphs. This study also proves that integration of GIS and remote sensing technologies is effective tool for urban planning and management. The quantification of LU/LC changes of the study area is very useful for environmental management groups, policy makers and for public to better understand the surrounding.

Acknowledgments: The authors acknowledge the management of Covenant University, Ota, Ogun State, Nigeria for publication support.

Interest: The authors declare no conflict of interest

References

[1] Parry ML, Carter TR, Konijn NT editors 2013 The impact of climatic variations on agriculture: : assessment in cool temperate and cold regions *Springer Science & Business Media,* **1**

[2] Bonan GB 2008 Forests and climate change: forcings, feedbacks, and the climate benefits of Forests, *Science* **20**(5882):1444-9.

[3] Olaniyi OA, Olutimehin IO, Funmilayo OA 2019 Review of climate change and its effect on Nigeria ecosystem. *International journal of Rural Development, Environment and Health Research,* **3**(3)

[4] Akpodiogaga-a P, Odjugo O. 2010 General overview of climate change impacts in Nigeria. *Journal of human ecology* **29**(1):47-55.

[5] Singh A, Singh S, Garga PK, Khanduri K 2013 Land Use and Land Cover Change Detection: A Comparative Approach using Post Classification Change Matrix Function Change Detection Methodology of Allahabad City. *Int. J. Curr. Eng. Technol.* **3**:142-8.

[6] Halmy MW, Gessler PE, Hicke JA, Salem BB 2015 Land use/land cover change detection and prediction in the north-western coastal desert of Egypt using Markov-CA. *Applied Geography.* **63**:101-12.

[7] Selçuk RE, Nisanci R, Uzun B, Yalcin A, Inan H, Yomralioglu T 2003 Monitoring land-use changes by GIS and remote sensing techniques: case study of Trabzon. *In Proceedings of 2nd FIG Regional Conference, Morocco* 2003 Dec 25, 1-11

[8] Star JL, Estes JE, McGwire KC, Arvidson RE, Rycroft MJ, editors 1997 Integration of geographic information systems and remote sensing. *Cambridge University Press*

[9] Cihlar J 2000 Land cover mapping of large areas from satellites: status and research priorities. *International journal of remote sensing*. **21**(6-7):1093-114.
[10] Immerzeel WW, Droogers P, De Jong SM, Bierkens MF 2009 Large-scale monitoring of snow cover and runoff simulation in Himalayan river basins using remote sensing. *Remote sensing of Environment*.**113**(1):40-9
[11] Olukanni DO, Alatise MO 2008 Rainfall-Runoff Relationships and flow forecasting, Ogun River Nigeria. *Journal of Environmental Hydrology*.**16**
[12] Meiyappan P, Dalton M, O'Neill BC, Jain AK 2014 Spatial modeling of agricultural land use change at global scale, *Ecological Modelling*, **291**:152-74.
[13] Lambin EF 1997 Modelling and monitoring land-cover change processes in tropical regions, *Progress in physical geography*. **21**(3):375-93.
[14] Ojo SA, Olusina JO, Ngene BU, Busari AA, Adediran J, Eletu A 2019 Assessment of road infrastructure using remote sensing and GIS methodology for monitoring the condition of paved and unpaved roads. *Materials Science and Engineering* **640** (1) 102099
[15] Sobrino JA, Raissouni N 2000 Toward remote sensing methods for land cover dynamic monitoring: Application to Morocco. *International journal of remote sensing*, **21**(2):353-66.
[16] Ogunyemi SA, Muibi KH, Eguaroje OE, Fabiyi OO, Halilu AS 2014 A geospatial approach to evaluation of accessibility to secondary educational institution in Ogun State, Nigeria, *Earth and Environmental Science*, **20**(1), 012045
[17] Adeoye NO, Abegunde AA, Adeyinka S 2012 Geospatial analysis of deforestation and land use dynamics in a region of Southwestern Nigeria. *In Deforestation Around the World 2012 Mar 30. Paulo Moutinho, Intech Open,* DOI: 10.5772/35117
[18] Al-Husban Y 2017 Inverse Distance Weighting (IDW) for Estimating Spatial Variation of Monthly and Annually Rainfall in Azraq Basin during the monitor Period (1980-2016), *Journal A l-Hussein ibn Tallal*. **3**(2).
[19] Wahab MA. Interpolation and Extrapolation" A Seminar presentation to *the Department of Electrical Systems Engineering at the University of Paderborn*.
[20] Raj EE, Ramesh KV, Rajkumar R 2019 Modelling the impact of agrometeorological variables on regional tea yield variability in South Indian tea-growing regions: 1981-2015. *Cogent Food & Agriculture*. **5**(1):1581457.

Geographic Information System application in flood risks prevention, hazards reduction and planning

G A Adeyemi[1]*, F Siyanbola[2], B I Oniemayin[1] G O Ode[2] and G O Bamigboye[1]
[1]Department of Civil Engineering, Covenant University, Ota, Nigeria
[2]Department of Surveying & Geoinformatics, Bells University, Ota, Nigeria
*gideon.adewale@cu.edu.ng

Abstract. The effects of Flood hazard mapping and flood inundation modelling in saving lives and properties in our societies cannot be overemphasized. Haphazard land use planning and lack of compliance to rules and regulations concerning setbacks have been observed to worsen the impacts of flooding and equally pose as major contributing factors to the exposure of people to flood risks in the study area. The study aimed at determining the built areas within the floodable area. Integrated Geographic Information System (GIS) and Remote Sensing were utilized to analyze the topographic data and the land use data generated from the satellite imagery of the area. The study revealed that nearly half of the buildings and infrastructures in the study area lie within the flood plain. Applications of GIS and Satellite Imagery analysis techniques were used to estimate the population at risk of flooding in the study area. We submit that space based and Geospatial data analysis and applications in flood risks assessment and planning prior to infrastructural development would go a long way in the preservation of lives and properties and prevent great economic loss as a result of flooding.

Keywords: Geographic Information System (GIS), flooding, risk prevention, planning, and sustainable infrastructure.

1. Introduction

The negative effects of flooding on lives and properties cannot be over emphasized and can only be imagined than experienced, these include loss of lives and properties, inconveniences of relocation and recoveries [1]. The increase in global temperature that has led to heavy rainfall, polar ice melting, rise in sea level and other environmental challenges call for concern. Flooding has been one of the major problems causing havocs in many parts of the globe. These havocs need to be tackled with effective planning, and respect for setbacks. Geographic Information Systems can provide a platform for planning and flood risks prevention mechanism. The efficacy of Information Technology in Risks Assessment and analysis is of great importance to enhance qualitative decision making in order to save lives and properties.

IOP Conf. Series: Materials Science and Engineering 1036 (2021) 012004 doi:10.1088/1757-899X/1036/1/012004

Worldwide, flooding poses a significant danger to lives and property. It is also responsible for about a third of all fatalities, injuries and damages from natural disasters [2]. Flood hazard mapping and flood inundation modelling are the crucial parts in flood mitigation measures and land use planning. Progresses in geospatial technologies (GPS, Remote Sensing and GIS) has empowered the procurement of data and analysis of the river basin for flood mitigation ventures for the safety of structural or non-structural features in a speedier and more precise ways. GIS encourages incorporation of spatial and non-spatial data for adequate decision making in a problem solving environment. Flood maps assist in critical decision-making in flood mitigation, land use planning, emergency management and general public awareness mechanism. GIS can combine and analyze data like precipitation, land use, elevation, social and economic information. These could be inventoried for some time later. Flood maps are helpful in flood damage appraisal, future flood mitigation planning and flood control. Causes of flooding: [1] opined that causes of flood in Nigeria could be either natural or man-induced. Natural causes can arise from heavy or torrential rains, rainstorm, ocean storms and tidal waves usually along the coast. [2] observed that man-induced causes include burst of main water pipes, dam burst, levee failures and dam spills. Flooding occurs throughout Nigeria in the following forms; coastal flooding, river flooding, flash floods, urban flooding, Dam bursts and spills and levee failures. The built environment is quickly changing. Fresh advances might influence vulnerability, and in fact hazards, significantly. Therefore, the map apprising should progress at similar pace as hazard map or any significant alterations in the flood plain. The United Nations Sustainable Development Goal 11 laid emphasis on making human settlements inclusive, save, resilient and sustainable. Applications of Geographic Information System and remote sensing in planning would support in the achievement of this goal.

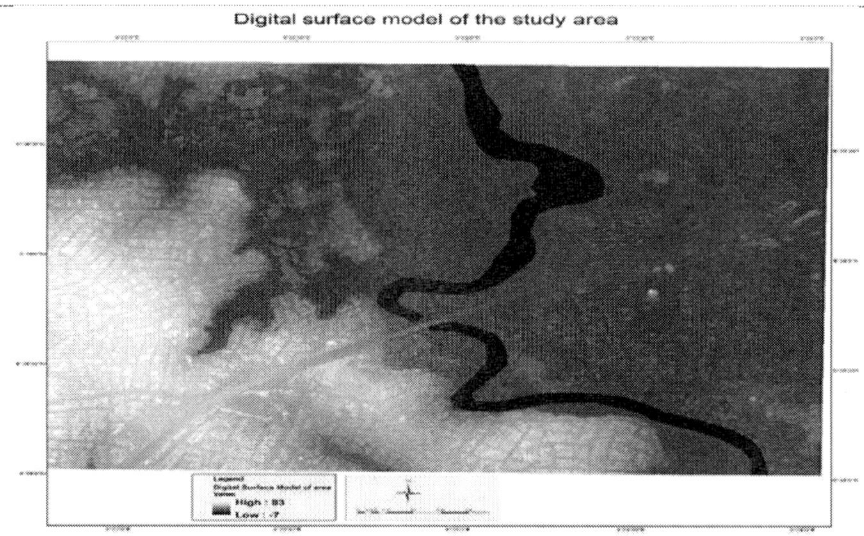

Figure 1. Digital Surface Model of the study area

2. Methodology

The satellite image and the elevation data of the study area were obtained. The hill shade map of the area was produced from the elevation data using ArcGIS 10.5. Land use map of the area was produced by feature extraction and digitization of the satellite imagery. Five classifications were made from the base map which include; built-up areas, roads, water body, grassland, and vegetation. The land use map was overlaid on the hill shade map to determine the land use features in the floodable zones of the hill shade and the areas in the high elevation zones. Geo-processing analysis of intersection was carried out to intersect the floodable layer and the residential area so as to determine the residential areas that falls within the floodable zones.

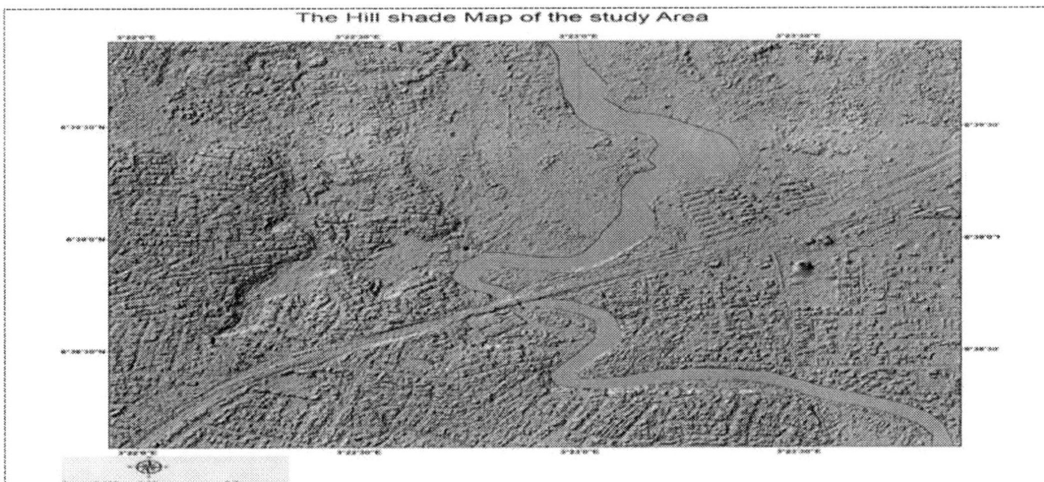

Figure 2. Hill shade map of the study area

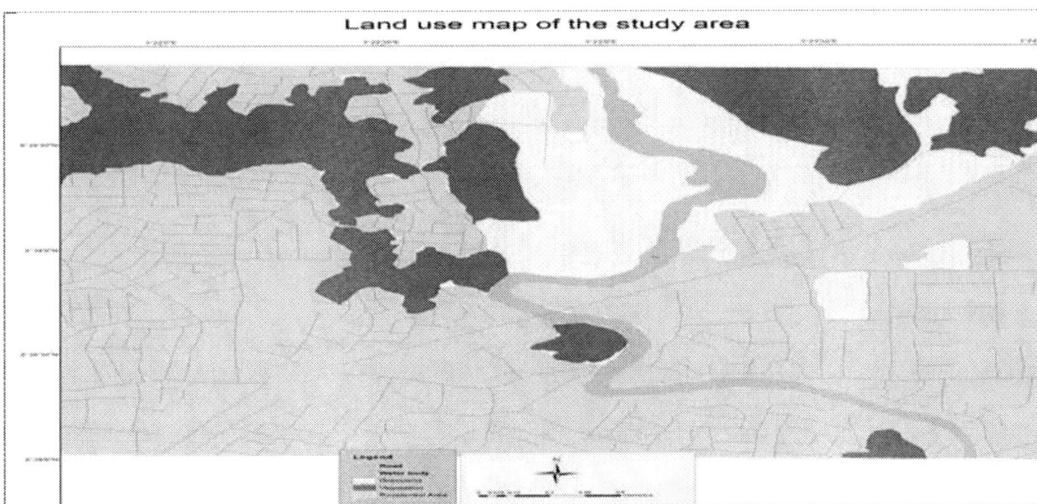

Figure 3. Land use map of the study area

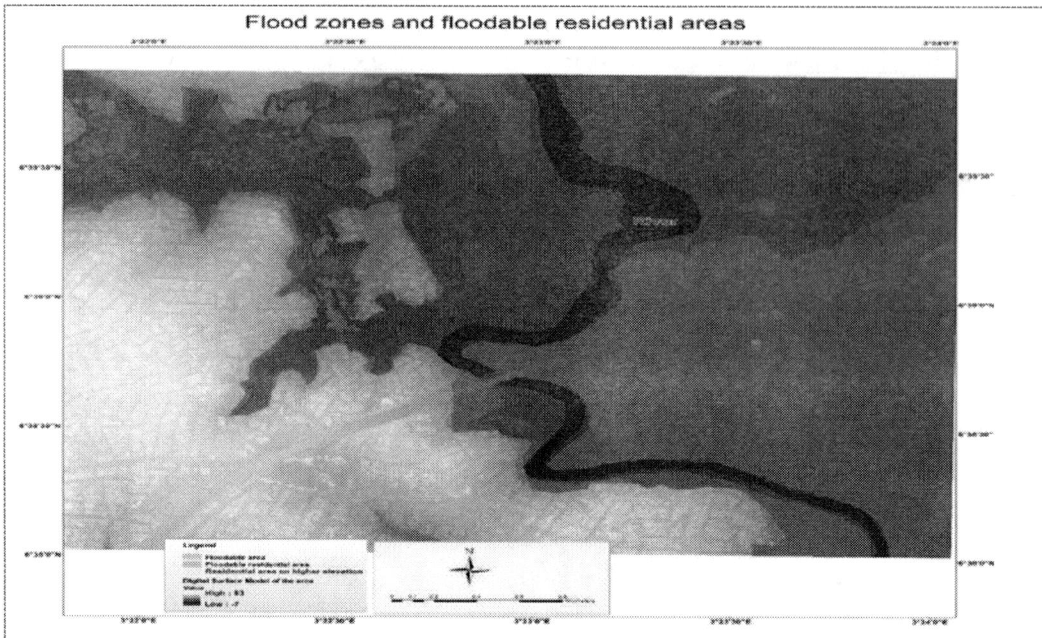

Figure 4. Vulnerable residential areas.

3. Results

It was observed that nearly half of the floodable zones in the area are residential, while more than half of the study area falls within the low elevation area. Roads and vegetation including farmland in the area are vulnerable to the effects of flooding. The residential locations that fall within the floodable zone may be submerged in case of heavy rainfalls due to the river overflowing its boundaries and erosion from the high elevation areas. Figure 1 is the satellite imagery of the study area. Different land use could be inferred from the imagery. Figure 2 shows the digital terrain model of the study area. The brighter areas are the regions with high elevation while the darker areas have lower elevations. It can be observed that there are patches of built-up areas in the region with low elevation. Figure 3 is the hill shade map of the study area. 3D view of the areas can be observed; the extent of depression and elevations can be clearly distinguished. Visual deduction of floodable areas can also be made. The extraction of the floodable areas from the hill shade map was done and shown. Figure 4 is the land use classification map of the study area, overlaying the land-use map and the floodable region.

4. Discussion

Flood map has the capacity of empowering residents and property owners with useful information that allows them to make informed decisions about the imminent flood risks and this would guide their levels of emergencies preparedness. [3] opined that recent cases of major flooding across the globe has further spiked the calls for urgent and thorough flood risk assessments in both rural and urban settings: Pakistan had a major flooding incident in 2010; Australia experienced one between 2010 and 2011, while China, Thailand, Laos, Nigeria and United Kingdom all endured devastating flooding incidents between 2011 and 2012. The need for attitudinal change and adequate land use planning is of great importance in flood control and management especially in the developing world. In some instances, drainages have been observed to be blocked by solid wastes thereby hindering the free flow of water.[4]opined that increasing population growth, lack of suitable waste treatment and disposal sites and inadequate town planning have been observed as palpable causes of poor waste management in our societies. [5] The landfill citing process requires the evaluation of several criteria such as land use, vegetation classification, road networks, hydrological and elevation data in addition to other socio-economic data. [6] observed that "flooding is the temporary inundation by water of normally dry land, and it can occur in coastal and lake areas, along rivers, from stream blockages and failure of engineering works including dams". [3] opined that Myriads of approaches and methods exist to determine flood hydrographs and inundation areas. Their selection hinges on the intended accuracy of the maps and the accessible data and resources.

5. Data availability

Flood mapping majorly depends on data availability. The need for accurate and reliable data is of high importance in flood modelling and predictions. Various current flood hazard maps require revision since they are outdated [7][8][9]. The concerted and coordination efforts of UNOOSA, UNSPIDER, USGS and other geospatial data providers have made it possible to obtain accurate and usable digital data for hazards control and mapping. This has helped to ameliorate the problems of certain geospatial data availability.

6. Conclusion

Efficacies of Geographic Information Systems in planning and hazard reduction should constantly be employed in order to save lives and property. Detailed environmental analysis of built environment, should be done before building approval is given by the authorities and agencies concerned. Similarly, in the larger scheme of global environmental management, GIS can be a viable tool in pattern delineation and assessment.

Acknowledgements

The authors wish to appreciate the Covenant University Centre for Research, Innovation and Discovery for the publication supports.

References

[1] Adeyemi GA, Onimowo D, Oniemayin BI and Eletu AA 2019 The place of geoinformatics in waste management, sanitation and health risks prevention in Ado-Odo/Ota local government area. *IOP Conference Series: Materials Science and Engineering*.

[2] Askew AJ 1999 Water in the international decade for natural disaster reduction. In Leavesley *et al.*(eds) Destructive Water: water-caused natural disasters, their abatement and control. *IAHS*. **239**.

[3] Angela KE 2011 The devastating effect of flooding in Nigeria: bridging the gap between cultures, FIG working week, Marrakech, Morocco.

[4] Douben JK 2006 Characteristics of river floods and flooding: a global overview, 1985-2003.*Irrigation and Drainage*. **55**(1): 9-21

[5] APFM 2013. The associated programme on flood management (APFM). Available at: www.apfm.info

[6] Adeyemi GA, Markus M, Gbolahan OG and Edeki SO 2018. Landfill site selection in Ado-Odo Ota L.G.A. using geographic information systems. *International journal of civil engineering and technology* **9** (8): 1434-45

[7] FFMGS 2018. Federal flood mapping framework Version 2.0. Available at: https://www.publicsafety.gc.ca/cnt/mrgnc-mngmnt/dsstr-prvntn-mtgtn/ndmp/fldpln

[8] Vahdettin D and Ozgur K 2016 Flood hazard mapping by using geographic information system and hydraulic model: Mert river, Samsun, Turkey. *Advances in Meteorology*. **2677**: 1- 9. http://dx.doi.org/10.1155/2016/4891015

[9] David OO, Favour P and Sunday OJ 2020. Public Perception of Solid Waste Management Practices in Nigeria: Ogun State Experience, *Recycling* **5**(2):8

A study on physicochemical parameters of fish pond effluents: A case study of Umudibia fish farm

H O Ibearugbulam[1], E I Ugwu[2]*, A C Ekeleme[3], C E Njoku[4], E C Amanamba[3], V Ezebuiro[5], O P Ibe[1] and EW Igwegbe[1]

[1] Department of Civil Engineering, Federal, Polytechnics Nekede, P.M.B 1036 Imo State, Nigeria
[2] Department of Civil Engineering, Michael Okpara University of Agriculture Umudike, Abia State Nigeria
[3]Department of Civil Engineering, Abia State University Uturu, P.M.B 2000, Abia State, Nigeria
[4]Department of Metallurgical and Materials Engineering, Federal University of Technology, Owerri, Imo State Nigeria
[5] World Bank African Centre of Excellence in Oil Field Chemicals Research, University of Port Harcourt, Nigeria.
*emmanuelugwu@mouau.edu.ng

Abstract. This study investigated the physicochemical characteristics of fish pond effluents. Four (4) effluent samples were collected at two-day intervals from a fish pond located at Umudiba Nekede, Owerri, and analyzed to ascertain their characteristics. The physiochemical parameters assessed are the temperature, biochemical oxygen demand, chemical oxygen demand, total dissolved solids, turbidity, ammonia, total hardness, alkalinity, pH, and electrical conductivity. The result revealed the following ranges for the physicochemical characteristics: temperature (23.0 to 25.9°C), pH (6.24 to 3.1), total alkalinity (43.1 to 50.4mg/l), total dissolved solid (27.9 to 95.2 mg/l), total hardness (19.7 to 21.5mg/l), turbidity (12 to 170 NTU), and electrical conductivity (137.6 to 144.3 μmhos/cm). The result of the study indicated that the effluents from the fish pond could constitute a threat to the ecology of the aquatic water bodies if not properly treated before discharge.

Keywords: Physicochemical characteristics; fish pond effluents; fish farm; Umudibia

1. Introduction

The most abundant compound on earth is water, and it covers approximately seventyfive percent of the earth's surface. Most of the water on the earth crust is saline in nature, with just a little amount existing as freshwater. Overexploitation and pollution have made freshwater a scarce commodity [1]. Water is becoming more unfit every day to mankind due to unwise use, neglect, and mismanagement. The entry of toxic substances into water bodies results in dissolution, suspension, and deposition on the bed of the water body, resulting in water pollution. This leads to the deterioration of water quality and negative effects on the aquatic ecosystem. When the wastewater in aquaculture is discharged, the pollutants seep down and affect the groundwater sources. If the concentration of nutrients, organic matter, and suspended solids in fish ponds rise, it results in a direct increase in turbidity, oxygen demand, and eutrophication in receiving water. Because of these problems, it becomes imperative to monitor water bodies' physicochemical properties to ensure that the quality is good for whatever purpose it is meant for. Some of the physicochemical properties regularly monitored for water quality include pH, dissolved oxygen, biochemical oxygen demand, temperature, total hardness, and chemical oxygen demand. However, the characteristics of the water bodies are influenced by seasonal variations [2].

Aquaculture consumes a large amount of water and this large amount of water is usually required in high quality. According to Boyd and Gross [3], similar principles are applied in the execution of the different forms of aquaculture. Water chemistry and the net results are usually determined by these principles [4]. Studies have shown that when water quality and quantity are both insufficient, profitable fish farming may be seriously threatened. Physicochemical and biological features of water used in fish ponds greatly define successful fishing farming and the survival of other aquatic biotas that form important food chain components [5]. Thus, maintaining good water quality is critical for optimum growth and sustenance of the aquatic organisms. Furthermore, information on the physicochemical qualities of raw water is essential to determining its appropriateness for use. Physicochemical parameters may also give insight into the various metabolic occurrences in the aquatic ecosystem. This research is therefore aimed at determining the physicochemical characteristics of wastewater from a fish pond located at UmudibiaNekede, Owerri, Imo State Nigeria as well as their effect on water quality in Materials and Methods

1.1. Study area
Umudibia Nekede is situated in Owerri West Imo State, South East, Nigeria. There are three Local Government Areas in Owerri; the Imo state capital; Owerri Municipal, Owerri West and Owerri North. Umudiba Nekede is bounded by latitude 5.4266° N and longitude 7.0176° E. The annual rainfall is between 2250 – 2500 mm. The study area has an average temperature of 25-27°C and a relative humidity of 80%, and it is located within the vicinity of Otamiri River which links Nworie River Owerri [6]. Figure 1 shows the map of the study area.

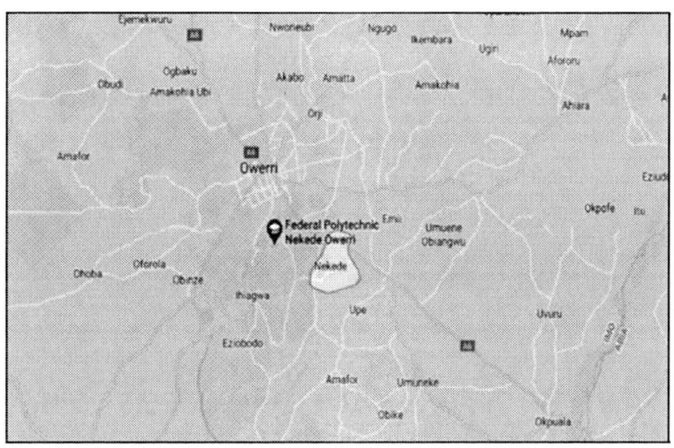

Figure 1: Map of the study area.

Legend: The black spot indicates the sampling area

1.2. Sample collection and analysis
Wastewater samples were collected from Umudibia Nekede fish farm. Plastic and glass screw-capped bottles were deployed in the sample collection for physicochemical factors. The bottles were washed with non-ionic detergents and rinsed with tap water as a pre-cleaning measure. Thereafter, the bottles were rinsed three times, first with distilled water and then with the sample before being filled with the sample. The sample containers were labelled and transported to the laboratory immediately in an ice pack. The temperature of the water was measured using a thermometer and electrical conductivity with the conductivity meter. Sample pH was measured using a pH meter (Q150 pH meter); measurement of total

hardness was performed using the ASTM-2340C method (EDTA titration method). Total dissolved solids (TDS) was determined using ASTM-D5907 method; ASTM D6238-98 determined dissolved oxygen (DO) and Biochemical oxygen demand (BOD), turbidity was measured using a hatch turbidity meter, ammonia by ASTM D1426-08 method, and total alkalinity was determined using the method consisting of methyl orange, Erlenmeyer flask and sulphuric acid.

2. Results and discussion

A total of ten (10) different physicochemical parameters were analyzed. The results obtained from the study are presented in Table 1 and Figure 2.

Table 1: Comparison of Results with other Standards Values

Parameters	Sample 1	Sample 2	Sample 3	Sample 4	WHO	FEPA	Acceptable range BD
Temperature (°C)	24.9±1.5^{a}	25.30±2.1^{b}	23.00±1.03^{c}	25.10±0.8^{d}	<35	27	-
pH	6.24±0.2^{a}	6.28±0.19^{b}	6.58±0.19^{c}	6.68±0.3^{d}	6.5-8.5	6-9	7-9.5
Ammonia (mg/l)	0.90±0.05^{abcd}	0.10±0.0^{ab}	0.13±0.0^{ac}	0.15±0.0^{ad}	50	20	0-1
Dissolved Oxygen (mg/l)	13.30±0.8^{acd}	11.90±0.9^{b}	10.30±0.4^{ac}	9.30±0.24^{abd}	6	8-10	3-5
BOD (mg/l)	2.90±0.02^{ad}	3.00±0.02^{bd}	3.18±0.18^{c}	3.53±0.1^{abd}	6	10	3-6
Total Alkalinity (mg/l)	43.10±2.09^{a}	45.20±1.9^{b}	47.60±2.09^{c}	50.40±2.2^{d}	600	-	50-200
Turbidity (NTU)	125.00±6.7^{ad}	135.00±1.6^{bd}	145.0±7.1^{c}	170.0±8.1^{abd}	10	<	-
TDS (mg/l)	27.90±1.4^{abcd}	49.80±0.21^{abd}	55.20±2.4^{acd}	95.20±3.8^{abcd}	500	500	-
Total hardness (mg/l)	19.70±1.1^{a}	20.10±0.8^{b}	20.70±1.1^{a}	21.50±0.95^{a}	600	-	>20
Electrical conductivity (μmhos/cm)	137.6±7.04^{a}	128.50±6.4^{b}	117.30±5.1^{c}	144.30±6.9^{d}	-	200	20-1500

The values represent mean and standard error of triplicate measurements. a: sample 1; b: sample 2; c: sample 3; d: sample. Note: For a given row, all values containing similar alphabets are statistically significant with the given sample. For example, all values containing "a" are statistically significant with sample 1; all values containing "b" are statistically significant with sample 2; all values containing "c" are statistically significant with sample 3 and all values containing "d" are statistically significant with sample 4. WHO-World Health Organization [7], FEPA-Federal Environmental Protection Agency [8], BD – Bhatnagar and Devi [9].

Temperature

Temperature is a key physiological controller and primarily influences the digestive system of aquatic organisms [10]. A change in temperature affects the physiology and metabolism and productivity of fishes. An optimum temperature range of 31°C to 36°C is required for the growth and sustenance of fishes [11]. In this study, the temperature range was between 23°C to 25.3°C (Table 1), which implies that the temperature of the water in the pond is not favourable for the fishes.

pH

This is a measure of hydrogen ion concentration. In this study, the pH values ranged from 6.24 to 6.68 (Table 1), which is a favourable condition for fish production. The optimal water pH range for fish farming is usually 6.5 to 9. Typically, pH ranges 4 to 6 or 9 to 10 reduce the growth rate of the fishes, although the majority of the fishes will still be alive. The pH of less than six results in a decrease in regeneration. The mortality of fish starts at pH less than 4 (acidic) or more than 11(fundamental) in the fish farm [12]. The pH value also agrees with the results of Bisht et al. [13] and falls within the standard limit of WHO and FEPA. Thus, it is favourable for fish life.

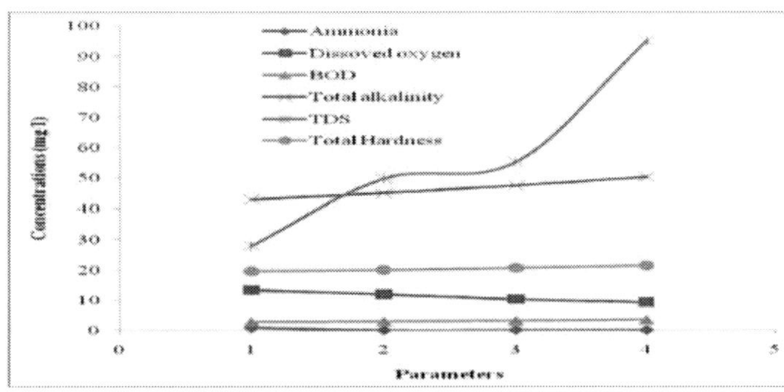

Figure 2: Result of selected physico-chemical parameters

Ammonia

Ammonia enters the ponds through uneaten feeds, phytoplankton, decaying, and dead organics. It is a bye-product from the fish excreta produced during the organic matter degradation [9]. The values recorded ranged between 0.9 and 0.15 mg/l, thus are beyond the acceptable range (Table 1 and Figure 2). Values of ammonia greater than 0.1 mg/L could cause damages to the gills of fishes as well as other adverse effects such as reduced immune system; poor bolster change, decreased growth and crush mucous creating membrane [9].

Turbidity

Turbidity in water refers to the intensity of the cloudiness of the sample due to the suspended particles [14]. In this study, the turbidity ranged from 125 to 170 NTU. Bhavimani and Puttaiah [15] reported a turbidity range of 20 to 72 NTU. The value of the turbidity obtained in this study (Table 1) is higher than both the FEPA [7] and WHO [8] stipulated limits. Excess turbidity leads in the reduction of sunlight absorbed in the pond which causes death of fishes [16]. This high turbidity could have arisen from the poor facility management, which allowed the influx of unwanted debris into the fish pond. Another reason may be due to the infrequent replacement of the source of water for the fish which causes the accumulation of wastes from the fishes. Turbidity arising from effluents when discharged into water bodies may hinder photosynthesis of organisms that usually participate in the phytoremediation of contaminated water [17]. However, according to WHO [18] it may form a shield against the hazardous effect of disinfectants on microorganisms.

Total Dissolved Solids (TDS)

The TDS is a measure of the dissolved organics and inorganics in the given water sample. The values of the TDS obtained in the current study are within the range of 27.9 to 95.2 mg/l (Table 1 and Figure 2).

Thus, the values were within the WHO and Federal Environmental pollution authority values, which shows a propensity to increase as the harvesting of fishes progresses [19].

Dissolved Oxygen (DO)
DO is a measure of the amount of dissolved oxygen in the aquatic ecosystem. The sustenance, development, mental soundness, behaviour and efficiency of living organisms are influenced by a change in dissolved oxygen [20], and it is used up usually during microbial decomposition of organic matter. In this study, the results of DO obtained ranged from 9.3 to 13.3 mg/l (Table 1). These values exceeded the limits set by WHO and FEPA. The higher values of DO obtained is an indication that the wastewater samples used in this study were polluted.

Biochemical Oxygen Demand (BOD)
According to Bhatnagar and Devi [9], BOD is the volume of oxygen microorganisms take up for decomposition of organics in water. Usually, the BOD value above 5mg/l is an indication of pollution of the water sample [21]. In this study, the value of BOD ranged from 2.9 to 3.53 mg/l (Table 1 and Figure 2). Previous studies reported that the levels of BOD range from 3.0 to 6.0 mg/l enhance fish activities [22]. The values obtained from the study were within the range of WHO and FEPA limits.

Alkalinity
This shows the buffering capacity of water and it has a strong impact on aquatic life. According to Bhatnagar and Devi [9] alkalinity equilibrates the changes in pH occurring naturally due to the photosynthetic activity exhibited by the phytoplankton. Fishes produce optimally at alkalinity between 50 and 200 mg/l. The alkalinity of unpolluted pond ranges from 171.2 to 235.5mg/l [23]. Therefore, the result obtained in this study which had alkalinity range from 43.10 to 50.40 mg/l is suitable for fish farming in the pond (Table 1).
Total Hardness
This is the sum of calcium and magnesium concentration in aquatic bodies usually present in the form of a combination of carbonates and bicarbonates which result in temporary hardness. Magnessium and calcium are essential nutrients needed for the formation of scale and bones in fishes [9]. Stoskopf [24] defined the total hardness as the minimum amount of hardness that is proper for fish farming is 100 mg/l [25]. The results of the total hardness obtained in the current study ranges from 19.7 and 22.5 mg/l which falls within the specified WHO and FEPA limits (Table 1 and Figure 2).

Electrical Conductivity (EC)
Electrical conductivity is a measure of the capacity of water to allow the passage of electricity through it [26]. The results of the electrical conductivity obtained in this study ranged from 117.3 to 144.3μmhos/cm (Table 1). Zimmermann [27] reported that the electrical conductivity values recommended for fish farming range from 20 to 150 μS/cm. They stressed that EC recognizes polluting sources and linearly increase with increasing concentration of salts within an aquatic setting.

3 Conclusions
From the results, the wastewater from Umudibia fish farm was polluted and could harm water bodies if not properly treated before discharge. Therefore, the fish pond needs to be checked at regular intervals for quality assurance procedure to guarantee there are no harmful substances in the ponds prompting a conceivable bio-aggregation. This will guarantee the great wellbeing of the aquatic ecosystem, man, and the environment. It is recommended that farmers should be acquainted with the standard practices with respect to waste minimization, pond management, feeding and the environmental protection.

References

[1] Basavaraja SM Simpi KNS Hiremath KN Murthy AN and Chandrashekarappa ETP 2011 *G. J. Sc. Front. Res.* 1 **3** 31- 34
[2] Shinde SE Pathan TS Raut KS Sonawane DL 2011 Middle-East *J. Sc. Res.* **8** 3 544-554
[3] Boyd CE and Gross A 2000 *Fisheries Manag. Ecol.* 7 **1** 2 55-63
[4] Boyd CE 1990 *Agriculture Experiment Station (*Auburn University, Alabama press) p 252
[5] Gupta S and P. Gupta P 2006 *General and Applied Technology (Fish and Fisheries)* (S. Chand and Company, New Delhi) p 1130
[6] Njoku–Tony RF Ebe TE Ihejirika CE and Ejiogu CC and Uyo N 2016 *IOSR J. Environ. Sc. Toxicol. Food Technol.* ***10*** *6 pp 67-75*
[7] World Health Organization (WHO) 2009 *Guidelines for Drinking Water Quality* (World Health Organization, Geneva)
[8] Federal Environmental Protection Agency (FEPA) 1991 *Guidelines and Standard for Environmental Pollution Control (*Federal Environmental Protection Agency) 27: 20.
[9] Bhatnagar A and Devi P 2013 *Int. J. Environ. Sci.* **3** 6 1980-1993
[10] Zutshi B Singh A and Dasgupta P 2020 *J. Basic Applied Zool.* **81** 4
[11] Food and Agriculture Organization (FAO) 2011 *Fisheries and Aquaculture Department* (Available at http://www.fao.org/fishery/culturedspecies/Oreochromis_niloticus/en Accessed on 29.08.2017)
[12] Lawson TB 2001 *Fundamentals of Aquacultural Engineering* (Chapman and Hall, New York) p 364
[13] Bisht, AS Gulam A Rawat DS and Pandey NN 2013 J. Ecol. Natural Environ. **5** 12 387 – 95
[14] Ugwu EI Uzoma NE Ikechukwu EL 2016 *South Asian J. Eng. Technol.* **2** 25 1–8
[15] Bhavimani H and Puttaiah ET 2014 *Fish Culture and Physico-chemical Characteristics of Madikoppa Pond, Dharwad Tq/Dist, Karnatak* (Hydrology: Current Research)
[16] Srinivasa A V Sai A V Lakshmi D V Saipriya P 2018 *Int. J. Recent Technol. Eng.* **7** 4
[17] Elango G Rathika G and Elango S 2017 *Int. J. Res. Chem. Environ.* **7** 1 17–24
[18] World Health Organization (WHO) 2017 *Guidelines for drinking-water quality* (World Health Organization Geneva) 4th edition
[19] Coldebella A Gentelini A L Piana P A Coldebella P F Boscolo W R Feiden A 2017 *Sustainability* **10** 3
[20] Garcia M Trambulo E M Pajarillo J M Apsay R B Tenorio J E Chua M G 2013 IEEE 3rd *International Conference on System Engineering and Technology*, Shah Alam p 83-88
[21] Clerk R.B 1996 *Marine Pollution* (Clarandon Press Oxford) p 256
[22] Bhatnagar A Jana SN Garg SK Patra BC Singh G and Barman UK 2004 *Water Quality Management in Aquaculture., In: Course Manuel of Summer school on Development of Sustainable Aquaculture Technology in Fresh and Saline Waters* (CS Hargyana Agricultural, Hisar, India) p 203 – 10
[23] Shastress A 1991 *Characterization of the Effluent from Commercial Catfish Pond*
[24] Stoskopf M 1992 *Fish Medicine* (W.B. Saunders Company) p 333-337
[25] Wedemeyer GA 1996 *Physiology of Fish in Intensive Culture Systems (*Chapman and Hall, New York) p 300
[26] Dahiya S and A. Kaur A 1999 *J. Environ. Pollution* **6** 4 281
[27] Zimmermann S Moreira H Vargas L Ribeiro R 2001 *Fundamentos da Moderna Aquicultura; ULBRA* (Canoas, Brazil)

Health Risk Assessment of Groundwater in Omu-Aran, Nigeria

O O Elemile[1], E M Ibitogbe[1], J R Adewumi[1, 2], O P Folorunso[3] and C O Osueke[4]

[1]Department of Civil Engineering, College of Engineering, Landmark University, Omu-Aran, Nigeria
[2]Department of Civil Engineering, School of Engineering and Engineering Technology, Federal University of Technology, Akure, Nigeria
[3]Department of Civil Engineering, Faculty of Engineering, Ekiti State University, Ado-Ekiti Nigeria
[4]Department of Mechanical Engineering, College of Engineering, Landmark University, Omu-Aran, Nigeria

*Correspondence Author: ibitogbe.enoch@lmu.edu.ng

Abstract. Pollution of groundwater by bathroom, laundry and kitchen effluent has been documented to have non-desirable health and environmental effects. There is little information about the quality of groundwater and associated health effects in Omu-Aran. The present research therefore meets the need to investigate the levels of heavy metals, the pollution effects, and the long term human health risk effect on local inhabitants in Omu-Aran. Thirty water samples were collected from five hand dug wells (HWs) in different parts of Omu-Aran spanning a period of three months and the heavy metal parameters were determined using standard protocol. The non-carcinogenic risk was evaluated using hazard index (HI) according to United States Environmental Protection Agency (USEPA) risk assessment method. Data analysis were carried out by the use of descriptive statistics. From the result, mean values and SD of heavy metals in the water samples obtained from selected HWs ranged from 0.16 ± 0.13 to 0.29 ± 0.18 mg/L, 0.29 ± 0.02 to 1.36 ± 0.07 mg/L, 0.00 ± 0.00 to 0.72 ± 0.04 mg/L, 0.00 ± 0.00 to 0.51 ± 0.09 mg/L, 0.01 ± 0.01 to 5.26 ± 0.07 mg/L, and from 0.03 ± 0.01 to 12.96 ± 2.82 mg/L for Al, Fe, Mn, Cr, Cu, and Zn respectively. The heavy metals were all above the WHO limits except Pb and Fe. The hazard index values ranged from 1.406 to 10.075 indicating significant non-carcinogenic risk. The findings of the study, has shown that there is considerable amount of heavy metal pollution which stems from anthropogenic sources. There was no observable risk via dermal pathways based on hazard quotient values while non-carcinogenic risk via ingestion was observed for all the locations. There is therefore a need to bring awareness to the community for the purpose of ensuring safety of its active water sources.

1. Introduction

Heavy metals generally exist as trace substances usually in quantities that do not pose any health risk. There is, however, a high risk that toxic and recurrent waste will be disposed of in the surroundings due to anthropogenic activities in improperly protected landfills, atmospheric deposition, grey water and kitchen effluents [10, 14]. The pristine nature and the availability of groundwater for different purposes makes it a very reliable source in comparison to water obtained from the surface [5]. Although natural features and anthropogenic activities play a role in the quality of groundwater [1], scientists have found that heavy metal pollution in the environment is significantly caused by the latter [9]. In addition, this progressive decline in groundwater quality and purity can be attributable to human activities, particularly in view of the rapid growth, farm activities, urbanization and industrial development [18]. Heavy metals are a known set of pollutants impacting the quality of groundwater, which presents a safety threat to living organisms. [1] suggest that their being highly persistent is because they bio-accumulate in the environment. Toxic heavy metals such as lead, arsenic and mercury retain harmful effects on plants, animals and human health. In the same way, chromium can be ingested through food and water. High exposure can lead to chronic neurological effects of the gastrointestinal system [3, 17]. If the heavy metal levels surpass the prescribed limit, they may lead to chronic disorders, such as liver conditions, high blood pressure, kidney problems, skin irritation, etc. [19, 20]. This makes all the more important to identify and characterize the nature of the risk and the dangers it poses to human health [15]. Risk assessment refers to the systematic detection and characterisation of the toxicity of a chemical. This will help to generate a qualitative (descriptive) risk analysis focused on the negative effects reported (chronic and carcinogenic) and the numerical risk level (calculation of health risk). Over the years, Omu-Aran has developed into a peri-urban region witnessing incredible expansion of human, businesses and economic activities with a consequent increase in the production of wastewater. Unfortunately, this community lacks the potential for wastewater treatment and segregation before disposal thus, wastes from the residential areas, schools, hotels, hospitals, and abattoirs in the area are handled by the residents since they have no central systems. The result of these practices make it almost inevitable for uncontrolled levels and contents of heavy metals in the immediate vicinity. A study by [7] showed that soil was contaminated as a result of abattoir pollution. However, there has been no documented study in terms of establishing the heavy metal pollution status to the immediate environment of the residential areas. This research therefore looked at the levels of heavy metals, the effects of contamination and the long-term consequences of human health threats on the local inhabitants of Omu-Aran.

2. Study Area

Omu-Aran is positioned in the southern part of Kwara State as shown in Figure 1. It occupies a land area of 73.7 km^2 and located on latitude 8°08′00″N and longitude 5°06′00″E with an average elevation of 564m above sea level [6]. Having an estimated 150,000 persons, (Bureau office of Statistic, 2018), it is the third largest town in Kwara state. The vegetation is largely guinea savannah while the climate is tropical maritime monsoon [15]. The average annual rainfall is 1262.8 mm while the average annual maximum temperature 35.8 °C.

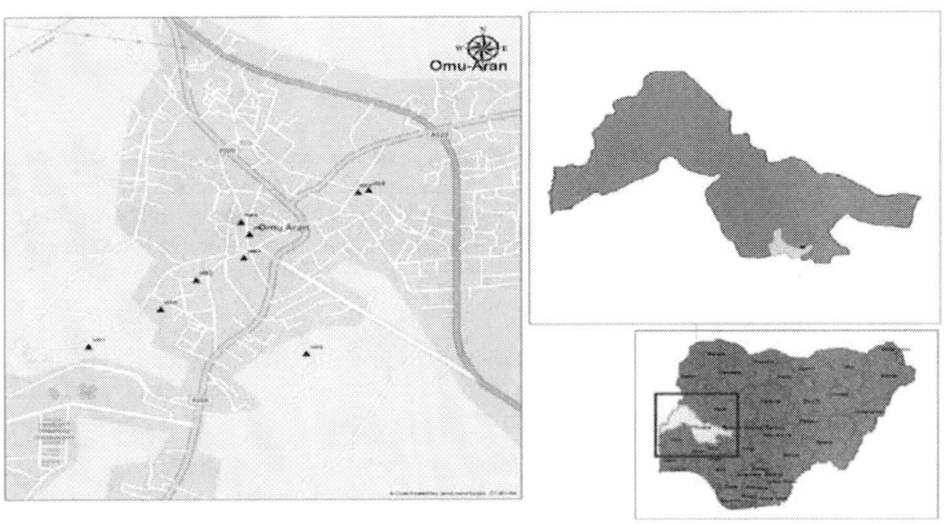

Figure 1: Location map showing collected samples.

3. Sampling Method and Analysis

Figure 2 shows one of the hand dug wells where groundwater samples were obtained between October and December in compliance with the protocol of [19] while Figure 3 shows an open dump which is one of the point source of pollution to the hand dug well at Latinwo area, Omu-Aran. Thirty (30) samples were collected bi-monthly across five locations from five (5) hand dug wells. The Hand dug wells are designated as follows HW_1 – Hand dug well at Aperan, HW_2 – Hand dug well at Egbe garage, HW_3 – Hand dug well at Latinwo, HW_4 – Hand dug well at Mode, HW_5 – Hand dug well at Igangu. Prewashed polyethylene bottles were used in collecting according to the method adopted by [18]. Sample preparation was done prior to analysis using the method prescribed by [2]. Heavy metal (Fe, Mn, Cu, Zn, Pb, Al, Cr) concentrations were then determined by using bulk scientific (AAS.AA 320 N).

Figure 2: Resident using dug well at Igangu, Omu-Aran well site

Figure 3: Point source pollution at

4. Risk Assessment

Risk assessment is a tool developed in order to measure and evaluate likely health risk associated with the exposure to certain elements such as heavy metals which could be injurious to the human health. Contamination of drinking water gives rise to many health challenges. Water-invasive toxic chemicals may have well-advanced health risks resulting its frequent exposure. This segment was used to evaluate the overall likely, non-carcinogenic effect of metals using the Hazard Index (HI) method. According to [12], risk assessment entails hazard tracing, exposure valuation, dose response and estimating the risk effect. Generally, exposure could happen through dermal pathways or by ingestion. In estimating the risk related with exposure to heavy metals, chronic daily intake (CDI) ($mgkg^{-1}day^{-1}$) via ingestion (CDI_{ingest}) and dermal contact (CDI_{dermal}) were evaluated by the equation as shown in equations 1 and 2 obtained from [19];

$$CDI_{ingest} = \frac{(Ci\ x\ IR\ x\ CF\ x\ FI\ x\ ED\ x\ EF)}{(BW\ x\ AT)} \quad (1)$$

$$CDI_{dermal} = \frac{(Ci\ x\ CF\ x\ SA\ x\ AF\ x\ ABS\ x\ ED\ x\ EF)}{(BW\ x\ AT)} \quad (2)$$

where Ci (mg / L) is level of chemical impurity in potable water; IR (L/day) is daily rate of ingestion of water; EF (days/year) is the frequency of exposure; ED (years) is the length of exposure; FI is the fraction ingested and is usually 1; CF (L/cm^3) is the converting factor; BW (kg) is the bodyweight; AT (day) is the average exposure period; SA (cm^2) refers to area of exposed skin surface; AF is the skin adherence factor.

IOP Conf. Series: Materials Science and Engineering 1036 (2021) 012006 doi:10.1088/1757-899X/1036/1/012006

$$\text{Hazard quotient (HQ)} = \frac{\text{CDI}}{RfD} \quad (3)$$

$$\text{Hazard Index (HI)} = \sum HQ = \text{HQ}_{\text{ingest}} + \text{HQ}_{\text{dermal}} \quad (4)$$

Hazard quotient, HQ is the ratio of chronic daily intake (CDI) to the corresponding reference dose (RfD). The RfD values for Cr, Mn, Cu, Zn, Pb and Fe are 0.003, 0.14, 0.0371, 0.30, 0.0035 and 0.70 $mg.kg^{-1}day^{-1}$. The hazard index (HI) is evaluated by the summation of HQs. HQ and HI are described in equations 3 and 4 respectively. HI value below 1 shows no significant non-carcinogenic risk while the likely non-carcinogenic risk would be above 1 [19]. The summary of assumed values is illustrated in Table 1.

5. Results and Discussion

5.1 Heavy metal evaluation

Water quality is an entire summary of its constituent parameters in terms of Physical, Microbial and Chemical (heavy metals). The factors that affect these parameters do not depend only on natural characteristics of reservoirs but also on human related discharges. The statistical summary and limit values of heavy metal parameters according to [19] for drinking waters is presented in Table 2. Concentrations for Al, Fe, Mn, Cr, Cu and Zn ranged from 0.16 ± 0.03 to 0.29 ± 0.08, 0.29 ± 0.02 to 1.36 ± 0.07, 0.00 ± 0.00 to 0.72 ± 0.04, 0.00 ± 0.00 to 0.51 ± 0.09, 0.01 ± 0.01 to 5.26 ± 0.07, 0.03 ± 0.01 to 12.96 ± 2.82 mg/L respectively while lead was not undetected as there was no anthropogenic activity causing lead contamination around the wells. This situation agrees with [7] who reported undetected Cu and Cd values for groundwater samples.

Table 1: Input parameters to calculate chronic daily intake.

Exposure parameters	Symbols	Unit	Value
Concentration	Ci		from analysis
Ingestion rate	IR	liters/day	2
Conversion factor	CF	liter/cm^3	1 x 10^{-3}
Fraction ingested	FI		1
Exposure frequency	EF	days/year	365
Exposure duration	ED	Years	70
Body weight	BW	Kg	70
Average time	AT	days/years	25550
Surface area	SA	cm^2	5700
Skin adherence factor	AF		0.07
Adsorption factor	ABS		0.001

The mean concentration of Al, Mn, Cr, Cu, Zn were above standard WHO limit of 0.1, 0.2, 0.05, 2.0 and 5.0 mg/L, respectively. The mean values of Pb and Fe did not exceed WHO permissible values of 0.01 mg/L and 3 mg/L respectively. The value of Al at all the HWs were above standard limit. Moreover, the concentrations in this study were much lower than 0.4310 mg/L reported by [15]. Aluminum is introduced into groundwater via interaction with rocks and leaching from soil [20]. The high values of Fe at HW_1 could be due to dissolution of iron minerals from water-rock interaction [15]. However, the values are higher than values (0.002 to 0.568 mg/L) from a study carried out previously [4]. The values of Mn fell within permissible limit set by WHO (0.2 mg/L) with the exception of HW_5. This could be due to close proximity to the road as well as poor maintenance practice as the well has no cover giving enough room for exposure. The observed values of Cu in this study were higher compared to 1.984 ± 0.066 mg/L while Cr was lower than reported values reported by [11]. Zn was observed to be higher than very high across all the sampled wells apart from HW_2. This suggest percolation of rainwater from the roof top of houses in the surrounding [6]. The reported values in this study exceed results reported by [13].

Table 2: Statistical summary of heavy metal concentration in groundwater samples collected from the study area.

Parameters (mg/L)	HW_1		HW_2		HW_3		HW_4		HW_5		Standard
	Mean	Std. dev	Mean	Std. dev	Mean	Std. dev	Mean	Std. dev	Mean	Std. dev	WHO
Pb	-	-	-	-	-	-	-	-	-	-	0.01
Al	0.22	0.01	0.29	0.08	0.16	0.03	0.23	0.02	0.18	0.04	0.10
Fe	1.36	0.07	0.44	0.02	0.48	0.05	0.54	0.07	0.29	0.02	3.00
Mn	0.00	0.00	0.11	0.08	0.00	0.00	0.00	0.00	0.72	0.04	0.20
Cr	0.00	0.00	0.02	0.01	0.31	0.18	0.02	0.02	0.51	0.09	0.05
Cu	0.12	0.01	0.01	0.01	0.13	0.03	0.15	0.02	5.26	0.07	2.00
Zn	12.96	2.82	0.03	0.01	12.03	1.56	12.12	2.40	10.03	1.40	5.00

5.2 Health Risk Assessment

Table 3 shows HQ and HI values for the oral and dermal pathways as it relates to the health of consumers. The HQ_{dermal} (hazard quotient by dermal pathway) of all trace elements at all the sampling locations were less than one except for Pb which was not detected for any of the samples. Cr was not detected for at HW_1 hence the nil value. Likewise, Mn was not detected at HW_3 and HW_4 locations. This indicates that these metals posed little danger by dermal absorption. However, in the case of $HQ_{ingestion}$ (hazard index by oral ingestion), Cr was above one at HW_3 and HW_5 whereas Cu is greater than one at HW_5 which suggest non-acceptable health risk at these locations. Zn poses health risk since it exceeds unity for all locations with the exception of HW_5. All other $HQ_{ingestion}$ values are < 1, which suggests no adverse non-carcinogenic health effect is posed even if the water is used for drinking. However, health effects could arise as a result of the synergistic effect of these trace metals. Conversely, the HI values are all above one with a range of 1.41 to 10.08 and are relatively higher as compared to values reported by [11]. The HI value at HW_5 was the highest, which denotes a relatively higher risk than other wells as shown in Figure 4.

6 Conclusion

This study was conducted to evaluate the associated health risk with exposure to heavy metal in the study area of Omu-Aran due to improper waste disposal methods. Risk assessment via dermal and ingestion pathways were investigated for water samples taken from majorly used HDWs by the residents. The average values of heavy metals (Al, Mn, Cr, Cu and Zn) across the wells exceed permissible values given by WHO except for Pb and Fe which was within limit. The order of toxicity of heavy metals based on mean concentrations found in drinking water in the region under review: Zn > Cu > Fe > Mn > Cr > Al > Pb. Zn had the highest concentration of 12.96 mg/L while Pb had the lowest value of 0.00 mg/L. This result is helpful in understanding the state of the drinking water source. It also provides relevant information such as is required by stakeholders to respond adequately to the current pollution in the area. It is therefore recommended that protective and preventive measures are to be carried out by residents to keep their waters close to pristine conditions.

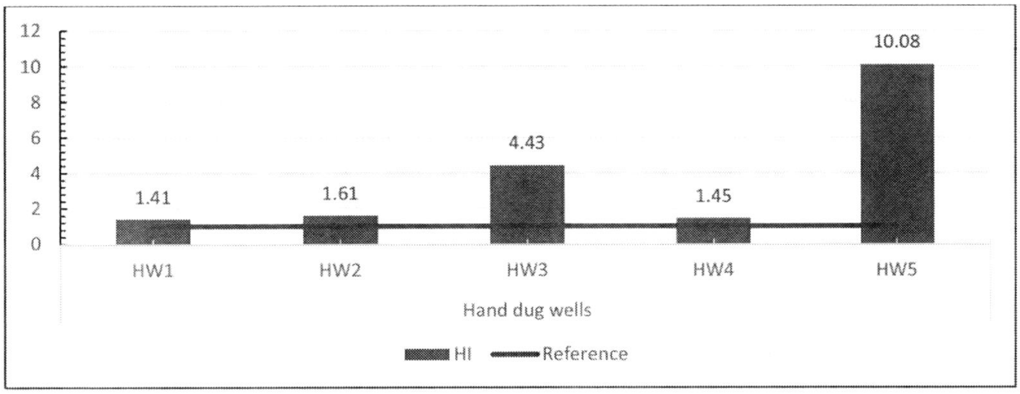

Figure 4: Summary of HI values for each hand dug well.

Table 3: HQ and HI values for oral and dermal pathways as it relates to health effect.

	Location									
	HW_1		**HW_2**		**HW_3**		**HW_4**		**HW_5**	
	HQ ingest	**HQ dermal**	**HQ ingest**	**HQ dermal**	**HQ ingest**	**HQ dermal**	**HQ ingest**	**HQ dermal**	**HQ ingest**	**HQ dermal**
Cr	0.00	0.00	2.18E-01	4.36E-05	3.35	6.69E-04	1.6E-01	3.2E-05	4.90	9.78E-05
Mn	4.07E-04	8.00E-08	2.31E-02	4.61E-06	0.00	0.00	0.00	0.00	1.47E-01	2.93E-05
Cu	8.76E-02	1.75E-05	8.76 E-02	1.75E-05	1E-01	1.99E-04	1.12E-01	2.23E-05	4.05	8.09E-04
Zn	1.25	2.51E-04	1.25	2.51E-04	9.55E-02	2.28E-04	1.146	2.30E-04	9.55E-01	1.90E-04
Pb	0.00	0.00	0.00	0.00	0.00	0.00	0.00	0.00	0.00	0.00
Fe	5.54E-02	1.11E-05	1.79 E-02	3.57E-06	1.96E-02	3.91E-06	2.21E-02	4.41E-06	1.18E-02	2.35E-06
Al	6.24E-03	1.25E-06	8.37 E-03	1.67E-06	4.59E-03	9.15E-07	6.68E-03	1.33E-06	5.14E-03	1.03E-06
HI=ΣHQs	**1.41**		**1.61**		**4.43**		**1.45**		**10.08**	

Reference

[1] Adewoyin O O, Kayode O T, Omeje O, Odetunmibi O A and Shukla S K 2019 Risk assessment of heavy metal and trace elements contamination in groundwater in some parts of Ogun State *Cogent Engineering* **6**(1) 1–12. https://doi.org/10.1080/23311916.2019.1632555

[2] American public association (APHA) 1992 Standard Methods for Examination of Water and waste 18th ed. American Public Health Association, Washing ton D.C

[3] ATSDRA Agency for Toxic Substances and Disease Registry 2007 Toxicological Profile for Lead ATSDR Atlanta Georgia USA

[4] Boateng T K, Opoku F, Acquaah S O and Akoto O 2015. Pollution evaluation, sources and risk assessment of heavy metals in hand - dug wells from Ejisu - Juaben Municipality Ghana. *Environmental Systems Research* https://doi.org/10.1186/s40068-015-0045-y

[5] Chabukdhara M, Gupta S K, Kotecha Y and Nema A K 2017 Groundwater quality in Ghaziabad district, Uttar Pradesh, India: Multivariate and health risk assessment. *Chemosphere* **179** 167–178. doi: 10.1016/j.chemosphere.2017.03.086

[6] Elemile O O, Raphael D O, Omole D O, Oloruntoba E O, Ajayi E O and Ohwavborua N A 2019 Assessment of the impact of abattoir effluent on the quality of groundwater in a residential area of Omu - Aran Nigeria *Environmental Sciences Europe* https://doi.org/10.1186/s12302-019-0201-5

[7] Elemile O O, Raphael O D, Omole D O, Oluwatuyi O E, Ajayi E O, Umukoro O and Elemile M G 2019 Assessment of the impact of abattoir activities on the physicochemical properties of soils within a residential area of Omu-Aran, Nigeria IOP Conference Series: M*aterials Science and Engineering* ***640***(1) https://doi.org/10.1088/1757-899X/640/1/012083

[8] Emenike P C, Tenebe I, Ogarekpe N, Omole D and Nnaji C 2019 Probabilistic risk assessment and spatial distribution of potentially toxic elements in groundwater sources in Southwestern Nigeria. *Scientific Reports* *9*(1) 1–15 https://doi.org/10.1038/s41598-019-52325-z

[9] Kasassi A, Rakimbei P, Karagiannidis A, Zabaniotou A, Tsiouvaras K, Nastis A and Tzafeiropoulou K 2008 Soil contamination by heavy metals: Measurements from a closed unlined landfill. *Bioresource Technology* **99**(18) 8578–8584. https://doi.org/10.1016/J.BIORTECH.2008.04.010

[10] Khan S, Cao Q, Zheng Y M, Huang Y Z and Zhu Y G 2008 Health risks of heavy metals in contaminated soils and food crops irrigated with wastewater in Beijing China. *Environmental Pollution* **152**, 686–692

[11] Kwame T, Francis B and Akoto O 2019 Heavy metal contamination assessment of groundwater quality : a case study of Oti landfill site, Kumasi *Applied Water Science* **9**(2) 1–15 https://doi.org/10.1007/s13201-019-0915-y

[12] Lee J S, Chon H T and Kim K W 2005 Human risk assessment of As, Cd, Cu and Zn in the abandoned metal mine site. *Environmental Geochemistry and Health* **27**(2) 185–191 https://doi.org/10.1007/s10653-005-0131-6

[13] Oyatayo K T, Songu G, Amos G and Ndabula C 2015. Assessment of Heavy Metal Concentration in Hand Dug Well Water from Selected Land Uses in Wukari Town , (November), 1–10.

[14] Ali S S, Anwar Z and Khattak J Z K 2012. Microbial analysis of drinking water and water distribution system in new urban Peshawar. *Current Research Journal of Biological Sciences* *4*(6) 731-737

[15] Şener Ş, Şener E and Davraz A 2017 Assessment of groundwater quality and health risk in drinking water basin using GIS. *Journal of water and health* **15**(1) 112-132

[16] Sojobi A O 2016 Evaluation of groundwater quality in a rural community in North Central of Nigeria. *Environmental monitoring and assessment* **188**(3) 192

[17] Samantara M K, Padhi R K, Sowmya M, Kumaran P and Satpathy K K 2017 Heavy metal contaminations major ion chemistry and appraisal of the groundwater status in coastal aquifer, Kalpakkam Tamil Nadu *India Groundwater for Sustainable Development* **5**(49–58), 49– 58 doi: 10.1016/j.gsd.2017.04.001

[18] Tirkey P, Bhattacharya T, Chakraborty S and Baraik S 2017 Assessment of groundwater quality and associated health risks: A case study of Ranchi Jharkhand India *Groundwater for Sustainable Development* **5** 85–100. doi: 10.1016/j.gsd.2017.05.002

[19] United States Environmental Protection Agency 2000 EPA Drinking Water Standards and Health Advisories Office of Water Washington, D.C., (June) 1–30 Retrieved from https://web.ics.purdue.edu/~peters/HTML/docs/drinking-water-standards.pdf%0Ahttp://www3.epa.gov/region09/water/drinking/files/dwsha_0607.pdf

[20] WHO 2011 Guidelines for drinking-water quality (4th ed.) Geneva Switzerland: environmental health cri- teria. http://www.who.int

Effectiveness of *Moringa Oleifera* and Blends of Both Alum and *Moringa* as Coagulant in the Treatment of Dairy Wastewater.

O O Elemile[1]*, N E Eze[2], and K Ogedengbe[1,2]

[1]Department of Civil Engineering, College of Engineering, Landmark University, Omu-Aran, Nigeria
[2] Department of Agricultural and Environmental Engineering Faculty of Technology, University of Ibadan, Nigeria
* corresponding author:elemile.olugbenga@lmu.edu.ng

Abstract. Food processing industries can cause some serious problems for the environment through the discharge of wastewater. There is little information on the effectiveness of *Moringa Oleifera* (*MO*)seeds on the treatment of dairy wastewater, this study therefore evaluated the effectiveness of *MO* seeds and blends of both alum and *MO* in the coagulation of diary wastewater. Wastewater collected from oriental food industry limited was kept in a 10L capacity plastic container covered using a black polyethene bag containing ice blocks and conveyed to the laboratory where it was kept in a refrigerator. Six different beakers (Nos 0-5) were filled with 200 mL of wastewater and different doses of the selected coagulant (alum, *MO* and a blend of both) was added to the six beakers ranging from 0-10mL at an increment of 2mL to treat the wastewater. The beaker with number 0 served as the control (no dosage) and this whole process was applied to both coagulants. This test was carried out at a settling time of 1 hour. The physicochemical characteristics of the treated dairy wastewater were determined using the standard methods. Data were analyzed using descriptive statistics and ANOVA. The analysis indicated that the raw wastewater had values of 932 NTU, 1690.99 mg/L, 7.2 and 685 mg/L for turbidity, BOD, pH and TSS respectively. The results revealed that at an optimum dosage and settling time of 10mL and 1hr respectively, the TSS removal efficiencies of 88.76, 82.63 and 86.42 % were obtained for blend of moringa and alum, alum and moringa respectively. The three types of coagulants did not have much effects on the BOD and there were no significant differences in the values. The use of the blend of moringa and alum is effective in the treatment of dairy wastewater.
KEYWORDS: *Moringa Oleifera,* Blend of coagulants, diary wastewater

1. Introduction

Speedy growth of human population and industrial development in the world's nations are destroying the environment through unchecked urbanization growth and rapid industrialization, and leading to the loss of natural habitats. The rising population and the degradation of the climate face the challenge of sustainable growth without harming the ecosystem. Chemical compounds are increasingly being used in industrial production today. Many of these chemicals have made their way into nature with harmful effects that can slowly impact the environment negatively [1]. The dairy industry is generally considered in many countries to be the main source of wastewater for food processing [2]. These dairies extract milk from the products and either simply bottle it for marketing purposes or manufacture specific milk products depending on their capabilities. Due to various processes in food production, large volume of effluents is produced. The biodegradable components in the effluents are available in the state they are present in the milk or in a deteriorated state due to further processing. Therefore, the effluents although organic are very persistent in nature [3]. The characteristics of the effluents are dependent on the by-products and plant processing capabilities. It comprises of a number of disinfecting substances and numerous acidic and basic cleansing agents. Therefore, the pH of effluents can differ significantly depending on the scrubbing method used. Dairy effluents are categorised by high biochemical oxygen demand (BOD) and chemical oxygen demand (COD) concentrations [4] Coagulation process can effectively be used in reducing some pollutants such as turbidity, colour and organic matter present in water. This process is aided by sedimentation employed for the removal of the formed flocs. Some of these methods involve the use of chemicals which are relatively expensive and may have immediate or later effect on the environment. Thus, there is the need to actually look into the use of natural method for the reclamation of industrial effluent. The two main coagulants most widely used in the past are salts of aluminum and iron [5] A lot of less costly materials have been used, these include application of industrial and agricultural wastes in the removal of different contaminants from the industrial wastewaters for their safe dumping into the environment [6]. Several plants have been applied in the clarification of water and these include *Moringa Oleifera, Moringa stenopetala, Vicia faba* [6]. Recently, more attention has been focused in the advancement of natural coagulants such as *Moringa Oleifera* (MO) and Chitosan. Naturally occurring coagulants are usually considered safe for man's wellbeing. While several preceding papers had examined the application of MO as a natural adsorbent for the special removal of contaminants [7], little has been done in the utilization of MO and Alum as a blend. This paper therefore assessed the effectiveness of MO and blends of both alum and *moringa* as coagulant in the treatment of dairy wastewater.

2. Materials and methods

2.1. *Collection of dairy effluent*

The dairy wastewater used for this study was collected from the Fan Milk PLC located at Eleyele, Ibadan. The wastewater sample was collected in a 10L capacity plastic container and conveyed immediately in a bag containing ice blocks to the Environmental Engineering Laboratory of the University Ibadan and kept in a refrigerator.

2.2. *Moringa Preparation*

The MO seeds used for this experimental work were obtained from the Bodija market, Ibadan. The seeds were shelled while seeds showing signs of deterioration were removed after the shelling process. The good ones were dried in an oven at 65 ^{0}C for 20 hours. The dried seeds were milled and oil extraction process was carried out on the powdered MO seeds in order to increase its coagulation properties as recommended by other Researchers [1]. The oil extraction process was achieved using a solvent called n-hexane in a soxhlet extractor. In this process, 20g of the seed powder was wrapped with a white cloth and placed in the soxhlet extractor chamber, and then a conical flask containing 250mL of n-hexane was placed in a water bath which was the heating chamber

2.3. *Preparation of Stock Solution (MO and alum)*

Stock solution of both coagulants (Moringa Oleifera and alum) was prepared by dissolving 15g of the coagulants at the ratio of 2: 1 in 100mL of distilled water and each was blended using the stirrer for 5min to extract the active ingredients after which the solution was filtered using Whatmans filter paper of 12.5mm in order to remove some of the suspended or undissolved solids. The alum was grounded into powder using the laboratory mortar and pestle. Each stock solution formed after filtration was 100,000mg/L.

2.4. *Treatment of Dairy waste water with the coagulants*

A stirrer having 4 paddles with varying rotational speed was used. Six 250 mL different beakers (Numbered 0-5) were washed and rinsed with distilled water after which they were filled with 200mL of wastewater. Different doses of the selected coagulant were added to the six beakers ranging from 0-10mL at an increment of 2mL to treat the wastewater. The beaker with number 0 served as the control (no dosage) and this whole process was applied to both coagulants. After dosing each of the wastewater with different concentrations of the coagulants, the solution was stirred rapidly for 5min at 80 rpm in order to achieve thorough mixture. This was then followed by 10min of gentle stirring at 20 rpm. Gentle stirring was to aid the formation of flocs through the collision of particles in the wastewater. As for the blends, moringa was first added to the wastewater and stirred for 2min before alum solution was then added and stirred for another 10min so as to achieve proper mixing. Settling phase was initiated immediately after stirring by allowing the suspension in a quiescent condition. After the settling phase, the supernatants formed in the beakers were extracted using a pipette at a depth of 2cm below the water surface for analysis of the physico-chemical parameters as per standard methods [8]. Table 1 shows the coagulant doses for the treatment of wastewater, while Figures 1 and 2 shows the treatment process using jar test apparatus and coagulants.

2.5. *Data Analysis and Management*

The data analysis was carried out by means of the SPSS software Version 21.0 while ANOVA was used to compare physicochemical parameters from the treatment by different coagulants using a 5% significance level ($p<0.05$). Descriptive statistics was used to analyse the data which were associated with the standard limits as set by the World Health Organization [9].

Table 1: Coagulant dosage for Treatment of Wastewater

Blends (mL)	Alum (mL)	*Moringa Oleifera* (mL)
2	1	1
4	2	2
6	3	3
8	4	4
10	4	6

3. Results and Discussion

3.1. *Raw Dairy Wastewater Characteristics*

Table.2 shows the physicochemical characteristics of the wastewater from the yoghurt producing industry studied. From the Table the wastewater has a very high BOD content of 1690.99 mg/L which is greater than the discharge standard of 30 mg/L according to National Environmental Standard and Regulation Enforcement Agency (NESREA). From the Table, the wastewater was extremely turbid with a value of 932 NTU which does not meet the World Health Organization (WHO) discharge standard of 5 NTU [10]. The pH of the wastewater was 7.2 which fell within the above stated guideline of 6.0 – 9.0. The total suspended solid of 685 mg/L was also above NESREA standard of 100 mg/L. The value of the total dissolved solids was 810 mg/L which met the NESREA standard of 2,000mg/L but the alkalinity was 308 mg/L, which is above the WHO standards of 120 mg/L. the results are consistent with the findings of [11]. The values of the Total dissolved solids and total suspended solids are in line with that of [12].

Figure 1: Coagulation/flocculation using the Jar test apparatus

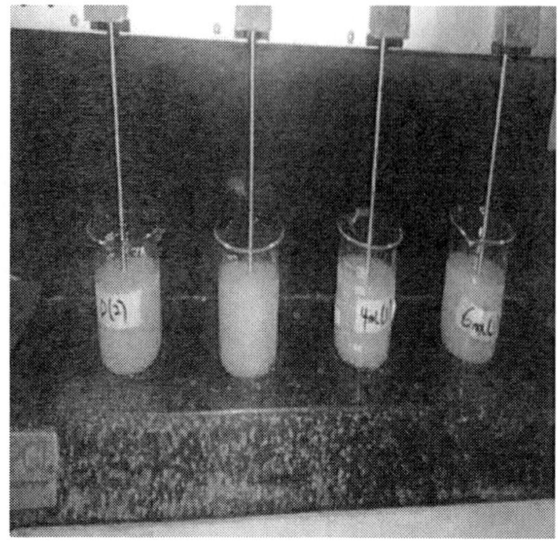

Figure 2: Treatment process using alum

Table 2: Physiochemical analysis of the wastewater sample

Parameters	Concentration	NESREA	WHO
pH	7.2	6-9	6.5-8.5
TDS (mg/L)	810	2000	500
TSS (mg/L)	685	30	<10
Conductivity (μS/cm)	1161	-	-
Turbidity (NTU)	932	5	5
Alkalinity (mg/L)	308	120	120
DO (mg/L)	4.83	8-10	8-10
BOD (mg/L)	1690.99	30	30

NESREA (2000) - National Environmental Standard Regulation and Enforcement Agency

3.2. *Treatment of dairy effluent by various coagulants*

Table 3 shows the values obtained of selected physicochemical parameters after treatment of the wastewater using Alum, Moringa oleifera and Moringa oleifera blends (alum + Moringa oleifera) at different concentrations after 2 hours retention time.

3.2.1. pH

The values of pH ranged between 6.6± 0.19 to 6.0±0.29; 7.2± 0.26 to 7.0± 0.51 and 6.8±0.56 to 6.2± 0.07 at dosages of 2mL to 10 mL using Alum, Moringa oleifera and Moringa oleifera blends (alum + Moringa oleifera) respectively. It was found that after treatment of the wastewater with all the coagulants, the pH was found to decrease. The results were consistent with [10] but at variance with some research works conducted previously [13]. The pH of the wastewater was also more acidic at 10 mL dosage for all the coagulants and did not meet NESREA and WHO standard at that dosage for 2-hour retention time. There is no need to further treat the wastewater because the values of pH fall within the acceptable values of pH [14].

3.2.2. *Total Dissolved Solids (TDS) and Total Suspended Solids (TSS)*

The values of TDS ranged between 226.00± 10.00 to175± 0.42; 209± 5.00 to 140 ± 7.52 and 199 ± 4.10 to 123± 5.64 mg/L at dosages of 2 mL to 10 mL using Alum, Moringa oleifera and Moringa oleifera blends (alum + Moringa oleifera) respectively. The result agrees with [6] that all the types of coagulants have an effect on the TDS with the best performance from the blend of Alum and Moringa oleifera. The TDS of the wastewater was lowest at 10 mL dosage but did not meet NESREA and WHO standard at that dosage for 2-hour retention time. The values of TSS ranged between 177± 3.50 to 135± 10.12; 191± 5.86 to 94.64 ± 7.92 and 169.10 ± 10.15 to 82.54 ± 6.18 mg/L at dosages of 2 mL to 10 mL using Alum, Moringa oleifera and Moringa Oleifera blends (alum + Moringa oleifera) respectively. The result agrees with [15] that all the types of coagulants have an effect on the TDS with the best performance from the blend of Alum and Moringa oleifera. The TSS of the treated wastewater was lowest at 10 mL dosage but did not meet NESREA and WHO standard at that dosage for 2-hour retention time. The high values of suspended and dissolved solids available in raw effluent could be adduced to the availability of biodegradable organic matter in high concentrations in the dairy waste water [12]. There is therefore the need for further treatment of the dairy effluent before final disposal.

3.2.3. *Biochemical Oxidation Demand (BOD)*

The BOD content of raw dairy waste water was 1690.99 mg/L. The values ranged from 1216.26± 24.24 to 1317.60± 25.13; 1236.52 ± to 1439.25± 42.58 and 1175.71± 37.56 to 1033.81±27.65 mg/L at dosages of 2 mL to 10 mL using Alum, Moringa oleifera and Moringa oleifera blends (alum + Moringa oleifera) respectively. The result agrees with [15] that all the types of coagulants have an effect on the BOD with the blend of Alum and Moringa oleifera having the best result. The BOD of the wastewater was lowest at 10 mL and did not meet NESREA and WHO standards at that dosage for 2- hour retention time. From the results it could be seen that the application of the blend of Moringa oleifera with Alum coagulant is more effective than the other coagulants. It is also seen that as the quantity of dosage is increased, there is greater efficiency in the rate of removal. The implication of this is that with the high after-treatment BOD, there are still organic materials and active microorganisms therefore there is need for further treatment of the wastewater before final disposal of the dairy wastewater [16][17]

Table 3: Treatment of dairy effluent with *Moringa Oleifera and Moringa Oleifera* blends (Alum + *Moringa O*) at different concentrations

S/No	Parameter	Coagulant	Dosage (mL)					
			0	2	4	6	8	10
1	pH	Alum	7.0	6.6± 0.19^{a}	6.3± 0.34^{a}	6.1± 0.27^{a}	6.1± 0.25^{a}	6.0±0.29^{a}
		Monringa		7.2± 0.08^{b}	7.2± 0.26^{b}	7.1± 0.36^{b}	7.1± 0.58^{b}	7.0± 0.51^{b}
		Moringa + Alum		6.8± 0.56^{a}	6.5± 0.06^{a}	6.4± 0.21^{c}	6.3± 0.20^{a}	6.2± 0.07^{a}
2	TDS (mg/L)	Alum	724	226.00±10.00^{c}	220.00± 6.75^{c}	190.00± 9.82^{c}	184.00± 10.25^{c}	175.00± 9.42^{c}
		Monringa		209.00± 5.00^{b}	186.00± 5.52^{b}	173.00± 4.21^{b}	160.00± 5.92^{b}	140.00± 7.52^{b}
		Moringa + Alum		199.00± 4.10^{a}	167.00± 6.72^{c}	150.00± 8.25^{a}	125.00 ± 7.89^{a}	123.00± 5.64^{a}
3	TSS (mg/L)	Alum	194	177.00± 3.50^{a}	131.00± 7.34^{a}	132.00± 5.45^{b}	134.00± 3.72^{b}	135.00± 10.12^{c}
		Monringa		191.00± 5.86^{b}	180.00± 10.21$_{c}$	116.00± 7.52^{a}	97.00± 6.12^{a}	94.64± 7.92^{b}
		Moringa + Alum		169.10± 10.15c	145.32± 6.75^{b}	129.25± 4.27^{b}	99.15± 5.17^{a}	82.54± 6.18^{a}
4	BOD (mg/L)	Alum		1317.60±25.13^{b}	1277.07±16.02^{b}	1216.26±25.12^{b}	1216.26±24.24^{b}	1094.61± 32.10^{b}
		Monringa	1470	1439.25± 42.58^{c}	1358.16± 6.25^{c}	1337.99± 26.92^{c}	1256.79± 45.21^{c}	1236.52± 32.23^{c}
		Moringa + Alum		1195.98± 19.70^{a}	1175.73± 16.20^{a}	1175.71± 37.56 ^{a}	1137.88± 23.78^{a}	1033.81± 27.65^{a}

For each parameter, means with the different letters (superscripts) are significantly different ($p < 0.05$), using Duncan's multiple range test.

3.3. **Optimum dosage and reduction efficiencies for turbidity**

3.3.1. Optimum dosage and reduction efficiencies for turbidity by the coagulants

Figures 3 and 4 showed the turbidity removal in the wastewater using Alum, *Moringa Oleifera* and blend at 1 hour and 2 hours retention time. It was shown that the initial turbidity of the raw wastewater was 932 NTU. The effects of coagulant dosage for the three coagulants at different settling time are shown in the figures. These figures show that turbidity was reduced as the coagulant's dosage was increased at both settling time of 1 and 2 hours. The results indicated that the optimum dosages in the reduction of turbidity of the wastewater using alum, *moringa* and blends were achieved at 8, 10 and 10 mL respectively at both I hour and 2 hours settling time respectively. Figure 5 showed the removal efficiencies for turbidity at optimum dosage of each coagulant in different retention time. At those optimum dosages, the turbidity was reduced to 117, 107 and 95 NTU for alum, *moringa* and blends respectively. The removal efficiencies of 87.45, 88.52 and 89.81% for alum, *moringa* and blends. respectively at a settling time of 1 hour. At a settling time of 2 hours, the optimum dosage reduced the initial turbidity of the wastewater to 131, 102 and 102 NTU for alum, *Moringa* and blend respectively representing a removal efficiency of 85.94, 89.06 and 89.06 % alum, *moringa* and blends respectively. Removal efficiencies for turbidity at these optimum dosages for both times are represented in Figure 3 and it can be seen from the results that the blend of Alum and *Moringa Olifera* gave the best removal efficiency of 89.81 % at 10 mL in a settling time of 1 hour. The result was in line with the reports of [1]; [18] who stated that all the coagulants removal values of turbidity were high and greater than 60%. The implication of this is that since the removal rate of turbidity is very high, there is no need for further treatment.

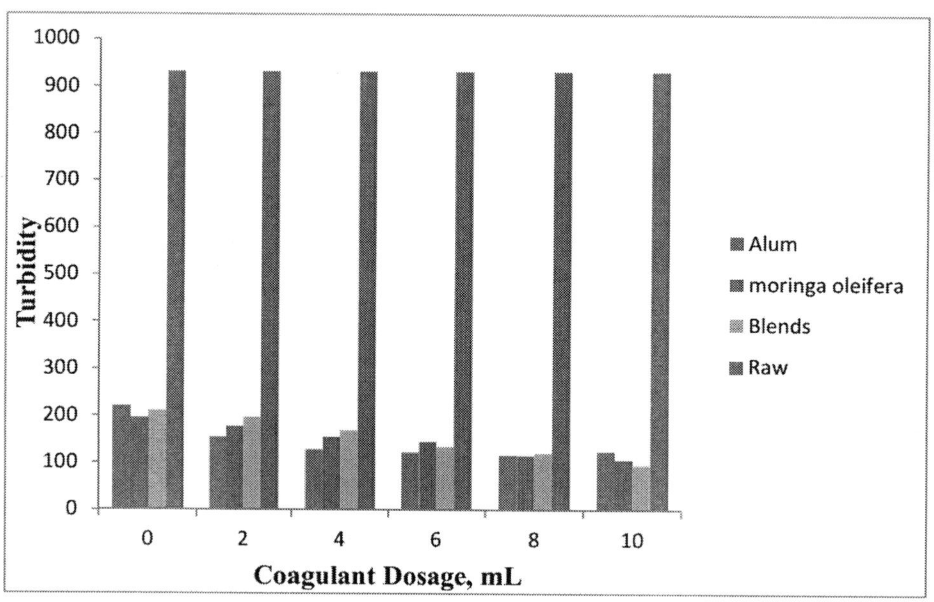

Figure 3: Turbidity removal in the wastewater using Alum, *Moringa Oleifera* and blend at 1hour retention time.

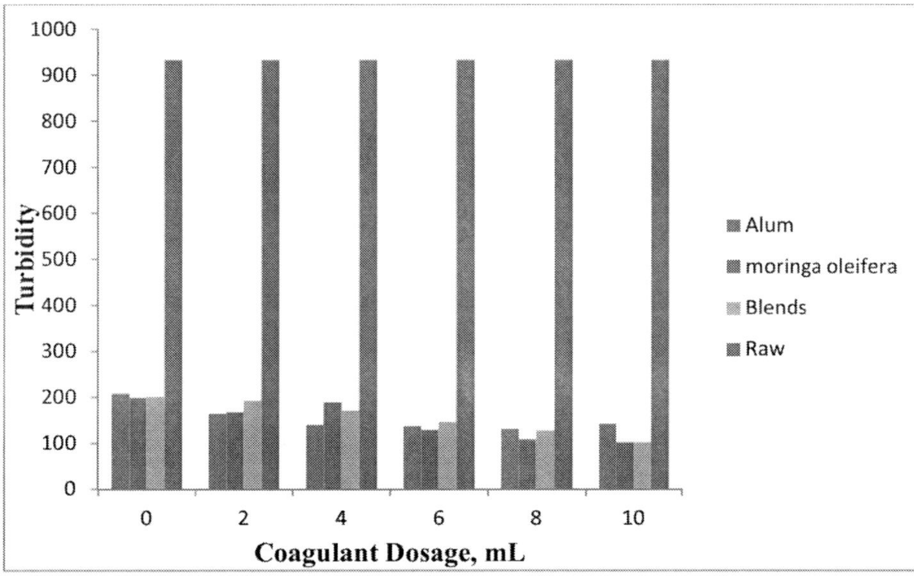

Figure 4: Turbidity removal in the wastewater using Alum, *Moringa Oleifera* and blend at 2 hours retention time.

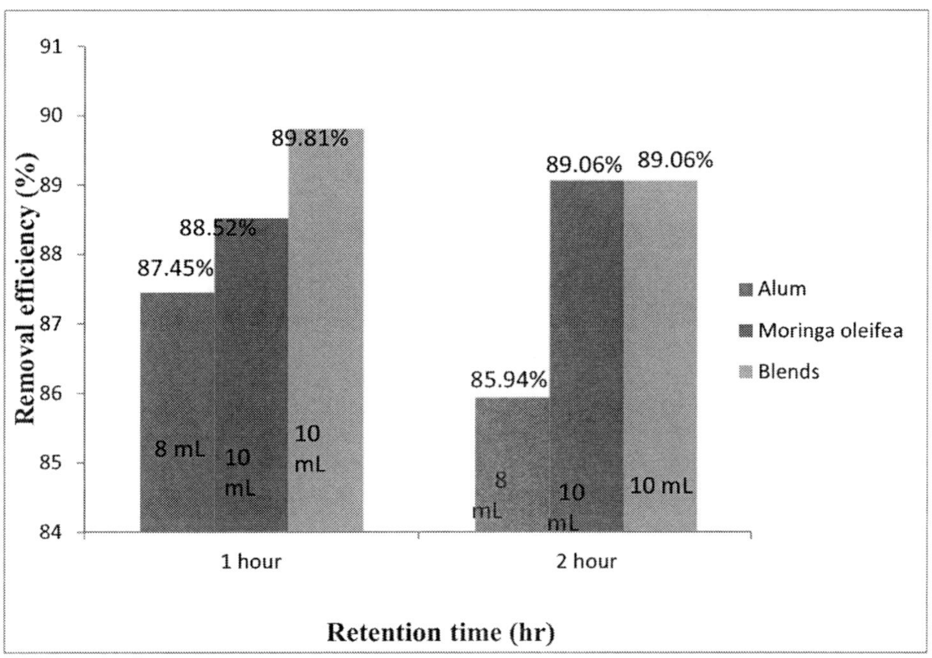

Figure 5: Removal efficiencies for turbidity at optimum dosage of each coagulant in different retention time

4. Conclusion

From the study it is shown that all coagulants possess properties for wastewater treatment. The treatment of dairy wastewater shows significant reduction in values of parameters like TDS and TSS. The values for TDS and TSS were 824, 140, 135 and 123 mg/L for control, Alum, *Moringa Oleifera* and Blend respectively and 194, 135, 94 and 82mg/L for control, Alum, *Moringa Oleifera* and Blend respectively. The study also showed that these coagulants are not very effective in BOD (Biochemical oxygen demand) removal. They were very useful in turbidity removal as well as reduction in the total suspended solids. *Moringa* and blend (alum + *moringa*) at a dosage of 8 mL and 10mL gave the best turbidity removal efficiencies of 89.06 % and 89.80 % making the blend the best. It can therefore be concluded that blending alum and *M. oleifera* seed powder showed characteristic synergies in water treatment than using alum alone or Moringa alone. The turbid and polluted water can also both be treated with *moringa* extract and blend of both alum and *moringa* but the best result in terms of turbidity removal are achieved using blends.

References

[1] Dehghani, M & Alizadeh, MH 2016 The effects of the natural coagulant *Moringa Oleifera* and alum in wastewater treatment at the Bandar Abbas Oil Refinery *Environmental Health Engineering and Management Journal* **3**(4):225–230

[2] Elemile OO, Raphael OD, Omole DO, Oluwatuyi, OE, Ajayi EO, Umukoro O and Elemile MG 2019 Assessment of the impact of abattoir activities on the physicochemical properties of soils within a residential area of Omu-Aran, Nigeria *IOP Conf. Series: Materials Science and Engineering* **640** 012083IOP Publishing doi:10.1088/1757-899X/640/1/012083

[3] Rao MN and Datta AK 1999 "wastewater treatment, Rational Methods of design an industrial practices", Oxford & IBH Publishing Co. Pvt. Ltd, 3rd Edition.

[4] Shivsharan VS, Kulkarni SW and Wani M (2013) Physicochemical Characterization of Dairy Effluents *International Journal of Life Sciences Biotechnology & Pharmaceutical Research* **2**(2):182-191

[5] Okuda T, Baes AU, Nishijima W and Okada M (1999). Improvement of extraction method of coagulation active components from *Moringa Oleifera* seed. *Water Resources* **33**(15): 3373–3378

[6] Chonde S and Raut P 2017 Treatment of dairy wastewater by *Moringa Oleifera* Seeds *World Journal of Pharmaceutical Research* **6**(8):1484-1493 doi: 10.20959/wjpr20178-9015

[7] Ghebremichael K, Gunaratna K, Henrikson H, Burmer H, Dalhammar G 2005 A simple purification and activity assay for the coagulant protein from *Moringa Oleifera* seed. *Water Resources* **39**(11):2338-44. doi: 10.1016/j.watres.2005.04.012

[8] Maithi SK 2004 Handbook of methods in environmental studies vol.1: water and waste water analysis, second edition, 2004, published by ABD publisher Jaipur – 302015

[9] APHA (2005) Standard methods of examination of water and wastewater, 20th edition, IWA Publishing Washington DC, American public health association

[10] WHO (2006) Guideline for drinking water quality incorporation first addendum **1** Recommendations 3th ed. ttp://www.who.int/water.sanitation.health/dwh/qdwa0506.pdf Accessed 15th January 2019

[11] Mortula M and Shabani S 2012 Removal of TDS and BOD from Synthetic Industrial Wastewater via Adsorption International Conference on Environmental, Biomedical and Biotechnology IPCBEE 41:166-170 *IACSIT Press, Singapore*

[12] Garg AR, Garcha S, Singh AK, Kamaldeep K and Magotra, A 2017 Characterization and Optimization of Physicochemical Properties of Dairy Effluents *Poll Research* **36**(2):307-312

[13] Bhuptawat H, Folkard GK and Chaudhari S 2007 Innovative physico chemical treatment of wastewater incorporating *Moringa Oleifera* seed coagulant *Journal of Hazardous Materials* **142**(1-2): 477-82. doi: 10.1016/j.jhazmat.2006.08.044

[14] Tetrapak 1995 TetraPak Dairy Processing Handbook TetraPak Printers: London, UK
[15] Kaushal RK and Goyal H 2019 Treatment of Waste Water Using Natural Coagulants International conference on "Recent Advances in Interdisciplinary Trends in Engineering and Applications
[16] Kulkarni PG and Dharwadkar SM 1998 Effect of dairy effluent on biochemical parameters of wheat seeds and fish *Proc. Acad. Environ. Biol.* **7**(1):57-60
[17] Choudhary M and Neogi S (2017) A Natural Coagulant Protein from *Moringa Oleifera*: Isolation, Characterization, and Potential Use for Water Treatment *Material Research Express* in press https://doi.org/10.1088/2053-1591/aa8b8c
[18] Bhutada DS, Datar MT and Kaul SN 2006 Use of Herbal Coagulant for Primary Treatment of Dairy Wastewater *Journal of Industrial Pollution Control* **22**(1): 139-148

Management of Infrastructures in Water Sector: A Veritable Tool for Healthcare and Sustainable Development in Nigeria

O J Oyebode[1]*, **A O Coker**[2]
Civil and Environmental Engineering Department, College of Engineering,
Afe Babalola University, Ado-Ekiti (ABUAD), Ekiti State, Nigeria
[2]Department of Civil Engineering, Faculty of Technology, University of Ibadan, Ibadan.
Corresponding Author: oyebodedare@yahoo.com

Abstract. There are a lot of health issues traceable to water sector in most developing nation of the world. Despite water rights and lots of researches in water resources engineering, strategic management of infrastructures still requires urgent and adequate attention for better healthcare delivery and sustainable development in Nigerian water sector. This paper assessed the operations of government, private and agencies involved in the supply and management of water resources. Data were collected through reconnaissance survey, literature review and interview of stakeholders in selected establishments. Challenges of this sector were checked and appraised with special focus on health implications, adequacy of supply, management and maintenance of infrastructures. Current trends within the administration of development comes were checked. Strategic management tools were place in place for effective handling of care, adequate funding, capability building, technological advancement and innovation in water sector. The study concluded that adequate attention should be given to strategic management of water sector to tackle healthcare issues and water challenges for accomplishment of infrastructural development in Nigeria. It was recommended that government; engineers, researchers and all stakeholders should embrace productive activity, viable policies, pollution management, innovative analysis, acceptable funding, technological advancement, manpower development and sanitation of water sector in all ramifications.

Key words: Strategic Management, Health Issues, Water Resources Engineering, Sustainable Infrastructures, Manpower Development

1. Introduction

Water infrastructure requires strategic management in all ramifications for the achievement of sustainable development goals and for better healthcare delivery. Amidst various healthcare interventions, national issues, engineering inputs, agenda, policies, development plans, sustainable development goals (SDG) and a host of others, Nigerian government has not achieved the desired goal in water sector. Developmental goals for developing nations includes adequate water supply, poverty eradication, unemployment; educational advancement, excellent healthcare delivery, socio-economical development, housing and food to every citizen.

Strategic management is highly required in water development organizations. Strategy should be given priority in the modification of previous management activities for better outcome [39-41]. There are copious challenges facing public water supply in Nigeria such as inadequate data, funds, manpower development and poor maintenance of infrastructures [2-8].

Public water utilities issues include legal framework, environmental issues, cultural and other technical and socio-cultural issues [9]. Urbanization aspects, insufficient assets, mismanagement and regulative framework are major challenge in the country [10].
The essential economic problems concern the requirement to foster property speedy economic process will assist Water Corporation. [11-16]
The housing demand for water in the metropolitan areas (120–150 liter per capita per day (lpcd)) complied with 120 lpcd national normal water demand [17,18, 22].
Wastewater management is in forefront of advancing health, economy, urbanization and temperature change, infrastructural growth. [23,24,25].
Civil Engineering infrastructures have a lot of impacts on the environment and sustainable development. Impacts of infrastructures should be evaluated with special focus on the planning, maintenance and implementation of civil infrastructures that preserve natural environment [1, 27, 28, 29].
Water transport in Nigeria is proscribed to the areas enclosed by water. It is oftentimes employed by those living in the water-side and features a mean share of regarding one.6 % of Nigeria's gross domestic product. Water transport involves each native and international movement [30,31].
Individuals will determine the way to create property lifestyles decisions, employment and creative development especially in water sector [32].
Environmental property means the climate system is stable, variety is preserved, ecosystems operate well, fresh is secured, and rural and concrete settlements area unit protected against pollution and area unit resilient to climate shocks.

2. Healthcare Provision System in Nigeria

Health of Nigerian must be protected in all areas especially in water resources aspect. Water provision has a long way to go in health care and infrastructural preservation. Health and national security of populace should not be endangered [33-36].
Figure 1 presented National Health Act [37-38].

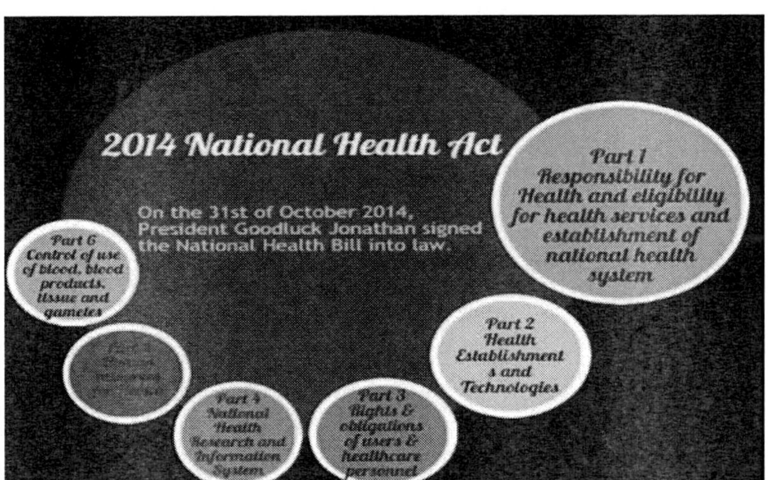

Figure 1: 2014 National Health Act

IOP Conf. Series: Materials Science and Engineering 1036 (2021) 012008 doi:10.1088/1757-899X/1036/1/012008

3. Methodology

Operations of government, private and agencies involved in the supply and management of water resources were assessed. Data were collected through reconnaissance survey, literature review and interview of stakeholders in selected establishments. Challenges of this sector were checked and appraised with special focus on health implications, adequacy of supply, management and maintenance of infrastructures. Figure 2 indicates the map of Nigeria based on geopolitical zone.

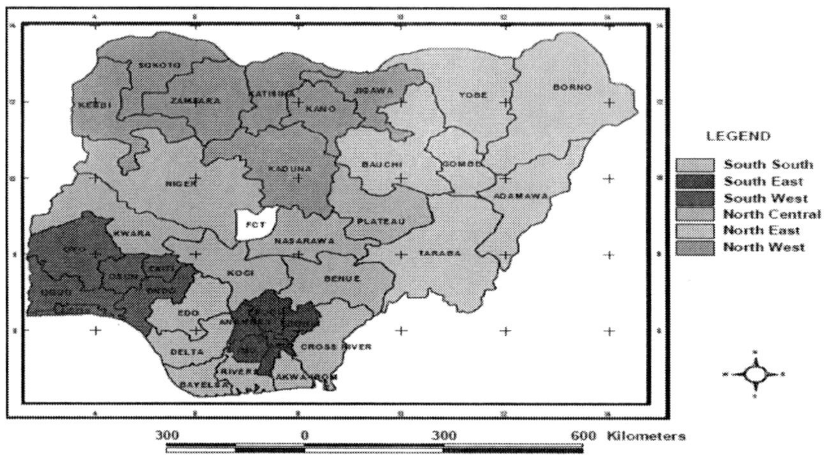

Figure 2: Nigeria's Geopolitical Zones

Figure 3 indicates the relevance of security-sustainability assessment framework in healthcare and water sector.

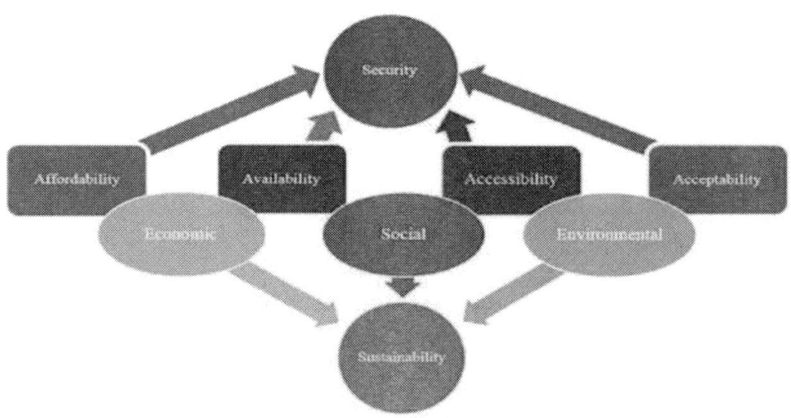

Figure 3: Relevance of Security-sustainability assessment framework

River basins authority in Nigeria such as Ogun – Osun and other agencies are trying their best in Nigeria.

Figure 4 indicated links between climate, agriculture, management and sustainability.

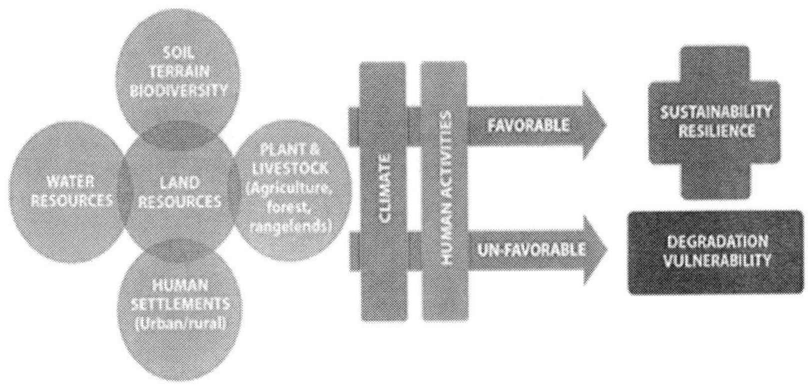

Figure 4: Links between Climate, Agriculture, Management and Sustainability

4. Discussion of Findings

Water sector and healthcare delivery are linked with economic, social, ecological and environmental sustainability as indicated by Figure 5. The health and well-being of individuals and populations is suffering from many wider environmental determinants. Figure 6 presented recent Water Supply Projects in Nigeria.

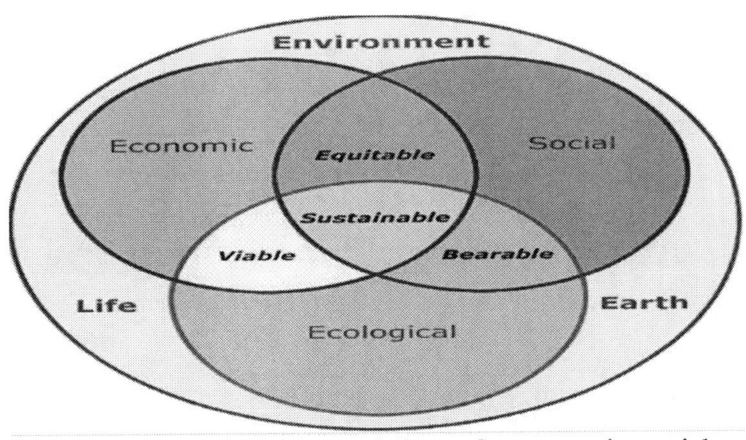

Figure 5: Water sector and healthcare delivery for economic, social, ecological and environmental sustainability

There is need to strengthen water supply programmes and projects in order to enhance their performance.

Safe drinking water is fundamental to health, survival, growth and development. While the goal of improved management of the water sector is no doubt a worthy goal, experiences from other parts of the world where the World Bank has pushed through the privatization of water reveals at best a mixed result with the common fallout being the systematic denial of the poor who cannot afford the exorbitant rates imposed by private water companies of the right to safe water. For new approaches to financing water supplies needs to be checked before implementation.

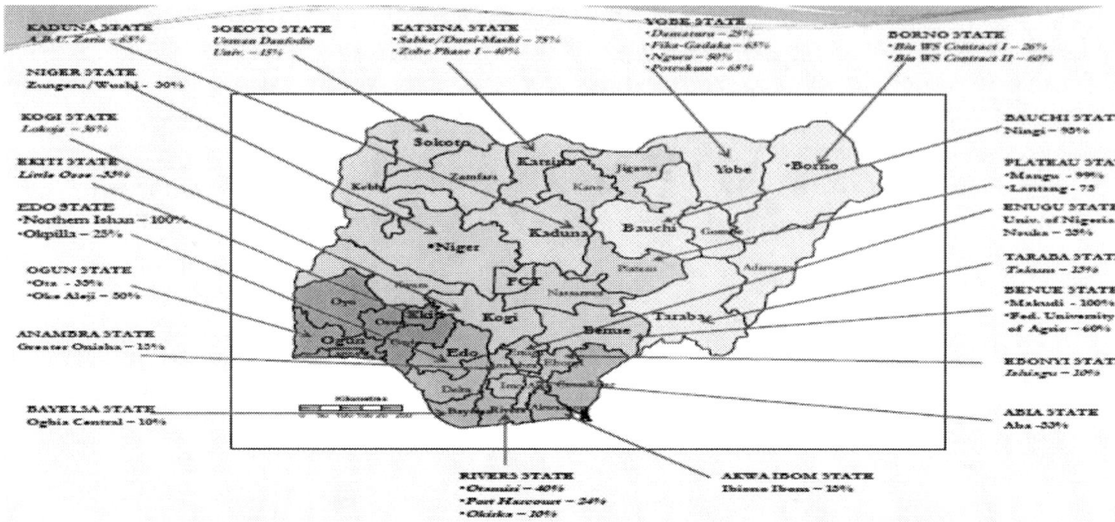

Figure 6: Developmental Projects related to Supply of Water

Figure 7 gives an illustration of Transformation for 2050 projection conceptual framework.

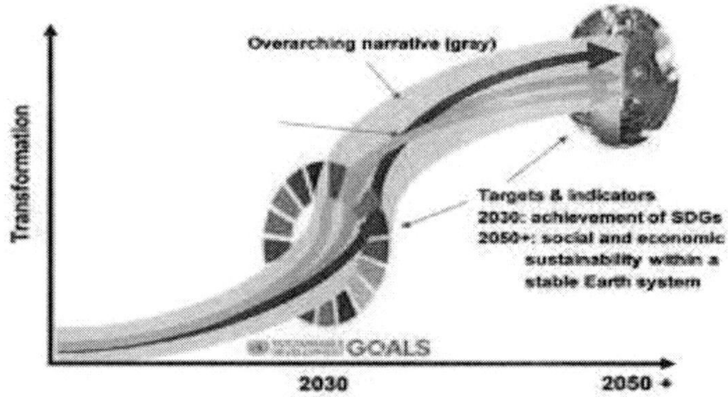

Figure 7: Transformation for 2050 projection conceptual framework

5. Challenges in water resource management in Nigeria

There are a lot of challenges facing our healthcare delivery which can be traced to water sector. Some of the challenges are highlighted as follows:

i. High water stress as a result of Increase in Water demands far above population growth in Nigeria.
ii. Regulation of water and processing of water licenses to qualified service provides by putting necessary law in place.
iii. Increase in climate change due to frequency of flooding, drought, groundwater pollution, groundwater overdraft, groundwater contamination, degradation of land and water.
iv. Inadequate attention to the link between water sector and healthcare delivery system.
v. Lack of cooperation for financing for sustainability; local, and International and regional basis.

vi. Infrastructures that can attract productive engagement of local and foreign investors should be encouraged.
vii. Government should bring strategic management into the financing and operation of water sector administration.

6. Strategic Management Approach of Infrastructure in Water Sector

The strategic management approaches for sustainable development of water sector in Nigeria are:

i. Adequate and planned provision of water infrastructure that can enhance sustainable development in the ministry of water resources.
ii. Good Governance in the administration of water sector.
iii. Effective communication among government, populace and stakeholders.
iv. Sustainable infrastructure therefore needs to be planned, designed, delivered, managed and decommissioned to minimize its negative impacts and maximize its positive impacts.
v. Adoption of integrated water resource management to execute the plan and review progress on a regular basis.

7. Conclusions

The following conclusions were made as a veritable tool to healthcare and water sector:

i. Appropriate attention must be given to strategic management of water sector to tackle healthcare issues and water challenges.
ii. The sustenance of the sustainability concept is hinged and underpinned by the integrated water management concept for infrastructural development. Strategic management approaches for sustainable development of water sector will improve public health in Nigeria.
iii. This strategy must be communicated through the organization to form broad support in private and public industries.
iv. Whenever a strategic management of infrastructure is adopted in water sector, it will revive the healthcare and enhance sustainable development goals.
v. It has been observed that only a few companies are clearly managed strategically, and all of them are multinational, diversified, manufacturer concerns.

8. Recommendations

Specific recommendations for strategic management and resolution of issues in water sector situation in Nigeria include:

i. Government; engineers, researchers and all stakeholders must embrace productive activity, viable policies, pollution control, innovative research, appropriate funding, technological advancement, manpower development and sanitation in all ramifications of health and water sector.
ii. Supervise the day-to-day operations by the various departments to ensure their efficiency and effectiveness;
iii. The Infrastructure Concession Regulatory Commission (ICRC) should be drastically supported and improved for optimal benefit of water sector.
iv. Management of infrastructures should be given adequate priority for better healthcare delivery and sustainable development in developed and developing countries.
v. Future policies, legal framework and water projects should take cognizance of strategic

management, capacity building, productivity and financing of water related research into consideration.

Acknowledgements
We extend my profound appreciation to professionals, stakeholders and engineers that have contributed to this research work. We also thank my reviewers, International Conference on Sustainable Infrastructural Development planning and scientific advisory committee for their thoroughness and unrelenting efforts.

References

[1] Oyebode OJ 2019 Impacts of Civil Engineering Infrastructures In the Sustainability of the Environment. a paper published in the *conference proceedings of 17th International Conference and AGM tagged "PEACE 2019" of Nigerian Institution of Civil Engineers (NICE)* held from October 30 – 1st November, 2019 at the Multipurpose Hall, Government House, Maiduguri, Borno State, Nigeria
[2] Adams EA, & Zulu LC 2015 Participant or customers in water governance? Community-public partnerships for peri-urban water supply. *Geoforum,* **65**, 112–124. https://doi.org/10.1016/j.geoforum.2015.07.017
[3] Adnan MA 2013 Drinking water quality assessment of the middle governorate in the Gaza strip Palestine. *Water Resources and Industry*, **4,** 13–20.
[4] Adnan S and Iqbal J 2014 Spatial analysis of the groundwater quality in the Peshawar District, Pakistan. *Procedia Engineering*, **70**, 14–22
[5] Cohen B 2006 Urbanization in developing countries: Current trends, future projections, and key challenges for sustainability. *Technology in Society*, **28**, 63–80 https://doi.org/10.1016/j.techsoc.2005.10.005
[6] Haider H, Sadiq R and Tesfamariam S 2014 Perfromance indicators for small and medium-sized water supply systems: A review. *Environmental Reviews,* **22**, 1–40. https://doi.org/10.1139/er-2013-0013
[7] Saravanan VS and Gondhalekar D 2013 Water supply and sanitation as a 'preventive medicine': Challenges in rapidly growing economies. *Water International,* **38**, 867–874 https://doi.org/10.1080/02508060.2013.857142
[8] Zérah MH 2000 Household strategies for coping with unreliable water supplies: The case of Delhi *Habitat International,* **24**, 295–307. https://doi.org/10.1016/S0197-3975(99)00045-4
[9] Jideonwo JA 2014 Ensuring sustainable water supply in Lagos, Nigeria (MSc Thesis), University of Pennsylvania, USA
[10] Sojobi AO 2016 Evaluation of groundwater quality in a rural community in North Central of Nigeria. *Environmental Monitoring and Assessment,* **188**(3), 1–17. doi:10.1007/s10661-016-5549-z
[11] Varis O 2006 Megacities, development and water. *Water Resources Development,* **22**, 199–225.https://doi.org/10.1080/07900620600648399
[12] Longe EO Kehinde MO and Olajide CO 2015 Private sector participation in water supply: Prospects and challenges in Developing Economies. XVth IWRA (*International Water Resources Association) World Water on Global wate*r, a resource for development: Opportunities, challenges and constraints, Edinburgh, Scotland, May 25–29, 2015. Retrieved 2016, November 14 from www.iwra.org/congress/resource/abs155_article.doc
[13] Olajuyigbe AE, Rotowa OO Adewumi IJ 2012 Water vending in Nigeria-A case study of Festac Town, Lagos, Nigeria. *Mediterranean Journal of Social Sciences*, **3**, 229–239
[14] Omole DO, Ndambuki J M, Badejo AA, Oyewo DO Soyemi TO 2016 Public feedback on state of domestic water supply in Lagos: Implications for public health. *Indian Journal of Traditional Knowledge,* **15**, 245–253
[15] Olukanni DO, Ajetomobi MO, Tebowei SO, Ologun OO and Kayode OM 2014 Water supply and

sanitation challenges in an urban setting: A case study *International Journal of Engineering and Applied Sciences*, **1**, 34–38
[16] Oyegoke SO Sojobi AO 2012 Developing appropriate technique to alleviate the Ogun River network annual flooding problems. *International Journal of Scientific and Engineering Research,* **3**,2 1-7
[17] Ezenwaji EE, Eduputa BM Okoye CO 2016 Investigations into the residential water demand and supply in Enugu Metropolitan Area. Nigeria *American Journal of Water Resources*, **4**, 22–29
[18] Martins O 2001 Water resources management and development in Nigeria -Issues and challenges in a new millennium. *An inaugural lecture delivered at the University of Agriculture, Abeokuta* Retrieved 2001, August 22 from https://www.unaab.edu.ng/attachments/ Water%20Resources%20Management%20And%20Development%20In%20Nigeria%20Issues% 20 And%20 Challenges%20In%20A%20New%20
[19] Samuel N 1986 Domestic water consumption patterns in selected areas in Nairobi (MSc Thesis) Tempere University, Nairobi

[20] UKEA 2008 Water and the environment, International comparisons of domestic per capita consumption. Prepared for the *United Kingdom Environment Agency by Aquaterra*
[22] WHO 2003 Domestic water quantity, service, level and health. Geneva: *World Health Organization. Retrieved* from http://www.who.int/water_sanitation_health/ diseases/WSH03.02.pdf
[23] Oyebode OJ 2019 Critical appraisal of options for removing various Contaminants from wastewater for economic sustainability. A paper in the *proceeding of 2019 Nigerian Society of Engineers Annual Engineering Conference*, Ado-Ekiti Branch, Ado-Ekiti
[24] Abubakar IR 2016 Quality dimensions of public water services in Abuja, *Nigeria. Utilities Policy,* **38**, 43–51. https://doi.org/10.1016/j.jup.2015.12.003
[25] Enwegbara B 2005 "How to Develop Our Economy." *This Day*, Vol.**10**, No 3711, p.14
[26] National Population Commission and ICF Macro 2009 *Nigeria Demographic and Health Survey*: 2008
[27] Federal Ministry of Environment 2008 Managing the Nigeria Environment for Sustainable Development: Challenges and Opportunities
[28] ECOSOC (United Nations Economic and Social Council) 2016 Progress towards the Sustainable Development Goals: *Report of the Secretary-General,* ECOSOC, New York, NY
[29] WHO (World Health Organization) 2012 UN-Water Global Analysis and Assessment of Sanitation and Drinking Water 2012 Report: The Challenge of Extending and Sustaining Services, *WHO*, Geneva
[30] Mosley LD Sharp and Singh S 2004 'Effects of a tropical cyclone on the drinking-water quality of a remote Pacific island', *Disasters*, **28**, No. 4, 405–17
[31] Nauges C and Strand J 2013 'Water-hauling and girls' school attendance: some new evidence from Ghana', World Bank Policy Research Working Paper 6443, *World Bank Group,* Washington, DC
[32] UNESCO-UNEVOC International Centre for Technical and Vocational Education and Training 2012 Skills Challenges in the Water and Wastewater Industry: Contemporary Issues and Practical Approaches in TVET, *United Nations Educational, Scientific and Cultural Organization,* Paris
[33] Moe JL, Pappas G, Murray A. Transformational leadership, transnational culture and political competence in globalizing health care services: A case study of Jordan's King Hussein Cancer Center.]; Globalization and Health. 2007 3:11. [Last accessed on 2019 Oct12] Available from: http://www.globalizationandhealth.com/content/3/1/11
[34] The Library of Congress Country Studies CIA World Factbook. [Last accessed on 2019 June 21] Available from: http://www.cia.gov
[35] Taylor CA, Taylor JC, Stoller JK 2008 Exploring leadership competencies in established and aspiring physician leaders: An interview-based study. *J Gen Intern Med.* **23**:748–54
[36] Jones DS, Tshimanga M, Woelk G, Nsubuga P, Sunderland NL, Hader SL, et al 2009 Increasing leadership capacity for HIV/AIDS programmes by strengthening public health epidemiology and

management training in Zimbabwe. [Last accessed on 2019 Nov 10Available from: http://www.human-resources-health.com/content/7/1/69
[37] Omoruan AI, Bamidele AP, Phillips OF 2009 Social health insurance and sustainable healthcare reform in Nigeria. *Ethno Med.* **3**:105–10
[38] National Health Insurance Scheme Decree No 35 of 1999 *Laws of the Federation of Nigeria.* [Last accessed on 2019 Nov 6]. Available from: http://www.nigeria-law.org/National%20Health %20Insurance %20Scheme%20Decree.htm
[39] Carte C 2013 "The age of strategy: strategy, organizations and society," *Business History*, **55**, no. 7, pp. 1047–1057, 2013
[40] Mackay D and Zundel M "Recovering the divide: a review of strategy and tactics in business and management," *International Journal of Management Reviews*, **19**, no. 2, pp. 175–194, 2016
[41] Cattani G, Porac FJ and Thomas H 2017 "Categories and competition," *Strategic Management Journal*, **38**, no. 1, pp. 64–92

Comparative study on the effects of guar gum nanocomposite and guar gum polymer solutions on enhanced oil recovery

A Sowunmi[1,3], V E Efeovbokhan[1], O D Orodu[2] and T Ojo[2]

[1]Department of Chemical Engineering, Covenant University, Ota, Ogun State, Nigeria
[2]Department of Petroleum Engineering, Covenant University, Ota, Ogun State, Nigeria
[3]Corresponding author – akinleye.sowunmi@yahoo.com

Abstract. Guar gum is a polysaccharide that occurs naturally, and has useful properties in thickening and stabilizing. This makes it of interest in enhanced oil recovery, because of its ability to increase the sweep efficiency of the recovery process. Also, guar gum has an economical and reliable supply as well as flexible chemistry. Guar gum, however, cannot interact with oil-rock surface or the oil-water interface, thereby limiting its capacity to recover crude oil. Nanoparticles can, therefore, be mixed with guar gum to increase its recovery potential. Silicon oxide nanoparticles can be combined with a polymer to form a polymer nanocomposite, which can then be used to increase the overall efficiency of the crude oil recovery process. In this study, the viscosity profile of guar gum was investigated under different conditions of temperature and weight percentages. Four temperatures were investigated: 30, 50, 75, and 90°C. Six weight percentages were also investigated in the viscosity tests: 0.1, 0.2, 0.3, 0.4, 0.5, and 1. The oil recovery potential of guar gum and guar gum nanocomposite was also investigated at different weight percentages. Across the six weight percentages used in the viscosity study, it was observed that there was an average percentage loss in viscosity of 33 % as the temperature was increased from 30°C to 90°C at the highest shear rate of 1021 s-1. At 30°C, the viscosity of guar gum was seen to increase by close to 1000 % as weight percentage was increased from 0.1 to 1. The recovery factor observed during core flooding tests utilizing guar gum at three different concentrations indicated an increment of 48, 51, and 54% respectively. On addition of silicon oxide nanoparticles at different concentrations, oil recovery was enhanced by 54 % to 67.2 % OOIP. This validates the recovery potential of guar gum for enhanced oil recovery.

Keywords: guar gum, enhanced oil recovery, viscosity, silicon nanoparticles, nanocomposite

1. Introduction

Several challenges exist in the process of oil recovery. Such challenges include the presence of an interface between the displacing fluid and the crude oil, mobility level of displacing fluid, heterogeneity of reservoir structure, wettability of rock surface, etc. [1]. The use of chemicals in enhanced oil recovery (EOR) has been found to reduce the impact of some of these challenges and increase oil recovery.

Chemical flooding involves the injection of solutions such as polymer, surfactant, or alkali into the reservoir, usually alongside water, to increase the sweep efficiency of the process. A combination of chemicals may also be used in the flooding process to achieve the targeted effect. Several types of biopolymers and synthetic polymers have been studied in the literature for use in the EOR process. Some synthetic polymers that have been studied include acrylic acid, comb polymers, associative polymer, polyacrylamide and NVP (N-Vinylpyrrolidone) [2]. Some biopolymers that have been tested include scleroglucan, guar gum, xanthan gum, schizophyllan, etc. Biopolymers have been found to have a higher tolerance for shear and salinity than synthetic polymers, but are more prone to biological degradation during the flooding process [3]. Nonetheless, synthetic polymers that contain ATBS monomer have been found to show the following characteristics: higher resistance to shear and salinity; lower adsorption of the polymer to reservoir rock during the flooding process, and reduced polymer precipitation in reservoirs with high temperatures. On the other hand, NVP monomers give polymers greater resistance to hydrolysis that the ATBS monomer. Guar gum polymer has been found to have good viscosifying properties which makes it a potentially good agent for chemical flooding in EOR.

Guar gum or guaran is a biopolymer produced from the plant *Cyamopsis tetragonoloba.* Guar gum is composed mostly of mannose and galactose; and has been used extensively as thickener, stabilizer, and emulsifier, properties that make it potentially useful in EOR. It has been found to show pseudoplastic behaviour when an external force is applied to it, and this behaviour shows a pronounced effect with molecular weight [4]. Guar gum has been used in the oil industry due to its relative cost, resistance to thermal and mechanical degradation, and its environmental biodegradability. It has been used as a fracturing fluid, in formulating drilling muds and now in chemical flooding process [5, 6]. The use of guar gum polymer for chemical flooding can produce up to 35% of additional oil from a reservoir after water-flooding [7-9]. The polymer achieves oil recovery by reducing the mobility of the displacing fluid. It is however believed that if guar gum is combined with a nanoparticle, the combination or nanocomposite would be able to achieve greater recovery by combining the macroscopic efficiency of the polymer with the microscopic efficiency of the nanoparticle to achieve greater overall efficiency; hence, the use of guar gum nanocomposite.

The application of guar gum in combination with nanoparticles for EOR has not been studied extensively. Nanoparticles range between 1-100 nm in size and are composed of a core and thin shell [10]. The core controls the physicochemical properties of the nanoparticle, while the shell is made up of inorganic material [11]. Nanoparticles that have been commonly used in the flooding process are silica and alumina. Others include oxides of copper, magnesium, nickel, zinc, iron tin, etc. Nanoparticles have the potential to change oil wettability, which is an important mechanism for increased oil recovery [12-14]. Silicon oxide nanoparticles have been found to enhance oil recovery when dispersed in ethanol [15]. Aluminium oxide nanoparticles are effective in increasing oil recovery when used in heavy oil fields because aluminium oxide can reduce the viscosity of the oil, therefore leading to increased mobility of the oil [16]. Magnesium oxide nanoparticle has been reported to control fines migration in reservoirs [17-19]. Other nanoparticles that can control fines migration have not been studied in detail. Silicon oxide was chosen in this work because of its potential in EOR, and its cost and availability. It was combined with guar gum to harness the strength of both to enhance oil recovery. Silicon oxide nanoparticle interacts with the oil-rock and oil-water interface to enhance oil recovery, while guar gum reduces the mobility of the displacing fluid. In this work, the viscosity of guar gum under different conditions of weight percentage, temperature and silica concentration was studied. The capacity of guar gum and its nanocomposite for enhanced oil recovery was also studied.

2. Experimental description

2.1. Materials

Guar gum polymer was purchased from a local dealer in Lagos Nigeria. Sigma Aldrich silicon oxide nanoparticles with the purity of 98+% and an average size of 30-50 nm were used for the experiments. Guar gum and silicon oxide nanoparticles were used in viscosity measurement and the core flooding process. Other materials used include; crude oil with API 24.8^0 gotten from Izombe Oilfield, Niger Delta region of Nigeria, and reservoir core samples.

2.1.1. Equipment

The viscosities of all samples were measured using an OFITE® 800 Model Viscometer. The enhanced oil recovery process was initiated and actualized with the use of OFITE® Reservoir Permeability Tester (RPT), and all flooding was done at 28°C. The saturation of core samples for porosity determination was done using Vinci Equipment at high-pressure.

2.2. Preparation of core samples used

Four core samples representing the porous medium were used for the core flooding equipment in the laboratory to measure the effects of testing fluids for this enhanced oil recovery process. For this experimental work, the core samples were acquired from Offshore Depobelt – Niger Delta. The core samples were cleaned in a soxhlet extractor using toluene as the cleaning solvent. Density and fluid viscosity were determined using a pycnometer and glass capillary viscometer, respectively. The core sample was saturated using Vinci Equipment's high-pressure core saturator for Porosity Studies. Table 1 shows the core sample properties.

Table 1: Petro-physical properties of core sample used

Sample	Pore Volume (cc)	Porosity	Permeability (mD)
Core 1	12	0.27	248.15
Core 2	21.8	0.31	228.4
Core 3	20.8	0.39	205.7
Core 4	5.8	0.15	215.3

2.3. Preparation of polymer fluids and polymer nanocomposites

2.3.1. Formulation of guar gum polymer

Plastic bottles were filled to 500 mL with deionized water. Respective quantities of guar gum were then measured and added to the deionized water. Guar gum was prepared to the following weight percentages for viscosity measurements: 0.1,0.2, 0.3, 0.5, and 1. These weight percentages were selected based on literature and pre-tests conducted in the laboratory. Core flooding was done using weight percentages of

0.1, 0.3 and 0.5. These weight percentages were chosen based on the requirement of the OFITE RPT machine (see table 2).

Table 2: Guar – gum polymer composition for flooding

S/N	Weight (%)	Weight (g)	Concentrations
1	0.1	0.5	GGC1
2	0.3	1.5	GGC2
3	0.5	2.5	GGC3

2.3.2. Formulation of guar gum polymer and silicon oxide nanocomposites

Guar gum nanocomposite was prepared using a similar method as the guar gum polymer. The nanocomposite is a mixture of guar gum and silicon oxide nanoparticles in deionized water. For the core flooding experiment, three weight percentages of silicon oxide nanoparticle were mixed with three different weight percentages of guar gum polymer. The three weight percentages of silicon oxide nanoparticle used are 0.1, 0.3 and 0.5 while the three weight percentages used for guar gum are 0.1, 0.3, and 0.5. In this case, silicon oxide nanoparticle was first added to deionized water before the addition of guar gum polymer. Table 3 shows the composition of guar-gum and silicon oxide nanoparticles.

Table 3: Composition of guar-gum and silicon oxide nanoparticle

S/N	Guar-gum concentration [g,%]	Silicon oxide concentrations [g,%]
1	[0.5, 0.1] ~ GGC1	[0.5, 0.1] ~ S1
2	[1.5, 0.3] ~ GGC2	[1.5, 0.3] ~ S2
3	[2.5, 0.5] ~ GGC3	[2.5, 0.5] ~ S3

2.4. Viscosity measurements

Viscosity was measured using OFITE 800 Model Viscometer. The mixture to be measured was filled into the sample cup of the viscometer. The rotor sleeve of the viscometer was immersed to the fill line of the sample cup. The viscometer was set to the required shear rate, and measurement was taken from the dial reading of the viscometer. In the case of higher temperatures, the heating mantle of the viscometer was used to achieve the desired temperature before the measurement was taken.

2.5. Core flooding

Guar-gum and Guar-gum nanocomposite flooding were implemented successively using the OFITE® Reservoir Permeability Tester. All core flooding experiments were done at 28^0C.

This setup comprises three different compartments that are either filled up with crude oil, brine or the polymer solution. For regulation of the flow of fluid, each compartment is equipped with inlet and outlet valves. The flood operates using a high precision pump to inject into one of the compartments at that instant, then acting as a driving force pushing a piston plate inside it, which drives the liquid contained in the compartment at the same time isolating the driving fluid from the displacing reservoir fluid. Figure 1 illustrates the experimental flooding setup. It shows the following parts of the core flooding equipment: pump fluid, pump, valves, displacing reservoir fluid, piston to separate the oils, crude oil, brine, polymer (Treating fluid), pressure gauge, bypass valve, hassler cell holder with core, sleeve pressure, effluent into test tubes.

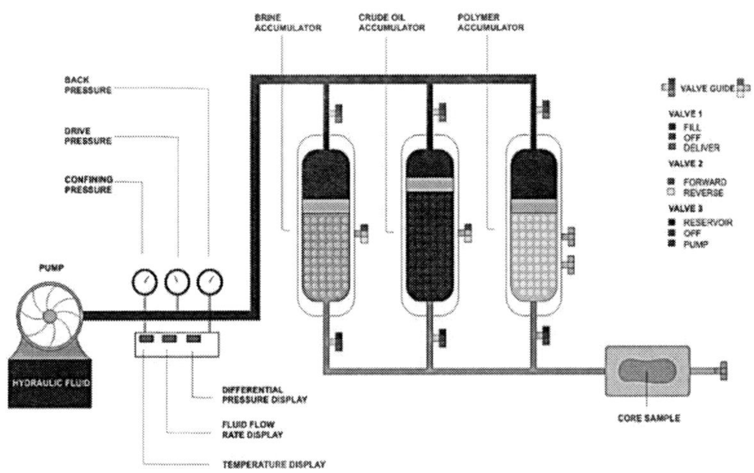

Figure 1: Schematic of the Experimental setup for core flood apparatus (Ojo et al., 2019)

2.5.1. Data collection using core flooding equipment

In carrying out this experiment, the actual reservoir system has to be reproduced at the laboratory scale. The core samples used for this experiment were already saturated 100 % with brine after the cleaning procedure; therefore, there was the need to introduce oil into the core plugs. At a flow rate of 3cc/min, oil was injected into the core plugs draining out the initial brine in the core sample until no more brine was produced. This process enabled the establishment of the initial water saturation, S_{wi}. After injecting oil into the core plugs and establishing an initial water saturation value, at a rate of 3 cc/min water was injected into the core plugs. This was done in the representation of the secondary recovery process.

However, Polymer fluids and polymer nanocomposites Flooding were conducted after water flooding considering the concentration of the recovery agents, and each value was collected every 4mins and recorded. To measure the oil recovery, the experiment was monitored closely all through the duration of each flooding and manually recording the amount of fluid produced at various time intervals.

2.5.2. Polymer fluids and polymer nanocomposites flooding analysis

All individual core plugs were flooded using the same method. Core plug was first injected with oil at a flow rate of 3 cc/min, displacing the brine already 100 % saturated in the core sample. This drainage process was done to attain initial water saturation and oil saturation values. This was followed by the injection of water at 3 cc/min to displace the oil in the core sample till no more oil can be produced via the water being injected. The back pressure set at 200 psi and the confining pressure having built to 700 psi at the end of the flooding.

The production of oil from the core sample into the receiving tube was monitored by observing the level of the oil-brine interface. Movement of the interface level indicated the continuous production of fluid. When oil production stopped, it implied that the core content has reached its residual oil saturation. The volume of the produced oil was oil was recorded in cc. This concluded the secondary recovery process for the core sample, thereby making it viable for tertiary recovery.

After water flooding, the core sample was injected with guar-gum polymer at different concentrations, and the flooding continued until no more oil was produced from the core sample. This was done to investigate the capacity of the guar-gum polymer to recover the residual oil left in the reservoir after water flooding. The flooding process recovered oil continuously till only brine was being produced.

3. Results

3.1. Viscosity profile of guar gum polymer

3.1.1. Effect of weight percent on the viscosity of guar gum

Figure 2 shows the viscosity of guar gum polymer at different weight percentages of guar gum, ranging from 0.1 to 1.0.

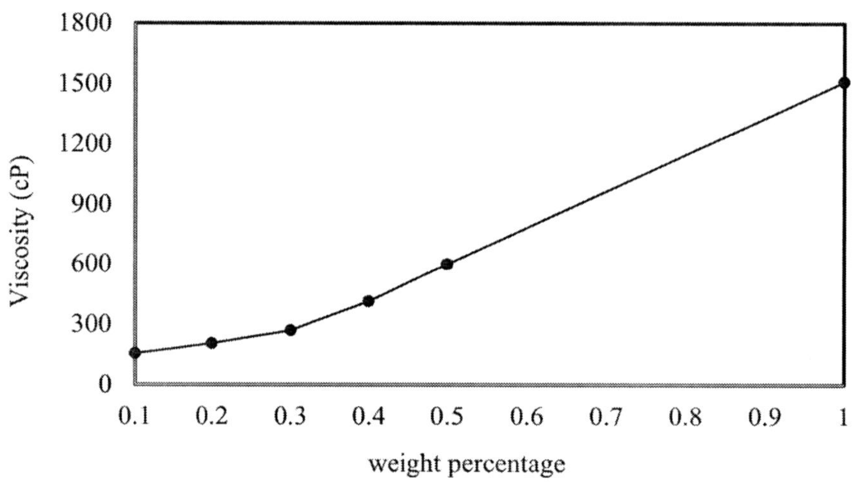

Figure 2: Viscosity of guar gum polymer at different weight percentages

3.1.2. Effect of temperature on viscosity of guar gum

Figure 3 shows the effect of temperature on the viscosity of 1 wt% guar gum at different weight percentages.

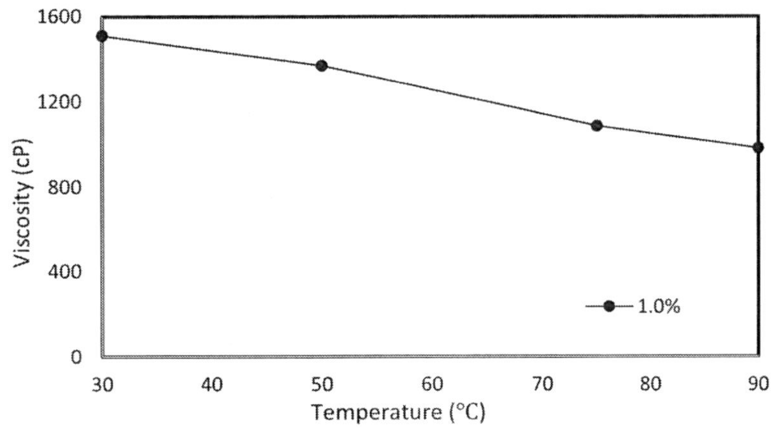

Figure 3: Temperature effect on viscosity of guar gum polymer

3.1.3. Effect of silicon oxide nanoparticle on the viscosity of guar gum

Figure 4 shows the effect silica nanoparticle on the viscosity of guar gum polymer at different weight percentages of silica.

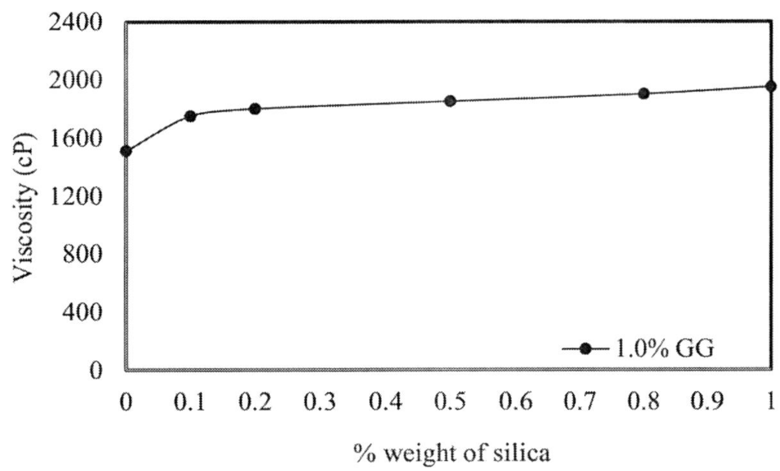

Figure 4: Viscosity profile of guar gum nanocomposite

3.2. Crude oil recovery potential of polymer and nanocomposite

3.2.1 Guar-gum polymer flooding analysis

Figure 5 shows the result for the core flooding experiment conducted using three weight percentages of guar gum polymer. The core flooding of the three weight percentages was done sequentially after water flooding, and result shown in the plot.

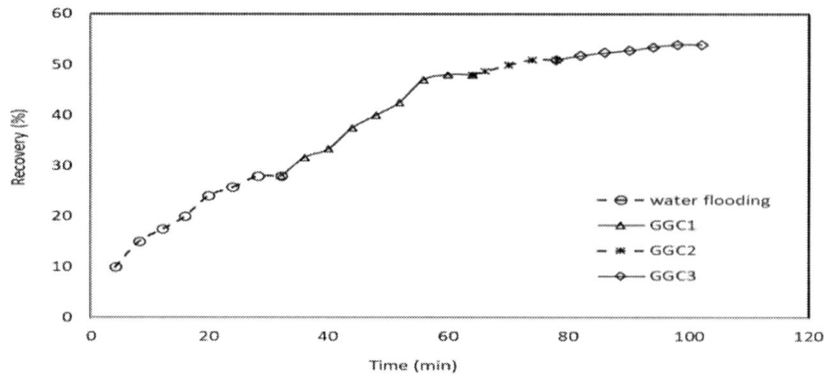

Figure 5: Recovery for guar gum on core sample 1

3.2.2. Guar-gum nanocomposites flooding analysis.

To enhance polymer flooding, nanoparticles were added to the polymer solution to test for improved recovery. The results are presented in Figures 6-8. The lowest weight percentage of guar gum (GGC1) was first used in combination with three weight percentages of silicon oxide nanoparticles incrementally (Table 3). The result is illustrated in Figure 6. Table 4 shows the properties of the core plugs used.

Table 4: Petro-physical properties of core sample 2 used for a new solution (GGC1 and silicon oxide)

Permeability (mD)	228.4	Porosity	0.31
Pore volume	21.8	Length	4.7
Core sample	2	Diameter	2.2

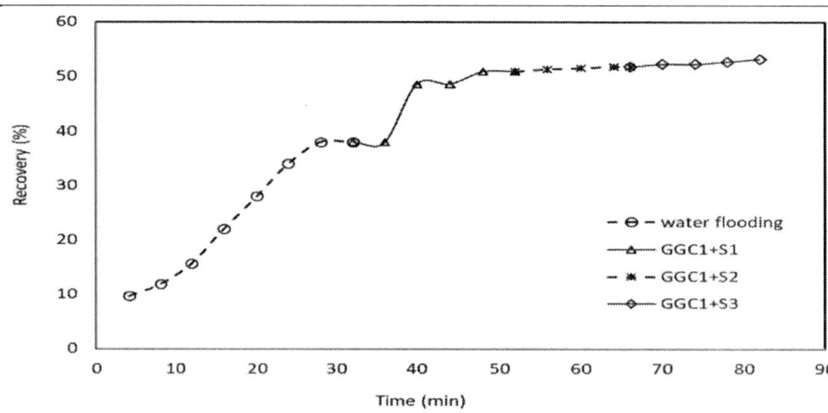

Figure 6: Flooding result of guar gum (GGC1) and silicon oxide nanoparticle

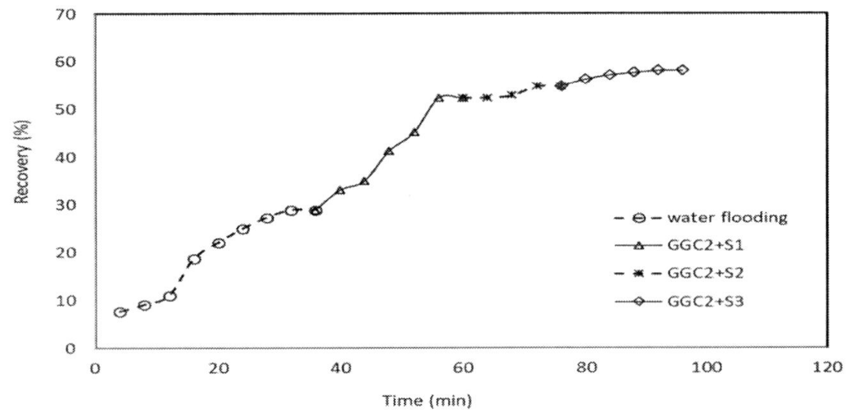

Figure 7: Flooding result of guar gum (GGC2) and silicon oxide nanoparticle

The third concentration of polymer solution from table 2 was mixed with different weight increments of silicon oxide nanoparticle. Table 5 shows the properties of the core plug used and Figure 8 shows the result obtained from polymer (GGC3) and silicon oxide nanoparticle.

Table 5: Petro-physical properties of core sample 4 used for guar gum (GGC3) and silicon oxide nanoparticle

Permeability (mD)	215.3	Porosity (%)	0.15
Pore volume	5.8	Length (cm)	4.5
Core sample	4	Diameter (cm)	2.2

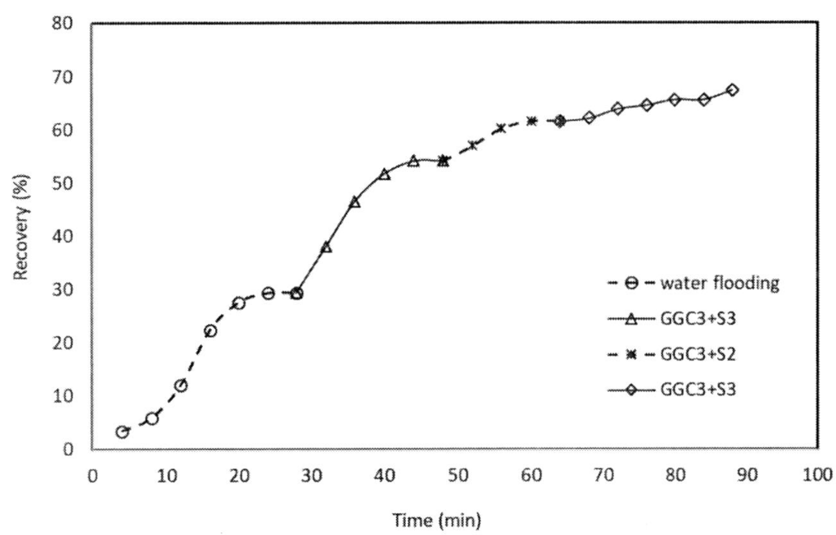

Figure 8: Flooding result of guar gum (GGC3) and silicon oxide nanoparticle

4. Discussion

4.1. Result for viscosity tests of guar gum polymer

The viscosity of guar gum showed a proportionate increase with weight percentage, up to about 1600 cP at 1 weight percent. Also, as the weight percentage increased tenfold, the viscosity was observed to increase by about ten times. From figure 3, it is observed that the viscosity of guar gum reduces with temperature. This means that as the temperature increases, its intermolecular entanglement loosens up and reduces. The viscosity of guar gum dropped by about 25 percent at the temperature was increased from 30 to 90°C. Silicon oxide nanoparticle was observed to have produced a slight increase in the viscosity of guar gum polymer. The viscosity of guar gum increased by about 29 % as the weight percentage of silica was increased to 1 wt%.

4.2. Result analysis for guar-gum polymer flooding

Guar gum polymer flooding was successfully conducted on core sample 1 at different concentrations. The values obtained from the core flooding experiment show some level of recovered oil increment after each flooding. The different concentrations used for guar gum polymer were GG conc 1, GG conc 2 and GG conc 3 selected from table 1.

However, the effectiveness of each concentration was monitored, and the amount of oil recovered, and the recovery was plotted against time, as illustrated from figures 5 to 8. Figure 5 shows a plot of crude oil recovery considering guar gum polymer (Conc 1, Conc 2, Con 3) flooding against the time of core contacted by the solution at 28°C on core plug.

For the lowest concentration of GG conc 1 in figure 5, water flooding performance gave a recovery of approximately 28 % OOIP leaving a huge amount of oil in the core. Sequel to water flooding at figure 2, GG conc 1 was then introduced, and it recovered additional oil up to 48 % accumulated recovery.

When GG, conc 2 was then introduced, figure 5 shows that it recovered additional oil up to 51% OOIP accumulated recovery.

Finally, in figure 5, GG conc 3 was later used for the flooding to check for more oil and it recovered an additional 54 % OOIP. This gave the optimum recovery since any further increment in polymer concentration did not recover more oil.

4.3. Result analysis for guar-gum nanocomposite flooding

To enhance the recovery of polymer flooding, nanoparticle was introduced to polymer flooding. The addition of silicon nanoparticle to each guar gum polymer was done on three different sandstone core plugs. The flooding process was conducted at 28°C, and the pressure was a buildup of 700 psi. Table 2 shows the composition of each nanoparticle mixed with polymer solutions.

The lowest concentration of guar gum (GG conc 1) was mixed with different concentrations of silicon oxide nanoparticle i.e. 0.1, 0.3, and 0.5 wt%. The effectiveness of this new mixture was tested on core sample 2, and the amount of oil recovered was improved and illustrated in figure 6. The recovery increased gradually with each weight percentage achieving 50, 51 and 53 % recovery of OOIP respectively.

Figure 7 shows the result for GG conc 2 mixed with three weight percentages of silicon oxide nanoparticle respectively. Recoveries of 52, 55 and 58 % OOIP were achieved respectively for the three weight percentages. The improvement in mobility ratio and reduction in the interfacial tension between the reservoir fluids added this additional increment.

The highest concentration of guar gum (GG conc 3) was as well mixed with three concentrations of silicon oxide nanoparticle. The effectiveness of this new mixture was tested on the core sample 4 and the amount of oil recovered was improved and illustrated in figure 8. The recovery increases gradually at each weight increment achieving the highest recovery of 67.22 % at the highest concentration of silicon oxide nanoparticle.

5. Conclusion

Polymer nanocomposite worked as a dual oil recovery agent with good rheological behaviour. Several experimental runs were performed to examine the activity of each solution. A rheology device and core flooding equipment were used to obtain results adopted in this article to investigate the EOR potential and rheological behaviours of each solution. Several conclusions are obtained as follows:

(1) The viscosity of guar gum polymer increased proportionately with weight percentage and it reduced by 25 % when the temperature was increased from 30 to 90°C.
(2) Silicon oxide nanoparticle can increase the viscosity of 1 wt% of guar gum by 29 %. The selected system can provide resistance for the polymer under harsh reservoir conditions
(3) The optimized formulation of the guar gum nanocomposite flooding system at 28°C achieved enhanced oil recovery of 67.22 % OOIP while the guar gum polymer of equivalent weight percentage achieved a recovery of 54 %. The improved recovery is attributed to an improvement in mobility ratio and a reduction in the interfacial tension between the reservoir fluids.

As more and more guar gum molecules are trapped in the throat, the intramolecular association of polymers progressively alter into intermolecular condensation, creating a denser network structure. The network structure in the throat plays an important role in increasing flow resistance and changing the direction of liquid flow, and its viscoelasticity contributes to the discharge of residual oil.

Under the synergistic effect of silicon oxide nanoparticle and guar gum polymer, hydrodynamic residual oil, membrane residual oil, and residual oil droplets can be mobilized, thus helping to enhance oil recovery.

References

[1] Fairhurst C 2017 Some Challenges of Deep Mining This paper was written shortly after this author had presented a video lecture titled “Why rock mechanics and rock engineering?” at the invitation of the ISRM. The lecture addressed the more general topic of the importance of rock engineering in the 21st century, and may be of interest as complementary to this article. The lecture (~ 45min) may be heard at https://www.isrm.net/gca/?id=1309. Engineering. 3(4):527-37.
[2] Eiroboyi I, Ikiensikimama S S, Oriji B A, Okoye I P 2019 Experimental Investigation of the Macroscopic Displacement Efficiency of Locally Sourced Locust Bean Gum and Gum Arabic. *SPE Nigeria Annual International Conference and Exhibition;* 2019/8/5/; Lagos, Nigeria. SPE: Society of Petroleum Engineers; p. 13
[3] Reksidler R, Vieira R A M, Orlando A E, Costa B R S, Pereira L S 2015 Offshore Chemical Enhanced Oil Recovery OTC Brasil; 2015/10/27/; Rio de Janeiro, Brazil *OTC: Offshore Technology Conference*
[4] Mudgil D, Barak S, Khatkar B 2011 Guar gum: Processing, properties and food applications-A Review. *J Food Sci Technol.* **51**(3):409-18
[5] Hasan A, Abdel-Raouf M 2018 Applications of guar gum and its derivatives in petroleum

IOP Conf. Series: Materials Science and Engineering 1036 (2021) 012009 doi:10.1088/1757-899X/1036/1/012009

industry: A review *Egyptian Journal of Petroleum* **27**(4):1043-50

[6] Amundarain J L, Castro L J, Rojas M R, Siquier S, Ramírez N, Müller A J, et al 2009 Solutions of xanthan gum/guar gum mixtures: shear rheology, porous media flow, and solids transport in annular flow *Rheologica Acta* **48**(5):491-8

[7] Mandal A 2015 Chemical flood enhanced oil recovery: A review. *International Journal of Oil, Gas and Coal Technology* **9**:241

[8] Ghoumrassi- Barr S, Aliouche D 2015 Characterisation and Rheological Study of Xanthan Polymer for Enhanced Oil Recovery (EOR) Application *Offshore Mediterranean Conference and Exhibition;* 2015/6/1/; Ravenna Italy OMC: Offshore Mediterranean Conference

[9] Su S, Giddins M A, Naccache P, Clarke A, M Howe A 2015 Accurate Modeling of Polymer Enhanced Oil Recovery Corefloods by Reservoir Simulation SPE Reservoir *Characterisation and Simulation Conference and Exhibition* 2015/9/14/; Abu Dhabi, UAE SPE: Society of Petroleum Engineers

[10] Ojo Temiloluwa Ibukun O O, Oguntade Tomiwa 2018 Nano Augumented Biosurfactant Formulation for Oil Recovery in Medium Oil Reservoirs. SPE, NAICE Conference Lagos: SPE

[11] Virkutyte J, and Varma R 2013 Environmentally Friendly Preparation of Metal Nanoparticles. p. 7-33

[12] Ju B, Fan T, Ma M. (2006). Enhanced oil recovery by flooding with hydrophilic nanoparticles *China Particuology* **4**(1):41-6

[13] Di Q, Shen C, Wang Z, Jing B, C.Y G, Qian Y 2010 Innovative drag reduction of flow in rock's micro-channels using nano particles adsorbing method. Society of Petroleum Engineers – *International Oil and Gas Conference and Exhibition in China 2010*, IOGCEC. 2:1152-62.

[14] Ju B and Fan T 2009 Experimental study and mathematical model of nanoparticle transport in porous media. *Powder Technol.* **192**(2):195-202

[15] Onyekonwu M O and Ogolo N A 2010 Investigating the use of nanoparticles in enhancing oil recovery *Nigeria Annual International Conference and Exhibition*; 2010/9/7/; Calabar, Nigeria Society of Petroleum Engineers

[16] Nares H R, Schachat P, Ramirez-Garnica M A, Cabrera M and Noe-Valencia L 2007 Heavy-Crude-Oil Upgrading With Transition Metals. *Latin American & Caribbean Petroleum Engineering Conference;* 2007/1/1/; Buenos Aires, Argentina. SPE: Society of Petroleum Engineers; p. 10.

[17] Huang T, Evans B A, Crews J B and Belcher C K 2010 Field case study on formation fines control with nanoparticles in offshore applications. *SPE Annual Technical Conference and Exhibition;* 19-22 September; Florence Italy: Society of Petroleum Engineers Journal

[18] Huang T, Clark D E 2015 Enhancing Oil Recovery With Specialized Nanoparticles by Controlling Formation-Fines Migration at Their Sources in Waterflooding Reservoirs

[19] Belcher C K, Seth K, Hollier R, Paternostro B P 2010 Maximizing Production Life With the Use of Nanotechnology to Prevent Fines Migration *International Oil and Gas Conference and Exhibition in China;* 2010/1/1/; Beijing, China. SPE: Society of Petroleum Engineers; p. 6

Evaluating solid waste storage and collection practices in Traditional Residential Area of Abeokuta Metropolis, Ogun State, Nigeria

O P Akinpelu, R F Simon, M A Owolabi, and I L Sotayo
Department of Urban and Regional Planning, College of Environmental Sciences, Bells University of Technology, Ota, Ogun State, Nigeria
olusegun_akinpelu@yahoo.com

Abstract. This study evaluated solid waste storage and collection practices in traditional residential area of Abeokuta Metropolis, Ogun State, Nigeria. This is with a view to providing information for policy intervention. The study area consists of seven political wards, stratified into three residential zones, namely; core, transition and sub-urban zones. Each of the zones comprises of neighbourhoods with a number of residential buildings. The three zones consist of 408 houses. Using systematic sampling technique, one of every five building was surveyed, thus, a total of 81 (20%) houses were sampled. Data collected were analysed using descriptive and inferential statistics. The study established a significance difference in income, occupation and tenancy status, while no significant difference exist in residents' education, family size and type of house occupied ($p < 0.05$). The study found that storage receptacles used in the three residential zones were similar and majority (70.4%) was not ecofriendly. Substantial proportion (49.4%) of the residents had no access to waste collection truck. Results of ANOVA computed ($F = 6.721$and $p < 0.05$) established that significant difference exist in the frequency of waste collection. It is recommended that public awareness on environment-friendly waste storage receptacle should be given priority; and provision of adequate equipment and personnel for sustainable waste collection through public-private partnership.

Key words: Storage receptacle, solid waste, waste collection, residents and residential area

1. Introduction

Managing solid waste in urban areas of developing countries poses serious challenges to municipal authorities [1] [2]. Inadequate allocation of funds has been a major factor responsible. Governments efforts at enhancing urban landscape that is aesthetically pleasing and functionally efficient were often thwarted due to inadequate solid waste management. Perpetual heaps of waste in every hook and cranny of most Nigerian cities is a source of concern and attention [3]. However, in order to ensure a sustainable

IOP Conf. Series: Materials Science and Engineering 1036 (2021) 012010 doi:10.1088/1757-899X/1036/1/012010

management of solid waste, priority must be given to every stage involved in its management activities. Solid waste management is a function of combination of various activities. This includes generation, sorting, storage and collection. Others are transportation, processing, treatment and disposal [4] [5] [6]. Attempt to compromise at any stage of the waste management practices, could jeopardise the whole process. Besides, empirical studies on integral of waste management practices would go a long way to provide much needed data that will foster an informed and comprehensive policy on urban solid waste management especially in developing countries such as Nigeria.

The current study is therefore set to investigate empirically solid waste storage and collection practices in the traditional residential area of Abeokuta Metropolis, Ogun State, Nigeria. Traditional residential areas in most Nigerian cities are zone occupied by early settlers in the cities. Regardless of the location, traditional residential areas in the urban centres of Nigeria have similar socio-economic and physical characteristics [7]. The most distinguishing physical feature of these residential districts is that most of them have never been planned [8]. Consequently, houses are built without any reference to a street network. Most of the houses were constructed with the use of locally made building materials, such as sunbaked mud and bricks [8]. However, much of the housing is overcrowded, insanitary and dilapidated. The districts are also without adequate vehicular access. The vast majority of people living in the zone belong to the low socio-economic stratum.

Studies have shown a significance relationship between socioeconomic background of residents and different aspects of their life. For instance, relationship has been established between socioeconomic attributes and housing modernization {9]; transport modal choice [10]; patronage of health facilities [11] solid waste generation [12]; perception of environmental problems and hazards [13] and variation in crime occurrence [14]. The issue of solid waste storage and collection might not be an exception when it is related to socioeconomic background of the residents. However, specific studies that scientifically documented the relationship between solid waste storage and collection practices on one hand; and the residents' socio-economic attributes on the other hand is scanty in literature. The reason for establishing this is predicated on the fact that information on storage receptacle usage across different residential zones will be brought to fore.

This study is therefore an attempt to answer the following salient questions: Are there significant variation in the socioeconomic attributes of the residents in the traditional residential areas? How this variation affects solid waste storage and collection practices? And what are the implication of this on environmental education and public enlightenment? Thus, the hypothesis that, while significant variation exist in the socioeconomic characteristics of the residents in the traditional residential zones, no significant variation exist in the solid waste storage and collection practices in the study area. The findings of the study is expected to provide information for the governments and relevant stakeholders on

IOP Conf. Series: Materials Science and Engineering 1036 (2021) 012010 doi:10.1088/1757-899X/1036/1/012010

how to adequately tackle the issue of solid waste storage and collection that has become intractable in most urban areas of developing countries, Nigeria inclusive.

2. Literature review

Solid waste storage and collection are important components of solid waste management. Because of its importance to successful implementation of sustainable waste management, a number of studies have been conducted on this issue. The work of [15] for instance, reported that some people use the appropriate bins, while others use other inappropriate means such as polythene bags, rice sacks, buckets and open dumps in a study carried out in Abuja, Nigeria. It is pertinent to note that Abuja Environmental Protection Board (AEPB) has a standard for the type of waste-bin that should be used at different land-uses within the city. The size of waste storage container for residential area is expected to be from 120ltr, 240ltr, 1100ltr, 13,000ltr and more. The smaller sizes come in hard plastic while the larger storage containers are made of metal. The location of the storage container is based on building layout and type of housing unit. A study of two local governments in Lagos State reveals that residents store waste in small containers [16]. The study further shows that some of the containers include: plastic and metal; unused buckets, bowls; and polythene and nylon bags. The use of such small storage containers helps residents' to get a good bargain by paying less to informal waste collectors for the disposal of their waste. He added that this storage system attracts disease carrying vectors such as flies and rats. [17] reported that there was no uniformity in waste containers used for storage prior to disposal. The different types of containers used include: plastic waste containers (about $0.4m^3$), waste bags and old/used bucket and were mostly kept outside the house.

It has been observed that between one-third and one-half of the solid waste within most cities in low and middle-income countries are not collected [18] (Ajani, 2007). Similarly, [19] Ogu (2000) interviewed 591 households in Benin-City, Nigeria and found out that three-fifth of the respondents had no solid waste collection service. This is attributed to inadequate resources, and the privatization scheme set up in 1995 to address the environmental issues. The study stressed the need for private partnership with government in providing adequate delivery services to the public. Similarly, the efficiency of the solid wastes collection services in Owerri municipality was assessed and observed that the level of efficiency in waste collection was only about 61%, a situation they attributed to a wide range of socio-economic and technical factors. [20] provided an overview of the state of municipal solid waste (MSW) management in the capital city of Cameroon, Yaoundé. Their study revealed that remoteness and inefficient infrastructure have a major impact on waste collection. In another study, it was observed that waste segregation at source is nonexistent in the Federal Capital Territory (FCT)

Abuja, the waste collection vehicles are not compartmentalized [15]. They noted that waste is collected weekly or as arranged with the AEPB. Waste collection constraints such as waste generation rates, physical development, changes in road networks, illegal structures, traffic conditions, insufficient collection vehicles, traffic density, vehicle conditions, hauling distance, collection time and collection routes are equally observed. [21] observed in a research conducted in Churriana de la Vega (Granada, Spain) that the town had an excessive number of containers for organic matter and rest-waste fraction, reducing the efficiency of waste collection and raising the costs related to the purchase of containers, collection time, personnel costs, collection route length, and vehicle maintenance. However, they asserted that optimization of solid waste collection can reduce the management costs as well the negative impacts on the environment.

3. Methodology

The study area, Abeokuta Metropolis consists of seven political wards (Ogun State Electoral Commission, 2015). Each of the wards comprises of neighbourhoods with a number of residential buildings, which can be stratified into core, transition and sub-urban zones. Using systematic sampling technique, one of every two wards (50%) was selected. This translates to three wards, representing the core, transition and sub-urban areas. Areas selected consist of varying number of neighbourhoods. For example, core zone has five neighbourhoods, namely: Fajol Estate, Kotogbo, Ebenezer and Oke Aregba. There are seven neighbourhoods in transition zone. These include: Labaiwa, Saje, Oke Efon, Kugba and Isoka. Others are Oke Saje and Arinlese. Sur-urban comprises five neighbourhoods. These are Afo, Emere, Itoko, Itiki-Titun and Oke Aleji. In order to arrive at sample size that is representative, one of every two neighbourhoods were selected in each of the wards. This added up to sampling of two, five and two neighbourhoods in Wards II, IV and VI respectively.

The two neighbouhoods selected in core area consists of 161 residential buildings. Whereas, there were 151 buildings in the five neighbourhoods sampled in transition zone, while two neighbourhoods picked in sub-urban had 96 buildings. Total number of buildings in all the neighbourhoods selected added up to 408 buildings. Using systematic sampling technique, one of every five building was surveyed. Thus, a total of 81 buildings were sampled. The questionnaire was targeted at the adult female members of the household. This is because the focus of the study is more or less a woman's business in Yorubaland. However, where this is not applicable or impossible such as the cases of bachelors, spinsters or widower, respondent would either be a male or female adult. Information to be obtained would include: households' social, economic and demographic attributes and how these would affect solid waste generation, storage and disposal activities; and willingness and affordability to pay management costs among other things.

Table 1: Residential zone, neighbourhoods and buildings selected

Residential zone	No. of Neighbourhood	Selected Neighbourhood	No. of Residential Building	Selected Residential Building
Core	5	2	161	32
Transition	7	3	151	30
Sub-urban	5	2	96	19
Total	17	7	408	81

Source: Ogun State Electoral Commission, 2015

Data collected were analysed using descriptive and inferential statistics. The descriptive statistics used were cross-tabulation, frequency distribution, mean and percentage. Analysis of Variance (ANOVA) was used to see if there is a significant difference between socioeconomic variables of the residents and solid waste storage and collection practices in the study area.

4. Results and discussion

The focus of this section is to discuss findings on residents' socioeconomic characteristics, solid waste storage and collection practices in the study area.

4.1 Residents' socioeconomic background

The residents' socioeconomic characteristics investigated include education, income and occupation. Others are tenancy status and family size. Findings in the different zone of the residential area are as summarized in Table 2. Majority of the residents were holders of either Ordinary National Diploma (OND) certificate/Nigerian Certificate in Education (NCE) or Higher National Diploma (HND) certificate. Their proportions are 32.1% and 48.1% respectively. Residents with secondary school certificate accounts for 11.1%, while those having technical certificate represent 6.2%. The least (2.5%) are those with primary school certificate. The preponderance of HND or First Degree holders in the study area might be attributable to early awareness of the importance of education through the missionary efforts at establishing first set of formal education in the area. All things being equal, the level of education in the area is expected to influence the storage and collection practices positively. However, there is no significant difference in the educational background of the residents across different residential zones with ANOVA results of $F = 0.257$ and $p > 0.05$. Results of monthly income reveal that almost fifty percent of the residents are within the low income group. For instance, 24.7% and 23.5% of the residents earned less than #18,000 and between #18,000 and #37,000 respectively, while those that earned above #57,000 accounts for 51.8%. The computed ANOVA ($F = 4.394$ and $p < 0.05$) established that a significance difference exits in the monthly income of residents in the surveyed wards.

Table 2: Residents' socioeconomic background

Parametres	Residential Zone			Total
	Core N (%)	Transition N (%)	Sub-urban N (%)	N (%)
Education background				
Primary	1 (3.1)	1 (3.3)	0 (0.0)	2 (2.5)
Secondary	4 (12.5)	2 (6.7)	3 (15.8)	9 (11.1)
ND/NCE	11 (34.4)	8 (26.7)	7 (36.8)	26 (32.1)
HND/First Degree	13 (40.6)	18 (60.0)	8 (42.1)	39 (48.1)
Technical	3 (9.4)	1 (3.3)	1 (5.3)	5 (6.2)
Total	**32 (100)**	**30 (100)**	**19 (100)**	**81 (100)**
Monthly income				
Less than #18,000	12 (37.5)	4 (13.3)	4 (21.1)	20 (24.7)
#18,000-#37,000	6 (18.8)	4 (13.3)	9 (47.4)	19 (23.5)
#38,000-57,000	8 (25.0)	9 (30.0)	2 (10.5)	19 (23.5)
#58,000-#77,000	1 (3.1)	2 (6.7)	2 (10.5)	5 (6.2)
#78,000-#97,000	2 (6.3)	4 (13.3)	0 (0.0)	6 (7.4)
Above #97,000	3 (9.4)	7 (23.3)	2 (10.5)	12 (14.8)
Total	32 (100)	30 (100)	19 (100)	81 (100)
Occupational background				
Farming	4 (12.5)	0 (0.0)	1 (5.3)	5 (6.2)
Trading	7 (21.9)	4 (13.3)	5 (26.3)	16 (19.8)
Artisan	3 (9.4)	1 (3.3)	1 (5.3)	5 (6.2)
Civil service	13 (40.6)	15 (50.0)	7 (36.8)	35 (43.2)
Professional	4 (12.5)	8 (26.7)	5 (26.3)	17 (21.0)
Others	1 (3.1)	2 (6.7)	0 (0.0)	3 (3.7)
Total	**32 (100)**	**30 (100)**	**19 (100)**	**81 (100)**
Tenancy status				
Landlord	8 (25.0)	17 (56.7)	3 (15.8)	28 (34.6)
Tenant	23 (71.9)	12 (40.0)	16 (84.2)	51 (63.0)
Squatter	1 (3.1)	1 (3.3)	0 (0.0)	2 (2.5)
Total	**32 (100)**	**30 (100)**	**19 (100)**	**81 (100)**
Family size				
Less than 2 people	1 (3.1)	0 (0.0)	2 (10.5)	3 (3.7)
2 – 5 people	17 (53.1)	16 (53.3)	8 (42.1)	41 (50.6)
6 - 9 people	10 (31.3)	10 (33.3)	8 (42.1)	28 (34.6)
10 – 13 people	3 (9.4)	4 (13.3)	1 (5.3)	8 (9.9)
More than 13 people	1 (3.1)	0 (0.0)	0 (0.0)	1 (1.2)
Total	**32 (100)**	**30 (100)**	**19 (100)**	**81 (100)**
House type				
Self contain	18 (56.3)	23 (76.7)	6 (31.6)	47 (58.0)
Face-me-I-face-you	10 (31.3)	3 (10.0)	13 (68.4)	26 (32.1)
Traditional courtyard	4 (12.5)	4 (13.3)	0 (0.0)	8 (9.9)
Total	**32 (100)**	**30 (100)**	**19 (100)**	**81 (100)**

Source: Author's field survey, 2018

Occupational background of the residents shows that majority (43.2%) were civil servants. The proportion of the residents that were professionals and traders represent 21.0% and 19.8% respectively. Those that were farmers, artisans and others (such as driver, cleaner, gatekeeper, conductor, etc.)

constitute 6.2%, 6.2% and 3.7% respectively. There is a statistically significant difference existed in the occupation of the residents as F value of computed ANOVA is 3.427 at $p < 0.05$. The residents' tenancy status was group into three. Of the three groups, tenants are the majority. Fifty One (51) of the 81 respondents were tenants. This accounts for 63.0%. Its distribution shows that 71.9%, 40.0% and 84.2% of the residents in core, transition and sub-urban zones were tenants respectively. While landlords constitute 34.6%, squatter only accounts for 2.5%. There is a statistical significance difference in the tenancy status of the residents across the surveyed zones with computed ANOVA ($F = 4.394$ and $p < 0.05$). Family size of between 2 to 5 people accounts for the majority (50.6%). Residents with family size of between 6 and 9 people represents 34.6%, While those that 10 to 13 people and less than 2 people constitute 9.9% and 3.7% respectively, residents with family size that is above 13 people was the least (1.2%). No statistical significant difference in the family size of the residents across the sampled zones as F value of the computed ANOVA is 2.825 at $p > 0.05$. The building type in the study area was majorly self-contained, representing 58.0% of the type of house residents occupied. While face-me-I-face-you constitute 32.1%, traditional courtyard accounts for 9.9%. However, the computed ANOVA established that there is no significance difference exists in the type of house residents occupied with $F = 0.321$ and $p > 0.05$.

4.2 Solid waste storage practice

As presented in Table 3, different waste storage receptacles were used by the residents. According to the findings, these include: paper carton, sack, dustbin, polythene bag, bucket and basket among others. Most of the residents (29.6%) used dustbin as a storage receptacle. While 24.7% of the residents used polythene bag, those that use sack represent 21.0%. The proportion of residents that employed the use of basket, bucket and others (such as nylon bag) represent 14.8%, 4.9% and 3.7% respectively. Paper carton accounts for the least (1.2%) as storage receptacle in the study area. The computed ANOVA ($F = 0.703$ and $p > 0.05$) established that there is no statistically significance difference in the waste storage receptacles used in the different ward of the study area.

Table 3: Type of storage receptacle

Residential zone	Storage receptacle							Total (%)
	Paper carton (%)	Sack (%)	Dustbin (%)	Polythene bag (%)	Bucket (%)	Basket (%)	Others (%)	
Core	1 (3.1)	9 (28.1)	12 (37.5)	2 (6.3)	1 (3.1)	6 (18.8)	1 (3.1)	32 (100)
Transition	0 (0.0)	4 (13.3)	8 (26.7)	11 (36.7)	3 (10.0)	2 (6.7)	2 (6.7)	30 (100)
Sub-urban	0 (0.0)	4 (21.1)	4 (21.1)	7 (36.8)	0 (0.0)	4 (21.1)	0 (0.0)	19 (100)
Total	**1 (1.2)**	**17 (21.0)**	**24 (29.6)**	**20 (24.7)**	**4 (4.9)**	**12 (14.8)**	**3 (3.7)**	**81 (100)**

Source: Authors' field survey, 2019

This implies that irrespective of the residential zone, the type of solid waste storage receptacle used by most of the residents is similar. It is noteworthy that of all the storage receptacle identified in the study area, only dustbin (29.6%) can be considered to be ecofriendly, other receptacle pose serious danger to the environment and residents' health.

4.3 *Accessibility to waste collection truck*

Findings on accessibility of waste collection truck in the study area reveal that over fifty percent (50.6%) of residents indicated that they have access to collection truck, while almost fifty percent (49.4%) confirmed that waste collection truck is inaccessible to them. The finding of this study is similarly to the work of [19]. He documented that three-fifth of the respondents had no solid waste collection service in his study of 591 households in Benin-City, Nigeria. This might be one of the reasons why road divides and road sides of most urban areas in Nigeria are often littered with waste.

Table 4: Accessibility to waste collection truck

Residential zone	Accessibility to truck		Total (%)
	Yes (%)	No (%)	
Core	22 (68.8)	10 (31.3)	32 (100)
Transition	13 (43.3)	17 (56.7)	30 (100)
Sub-urban	6 (31.6)	13 (68.4)	19 (100)
Total	**41 (50.6)**	**40 (49.4)**	**81 (100)**

Source: Authors' field survey, 2019

4.4 *Solid waste collection practice*

The frequency of solid waste collection is very important to its sustainable management. Uncollected waste has been known to serve as haven of rodents and insects, creating an eye sore, offensive odours and environmental contamination through production of leachate that could have detrimental effects. As observed from Table 5, residents that affirmed that their waste is collected once in a week constitute the majority (29.6%). Those that indicated that their waste is picked fortnightly represent 24.7%. While those that confirmed that their waste is collected once every two days accounts for 21.0%, residents that indicated that they cannot remember constitute 14.8%.

Table 5: Frequency of solid waste collection

Residential zone	Frequency of solid waste collection						Total (%)
	Once daily (%)	Once every two days (%)	Once in a week (%)	Once in a fortnight (%)	Randomly (%)	I cannot remember (%)	
Core	1 (3.1)	9 (28.1)	12 (37.5)	2 (6.3)	1 (3.1)	6 (18.8)	32 (100)
Transition	0 (0.0)	4 (13.3)	8 (26.7)	11 (36.7)	3 (10.0)	2 (6.7)	30 (100)
Sub-urban	0 (0.0)	4 (21.1)	4 (21.1)	7 (36.8)	0 (0.0)	4 (21.1)	19 (100)
Total	**1 (1.2)**	**17 (21.0)**	**24 (29.6)**	**20 (24.7)**	**4 (4.9)**	**12 (14.8)**	**81 (100)**

Source: Authors' field survey, 2019

Residents that enjoyed waste collection service every day represent 1.2%. The implication of this finding is that the residents whose waste is collected within a week might suffer offensive odour and breeding of mosquitoes from such uncollected waste. This might later result into infection of diseases such as malaria, diarrhea, dysentery and lassar fever among others. However, the computed ANOVA established that statistically significance difference exists between the frequency of solid waste collection and residential zones with $F = 6.721$ and $p < 0.05$. The variation might be as a result of inaccessibility of some houses especially in core area.

5. Conclusion

The study revealed that while there is differential in residents' monthly income, occupation and tenancy status, however, the storage receptacles employed across the different residential zones were similar. Most of the storage receptacles in use in the study area were observed to be environmentally-unfriendly. The frequency of solid waste collection varied across the residential zones. Some of the residents in the study area did not have access to waste collection truck. This might promote indiscriminate disposal of waste, with attendant consequences such as outbreak of disease, flooding and unsightliness among others. It is therefore suggested that urgent step should be taken at creating awareness on the need for adoption of ecofriendly waste storage receptacle on one hand, and provision of adequate waste collection equipment and personnel for sustainable solid waste collection service delivery through public-private partnership. With the continual geometric increase in urban population, governments alone could not sufficiently shoulder the responsibility of providing adequate urban social services. It is therefore, imperative that the governments especially municipal authority must see the need to collaborate with private organization in the area of waste management and by extension other urban social services.

The need to pay more attention to cleanliness of urban environment becomes more important with the advent of global pandemic (Covid-19), which is believed to spread faster in an untidy environment. Both residents and municipal authority must be on the same page towards addressing the issue of waste storage in a manner that would foster conducive, functional and aesthetically pleasing urban settlements. Nigerian urban areas cannot afford to be behind in the global trend in relation to sustainable waste management. Considering the cardinal role urban areas play in economic development of a nation, any attempt to downplay the required efforts at ensuring adequate solid waste storage and collection might spell doom for the entire process of waste management. Consequently, socio-economic development might be greatly hampered.

References

[1] Adekunle IM, Adebola AA, Aderonke KA, Pius OA and Toyin AA 2011 Recycling of organic wastes through compos ting for land applications: A Nigerian experience, *Waste Manag. Res.*, 29, **6**: 582-93

[2] Jalil A 2010 Sustainable development in Malaysia: A case study on household waste management. *Sustainab. Dev.*, 3, **3**: 91-102
{3] Imam A, Mohammed B, Wilson DC and Cheeseman CR 2008 Solid waste management in Abuja, Nigeria, *Waste Management*, **28,** 468-472
[4] Robinson WD 1986 *The solid waste handbook: A practical guide,* Chichester, John Wiley & Sons, pp. 7
[5] Asanuma M and Ariyama T 2004 Recycling of waste plastics in blast furnace. *Journal of Japan Institute of Energy*, **83**, 4: 252-256
[6] Mastral FJ, Esperanza E, Berrueco C, Juste M and Ceamanos J 2003 Fluidized bed thermal degradation products of HDPE in an inert atmosphere and in air–nitrogen mixtures. *Journal of Analytical and Applied Pyrolysis*, **70,** 1, 1-17
[7] Onokerhoraye AG 1994 *Benin: A traditional African city in transition,* Benin: Benin Social Science Series, University of Benin
[8] Onokerhoraye AG 1977 "The spatial pattern of residential districts in Benin, Nigeria" *Urban Studies*, **14**, 292-302
[9] Adedibu AA 1989 Housing modernization and socio-economic area in Ilorin, *Africa Urban Quarterly*, 1&2, 30-37
[10] Fadare SO and Wojuade CO 2007 Study of traffic delay situation in Ibadan, Nigeria. *Planning Journal*, **3**, 1, 1-16
[11] Adeyinka SA 2007 Multivariate analysis of health care service utilization factors in Ife region, Nigeria, *Ife Planning Journal*, **3**, 1
[12] Afon AO and Okewole IA 2007 Estimating ihe quantity of solid waste generation in Oyo, Nigeria, *Waste Manag. Res.* **25**, 371–379
[13] Sjoberg L 2000 *Factors in risk perception, Risk analysis*, **20**: 1, 1-12
[14] Adedibu and Abiodun FO 2005 Intra-urban variations in crime occurrence: case of Ogbomoso, Oyo State, Nigeria. *Journal of Nigerian Institute of Town Planners*, **13**, 31-38
[15] Njoku CG and Okeniyi OO 2014 The survey of waste-Bins and collection methodology In households of Federal Capital City, Abuja, Nigeria, *International Journal of Scientific and Technological Research*, **3**
[16] Afon AO 2007 Informal sector initiative in the primary sub-system of urban solid waste management in Lagos, Nigeria, *Habitat Int.* **31**, 193–204
[17] Ibrahim SI, Awomuti AA and Ajibade LT 2012 Storage and treatment system of solid waste as strategy for sustainable environmental development in Ilorin Metropolis, Nigeria *Applied Science Research,* 3, **2**, 801-808
[18] Ajani OIN 2007 Determinants of an effective solid waste management in Ibadan Metropolis, Oyo State, Nigeria, *Journal of Food, Agriculture and Environment*, **6**:1, 152-157
[19] Ogu V 2000 Stakeholders' partnership approach to infrastructure provision and management in developing world cities: lessons from sustainable Ibadan project, *Habitat International*, **24**: 4, 517-523
[20] Parrot L, Sotamenou J and Kamgnia Dia B 2009 Municipal solid waste management in Africa: Strategies and livelihoods in Yaoundé, Cameroon. *Waste Manage*., **29:** 2, 986-995, ISSN 0956-053X, http://dx.doi.org/10.1016/j.wasman.2008.05.005
[21] Zamorano M, Molero E, Grindlay A, Rodríguez MI Hurtado A and Calvo FJ 2009 A planning scenario for the application of geographical information systems in municipal waste collection: A case of Churriana de la Vega (Granada, Spain). *Resour. Conserv. Recy.* **54**(2), 123-133, ISSN 0921-3449, http://dx.doi.org/10.1016/j.resconrec.2009.07.001

IOP Conf. Series: Materials Science and Engineering 1036 (2021) 012011 doi:10.1088/1757-899X/1036/1/012011

Water Quality of Selected Sachet Water Brands Sold in Sango-Ota, Nigeria

K S Ayegbo[1], K E Ogundipe[2], G Olimaro[1], A D Ajimalofin[1] and I I Akinwumi[1]
[1]Department of Civil Engineering, Covenant University, Ota, Ogun State, Nigeria.
[2]Department of Building Technology, Covenant University, Ota, Ogun State, Nigeria.
Corresponding Author: korede.ayegbo@covenantuniversity.edu.ng

Abstract. Clean and safe water drinking is essential for the attainment of the United Nations sustainable development goals. The study assessed the physical and chemical parameters of ten (10) brands of sachet water sold at Sango-Ota in Ogun State, Nigeria. The physical parameters (temperature, electrical conductivity (EC), total dissolved solids (TDS), total suspended solids (TSS), power of hydrogen (pH) and turbidity) and chemical parameters ((calcium (Ca), potassium (K), copper (Cu), manganese (Mn), zinc (Zn), magnesium (Mg), iron (Fe) and chloride (Cl)) of the sachet water samples were determined. The physical parameter tests were analysed using Hanna Instruments (Model HI 98130). The compliance of the physico-chemical parameters of the samples with the drinking water standard requirements of the World Health Organisation (WHO) and Nigeria Standard Drinking Water Quality (NSDWQ) was analysed. The pH values of the samples were within the range 6.5-8.5, in conformity with the WHO and NSDWQ standard for drinking water. The electrical conductivity ranged from 0.02 - 0.06 μS/cm, falling within the permissible limit (1000 μS/cm) of WHO standard. The TDS of 0.01 - 0.03 mg/l were within the recommended permissible limit (1000mg/l) recommended by WHO. Similarly, the TDS, TSS and Turbidity were within the recommended permissible limits of the WHO. For the chemical parameters, the concentrations of K, Cu, Mn, Zn, Mg, Fe and Cl in the sachet water samples were within the WHO permissible limits of drinking water. However, some brands had their Ca concentrations above the WHO drinking water permissible limit. Generally, all the brands of sachet water samples studied were considered safe for drinking.

1. Introduction

Water is one of the major natural resources endowed mankind for the continued existence [1, 2]. However, access to a portable watercourse remained the basic need for all human beings [3]. It is needed to maintain the proper functioning of the human body. Little wonder, when ill persons get admitted in hospitals, normal saline solution (sterile water with sodium chloride) is usually administered to those who cannot take fluids orally and are in danger of being dehydrated. Water is also vital for personal hygiene.

Judging by the recent report from the World Health Organization and UNICEF[4], about 1 in 3 persons globally lack access to portable water. This lack of clean and safe drinking water is more challenging in many developing countries like Nigeria. The government water supply schemes serve only a few. People are, therefore, left with no other option than to source their water themselves. Some people source their water from streams, rivers, hand-dug wells or drilled borehole wells. Others purchased their drinking water from water vendors. Some of these vendors sell water using different sealed and unsealed containers. Cart-pushing vendors sell unsealed water in 25 litre or 50 litre containers, while some water vendors sell water in sealed plastic bottles and sachet (plastic) bags. This research work aims at assessing the quality of some of the selected brands of sachet water sold in Sango-Ota metropolis of

Ogun State, Southwest, Nigeria using Hanna Instruments (Model HI 98130) in analysing the physical parameters. While Palintest Photometer 8000 was used in checking the level of chemical pollutants in the water samples.

2. Materials and Methods

This research obtained two samples of sachet water procured from ten different sources across Sango-Ota in Ado-Odo Ota Local Government Area, Ogun State, Nigeria, in order to assess the quality of the sachet water consumed in the study area. All the sachet water brands selected were registered under the National Agency for Food and Drug Administration and Control (NAFDAC) [5]. Their consumption is not limited to the study area.
The samples were kept in a refrigerator at 4°C for twenty-four hours prior to the analysis at the Environmental Laboratory of Covenant University. The brands were denoted by A, B, C, D, E, F, G, H, I and J, while 1 and 2 were used to indicate the two different samples. Physical analysis like odour, taste and colour were determined using physical observations. The temperature, electrical conductivity (EC), total dissolved solids (TDS) and pH of the water samples were determined using the Hanna Multiparameter Instrument (model HI 98130) under room temperature (27°C), while the turbidity was determined using a turbidity meter. Filtration method was used to determine the total Suspended Solids (TSS) present in the samples. Calcium (Ca), Potassium (K), Copper (Cu), Manganese (Mn), Zinc (Zn), Magnesium (Mg), Chloride (Cl) and Iron (Fe) were analysed using a Palintest Photometer 8000.

A part of each of the samples was poured into different beakers and a well-calibrated pen-type Hanna multiparameter instrument was dipped into the samples. The device was suspended for the tip not to touch the base of the apparatus for an accurate reading. When the device reading became stable, it was recorded. The TSS was determined by filtration method. A little portion of each sample was poured into a beaker and weighed (initial reading, D_0). Filter paper of size 110 mm diameter was folded properly into a conical flask. The sample was then poured into the filter paper. When the filtration process was completed, the filtrate was poured back into the beaker and weighed. This last reading was recorded as the final reading (D_1). The difference in initial (D_0) and final reading (D_1) was taken as the TSS (see Equation 1).

$$D_0 - D_1 = TSS \quad (1)$$

The turbidity test was carried out using a turbidity meter. A turbidity test tube was selected, filled with the sample to the 10 ml mark and inserted into the turbidity meter. When the reading became stable, the result was recorded.

Using the Palintest Photometer 8000, calcium (Ca), potassium (K), copper (Cu), manganese (Mn), zinc (Zn), magnesium (Mg), chloride (Cl) and iron (Fe) were analysed. The water samples were filtered to obtain a clear solution. For each of the elements analysed, two (2) 10 ml test tubes were rinsed with distilled water, filled with the water sample and the palintest water test tablets or reagents was added. For the determination of the concentrations of calcium, potassium, copper, manganese, zinc, and chloride, readings were taken directly from the photometer. In contrast, those of magnesium and iron were determined after their solutions were left to stand for the specified time (using a stopwatch) to allow for full colour development. Also, the safety precautions observed during the experimental work was in line with [6].

3. Results and Discussion

The results for the determination of the odour, taste, colour, pH, TDS, EC, temperature, TSS and turbidity of the selected sachet water samples are shown in Table 1.

Table 1. Physico-chemical parameters of selected sachet water brands in Sango-Ota

ID	Odour	Taste	Colour	pH	TDS (mg/l)	EC (μS/cm)	Temperature (°C)	TSS (mg/l)	Turbidity (NTU)
A1	Odourless	Tasteless	Colourless	8.08	0.03	0.06	25.5	0.01	0.35
A2	Odourless	Tasteless	Colourless	8.05	0.03	0.06	28.8	0	0.35
B1	Odourless	Tasteless	Colourless	7.71	0.02	0.05	25.3	0	0.31
B2	Odourless	Tasteless	Colourless	7.64	0.02	0.05	28.9	0.01	0.25
C1	Odourless	Tasteless	Colourless	7.31	0.01	0.05	30.3	0.01	0.16
C2	Odourless	Tasteless	Colourless	6.98	0.01	0.03	30.5	0.01	0.29
D1	Odourless	Tasteless	Colourless	6.62	0.01	0.03	25.9	0	0.23
D2	Odourless	Tasteless	Colourless	6.65	0.01	0.03	25.5	0	0.24
E1	Odourless	Tasteless	Colourless	6.42	0.02	0.03	25.7	0.01	0.18
E2	Odourless	Tasteless	Colourless	5.97	0.01	0.02	25.4	0.01	0.28
F1	Odourless	Tasteless	Colourless	6.06	0.02	0.04	26	0	0.25
F2	Odourless	Tasteless	Colourless	6.07	0.02	0.04	26	0	0.28
G1	Odourless	Tasteless	Colourless	6.74	0.01	0.02	26.6	0	0.25
G2	Odourless	Tasteless	Colourless	6.8	0.01	0.03	26.6	0	0.16
H1	Odourless	Tasteless	Colourless	7.07	0.03	0.05	26.3	0.01	0.25
H2	Odourless	Tasteless	Colourless	7.08	0.03	0.05	26.4	0.01	0.22
I1	Odourless	Tasteless	Colourless	5.74	0.02	0.05	23.5	0.01	0.02
I2	Odourless	Tasteless	Colourless	5.84	0.02	0.05	23.5	0.01	0.02
J1	Odourless	Tasteless	Colourless	6.6	0.02	0.04	22.9	0	0.11
J2	Odourless	Tasteless	Colourless	6.54	0.02	0.05	29.6	0	0.06
WHO	Odourless	Tasteless	Colourless	6.5-8.5	1000	1000	-	-	5
NSDWQ	Odourless	Tasteless	Colourless	6.5-8.5	1000	1000	-	25	-

A1-J2 was used to denote sample ID of the selected sachet water brands.

The physical analysis of the sachet water samples studied showed that all the brands were colour, odour and taste free, which conforms with the World Health Organisation [7] and Nigeria Standard Drinking Water Quality (NSDWQ) [8] standards for drinking water.

The pH of a given water sample is the measure of hydrogen ions (H^+) in the water, and it is measured on a scale of 1-14. If the pH of the sample is between 0 - 7, the water is said to be acidic. However, if the pH is at 7, it is neutral. A pH between 7 - 14 is considered basic. The pH of the brands D, E, F, G, I and J can be described as being slightly acidic (<7), while the pH of the brand H is neutral. The pH of the brands A, B and C can be described as slightly basic. However, the pH of all the brands was within the range of acceptable pH (6.5 – 8.5) recommended by the WHO [7] and NSDWQ [8] drinking water quality standards.

The TDS of the brands ranged from 0.01-0.03 mg/l and are within the recommended permissible limits specified by the WHO [7]. This also indicates that all the values are within the acceptable limit of 500 mg/l recommended by the NSDWQ [8].

The figures of Electrical conductivity (EC) of the samples tested ranges from 0.02 - 0.06 μs/cm, these were is inline with [8-10]. Using physical testing method, the EC of the given samples also gave a clarity, which mean the water is portable for comsuption. The EC in all the sachet water samples analysed were below the maximum limits of 1000 μs/cm stated by NSDWQ [8].

The TSS figures recorded ranges from 0 – 0.01 mg/l, these falls within the maximum limits of NSDWQ [8] of 25 mg/l. This indicates that all the TSS values of the samples were within the recommended limits of WHO [7] and NSDWQ [8].
Turbidity is attributed to cloudiness of water due to variety of particles present in drinking water. The turbidity of the twenty (20) samples under investigation fall within the maximum limits of WHO turbidity standard of portable water of 5 Nephelometric Turbidity Unit (NTU). According to [9-11] the sources of turbidity includes; sewage solids, silt and sand washings, clay particles, organic and biological sludges or industrial discharges.

The results for the determination of the chemical constituents in the selected sachet water samples are presented in Table 2.

Table 2. Elemental test of the selected sachet water brands in Sango-Ota

ID	Calcium (Ca) (mg/l)	Potassium (K) (mg/l)	Copper (Cu) (mg/l)	Manganese (Mn) (mg/l)	Zinc (Zn) (mg/l)	Magnesium (Mg) (mg/l)	Iron (Fe) (mg/l)	Chloride (Cl) (mg/l)
A1	124	2.1	<0.00	<0.000	0	<0.0	<0.00	21
A2	231	1.8	0.88	0.003	<0.00	<0.0	0.01	3.4
B1	342	1.6	0.06	0.001	0.38	<0.0	<0.00	0
B2	349	2.2	0.34	<0.000	<0.00	<0.0	<0.00	5.4
C1	270	1.2	<0.00	<0.000	<0.00	<0.0	<0.00	<0.0
C2	153	2.3	<0.00	<0.000	<0.00	<0.0	<0.00	<0.0
D1	269	<0.0	<0.00	0	<0.00	<0.0	<0.00	22.5
D2	69	6.7	<0.00	<0.000	<0.00	<0.0	<0.00	0
E1	34	9	0.58	0	0.05	<0.0	0.06	25.5
E2	99	2.5	<0.00	0.003	0.01	<0.0	<0.00	5.7
F1	63	1.9	0	<0.000	0.05	0	0.02	42
F2	122	4	<0.00	0.001	<0.00	<0.0	<0.00	32
G1	188	1.8	<0.00	0.001	<0.00	<0.0	0.02	32
G2	79	1.3	<0.00	0.002	<0.00	<0.0	<0.00	14.5
H1	*ND	5	0.02	0	<0.00	<0.0	<0.00	2.2
H2	40	7	<0.00	<0.000	0.21	<0.0	0.01	20
I1.	83	5	0.42	<0.000	<0.00	<0.0	<0.00	1.5
I2	85	2	0.72	0	0.03	<0.0	<0.00	15.5
J1	81	2	0.38	<0.000	<0.00	<0.0	<0.00	6.7
J2	44	ND	0.94	0.003	<0.00	<0.0	<0.00	1.1
WHO	200	12	5	0.03	5	100	1	250
NSDWQ	200	12	5	-	3	-	0.3	250

*ND- Not Detected
A1-J2 was used to denote sample ID of the selected sachet water brands.

Elements tested are; Calcium, Potassium, Copper, Manganese, Zinc, Magnesium, Iron and Chloride. The calcium concentrations in the water samples from Table 2 ranges from 34 – 349 mg/l. However, the permissible limits for calcium in drinking water is 200 mg/l, according to the WHO [7]. The presence of calcium ions in water samples tend to hardened the water and may not be suitable for consumption. Because the calcium salts precipitates from water causes high levels of calcium ions which tend to promote scale formation in water systems (Palintest Photometer 8000 manual). Meanwhile, brand B has its calcium concentration exceeding the permissible limit.

The potassium concentrations in the sachet water samples ranges from 0 - 9.0 mg/l, which were within the permissible limit of 12 mg/l stated by the WHO [7]. The daily comsuption of potassium is a vital element for human growth. As such, the daily recommended requirement of potassium should be greater than 3000 mg WHO [7].
The concentration of copper in the water samples ranges from 0 - 0.94 mg/l, these were within the stipulated permissible limits of 5 mg/l specified by the WHO [7]. Therefore, the presence of copper occurs naturally, only in minute quantity (few micrograms per litre) in drinking water [12].
The concentration of manganese in the samples ranges from 0 - 0.003 mg/l and it complies with the WHO [7] standard of drinking water quality. Manganese occurs mostly due to distributions of ores and rocks. It has no particular toxicological effect [7]. Similarly, the concentrations of zinc (0 - 0.38 mg/l), magnesium (0.0 mg/l), iron (0 - 0.06 mg/l), and chloride (0 - 42.0 mg/l) were all far less than the permissible limits of the WHO [7] and NSDWQ [8] standard of portable water quality.

4. Conclusions

The results of the physical properties tests on the various samples indicate that all the brands of water tested were colour, odour and taste free. Also, the pH, TDS, EC, temperature, TSS and turbidity of all the brands were within the permissible limits of the WHO (2017) drinking water. The chemical properties of the brands were within the permissible limits of WHO except samples A2, B1, B2, C1 and D1 that had calcium concentration exceeding the WHO permissible limits, the conceentrations of potassium (K), copper (Cu), manganese (Mn), zinc (Zn), magnesium (M), iron (Fe) and chloride (Cl) were within the WHO [7] drinking water standards requirements. Consequently, we recommend that the ten (10) brands of sachet water studied were safe for public consumption.

Acknowledgements

The authors appreciates Covenant University Centre for Research Innovation and Discovery (CUCRID), for securing open acess for this article.

References

[1] Agboli E, Egu E, Kweku M, Cham M, Zotor F and Binka F 2017 Assessment of Some Physico-chemical Properties and Bacteriological Status of Sachet Water Consumed in the Hohoe Municipality, Ghana. *International Journal of Tropical Disease & Health*, **28**(2): 1–8
[2] Shyamala G, Jeyanthi J, Gobinath R, Akinwumi I I and Maheswari M 2017 Assessment of groundwater quality using spatial variation technique. *Journal of Chemical and Pharmaceutical Sciences*, **10**(4): 9–15
[3] Edema M O, Atayese A O, and Bankole M. O 2011 Pure Water Syndrome: Bacteriological quality of sachet-packed drinking water sold in Nigeria. *African Journal of Food, Agriculture, Nutrition and Development*, **11**(1): 1–8
[4] UNICEF and World Health Organization 2019 1 in 3 people globally do not have access to safe drinking water – UNICEF, WHO. Available on https://www.who.int/news-room/detail/18-06-2019-1-in-3-people-globally-do-not-have-access-to-safe-drinking-water-unicef-who [accessed: 5 May 2020]
[5] Yusuf Y O, Jimoh A I, Onaolapo E O and Dabo Y 2015 An assessment of sachet water quality in Zaria Area of Kaduna State, Nigeria. *Journal of Geography and Regional Planning*, **8**(7): 174–180
[6] Ogundipe, K.E., Ogunde, A., Olaniran, H.F., Ajao, A.M., Ogunbayo, B.F. and Ogundipe, J.A., 2018. Missing gaps in safety education and practices: academia perspectives. *International Journal of Civil Engineering and Technology (IJCIET)*, *9*(1), pp.273-289
[7] World Health Organisation 2017 Guidelines for drinking-water quality, fourth edition. World Health Organisation
[8] Nigerian Industrial Standards 2015 Nigerian Standard for Drinking Water Quality, Standards Organisation of Nigeria (SON)
[9] Rahmani N, Ali S H B, Homayoonfard M, Ali N J, Rehan M, Sadef Y and Nizami A S 2015 Analysis of Physiochemical Parameters to Evaluate the Drinking Water Quality in the

State of Perak, Malaysia. *Journal of Chemistry*, **2015**(Article ID 716125), 1–10
[10] Chindo I Y, Karu E, Ziyok I and Amanki E D 2013 Physicochemical Analysis of Ground Water of Selected Areas of Dass and Ganjuwa Local Government Areas, Bauchi State, Nigeria. *World Journal of Analytical Chemistry*, **1**(4), 73–79
[11] Durotoye, T O, Aderonke A A, David O. O, and Olumuyiwa O. 2018 "Impact assessment of wastewater discharge from a textile industry in Lagos, Nigeria." *Cogent Engineering* 5, no. 1 1531687
[12] Airaodion A I, Airaodion E O, Osemwowa E U, Ogbuagu E O and Ogbuagu U 2019 Quality Assessment of Sachet and Bottled Water in Ogbomoso Metropolis, Nigeria. *Asian Food Science Journal*, **9**(2), 1–15

IOP Conf. Series: Materials Science and Engineering 1036 (2021) 012012 doi:10.1088/1757-899X/1036/1/012012

Remediation of Emerging Pollutants in Industrial Contaminated Water using Oxytenanthera abyssinica and Bambusa vulgaris in a Treatment Media

O O Ijaola[1,3] and AY Sangodoyin[2,4]

[1]Department of Civil Engineering, Faculty of Engineering and Technology, Federal University Otuoke
[2] Department of Agricultural and Environmental Engineering, Faculty of Technology, University of Ibadan, Ibadan Nigeria.
[3]Telephones: 08061179584, 08125107697 Email opodengineer@yahoo.com,
[4] Telephone:08033715522Email:ay.sangodoyin@mail.ui.edu.ng,

Abstract. Remediation of emerging pollutants is an aspect of environmental research with sparse information and knowledge, and the disposal of, these pollutants to the environment are mostly unregulated and on the increase. In view of this, the study examined the adsorptive capacity of adsorbent produced from *Oxytenanthera abyssinica* (COA 350°C KCl) and *Bambusa vulgaris* (CBV 350°C H_3PO_4) in remediating Pharmaceutical actives contaminants (PhACs) and polycyclic aromatic hydrocarbons (PAHs). Fourier Transform Infrared Spectroscopy, Scanning Electron Microscopy, Energy Dispersive X-ray were used in characterizing adsorbents. Derivative form of Solid Phase Extraction (SPE) and Liquid-Liquid extraction procedures were used for extracting selected PhACs and PAHs from sampled solutions respectively. Batch experiments were performed to study the adsorption capacity of adsorbents at varying contact time of 30min, 2hrs and 12hrs for PAHs and 30, 180, 360, 720 and 1440 mins for PhACs. Analysis of PhACs and PAHs concentrations for each sampling were determined by UV spectrophotometer and GC-MS respectively. The samples showed sharp absorption peaks at 3432 and 1632 cm^{-1} indicating O-H stretched and N-H stretched. The adsorbents showed both open and close pores. Carbon, Oxygen and Nitrogen in the adsorbent for CBV were 62.00, 33.53 and 3.98 and COA were 71.00, 22.80 weight percent, respectively. The adsorbents had good surface areas, pore volumes and pore sizes. Removal efficiency of COA showed the highest efficiency for PhACs removal at (73.3%, 78.1% and 86.2%) while CBV (63.9 %, 66.7% and 82.2%) for paracetamol, salbutamol and chlorpheniramine, respectively. CBV and COA ranged from 42.5-81.2% and 8.9-65.5% for PAHs removal respectively. The experimental result showed that adsorbents made from *Oxytenanthera abyssinaca* and *Bambusa vulgaris* can efficiently adsorb selected PhACs and PAHs in industrial Contaminated Water.

Keywords: Pharmaceutical actives contaminants, polycyclic aromatic hydrocarbons, adsorbent, *Oxytenanthera abyssinaca*, *Bambusa vulgaris*

1. Introduction

Lately, most research that involves the remediation of water contaminates focuses on the use of agricultural products as adsorbent. In Nigeria the use of agricultural products as adsorbents in the removal of heavy metals, trace metals, anions and cations as well as biological contaminates from water

and wastewater had been floored over flogged [1,2,3,4]. The tremendous advancement in research on environmental pollutants has led to various discoveries of contaminants in environmental bodies such as water, soil, and air. The twelve initial persistent organic pollutants (POPs) were first identified as the "dirty dozen", At the Stockholm Convention of 2001 and include: aldrin, chlordane, dichloro diphenyl trichloroethane (DDT), dieldrin, endrin, heptachlor. Mirex and Toxaphene, Hexa-chlorobenzene (HCB), Polychlorinated Biphenyls (PCB), Chlorinated Dioxins and Chlorinated Furans [5]. Other substances such as polybrominated diphenyl ethers (PBDEs), unintentional by-products such as dioxin furans and polycyclic aromatic hydrocarbons (PAHs) were added later at the Conference of the Parties (COP 4) of the Stockholm's Convention on POPs in 2009 and are called the "Nasty nine". These groups of organic contaminants are persistent in the environment with the ability to travel far across the atmosphere or move through waters to settle in areas far from where they were originated. They bio-accumulate in fat tissues and are toxic to wildlife and human beings. They are difficult to biodegrade due to their chemical persistence and semi-volatile nature [5].
Aside the toxic POPs, other organic contaminants may produce subtle ecological effect but there are uncertainties surrounding their adverse effects on the environment. These are called emerging organic contaminants (EOCs). According to [6], EOCs are any synthetic or naturally occurring organic chemical(s) or microorganisms that are not commonly monitored but have the potential to enter the aquatic environment and impair the quality of raw water or cause known or suspected adverse human health or ecological effects. These EOCs may not be seen as active or passive chemical only but also include their metabolites and other transformation products and chemical by-products generated during production [7]. Their introduction into the environment is continuous because of their wide usage, and they are seen as micro-pollutants which include endocrine-disrupting compounds (EDCs), Pharmaceutically Active Compounds (PhACs), Personal Care Products (PCPs), Additives, Preservatives etc. [7, 8,9,10,11].
Emerging contaminants need not to be persistent in the environment to cause negative effects since their high transformation and removal rates can be offset by their continuous introduction into the environment. [6, 12, 13, 14] stated that many of these organic contaminants can cause cancers, neuro-behavioral deficits, reproductive disorders effects, endocrine disruption, allergies, damage of the central and peripheral nervous system, disruption to immune system and even death. However, [11] reported that there are no accurate data concerning the occurrences, eco-toxicity and risk assessment for most of the emerging contaminants, hence it is difficult to predict their effects on the environment and health, and they are therefore currently unregulated in most countries. Among many industries in Nigeria that contribute to organic contamination of water resources are the petroleum and pharmaceutical industries. The products from these industries are ubiquitous because of their daily usage and the pollutants that result from them are persistent as well as emerging [8].
Most of the organic contaminants in water and wastewater associated with the petroleum and pharmaceutical industries have been seen to generate an adverse effect on surface water, which is a source of livelihood. Identification and remediation processes for these contaminants are of a necessity if portable water will be available to everyone. Various analytical techniques such as gravimetric, UV-spectrophotometerys, gas and liquid chromatography have been developed and employed in several literatures to identify the various organic contaminants [11]. The most effective methods so far that identified organic contaminants in their trace amount (or micro-pollutant) is gas chromatography or liquid chromatography with mass spectrophotometer followed by a clean-up method such as solid-phase extraction (SPE) and solid-phase micro-extraction (SPME) [11].
The petroleum industry makes use of conventional treatment methods such as the Induced Gas Flotation (IGF) or the Induced Air Flotation cells usually known as WEMCO with the Enviro-cell as the most recent technology in the class [15]. The conventional separation method employs the principle of gravity with differences in density between the oil and water. Similarly, the conventional wastewater treatment plants (WWTPs) are not always effective for the removal of these huge classes of pollutants and so further water treatments are necessary [16]. An alternative method of treatment that has been researched and still in research is the use of adsorption mechanism [11]. Activated carbon is one of the most popular adsorbents used in numerous industries for the removal and recovery of organic and inorganic compounds from gaseous and liquid streams [17]. It has high adsorption capability due to its high

internal surface area and porosity formed during carbonization process. The presence of activating agents and carbonization conditions influences the development of pore structures. However, the use of activated carbon is limited by its high cost which is a consequence of its production. Agricultural wastes or materials have been observed to be potential precursors in activated carbon production because of the abundant supply and low cost. Adsorption method of bioremediation had proven to be the choice treatment option for PAHs and PhACs and other micro-pollutant in aqueous media because its convenience advantages; efficiency, effectiveness and simplicity of design when compared to other form of treatment. Aside of these attributes, it adds no undesirable by-products or degradation metabolites which could be harmful [6, 10].

Adsorption is superior to any other wastewater treatment techniques because of its insensitivity to toxic substances and inexpensive nature of materials used as adsorbents [18]. However, full utilization of adsorption processes for water purification is impeded by inadequacies of commercial adsorbents like activated carbon and synthetic polymer resins. Therefore, the development of a low cost adsorbent is of necessity in environmental researches. Adsorbent produced from natural agricultural products have been the popular substitutes for commercial and synthetic adsorbents because of hydrophobic-oleophilic potential which is necessary in bioremediation processes [18, 19].

These views prompted the research designed to investigate the adsorptive capacity of a low-cost adsorbent (Bamboo) and examine the extent to which different modified Nigerian bamboo species can absorb PAHs and PhACs organic contaminants found in petroleum and pharmaceutical liquid waste respectively. The study also aims at evaluating the effectiveness of the most efficient modified bamboo species in a prototype developed treatment system

2. Materials & Methods

The Bamboos (*Oxytenanthera abyssinaca and Bambusa vulgaris)* were carbonized at temperature of 350°C in a Gallen kamp muffle furnace with aluminum foil as an oxidizing agent. After cutting at 20cm above the ground levels and processing them. The yield percentage of each carbonized bamboo species were then calculated thus:

$$Yield\ (\%) \quad = \frac{w_2 * 100}{w_1} \qquad (1)$$

where:

$w_1 = initialweightofbamboospeciebeforecharring\ (g)$

$w_2 = finalweightofbamboospecie\ aftercharring\ (g)$

Activation was done with Phosphoric acid (H_3PO_4) and Potassium chloride (KCl) as dehydrating agent. About 26.25w/w of activator was used in activation of the carbon samples. Characterization was chemically done using Fourier Transform Infrared Spectroscopy (FTIR), to determine the surface functional groups. The Scanning Electron Microscopy (SEM) was used to view the surface structural of the samples at magnification of 100, 300, 500, 2000 and 5000 times the original size in order to view the pore space development and reveal other information such as texture (external morphology) and structural orientation. EDS in conjunction with SEM was used to determine the chemical composition or elemental analysis of the samples.

2.1. Adsorption Experiments

Adsorption behavior of PhACs in pharmaceutical effluents and Polycyclic Aromatic Hydrocarbon (PAHs) in petroleum wastewater onto bamboo activated carbon was studied in batch process. . Experiments were carried out in ambient temperature and adsorption capacity of activated carbon from bamboo was tested on the basis of contact time. Half liter of pharmaceutical effluent was put into conical flasks of 600ml capacity. Two grammes of each selected bamboo activated carbons were weighed into conical flasks to from an adsorbent/solute solution. Solutions were agitated at stirring speed of 160rpm to ensure intimate contact of the adsorbent and solute in solution. Each solution was observed for 6hrs contact time at\which it attains dynamic equilibrium. Thereafter, solutions were filtered with filter paper

0.45μm size. The filtrate of 300ml was poured into sampling bottles with tie cap sealed with aluminum foils and kept at temperature of 4^{0}C for further analysis of extraction, clean up and Vis-UV. To obtain accuracy, all experimental analysis was duplicated.
200ml of simulated petroleum wastewater was sampled into different conical flasks of 250ml capacity. 1g of each selected bamboo activated carbon was weighed into the conical flasks to form an adsorbent/solute solution. Solutions were agitated at a stirring speed of 160 rpm to ensure intimate contact of the adsorbent and solute in solution, while observing each solution in equilibrium for 5hrs contact time to attain dynamic equilibrium. After 5 hrs solutions were filtered with filter paper 0.45μm and the filtrate of 150ml were poured into amber bottles with Teflon cap and kept at temperature of 40C for further analysis of extraction, clean up and GC-MS. The 5hrs contact time was chosen based on the experiment performed at the terminal station of the oil and gas industry. To obtain accuracy, all experimental analysis was duplicated.
The amount of PhACs (q_e) and PAHs (q_e) adsorbed by bamboo activated carbons can be expressed mathematically as:

$$q_e = \frac{C_o - C_e}{M} \times v \qquad (2)$$

The percentage removal is evaluated using

$$\% \, Removal \; = \frac{C_o - C_e}{C_o} \times 100 \ldots\ldots\ldots\ldots\ldots\ldots \qquad (3)$$

where
V is the volume of PhACs and PAHs in solution (L)
Co is initial concentrations of PhACs and PAHs (mg/l)
Ce is equilibrium concentrations of PhACs and PAHs (mg/l)
M is the mass of the adsorbent (g)

2.2 Extraction techniques and Quantification of contaminants:

Extraction techniques and Quantification of PhACs and PAHs were examined using different techniques respectively. Pure standards of all active ingredients observed with minimum of 98.5-99% purity were used for this study. Derivative form of Solid Phase Extraction SPE extraction procedure was used for extracting PhACs from sampled solutions.1cm of moderate packed cotton wool was placed at the bottom of each 10mm ID.250mm Loup chromatographic column used. 2g of activated silica gel 10ml of 1:1 acetonitrile and methanol was prepared and placed into the chromatographic column. To the top of the column was added 0.8cm of anhydrous sodium sulphate. The column was rinsed with additional 3ml of acetonitrile followed by 3ml of methanol and 3ml acidify ultra-pure milli-Q water to pre-elute the column. Elute was allowed to flow though the column at the rate of about 1ml/min until the liquid in the column was just above the sulphate layer and immediately 300ml untreated or treated sampled effluents were transferred into each prepared column. The sample bottles were rinsed with 1ml methanol eluent and added to the column as well. The eluent was collected into sampling bottles, after which acidified methanol pH 2 was immediately used to extract compounds adsorbed to the silica sorbent at a flow rate of 0.1ml/sec. Samples were collected in a graduated cylinder each and allowed to concentrate to 10ml under air vacuum after which sample was increased to 20ml by adding 10ml of methanol and kept at 4^{0}C temperature before Vis-UV analysis.
All solvents and chemicals used are of analytical grade with 99.0% purity. A standard mixture of 16 priority PAHs of 2.0 mg/ml each was used to calibrate the GC-MS used for PAHs analysis. Liquid-Liquid extraction procedures were used for extracting selected PAHs from sampled solutions. Mixture of dichloromethane (DCM) and n-Hexane in ratio 3:1 was used as extractant. The separatory funnels were shaken vigorously for at least 5min and the organic layers were allowed to separate clearly from the aqueous phase by gravitation technique for a period of 5-10 mins. Thereafter, the organic layers were collected in separate glass bottles of about 50ml that was well capped. The extraction processes were repeated twice for each sample. Cleanup of selected PAHs was carried out with 1cm of moderate packed cotton wool placed at the bottom of each 10mm ID 250mm loup chromatographic column used. Two

grammes of activated silica gel and 10ml of 3:1 (DCM and hexane) was prepared and placed in the chromatographic column. To the top of the column was added 0.8cm of anhydrous sodium sulphate. The column was rinsed with additional 20ml mixture of 3:1 (DCM andHexane) to pre-elute the column. Elute was allowed to flow though the column for about 2min or until the liquid in the column was just above the sulphate layer. Thereafter, 3ml of extracted samples was transferred into each prepared column. The extraction bottles were rinsed with 1ml of DCM (3) and Hexane (1)/eluent and added to the column as well. The stop cork of the column was opened and the aromatic extract collected into 10ml graduated cylinder each was allowed to concentrate and evaporate under oxygenated conditions and later kept at 4^0C before GC-MS analysis.
Quantification of PhACs and of PAHs was by an external standard method, which relies on the reproducibility of the standard preparation. The linearity of external calibration was done by preparing different concentration of Paracetamol, Salbutamol and Chlorpheniramine standards at different dilution rate of between 0.2to3.5ppm. Similarly, the linearity of external calibration of PAHs were done by preparing different concentration of PAHs mix standards at dilution rate of 12.5,25,50,100 and 200ppm according to the specification given by the International Conference for Harmonization [11] that requires a 5 point calibration curve serial dilution. The linearity was evaluated at different concentration using response factors and all showed good linearity with regression coefficients ranging from 0.972 to 1.000 for all PAHs. USEPA method 8270 was employed in GC-MS analysis.
Visible Ultra-violet spectrophotometry, UV-Vis spectrophotometer S/N 18-1901-01-0243 was further used to analysis the pharmaceutical effluents in order to ascertain concentration of PhACs of interest in the simulated pharmaceutical effluents. The software used in interpreting concentration of pollutant or chromatogram was UV win5 spectrophotometer version 5.2.0. Extracted concentrated analytes were further diluted with methanol and scanned with visible-UV to determine the wavelength of each PhACs in the samples. The wavelength of Paracetamol, Salbutamol and Chlorpheniramine was set at 257, 278 and 262nm respectively. The concentrated analytes were transferred into 10ml cuvette to read each concentration of PhACs with UV win5 spectrophotometer version 5.2.0 at different determined wavelength. Prior to each UV reading, the instrument was blanked with methanol in order to set a new baseline.
The gas chromatography that was used was Agilent Technologies 7890-A GC system G-3440-A A.01.14 series, coupled with Mass spectrometry detector (MS) Agilent Technologies 5975-C VL MS D series with Agilent Technologies 7683-B series injector powered with MSD chem-station G1701-EA E.02.10.1177 software to identify and quantify analytes in compounds. Selected hydrocarbon compounds were identified on the basis of their retention time and by comparing them to those of analytical standards. Matrix spikes, duplicates, solvent, and method blanks were also analyzed as quality control samples.

3. Results & Discussion

Figure 1(a & b) shows the FTIR spectra of CBV 350°C H_3PO_4 and COA 350°C KCl. The modified bamboo species revealed large number of functional groups, the spectra also indicates that the surface functional groups of the adsorbents do not exhibit significant difference irrespective of the activating agent used. Differences were slightly noticed on the intensity of the bands and spectrum shapes, with some functional groups shifted to different frequency levels. For the activated carbons analyzed the spectra shows a broad absorption peak at 372 - 3424cm^{-1}. The 3424cm^{-1} wave length was associated to OH hydrogen bonded phenol, alcohol hydroxyl group with H bonded to OH stretch was seen as most predominate in the adsorbent. The results for the carbon tested show similar functional groups but different structural shapes. This can be linked to the different activating salt.

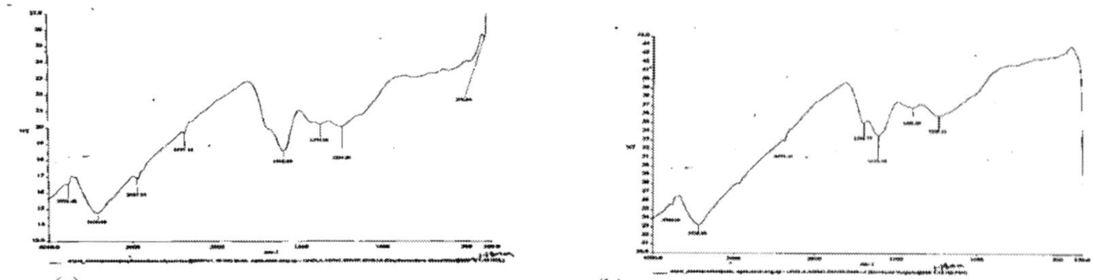

(a) Fourier Transform Infrared Spectrometer of COA 350°C KCl

(b) Fourier Transform Infrared Spectrometer of CBV 350°C H_3PO_4

(Parkin Elmer FT-IR system BX spectrum v5.3.1 Multidisciplinary Central Research Laboratory (MCRL) of the University of Ibadan, Ibadan, Nigeria.)

Figure 1 Fourier Transform Infrared Spectrometer of the absorbent used

The SEM-EDS analysis reveals the porous structure of the samples, the amount of pores within each species for sorption and the agglomeration of particles within the structures in distinctive irregular shapes. Both species show layers of micro pore material clustered and woven together with a semi-permeable bio-membrane indicating that bamboo can undergo both filtration and adsorption processes progressively. The SEM image CBV 350°C H_3PO_4 seem to have more open pores when compared with COA 350°C KCl which has a close pore with a thread-like surface. This indicates that CBV 350°C H_3PO_4 may be better in adsorption or sorption processes when compared with COA 350°C KCl,

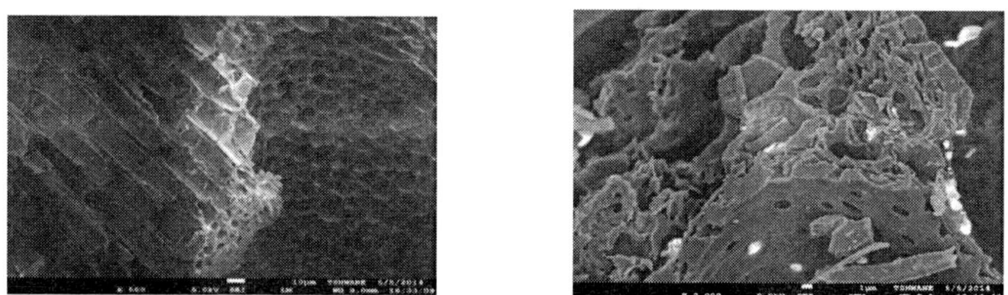

(a) COA 350°C KCl (B) CBV 350°C H_3PO_4

Figure 2 SEM images of Bambusa *vulgaris and* Oxytenanthera *abyssinaca*

The elemental composition of elements in the bamboo species, revealed the electron dispersion spectral of the selected areas. The SEM-EDS monograph of the modified bamboo species reveals that the chemical composition on the surface of each mapped spectrum include carbon, oxygen, nitrogen, silicon chlorine, zinc, potassium, phosphorus, magnesium and titanium. While table 1 shows the elemental constituents of COA 350°C KCl and CBV 350°C H_3PO_4

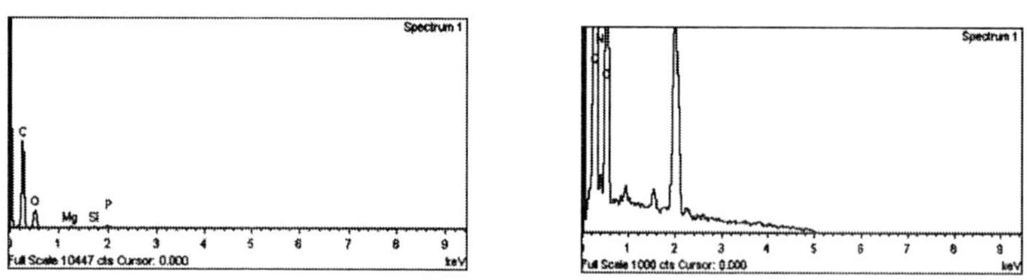

(a) COA 350°C KCl (b) CBV 350 °C H_3PO_4

(EDX: CSIR Nanotech at Tshwane in Pretoria, South Africa)

Figure 3: Elemental dispersion spectral of selected areas on SEM-EDS for modified Absorbents

Table 1: Elemental Composition of Modified Bamboo Species

Modified Bamboo Activated Carbon	Mapped Spectrum Area	Elements	Weight (%)	Atomic (%)
Bambusa Vulgaris(CBV) 350°C H_3PO_4	Spectrum 1	C	62.00	68.62
		N	3.98	3.74
		O	33.53	27.64
		Ti	34.48	14.80
Oxytenanthera *abyssinaca*(COA) 350°C KCl	Spectrum1	C	71.07	78.38
		O	22.80	18.88
		P	4.43	1.90
		Mg	0.56	0.30
		Si	1.14	0.54

The results in table 1 revealed high intensity of carbon and oxygen, which can be deduced that modified bamboo species, has good adsorptive properties. The rather high concentration of silicon in Oxytenanthera *abyssinaca* implies that the adsorption processes will be rapid as confirmed by [19] that incorporation of silicon in a solid will dramatically increase the solid binding strength for some pollutants. Some of the elements introduced resulted from the dehydrating salt used while others are component of the bamboo species.

3.1. Effect of contact time on Adsorption Behavior

3.1.1. Effect of contact time on adsorption of PAHs

Figure 4 shows the Spectra of 16 priority PAHs in Simulated Petroleum wastewater. The effect of contact time in removal of PAHs from simulated petroleum wastewater by COA KCl and CBV H_3PO_4 are detailed in Figures 5 & 6. The varying contact times used were 30min, 2hrs and 12hrs and there were no much changes or differences in the adsorption rate per time.

It was revealed that the percentage removal efficiency of PAHs with time by COA KCl was not consistent. At 30min of adsorption rate; the percentage removal efficiency was 49.8% of total PAHs, while 39.1% of total PAHs was adsorbed at 2hrs and about 72.3% of total PAHs was adsorbed at 12hrs contact time. Similarly, the adsorption rate of PAHs by CBV H_3PO_4 was observed not to follow the adsorption pattern wherein adsorption rate increased with time. At 30min, about 85.1% was adsorbed; an increase in contact time to 2hrs shows a reduction in adsorption rate to 25.1% but further increase in time to 12hrs further increases adsorption efficiency to 87.7% which is the optimal contact time for CBV H_3PO_4 in adsorbing PAHs.

Since the pattern of adsorption does not follow the norms of adsorption, it could be deduced that adsorption of PAHs is in two stages. At the first stage the PAHs were adsorbed easily unto the accessible hydrophobic site within the adsorbent or granular activated carbon matrix for the first 30min. These could have resulted from the chemical interaction between the PAHs and the adsorbent surface. The reduction in adsorption rate implies that in the second stage, adsorption rate may be restricted by slow movement of PAHs to less available site associated with the micropores within the adsorbents matrix which could take hours. Same observation was deduced by [20, 21, 22]. [23] Further explained that the sorption of PAHs onto porous adsorbent such as GAC undergoes three consecutive stages. The first stage is assumed to occur rapidly without a rate-limiting stage in the adsorption of organic compounds on the adsorbent, but at the second stage the mass transfer proposed a main resistance during the diffusion of PAHs into the internal structure of the adsorbent. It can then be implied that the pore structure of the adsorbents used for these experiment consists of macropores, transitional pore (mesopore) and micropores as early deduced by SEM and EDX analysis.

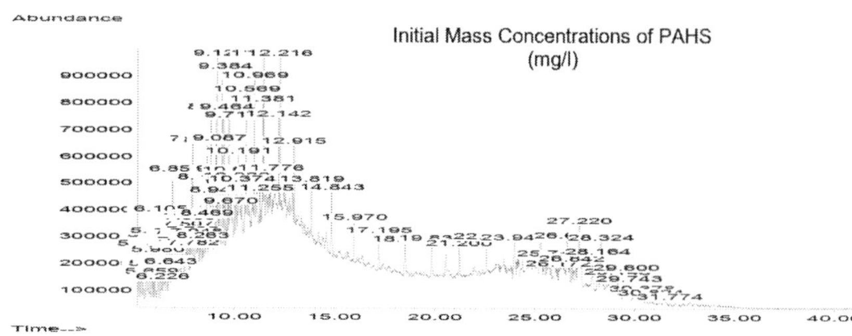

Central Research Laboratory, University of Lagos

Figure 4: Spectra of Simulated Petroleum wastewater indicating 16 priority PAHs,

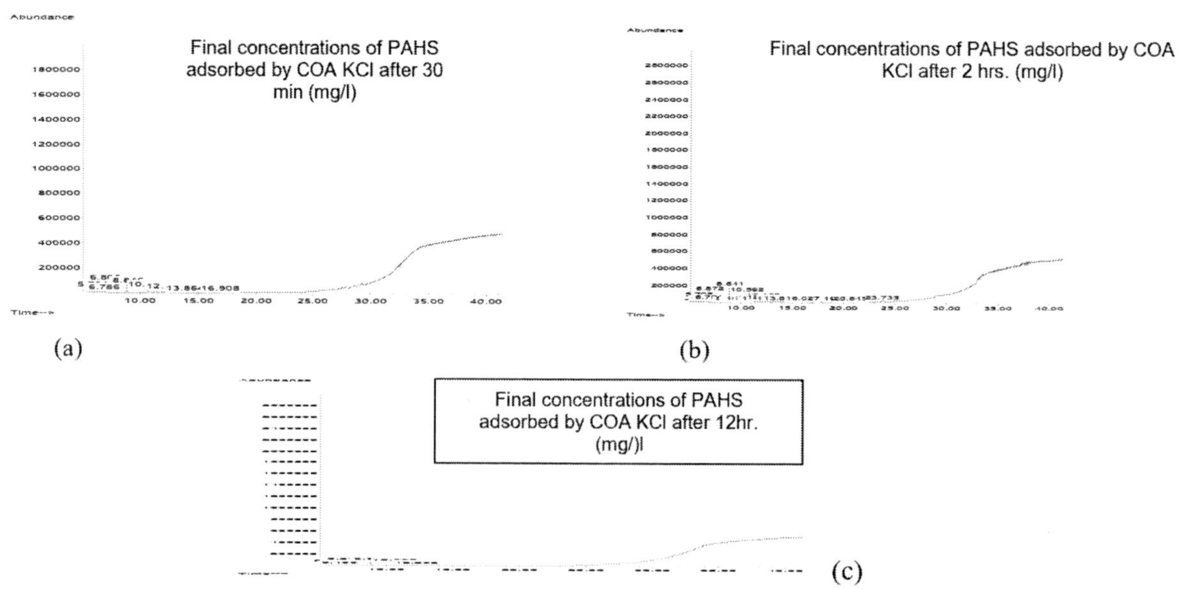

Central Research Laboratory, University of Lagos

Figure 5: Effect of contact time on adsorption rate of PAHs by COA KCl at (a) 30mins,(b) 2hrs and(c) 12hrs

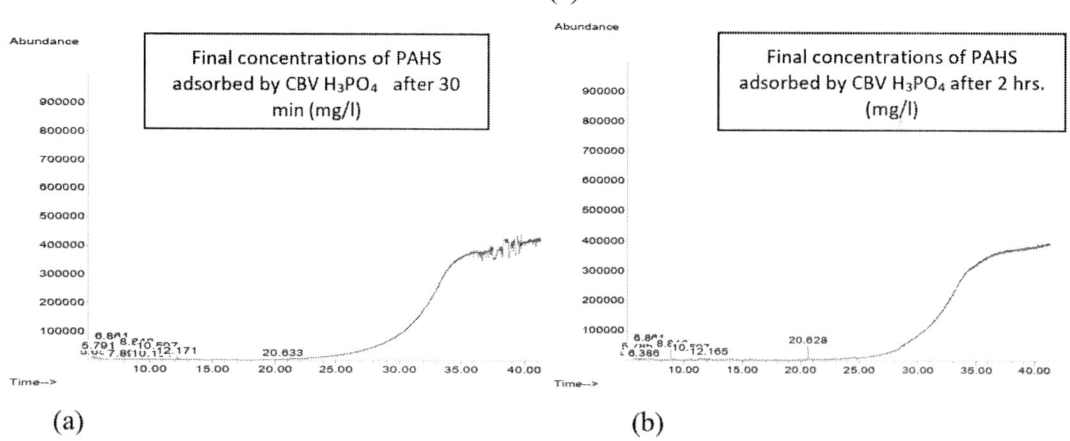

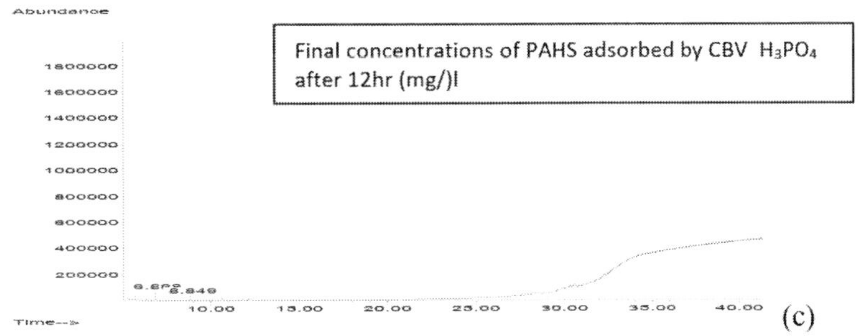

(c)

Central Research Laboratory, University of Lagos

Figure 6: Effect of contact time on adsorption rate of PAHs by CBVH_3PO_4 at (a) 30mins,(b) 2hrs and(c) 12hrs.

3.1.2 Effect of contact time on adsorption of PhACs

Figures 7 a-c shows the adsorption rate and removal efficiency of PhACs by COA KCl and CBV H_3PO_4 at varying contact time of 30, 180, 360, 720 and 1440 mins.

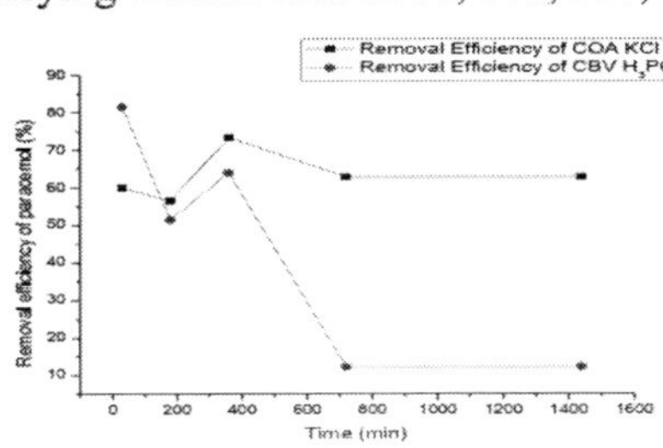

Figure 7a

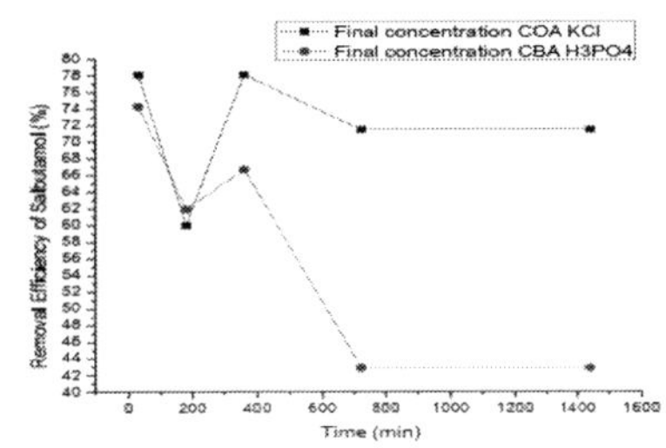

Figure 7b

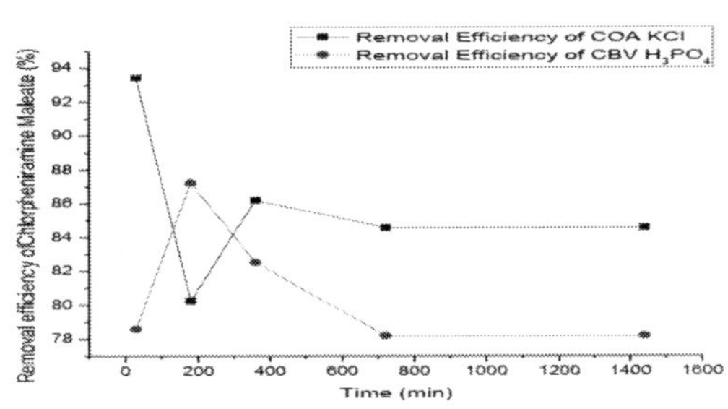

Figure 7c

Figures 7a-c :Effect of contact time on adsorption rate of PhACs by COA KCl and CBVH_3PO_4 at 30, 180, 360, 720 and 1440 mins

The trend of adsorption in Figs. 6a-c revealed a slight reduction of adsorption rate before an increase after which equilibrium was observed for COA KCl adsorbent. Adsorption trend of CBV H_3PO_4 revealed a sharp reduction to a level and there after an increase was observed before equilibrium point was reached. These observations are consistent with [24,25&,26] findings. It can be explained that absorption of PhACs with COA KCl and CBV H_3PO_4 occurred in two different stages. The first stage occurred during the first 30-360mins contact time, with high number of active binding sites on the adsorbents surfaces. Adsorption rate is rapid in this stage and points to adsorption being controlled by diffusion processes of paracetamol, salbutamol and chlorpheniramine molecule from the bulk phase to the adsorbent surface. The second stage of adsorption is an attachment-controlled processes due to decrease in the number of active site available for Paracetamol, salbutamol and chlorpheniramine

IOP Conf. Series: Materials Science and Engineering 1036 (2021) 012012 doi:10.1088/1757-899X/1036/1/012012

molecule onto the adsorbents surfaces. Slow uptake of adsorbate and establishments of equilibrium over a longer period indicate strong chemical binding of adsorbate with adsorbent.

3.2. Compatibility test of COA KCl and CBV H_3PO_4 adsorbent with developed wastewater system

In order to test for the compatibility of the activated carbon alongside other treatment section as well as the efficiency of the wastewater treatment system. 15.9L of simulated wastewaters of each pollutant was used alongside 78.5g of selected activated carbon.
The volume of simulated wastewater used was calculated thus:
Height of developed system = 900mm or 0.9m
Diameter of developed system = 150mm or 0.15m
Volume of a cylinder = $\pi r^2 h$
$\rightarrow \frac{22}{7} \times (0.075^2) \times 0.9 \rightarrow 0.015911m^3 = 15.911L$
For carbon
If 200ml→1g
15900ml→x
Implying that 79.5g of selected adsorbent.
Tables 2 and 3 showed the initial concentration, final concentration and percentage removal efficiency of PhACs and PAHs in simulated wastewater by developed prototype wastewater system.

Table 2 Remediation of PAHs by Developed Prototype wastewater Treatment System

Compounds	Mass Concentration mg/l	Final concentrations of PAHS adsorbed by AC (KCl) mg/l	Removal Efficiency AC (KCl) %	Final concentrations of PAHS adsorbed by AC (H_3PO_4) mg/l	Removal Efficiency AC (H_3PO_4) %
Naphthalene	BC	BC	BC	BC	BC
Acenaphthylene	BC	BC	BC	BC	BC
Acenaphthene	BC	BC	BC	BC	BC
Fluorene	0.66	0.23	65.15	ND	ND
Phenanthrene	5.379	3.275	39.11	2.56	52.40
Anthracene	4.227	2.1	50.31	1.27	69.95
Flouranthene	6.192	3.805	38.54	ND	ND
Pyrene	6.57	4.23	35.61	ND	ND
Benzo(a)anthracene	16.2	12.369	23.64	ND	ND
Chrysene	ND	ND	ND	ND	ND
Benzo(b)fluoranthene	ND	ND	ND	ND	ND
Benzo(k)fluoranthene	23.784	20.34	14.48	20.61276	13.33
Benzo(a)pyrene	ND	ND	ND	ND	ND
Dibenzo(a,h)anthracene	ND	ND	ND	ND	ND
Indeno(1,2,3-cd)pyrene	ND	ND	ND	ND	ND
Benz(e)acephenanthrylene	17.094	14.056	17.77	13.567	20.63
Benzo(e)pyrene	16.674	12.476	25.17	11.457	31.288
Benzo(g,h,i)perylene	0.021	0.005247	75.01	ND	ND
Total PAHS	96.801	72.88625	24.70	49.46676	48.898

Below calibration; ND: Not detected, COA KCl: oxytenanthera abyssinaca activated with KCl; CBV H_3PO_4: Bambusa vulgaris activated with H3PO4; CA: Oclansorb from peat moss.

Table 3 Remediation of PhACs by Developed Prototype wastewater Treatment system

Compounds (PhACs)	Initial mass concentrat. (mg/l)	Final concentrate. COA KCl (mg/l)	Removal Efficiency COA KCl (%)	Final concentrate. CBV H_3PO_4 (Mg/l)	Removal Efficien. CBV H_3PO_4(%)	Final concentrate. CA (mg/l)	Removal Efficien. CA (%)
Paracetamol	55.968	14.946	73.3	20.193	63.921	45.633	18.466
Salbutamol	16.695	3.657	78.1	5.565	66.667	10.971	34.286
Chlorphenirami.	77.274	10.653	86.214	13.515	82.511	37.842	51.029

4. Conclusion

The experimental result in this study showed that adsorbents made from *Oxytenanthera abyssinaca* and *Bambusa vulgaris* can efficiently adsorb selected PhACs and PAHs in industrial Contaminated Water. The adsorption mechanism of trace organics followed same pattern though with little differences. For all organic pollutants, adsorption rate are in two stages viz: optimization and reduction followed by equilibrium. The removal efficiency of the developed prototype system highly depends on the tertiary treatment part which is the adsorbent used.

5. Acknowledgement

The authors are grateful to the following organizations in Nigeria that afforded them the opportunity to use their facilities and contributed to this study; National Petroleum Investment Management Services (NAPIMS): Department of Health, Safety, Security and Environment (HSSED), Chevron Nigeria Limited, Health, Safety and Environment unit (HSE), The Shell Petroleum Development Company of Nigeria Limited, Remediation Unit. Members of Analytical unit of Central Research Laboratory, University of Lagos, Akoka, Nigeria, Fidson Healthcare Plc, Factory unit, Mospson Pharmaceutical Limited, Emezor Pharmaceutical Limited, Drug laboratory, Lagos State QA/QC for Pharmaceutical Companies and National Agency for Food and Drugs Administration Control (NAFDAC),

References

[1] Nwabanne, J.T. and Igbokwe, P. K. 2012. Mechanism of Copper (II) Removal from Aqueous Solution Using Activated Carbon Prepared from Different Agricultural Materials. *International journal of multidisciplinary sciences and engineering.*3(7) 46-52.

[2] Ademiluyi FT and Ujile A A 2013 Kinetics of Batch Adsorption of Iron II Ions from Aqueous Solution using Activated Carbon from Nigerian Bamboo *International Journal of Engineering and Technology* **3**(6)18-25

[3] Ijaola OO, Ogedengbe K and Sangodoyin AY 2013 On the Efficacy of Activated Carbon Derived from Bamboo in the Adsorption of Water Contaminants *International Journal of Engineering Inventions* **2** (4)29-34

[4] Ajayi-Banji A, Sangodoyin A and Ijaola O 2015 Coconut husk Char Biosorptivity in Heavy Metal Diminution from Contaminated Surface Water *Journal of Engineering Studies and Research* **21**(4) 7-13

[5] UNEP 2001 Stockholm Convention on Persistent Organic Pollutants Geneva Switzerland: United Nations Environmental Programme

[6] Daughton GC 2004 Non-regulated water contaminants: emerging research *Journal of Environmental Research Imp. Ass.*, **24**: 711-734

[7] Lindqvist N, Tuhkanen T and Kronberg L 2005 Occurrence of acidic in phamaceutical raw and treated sewages and receiving water *Water Res.***39**:2219–2228

[8] Wennmalm A and Gunnarsson B 2005 Public health care management of water pollution with pharmaceuticals: environmental classification and analysis of pharmaceutical residues in sewage water *Drug Inf. J.***39**: 291–297

[9] Lishman L 2006 Occurrence and reductions of pharmaceuticals and personal care products and estrogens by municipal wastewater treatment plants in Ontario *Canada. Sci. Total Environ.* **367**:544–558

[10] Esper J N 2009 Selection and prioritization of organi contaminants for monitoring in the drinking water value chain A thesis submitted in fulfillment of the requirements for the degreeof philosophiae doctor in public health in the School of Health Systems and Public Health Faculty of Health Sciences university of Pretoria Pp 1-268.

[11] Noelia RG 2011 Organic contaminants in environmental atmospheres and waters Doctoral thesis Departament de Química Analítica i Química Orgànica. Universitat Rovira I VIRGILI. Pp 1-399

[12] Hsu CH, Jeng WL, Chang RM, Chen LC and Han BC 2001 Estimation of potential lifetime cancer risks for trihalomethanes from consuming chlorinated

[13] Ying GG, Kookana R, and Waite TD 2004 Endocrine disrupting chemicals (EDCs) and Pharmaceuticals and Personal care products (PPCPs) in Reclaimed Water inAustralia. AWA

[14] Armbrurst KL and Wook-Kwon J 2005 Occurrence environmental fate and exposure assessment of SSRIs in aquatic environments.http://www.mscl.msstate.edu last visited 14/3/2020.

[15] Jeffrey L 2010 Towards a Novel Methodology for Environmental Remediation of Oil-polluted Aqueous Systems A PhD Thesis University of Aberdeen

[16] John AV, Markus GP, Deborah E, and Robert JR 2004 A White Paper Describing Production of Crude oil Natural gas and Coal Bed Methane US Department of Energy National Energy Technology Laboratory prepared under contact W-31-109-Eng 38.

[17] Foo PYL and Lee LY 2010 Preparation of activated carbon from *Parkia Speciosa* pod by chemical activation. In *Proceedings of the World Congress on Engineering andComputer Science*, 2.San Francisco USA. WCECS,October,20/22/2010.

[18] Chang EE, Wan J, Kim H, Lian C, Dai Y and Chiang PC 2015 Adsorptionof selected pharmaceutical compounds onto activated carbon in dilute aqueous solutions exemplified by acetaminophen, diclofenac, and sulfamethoxazole *Scientific World Journal* **186**(501): 1-11.

[19] Paul RA and Mark MB 1985 Effects of Silicon on Crystallization and Adsorption Properties of Ferric Oxides *Environ. Sci. Technology*. **19** (11)1048-1053

[20] Valderramaa C, Cortina JL, Farran A, Gamisans X and Laoa C 2007 Kinetics of sorption of polyaromatic hydrocarbons onto granular activated carbon and Macronet hyper-cross-linked polymers (MN200) *Journal of Colloid and Interface Science* **310**(1) 35–46

[21] Mouhamed KS, Ramzi K, Elimame E and Younes M 2014 Adsorption of anthracene using activated carbon and Posidonia oceanica *Arabian Journal of Chemistry* **7**(1) 109–113

[22] Xuemin X, Fei T, Yujun Y, Zhilin W, Zhansheng W and Giancarlo C 2015 Adsorption behavior of phenanthrene onto coal-based activated carbon prepared by microwave activation. *Korean Journal of Chemical Engineering* **32**(6) 1129-1136

[23] Valderrama C, Cortina JL, Farran A, Gamisans X and de las Heras FX 2008 Evaluation of Hyper-cross-linked Polymeric sorbents (Macronet MN200 and MN300) on Dye (Acid red 14) Removal Process *Reactive and functional Polymers* **68**(3) 679-691

[24] Vergilli I and Barlas H 2009 Removal of Selected Pharmaceutical compounds from Water by an Organic Resin *Journal of scientific and Industrial Research* **68**(2):17-425

[25] Meenakshisundaram M, Srinivasagan G and Rejinis J 2011 Novel Eco-Friendly Adsorbents for the Removal of Victoria Blue Dye *Journal of Chemical and Pharmaceutical Research* 3(6):584-594

[26] Lateefa A, Al-Khateeb A, Sitah A and Mohamad AS 2014 Adsorption of pharmaceutical pollutants onto graphene nanoplatelets *Chemical Engineering Journal* **248**: 91–199

Vulnerability Assessment of Groundwater Contamination from an Open dumpsite: Labete Dumpsite as a Case study

R A Olaoye [1], O D Afolayan [2*], R Adewale [1], T O Ogundairo[2]
[1] Civil Engineering Department, Ladoke Akintola University of Technology (LAUTECH), Ogbomoso, Nigeria.
[2] Civil Engineering Department, Covenant University, Ota, Nigeria.
Corresponding author: olaniyi.afolayan@covenantuniversity.edu.ng

Abstract. The assessment of leachate impact on groundwater is very crucial to human health and its environment. This research work aimed at assessing Labete dumpsite and its effects on groundwater integrity in the vicinity of the dumpsite. Soil samples were collected for leachate analysis at a depth of 0.5m from the earth's surface within the proximity of the dumpsite. Four different groundwater samples (GW1-GW4) were collected at radial distances from the dumpsite. Collected water samples were analyzed to examine their Cr, Fe, Cd, Pb, Zn, and Cu ions content. Physicochemical and bacteriological analysis were also considered and measured against NSDWQ and WHO standards for drinking water. The results showed that the level of groundwater contamination is directly proportional to the depth of the water table (collection) and the distance of the dumpsite from the shallow wells. GW1 and GW2 groundwater samples were not fit for consumption unless adequately treated.
Keywords: Dumpsite; leachate; groundwater; contaminant; electrical conductivity

1. Introduction

In developing countries, solid wastes are disposed of without effective management and consideration for groundwater contamination through leaching. In an engineered waste disposal system, groundwater contamination is inhibited by the introduction of a liner system [1]. Thus helps in the provision of safe drinking water. Globally, one of the primary human needs is safe drinking water. Most communities tap into the closest water source in their area, whether it is groundwater via a borehole or surface water such as a river [2]. The safety of drinking water is thus paramount and needs to be adequately monitored. In most developing countries, surface and groundwater are the primary sources of water for all basic needs. These water sources are also easily susceptible to contamination by human activities and, the degree of pollution from any of these sources is dependent on environmental control, waste management practices, and environmental protection laws in each area [3]. However, owing to several anthropogenic activities, land use [4], and consequently, high cost involved in its purification and distribution, the reliance of rural dwellers in developing nations on these water sources is on its lowest ebb. These reductions in consumption of surface water leading to increased dependency on the groundwater.

Groundwater, as defined by [5], is the water found in soil and rock cracks and crevices. It is stored in and moves slowly through geologic formations of clay, sand, and rocks called aquifers. It is everywhere, however, at varying depths, and its supplies are replenished by precipitation, which infiltrates and percolates through the cracks and crevices in soil and rock into the aquifer [5]. However, when permeable materials overlay the aquifer, it can be easily contaminated and polluted from the myriads of human activities [5, 6]. Of particular note is the contribution of waste management and disposal to groundwater contamination. In developing nations, un-engineered landfills and dumpsites are ubiquitous [7, 8]. These systems, if engineered, will help urban centers in effectively and efficiently managing the colossal waste generated through the myriad of activities taking place. However, engineered landfill adoption in developing countries is still lagging, hence contamination the groundwater resources [7, 9-11]. Mor et al. [12] explained that areas surrounding landfills and municipal dumpsites are at risk of groundwater contamination owing to the decomposition of different types of wastes; these wastes percolate into the aquifer during groundwater recharge or on gravity.

Bishop [13] defined municipal solid waste as waste generated from single-family and multifamily residences and hotels. In developing nations, due to urbanization and industrialization, there arose significant accumulation of municipal solid waste [14-16]. In these nations, dump or tip, open placement of solid waste, or incineration is the usual practice due to the low budget for waste disposal and non-availability of trained human resources [15, 17]. Ali et al. [15], Kanmani and Gandhimathi [17], and Papadopoulou et al. [6] in their research identified the dangers of open dumpsites and environmental contamination inherent in its adoption. Management of solid waste and its disposal through various means such as landfills, open dumps, or incinerations release a significant quantity of metals into the environment and groundwater [17-22]. Samuding [23], in his research, stated that leachate, which contains organic compounds, inorganic compounds, and heavy metals in high quantities, is produced when precipitation comes in contact with the decomposed solid waste. These leachates infiltrate and percolate into the shallow aquifer and pollute it [24]. Due to leachate migration, soil-water becomes contaminated with heavy metals. These metals, due to their non-biodegradable properties, lead to health problems when consumed [25] and cause taste and odor problems and oxygen depletion in groundwater [25]. Hence, the need to assess the effect and contamination level of wells built in the circumference of a dumpsite. This study is unique in its focus on usable water and in its approach of considering water sources from the homes of community members close to pollutant sites.

2. Materials and Methods

2.1 Description of the Study Area

The Labete dumping site is located at Kajola, Oshogbo, Osun State, Nigeria, as shown in Figure 1. The dumpsite is about 17.5 meters square in size and bounded by Olodo, Ile sango, and Labete communities. This dumpsite is an open dumping site and has neither a leachate collection system nor an impermeable liner system. Due to the lack of or inadequate supplies of public water in the vicinity of the dumpsite, the majority of the inhabitants rely on shallow wells as their sources of water for domestic uses. From visual inspection, wastes deposited in this dumpsite includes; waste from single-family and multifamily residences, hotel. These consist of residential wastes, commercial and industrial wastes, and agricultural waste, as shown in figure 2a and b.

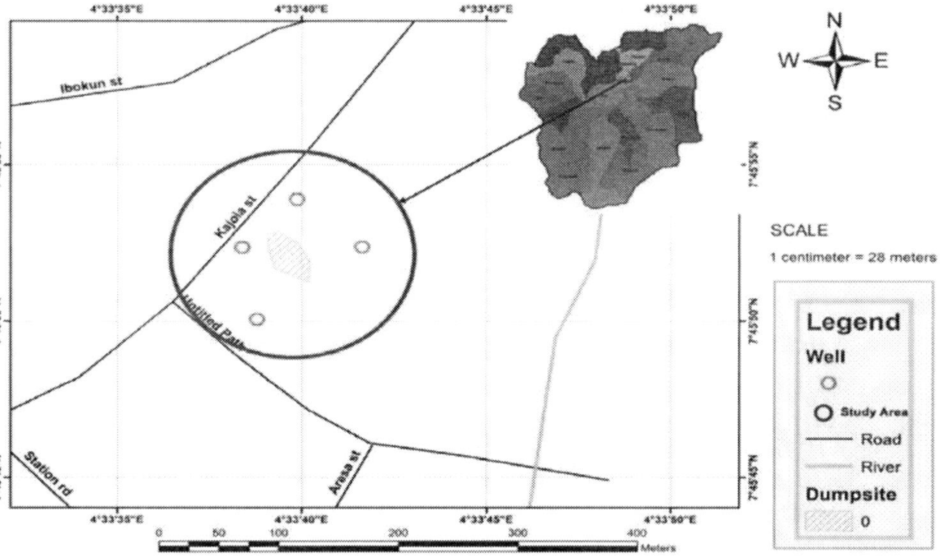

Figure 1. The layout of Kajola Osogbo (location of the dumpsite), Osun State

Figure 2 (a) Front view of the dumpsite (b) Transported waste to the downstream by erosion and wind

2.2 Field Survey

2.2.1 Determination of Well Distance and Depth

The distance of the wells was obtained by taking coordinate of the two points to be measured, using the GPS (global positioning system), from the geographical coordinate german 76 series handheld GPS. The GPS receiver obtained the longitude and latitude in World geodetic system datum, reference with UTM (Universal transverse Mercator) zone 31 north. The depth of the wells was determined by immersing a water supply pipe with a total length of 7.24m into the selected well. A measuring tape was used to measure the marked point on the pipe immediately before taking the next measurement.

2.2.2 Sampling of groundwater and soil for leachate test
Water samples were collected from four wells (GW1-GW4), as shown in table 1. Samples were collected in pre-treated 1L plastic bottles (pre-treatment of bottles was carried out by washing with distilled water) and rinsing thrice with the representative samples. The procedures followed the work of Aderemi and Falade [26]. Furthermore, soil samples were collected using digging set to excavate to a depth of 0.5m below the earth's surface around the dumpsite's vicinity during the dry season of January. The Samples were collected in a plastic bag and transported to the laboratory for leachate analysis. Leachate analysis was done in accordance with ASTM D4874.

2.2.3 Characterization of the selected samples.
Collected water samples were analyzed for metals, physicochemical, and microbial parameters, according to [27] and compared to Nigerian Standards for Drinking Water Quality (NSDWQ) and water quality parameters of World Health Organization (WHO). The concentrations of heavy metals were determined using an atomic absorption spectrophotometer (TAG 990 AS machine).
Parameters such as pH, electrical, dissolved oxygen (DO), chemical oxygen demand (COD), biochemical oxygen demand (BOD) turbidity, total alkalinity, magnesium (Mg^{2+}), calcium (Ca^{2+}), chromium (Cr^{2+}), copper (Cu^{2+}), iron (Fe^{2+}), zinc (Zn^{2+}), cadmium (Cd^{2+}), and lead (Pb^{2+}) were analyzed in the leachate and groundwater samples.

Table 1. Average depth and distance Groundwater from Dumpsite

Well	depth(m)	distance(m)	distance/depth(m)	GPS location
GW1	3.7	4.5	1.2	7.764667 N, 4.560783 E
GW2	4.9	18	3.7	7.764714 N, 4.561211 E
GW3	4.5	23.3	5.2	7.764380 N, 4.561376 E
GW4	4.7	19.6	4.2	7.764380 N, 4.561376 E

3. Result and Discussion
3.1 Heavy metal analysis for groundwater and leachate sample.
The groundwater parameters analyzed are all within the permissible limit provided by WHO and NSDWQ except for cadmium, which was in higher concentrations for GW1 and GW2 wells with values of 0.06 and 0.02mg/l respectively as shown in Table 2. These values depict that cadmium levels in GW1 and GW2 wells and its environs are at an alarming stage, with high concentrations. World Health Organization [28] recommended a concentration of Cd ions below 0.003 mg/l in drinking water. Still, the results of these two wells had levels of the toxic metal that exceeded the recommended threshold. Furthermore, the position of the well plays an essential role in its contamination as GW3 and GW4, which were at the downstream, and a considerable distance shows no Cd contamination as opposed to GW1 and GW2.

Cadmium (Cd) ions contamination in water according to research by [29-32] is associated with the presence of either electronic waste, metal industries, sewage sludge, mining wastes, open-air waste burning, coal combustion, steel industry, concrete production, or artificial phosphate fertilizers in the dumpsite which thereby leach into the subsoil. Also, aquifer type contributes to the presence of these ions in a significant quantity [29]. Consequently, water with a high-level of Cd ions is carcinogenic [33, 34]. Hence, it adversely affects the vital organs in the human body. The presence of iron, lead, and zinc ion in the leachate signifies the presence of metal scraps, batteries, lead-based paints and lead pipes, fluorescent tubes, batteries, food wastes, and burning of tires at the dumpsite [35].Furthermore, the presence of lead and cadmium in the samples can be attributed to the sites' sub-surface geology [36, 37].

Table 2: Characteristics of heavy metal in leachate and groundwater samples

Sample	Cr	Fe	Cd	Pb	Zn	Cu
GW1	0.04	0.26	0.06	ND	0.53	0.96
GW2	0.02	0.10	0.02	ND	0.34	0.87
GW3	0.03	0.03	ND	ND	0.26	0.69
GW4	0.03	0.17	ND	ND	0.22	0.64
LS	2.67	10.62	3.045	2.39	10.61	6.91
NSDWQ	0.05	0.3	0.003	0.01	3.0	1.0
WHO	0.05	0.3	0.003	0.01	NS	2.0

Note. All heavy metal values are given in mg/l
NSDWQ Values are the maximum permitted levels in the Nigerian Standards for Drinking Water Quality, ND- Not detected, NS-Not supplied and LS-leachate sample. WHO values are the maximum permitted levels in the WHO Drinking Water Quality Guideline

3.2 *Physicochemical analysis*

pH, a measure of acidity and alkalinity of a solid/liquid, it is an important parameter as it influences the rate of contaminants transportation, dissolution of minerals in the soil and the growth of microbes. A low and high pH value indicates acidic and alkaline water, respectively. As seen in Table 3, the water samples and the leachate had an acceptable pH value. Furthermore, calcium and magnesium ions, turbidity, dissolved oxygen, color, and electrical conductivity of the groundwater samples collected conformed to WHO water quality standards.

BOD and COD indicators in environmental studies indicate the pollutant strength of polluted water [38, 39]. Table 3 showed that the samples tested had high BOD, high COD, and alkalinity values, which is an indication of leachate infiltration into the subsoil.

3.3 Microbial analysis

Coliform is naturally present in the environment [11, 40, 41], fecal coliform and Escherichia-Coli (E. coli) are derived from human and animal waste [42]. As shown in Table 4, the presence of fecal coliform and E. coli in water the presence of human or animal waste. The presence of E. coli in groundwater is influenced by groundwater flow based on the topography of the area [42]. In groundwater, sources of total and fecal coliform may include run-offs from decomposed agricultural waste, effluents from a septic tank or sewage disposal, and infiltration of animal fecal

IOP Conf. Series: Materials Science and Engineering 1036 (2021) 012013 doi:10.1088/1757-899X/1036/1/012013

matter [43]. E-coli and total coliform were counted in the four wells, and the result presented in Table 4.

Table 3. Comparison of the physicochemical result of leachate and groundwater samples with WHO and NSDWQ standard

Sample	GW1	GW2	GW3	GW4	LS	WHO	NSDWQ
pH	7.90	6.78	6.90	7.72	6.45	6.5-9.5	6.5-8.5
EC (ms/cm)	-	-	-	-	20.24	<5	
BOD5	8.50	8.25	7.75	7.00		6	
Mg (mg/l)	3.50	3.00	2.45	3.10	3.55	50	
DO (mg/l)	0.32	0.26	0.23	0.16	4.11	7.27	13.05
Ca (mg/l)	23.52	22.55	19.24	17.45	32.02	75	75
COD	5.30	6.01	5.10	5.49	7856.56	1000	
Alkalinity	178	178	178	178	356	200	
Turbidity (NTU)	2.50	3.00	2.01	3.50	38	5	
Color	Colorless	Colorless	Colorless	Colorless			

Table 4. Microbial analysis of the collected water samples

Samples	10^{-5} (CFU/ml) Coliform	10^{-5}(CFU/ml) General Bacterial
GW1	3.57	6.75
GW 2	2.62	2.12
GW3	2.62	2.17
GW4	1.55	1.75
WHO	0	0

IOP Conf. Series: Materials Science and Engineering 1036 (2021) 012013 doi:10.1088/1757-899X/1036/1/012013

This result is expressed as (Coliform Microbial Unit) and indicates the extent of fecal matter present in it. Table 4 showed that the four wells had been contaminated with fecal material based on the coliform and bacterial content in comparison to the WHO standard; these microbes can cause a myriad of disease to the consumer if the water is not properly treated before consumption as [28] specified for drinking water quality, and water must utterly free from any colony.

In summary, based on the obtained results from the physicochemical, heavy metals and microbial analysis of the water samples and leachate, it can be deduced that the concentrations of the analyzed parameters have decreased with the distance from the landfill, which can be linked to biodegradation and dilution [44, 45].

4. Conclusion.

From the study conducted, the followings are the conclusions.

i The distance and depth of the wells from the source of leachate have a more significant impact on the degree and extent of contamination of groundwater and surface water. Therefore, wells with longer distances away from the dumpsite boundary and appreciable varying depth difference should be considered when planning for good construction in the future.

ii Groundwater samples from all the wells are unsuitable for drinking purposes owing to the elevated levels of cadmium and fecal waste in them

iii Therefore, there is an urgent need to improve the waste management practices in the study area through the construction of proper engineered sanitary landfill sites to curtail the pollution of groundwater.

Acknowledgments

The authors wish to acknowledge the financial support offered by Covenant University CUCRID in the actualization of this research work for publication.

References

[1] Olaoye R A, Afolayan O D, Oladeji V O and Sani R O 2019 Influence of bentonite on clayey soil as landfill baseliner materials *IOP Conf. Series: Materials Science and Engineering* **640** 012107 DOI:10.1088/1757-899X/640/1/012107

[2] Odukoya A M and Abimbola A F 2010 Contamination assessment of surface and groundwater within and around two dumpsites. *International Journal of Environmental Science and Technology*, **7**(2): 367-376, https://doi.org/10.1007/BF03326146

[3] Han D M, Long X X, Jin M G, Emily H, Tong C S and Song X F 2013 Evaluation of organic contamination in urban groundwater surrounding a municipal landfill, Zhoukou, China. *Environmental Monitoring & Assessment,* **755**(1): 3413-3444, DOI 10.1007/s10661-012-2801-z

[4] Emenike P C, Tenebe I T, Neris J B, Omole D O, Afolayan O, Okeke C and Emenike I K 2020 An integrated assessment of land-use change impact, seasonal variation of pollution indices and human health risk of selected toxic elements in sediments of River Atuwara, Nigeria *Environmental Pollution* **265**: 114795

[5] The Groundwater foundation 2019 3201 Pioneers Blvd Suite 105 Lincoln, NE 68502.accessed from: https:// www.groundwater.org/get-informed/basics/whatis.html

[6] Papadopoulou M P, Karatzas G P, and Bougioukou G G 2007 Numerical modelling of the environmental impacts of landfill leakage on groundwater quality – a field application. *Environmental Modelling Assessment.* **12:**43–54, https://doi.org/10.1007/s10666-006-9050-x

[7] Longe E O and Balogun M 2010 Groundwater quality assessment near a municipal landfill, Lagos, Nigeria *Research Journal of Applied Sciences, Engineering and Technology,* **2**(l): 39-44

[8] Jhamnani B and Singh S K 2009 Groundwater contamination due to Bhalaswa landfill Site in New Delhi. *International Journal of Civil and Environmental Engineering,* **1**(3): 121-5

[9] Sabahi E A, Rahim S A, Zuhairi W Y and Nozaily F A 2009 Assessment of groundwater pollution at municipal solid waste of a landfill in Yemen. *Bulletin of the Geological Society of Malaysia,* **55**(1): 21-26. DOI: 10.7186/bgsm55200904

[10] Fatta D, Padadopoulos A and Loizidou M 1999 A study on the landfill leachate and its impact on the groundwater quality of the greater area. *Environmental Geochemical Health,* **21** (2): 175 – 190. https://doi.org/10.1023/A:1006613530137

[11] United States Environmental Protection Agency 2009 *National Primary Drinking Water Regulations* (https://www.epa.gov/ground-water-and-drinking-water/national primarydrinking- water-regulations #Microorganisms)

[12] Mor S, Ravindra K, Dahiya R P and Chandra A 2006 Leachate characterization and assessment of groundwater pollution near municipal solid waste landfill site. *Environmental Monitoring and Assessment,* **118**: 435-456. DOI: 10.1007/s10661-006-1505-7

[13] Bishop P L 2000 *Pollution prevention: Fundamentals and Practice.* McGraw-Hill, Companies Inc., USA

[14] Yoada R M, Chirawurah D and Adongo P B 2014 Domestic Waste Disposal Practice and Perceptions of Private Sector Waste Management in Urban Accra. *BMC Public Health,* **14**, 697. https://doi.org/10.1186/1471-2458-14-697

[15] Ali S M, Pervaiz A, Afzal B., Hamid N., and Yasmin A 2014 Open dumping of municipal solid waste and its hazardous impacts on soil and vegetation diversity at waste dumping sites of Islamabad city Syeda *Journal of King Saud University – Science,* **26**:59–65, http://dx.doi.org/10.1016/j.jksus.2013.08.003

[16] Ziadat A H and Mott H 2005 Assessing solid waste recycling opportunities for closed campuses." *Managing Environmental Quality,* **16**:250–6 https://doi.org/10.1108/14777830510591679

[17] Kanmani S and Gandhimathi R 2012 Assessment of heavy metal contamination in soil due to leachate migration from an open dumping site *Applied Water Science,* **3**:193–205. DOI: 10.1007/s13201-012-0072-z

[18] Rizo O D, Merlo M H, Castillo F E and Lopez J A O 2012 Assessment of metal pollution in soils from a former Havana (Cuba) solid waste open dump. *Bulletin on Environmental Contamination and Toxicology* **88**:182–86, https://doi.org/10.1007/s00128-011-0505-7

[19] Bretzel F C and Calderisi M 2011 Contribution of a municipal solid waste incinerator to the trace metals in the surrounding soil. *Environ Monitoring Assessment,* **182**:523–533. DOI: 10.1007/s10661-011-1894-0

[20] Iwegbue C M A, Nwajei G E, Ogala J E and Overah C L 2010 Determination of trace metal concentrations in soil profiles of municipal waste dumps in Nigeria. *Environ Geochem. Health*, **32**(5):415–430. DOI: 10.1007/s10653-010-9285-y
[21] Waheed S, Siddique N, Hamid Q and Chaudhry M M 2010 Assessing soil pollution from a municipal waste dump in Islamabad, Pakistan: a study by INAA and AAS. *Journal of Radiational and Nuclear Chemistry,* **285**:723–732. https://doi.org/10.1007/s10967-010-0623-4
[22] Yarlagadda P S, Matsumoto M R, Van-Benschoten J E and Kathuria A 1995 Characteristics of heavy metals in contaminated soils. *Journal of Environmental Engineering*, **121**(4):276–286, DOI: 10.1061/(ASCE)0733-9372(1995)
[23] Samuding K 2009 Distribution of heavy metals profile in groundwater system at a solid waste disposal site, *European Journal of Science and Research,* **37**:58–66
[24] Bhalla G, Swamee P K, Kumar A and Bansal A 2012 Assessment of groundwater quality near municipal solid waste landfill by an Aggregate Index Method. *International Journal of Environmental Science,* **2**(2):1492 DOI: 10.6088/ijes.002020300034
[25] Hong K J, Tokunaga S and Kajiuchi T 2002 Evaluation of remediation process with plant-derived biosurfactant for recovery of heavy metals from contaminated soils. *Chemosphere* **49**:379–387
[26] Aderemi A O and Falade T C 2012 Environmental and health concerns associated with the open dumping of municipal solid waste: a Lagos, Nigeria experience. *American Journal of Environmental Engineering*, **2**(6):160–165, DOI: 10.5923/j.ajee.20120206.03
[27] APHA 2005 *Standard Method for the Examination of Water and Wastewater*, 21st ed. American Public Health Association, Washington DC
[28] World Health Organization 2011 Guideline for drinking-water Quality. 4th Edition. pp.541http://apps.who.int/iris/bitstream/handle/10665/44584/9789241548151_eng.pdf;jsessionid=E08724B01DEBF746D639045830AB198C?sequence=1>
[29] Minnesota Pollution Control Agency 1999 *Cadmium, Lead and Mercury in Minnesota's Groundwater*, 520 Lafayette Rd. N., Saint Paul, MN 55155-4194(651)
[30] Ramachandra T V and Saira V K, 2004 An environmentally sound option for waste management. *Envis. J. Human Settlements* (March: 1-10)
[31] Sivakumar T, Vijayaraghavan G and Kumar AV 2011 Global Challenges in E-waste management: Indian Scenario. *International Journal on Advancement of Engineering and Technology* **2**:10–15
[32] Idrees N, Tabassum B, Abd-Allah E F, Hashem A, Sarah R and Hashim M 2018 Groundwater contamination with cadmium concentrations in some West U.P. Regions, India, *Saudi Journal of Biological Sciences,* **25**: 1365–1368. https://doi.org/10.1016/j.sjbs.2018.07.005
[33] Tabassum B, Bajaj P and Azeem I 2016 Groundwater contamination due to uncontrolled disposal of E-waste in some regions of western UP. *EJBPS* **3** (8), 360–364
[34] Stewart B W and Kliehues P 2003 *World Cancer Report, Lyon*: IARC Press, Lyon
[35] Moturi M C Z, Rawat M and Subramanian V 2004 Distribution and fractionation of heavy metals in solid waste from selected sites in the industrial belt of Delhi, India *Environ. Monitoring and Assessment,* **95**:183–199, https://doi.org/10.1023/B:EMAS.0000029900.86810.85
[36] Mor S, Ravindra K, Dahiya R P and Chandra A 2006 Leachate characterization and assessment of groundwater pollution near municipal solid waste landfill site.

Environmental Monitoring and Assessment, **118**: 435-456 DOI: 10.1007/s10661-006-1505-7

[37] Longe E O and Enekwechi L O 2007 Investigation on potential groundwater impacts and influence of local hydrogeology on natural attenuation of leachate at a municipal landfill. *International Journal of Environmental Science & Technology,* **4** (1), 133-140, https://doi.org/10.1007/BF03325971

[38] Kalenik M 2014 Sewage Treatment Efficacy of Sandy Soil Bed with Natural Clinoptilolite Assist Layer. *Ochr. 'Srodowiska*, **36**:43–48

[39] Ngang B U and Agbazue A V E 2016 Seasonal Assessment of Groundwater Pollution due to Biochemical Oxygen Demand, Chemical Oxygen Demand and Elevated Temperatures in Enugu Northern Senatorial District, South East Nigeria. *IOSR Journal of Applied Chemistry (IOSR-JAC) e-ISSN: 2278-5736.Volume 9, Issue 7 Ver. I (July 2016), pp. 66-73*

[40] Pal A Gin K Y H Lin A Y C and Reinhard M 2010 Impacts of Emerging Organic Contaminants on Freshwater Resources: Review of Recent Occurrences, Sources, Fate and Effects. *Science of the Total Environment,* **408**, 6062-6069, DOI:10.1016/j.scitotenv.2010.09.026

[41] Nicholson K N, Neumann K, Dowling C and Sharma S 2017 E. coli and Coliform Bacteria as Indicators for Drinking Water Quality and Handling of Drinking Water in the Sagarmatha National Park, Nepal. *Environmental Management and Sustainable Development*, **6** (2) 2164-7682, DOI: 10.5296/emsd.v6i2.11982

[42] Rohmah Y, Rinanti A and Hendrawan D I 2018 The determination of groundwater quality based on the presence of *Escherichia coli* on populated area (a case study: Pasar Minggu, South Jakarta). *The 4th International Seminar on Sustainable Urban Development. IOP Conf. Series: Earth and Environmental Science* **106** 012079. DOI :10.1088/1755-1315/106/1/012079

[43] Water Stewardship Information Series 2007 *Total, Fecal & E. coli Bacteria in Groundwater.* Retrieved on 3/13/2020

[44] Koda E, Miszkowska A and Sieczka A 2017 Levels of Organic Pollution Indicators in Groundwater at the Old Landfill and Waste Management Site. *Applied Sciences*, *7*(6), 638; DOI: 10.3390/app7060638

[45] Nyenje P M, Foppen J W, Muwanga R and Uhlenbrook S 2013 Nutrient Pollution in shallow aquifers underlying pit latrines and domestic solid waste dumps in urban slums. *Journal of Environmental Management.* **122**: 15-24, DOI: 10.1016/j.jenvman.2013.02.040

IOP Conf. Series: Materials Science and Engineering 1036 (2021) 012014 doi:10.1088/1757-899X/1036/1/012014

Computational forecast of municipal waste in Lagos: What may happen in 2025?

M E Emetere[1,2] and C O Iroham[3]
[1] Department of Physics, Covenant University Canaan land, P.M.B 1023, Ota, Nigeria.
[2] Department of Mechanical Engineering Science, University of Johannesburg, South Africa.
[3] Department of Estate Management, Covenant University Canaanland, Ota, Nigeria.

Abstract. Lagos is the commercial hub of Nigeria with almost the highest population in Nigeria. The municipal waste of Lagos is currently huge with existing health threats to residence around the landfills. Thirty-seven years (1980-2017) dataset for municipal waste in Lagos was projected to 2025 using computational analysis i.e., moving averages and excel forecast tool. It was discovered that the averages of all the forecast show that the municipal waste in Lagos will be fairly stable between 2020-2025 with an addition of 12 million tonnes by 2025. It is recommended that waste management authorities in Lagos should step-up with the establishment of municipal waste processing companies in all the five landfill sites because the land fill leachates is expected to pollute more ground water sources in Lagos metropolis.

1. Introduction

In the country Nigeria, the law forbids indiscriminate dumping of plastic waste, e-waste and other forms of waste [1]. However, the enforcement of the law is overwhelmingly impossible because of the inadequacies of regulating bodies. Public awareness on the dangers of indiscriminate dumping of waste is very low, thereby exposing over 100 million people to dangerous substances that cause defects in human organs such as the skin, nose lungs, eye and brain [2-6]. Ultimately, this danger leads to loss of lives. Due to the high level of poverty, the importation of used goods from developed world is a big business in Africa and beyond [7]. Due to the Nigeria population the patronage of imported used goods is high, thereby necessitating her ranking as the top for waste generation [8]. Still in the sub-Saharan Africa, Ghana is also recognized for its huge importation of used goods. Most of the used-goods are electronics, clothes and automobiles. Unfortunately, most of the imported goods are near expiry and becomes waste. Out of this wastes, the electronics waste is more worrisome because high level of heavy metals such as copper, lead. The e-waste also is made-up of plastic, which is an additional cause for caution to marine lives [9].

Lagos state, Nigeria, is believed to have large amount of e-waste because it is the commercial capital of Nigeria. The heap of wastes in landfill across is unimaginably large as presented in Figure 1. With an estimated 21 million residents, waste management is a huge task, hence the dumping of refuse along water bodies may not be eradicated. There are 3 major landfills and 2 temporary sites serving Lagos State. The two largest land fill i.e. Olushosun Landfill and Abule-Egba Landfill sites. Olushosun Landfill receives approximately 40% of the total waste deposits from Lagos [10]. The size is 42.7 hectares and it has a residual life span of 20 years. Abule-Egba Landfill occupies a land of about 10.2 hectares and receives waste from the densely populated area. The residual life span is approximately 8 years.

Figure 1: A typical landfill location in Lagos Nigeria

Basel Action Network (BAN) conducted a research that maps the distribution of e-waste over Lagos State. The research shows that there are huge e-waste in Lagos, which includes central processing units (CPU), Dell CPU, Dell computer (made in Ireland), Pentium II computer, Toshiba computer, gateway computer, computer monitor, TV monitor, used computers, CTX computer monitor, Hewlett Packard scanner, Compaq CPU, electronic equipment, Hewlett Packard printer, Hewlett Packard fax, Hewlett Packard copier, OPUS technology computer equipment, old computer, computer keyboard, UTAX C123, IBM computer, used printerPanafax UF-160, refrigerator, NEC computer, AGFA copy machine, XEROX copy machine, RICOH copy machine, EPSON, and Macintosh CPU (BAN, 2005) [11].

2. Methodology

Lagos State is located on longitude 2° 42'E and 3° 22'E and between latitude 6° 22'N and 6° 42' N [12-13]. It is bounded in the south by Atlantic Ocean. This research thirty-five years dataset was used (1980-2014) was used to forecast the volume of municipal waste by 2025. The municipal waste dataset for all the landfill in Lagos for 2012 was gotten from Lagos Waste Management Authority (LAWMA) [14]. The dataset covers the refuse dumping of private and government operators in the State. The cubic meters of refuse deposited was calculated according to the number of trucks that deposits waste per day. The source error may likely occur via manual counting system. Constructing computational systems that automatically counts truck coming in and out of the dumpsite can minimize this error. Opejin [15] had reported that organic waste accounts for 45% of the total wastes in Lagos. The computational tool that was moving averages and excel forecast tool. The techniques have been reported to be very reliable in understanding both the trend and anomalies of dataset [15-16]. The excel forecast was particularly chosen because most professionals and students appreciate the tool.

The dataset ranges from 1980-2017. The 92% of the dataset was used to forecast and 8% was used to test the validity of the dataset. The forecast technique are two moving averages, excel forecast tool and the averages of all forecast.

3. Results and Discussion

The municipal waste in Lagos generally dwindled between two highest outputs i.e., 1991 and 2010. Though the significant decrease between 1991 and 2010 is quite questionable since it is believed that the population over Lagos has been increasing with respect to opportunities that abound in the city. World population review believes that as of 2012, about 17.5 million people live in Lagos [17]. Beyond 2010, the municipal waste keeps reducing even when there are no clear evidence that population had decreased. There are two broad explanations, i.e., first, some people have their private landfills; second, residence living at the outskirt of the state patronize the waste management system of neighboring states. The forecast of the municipal waste in Lagos is presented in Figure 2.

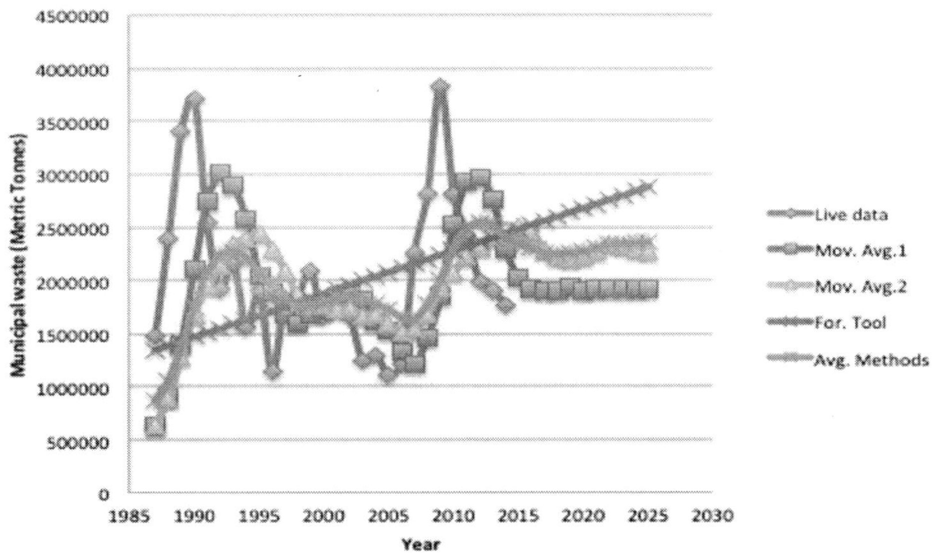

Figure 2: Forecast of municipal waste (2020-2025)

The forecast extended in 2025. 92% of the dataset was used for the prediction while the remaining 8% of the dataset was used to test the accuracy of the forecast. The prediction of the moving averages show that the annual municipal waste will be fairly stable and may not go beyond 2.2 million tons. The excel forecast tool predicted that the municipal waste would linearly increase to 2880285.15 in 2025. The averages of all the forecast show that the municipal waste in Lagos will be fairly stable and will not go beyond 2.3 million tons in 2025.

The validity of the prediction was carried out using 8% of the dataset i.e., 2015-2017 with a degree of confidence given as 95%. It was observed that the outcome shows that the moving averages were near accurate in that the dataset was fairly stable. The correlation between the forecast and the test dataset is shown in Table 1 below. Corrl. 1 is the correlation between the first moving average and the test dataset; Corrl. 2 is the correlation between the first moving average and the test dataset; Corrl. 3 is the correlation between the excel forecast tool and the test dataset; and Corrl. 4 is the correlation between the averages of all forecast and the test dataset. The second moving average showed high correlation while the excel forecast tool showed a negative correlation, i.e., meaning that there is an inverse relationship between two variables - when one variable decreases, the other increases.

Table 1: correlation between forecast and 8% test dataset

Corrl. 1	Corrl. 2	Corrl. 3	Corrl. 4
0.736	0.999	-0.941	0.972

The inter-variable covariance of the two averages and the original dataset is presented in Table 2. Also, the inter-variable covariance of the excel forecast tool, averages and the original dataset is presented in Table 3.

Table 2: Inter-Variable Covariance for moving averages

Original Data		Moving average 1	Moving average 2
X:	509596226721.99	48651054866.189	-20930871482.474
Y:		397977001951.01	229187557686.53
Z:			195821000780.64

The inter variable covariance, which describes the extent random variables vary together show that original data/moving average 1 had the highest covariance while the original data/moving average 1 had the lowest covariance. Hence, though the moving average 2 showed the highest correlation for 8% test dataset (Table 1), the order of accuracies using the trained dataset i.e., 92% follow in descending order i.e., moving average 1, forecast average, moving average 2 and excel forecast tool.

Table 3: Inter-Variable Covariance excel forecast

Original Data		Excel forecast Tool	Averages of all forecast
X:	509596226721.99	-16206994259.884	3837729707.9433
Y:		105666088951.68	71473854302.215
Z:			152816683885.15

This result show that there will an addition of over 12 million metric tons of municipal waste will be added to the Lagos landfill between 2020 and 2025. This result means the estimated e-waste may also rise to about 40%. Hence, the local waste management authorities in Lagos have a herculean task to prevent high heavy metals contaminations that can increase diseases to dumpsite workers and neighbouring communities. It is recommended that waste management authorities in Lagos should step-up with the establishment of municipal waste processing companies in all the five landfill sites, as the land leachates is expected to pollute more ground water sources in Lagos metropolis. In other words, public awareness is important to educate populace on waste management [18]

4. Conclusion

The study reveals that the municipal waste would increase between 2020 and 2025 by 12 million tons. The undocumented municipal waste would also increase within the city with high risk of outbreak of airborne diseases. It is expedient that the waste management authorities in Lagos should urgently incorporate waste processing plants that would help reduce the weight of the waste on the surface geology of the landfill. When the depth of the leachate increases, there is a high possibility that deep ground water sources around the landfill will be polluted. It is recommended that the waste disposal laws be enforced via large campaign awareness and stricter law concerning the disposal of waste.

Acknowledgement

The authors appreciate Covenant University for partial sponsorship.

5. References

[1]. Nwufo CC 2010 Legal Framework for the Regulation of Waste in Nigeria, *African research review*, **4** (2) 491-501

[2]. Emetere ME and Jumbo H 2019 Computational extrapolation of municipal waste generation in Nigeria, *Journal of Physics: Conference Series* **1299** (1), 012032

[3]. Smyth DP, Fredeen AL Booth AL 2010 Reducing solid waste in higher education: The first step towards 'greening' a university campus. *Resour. Conserv. Recycl.*, ***54***, 1007–1016
[4]. Hamer G 2003 Solid waste treatment and disposal: effects on public health and environmental safety, *Biotechnology Advance;* **22**, 71-79
[5]. An H, Englehardt J, Fleming L and Bean J, "Occupational health and safety amongst municipal solid waste workers in Florida," *Waste Management and Research*, **17**, 5, pp. 369–377, 1999
[6]. Ray MR, Roychoudhury S, Mukherjee G, Roy S and Lahiri T, "Respiratory and general health impairments of workers employed in a municipal solid waste disposal at an open landfill site in Delhi," *International Journal of Hygiene and Environmental Health,* **208**, 4, pp. 255–262, 2005
[7]. Jutta G 2009 Waste, poverty and recycling, *Waste Management* **30**(2):171-173
[8]. Alamgir M, McDonald C, Roehl KE, and Ahsan A, Eds., (2005). Final Report of the EU Asia Pro Eco Project "WasteSafe", *Khulna University of Engineering & Technology* (KUET), **398** p.; Khulna, Bangladesh (2005) - ISBN 984-32-3003-5
[9]. Mohiuddin KM, Alamgir M and Roehl KE 2005 Present scenario of solid wastes disposal sites in Khulna city of Bangladesh," *in Proceedings of the 3rd Annual Paper Meet and International Conference on Civil Engineering*, pp. 429–438, IEB, Dhaka, Bangladesh,
[10]. LAWNA, (2020) https://www.lawma.gov.ng/ (Accessed 8/4/2020)
[11]. Basel Action Network (2005), The digital dump: Exporting re-use and abuse to Africa https://static1.squarespace.com/static/558f1c27e4b0927589e0edad/t/55d79038e4b069c9055c8720/1440190520196/BANsDigitalDump-2005.pdf (Acessed 8/04/2020)
[12]. Emetere ME, Emetere JM, and Dania OE 2019 Modified Angstrom Solar Radiation Model for Lagos Nigeria, *IOP Conference Series: Earth and Environmental Science* **331** (1), 012054 (2019)
[13]. Emetere ME, Akinyemi ML and Akinojo O2015 Theoretical Aid for Improving Measuring Instruments Efficiency within the Meteorological Space of Lagos-Nigeria, 2015 PIAMSEE: *AIP Conference Proceedings,* **1705** (1), 020054
[14]. LAWNA, (2020) https://www.lawma.gov.ng/inside-lawma/departments/landfill/ (Accessed 8/4/2020)
[15]. Murat M, Malinowska I, Gos M, Krzyszczak J, 2018. Forecasting daily meteorological time series using ARIMA and regression models. *Int. Agrophys.,* **32**, 253-264, doi:10.1515/intag-2017-0007
[16]. Ye LM, Yang GX, Van Ranst E, and Tang HJ 2013 Time-Series modeling and prediction of global monthly absolute temperature for environmental decision making. *Adv. Atmos. Sci,* **30**, No.2, pp. 382-396
[17]. Unknown (2020), Lagos population 2020, https://worldpopulationreview.com/world-cities/Lagos-population/ (Accessed 9/4/2020)
[18]. Rajamanikam R and Poyyamoli G, 2014 Towards Zero-Waste Campus: Compositional Analysis of Solid Waste at the Staff Quarters to frame Inclusive Sustainable Campus. Waste Manag. Syst. *3*, 11255–11264

IOP Conf. Series: Materials Science and Engineering 1036 (2021) 012015 doi:10.1088/1757-899X/1036/1/012015

Effect of varying concentrations on the yield of activated carbon produced from African star apple

G Abatan[1]* A Busari[2] O Odunlami[1] G Alao[3], B A Oni[1] A Ayoola[1] and D Abraham[1]

[1]Department of Chemical Engineering, Covenant University, Ota, Nigeria
[2]Department of Civil engineering, Covenant University, Ota, Nigeria
[3]Department of Mathematics, Covenant University, Ota, Nigeria
Corresponding author: olubunmi.abatan@covenantuniversity.edu.ng

Abstract. Activated carbon has been developed using different techniques and activating agents. The activation of carbon is done using different chemical, physical, and bio activating agents. This activating agent gives various outcomes of the activated carbon depending on the structure of the cursor used. The increase in population has also brought about a significant increase in waste generated. In this work, we were able to produce activated carbon from African star apple seed using the seed, husk, and shell. This work was carried out using a furnace at 500 C. The carbon produced was then activated using different concentrations of Phosphoric acid (H_3PO_4), which are 0.2M, 0.4 M, 0.6 M, 0.8 M and 1.0 M. Whole seed which was impregnated with 0.2 M gave the highest yield of 87.192%. The amount of activating agent used did not determine the yield.

1. Introduction

According to World bank, the world population is increasing daily, so also do we have the waste on the increase [1]. These wastes are generated from households, industries, religious centers, etc. Some agricultural waste can be converted into activated carbon in order to purify other waste for usage, especially waste water from different sectors of industries and households [2]. Water pollution is one of the key problems confronted internationally due to the unsystematic discharge of domestic and urban polluted waters, agricultural wastes and industrial wastes into water bodies. Food industries produce a lot of waste water and if not well treated and disposed can pollute the environment (Mohamed 2016). It was also stated that the quick industrialization rate has caused discharge of wastes encumbered with pollutants into our water bodies which has affected humans, marine life and the environs at large [4]. This pollution can also be related to the cost incurred in treating the waste water. Producing activated carbon from other waste can be a source of reduction in cost if the field is high. In this present day several activated carbon has been produced from different materials like rice husk, Moringa seed husks, palm shells, coconut Shells etc, [2] The manufacture of activated carbon by means of agricultural waste is more cost-effective likened to commercially activated carbon [3].

Activated carbon (AC) was defined as a form of carbon that was primed to make it astonishingly porous to achieve a widespread surface area available for adsorption responses [6]. It was also defined it as a non-graphitizable carbon, denoting that it cannot be converted into graphite just by extraordinary heat treatment [7]. Arvanitoyannis has made it known that the surface region of Activated carbon can be

IOP Conf. Series: Materials Science and Engineering 1036 (2021) 012015 doi:10.1088/1757-899X/1036/1/012015

larger than 1000 m^2/g [8]. This implies that it is possible for 5 g of activated carbon to have a surface area which is larger than of a football field. The process of preparing activated carbon are carbonization of raw materials and carbon activation [9]. Carbonization is done when the raw material is decomposed thermally which removes non-carbon species. This helps to create fundamental pore structure which are small in size. Activating the carbon helps to increase the pores largely. Activation can be done chemically or physically. Chemical activation is done using activating chemical agents like Sulphuric acids, Phosphoric acids, Zinc Chloride, Nitric acids, Sodium Chloride, etc. [10]. Bio-activators can also be used to manufacture bio centered activated carbon particularly in the medical field, food and pharmaceuticals industry [11]. Yield of activated carbon will be very important to the economic value. Table 1 below shows the yield from different activate carbon which were activated using Phosphoric acid.

Table 1: Yield of activated carbon gotten when phosphoric acid is used.

	Raw Material	Molarity of Phosphoric acid (M)	% yield	Source
1	Pineapple peel	1.5	58.15	[12]
2	Pineapple crown	1.5	15.24	[12]
3	African Star Apple seed husk	0.008	55.3	[2]
4	African Star Apple oil seed	0.008	54.3	[2]
5	African Star Apple whole seed	0.008	41.6	[2]
6	Coconut shell		77.74	[13]
7	Orange peel		72.9	[13]
8	Banana peel		60.8	[13]

The study of Ref [4] was done using a single molarity. In this work further steps have been carried out to vary the concentration of phosphoric acid in activating the carbon. The carbon was activated using various concentrations of phosphoric acid.

2. Materials and methods

The ripe African Star Apple fruit was gotten from Tejuosho Market, Yaba, Lagos Nigeria and Industrial Waste Water from Honeywell Industries, Nigeria. De-ionized water available in the laboratory Covenant University and Phosphoric acid (J.T.Baker Deventer- Holland). Digital Weighing Balance (OHAUS Scout pro 2000g); Mechanical Stirrer (JENWAY 1000 Hot Plate & Stirrer ST15 OSA, UK); SVAC2 Compact Vacuum Drying oven, 1.7 Cubic Foot, 240°C, Model SVAC2; Model F6018, Thermo Scientific Thermolyne A1 Premium Muffle Furnace - 208V; Mercury-in-glass Thermometer (GH Zeal LTD, 76mm 0-360°C); Conical Flask (250 ml); Measuring Cylinder, Beaker (500 mL); Plastic containers for samples; Mortar pestle; Spatula and Stopwatch

2.1 Removal and drying of seed husks:

The seed kernels of the African star apple were physically torn apart from fleshy fruit, retrieved and washed to remove grease and debris and then dried for 3 days in the sunlight after which it was dehydrated in an oven at temp 105 °C for 3 hours to ensure all effective removal of all water residue in the seeds. The dried precursor was separated into 3 samples (seed, husk and shell) and were ground using mortar and pestle and then the ground kernels were ground further to make particle size of -6 to +40 mesh.

IOP Conf. Series: Materials Science and Engineering 1036 (2021) 012015 doi:10.1088/1757-899X/1036/1/012015

2.2 Preparation of Activated Carbon:

Precisely measured samples of African Star Apple seeds (husk, whole seed and shell) that was dried, were carbonized in crucibles at 500°C in furnace. The charcoal produce obtained was milled and sieved to 2mm size. Five solutions of known concentrations; 0.2 M, 0.4 M, 0.6 M, 0.8 M, and 1 M of the Phosphoric acid (H_3PO_4) were made. Three separate 10 g servings of the African Star apple samples were stirred and left to be soaked distinctly in 100 mL of 0.2 M, 0.4 M, 0.6 M, 0.8 M and 1 M H_3PO_4 respectively for 24 hours. The samples were the labelled ABD1, ABD2, ABD3, and ABD to ABD15 respectively before being heated in the oven at a temperature of 105°C for 2 hours. The activated carbon was then cooled to 27 °C and washed with warm distilled water until it reached a pH of 7 to remove any undiluted residue of phosphoric acid after which the precursor was dried in an oven at a temperature of 105 °C for 1 hour. Finally, the dried precursor was milled and sieved to get the molecule size of 125-150μm. All proximate analysis was done on the activated carbon. It was kept in plastic containers for further use and the results were tabulated. Table 2 shows the activated carbon sample at different concentrations.

Table 2: Activated Carbon samples

Sample Name	Precursor	H_3PO_4 Concentration(M)
ABD1		0.2
ABD2		0.4
ABD3	Seed husks	0.6
ABD4		0.8
ABD5		1
ABD6		0.2
ABD7		0.4
ABD8	Oil seed	0.6
ABD9		0.8
ABD10		1
ABD11		0.2
ABD12		0.4
ABD13	Whole Seed	0.6
ABD14		0.8
ABD15		1

3. Result and discussion

The yield of each activated carbon sample produced by chemical activation with phosphoric acid was obtained from calculations and is being summarized in Figure 1 It is calculated as a function of the ratio of the Mass of carbonized sample to Mass of activated sample after washing and drying.

$$Yield\ (\%) = = \frac{Wo}{Wc} X\ 100$$

Wo = Mass of carbonized sample
Wc = Mass of activated sample after washing and drying

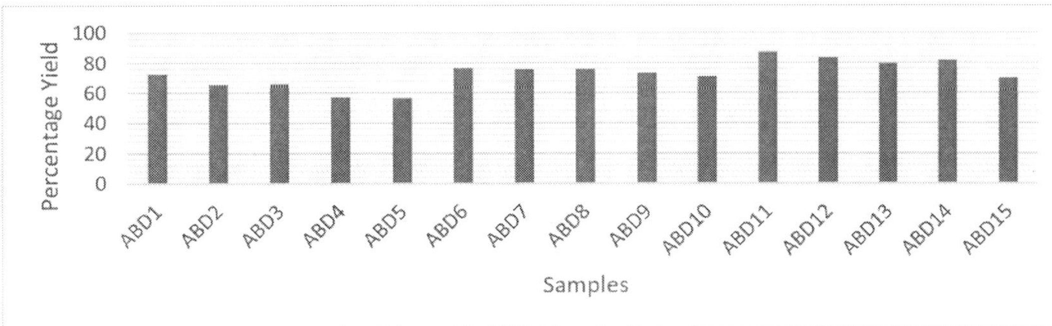

Figure 1 Percentage yield of each activated carbon

From the results, it is clearly shown that as the concentrations of each precursor increases, the resulting yield exhibits a form of decreasing trend. Concentration levels of 0.2N for seed husks, seed shells and whole seeds (ABD1-72.56%, ABD6- 76.471% and ABD11- 87.192%) respectively being the smallest concentration levels displayed the highest amount of yields. In particular, that of whole seed at 0.2 M gave the highest yield.

The yield for seed husk (ABD1, ABD2, ABD3, ABD4, and ABD5) decreased as the concentration increased. The same trend was seen in the yield for the oil seed (ABD6, ABD7, ABD8, ABD9, and ABD10). The trend for the whole seed (ABD11, ABD12, ABD13, ABD14, and ABD15) was similar, although the yield at 0.8 M was slightly above that of 0.6 M. The combination of the seed husk and the oil seed could be the reason while it had the highest yield suspecting that the structure will definitely be different from when it was treated separately and also because of its non-volatile structure, the chemical transformations which include degradation and dehydration of molecules achieved mainly by impregnation with the phosphoric acid did not seem to evolve gaseous products from its well-developed char structure which could lead to decreasing yield of activated carbon [14].

It also stated that the yield of the raw materials could be dependent on its original structure before activating it [3]. The bulk mass of the whole seed gives it a more weight retentive ability and the less concentration levels of the activating agent favours its yield as well because of less gasification of char and reduced loss in weight of carbon. Higher concentrations of acid in turn leads to a greater gasification level of char achieved resulting in a decreased final mass and yield of activated carbon.

4. Conclusion

The concentration of activating agent used can also be a major determining factor to the yield of activated carbon produced from waste generated in the agricultural sector. Phosphoric acid is a good activating agent but it best of it is gotten when the right concentration is used. The yield as relating to the concentration does not depend on the amount used but more on the type of carbon been activated.

Acknowledgement

Our appreciation goes to Covenant University, Ota, Nigeria for assistance give to the success of this studies.

References

[1].https://www.worldbank.org/en/topic/urbandevelopment/brief/solid-waste-management 25/07/2020 4:50 am

[2]. Efeovbokhan VE, Olurotimi OO, Yusuf EO, Abatan OG and AlagbeEE 2019a Production of Clay Filters for Waste Water Treatment. International Conference on Engineering for Sustainable World *Journal of Physics: Conference Series* **1378** (2019) 032028

[3] Mohamed NA, Walid SA and Hisham SA 2016 Industrial Wastewater Treatment of Food Industry Using Best Techniques, *International Journal of Engineering Science Invention* **2016** pp15-28

[4]. Abatan GO, Oni BA, Agboola O, Efevbokhan V and Odedoyin OA 2019 Production of activated carbon from African star apple seed husks, oil seed and whole seed for wastewater treatment. *Journal of Cleaner Production* 441-450

[5]. Adib M R M, Attahirah M H M N and Amirza A R M (2018) Phosphoric acid activation of sugarcane bagasse for removalof o-toluidine and benzidine *Earth and Environmental Science* **140** (2018) 012029

[6]. Mattson EA 1971 Activated carbon: surface chemistry and adsorption from solution. Resour. Conserv. *Recycl.* **52**, 13e27

[7]. Marsh H and Rodríguez-Reinoso F 2006 Production and Reference Material, Activated Carbon,Chapter 9 - *Elsevier Science Ltd*, 2006, Pages 454-508, ISBN 9780080444635, https://doi.org/10.1016/B978-008044463-5/50023-6

[8] Arvanitoyannis IS, Kassaveti A and Ladas D 2008 Food Waste Treatment Methodologies, in Food Science and Technology, Waste Management for the Food Industries, *Academic Press,* Pages 345-410, ISBN 9780123736543, https://doi.org/10.1016/B978-012373654-3.50009-2

[9] Skouteris G, Saroj D, Melidis P, Hai FI and Ouki S 2015 The effect of activated carbon addition on membrane bioreactor processes for wastewater treatment and reclamation – A critical review, *Bioresource Technology* **185** 399–410

[10]. Mohammadi SZ, Karimi MA, Afzali D and Mansouri F 2010 Removal of Pb(II) from aqueous solutions using activated carbon from Sea-buckthorn stones by chemical activation. *Desalination* **262**, 86e93.

[11]. Efeovbokhan VE, Alagbe EE, Odika B, Babalola R, Oladimeji TE, Abatan OG, and Yusuf EO, 2019b Preparation and characterization of activated carbon from plantain peel and coconut shell using biological activators International Conference on Engineering for Sustainable World *Journal of Physics: Conference Series* **1378** (2019) 03203

[12]. Oni BA, Abatan OG, Busari A, Odunlami O and Nweke C 2020 Production and characterization of activated carbon from pineapple waste for treatment of kitchen wastewater *Desalination and Water Treatment* 413–424

[13]. Shukla SK, Al Mushaiqri NRS Al Subhi HM, Yoo K and Al Sadeq H 2020 Low-cost activated carbon production from organic waste and its utilization for wastewater treatment *Appl Water Sci* **10,** 62 (2020). https://doi.org/10.1007/s13201-020-1145-z

[14]. Timur SK 2006 Preparation of activated carbon from organum stalks by chemical activation. *Energy Fuel*, 2636-2641

IOP Conf. Series: Materials Science and Engineering 1036 (2021) 012016 doi:10.1088/1757-899X/1036/1/012016

Evaluation of Solid Waste Storage and Disposal Practices in Nsukka, Enugu State

*C C Nnaji[1,4], B C Afangideh[2], U U Udokpoh[1], J P Nnam[3]

[1] Civil Engineering Department, University of Nigeria, Nsukka, Enugu State, Nigeria
[2]Civil Engineering Department, Akwa Ibom State University, Ikot Akpaden, Nigeria
[3]Departamento de Ingenierı´a Quı´mica y Tecnologı´a del Medio Ambiente, Universidad de Valladolid, Paseo del Prado de la Magdalena, s/n, Valladolid, Spain
[4] Faculty of Engineering and Built Environment, University of Johannesburg, South Africa

Abstract. Poor waste management has remained a nagging environmental issue for many developing countries. This paper examined the mode of solid waste storage and disposal in Nsukka, with the aim of providing and synthesizing data needed for waste management planning and policy making. A total of three hundred (300) well-structured questionnaires were administered to households cutting across ten communities to generate quantitative data on demographics as well as waste storage and disposal practices. The qualitative methods such as oral interviews, observation surveys, and secondary data were also employed. An initial environmental quality assessment of the selected households showed that 58.5% of houses surveyed were littered with waste and 71% had bushy surrounding. The study reveals that the distribution of waste storage container sizes are: 0 – 5 L (4%), 6 – 10 L (43%), 11 – 20 L (51%) and > 21 L (2%). It was also observed that the prevalent types of waste storage containers in Nsukka are: plastic bins (49%), polythene bags (15%) and cartons (10%) with volume ranging from 0.1 to 0.625m^3. The remaining twenty-six (26%) was distributed among other unconventional waste storage types. Eighty-eight percent (88%) of the residents in the study area can access waste dumps within a radius of 1 km or five minutes walking distance. The waste storage containers in the study area were of varying types and capacities which makes waste management more tasking. Indiscriminate solid waste disposal is necessitated by perceived non proximity of waste dumps to residents. Door to door services should be provided to residents by integrating private sector into waste management. Material such as aluminum should be used in manufacturing of waste containers because of it lightweight and durability.

1. Introduction

Solid waste is the end product of human daily activities. They are residues from domestic, commercial, industrial, institutional and agricultural activities [1,2]. These wastes that are generated on daily basis include polythene, paper, food waste, bottles, plastics, metal scraps, aluminum, rubber, bones, saw dust, and fabrics. An evolving and occasional waste such as mobile phones, computers and other kinds of electronic gadgets are also generated [3]. Solid waste generation in Nigerian cites have been estimated to be an average value of 0.49 kg/capita/day (Table 1) with over 25 million metric tonnes generated annually [4]. The generation rate varies widely depending on the peculiar features of the city [5]. Solid waste is generated largely from households followed by urban markets and stores. In Nigeria, household waste generated ranges from 49 percent to 78.9 percent and is greatly influenced by the economic status, lifestyle/habits, gross domestic product, sensitization on environmental sanitation, and population density as shown in Table 2 [6, 7]. According to [8], China is the leading solid waste generator in the world with an annual waste generation of 190 million tonnes in 2004 and 480 million tonnes was projected by the year 2030.

IOP Conf. Series: Materials Science and Engineering 1036 (2021) 012016 doi:10.1088/1757-899X/1036/1/012016

Between 35 and 50 percent of the solid waste generated in most developing countries are not collected. They usually end up in illicit dump sites, open spaces, waste land and drainages [9]. The percentage of solid waste collected and disposed ranges between 25 and 60 percent in Tanzania, Pakistan and Indonesia. It has been suggested that 80 to 90 percent of wastes generated in countries within the sub Saharan Africa are not collected for safe disposal [10].
In Nigeria about 30-60 percent of solid waste generated is not collected owing to low accessibility resulting from poor road networks, technical, economic and management failures on the part of waste management authorities, inadequate facilities, refusal to pay collection bills and non-implementation of route optimization [4]. About 80 percent of residents in most cities of Nigeria get rid of their waste by open air burning, unauthorized open dumping, burial and other unconventional means. Solid waste collection, storage and disposal are the most important aspects of municipal waste management. Wastes generated in urban areas in Nigeria are collected and disposed by the municipal waste authorities. Municipal waste collection in Nigeria is characterised by routine primness, bureaucracy and intriguing politics, unsuitable and unsustainable solid waste management techniques [5]. Waste collection and disposal services are offered mainly by the public sector through the State governments and involvement of formal public private participation (PPP) [11].

Waste containers are used for temporal storage of waste. It can be made of plastic, metal, wood and polythene materials of different sizes and shapes. Plastic bins and polythene bags are the most common waste storage containers found in most household in Nigeria with paper bags, used drums and sacks as unconventional waste storage containers. Polythene bags are usually provided by municipal waste management authorities to encourage bagging of waste. Bagging makes waste collection less chaotic compared to plastic bins though not durable [3]. The choice of waste storage containers used depends on the nature of waste to be stored, conveyance, durability and affordability. Storage and collection form the final link between waste generators and waste managers. Proper waste storage makes for easy collection and disposal of waste. In cities where there is effective waste management, large and more durable metal containers are used in order to make waste collection and disposal more expedient.

Table 1. Waste Generation and Composition in some Cities in Nigeria.

City	Population	Tonnes/month	Density (kg/m3)	kg/capita/day
Lagos	8,029,200	255,556	294	0.63
Kano	3,228,700	156,676	290	0.56
Kaduna	1,458,900	114,433	320	0.58
Port Harcourt	1,053,900	117,825	300	0.6
Onitsha	509,500	84,137	310	0.53
Ibadan	307,840	135,391	330	
Makurdi	249,500	24,242	340	0.48
Abuja	159,900	14,785	280	0.66
Nsukka	100,700	12,000	370	0.44

Source: [11]

Solid waste management has posed enormous challenge in most developing countries of the world due to rapid increase in waste generation and laxity in effective waste management [12]. The volume of solid waste generation globally in the past few years has increased resulting in overstressing of waste management facilities (such as storage containers, collection trucks etc.) and ineffective waste management by authorities concerned. Serious challenges that are being faced by municipal solid waste management in various cities of the world includes urbanisation, population growth, lifestyles, rapid industrialisation and consumption of foods with materials that are less decomposable [13]. The issue of effective waste management is not only the obligation of government but is a shared responsibility that includes the citizens and households, who are the main end-users of waste management facilities and services. Some of the nagging challenges associated with solid waste management are: quantity of waste generated, low technical experience, inadequacy and frequent breakdown of collection vehicles, low level of manpower and low financial resources which often cover only collection and transfer costs, leaving no resources for safe final disposal [14].

Table 2. The Percentage Generation of Solid Waste by Different Sector in Nigeria.

City	Domestic	Industrial	Commercial	Institutional	Hospital	Agricultural
Abeokuta	73.9	0.5	17.5	-	-	8.2
Ado-Ekiti	78.9	2.7	14.3	-	-	4.1
Akure	70.3	4.8	18.6	-	-	6.3
Ibadan	66.1	11.4	20.3	-	-	-
Osogbo	68.2	6.2	23.5	-	-	2.1
Umuahia	8	12	80	-	-	-
Zaria	49	14	23	10	4	-

Source: [3]

It was observed by [3] that with a population of about 97,000 residents in Nsukka, Enugu State Waste Management Authority (ESWAMA) uses only two collection trucks to service the metropolis and its environs. As at 2012, only one of the trucks was functional. Poor services from this waste management authority could be attributed to none payment for waste collection by residents. The inability of waste management authorities to cope with the volume of waste generated has left residents of most cities in Nigeria with no option than to discard their waste in approved open waste dump sites. These wastes are evacuated in turns by collection vehicles. Involving private firms in waste management of waste makes it more efficient as attested by [13], that over 80 percent of waste generated is adequately collected for disposal.

Municipal solid waste management is an important aspect of urban infrastructure that ensures the protection of environment and human health [15, 16]. Where there is no effective waste management, waste generated will be indiscriminately dumped in open spaces, access roads, along water courses or burnt. These dumps are invaded by scavengers and animals thereby constituting a nuisance to the environment. Open dumps also serve as breeding grounds for disease vectors, primarily flies and rodents. Leachate from decomposing garbage contaminate the soil, groundwater and surface water sources. Uncollected wastes may also find their way into open drains, causing blockage and dammed-up stagnant water which encourages the breeding of mosquitoes. Some of the health problems associated with improper collection storage and disposal of solid wastes include typhoid fever, cholera, dysentery, skin disease, malaria, hepatitis and guinea horn disease [17]. Also some chemical waste when touched or inhaled in excess can cause outbreak of epidemic. Hence, there is a growing global concern about the sterility of our environment and the need for proper handling and disposal of solid waste [1]. Burning of municipal solid waste has severe impact on public health resulting in the release of toxic substances [17].

Solid waste management has become an area of major concern in Nsukka today. It was observed by [5, 12] that 87% of Nigerians practice unhygienic methods of waste storage and disposal. Waste storage and disposal in developing countries is still largely indiscriminate and uncontrolled. Gharaibeh et al. [18] observed that in Saudi Arabia, the municipal solid waste collection services are faced with an increasing number of problems such as increasing population growth, changes in habits and lack of awareness of the impact of solid waste on the environment. Proper solid waste management (i.e., storage, collection and disposal) requires accurate information regarding waste generation rates and quantities, mode of generation, sources, storage containers and locations of waste [19]. The lack of information on waste storage containers is partly responsible for inefficient waste collection and disposal practices in many Nigerian cities. Without this information, it would be difficult to effectively plan for solid waste collection. Moreover, there is no reliable database indicating the amount of various types of waste collected and the volumes of waste generated per capita. It is against this backdrop that this study seeks to evaluate the nature and types of waste storage containers that are prevalent in the study area. Therefore, it is the aim of this paper is to evaluate the types and capacity of waste storage containers, distance from households to waste disposal points and common solid waste disposal practices in Nsukka, Enugu State.

2 Methodology

2.1 Description of Study Area

This study was carried out in Nsukka Local Government Area, Enugu State. Nsukka is bounded by Igbo-Eze South, Igbo-Eze North, Enugu-Ezike and Uzo-Uwani Local Government Areas. It has an average temperature of about 27°C (80°F) and a typical vegetation of rain forest. It lies between longitude 6°51′24″N and 7°23′45″E with an elevation of 550 m above sea level (Figure 1). Nsukka had a population of 309,633 as at 2006 census and an area of 5,545.38 square kilometres. Its inhabitants are predominantly civil servants, traders and farmers. The main campus of University of Nigeria is located in the study area.

2.2 Data Collection

Data was collected through the use of well-structured questionnaire, oral interview, and physical observation. A total number of 300 questionnaires were administered to households in Nsukka communities, out of which 234 were retrieved and analysed, giving a response rate of 78%. The questionnaires were constructed to achieve the aim and objectives of the study. It was categorised into three sections. Section A consisted of bio-data of the respondents, Section B focused on general waste-related practices such as mode of waste generation, frequency of waste collection, waste storage, waste transportation and waste disposal and Section C was designed to assess the attitude of respondents towards waste management. Oral interview was introduced to give clarification of question to respondents and also obtain additional information that was not captured in questionnaire. The physical observation was employed to get more information about the environment of the respondents. Measuring tape was used to measure the dimensions of waste storage containers in order to determine the volume of waste generated in households.

2.3 Sampling Site and Size

The sampling site was divided into 10 sections spreading across the entire city. Ten (10) communities were carefully selected to represent the sections (Table 3). Stratified random sampling at each of the section was used to ensure appropriate representation of the entire classes in the population. The first household in each residential area was selected and the households were subsequently followed alternatively until all the households were exhausted. The distribution of the sample size among the ten communities and the number of respondents are presented in Table 3. Relevant Statistical tools in Microsoft excel were used for data analyses. The use of tables, pie chart, and bar charts was employed for the data analysis and presentation. Both the descriptive and inferential techniques were applied in the data analysis.

IOP Conf. Series: Materials Science and Engineering 1036 (2021) 012016 doi:10.1088/1757-899X/1036/1/012016

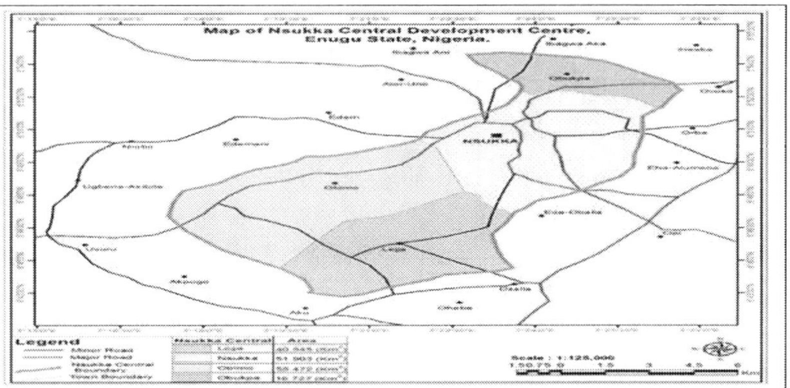

Figure 1: Map of Nsukka

Table 3. Communities in Nsukka and Questionnaires response data.

S/N	Zone	No. Administered	No. of Respondents
1	UNN	60	50
2	Obukpa	30	21
3	Amaezeani	30	21
4	Onuiyi	25	20
5	Leija	20	15
6	Alor-Uno	30	24
7	Ede-Oballa	25	20
8	Nru	30	26
9	Eha-Alumona	30	27
10	Ede-Ani	20	10
Total		300	234

3 Results and Discussion

3.1 Environmental Quality Assessment

The on-the-spot assessment are presented in Figure 2. The environment of the respondents was assessed in order to evaluate the hygienic status of the premises of respondents. It was observed that a greater number of 137 households dispose their waste around their premises despite the fact that 147 houses placed waste bin in front of their building.

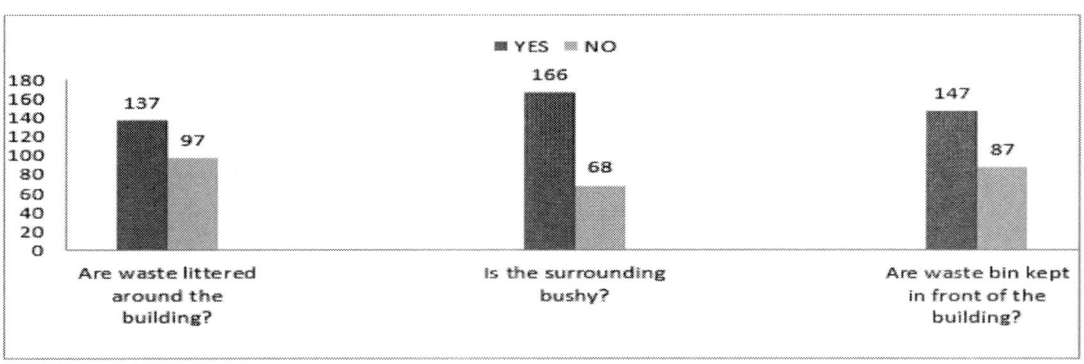

Figure 2. On the Spot Assessment of Respondents Environment.

The culture of disposing wastes in waste bin is not practiced in most homes in Nsukka. Another useful information gathered from this assessment was the nature of development of the respondent's environment. It was also observed that a total of 166 houses were surrounded by bushes which entails that the study area is still undeveloped. It also indicated a low level of environmental sanitation and personal hygiene.

3.2 Personal Information of the Respondents

Some of the personal information of respondents are presented in Figures 3 & 4. The results indicate that 53 percent of respondents are female while 56 percent are single. On level of education, it was recorded that a favourable percentage of respondents are learned. The result also showed that 118 houses within the study area has a range of 4 to 7 inhabitants. There seems to be a poor correlation between educational status and best waste management practices.

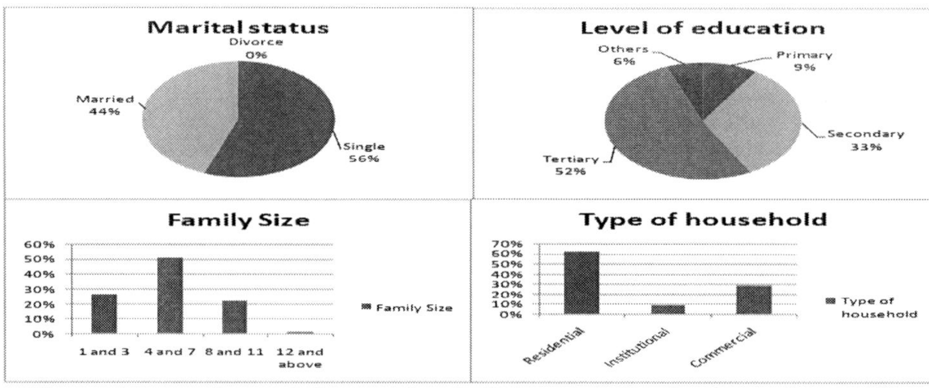

Figure 3. Charts showing the marital status, level of education, family size and type of household of the respondents.

The monthly income of respondents within the study area was low with 54 percent of respondents earning N20,000 or less monthly. All these features could be a strong factor affecting the rate of waste generation, storage and disposal pattern. For instance, a household of 1-3 inhabitants will not generate the same quantity of waste with a house of 8 to 11 inhabitants, owing to the fact that waste is as a result of human activities. Results from Figure 3 shows that 62 percent of household assessed were residential apartment, which 9 and 29 percent were institutional and commercial building respectively. About 45 percent of these buildings were blocks of flats and was situated within the urban centre of Nsukka and the university community. Fifty-nine percent of the houses have waste bin in their premises but a greater percentage of 53 choose to discard their waste within the surrounding and others in nearby river/stream.

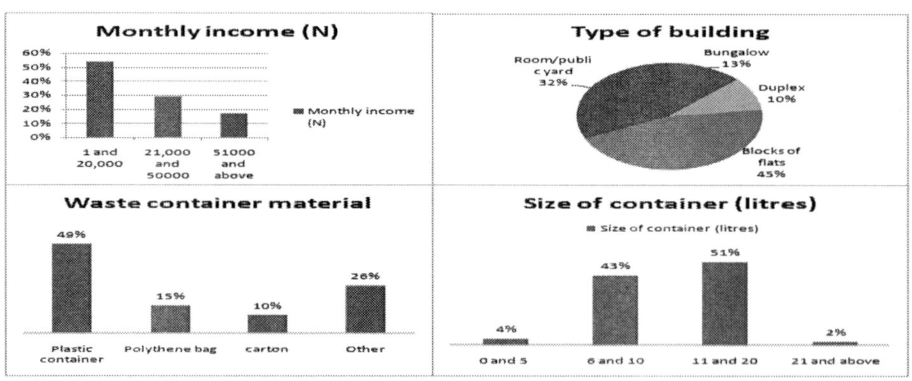

Figure 4: Charts showing the monthly income, type of building, waste container material and size of container of the respondents.

3.3 Waste Collection and Storage Containers

Solid waste storage container is a receptacle used for the temporary storage of solid waste while awaiting collection. The design of an efficient waste collection system requires careful consideration of the type, size and location of containers at the point of generation for storage of wastes until they are collected for disposal. The type and size of storage containers largely depends on the size of household and the purpose of the premises. Single-family households generally use smaller containers while residential units, commercial units and institutions require larger containers. Smaller containers are usually handled manually whereas the larger, heavier ones require mechanical handling. The desirable characteristics of a well-designed waste storage and collection containers are, low cost, weight, shape, resistance to corrosion, water tightness, strength and durability. Containers used in all the households in Nsukka are manually handled by one person. The weights of the containers including waste were between 5 and 15 kg. Households in the study area are very sensitive about the weight of their containers and wastes because emptying is done by members of the household. Heavy big heavy containers could lead to muscular strain. Larger containers are more suitable for waste dumps where several households deposit their wastes prior to collection. The container plus waste weighs up to 40kg, although emptying of wastes into collection vehicle is manually done by two or more crew members when it is filled. The process usually results in low collection efficiency and requires more manpower. Containers at collection points had handles and wheels to facilitate mobility. The body is made strong enough to resist and discourage wandering animals and scavengers from ripping it as well as withstand rough handling by the collection crew.

In Nsukka the type of waste containers used in most households are plastic containers, polythene bags, cartons and others such as metal containers, drums, etc. The plastics may be thin or heavy. On the average, 49% of Nsukka residents uses plastic containers, 15% uses polythene bags, 10% uses cartons, while 26% uses other means such as metal containers and drums (Table 4). The choice of plastic containers used by most households could be attributed to the fact that it is affordable, flexible and lightweight compared to metal container. Afullo and Frank [20] observed that in Nairobi, plastic bins were used by most households because it is relatively cheap and affordable compared to metal bins.

Table 4. Distribution of types of waste bin in various cities (%)

City	Abeokuta	Ibadan	Awka	Uyo	Nsukka (This study)
	Source: (Nnaji, 2015 and Okey et al., 2013)				
Plastic bin	48.89	40	81.5	-	49
Paper bags	6.53	-	-	71.4	-
Polythene bags	28.44	39.6	7.7	-	15
Carton	-	-	-	-	10
Used drums	15.15	28.4	-	-	-
Sacks	6.53	-	-	-	-
Metal tin	-	-	10.8	25.6	-
Others	-	-	-	3	26

[3, 21]

Some household used polythene bag for storage of waste, because it makes waste disposal less chaotic and flexible. Cartons were used mostly in commercial premises (such as stationaries stores, café,

IOP Conf. Series: Materials Science and Engineering 1036 (2021) 012016 doi:10.1088/1757-899X/1036/1/012016

computer business centres, etc) which generates mostly non-biodegradable waste and paper. Some households convert old metal and drums as waste bin. It was observed in this study that the size of solid waste containers mostly used ranged from 11 to 20 litres representing 51% of the total households, 6 to 10 litres representing 43% and 0 to 5 litres representing 4. Large containers greater than or equal to 21 litres were used by only 2% of Nsukka residents. Household with single family unit use storage containers ranging in size from 0 to 10 litres, while residential unit, commercial and institutional use larger storage containers > 11 litres containers. Generally, the size of storage containers used is a function of the anticipated rate of waste generation and frequency of disposal.

Solid waste storage containers are classified as stationary and hauled containers. Stationary containers are used for contents to be transferred to collection vehicles at the site of storage while the later are used for contents to be directly transferred to a processing plant, transfer station or disposal site for emptying before being returned to the storage site. Hauled containers are made of durable materials and are also made large enough to accommodate a reasonable amount of waste generated in order to reduce the cost of haulage. The containers in the study area were predominantly 1 m^3 stationary containers kept at a convenient positon in the occupant's residence. These containers are emptied once they are filled into larger stationary container placed at designated locations provided by Enugu State Waste Management Agency (ESWAMA). These containers are made of metal material and are coated regularly to avoid corrosion. When the container is full, the contents are transferred into collection vehicles for disposal at the municipal landfill. These containers are also fitted with lids to keep insects and other animals, and also to minimize infiltration of rainfall. An ideal container is shown in Figure 5, but open containers known as skips are often used (Figure 6).

Figure 5. Solid Waste Storage/Collection Containers with Lid

Figure 6. Solid Waste Storage/Collection Container without Lid.

IOP Conf. Series: Materials Science and Engineering 1036 (2021) 012016 doi:10.1088/1757-899X/1036/1/012016

3.4 Waste disposal Practices

It was observed that waste stored in waste bin was not disposed frequently where 46 percent does not empty their waste bin within one week. The result also revealed that 49 percent of respondent use plastic containers for waste storage while the remaining 51 percent use either polythene bag, carton or any other means. The volume of containers used in storing waste is reliable means of determining the quantity of waste generated in each household. The minimum volume of waste container with a uniform value of 0.01m3 was observed in all the communities while the maximum volume varies between 0.10 to 0.625 cubic metres. The sizes of waste containers available in most of the households are relatively small. Coupled with the habit of inhabitants not disposing of their waste within one week could be a major pointer as to why wastes were littered within the environment. Small waste storage containers necessitate frequent disposal Large containers will take a longer time to fill but will provide adequate residence time for the wastes to decompose and create unsightly and unhygienic conditions such as odour and leachate production.

Table 5. Descriptive statistics of volume of Solid Waste Storage Containers.

Zone	Mean	Standard Deviation	Maximum	Minimum
UNN	0.036	0.00227	0.27	0.01
Obukpa	0.044	0.0038	0.3	0.01
Amaeze-Ani	0.107	0.0113	0.41	0.01
Onuiyi	0.042	0.0016	0.041	0.01
Leija	0.048	0.00119	0.1	0.01
Alor-Uno	0.057	0.02699	0.45	0.01
Ede-Oballa	0.029	0.00099	0.1	0.01
Nru	0.092	0.01459	0.47	0.01
Eha-Alumona	0.063	0.00868	0.41	0.01
Ede-Ani	0.099	0.003428	0.625	0.01

Figure 7 present results of respondent's view on solid waste management by municipal waste management authorities. The result indicates that 69 percent of respondents are fully aware of municipal waste management agency. The poor management of waste by these agencies could be attributed to none payment for waste collection as reported by [3].

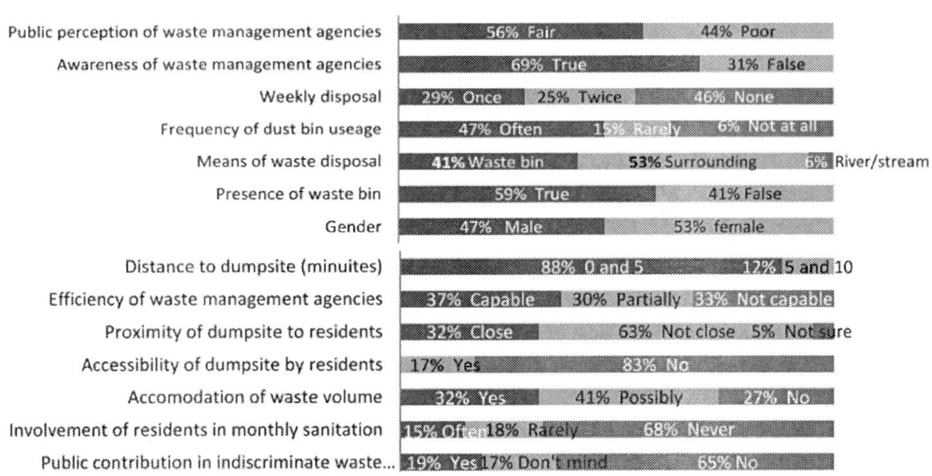

Figure 7. A chart showing results of responses about solid waste management in the study area.

Field survey showed that 88% of residents can access a waste dump within a radius of one kilometre or five minutes walking distance. However, 63% of respondents perceived the distance from their residence to waste dump as unfavourable. The perceived non-proximity of waste dumps coupled with unavailability of waste bins in open places is responsible for waste littering and indiscriminate dumping of refuse in places like farmlands, roadside drains and streets. It is also responsible for the creation of illegal dumping points, especially along access roads. Another intriguing aspect that was revealed by this research was the size of collection containers at approved dumpsites. The volume of collection containers at dumpsite provided by municipal agencies were not enough to accommodate the volume of wastes generated by some households. As a result, the containers at the waste dumps are always in perpetual state of overflow.

4 Conclusion

The type, size and weight of waste storage containers help in determining the volume of waste generated by households and their perception. The prevalent type of waste storage containers in Nsukka are, plastic bins, polythene bags, cartons, metal boxes and used drums with widely differing volumes. The preference for plastic waste bins is because of their flexibility, affordability and ease of disposal. The volume of containers used in most households were small owing to the fact that agency concerned doesn't provide door to door services, individuals are responsible for emptying their bins. Using containers of smaller volume will require frequent disposal, with collection points not close to most households will make waste disposal stressful. Aluminum containers of considerable size should be introduced because of its lightweight and durability. A knowledge of the volume of waste generated by households will help authorities in strategizing, planning and formulation of waste management policies. Indiscriminate disposal of solid waste can be mitigated if municipal waste agencies provide door to door services and individuals are willing to pay for such services. It was observed that private sector was not actively engaged in waste management, thereby stressing the agency available. More collection points should be established close to residents, services rendered should not be limited only to urban areas. Individuals should be enlightened on the adverse effect of improper waste storage and disposal. It was seen that households where collection point is close were still discarding their waste to the surrounding. Further study should be carried out on the cost and durability of the prevalent types of solid waste storage containers in households.

References

[1] Ikebude C 2017 Feasibility study on solid waste management in Port Harcourt metropolis: Causes, effect and possible solutions, *Niger. J. Tech*, **36**(1) 276 – 281

[2] Leton T G 2007 *Water and Waste Water Engineering*, The Pearls Publishers, Port Harcourt, Rivers State, Nigeria

[3] Nnaji C C 2015 Status of municipal solid waste generation and disposal in Nigeria, *Manag. Environ. Qual.: An International Journal* **26**(1) 53 -71

[4] Ogwueleka T C 2009 Municipal solid waste characteristics and management in Nigeria, *Iran J Environ. Health Sci. and Eng.* **6**(3) 173 – 180

[5] Ibrahim R A, Oluwagbemiga E A and Adeolu A 2014 Municipal household solid waste collection strategies in an African megacity: Analysis of public private partnership performance in Lagos, *Waste Manag. & Res.* **32**(9) 67 – 78

[6] Fakere A, Fadair G, and Oriye O 2012 Domestic waste management and urban residential environment: Focus on Akure, Nigeria, *Int. J. Eng. Tech.* **2**(5) 878 – 887

[7] Ukoje J 2011 Analysis of the determinants of participation of stakeholders in solid waste management in Zaria, Nigeria, PhD thesis, School of Postgraduate Studies, Ahmadu Bello University, Zaria

[8] Minghua Z, Xiumin F, Rovetta A, Qichang H, Vicentini F, Bingkail L, Giusti A and Yi L 2009 Municipal solid waste management in Pudong New Area, China, *Waste Manag.* **29** 1227 – 1233

IOP Conf. Series: Materials Science and Engineering 1036 (2021) 012016 doi:10.1088/1757-899X/1036/1/012016

[9] Bogoro A and Babanyara Y 2011 Evacuation of solid waste in residential areas of Bauchi Metropolis, Nigeria, *J. Environ. Sci. Resour. Manag.* **3**(1)

[10] Ogu V 2000 Private sector participation and municipal waste management in Benin City, Nigeria, *Environ. Urban.* **12**(2) 103 – 116

[11] Oyebode O 2018 Evaluation of municipal solid waste management for improved public health and environment in Nigeria, *Eur. J. Adv. Eng. Tech.* **5**(8) 525 – 534

[12] Longe E and Williams A 2006 A preliminary study of medical waste management in Lagos Metropolis, Nigeria, *Iran J. Environ. Health. Sci. Eng.* **3**(2): pp 133-139

[13] Asase M, Yanful E, Mensah M, Stanford J and Amponsah S 2009 Comparison of municipal solid waste management systems in Canada and Ghana: A case study of the cities of London, Ontario, and Kumasi, Ghana, *Waste Manag.* **29** 2779 – 2786

[14] Awosusi A (2010). Assessment of environmental problems and methods of waste management in Ado-Ekiti, Nigeria, *Afr. Res. Rev.* **4** (3b) 331 – 343

[15] Aliyu B 2010 An analysis of municipal solid waste in Kano Metropolis, Nigeria, *J. Hum. Ecol.* **31**(2) 111 – 119

[16] Ogbonna A C, Mikailu A and Jiya G 2018 municipal solid waste management in African cities: a case study of Lagos State, Nigeria *Malays. J. Civ. Eng.* **30**(1) 1 – 12

[17] Bassey B, Benka-Coker M and Aluyi H. S. 2006 Characterization and management of solid medical wastes in the Federal Capital Territory, Abuja Nigeria *Afr. Health Sci.* **6**(1) 58 – 63

[18] Gharaibeh E, Haimour N, and Akash B 2011 Evaluation of current municipal solid waste practice and management for Al-Ahsa, Saudi Arabia *Int. J. Sustain Water Environ. Syst.* **2**(2) 103 – 110

[19] Agarwal A, Singhmar A, Kulshrestha M and Mittal A 2005 Municipal solid waste recycling and associated markets in Delhi, India *Resour. Conserv. Recycl.* **44** 73 – 90

[20] Afullo A and Frank O 2009 The primary solid waste storage gaps experienced by Nairobi households *Ethiop. J. Environ.l Stud. Manag.* **2**(3) 38 – 43

[21] Okey E, Umana E, Markson A and Okey P 2013 municipal solid waste characterization and management in Uyo, AkwaIbom State, Nigeria in Brebbia, C.A. (Ed.), *Sustainable Development and Planning IV*, Wessex Institute of Technology, Southampton, 639 – 648

IOP Conf. Series: Materials Science and Engineering 1036 (2021) 012017 doi:10.1088/1757-899X/1036/1/012017

Plume Model: A Simple Approach to Air Quality Control

O O Ajayi[1], B U Ngene[1] and S A Ogbiye[1]

[1]Department of Civil Engineering, Covenant University, Ota Nigeria
Oluwatoyin.ajayi@stu.cu.edu.ng

Abstract. Air is an essential element in the life of humans, and everyone deserves to breathe in clean air. But anthropogenic activities from industrialization and modernization, as well as some natural disasters such as earth quake and volcanic eruption pollute the air. Consequently, air pollution threatens the health of people living in close proximity to the source of pollution by causing or aggravating sicknesses like, heart problem, cancer and respiratory disorder. The dispersion of air from a point source tends to follow the shape of Gaussian or normal statistical distribution. The present paper aims at examining Gaussian plume model as the easiest and simplest model among the lists of air dispersion models evaluated, stating the various issues from the source to the receptors. Gaussian Plume model is mostly used in the analysis of air dispersion with the concentration of a given place, having function of the distance from the source, emission rate and the meteorological data, mostly temperature, wind speed and wind direction, with the help of some various assumptions, in order to estimate how much air has reduced or travelled from the source and the concentration at the ground level.The model uses Briggs coefficients sigma y and z to get both vertical and horizontal dispersion. Plume rise determines dispersion and transportation levels, and thereby affect the distance to maximum ground level concentration and the maximum ground level concentration, the taller the stack or chimney (source), the larger the plume and the lower the concentration at the ground. A week analysis on pollutants generated from Larfarge cement factory Sagamu, Ogun State was carried out. The result shows mathematical simulation of Gaussian plume model of pollutants concentrations from the source and the applicability of the model.

Keywords: Stack height, air quality, cement factory, meteorology, models, dispersion

1. Introduction

The control of Air Quality is a challenge of the present world especially for the industrial cities and towns. [1] defined air pollution as the "presence especially in the outdoor environment of contaminants such as dust, fumes, gas, mist, odour, smoke or vapour in quantities or characteristics and of duration which can cause injurious to human, plant or animal life or to property or which unreasonably interferes with the comfortable enjoyment of life and property." Though the definition is outdoor specific, it is necessary to observe that indoor pollution from kitchen and faulty planning cannot be neglected. Air pollution importance cannot be over-emphasized because unlike others, an average adult consumes about 12 Kg of air per day which is about 12 times more than the food consumption [2]. A number of air pollution in the past such as London smog (1952) with about 4000 casualties and Bhopal India Dec. 1984, in Union Carbide industry where about 2500 people died when toxic gas methyl isocyanate leaked accidentally and over 100000 were severely affected. This defines the need for Air Pollution control especially in urban or Industrial cities [2].

Air pollution modeling is a computer/mathematical method for simulating air quality processes. It is the tool that describes the association among factors such as the atmospheric concentration, meteorological conditions, emissions and deposition. Estimation of air pollution gives necessary information about the air cluster, settling and it gives the description of air quality at particular times and position not stating the causes of the problem. Consequently, air pollution modeling gives sufficient description of the problem as well as the causes and factors responsible.
Air pollution modeling predicts both the past and the future consequences [3]. Subsequently, a simple model "The Gaussian Plume Model, is a reactive model developed to calculate the ground level and impact from the source. Gaussian model was derived practically from both horizontal and practical spread of the plume [4]. [5] and [6], concluded that understanding of the dispersion of plumes from stacks has been a challenge in air pollution modeling. Over the years, there have been several modeling used in solving air quality challenges, according to [7].
Air pollutants may be classified into Natural Contaminants such as pollen grains; Aerosols like dusts, smoke, mists, fogs and fumes; and Gases which include: Sulphur dioxide, Oxides of nitrogen, Carbon monoxide, Hydrogen sulphide, Hydrogen fluoride, Chlorine and hydrogen chloride, Ozone and Aldehydes. Also, Air pollutants are classified based on source as being stationary or mobile: Point source (large stationary source as power plants), Area source (small stationary sources like residential heating) and Mobile source (line source like highway vehicles or area source like Aircraft at airports) [2]. The aim of this work is to assess how to control point source pollutants such as in Cement factory with high Stack or Plume/Chimney and through that reduce air pollution.

1.1 Types of Air Quality Models:

The three types of models used in Air Quality are:

1.1.1 Photochemical Models

They are mathematical quality model in large scale used in simulating the changes in the atmosphere. This kind of model can be used at multiple spatial scales with local, region and global, it is also used in regulating policy and simulating the impacts from source by giving/circulating pollutant concentrations and deposition from inert and chemically reactive pollutants over large spatial scales. There are two of photochemical air quality models commonly used in air quality evaluation: the Eulerian grid model that makes use of a fixed coordinate system with respect to the ground and the Lagrangian trajectory model.
Community Multi-scale Air Quality (CMAQ)-Model is an active open source project form U.S. EPA that is used in determining the impact of air quality for multiple pollutants at multiple scales and to deduce and simulate physical interactions in the atmosphere [8].
Comprehensive Air quality Model with extensions (CAMX) - This is an open source computer comprehensive air quality model, it is for gaseous and particulate air pollution. The computer aid is easy to work with, simulates to use over geographical scales and it deals with inert and chemically active pollutants. It is also a plume-in-grid and source apportionment capabilities
Urban Airshed Model Variable Grid (UAM-V)-The UAM-V is the pioneering effort in photochemical modeling in the early 1970s in staging ozone, is the most used photochemical air quality model, that is technically okay and reliable, it is a three-dimensional grid model developed in calculating the inert and chemically reactive pollutants in order to simulate both physical and chemical processes in the atmosphere.
Regional Modeling System for Aerosol and Deposition (REMSAD)-REMSAD was developed to calculate the chemically and inert pollutants concentration by simulating the physical and chemical processes in the atmosphere that affect pollutant concentrations over regional scales. The processes include particulate matter, haze and airborne pollutants, like mercury and soluble acidic components [8].

1.1.2 Meteorological Models

Meteorology is important for the dispersion of pollutants because it is the primary factor that determines the diluting effect of the atmosphere, for the purpose of air quality, meteorological grid models are used with chemical interaction models. Meteorological grid models use mathematics to formulate and simulate processes like wind speed, wind direction and temperature.

IOP Conf. Series: Materials Science and Engineering 1036 (2021) 012017 doi:10.1088/1757-899X/1036/1/012017

California Meteorological model (CALMET)-This reconstruct the 3D wind and temperature fields; it also determines the 2D of micro meteorological variables that is used for dispersion simulation.
The Fifth Generation Penn State University/National Centre for Atmospheric Research mesoscale model(MM5) - This is a regional mesoscale model used in creating weather forecasts and climate projections, it can also be used in research by comparing it to other model.MM5 can also be used for air quality models. Plate 1 shows weather station used to obtain Data on wind speed and direction, rainfall volume, relative humidity and air pressure for weather forecasting.
Weather Research Forecasting model (WRF)-Numerical weather prediction system used for research and forecasting applications. WRF produces the actual atmospheric conditions for researchers (observation to analyses).

Plate 1: Airport weather station used to obtain Data on wind speed and direction, rainfall volume, relative humidity and air pressure.

1.1.3 Dispersion Models

Dispersion models are computer programs that use mathematical algorithms to simulate how pollutants disperse and react in the atmosphere, it estimates mainly based on Gaussian plume model which deals with the stable wind speed and turbulent and height [9]. Dispersion modeling allows estimating concentration of pollutants at specific ground level receptors surrounding the emission source; it is used for planning and managing existing emission. The most commonly used dispersion models are the steady state Gaussian plume models. Examples include AUSPLUME, ISCST3, AERMOD and CTDMPLUS [10].
America Meteorological Society/Environmental Protection Agency Regulatory Model (AERMOD) -A steady type of Gaussian plume model that uses wind in transmitting emission, it is been used to substitute ISCST3, the model is based on atmospheric boundary layer. It employs flat or complex, rural or urban terrain. Gaussian dispersion is used for stable conditions while non Gaussian is used for the unstable conditions.
California Puff Model (CALPUFF)-An advanced non steady state Lagrangian puff model that can simulate dispersion in quiet advanced all coastal terrain. It is used for long range coverage above 50miles. This takes care of visual impact plumes [11].
Industrial Source Complex for Short Term Version 3 (ISCST3)-This is a Gaussian model used in assessing pollutant concentration in sources from industrial complex. It takes care of the plume rise, point source, deposition of particles and settling.
Atmospheric Dispersion Modeling System (ADMS)-An advanced dispersion model designed for calculating concentrations continuous pollutants from line, point area sources from point source or discrete from point sources. Plate 2 shows the dispersion of pollutant through the point source of a Stack.

IOP Conf. Series: Materials Science and Engineering 1036 (2021) 012017 doi:10.1088/1757-899X/1036/1/012017

Plate 2: Cement factory Stack showing the discharge of pollutants into the atmosphere from waste generated by the plant following the Gaussian model.

2. Methodology

2.1 Nature and Causes of Air Pollution

There are many forms of environmental pollution, and the resultant health risks include diseases of all kinds affecting almost all organ systems of the body. According to [12], this variation in forms of environmental pollution makes each pollutant have its own health risk profile, which makes summarizing all relevant information into a difficult task. Air pollutants are usually categorized into suspended particulate matter (PM) (dusts, fumes, mists, and smokes); gaseous pollutants (gases and vapors); and odours. Outdoor air pollution is caused mainly by the combustion of petroleum products or coal by motor vehicles, industry such as cement factory, and power stations [12].

2.2 Location of Study

Sagamu is a city in Ogun State, South-west of Nigeria, Figure 1. It is located between the cities of Ibadan and Lagos and on Coordinates 6°50′N 3°39′E on Area of 614 km^2, with Population 253,412 [13]. Sagamu has many industries, of which the Larfage cement factory is the major in region.

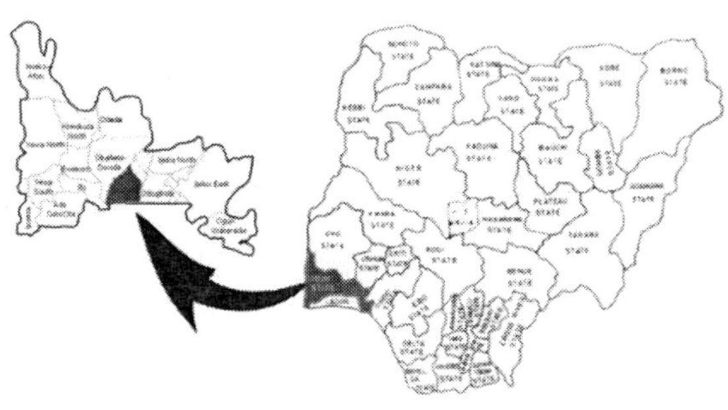

Figure 1: Map of Nigeria Showing Ogun State and Sagamu [14]

Larfarge cement factory is one of the major contributor of air pollution in Sagamu environs, thereby threating the health of people living or working around the factory. [15] established that there were contamination and impacts on air, land and noise pollution conducted at Ewekoro area of Ogun and through the study investigated on health and environmental challenges of the Cement Industry on Physical surrounding settlement of Ewekoro, Ogun State.

Data collection was done on the three (Ikorodu, Ijebu-ode and Abeokuta) major routes to Larfarge cement factory Sagamu using Aeroqual 500 series. Concentration of four different gases, CO2, SO2, VOC NO2 and particulate matters were collected on weekly basis over a period. Meteorological data such as Relative humidity, Temperature, Rainfall, Wind speed and

Wind direction were also determined. All data collected were used in calculation in other to obtain the concentration levels of the pollutants.

2.3 Controlling Pollutants with Plumes

The aim of air pollution control is to reduce to the minimum the release of pollutants to the atmosphere and hence promote sufficient dispersion and dilution of released pollutants within the atmosphere. Efficient Stack height determination ensures optimal release condition of pollutants that will limit ground level concentration [16]. Air pollution is therefore controlled in an Industrial plant by means such as the nature and height of Plumes, changing the raw material, the process conditions, procedure and by ensuring preventive maintenance. Pollutants are being controlled through different plume behaviours, and it depends on the degree of instability of the atmosphere and the wind turbulence in form of;

Looping: It is an unstable condition which occurs with heat and light conditions, it moderates wind speed during hot weather in the presence of eddies, Eddies carry plume to the ground level for a while by causing surface concentration of pollutants near the stack.

Coning: A neutral lapse rate condition which occurs within cloudy skies both in the day and night, shape of plume is vertically symmetrical, pollutants concentrations are fairly carried downwind before reaching the ground, it also works between moderate and high wind.

Fanning: This is an inversion condition which occurs during the dispersion of plume in the presence of light wind. If the density of the plume is different from that of the surrounding atmosphere, then the plume travels parallel to the ground when stable plume is within 10 to 20 km downwind.

2.4 Effect of Stack Height on Pollutant Removal

The height of the stack or chimney is governed by such factors as meteorological parameters, height of the nearest building or structure, prevailing wind direction, height of natural landforms, location of air intake, type of equipment draft (natural or forced), type of fuel and local and/or national ambient air quality requirements. Stack height is the most important parameter which highly affects transportation and dispersion of pollutants [17]. Stack height can affect pollutant removal by: Limiting pollution at ground level in order to make pollution stay within acceptable limits; Preventing downwash during periods of high wind velocity; Lowering ground level concentrations near the source; Increasing the amounts of pollutants emitted to the atmosphere.

2.5 Application of Gaussian Plume Model

The application and the use of the model under review are enormous,Gaussian plume can be used in the day to day activities.The model can be used in predicting the impacts of pollutants from the industry/factories, to disseminate pollen in agriculture.[18], it is also used in wind control (pheromones) during insect entomology behaviour[19]

3. **Result and Discussion**

Gaussian plume model is used in determining or predicted ground level concentration of particulate matters long downwind of the factory. Simulation of the effects of deposition and fumigation is best done by Gaussian plume model.

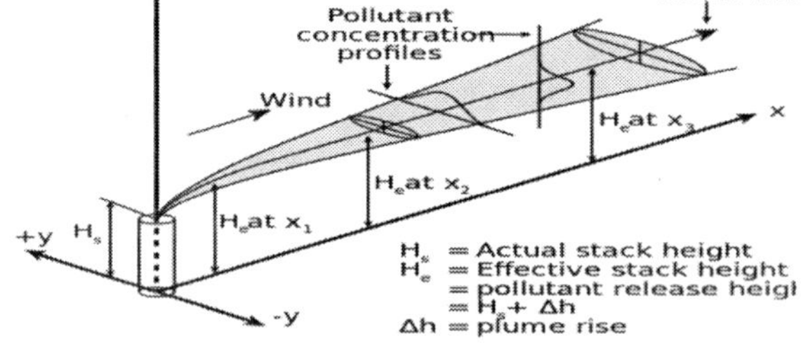

Figure 2: Schematic figure of a Gaussian plume. The *He* effective stack height and the crosswind and vertical deviation of the profile are the key parameters of the model [20]

IOP Conf. Series: Materials Science and Engineering 1036 (2021) 012017 doi:10.1088/1757-899X/1036/1/012017

The equation for a Gaussian plume is

$$c(x,y,z)=\frac{Q}{2\pi\sigma_y\sigma_z u}\exp\left(\frac{-y^2}{2\sigma_y^2}\right)\left(\exp\left(\frac{-(z-h)^2}{2\sigma_z^2}\right)+\exp\left(\frac{-(z+h)^2}{2\sigma_z^2}\right)\right)$$

where c is a concentration at a given position, Q is the source term, x is the downwind, y is the crosswind and z is the vertical direction and u is the wind speed at the h height of the release. The σy, σz deviations describe the crosswind and vertical mixing of the pollutant. The following assumptions are made to develop emission from cement factory [21]: (i) A continuous and point source emission (ii) Eddy diffusion, advection/diffusion equation (iii) Diffusion in the downwind direction is negligible, presence of vertical and crosswind only (iv) Wind speed from point source to receptor is constant (v) Atmospheric turbulence is constant (vi) Plume reflection

The Gaussian plume is derived by assuming steady state conditions. The formulae do not depend on time, the meteorological conditions are also assumed to remain constant from source to receptor instantaneously. Although, the plume characteristics will not change over time, and they depend on emission change and meteorological conditions. Gaussian plume model can be used when: (i) the pollutants are chemically inert (ii) the terrain is not complex (iii) the meteorology may be considered uniform spatially and (iv) there few periods of calm or light winds.

The concentrations of the pollutants and the meteorological data results for the two weeks period were obtained to calculate the emission rate, which were used as input data for the Gaussian plume modeling for the point source.

Table 1: Shows the typical parameters measured and used for the modeling of concentration of pollutants at the site of a Cement factory

Point	Land-mark/ Name	Locati-on/Cor-dinates	Time (Hr)	Temp °c	RH (%)	Light Ft-cd	Wind Speed- (m/s)	P.M$_{2.5}$ µg/m3	CO_2 ppm	SO_2 Ppm	VOC Ppm
1	Regal Heights	754285 569503	11:50	29.9	73.60	980.2	0.7	8.6	650	0.00	1104
2	UBA	755062 570785	12:28	30.3	69.40	1245.1	2.1	16.9	677	0.02	736.6
3	OPC-Signboard	756754 571896	12:46	29.6	71.50	824.5	2.2	62.7	687	0.00	1089
4	Terrazo House	758071 572669	13:07	30.5	72.30	723.3	0.4	37.5	665	0.00	624.1

4. Conclusion

This paper presents a description of Gaussian plume mathematical model as a flexible model applicable for dispersion from source to receptor; it is based mainly on assumptions developed for simulation. In the current study, wind speed controls the dispersion of air pollutants released from the point source as well as the vertical height above the eddy diffusion (vertical and horizontal). From the research it is established that Stack height affects the dispersion of pollutants in a Cement factory and hence optimally designed chimney height influence the air quality around the location of the factory.

Acknowledgements
The authors wish to thank the Chancellor and Management of Covenant University for the platform made available for this research. The authors also acknowledge the reviewers for their valuable comments.

References

[1] Perkins H C 1974 Air pollution *McGraw-Hill Higher Education*. New York: ISBN 10: 0070493022 ISBN 13: 9780070493025: https://trove.nla.gov.au/version/45416016
[2] Gaur RC 2008 Basic Environmental Engineering. *New Age Inter. Ltd, Publishers*, New Delhi, India. ISBN (13): 978-81-224-2701-1
[3] Buitjes PH 2003 Impact of the extreme meteorological conditions during the summer in Europe on particulate matter concentrations. Accessed from https://www.researchgate.net
[4] Boubel DL, Fox DB, Tumer DB and Stern AC 1994 Fundamentals of Air Pollution.3rd Edition. *Academic Press*, San Diego
[5] Sutton OG 1932 A Theory of eddy diffusion in atmosphere, https://doi.org/10.1098/rspa.1932.0025.
[6] Bosanquet LS 1936 Note on the Absolute Summability (C) of a Fourier Series. *Journal of the London Mathematical Society*
[7] Johnson WB 1980 Inter-regional exchange of air pollution model types and applications 10th ITM, 3 Amsterdam
[8] United States Environmental Protection Agency 2017 Air Quality Dispersion Modeling-Preferred and Recommended Models. Support Centre for Regulatory Atmospheric Modeling (SCRAM) US EPA
[9] Hinrichsen EL 1986 Environmental Ecology: The impact of Pollution and other https.//books.google.com.ng,books
[10] Perry DJ, Harperb PL, Fairham M and Carrell RW 1989 Geologic and environmental aspects of surface cementation, north coast, Yucatan, Mexico
[11] Scire JS, Robe FR, Fernau ME and Yamartino RJ 2000a A user Guide for the Calmet meteorological model Version5.Earth Tech. Inc, MA
[12] Kjellstrom T, Lodh M, McMichael T, Ranmuthugala G, Shrestha R and Kingsland S Jamison DT, Breman JG, Measham AR, et al., editors 2016 Air and Water Pollution: Burden and Strategies for Control In *Disease Control Priorities in Developing Countries*. 2nd edition. Washington (DC): The International Bank for Reconstruction and Development/ The World Bank; 2006. Chapter 43. NBK11769/ Available from: https://www.ncbi.nlm.nih.gov/books
[13] National census 2006. www.nigeriamasterweb.com/Nigeria06CensusFigs.html. Retrieved January 12, 2017.
[14] Ogun Geographical Information System. (Ogun GIS). Retrieved from https://allafrica.com/stories/201405192312.html on June 20, 2020.
[15] Aribigbola A, Fatusin AF, Fagbohunka, A 2012 Assessment of Health and Environmental Challenges of Cement Factory on Ewekoro Community Residents, Ogun State, Nigeria. *America Journal of Human Ecology* **1**:2, 51-57
[16] Kapadia J 2012 Stack height determination: GIL Power Generation Project, Kilkenny. https://www.scribd.com/document/434538450/Stack-height-determination
[17] Kumar SA 2012 Effect of Stack Height on Transport and Dispersion of Air Pollutants. ENVIS Newsletter. Bihar Envis Centre, India.www.bhenvis.nic.in
[18] Loos C, Seppelt R, Merier-Bethke S, Shiemann J and Richter O 2003 Spatially explicit modelling of transgenic maize pollen dispersal and cross-pollination. *J. Theor. Biol.* **255**, 241-255
[19] Murlins J, Elkinton JS and Carde RT 1992 Odor plumes and how insects use them *Amu. Rev. Entomol.* **37**, 505-532
[20] Dobbings RA 1979 Atmospheric Motion and Air Pollution, John Wiley and sons, New York
[21] Beychok MB 2005 Fundamentals of Stack Gas Dispersion: Fourth Edition

IOP Conf. Series: Materials Science and Engineering 1036 (2021) 012018 doi:10.1088/1757-899X/1036/1/012018

Toxicity potential of automobile carbon monoxide emission along a majour highway in Sango-Ota, Ogun State, Nigeria

F B Elehinafe[1]*, S D Uzuana[1], O A Odunlami[1], A O Mamudu[1], A O Ayeni[1] and O B Okedere[2]
[1]Department of Chemical Engineering, Covenant University, Ota, Ogun state, Nigeria
[2]Department of Civil Engineering, Osun State University, Osogbo, Nigeria
*Corresponding Author: francis.elehinafe@covenanuniversity.edu.ng

Abstract: This study investigated the toxicity potential of carbon monoxide along a major highway in Sango-Ota, an emerging city, in Nigeria. Ambient air carbon monoxide was monitored and measured at ten different locations using the GM8805 benetech carbon monoxide monitor. Coca-Cola junction was selected as the control location for this study. The 1-hr averaging period measured for carbon monoxide concentrations within the selected locations ranged between 25.2 ppm and 29.3 ppm while at the control point, it was measured to be 8 ppm. The extrapolated 8-hr averaging period for carbon monoxide concentration ranged between 14.076 ppm and 16.366 ppm with the concentration at the control point being at 4.4688 ppm. The extrapolated 24-hr averaging period for carbon monoxide concentration ranged between 10.357 ppm and 12.042 ppm with the concentration at the control point being 3.288 ppm. An assessment of toxicity potential of these locations based on Federal Environmental Protection Agency (FEPA), United States Environmental Protection Agency (USEPA) and World Health Organization (WHO) statutory limits were all above unity (one) except at the control location and this poses as a threat to human lives. These values showed unhealthy air conditions at these regions of the emerging city. This study establishes that traffic congestions contribute significantly to the ambient air thereby reducing its quality thus calls for the attention of regulatory bodies to rise to protection of commuters and businessmen along the major highways in emerging cities in Nigeria.
Keywords: Developing countries, emerging cities, highways, carbon monoxide emission, and toxicity potential

1. Introduction

Automobiles emit pollutants, such as carbon dioxide, that add to global climate and environmental changes, and are within the class of ozone harming substances. One gallon of gasoline makes 18 pounds of CO_2 [1]. The U.S. alone produces 22 billion tons a year of CO_2 [1]. The transportation area as of now represents over a fourth of all U.S. ozone harming substance discharges [2]. Global effects: Carbon monoxide (CO) being a poisonous gas that influences carbon dioxide (CO_2), methane (CH_4) and tropospheric (lower atmospheric) ozone (O_3). Thus, CO plays a role in air contamination and climatic change and so, consequently controlled in numerous parts of the globe (Matthew *et al.*, 2018). CO is exceptional among air pollutants in the tropopause (lower atmosphere) in that it persists for approximately a month, sufficiently long to be conveyed long distances yet not does not take time to becomes circulated uniformly [2]. [3] examined the long-lasting impact of greenhouse gas (GHG) emissions on health aftermaths in Nigeria using data of time series and reported that human undertakings increase GHG to the atmospheric ambience through burning of fuels from fossils.

Being an evolving country, Nigeria is threatened by pollution being emanated from traffic congestion which occurs separately from pollution arising from sources like wide-ranging industrial

IOP Conf. Series: Materials Science and Engineering 1036 (2021) 012018 doi:10.1088/1757-899X/1036/1/012018

pollution and pollution from oil industries [4]. This threat is an outcome of importation of old and over used automobiles, poorly constructed roads and inefficient agencies that do not ensure regulations are maintained nor enforce the implementation of pollutants level projected to be good to human health [5]. Carbon Monoxide is most toxic during power outages due to storms, inappropriate practice of using gasoline-powered portable generators to generate electricity and indoor use of charcoal briquettes for cooking and heating. Carbon Monoxide poisoning can also arise from sources like exhaust from generators and propulsion engines on houseboats [6].

Nigeria is considered as one of Africa's most populous country (170 million people as of 2012). Nigeria's population increased by 60% between 1990 and 2008 and it is also projected to fall within the range of 0.5 – 1 billion people by 2100 [7]. The GDP rate in Nigeria as at 2012 was supposed to be 7% every year, conjecture to proceed [8], is viewed as one of the most elevated on the planet. The Nigerian economy recently surpassed South Africa as the biggest on the continent. This quick growth in Nigeria causes a variety of ecological concerns concerns including air quality. Pollution from vehicular sources is worsened by inefficient automobiles, disrupted street systems, inadequately built streets, traffic clog, and polluted fuels [8]. [9] was able to establish that increase in the use of automobiles and establishment of industries contribute significantly to the high level of carbon monoxide concentration in Lagos State, Nigeria.

Sango-Ota is an emerging city in Ogun State, Nigeria and is experiencing an increasing population as more companies, houses are being built within the area. This situation is leading to slow flow of traffic and longer time spent in congestions, which in return increasing burning of fossil fuels, thereby polluting the air. Hence, measuring the concentration of CO being emitted and determining the toxicity potential of automobile carbon monoxide emission in the ambient air along Sango-Ota highway is of paramount importance.

2. Materials and methods

2.1 Description of the studied emerging city

The study area is Sango-Ota, an emerging city in Ogun State, Nigeria (Figure 1). It borders Lagos State to the south, Oyo and Osun States to the north, Ondo to the east and the Republic of Benin to the west. It lies on the geographical coordinates of 6° 42' 0" N, 3° 14' 0" E. it is located in Ado-odo/Ota local government area, Ogun state. It covers a land area of 878 km². Sango-Ota was separated into four expansive sub-areas based on the residential features and socio-economic dissimilarities. The sub-divisions are: Sango fly-over neighborhood and Oju-Ore being the two-conspicuous indigenous developed pieces of Sango-Ota; the new private neighborhood area of Iyana-Iyesi covering the two colleges and the Winners Church neighborhood; and the high-status neighborhood zone of Toll-gate possessing the south-eastern segment of the city [10].

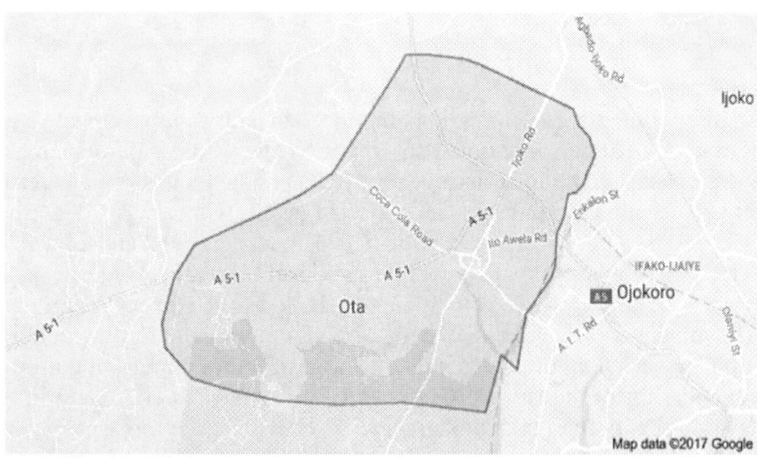

Figure 1: The highlighted region shows the studied emerging city, Sango-Ota.

2.2. Sampling of CO along Sango-Ota highway

Tours were made to identify different sampling locations at ten active junctions along Sango-Ota Highway. The choice of the different locations was based on the accessibility, central location and congestion in the areas. Ogun State, being a fast-rising industrious part of Nigeria has her highways with automobile activities and hence, automobile emissions in which CO emission is one. A user friendly GM8805 benetech CO monitor (Plate 1) was utilized to measure the CO concentrations at the selected sampling locations along Sango-Ota Highway. The CO monitor was put on and left to countdown for about l0 seconds so as to prepare internal systems ready. The monitor- GM8805 benetech CO - was positioned at an appropriate height, between l.5 m and 2.0 m to dodge measuring emission that could be fugitive. Readings were taken at 2 minutes interval for one hour at each of the identified sampling locations.

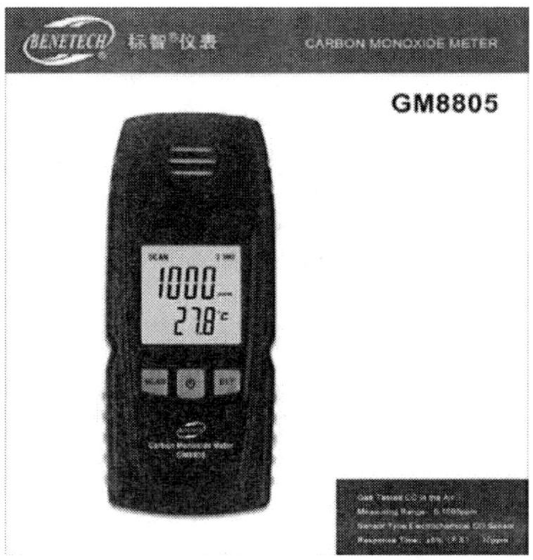

Plate 1: GM8805 benetech CO monitor

The toxicity potential of CO emission at each of the designated sampling locations was determined. To do this, first of all, the mean concentration of CO was determined on the averaging period of l-hour and extrapolated to 8-hour per day and 24-hour averaging time concentrations using the atmospheric stability formula [11] given in equation l. Then, the toxicity potential of CO emission was determined at each sampling location and on each averaging period using model equation 2 [12] mulling over the surrounding air quality statutory standards of CO by the FMENV, WB and that of the WHO as revealed by FEPA 1n 1991, WB in l998 and WHO in l979, separately.

$$C_o = C_l \text{ x } F \quad (1)$$

where,

C_0 = the concentration at the averaging period, t_o

C_l = the concentration at the averaging period t_l

F = factor to convert from the averaging period t_l to the $t_o = (t_l/t_o)^n$, n = 0.28 (the stability dependent exponent)

$$\text{Toxicity Potential (TP)} = \frac{M_{CO}}{S_{CO}} \quad (2)$$

where,

M_{CO} = measured CO
S_{CO} = statutory limit set for CO

IOP Conf. Series: Materials Science and Engineering 1036 (2021) 012018 doi:10.1088/1757-899X/1036/1/012018

3. Results and discussion

In this study, the impact of carbon monoxide on humans was explored. Furthermore, the different levels of exposure and toxicity potential of Carbon Monoxide based on the specific averaging periods were quantified. This approach allows a more accurate prediction of the effect of Carbon Monoxide emission on humans along this highway. Summarized in Table 1, are the mean values of the measured 1-hr averaging period of CO concentrations at the various selected sampling points at the junctions selected. With the average values calculated at each sampling location, the average concentration ranged from 25.20 ppm to 29.30 ppm with that of the control location being at 8.00 ppm. From Table 2, the l-hr averaging period when extrapolated to 8-hr averaging period, the concentration ranged from 14.076 ppm to 16.366 ppm while the concentration at the control location was 4.469 ppm. Again, as seen in Table 3, the 1-hr averaging period was extrapolated to 24-hr averaging period. The concentration ranged from 10.357 ppm to 12.042 ppm with the control location at 3.288 ppm. The means maximum and minimum Carbon Monoxide concentrations were recorded at Ifo Junction and Toll-Gate Junction correspondingly. This could be ascribed to the vehicular activities seen along this highway.

Table 1. Mean 1-hr Values from Measured CO concentration

S/N	Junction	CO (ppm)
1	Winners	25.40
2	Toll-Gate	29.30
3	Industrial Estate	25.47
4	Iyana-Iyesi	29.20
5	Oju-Ore	25.50
6	Idi-Iroko (Obasanjo)	27.70
7	Joju	25.73
8	Coca-Cola	8.00
9	Ifo	25.20
10	Iyana-Ilogbo	25.77

Table 2. Extrapolation from $t_1 = 1$hr to $t_0 = 8$hrs

S/N	Junction	C_1 (ppm)	$F = (t_1/t_0)^{0.28}$	C_0 (ppm)
1	Winners	25.40	0.5586	14.188
2	Toll-Gate	29.30	0.5586	16.366
3	Industrial Estate	25.47	0.5586	14.227
4	Iyana-Iyesi	29.20	0.5586	16.311
5	Oju-Ore	25.50	0.5586	14.244
6	Idi-iroko (Obasanjo)	27.70	0.5586	15.473
7	Joju	25.73	0.5586	14.372
8	Coca-Cola	8.00	0.5586	4.469
9	Ifo	25.20	0.5586	14.076
10	Iyana-Ilogbo	25.77	0.5586	14.395

IOP Conf. Series: Materials Science and Engineering 1036 (2021) 012018 doi:10.1088/1757-899X/1036/1/012018

Table 3. Extrapolation from t_1 = 1hr to t_0 = 24hrs

S/N	Junction	C_1 (ppm)	F= $(t_1/t_0)^{0.28}$	C_0 (ppm)
1	Winners	25.40	0.411	10.439
2	Toll-Gate	29.30	0.411	12.042
3	Industrial Estate	25.47	0.411	10.468
4	Iyana-Iyesi	29.20	0.411	12.001
5	Oju-Ore	25.50	0.411	10.480
6	Idi-iroko (Obasanjo)	27.70	0.411	11.384
7	Joju	25.73	0.411	10.575
8	Coca-Cola	8.00	0.411	3.288
9	Ifo	25.20	0.411	10.357
10	Iyana-Ilogbo	25.77	0.411	10.591

At Ifo Junction where the mean minimum carbon monoxide concentration was recorded could be as a result of lesser automobile movements in comparison to the rest of locations. In Ifo Junction, main activities that take place there include trading, car repairs etc. Important places which include gym and fitness club, schools, churches, can also be seen along this junction. Ifo junction is a developing area compared to other junctions along Sango-Ota highway and as such, less traffic congestion is seen. However, there are higher volumes of traffic at Toll-gate, when compared to other junctions due to how important a highway it is linking both Lagos State, commercial region of the country, and Ota, Ogun State, which contain factory plants to various industries. The automobile users reduce their speeds, always, around this location because of the poorly tarred road and the commercial buses that park along the highway negotiating with passengers. Although the activities at these two junctions are almost the same, the higher volume at this sample location could be ascribed to higher volumes of traffic experienced by Toll-gate junction than Ifo junction. Winners, Iyana-Iyesi, Joju, Idi-iroko, Iyana- Ilogbo, and Industrial Estate Junctions, all of which have lower number of automobiles than Toll-Gate junction but higher number of automobiles than Ifo Junction. This situation might be due to the lower mean CO concentrations in these sampling locations as compared to that of Toll-Gate Junction and the higher concentrations in comparison to Ifo Junction. The 8-hr and 24-hr averaging periods extrapolated CO concentrations followed the same trend in magnitude in all the sampling locations as the concentrations measured, where Ifo Junction had the least concentration and Toll-Gate Junction had maximum concentration.

From the Tables 4 to 6 above, it can be seen that the Toxicity Potential values gotten at 9 junctions after 1hr were between 2.52 and 3.525 across the regulating bodies, values gotten after 8hrs and 24hrs were above unity i.e 1. Toxicity Potential higher than unity engenders severe health issues to the inhabiting populace and the businessmen of the atmosphere where such is gotten. This shows the level of exposure commuters are exposed too. Minimum toxicity potential value is seen at Ifo Junction, this junction had least traffic congestion when compared with the others and fewer automobiles were seen at this junction. The toxicity potential ranged from 1.4076 to 1.564 for 8hrs and 1.0357 to 1.1507 for 24hrs. The cause for this tendency could be as earlier deduced for the measured levels of CO. At the selected control sampling point being Coca-cola Junction, the toxicity potential ranged from 0.44688 to 0.4965 for the 8-hr and 0.3288 to 0.3653 for 24hrs extrapolation. At this junction, the standards and regulations were maintained and toxicity potential values calculated at various times of the day were all below 1. Maximum toxicity Potential is seen at Toll Gate Junction where a large number of automobiles are seen regularly and regular traffic occurs. This is a majour worry for every road user along the particular road that passes through this axis to get to their various destinations. The regular traffic along this route shows more reason why attention should be given to this region in general.

Table 4. Toxicity Potential calculated for 1hrs based on FEPA, WHO, USEPA Statutory limits

S/N	Junction	TP_{FEPA}	TP_{WHO}	TP_{USEPA}
1	Winners	2.540	2.540	2.822
2	Toll Gate	2.930	2.930	3.255
3	Industrial Estate	2.547	2.547	2.830
4	Iyana-Iyesi	2.920	2.920	3.244
5	Oju-Ore	2.550	2.550	2.833
6	Idi-iroko (Obasanjo)	2.770	2.770	3.077
7	Joju	2.573	2.573	2.8588
8	Coca-Cola	0.800	0.800	0.888
9	Ifo	2.520	2.520	2.800
10	Iyana-Ilogbo	2.577	2.577	2.863

Table 5. Toxicity Potential calculated for 8hrs based on FEPA, WHO, USEPA Statutory limits

S/N	Junction	TP_{FEPA}	TP_{WHO}	TP_{USEPA}
1	Winners	1.4188	1.4188	1.5764
2	Toll Gate	1.6366	1.6366	1.8184
3	Industrial Estate	1.4227	1.4227	1.5800
4	Iyana-Iyesi	1.6311	1.6311	1.8123
5	Oju-Ore	1.4244	1.4244	1.5826
6	Idi-Iroko(Obasanjo)	1.5473	1.5473	1.7192
7	Joju	1.4372	1.4372	1.5968
8	Coca-Cola	0.4469	0.4469	0.4965
9	Ifo	1.4076	1.4076	1.5640
10	Iyana-Ilogbo	1.4395	1.4395	1.5994

Table 6. Toxicity Potential calculated for 24hrs based on FEPA, WHO, USEPA Statutory limits

S/N	Junction	TP_{FEPA}	TP_{WHO}	TP_{USEPA}
1	Winners	1.0439	1.0439	1.1598
2	Toll Gate	1.2042	1.2042	1.3380
3	Industrial Estate	1.0468	1.0468	1.1631
4	Iyana-Iyesi	1.2001	1.2001	1.3334
5	Oju-Ore	1.0480	1.0480	1.1640
6	Idi-Iroko(Obasanjo)	1.1384	1.1384	1.2648
7	Joju	1.0575	1. 0575	1.1750
8	Coca-Cola	0.3288	0.3288	0.3653
9	Ifo	1.0357	1.0357	1.1507
10	Iyana-Ilogbo	1.0591	1.0591	1.1767

The toxicity potential values ranged from 1.6366 to 1.8184 for 8-hrs and a range of 1.2042 to 1.338 for 24hrs. The next highest TP is Iyana-Iyesi Junction, with a range from 1.6311 to 1.81234 for 8hrs and a range from 1.2001 to 1.3334 for 24hrs. This is a route for a number of automobiles. This region is regularly filled with commuters and passengers.

It was observed form the field that the 24-hour averaging period, as extrapolated, of Carbon Monoxide concentration at Toll-Gate Junction breached the standard. So, did every other sampling location except that of Coca-Cola Junction. Since automobiles contribute considerably to ambient gaseous emissions (Xing et al., 2012) decrease in the traffic holdups in the vicinity of this segment of the high possibly will cut the Toxicity Potential. Likewise, at Iyana-iyesi Junction, the subsequent peak Toxicity Potential range in order of degree occurred here. This shows that road users along the road extend all the way to Iyana-Iyesi Junction. Shown in Figure 2 is the percentage impact of each of the selected sampling positions to the Carbon Monoxide level, overall, along Sango-Ota highway. This ranges between 10.357% and 12.042 %. Locations like Ifo, Winners and industrial estate are the least contributors (10.357%, 10.439 % and 10.4685%) (With exception to the control point) while Toll gate remains the highest contributor (12.042 %) with Iyana-Iyesi Junction as the next highest contributor (12.001). This ratifies the previous observation that traffic holdups around this segment of the highway should be minimalized if not absolutely eliminated in order reduce the exposure of road users (commuters) to high level of Carbon Monoxide on the highway. Also, Coca-cola Junction being at 3.288 remains the control point.

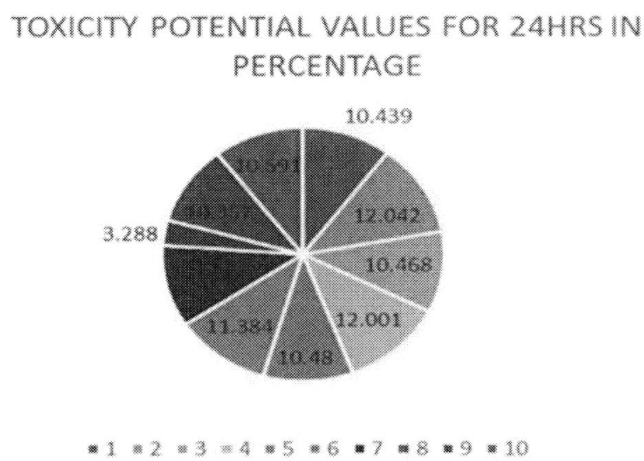

Figure 2: Pie-chart showing 24-hrs Toxicity Potential Values in Percentages

4. Conclusion

The outcome of the study showed that emissions from automobiles contribute considerably to CO levels in the outdoor air with great health impact on humans along the highways in evolving cities. The toxicity potential greater than unity at the selected locations, asides the control location, calls for attention most especially for allergic persons, aged persons, young ones as well as persons with respiratory diseases. Transport related emissions contribute significantly to the Carbon Monoxide present in ambient air along this highway. In emerging cities where people prefer commuting in cars, this work is of great value to change people's perception on the levels of exposure to carbon monoxide that are experienced in traffic congestions and the effect of intake of carbon monoxide Having known the toxicity potential of carbon monoxide from automobiles along Sango-Ota expressway through this study, appropriate agencies need to be made aware and road networks need to be intensified to decongest traffics along highways in emerging cities in developing countries.

Acknowledgement

The management of Covenant University, Nigeria is highly acknowledged for paying the publication fee of this manuscript.

References

[1] US Energy Information Administration 2014 How much carbon dioxide is produced by burning gasoline and diesel fuel? *Monthly report*
[2] Shindell D 2007 Understanding Carbon Monoxide as Pollutant and as Agent of Climate Change, National Aeronautics and space administration, Goddard Institute for Space Studies.

(Accessed on the 25th of September 2017). Available at: https://www.giss.nasa.gov/research/briefs/shindell_09/
[3] Matthew O, Osabohien R, Fasina F, Fasina A 2018 Greenhouse Gas Emissions and Health Outcomes in Nigeria: Empirical Insight from ARDL Technique, *International Journal of Energy Economics and Policy, 2018, 8(3), 43-50.* Available at http:www.econjournals.com
[4] Chidi C U (2013). Measurement of Carbon Monoxide Emissions in Some Selected Area in Lagos State, Journal of Scientific Research & Report, Vol 02, Issue-02, P 536- 539. Available at http://www.journalrepository.org/media/journals/JSRR_22/2013/Aug/1375590422-Uhuegbu222013JSRR3223.pdf
[5] Odunlami OA, Elehinafe FB, Oladimeji T E, Fajobi MA, Akeredolu FA, Okedere OB and Fakinle BS 2018 Implications of Lack of Maintenance of Vehicles on Ambient Air Quality, **9** (9), pp. 647 – 654
[6] Shochat G N and Lucchesi M 2016 Carbon Monoxide Toxicity (Online) Medscape (Accessed on the 25th of August 2017) Available at: http://emedicine.medscape.com/article/819987-overview
[7] National Population Commission (NPC) [Nigeria] and ICF International (2014). *Nigeria Demographic and Health Survey 2013.* Abuja, Nigeria, and Rockville, Maryland, USA: NPC and ICF International
[8] PricewaterhouseCoopers 2013 World in 2050 The BRICs and Beyond: Prospects, challenges and opportunities, is available at: http://www.pwc.com/world2050
[9] Oguntunde PE, Odetunmibi OA and Adejumo A O 2014 A Study of Probability Models in Monitoring Environmental Pollution in Nigeria, Volume 2014, Article ID 864965, 6 pages. Available at http://dx.doi.org/10.1155/2014/864965
[10] Oloyede S A, Ayedun C A, Durodola OD and Peter J N 2014 Residential Solid Waste Management in Sango-Ota, Ogun State: To Recycle or Not to Recycle? *Humanity and social sciences*, ISSN(paper) 2224-5766, ISSN(online)2225-0484, **4**, -26, P 190-191. Available at: http://eprints.covenantuniversity.edu.ng/3541/1/Pliagiarism%20Test%2018.pdf
[11] Bashar MA, Kamel K A and Khaldoun M S 2009 Assessment of Air Pollutants Emissions from a Cement Plant: A Case Study in Jordan, *Jordan Journal of Civil Engineering,* **3**(3), P 265– 282
[12] Sonibare JA, Akeredolu FA, Osibanjo O and Latinwo I 2005 ED-XRF Analysis of Total Suspended Particulates from Enamelware Manufacturing Industry, *American Journal of applied science*, **2**(2), 573-578
[13] Federal Environmental Protection Agency (FEPA) 1991 Guidelines and Standards for Environment Pollution Control in Nigeria, Lagos, P 250
[14] World Bank 1998 Pollution Prevention and Abatement Handbook, World Bank Group, Washington DC
[15] World Health Organization (WHO) 1979 Sulfur Oxides and Suspended Particulate Matter, *Environmental Health Criteria 8.* Geneva
[16] Xing W, Dane W, Jingnan H, Ye W, Hang Y, Xiaochuan P and Zhang K M 2012 On-road Diesel Vehicle Emission Factors for Nitrogen Oxides and Black Carbon in two Chinese Cities, *Atmospheric Environment*, **46**, 45-55

A review on plastic waste as sustainable resource in civil engineering applications

T O Ogundairo [1], D O Olukanni[1], I I Akinwumi[1] and D D Adegoke[1]

[1]Department of Civil Engineering, Covenant University, Ota, Ogun State, Nigeria.
*Corresponding Author e-mail: topepearse@yahoo.com

Abstract: Plastic wastes are non-biodegradable and can stay in the ecosystem for many decades. It can go through aging processes resulting from physical, chemical, and biological actions with the potential to harm habitats and weaken the life-supporting environment. The overwhelming environmental problem created by the indiscriminate disposal of plastic waste containers has led to a search for comprehensive research work seeking a viable alternative to manage, process and dispose of used plastic containers. Hence, researchers have found alternative use to its recovery. The waste can be recycled, reused or reprocessed as substitutes for construction materials because construction applications take up significant amounts of aggregates and cement. Its use is also applicable in bitumen modification, soil stabilization, geosynthetic materials, bricks, plastic reinforcement, and as natural aggregates, which all assist in reducing the quantity of natural aggregates that can be extracted for use in the construction industry. The review covers the collection of relevant information about plastics, the different types of plastics, their corresponding waste and its application in the construction industry from literature. It also examines plastic waste challenges and its sustainable utilization as a construction material. By changing the production design of plastics, it would proffer better ways to manage plastic waste, clean up and improve our natural environment. This review submits that there is an opportunity for the use of plastic waste as an innovative alternative in the construction industry which can stimulate economic growth and could boost the drive of government towards the achievement of some of the sustainable development goals.

Keywords: Municipal Solid Waste, Environmental Pollution, Environmental Sustainability, Construction Materials, Plastic Waste, Soil Stabilization

1. Introduction

The production of plastics and its use dates to 1950 [1]. They have become a crucial part of everyday living [2]. Nigeria is ranked ninth of the rundown of twenty nations, not managing plastic waste in a responsible manner [1]. In 2010, the worldwide volume of plastic waste not properly managed was 851,493 tons, and quantities are estimated to increase to 2,481,008 tons by 2020 [1]. Plastics consist of synthetic organic materials; this is easily shaped in the desired form when compressed then placed within slightly elastic and stiff shapes. The use of plastic is versatile, and examples are found in the medical industry, transportation, construction, household products such as electronics and food

packaging [3]. The consumer motivation to plastics is its economics, lightweight and chemical resistance, resilience to the natural deterioration. Its utilization has a direct connotation with the volume explosion of plastic waste for domestic and industrial use in recent times [4,5,6]. The increasing population in Nigeria is directly related to the increase in plastic waste. The average growth rate for the population between 2000 and 2017 in Nigeria was 2.37% [6]. It is relative to the growth of general municipal solid waste (MSW) as well as plastic waste. The global production of plastics was estimated at 260 million metric tonnes per annum in 2007, with 67% of the production been thermoplastic resins [5,6]. The estimated increase per annum in 2017 was 348 million metric tons [6]. However, single-use plastic such as consumer goods and packaging accounts for 50% of plastic produced [6]. The end of life management of plastics is a global issue, especially in developing countries that have non-existent or ineffective waste management practices. Consequently, the dumping of plastic waste in landfill sites and water bodies is widespread [3,5,6,7,8].
The bottled water industry in Nigeria is growing at an astonishing rate; it is a healthy alternative to carbonated drinks. It is a low-cost and convenient way of getting good quality, easy to open, safe drinking water. The Nigerian government has not been able to provide clean drinking water for its populace; as a result, bottled water is a trusted option [3]. Therefore, the Nigerian private sector has keyed into the United Nations Sustainable Development Goal; access to clean water and sanitation. The industry provides hygienically bottled, safe drinking water in polyethylene terephthalate (PET) containers (this is commonly called 'bottled water'), and it is affordable to the low and middle class in the population. The disposal of the empty bottle containers after use is a significant environmental challenge as it is dumped without prejudice. Plastics are versatile, not susceptible to biodegradation, and can stay in the soil without degradation for 450 years [9]. The large volume per unit mass poses disposal and environmental problem. Although plastics in municipal waste streams account for only 7-9% of the overall waste stream weight, this signifies 20-30% by volume [3,6,9,10,11,12].
It is distressing that in Nigeria and other developing countries, there is a lack of efficient means for dealing with the volume of waste generated. Many consumers are not aware that their choices influence the quantity of waste produced and managed [13]. The government in Nigeria at all levels–federal, state, and local–is facing this massive task in the management of nylon and plastic waste [14]. A popular sight on the highway is bottled water containers, and this blocks drainage and obstructs the free flow of stormwater [15]. Figure1 shows plastic waste blocking the drainage system. Also, when disposed into natural water bodies, they cause an obstruction and endanger aquatic life. Plastic waste hinders the environment remarkably, triggering a reduction in ecological diversity and the aesthetic quality of urban parks and beaches [16,17]. Therefore, proper management of plastic waste assists in improving the wellbeing of society [6,18]. The overwhelming environmental problem created by the indiscriminate disposal of plastic waste containers has led to a search for comprehensive research work seeking a viable alternative to manage, process and dispose of used plastic containers.

Figure 1. Plastic waste in a drain in Olowu Street Ikeja, Lagos

2. Plastic

In our society, plastics are generally utilized in different categories of daily applications. Plastics are universal materials used in the manufacturing and food industry, transportation, medical, and in households to solve a variety of challenges. Conventional plastic packaging shields our food and goods from pollution [19]. Furthermore, clean water can be provided using plastic water distribution systems and storage containers. Also, plastic materials are used for protective apparels, safety and security apparatus (e.g., textile impervious to fire, head protectors such as helmets, airbags) avert injuries. In the same way, plastic products for medical purposes lead to enhanced health conditions (e.g., tubing, blood bags, prosthesis, and disposable syringes). Equally, low-density plastic materials are used as alternatives for metals and ceramics in airplanes and motor vehicles. The use of plastic in different ways creates a variety of waste streams. Plastic packaging is produced in substantial quantities and usually discarded in the waste stream typically when used once within a limited time of purchase (about 40 percent of plastics products have an average life expectancy of under one month) [19]. This massive waste causes significant environmental management challenges.

2.1 Types of Plastic and Recycled Plastic

The fourth major category of civil engineering construction materials are polymers, and different types are manufactured for an array of uses [20]. Although new forms of polymeric materials are continually emerging on the scene, there are notable improvements in the properties of existing plastics. Recycled plastics are utilized in innovative structure and other construction implementations daily. These recycled plastics can be processed using new plastics to reduce costs without compromising quality. The recycled plastic materials are used to produce polymeric timbers used in fences to picnic tables, thereby assisting in the preservation of trees. Plastic from PET containers is convertible into fiber for the manufacturing of carpets. Plastic waste is utilized for waste-to-energy (WTE) systems, and the method includes converting and recovering the energy value from plastic waste material by applying elevated temperature and precise combustion. The heat recovered is then transformed to electrical energy in steam-driven turbine generators used for equipment, or it can also be sold directly either as steam or heated water within the manufacturing and heating procedures [21].

2.1.1 Thermoplastics

Upon application of heat, thermoplastics can potentially be dissolved into their liquid state and reshaped into the same or different form for the same or a variety of applications [6]. The gross plastic consumption of thermoplastics is about 80 % [6]. Examples of thermoplastics are Polystyrene, Polyvinyl chloride, Polyethylene terephthalate, Polymathic methacrylate, Polypropylene and Polyethylene [19].

2.1.2 Thermosets

These are plastics that are incapable of returning to their original shape once they have been cooled and hardened. They are manufactured through the process of chemical bonding by crosslinking comparable or diverse polymer chains, and they make up 20% of the total global production of plastics. These are tougher, stronger, and resilient in comparison to thermoplastic, although mostly non-recyclable. Recent studies have, however, revealed that some thermosets are recovered from previously cured thermosets by the exchange of bond reactions or as a result of additional reactions at active sites [6]. Epoxy resins, polyurethanes, phenolic, and polyester resins are examples of thermosets.

2.2 Uses of Plastic and Plastic Waste

With a growing dedication to environmental sustainability, the engineering sector is enthusiastic about reusing and recycling plastics and polymers considered as plastic waste. The construction industry is dependent on raw materials; therefore, the motivation for the utilization of plastic waste is an implication of the significant quantities of plastic used, and the exponential volume of plastic waste generated [6,19]. Likewise, the escalating cost of building materials also inspires recovering of plastic waste as a value-added product in construction applications. This emerges as an unconventional alternative solution for the utilization of plastic waste. Instead of littering the environment, plastic wastes have been used in civil engineering, as soil stabilizers and alternative aggregates for concrete and asphalt production. Plastic waste can replace a specific portion of aggregates, conventional stabilizers, or lightweight concrete pavement, reinforced geo-pavements, or can be used in low-volume asphalt pavements [19]. Consequently, the amount of waste going to the landfill reduces, thereby improving resource management and ultimately reducing the cost of construction. The use of these green substitutes has a positive influence on the environment. Moreover, the adaptability of plastic waste for homes, road infrastructure, or energy generation in the construction industry must be economical, efficient, and safe. Furthermore, some unconventional uses of plastic waste are high-grade polymers such as automobile parts, home equipment, fabrics, building insulation, and films that are manufactured from plastic waste. Consequently, plastic waste treatment and recycling procedures can be distributed into four main classes that are recycling, chemical, mechanical, and waste to energy recovery. Table 1 illustrates the diverse types of plastics and their recycling potential.

Table 1. Uses of Plastic and Recycled plastic [6,19].

Type of Plastic	Properties	Application of plastic	Application of recycled plastic
i. Polyethylene Terephthalate (PET)	Transparent solid plastic, suitable for fiber	Water bottle drink and minerals bottles	Transparent soft material for packing items, wrapping film, carpet fibers, rain jackets
ii. Low-density polyethylene (LDPE)	Soft, flexible plastic, cloudy white except the pigment is added	Closures of food containers, trash bins, and garbage bags	Wrapping business, soft cling films, for packing plants and nursery bags
iii. High density Polyethylene (HDPE)	Normally used plastic in coloured or white	Wrinkled bags for shops, freezer bags (for meat /milk storage)	Manure bins, soap bottles, crates, and mobile trash bins
iv. Plasticized Polyvinyl chloride (PPVC)	Flexible, transparent, elastic-plastic	Blood bags and tubing for the medical sector, soles for shoes, pipes and hose for gardens	Factory floors,
v. Polypropylene (PP)	Hard, but flexible plastic	Ice-cream- bowls/containers, potato crisp bags, chairs	Manure-bins, kerbside recycling bin/crates,
vi. Polystyrene (PS)	Stiff but brittle plastic. Transparent in nature and glossy	Cheap, transparent kitchenware, lightweight fittings, bottles, toys, and food containers	Laundry pegs, coat hangers, and video/CD boxes
vii. Polyester (EPS)	Foamed, lightweight, energy absorbing, and thermal insulation	Tea or coffee cups and food Takeaway bowls	Spoons, rulers and Movies/CD box sets
viii. Polyamides (PA)	Nylons	Fibers, toothbrush bristles, and fishing lines	Material for packing items, wrapping film,
ix. Polymethyl methacrylate (PMMA)	High UV light and chemical resistant, transparency, durability	Motorized lamp covers an alternative to glass in specific applications	Bins, shampoo bottles, crates
x. Polyethylene (PE)	Translucent, tough and semi-rigid, chemical resistance, waterproof and cheap	Packaging (plastic bottles, bags, wraps, and films), construction	Wrapping cling film

Table 1 shows the following plastics polyvinyl chloride (PVC), polyethylene terephthalate (PET), polyethylene (PE), polystyrene (PS) and polypropylene (PP) these are most recycled types of plastics [19] and are useful in the production of a wide variety of packaging applications, including, tubes, barrels, trays, bags, cups and packets, infant goods and protective packaging. Polymethyl Methacrylate (PMMA), Polyamides (PA), Polyester (EPS), Plasticized Polyvinyl chloride (PPVC) Low-density polyethylene (LDPE), High-density polyethylene (HDPE) are some of the other types of plastics in use. Figure 2 shows the processing of PET bottles for recycling use in a university community.

Figure 2. Processing of PET bottles for reuse in a University Community

2.3 Problems arising from Plastic Waste Pollution

Plastic pollution in the environment influence air quality as greenhouse gases are released during the production of plastic. Plastic waste burning discharges black carbon (soot); this adds to global warming and contamination of the air. The other toxic chemicals released from plastic waste burning are gases like polychlorinated biphenyls (BCPs), mercury, dioxins and furans. These gases present a hazard to human, animal wellbeing and vegetation when released into the air [22]. These toxic chemicals likewise impact water and soil. After plastic waste is dumped on the ground, it slowly releases phthalates and biphenyls into the soil, groundwater, and other water bodies [22,23]. The world's plastic production from the 1950s-2014 has increased by 200% [24]. The creation of disposable plastics is growing at an alarming rate, and the world is overwhelmed by its inability to address the problem. About 448 million tonnes of plastic was produced globally in 2015, 58 percent of waste plastic was thrown away or deposited, and about 18 percent recycled in 2015. Plastic production is expected to double by 2050 [25,26]. Every year about 8 million tonnes of plastic waste from activities on land enter the world's ocean, and this is equivalent to five full trash bags of rubbish along each foot of the global ocean [25]. Millions of birds, fish, and other marine organisms every year die as a result of plastic waste pollution [25].

Consequently, the most pressing environmental challenges in developing countries like Asian and African remain plastic pollution. Plastic pollution is significantly evident because waste collection programs are mostly ineffective or non-existent [14]. Notwithstanding, it is challenging to collect discarded plastics in Europe and America within areas of the country that have low recycling rates. The chemical industry contributes to around 15% to the global anthropogenic GHG (Green House Gas) emissions. The universal collective application of plastic GHG world life-cycle emissions in 2015 was 1.7 Gt (Gigatonne) of CO_2 -equivalent (CO_2e), increasing to a composite annual growth rate of 8.4% by 2050. The global emissions and negative environmental impacts associated with plastic production have triggered the United Nations to start plans to draw up a global treaty [25]. There is, therefore, a need to reduce and creatively discover ways to recycle and reuse plastics waste [26].

3. Plastic Waste as Construction Material

The review focuses on the utilization of recycled plastic waste as a construction material. The incorporation of plastic waste in concrete either as an aggregate or as cement-based materials such as cement mixtures or concrete mixes or as a soil stabilizer are better options for plastic waste disposal [19]. When plastic waste is substituted or replaced within a concrete mix, it has environmental and economic benefits. Also, plastic waste modification is utilized as a lightweight, low-strength concrete pavement, or low volume asphaltic pavements, and soil stabilizers in geo-pavements [19]. Numerous researchers have used plastic waste as cementitious materials or aggregates or as soil stabilizers in the construction industry. Plastic products such as polyethylene terephthalate (PET bottle), high-density polyethylene (HDPE),glass-reinforced plastic (GRP), expanded polystyrene foam (EPS), polyvinyl chloride (PVC) pipe, spent plastic waste, polypropylene fiber, polycarbonate, and thermoplastic recycled polystyrene [19].

3.1 Plastic Waste Materials Used in Civil Engineering Applications
In civil engineering applications, plastic waste is used as modified bitumen, fine and coarse aggregates in concrete, soil stabilization, geosynthetics, bricks, plastic waste reinforcement, and plastic waste cement.

3.1.1. Modified Bitumen
[27,28] investigated physical properties such as strength, durability using PET (polyethylene-tetra phthalate in bituminous roads at the various percentage (5%, 10%, 15%) and results reported were satisfactory. [29] evaluated the application of PE (polyethylene) plastic waste as a modifier in semi-dense bituminous concrete. In the research, shredded plastic waste was mixed in the hot aggregate, and the plastic was used to modify the mixture at 6%, 8%, 10%, 12%, and 14% plastic by weight of bitumen. The outcome observed that Marshall stability's highest value was at 12% plastic waste content. [30] investigated the inclusion of HDPE and LDPE (high-density polyethylene and low-density polyethylene) shredded plastic waste in the bituminous concrete, and the result shows a significant increase in the Marshall characteristics and stability value in the combined mixture. The study reported that the utilization of 8% HFPE and LDPE plastic waste in bituminous concrete is secure and suitable for light, medium, and heavy traffic roads. The application of polyethylene terephthalate (PET) as a strength modifier in asphalt road construction was examined [3]. The samples were collected and thoroughly cleaned, naturally dried, and shredded. The results from the sampling regime were compared with the ASTM, Nigeria Federal Ministry of Works, and AASHTO standards and the result showed that optimum binder content (OBC) by weight of aggregate was 1% [3]. The proportion of PET content was 15% by weight of bitumen, Bulk density (BD) 2.38 kg / m^3; Void in Total Mix (VTM) 3.33%; Void Filled with Bitumen (VFB) 82.20%; Marshall Flow (MF) 4.00 mm and Marshall Stability MS 17.01 kN [3]. The improved asphalt with 1% PET was observed to be useful for pavement construction and showed better binding properties, strength, density, and water resistance [3]. Utilizing the concrete modified blend alongside treated plastic waste of around 5–10% by weight of bitumen prompts a significant enhancement in the fatigue cycle, strength, and different characteristics necessary in bituminous concrete. As a result, it increases the service life cycle and road surface quality with small but substantial savings in bitumen constructed roads. Likewise, another study coated stone aggregates with molten plastic and concluded that coating aggregates with plastic decrease porosity, moisture absorption, and increased soundness. The combination of plastic waste with stone mastic bitumen demonstrated significant improvements in the mechanical and volumetric properties of the mix. It also revealed that the utilization of plastic waste is efficient and convenient in reducing plastic waste and the associated cost of highway construction [12,31,32,33]. Thus, the inclusion of plastic waste is feasible for the production of modified bitumen for road construction with its adequate binding properties and cost-saving potentials.

3.1.2. Fine and Coarse Aggregates in Concrete
Aggregate plays a vital role in the production of concrete mix and accounts for 65-85% of the standard concrete mix mass volume. The development of concrete strength is dependent on its aggregates and categorized by physical and chemical properties such as fresh properties of the mix, workability, slump test values, compressive strength, flexural strength, and the effects of sulfide attack. The substitution of aggregates with plastic waste either as part or complete replacement acts as an alternative solution. Some studies investigated partly replacing fine aggregates and coarse aggregate with Plastic fiber (PF), sliced polypropylene (PP) debris polyethylene (PE) pellets. Plastic aggregates (PA) have a lower bulk density than basalt, granite, or limestone. Therefore, utilized for lightweight concrete. The effect of fresh concrete density, workability, and hardened concrete density, compressive strength, water assimilation was evaluated [34]. The findings indicated that plastic waste has a beneficial impact on fresh concrete characteristics: the workability of the concrete improves and is lighter. Regrettably, the hardened concrete with high plastic waste content has an adverse effect: the compressive strength was altered and reduced with the increase of plastic waste in the concrete [35], mode of failure changes from brittle (rapid) failure to more ductile failure [36]. However, water

absorption, in turn, increased. However, the result observed, that maintaining the same value of concrete workability, there are opportunities to decrease water and cement ratio, and this, in turn, provide strength loss, and water absorption increase compensation [19,31,37,38,39]. Plastic waste was partly substituted as sand in an innovative mix for structural concrete. For the evaluation of the mix, eleven concrete mixes were assessed, measuring the aspect ratio, plastic content, and particle dimensions to entirely create a suitable material that can partially replace sand. The outcome of the research states that 10% substitution of plastic can replace sand by volume, and will save about 820milloin tons of sand yearly. However, plastic waste should be used with the right design mix considering the structure [40,41]. Vehicle plastic waste was studied for possible use in concrete as a substitute for sand. The study found concrete's impact strength properties in 2.5%, 5%, 10%, and 20%, respectively, of plastic waste particles as a substitute for sand at different content. The concrete mix showed the tendency to withstand a significant amount of energy from impact [42]. [37] evaluated the partial substitution of fine aggregates with polypropylene in the mixed model, and the compressive strength of concrete was tested by utilizing test regimes of six concrete mix samples prepared with substitutes of polypropylene varying from 0.5 to 3.0%. With the inclusion of plastic aggregate, the results showed a reduction in compressive strength, and this was regulated as plastic waste content increased the concrete mix. Subsequently, the results from two samples reported attaining the ST5 standard criteria at 28days for concrete mix compressive strength. The percentage proportion of plastic aggregates used was 1.5% and 2.5%; this confirms the possibility that recycled polypropylene can be used in structural applications as a part by mass of fine aggregate substitute in the concrete mix.
E-waste plastics are utilized to substitute fine and coarse aggregates in concrete. The compressive strength, tensile strength, and flexural strength of concrete are evaluated with replacement ranges from 0, 10, 20, and 30% and without E-waste plastic aggregates [43]. The results from the analysis reported that compressive strength at 10% optimum content of E-waste plastic in the concrete mix yielded stability and a recent compressive strength at grade 53 [44]. [45] examines the impact of plastic on various fresh and hard concrete properties as an alternative coarse aggregate. The thermoplastic polymer polyethylene terephthalate (PET) is evaluated as an artificial aggregate and replaced by natural coarse aggregate such as brick chips. The PET aggregate is acquired when the collected waste PET bottles are shredded, melted, and crushed. Figure 2 showed an example of the shredding process. The focus of the study is to examine PET aggregate concrete (PAC) compressive strength and unit weight along with its workability in comparison to concrete mixed with natural aggregate (NAC) [45]. PAC at 20 per cent PET substitution has a compressive strength at 0.42 w / c of 30.3 MPa; this is 9 per cent lower than the NAC. The observed slump rate of 1.8 cm, indicates PAC has substantially high workability. Low-density polyethylene (LDPE) plastic was utilized as a partial substitution for sand in cement mortar [46,47]. The sand was substituted with LDPE waste at different percentage ratios from 0, 5, 10, 20, 30, 40, 50, and 60%. The mortar mixture was one part of cement to three parts of plastic fine aggregate at water/cement ratio. The sample was then cast into cubes and prisms. The outcome of the experiments showed that performance decreases substantially as the percentage of LDPE in mortar increases, but mixtures containing 50–60% LDPE meet the performance requirements for mortar masonry [47]. However, a model for predicting the compressive strength of mortars containing waste plastics was proposed based on experimental data. The validation of the model generally provided accurate predictions of mortar strengths made from various types of plastics [47]. Subsequently, the addition of plastic waste is viable for the production of fine and coarse aggregates in concrete with satisfactory workability and energy on impact.

3.1.3. Soil Stabilization

For practical application of plastic waste to stabilize unstable and weak soils, the most important properties are shear strength, compaction (CBR; California bearing ratio, MDD; Maximum dry density, OMC; Optimum moisture content), relative density specific gravity, permeability, consolidation, consistency limits, particle size and resilient modulus (for road pavements) [48,49,50]. Several researchers have utilized different types of plastic waste as a soil conditioner, and the influence of compressive strength is that it increases to some extent [51,52]. [53,54] reported that bearing capacities of black cotton soil or silty soil increased with an increasing quantity of waste plastics. [53] studied the effect of plastic waste in black cotton soil, and the test reported a low specific gravity for plastic. [55] measured the specific gravity of dry soils using plastic fibers as an additive for the soils. The outcome of laboratory tests showed that there is an exceptional effect on the specific gravity of clayey soil with plastic fiber. [56] evaluated the properties of dune sand modified with

IOP Conf. Series: Materials Science and Engineering 1036 (2021) 012019 doi:10.1088/1757-899X/1036/1/012019

LDPE waste by Variable Head Permeability test. The outcome of the test demonstrated that the use of LDPE waste between 0.06 and 1% helps to improve the strength of dune sand and assists in altering the dune sand properties, thereby improving the subgrade soil. [57] tested four soils mixed with crushed waste ceramic (CWC) and crushed waste plastics (CWP) and subjected to laboratory conditions to test for subgrade stiffness. The sample soils are mixed with 10 to 120% by weight with these geomaterials. Lateral deformation from modified triaxial compression consistently decreased with the amplified CWC and CWP proportions. [55] studied utilizing plastic fiber as an additive to clay soil for soil stabilization. Atterberg limits tests were conducted on soft clayey soil with and without plastic fiber at 1, 2, and 4% by the proportion of stabilizing material to the net weight of soil. The laboratory tests showed a significant impact on the liquid limit and observed that plastic fiber content has a remarkable effect on the grain size distribution. The modulus (Mr) of clayey soil, and ordinary soil mixed with recycled polyethylene terephthalate (PET), and its impact as soil reinforcement exhibited significant improvement. The repeated-loading triaxial tests reveal Mr of 0.6% PET-reinforced samples increased by 58%. In contrast to Mr of the control sample (0) percent PET) [58]. Similarly, resilient modulus Mr, R-value, and stiffness of subgrade soils mixed with crushed waste ceramic (CWC) and crushed waste plastics (CWP) were investigated [54]. It was observed that the soil improved steadily with increased CWC and CWP. Therefore, the inclusion of plastic waste for soil stabilization improved the geotechnical properties of the soil considerably.

3.1.4. Geosynthetics

Plastic waste is used to make geocell and geogrid for soil reinforcement. The geocell and geogrid plastic waste were positioned from the top of the CBR mould at distinct depths. The CBR test was conducted for the unsoaked state of natural soil and reinforced soil, and load-versus-penetration behaviour was examined. Results show that by positioning the reinforcement of plastic waste at varying depths, the strength of the reinforced pavement soil in comparison to natural soil experienced a considerable increase. The confined geocell structure showed a significant improvement in strength compared to the geogrid of plastic waste. The findings indicate that at a certain depth, the considerable enhancement in soil strength using both geosynthetic materials has been successful. This study shows that it is an excellent option to use plastic waste to increase pavement soil capacity. For pavements, plastic waste can be used as an economical solution [59]. [60] investigated the reuse of plastic waste bottles to formulate reinforcement comparable to conventional geogrid for prospective application to advance load – settlement behaviour of soil. River sand was used to prepare soil beds at a relative density of 70%. The experiment was conducted using a load plate test compare unreinforced soil, conventional geogrid reinforced soil and plastic reinforced soil. The results in terms of cost efficiency between geogrid and plastic bottle reinforcement specified 60% savings when the plastic waste is utilized for soil reinforcement. The maximum bearing capacity for unreinforced soil is 254 kPa, geogrid reinforced soil 310 kPa (22% increase), and plastic reinforced soil 292 kPa (15% increase). The incorporation of plastic waste in geosynthetics records considerable increase in strength properties and significant cost savings.

3.1.5 Bricks

The research aimed to investigate the appropriateness of the production compressed earth bricks (CEBs) with a mix of soil and differing percentages of shredded plastic waste at (0, 1, 3, and 7%) [61]. Test on the technological properties of the soil, specific gravity, particle size distribution, Atterberg limits, and compaction experiments are conducted on the soil. The CEB's compressive strengths and erosion levels were established and calculated for the untreated soil and soil mixture at differing ratios of two-sizes (< 6.3 mm and > 9.6 mm) of the shredded plastic waste. Classification of the soil as clayey sand (SC). For the CEB, which contains one plastic waste of sizes < 6.3 mm, the highest compressive strength was obtained, and its compressive strength increase by 244.4% [61]. The specimen comprising 1% waste plastic with sizes < 6.3 mm also had the least erosion level of the CEB specimens stabilized with shredded plastic waste. Correspondingly, provided that the external wall surfaces created using the CEB were safe from erosion, it was advised to use 1% shredded plastic waste with particle sizes < 6.3 mm [61]. Plastic waste that would have been a threat to the atmosphere would create more reliable and more durable bricks to provide affordable housing [61,62]. Plastic waste is used to manufacture concrete paver blocks, and from the results obtained, it is to equal compressive strength of the conventional paver blocks. From the observation, they are used for footpaths and parks [63]. [64,65] studied three types of waste thermoplastics, namely, polystyrenes, mixed plastics, and polycarbonates to make eco-friendly bricks. Experiments were conducted by using different proportions of waste thermoplastics (0 – 10% by weight) and sand (60 – 70% by weight), keeping a constant 15% (by weight each) proportions for fly ash and ordinary Portland cement. Three separate sets of bricks were placed underwater for 28 days, some of the samples were positioned in the oven from temperatures ranging from 90 °C to 110 °C for two hours to allow the plastics melt and form structure voids. The observation from the experiments states that the thermal conductivity of the bricks reduced, and compressive strength increased sufficiently. Hence, the addition of plastic waste in bricks to produce eco-friendly bricks with acceptable compressive strength.

3.1.6. Plastic waste reinforcement

The principal disadvantage of steel fibers as concrete reinforcement is that it is prone to corrosion. There is the probability that seawater or saltwater can impact steel members if concrete surfaces are left unprotected. Foti [66] investigated the use of plastic reinforcement by conducting a series of pilot tests on concrete beam samples utilizing Polyethylene terephthalate (PET) waste and Carbon Fiber Reinforced Polymer (CFRP). The plastic waste in the form of PET and CFRP was made into strips and bars and arranged in a continuous position in the same manner steel reinforcement is arranged in a reinforced concrete structure. The results from the tests report that they both limit cracks and eliminate/ reduce corrosion in reinforced concrete structures. However, the CFRP samples exhibited improved behaviour. Similarly, plastic waste converted into Plastic fiber (PF) is utilized to reinforce structures by substituting it with traditional steel fibers enhanced structural stability and durability of the concrete element [19]. Therefore, the inclusion of plastic fiber seems to produce functional reinforcement in terms of strength, reduction of corrosion, and durability in concrete.

3.1.7. Plastic cement

High-density polyethylene wastes are combined with Portland cement to examine the prospect of plastic cement production and to test the impact of substituting sand with fine polyethylene wastes with the various proportion of product characteristics [34]. The tests were conducted utilizing the waste of polyethylene packages as a short reinforcement structure, including bottle and food crates in the span of 10% to 80% by volume as reinforcement with compact structure. The outcome of the test indicated that by using 60% and 40%, respectively, plastic cement can be made from polyethylene waste and Portland cement. Furthermore, the density of plastic cement reduced, there was an increase in ductility, and workability improved; hence, lightweight materials were produced [44]. [67] Formulated a cement that was made from recycled plastic waste. The recycled plastic bounded concretes (RPBCs) containing zero asphalt binder, and zero Portland cement was studied to evaluate the Mechanical properties of the RPBCs concerning Portland Cement Concretes (PCCs) and Asphalt Concretes (ACs). The compressive strength of some RPBCs is like ACs; however, some are like strong PCCs. The RPBCs samples showed significant moisture resistance compared to ACs. It observed that RPBCs shows improved stiffens and flexural strength, about three times compared to that of PCC and five times ACs. The RPBC's crack healing performance is 92 percent compared to 9 percent, according to AC's. The RPBCs establish much reduced thermal sensitivity and improved moisture resistance than the ACs. [68] examines the dry density characteristics of concrete utilizing plastic waste and polymer fiber to substitute cement. Tests conducted on concrete samples for compressive, tensile, and flexural strength, and lastly, a comparison reported on the concrete quality that utilized plastic waste and polymer fiber against the concrete with plastic waste. A detailed experimental study was conducted using plastic waste in concrete with a proportion range of 10 %, 20%, and 30% and polymer fiber proportion range of 2%, 4%, and 6%, respectively. The finding exhibited a lower density pattern in altered concrete in the polymer [67]. The utilization of plastic waste to replace cement decreases concrete compressive and flexural strength. This is possible as a result of fibers bridging activity that stores more energy and halts the concrete from sudden failure. However, low mechanical properties of the modified concrete that utilized plastic wastes and polymer fibers were documented [68]. Consequently, the production of plastic cement in concrete produces an effective, lightweight material with sufficient mechanical and strength properties.

Figure 3. Sea Turtle photographed eating Plastic Waste in the ocean. (Southey, 2019; Photo: Getty Images/Kwangmoozaa).

4. Conclusion

The use of plastic waste for construction applications in the engineering industry holds a high capability of decreasing worldwide ecological contamination and environmental pollution. Researchers generally are interested in discovering materials that can cut the budget of construction and increase the strength of engineering infrastructure. Plastic aggregates, cementitious materials, soil stabilization additives, and other construction resources obtained from plastic waste was examined. The examination demonstrated that the engineering properties of the construction materials improved. This creates an opportunity for recycling, reuse, and reprocessing of plastic waste materials. Given the review, the accompanying conclusions were drawn: The behaviour of plastic waste in construction materials is innately unique concerning the synthetic manufactured materials. Nevertheless, construction materials made from plastic waste can provide comparable or better outcomes than traditional construction materials. The use of plastic waste will gradually empower the reduction in the utilization of conventional construction materials, subsequently diminishing the carbon footprint associated with the production of such materials.

Several, researchers have used different compositions of plastic waste unaided, or as a combination with other materials; however, the use of plastic as a nanostructured material is not reported.

The non-biodegradable properties of plastic waste are able to assist the long-term workings of construction materials. It is appropriate for improving concrete mixes and soils of multiple classes. It is also suitable for different engineering design requirements, consequently upgrading the scope of concrete, aggregates, and stabilizers that can accomplish great products. However, these long-term effects of plastic waste in construction materials and its environmental impact after its service life and financial implications are in limited in the literature. There ought to be more examinations on the viability of plastic waste that produce high strength and load-bearing capacity in different types of structures to institute their appropriateness as good engineering construction materials. Plastic waste requires extensive research in their usage as construction materials because of the sheer volume of waste generated each year. Thus, more examination of the less investigated aspects of plastic waste and an introduction of engineering standards to evaluate its properties can open a broader use of waste in the construction industry.

Research should start now rather than later, so the issue of the control of plastic wastes are regulated by reasonable proportions. These proportions need to remain within applicable ranges for reuse as construction materials. Recycling of plastic waste in the manufacture of compressed earth bricks can give less expensive options to conventional blocks [34]. Plastic compressed earth bricks give strength characteristics that are similar to or higher than traditional bricks. Still, the research is inadequate as the investigation into the durability characteristics of such earth bricks has not been evaluated. Some other reuse applications of plastic waste areas strength modifiers in asphalt mix, and as part of flexible/ asphalt pavements, wood, plasters, plastic lumber (timber), in concrete production, plastic cement, sand, energy resource, soil backfill material and so on needs to be explored.

The utilization of plastic waste is anticipated to reduce the cost of pavement construction. This will improve the economic and environmental significance in underdeveloped and developing countries. In future research, construction applications with emphasis on economic valuation can be accomplished for pavement construction as well as for other engineering applications. The plastic waste construction materials should be examined in more detail under a wide variety of conditions to promote the acknowledgement of its use in industrialized civil engineering applications. The overall research activities should not exclusively be constrained to papers. However, pilot-scale undertakings should be organized and execution assessed to guarantee industry sustainability.

Acknowledgement: The management of Covenant University is appreciated for the platform provided.

References

[1] Geyer R, Jambeck J R and Law K L 2017 Production, use, and the fate of all plastics ever made *Science advances*, *3*(7), e1700782
[2] Hassan S and ul Haq I 2019 Pervasive Pollution Problems Caused by Plastics and its Degradation *International Journal of Online and Biomedical Engineering (IJOE)*, **15**(10), 29-39
[3] Badejo A A, Adekunle A A, Adekoya O O, Ndambuki J M, Kupolati K W, Bada B S and Omole DO 2017 Plastic waste as strength modifiers in asphalt for a sustainable environment *African Journal of Science, Technology, Innovation and Development*, *9*(2), 173-177
[4] Scalenghe R 2018 Resource or waste? A perspective of plastics degradation in soil with a focus on end-of-life options *Heliyon*, *4*(12), e00941
[5] Babayemi J O, Ogundiran M B, Weber R and Osibanjo O 2018 Initial Inventory of Plastics Imports in Nigeria as a Basis for More Sustainable Management Policies *Journal of Health and Pollution*, *8*(18), 180601
[6] Duru R U, Ikpeama E E and Ibekwe J A 2019 Challenges and prospects of plastic waste Management in Nigeria *Waste Disposal and Sustainable Energy*, 1-10
[7] Akinwumi I I, Booth C A, Ojuri O O, Ogbiye A S and Coker A O (2018) Containment of Pollution from Urban Waste Disposal Sites *Urban Pollution: Science and Management*, 223-234
[8] Olukanni D O, Adeleke J O, and Aremu D D 2016 A review of local factors affecting solid waste collection in Nigeria ("A review of local factors affecting solid waste collection in Nigeria (pp339-356)
[9] Pandi G P, Raghav S and Selvam D T 2017 Utilization of Plastic Waste in Construction of Roads *International Journal of Engineering Science*,**7**(03),5804-5806
[10] Singh T S 2016 Disposal of Plastic Waste and Stone Dust: A Sustainable Development Approach in *Environmental Science and Sustainable Development: International Conference on Environmental Science and Sustainable Development* 2015 (pp 26-31)
[11] Majeed A Z, Kurian T, Davis B, Alex S T, Fernandez K S and Mathew A V 2019 Partial Replacement of Aggregates with Granulated Waste Plastic in Solid Concrete Blocks—An Intensive Study in *Green Buildings and Sustainable Engineering* (pp 405-415) Springer, Singapore
[12] Duggal P, Shisodia A S, Havelia S and Jolly K 2020 Use of Waste Plastic in Wearing Course of Flexible Pavement in *Advances in Structural Engineering and Rehabilitation* pp177-187 Springer, Singapore
[13] Olukanni D O, Pius-Imue F B and Joseph S O 2020 Public Perception of Solid Waste Management Practices in Nigeria: Ogun State Experience *Recycling*, *5*(2)
[14] Olukanni D O and Oresanya O O 2018 Progression in Waste Management Processes in Lagos State, Nigeria in *International Journal of Engineering Research in Africa*, **35**, 11-23
[15] Olukanni D O, Adebayo R A and Tenebe I T 2014 Assessment of urban drainage and sanitation challenges in Nigeria *International Journal of Emerging Technology and Advanced Engineering*, *4*(12), 100-105
[16] Vethaak A D and Leslie H A 2016 Plastic debris is a human health issue *Environmental Science Technology*, **50** (13), 6825-6826
[17] Mrowiec B 2018 Plastics in the circular economy (CE) *Environmental Protection and Natural Resources;The Journal of Institute of Environmental Protection-National Research Institute*,

29(4), 16-19
[18] Adekomaya O and Ojo K 2016 Adaptation of plastic waste to energy development in Lagos: An Overview assessment *Nigerian Journal of Technology*, ***35***(4), 778-784
[19] Kamaruddin M A, Abdullah M M A, Zawawi M H and Zainol M A 2017 Potential use of plastic waste as construction materials: Recent progress and prospect in *Materials Science and Engineering Conference Series*, **267**(1), 012011
[20] Crawford RJ 2013 Plastics Engineering, Elsevier Science, ISBN 0080952100, 9780080952109, (pp1-364)
[21] Al-Sherrawi M H, Edaan I M, Al-Rumaithi A, Sotnik S, and Lyashenko V 2018 Features of plastics in modern construction use *International Journal of Civil Engineering and Technology*, *9*(4), 975-984
[22] Verma R, Vinoda K S, Papireddy M and Gowda A N S 2016 Toxic pollutants from plastic waste- a review *Procedia Environmental Sciences*, ***35***, 701-708
[23] Ren H, Zhou W, Guo Y, Huang L, Liu Y, Yu Y and Ma T 2020 A GIS-based green supply chain model for assessing the effects of carbon price uncertainty on plastic recycling. *International Journal of Production Research*, ***58***(6), 1705-1723
[24] Li W C, Tse H F, & Fok L 2016 Plastic waste in the marine environment: A review of sources, occurrence, and effects *Science of the Total Environment*, **566**, 333-349
[25] Parker L 2019 The world's plastic pollution explained, Retrieved October 18, 2019, from https://wwwnationalgeographiccom/environment/habitats/plastic-pollution/
[26] Zheng J and Suh S 2019 Strategies to reduce the global carbon footprint of plastics *Nature Climate Change*, **9**(5), 374
[27] Bhatnagar A, Kumar G, Pal G S, Singh P P and Gupta V 2019 To Enhance the Strength of Bituminous Roads Using Waste Plastic *International Research Journal of Engineering and Technology,* **6**(05), 4958-4961
[28] Singh M, Khan S, Sharma S, Tiwari H and Dohare D 2019 Improvising Pavement Strength using Plastic Waste *International Research Journal of Engineering and Technology,* **6**(4), 4628-4631
[29] Rajput P S and Yadav R K 2016 Use of plastic waste in bituminous road construction *IJSTE-International Journal of Science Technology and Engineering*, *2*(10), 509-513
[30] Tiwari A V and Rao,Y R M 2017 Study of Plastic Waste Bituminous Concrete Using Dry Process of Mixing for Road Construction *The Asian Review of Civil Engineering,***6**(2),1-6
[31] Sulyman M, Haponiuk J and Formela K 2016 Utilization of recycled polyethylene Terephthalate (PET) in engineering materials: a review *International Journal of Environmental Science and Development*, 7(02),100 -108
[32] Kader S A A, Jaya R P, Yaacob H, Hainin M R, Hassan N A, Ibrahim M H W and Mohamed A A 2017 Stability and Volumetric Properties of Asphalt Mixture Containing Waste Plastic In *MATEC Web of Conferences* **103**, p 09002 EDP Sciences
[33] Singh N, Hui D, Singh R, Ahuja I P S, Feo L and Fraternali F 2017 Recycling of plastic solid waste: A state of art review and future applications *Composites Part B: Engineering, 115*, 409- 422
[34] Almeshal I, Tayeh B A, Alyousef R, Alabduljabbar H, Mohamed A M and Alaskar A 2020 Use.of recycled plastic as fine aggregate in cementitious composites: A review *Construction and Building Materials*, ***253***, 119146
[35] Bahij S, Omary S, Feugeas F and Faqiri A 2020 Fresh and hardened properties of concrete containing different forms of plastic waste–A review *Waste Management*, ***113***, 157-175
[36] Jibrael M A and Peter F 2016 Strength and behavior of concrete contains waste plastic *Journal of Ecosystem & Ecography*, **6**(2), 2-5
[37] Bahadori-Jahromi A 2019 Sustainable development of medium strength concrete using polypropylene as aggregate replacement *Civil and Environmental Engineering*, ***13***(9), https://publicationswasetorg/10010763/pdf
[38] Gu L and Ozbakkaloglu T 2016 Use of recycled plastics in concrete: A critical review *Waste Management*, **51**, 19-42
[39] Skominas R, Zvinakevičius L, Gurskis V and Šadzevičius R 2017 Evaluation of suitability to use plastic waste in concrete production In *International scientific conference Rural Development 2017* (pp 428- 432)
[40] Thorneycroft J, Orr J, Savoikar P and Ball R J 2018 Performance of structural concrete with

recycled plastic waste as a partial replacement for sand *Construction and Building Materials*, **161**, 63-69

[41] Orr J 2019 India looks at using plastic waste as a partial replacement for sand in structural concrete In *Proceedings of the Institution of Civil Engineers-Civil Engineering*, **1729**(01) 1, 11-11, Thomas Telford Ltd

[42] Al-Tayeb M M 2017 Mechanical Propagates of Concrete with Partial Replacements of Sand by Plastic Waste of Vehicles under Impact Load *International Journal of Emerging Research in Management and Technology*, **6**(05), 19-25

[43] Kanwar V S and Shukla S K 2020 Sustainable Civil Engineering Practice, **72** ISBN: 978-981-15-3676-2

[44] Jassim A K 2017 Recycling of polyethylene waste to produce plastic cement *Procedia \ Manufacturing*, **8**, 635-642

[45] Islam M J, Meherier M S and Islam A R 2016 Effects of waste PET as coarse aggregate on the fresh and harden properties of concrete *Construction and Building materials*, **125**, 946-951

[46] Weimer N F 1987 Use of polyvinyl chloride liners for large irrigation canals in Alberta *Canadian Geotechnical Journal*, **24**(2), 252-259

[47] Ohemeng E A and Ekolu S O 2019 Strength prediction model for cement mortar made with waste LDPE plastic as fine aggregate *Journal of Sustainable Cement-Based Materials*, **8**(4), 228-243

[48] Chittoori B, Puppala A J, Reddy R and Marshall D 2012 Sustainable reutilization of excavated trench material In *GeoCongress 2012: State of the Art and Practice in Geotechnic Engineering* (pp 4280-4289)

[49] Rout R K, Ruttanapormakul P, Valluru S and Puppala A J 2012 Resilient moduli behavior of lime-cement treated subgrade soils in *GeoCongress 2012: State of the Art and Practice in Geotechnical Engineering* (pp 1428-1437)

[50] Farah R E and Nalbantoglu Z 2019 Performance of plastic waste for soil improvement *SN Applied Sciences*, **1**(11), 1340

[51] Singh K and Mittal A 2019 Soil Stabilisation Using Plastic Waste In *Recycled Waste Materials* (pp 91-96) Springer, Singapore

[52] Zhao J J, Lee M L, Lim S K and Tanaka Y 2015 Unconfined compressive strength of PET waste-mixed residual soils *Geomechanics and Engineering*, **8**(01), 53-66

[53] Kumar A and Vageesh HP 2017 Effect of Discarded Plastic Waste as Stabilizer on Engineering Properties of Cohesive Soil *International Journal of Engineering Technology Science and Research*, **4**(12), 779-786

[54] Peddaiah S, Burman A, and Sreedeep S 2018 Experimental Study on Effect of Waste Plastic Bottle Strips in Soil Improvement *Geotechnical and Geological Engineering*, **36**(5), 2907-2920

[55] Salim N M, AL-Soudany K Y and Ahmed A A 2018 The Impact of Using Recycled Plastic Fibers on the Geotechnical Properties of Soft Iraqi Soils in *IOP Conference Series:Materials Science and Engineering* 433 (01), 012017

[56] Kumar P, Das V, Gupta A and Ameta N K 2016 Stabilization of dune sand mixed with plastic (LDPE) waste strips for design of flexible pavement in construction of roads *Am J Eng Res (AJER)*, **5**(11), 315-320

[57] Onyelowe KC, Bui Van D, Ubachukwu O, Ezugwu C, Salahudeen B, Manh Nguyen Van M, Ikeagwuani C, Amhadi, T, Sosa F, Wu W, Ta Duc T, Eberemu A, Pham Duc T, Barah O, Ikpa C, Orji F, Alaneme G, Amanamba E, Ugwuanyi H, Sai V, Kadurumba C, Subburaj S and Ugorji B 2019 Recycling and reuse of solid wastes; a hub for ecofriendly, ecoefficient and sustainable soil, concrete, wastewater and pavement reengineering *International Journal of Low-Carbon Technologies, 14(03), 440-451* doi: https://doiorg/101093/ijlct/ctz028

[58] Hafez M, Mousa R, Awed A and El-Badawy S 2018 Soil Reinforcement Using Recycled Plastic Waste for Sustainable Pavements In *International Congress and Exhibition" Sustainable Civi Infrastructures: Innovative Infrastructure Geotechnology"* (pp 7-20) Springer, Cham

[59] Shah A and Modha H 2020 Improving the Soil Subgrade with Plastic Waste Reinforcement—An Experimental Study In *Advances in Sustainable Construction Materials*

(and Geotechnical Engineering (pp153-161) Springer, Singapore
[60] Dave T N and Thaker T P 2017 Reuse of plastic waste in foundation soil reinforcement application *Proc 19th ICSMGE*, 3369-3372
[61] Akinwumi I I, Domo-Spiff A H and Salami A 2019 Marine plastic pollution and affordable housing challenge: Shredded waste plastic stabilized soil for producing compressed earth bricks *Case Studies in Construction Materials*, *11*, https://doiorg/101016/jcscm2019e00241
[62] Jalaluddin M 2017 Use of Plastic Waste in Civil Constructions and Innovative Decorative Material (Eco-Friendly) *MOJ Civil Engineering,* **3**(05), 01-10
[63] Agarwal S and Gupta R K 2017 Plastics in buildings and construction In *Applied Plastics Engineering Handbook* (pp 635-649) William Andrew Publishing
[64] Ghuge J, Surale S, Patil B M and Bhutekar S B 2019 Utilization of Waste Plastic in Manufacturing of Paver Blocks *System*, *6*(04), 1967-1970
[65] Mondal M K, Bose B P and Bansal P 2019 Recycling waste thermoplastic for energy efficient construction materials: An experimental investigation *Journal of environmental management*, ***240***, 119-125
[66] Foti D 2016 Innovative techniques for concrete reinforcement with polymers *Construction and Building Materials*, ***112***, 202-209
[67] Dalhat M A and Wahhab H A A 2016 Cement-less and asphalt-less concrete bounded by recycled plastic *Construction and Building Materials*, ***119***, 206-214
[68] Ahmad M, Khalid L S, Manan A, Umar M and Ibrahim S 2019 The Effect of Cement Replacement with Plastic Bag Wastes (PE) and Polymer Fibre on the Concrete Performance *AJIRSET,* **4**(03), 01-06
[69] Ogundairo TO 2019 Photograph of Plastic Waste in a Drain at Olowu Street Ikeja, Lagos State
[70] Southey F 2019 Plastic waste 'state of play' reveals just 9 countries have specific prevention targets (Getty Images/Kwangmoozaa), Retrieved from October 10, 2019, https://wwwfoodnavigatorcom/Article/2019/06/06/Plastic-waste-state-of-play-reveals-just-9-countries have-specific-prevention-targets

Integration of Magnetic Data and Geo-Spatial Analysis for Characterising Aquiferious Yields in New Estate, Covenant University, Ota, Ogun State

M Omeje[1], O O Adewoyin[1], E S Joel[1], P C Emenike[2], U A Omeje[3], M O Johnson[1]
[1]Department of Physics, College of Science and Technology, Covenant University, Ota, Ogun State, Nigeria
[2]Department of Civil Engineering, College of Engineering, Covenant University, Ota, Ogun State, Nigeria
[3]Department of Public and Community Health Care, College of Medicine, University of Lagos, Idiaraba, Lagos State

Abstract This study was conducted to understand the possible causes of abortive boreholes along the New Estate in Covenant University. The conceptual model predicted the two groundwater characteristics of weathered and the unweathered zones. The V-shape curve revealed two groundwater anomalies with high and low curves. The peak anomaly indicates the higher resistivity, presumed to be the geological structures of unsaturated features in the area. The sink V-shape curve indicate an anomaly with groundwater features, presenting low resistivity values attributed to be water bearing formation that will be viable for groundwater potential. The Iso-resistivity contour map identified possible groundwater features. The first Iso-resistivity column is a region of aquitard with a semipervious in nature that could transmit water in a slower rate. The second groundwater feature in column 2 is a perch aquifer that could enhance the main aquifer due to interconnectivity and pore pressure of downward-upward migration in confined hydrogeological aquifer setting. Column 3 depicts a complete impermeable feature of aquifuge or aquiclude of subsurface structure with rich in clay sediments or stratum of sediments found at the depth range of 30 m to 270 m. The forth groundwater column indicates a promising water bearing formation, probably sandy in nature with a depth range of 210 to 300 m and thickness of about 90 m. The spatial analysis identified the two locations along the survey areas to be suitable for productive boreholes as well as the trending of the groundwater in the area to be Northwestern Southeast direction. Consequently, the model could be a guide to organize site data, understand site processes and screen for applicability of drilling. These multiple methods of geophysical, spatial analysis and conceptual model could be used a tool for mapping hydrocarbon bearing formation.
Keywords: Groundwater, GIS, Electromagnetic Method, Conceptual Model

1. Introduction

An Aquifer is a body of permeable rock that can contain and transmit groundwater. It is a geologic formation that bequests significant quantities of water. Groundwater is a source of fresh water, and it is the primary storage reservoir of freshwater. Groundwater is not just water under the ground; it is water held in rocks by specific forces, replenished by nature according to the climate and the local geology, and consequently variable in both amount and quality (Molle et al., 2017). Aquifers are essential to all life and beings (Omole et al., 2017). Aquifer behavior is related to many factors in the hydrological system. One of the essential ideas in hydrology, and in any water resource management, is the hydrologic cycle (also called the water cycle), this ideology has been stipulated on since ancient times (Maidment, 1993). The hydrological cycle gives a detailed description of water production and transportation in different ways from place to place (Rushton and Kruseman, 2004). Groundwater, a renewable and finite natural resource, vital for human life, for social and economic development and a valuable component of the ecosystem, is vulnerable to natural and human impacts (Vrba 2002). In Nigeria today, the unmanaged digging of wells and its utilization is also a source of hydrological stress (Ufoegbuna, 2016). The lack of proper laws on how people in residential areas dig wells, and even as they use these wells can cause severe damage to the aquifer. Lack of knowledge of how improper management of groundwater resources would continue to lead people to dig up wells in any available area continually. This study is aimed at using integrated geophysical, conceptual models and geospatial analysis to identifying the subsurface features that could cause the abortive boreholes along New Estate in Covenant University

2.0 Geology of the Study Area

Ogun State with other south-western Nigerian states including Oyo and Lagos States lies in the eastern Dahomey Basin. Its geology is composed of sedimentary and basement complex rocks that are Late Cretaceous to Early Tertiary (Olabode, 2006). Stratigraphically, the sedimentary rock of Ogun State consists of the Abeokuta group, Imo group, Ewekoro, Oshosun, Ilaro and Benin formations (Olabode, 2006; Omole, 2006). The Geology of the study area is shown in Figure 1 and the Google Earth map indicating the location is shown in Figure 2.

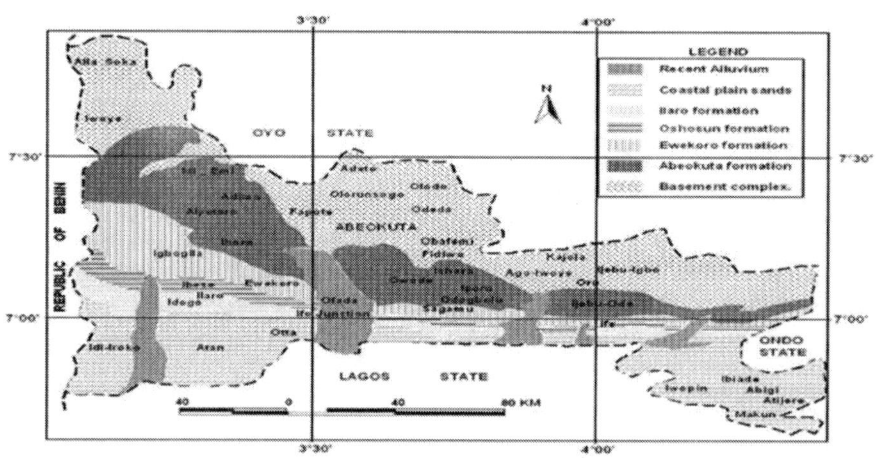

Figure 1: Geological Map of Ogun State (Olowofela et al., 2012)

3.0 Materials and Methods

TC-300 Series is a Hunan Puqi Geologic Exploration Equipment (Model No: PQWT-TC300) is an electromagnetic method used to study the subsurface geological features geological ore / mass bodies, groundwater and its abnormalities. These variations are useful in interpreting and solving the subsurface characteristics of the geological, geoengineering, environmental and disaster problems. The basic principle of the instrument is measuring the existing natural electric field of subsurface water / minerals /geological structures / ore bodies. The natural electric field geophysical instrument is one of the geophysical instruments, working with the source of natural electric field, measuring the subsurface rock / ore / groundwater, natural electric fields and variations at different frequencies and different depths. It detects the abnormality changes in the subsurface with a coupled 3 methods integrated and developed, consisting more and more data filters for analyzing data and filtering errors etc. It is easy operation, high quality touch screen and highly standard. The instrument profile map gives accurate mean depth, quantity, quality, shape of subsurface features and the geological structure. The instrument gives 100% results subsurface investigation and detection of bed rocks etc. The measurement is done by using tape to measure 10 meters distance and mark the starting point to be 0 meter. The configuration of M N two electrode bar equidistance is 10 meters, and subsequently both M N moved 1meter after measuring the first point placed at the position of 1.

3.1 Data Acquisition Using Magnetic Methods

Figure 2 presents the sampling points in the study area where the magnetic measurements and electromagnetic measurements were taken using Hand-held GEM Magnetometer. At each point, four different magnetic anomalous values in nano-tesla (nT) were measured and the average was taken to ensure accuracy. The measurements were carried out at an interval of 50 m from each other which was maintaianed all through the forthy nine (49) station points mapped. All the data were treated to ensure accuracy in georeferencing, capturing the subsurface geological structures that could be of interest for groundwater features according to Emenike et al., 2020 and Mama et al., 2020.

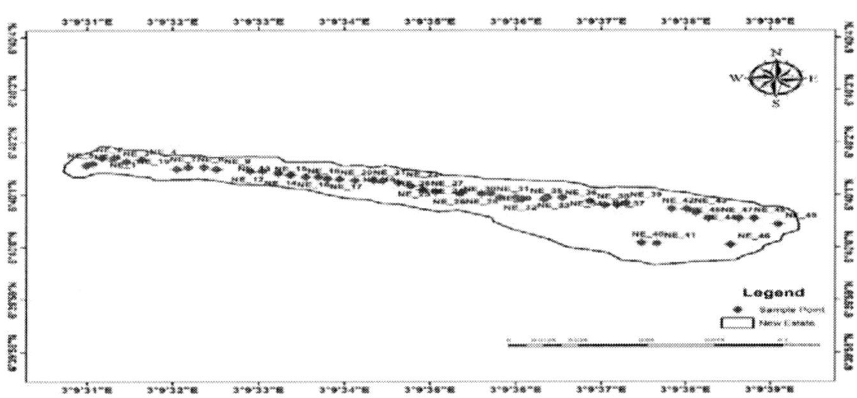

Figure 2: The Base Map of the Study Area Showing the Points Where the Magnetic Measurements were Carried Out

3.2. GIS Spatial Analysis of Magnetic Data from the Study Area

The magnetic data acquired from the study area were used to map the possible potential zones for water bearing formation. The acquired values were in nano-tesla (nT) with stations 10 and 48 identified to be the locations of water bearing zones as shown in Figure 6. The georeferencing and groundwater attributes pointed the area with magnetic values of 34331 to 34507 as the possible weathered or fractured zone that could trap large scale of groundwater. This permeable zones are within range of stations 10 – 11 and 48 – 49, respectively. The fissures and weathered zone picked from the magnetic values was used to justify the promising locations that form the leaky aquifers in New Estate.

3.3 Conceptual Model for Groundwater Flow in the Study Area

This conceptual model was developed to review the relationship between the hydraulic head, hydraulic gradient, hydraulic conductivity of productive and the unproductive boreholes in New Estate using two boreholes as a case study. This conceptual model is shown in Figure 3.

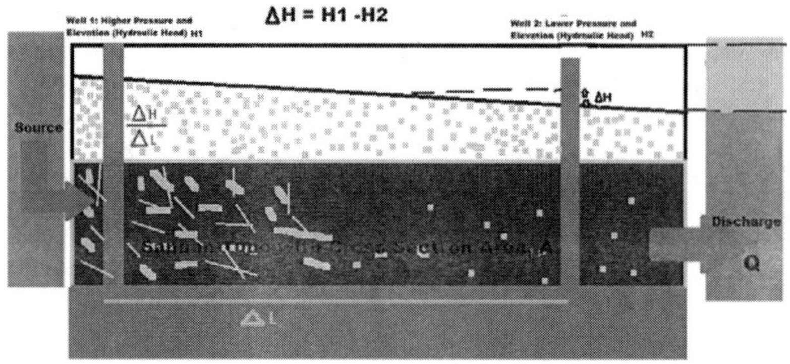

Figure 3: Conceptual Model for Prediction of Groundwater Flow in New Estate, Canaanland

3.4 GIS Spatial Analysis of Magnetic Data from the Study Area

The magnetic data acquired from the study area were used to map the possible potential zones for water bearing formation. The acquired values were in nano-tesla (nT) with stations 10 and 48 identified to be the locations of water bearing zones as shown in Figure 6. The georeferencing and groundwater attributes pointed the area with magnetic values of 34331 to 34507 as the possible weathered or fractured zone that could trap large scale of groundwater. This permeable zones are within range of stations 10 – 11 and 48 – 49, respectively. The fissures and weathered zone picked from the magnetic values was used to justify the promising locations that form the leaky aquifers in New Estate.

4.0 Results and Discussion

4.1 Interpretation of the Groundwater Conceptual Model of New Estate

The conceptual model developed in Figure 4 is an interpreted version of Figure 3 using selected values of the electromagnetic method acquired in the area. This model is used to infer the groundwater condition in New Estate, Covenant University. The prediction has revealed that any borehole sited within the left hand site of the model will be productive due to the possible large grain sizes, large pore spaces, less frictional resistance of flow and as well as higher hydraulic conductivity and hydraulic gradient which is geologically termed as the thick overburden. This region will be a suitable aquiferous zone with good yield of groundwater since the layers are weathered, hence presumed to be the saturated zone or water bearing formation. On the right hand side of the model, siting a borehole may constitute low to dry aquifer due to the smaller grain sizes, smaller pore spaces, more frictional resistance of the groundwater flow, lower hydraulic conductivity, and is termed as the unsaturated zone. Borehole 1 will give good yield in the study area whereas boreholes 1, 2, 3, and 4 will be highly unproductive due to the unweathered formation that may constitute clay and silt stone. The weathered zone is possibly interbedding of sandy-clay formation

IOP Conf. Series: Materials Science and Engineering 1036 (2021) 012020 doi:10.1088/1757-899X/1036/1/012020

presumed to be perch aquifer was noticed at the depth of 150 m at the left hand side of the model. The main water bearing formation from the model ranged from 190 m to 300 m deep along the saturated/water bearing zone. This prediction may not be precise but a tool to help understand the site processes and screen for groundwater flow. The validation of this model will be the field measurements along the zones of where both successful and unsuccessful boreholes were drilled in the study area.

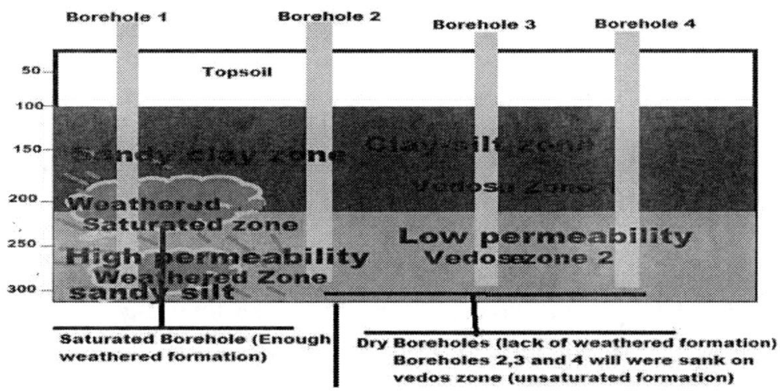

Figure 4: Conceptual Model for Prediction of Groundwater Flow in New Estate, Canaanland

4.2 The V-Shape Curve of the Measured Electromagnetic Methods for Groundwater Detection

The V-shape curve of the measured electromagnetic values were plotted in Figure 5 with several colour bands showing different frequency levels. The curve presented two different groundwater anomalies with high and low curves. The peak anomaly indicates the higher resistivity, presumed to be the geological structures of unsaturated features in the area. The sink V-shape curve indicate an anomaly with groundwater features, presenting low resistivity values pointing down. This sinking zone is termed the water bearing formation that will be viable for groundwater potential. At the same time, it could be a prone zone for groundwater contamination in the area.

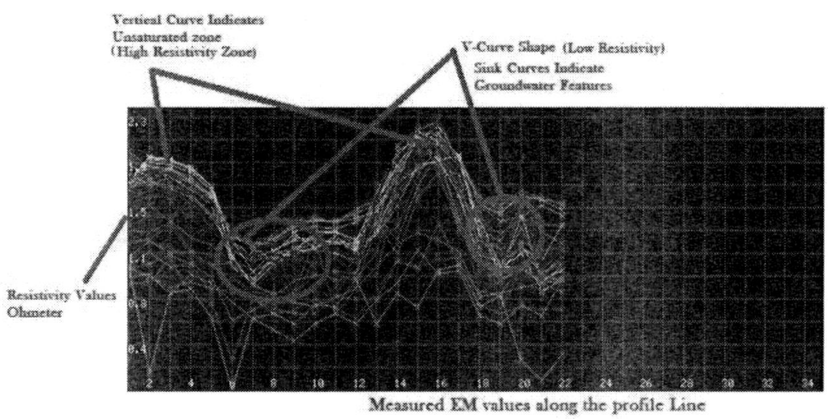

Figure 5: V-Curve Shape Profile Plot of Measured EM values and the Colour Band Frequencies in the Area

IOP Conf. Series: Materials Science and Engineering 1036 (2021) 012020 doi:10.1088/1757-899X/1036/1/012020

4.3 Iso-Contour Profile Map of the Measured Electromagnetic Method Values in New Estate

This Iso-resistivity profile map shows several variations in groundwater characteristics in the area. The region possesses 4 possible groundwater features. The first Iso-resistivity column which is labeled 1 is a region of aquitard that is semipervious in nature that could transmit water in a slower rate and possibly low hydraulic conductivity. The layers are compacted with probably clay or silt that retard groundwater flow; that is they act are groundwater barrier in Iso-resistivity column 1. The second groundwater feature in column 2 is the region of perch aquifer that could enhance the main aquifer due to interconnectivity and pore pressure of downward-upward migration experience in the aquifer hydrogeological setting. In the Iso-resistivity zone of column 3, it depicts a complete impermeable aquifuge or aquiclude of groundwater features with rich in clay sediments. This impermeable feature in column 3 may be attributed to the stratum of sediments found at the depth range of 30 m to 270 m. The forth groundwater column labeled column 4 indicates a promising water bearing formation, probably sandy in nature. This geological formation of groundwater potential found at this depth range between 210 and 300 m is inferred to be the aquiferous zone with a thickness of about 90 m. This layer seems to be made of denser geological materials that are completely weathered with fissures in alluvium aquifer which constitute high hydraulic conductivity.

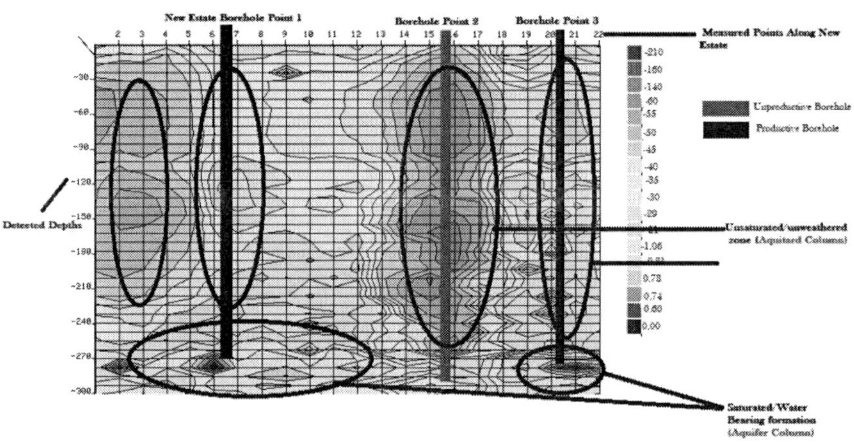

Figure 6: Iso-Resistivity Profile Map of the Study Area Showing 4 Different Groundwater Characteristics

4.4 GIS Spatial Analysis of Magnetic Data from the Study Area

The magnetic data acquired from the study area were used to map the possible potential zones for water bearing formation. The acquired values were in nano-tesla (nT) with stations 10 and 48 identified to be the locations of water bearing zones as shown in Figure 6. The georeferencing and groundwater attributes pointed the area with magnetic values of 34331 to 34507 as the possible weathered or fractured zone that could trap large scale of groundwater. This permeable zones are within range of stations 10 – 11 and 48 – 49, respectively. The fissures and weathered zone picked from the magnetic values was used to justify the promising locations that form the leaky aquifers in New Estate.

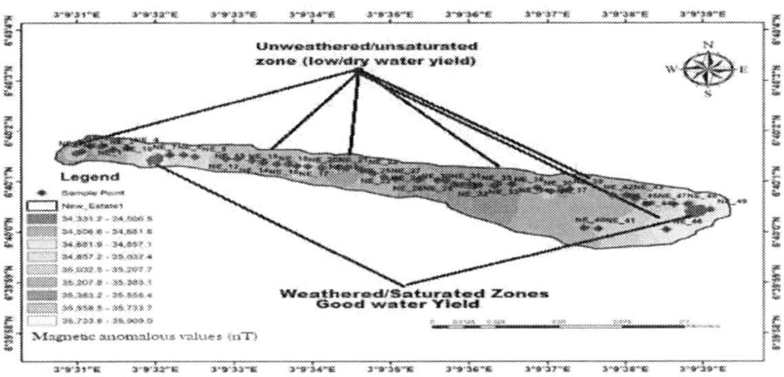

Figure 7: GIS Spatial Analysis of Magnetic Data Showing the Weathered and the Unweathered zones in the Study Area

4.5. Locating the Suitable Sites for Productive and Unproductive Boreholes along New Estate from Spatial Analysis Map

The successful siting of boreholes within the leaky aquifers was geologically inferred to be on dark green location of the spatial analysis map. This integrated map only identified two viable locations for successful boreholes, pointing to the extreme Northwestern part of the region and the other suitable site is trending towards the Eastern part of the region as shown in Figure 7. This spatial map revealed the trending of groundwater flow, with the source from NW trending towards SE direction.

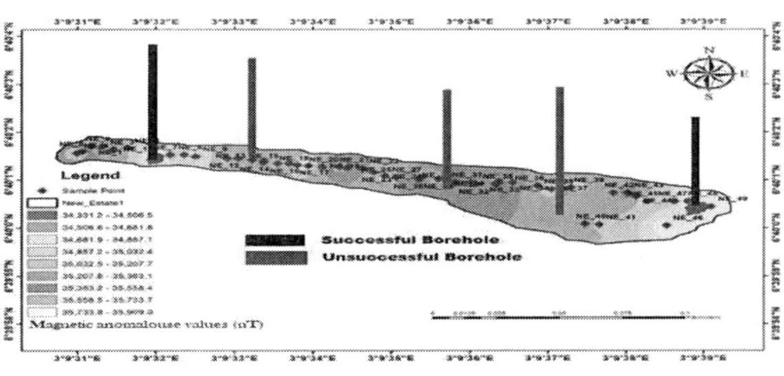

Figure 8: Spatial Analysis Map Showing the Successful and the Unsuccessful Zones for Borehole Siting or Drilling

5.0 Conclusion

This present study integrated conceptual model, geophysical methods and GIS spatial analysis to understand the subsurface structural formation that controls the groundwater yield at New Estate in Covenant University. The subsurface geology from the conceptual model identified the possible hydraulic head, hydraulic gradient and hydraulic conductivity. The V-shape curve from electromagnetic method revealed two groundwater potential points with low resistivity values at profile lines ranges between 6 – 8 and 18 – 20, respectively. The Iso-resistivity map of EM indicates the range of water bearing formation in the study area to be from 210 to 300 m with an aquifer thickness of about 90 m. The spatial analysis map identified measured points of 10 -11 and 48 -49, respectively as the viable zones for accessing productive boreholes in the area. The conceptual model could be a guide for organize site data, understand site processes and screen for applicability of drilling. These multiple methods of integrating groundwater Conceptual model, electromagnetic methods, magnetic methods and GIS spatial analysis for viable mapping of groundwater potential could serve as a tool for hydrocarbon mapping.

Acknowledgement
The authors would like to thank the management of Covenant University through Covenant University Center for Research Innovation and Discovery (CUCRID) for their 100% financial support

References

1. Molle F, Nassif M, Jaber B, Closas A and Baydoun S 2017 Groundwater governance in Lebanon: the case of Central Beqaa. A Policy White Paper
2. Olabode 2006 Siliciclastic slope deposits from the Cretaceous Abeokuta Group, Dahomey (Benin) Basin, southwestern Nigeria. *J. Afr. Earth Sci.* **46**: 187-200
3. Omole, D., Bamgbelu, O., Tenebe, I., Emenike, P., and Oniemayin, B. 2017. Analysis of Groundwater Quality in a Nigerian Community, *Journal of Water Resource and Hydraulic Engineering,* **6**, 2, PP. 22-26
4. Emenike PC, Jordan BN, Tenebe TI, Nnaji CC, Jarvis P 2020 Estimation of some trace metal pollutants in River Atuwara southwestern Nigeria and spatio-temporal human health risks assessment. *Chemosphere* **239**,124770 https://doi.org/10.1016/j.chemosphere.2019.124770
5. Mama C N, Nnaji CC, Emenike PC and Chibueze CV 2020 Potential environmental and human health risk of soil and roadside dust in a rapidly growing urban settlement, *International Journal of Environmental Science and Technology* https://doi.org/10.1007/s13762-020-02637-9
6. Ufoegbune GC, Atanley PA, Eruola AO, Makinde AA and Ojekunle ZO 2016 Geographical Information System (G.I.S.) Application for Planning and Improvement of Public Water Supply in Ota, Ogun State, Nigeria, *J. Appl. Sci. Environ. Manage.* **20** (2) 1105 – 1111

IOP Conf. Series: Materials Science and Engineering 1036 (2021) 012021 doi:10.1088/1757-899X/1036/1/012021

Building Collapse in Nigeria during recent years – Causes, effects and way forward

P O Awoyera, J Alfa, A Odetoyan* and I I Akinwumi

Department of Civil Engineering, Covenant University, Ota, Nigeria
*Corresponding Author: abimbola.odetoyan@covenantuniversity.edu.ng

Abstract. Building collapse in Nigeria in the last few decades is a growing concern for investors and the government alike. Many of the documented cases of building collapse in Nigeria are due to the use of defective or substandard building materials, no requisite technical knowledge, non-adherence to building codes and standards, the use of non-professionals and the high level of corruption which has ravaged every sphere of the construction industry including government and private parastatals. In addition to the established causes of the collapse of structures, empirical data from developed countries of the world has shown that many of the recorded cases in this climes are due to the fact that the current codes of practice do not make provisions for unexpected loads and an unexpected failure of a single member may lead to an all-round collapse of the entire structure. This study critically highlights the collapse cases of buildings in Nigeria and provides recommendations to halt this growing concern. It seeks to shed further light on these causes and the likelihood for the continued collapse of buildings in Nigeria if measures are not urgently and strategically put in place by all concerned stakeholders. It also seeks to reduce the impact of man-made phenomena on buildings through awareness and policy formulation by the government.

1. Introduction

Buildings are civil engineering structures, constructed in various forms, with different materials, and for specific uses, e.g. to provide shelter for lives and properties. A building is built with considerations for safety, serviceability and economy. This study provides critical reasoning and contributions regarding structural failures observed in Nigeria.

Building collapse is a failure in a structure due to its inability to serve the purpose for which it was built [1]. In 2013, collapse of Rana Plaza, an eighty storey commercial building in Savar, Dhaka, Bangladesh, resulted in over 1200 deaths and injury to an estimated 2500 other [2]. The 2014 Synagogue Church building collapse in Lagos, Nigeria with at least 116 deaths, the 2015 Mecca crane collapse in Mecca, Saudi Arabia claiming 184 lives and the Uyo Church collapse in Uyo, Nigeria causing 60 deaths and injuries to a lot more [3]. The 2019 Lagos School collapse in Lagos, Nigeria causing 20 deaths and over 60 injured. The 2015 collapse of the 19th century Tower, Dharahara building in Kathmandu, Nepal on May 25 caused by an earthquake resulted in 200 fatalities. These are just a few of the collapse cases around the world. These cases are more in developing cities as a result of urbanization and the need to provide shelter for the influx of people trooping into the cities every day. This has, in fact, led to an increase in 'building contractors' of zero training and doubtful competences. [4] researched the fatality rate in the forty-seven cases of building collapse from 2000 to 2010. He reported not less than 300 deaths and casualties in Nigeria's major cities alone with an enormous loss of investment. The increasing number of cases mandates the need for a wakeup call to all stakeholders in the industry.

Failure of a structure is not just a result of ultimate collapse but when any part of the structure or the whole structure becomes unfit for use for the intended loadings for which it was designed to carry. Failure of a structure then could be any of the three scenarios:

IOP Conf. Series: Materials Science and Engineering 1036 (2021) 012021 doi:10.1088/1757-899X/1036/1/012021

i. Serviceability limit state failure occurs when it becomes unserviceable by undergoing excessive deflection and cracking;
ii. Ultimate limit state failure happens when it fails by overturning, ultimate collapse, or wobbling of the columns. The structure is said to have failed in Ultimate limit state.
iii. Durability failure has to do with the weakening of the components beyond reparable limits.

The traditional materials used in previous centuries made buckling not a major structural problem and multi-storey structures were uncommon during these ages. There is a tendency to make sure important structures do not fail by using materials generously making many of the roman structures survived till date [5].

The standard of living has increased over the centuries bringing with it huge strides in technological advancements. Today, there are better procedures, more durable materials, better construction machinery and more advanced knowledge on construction. However, with all the advances in technology, there are still cases of collapse in countries of the world, though more prevalent in developing countries like Nigeria. Collapse in the developed countries is majorly due to natural causes or man-made causes like bombings like the world trade centre. However, collapse in developing countries are usually due to poor quality or substandard materials, poor supervision, non-adherence to standards and regulations, unqualified professional, overloading, no geotechnical/sub-soil investigation, poor construction procedure, illegal approval, wrong demolition process, lack of maintenance and numerous others [1, 8-12].

Table 1: Summary of collapsed buildings and fatalities between 2010 and 2019 in the 21st century in Nigeria.

References	Structure	Date	Location	Cause of Collapse	Fatality
[6]	Residential under construction	April 26, 2010	Isopakodowo street Cairo, Lagos.	Use of substandard building materials.	4
[7]	Residential under construction	June 2010	Oniru Estate, Victoria Island.	Substandard materials, haphazard works.	1
[7]	2-storey building under construction	June 2010	Nkwerre Street Garki, Abuja.	Non-compliance with building regulations	1
[8]	6 suspended floors for commercial purpose.	July 2010	Plot 702 Port-Harcourt crescent Garki Abuja	Substandard materials and unqualified professionals	11
[6]	A wall fence.	August 10, 2010	Aghaji crescent GRA. Enugu.	No proper drainage	1
[9]	Uncompleted 4-storey building	August 11, 2010	Ikole street, Abuja	Substandard materials & disregard for building regulations	23
[10]	4-storey building	September, 2010	28 Tinubu street VI Lagos.	Structural defect/overloading	3
[11]	Five-storey	June 2011	11 Aderibigbe Street, Maryland, Lagos.	No geotechnical investigation,	None
[7]	4-storey building	June 2011	Ndiagu Amechi Road Enugu	undersized reinforcement,	3

[7]	2-storey building	June 2011	Nyanya, Abuja	large span slab	4
[8]	2-storey building	June 29, 2011	Maraba (near Abuja)	No specific floor thickness on structural drawing	2
[7]	3-storey building	July 2011	Oloto Street, off Cemetery Road, Ebute Metta, Lagos	Non-adherence to building standards & regulations	10
[7]	3-storey building	August 2011	Orosanye Street Lagos.	Wrong Supervision	-
[8]	Two storey building.	January 28, 2012	Gwarinpa Estate Abuja	Structural defect and demolition operation	3
[7]	2-floor Commercial building.	Wednesday, June 13, 2012.	Apo Mechanic Village.	There was no building approval and no qualified on site.	14
[8]	101-year-old 3-storey commercial building.	July 2012	Hadeja Road by Ibrahim Taiwo Road Gombe	Building has passed its limit state.	1
[8]	2-storey building under construction	August 8, 2012	3, Ademola Awosike Road Kubwa Extension III, Abuja	Poor quality materials, poor workmanship, inadequate/weak foundation.	3
[8]	Building in use	November 20, 2012	Jakande estate Oke-Ake Afa, Isolo Lagos	Structural failure and occupants ignored the government's safety warning.	3
[12]	A twin four-storey duplex	November 3, 2013	Victoria Island, Lagos	Unknown	4
[8]	Abu Naimi school building.	September 2014	Bukuru Jos.	Structural defect and illegal conversion	10
[3]	Synagogue Church of All Nations (SCOAN)	September 12, 2014	Ikotun-Egbe area of Lagos state, Nigeria.	Structural failure	116
[8]	Synagogue Warehouse Church, Lagos.	September 12, 2014	Ikotun-Egbe area of Lagos state, Nigeria.	Demolition process	4
[8]	Liberty power bible church, Benin	September 30, 2014	Liberty power bible church, Benin	Structural defect and use of substandard material	-
[12]	4-storey building	March 13, 2015	6 Mogaji Street Idumota Lagos island	Unknown	1
[8]	3-storey building	July 15, 2015	Ebuta meta Lagos	Structural defects	-
[8]	3-storey building	October 21, 2015	Swamp street Odunfa Lagos island.	Structural defects	-

[13]	A five-story building under construction	March 9, 2016	Lekki, Lagos.	Addition to the approved number of floors.	34
[8]	Two storey building	March 19, 2016	Mile 12, Lagos	Structural defects	1
[8]	Residential building	April 2016.	Horizon 1, Lekki Garden, Ikate.	Structural defects	18
[8]	Commercial complex	May 13, 2016	Lafenwa Sapon Road, Itoku, Abeokuta.	Structural defects	2
[3]	The Reigners Bible Church.	December 10, 2016	Akwa Ibom State, Nigeria	Structural failure	100
[14]	4-storey building	July 25, 2017	3 Massey St, Lagos Island	Undisclosed	6
[14]	A storey building	July 8, 2017	Ulakwo junction, Owerri North LGA, Imo State	Undisclosed	3
[8]	Four storey building	August 18, 2017	Zulu Gambari Road, Ilorin	Undisclosed	-
[14]	4-storey building	July 17, 2018	Owelle Aja Layout, Obosi, Anambra.	Substandard materials with addition of two un-designed floors.	-
[14]	4-storey building	August 16, 2018	Jabi, FCT Abuja.	Substandard materials	2
[14]	3-storey building	October 18, 2018	Okpuno, otolo in Nnewi, Anambra.	Substandard materials.	-
[14]	3-storey building	October 2018	Ifite Awka, Anambra State	Substandard material	-
[8]	7-storey building	November 23, 2018	Woji road, GRA phase 2, Port Harcourt	Undisclosed	5
[8]	3-storey building	February 3, 2019	Lagos Island	Not reported	2
[8]	A three-storey building	March 13, 2019	Ita Faaji area of Lagos state, Nigeria	The change of use of the building from the intended purpose.	20
[8]	3-storey building under construction	March 15, 2019	Sogoye, Bode area of Ibadan	Concrete was not adequately cured during construction.	-

2. Research Methodology

To actualize the aim of this research data was collected using primary source of well-structured questionnaires distributed among the various professionals in the built industry as well as secondary sources from journal publications and conference proceedings, Nigerian dailies and technical reports from credible sources on the internet.

The questionnaire was designed using google form with the sole aim of reaching in time a wider range of the targeted professionals through the electronic mail. It captured the causes of collapse, roles of individuals, Natural and man-made conditions.

3. Results and Findings

From Figure 1, Man-made conditions are regarded as the major cause of building collapse over the decade. Whereas natural conditions contributes less to the cause of structural failure in most countries this present decade.

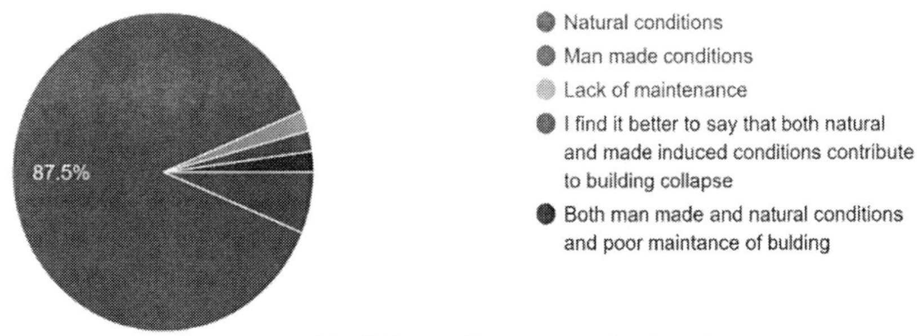

Figure 1: Major cause of building collapse over the decade

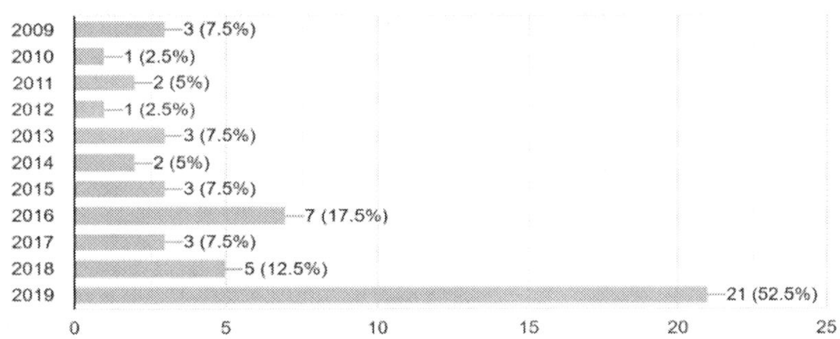

Figure 2: The most amount of building collapse per year.

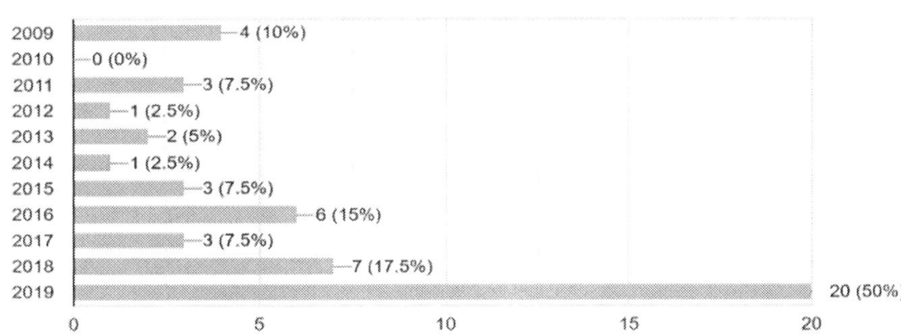

Figure 3: The fatality rate from building collapse per year.

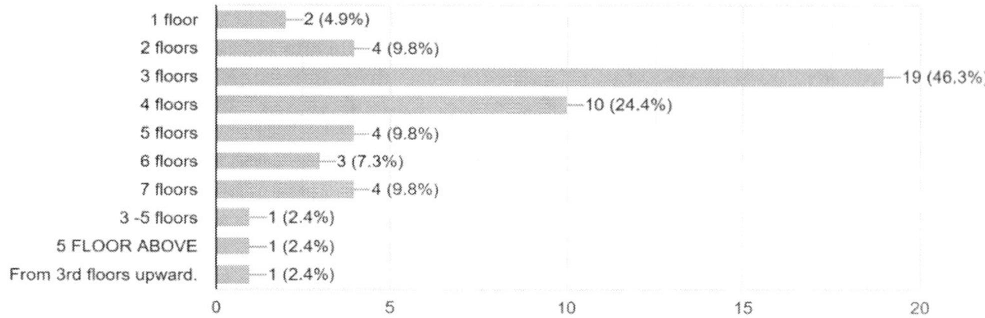

Figure 4: The height of mostly collapsed buildings.

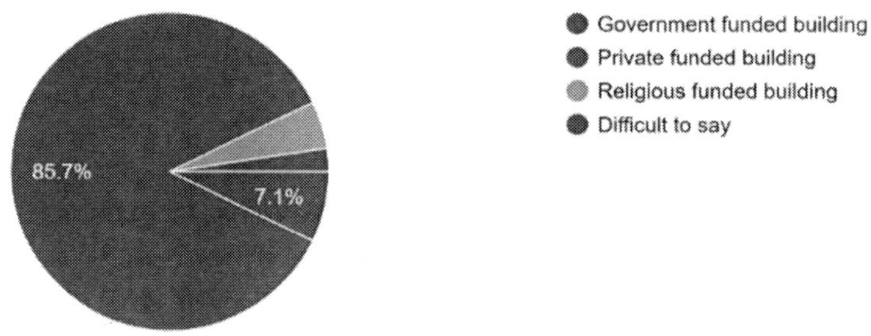

Figure 5: Funding of the collapsed buildings.

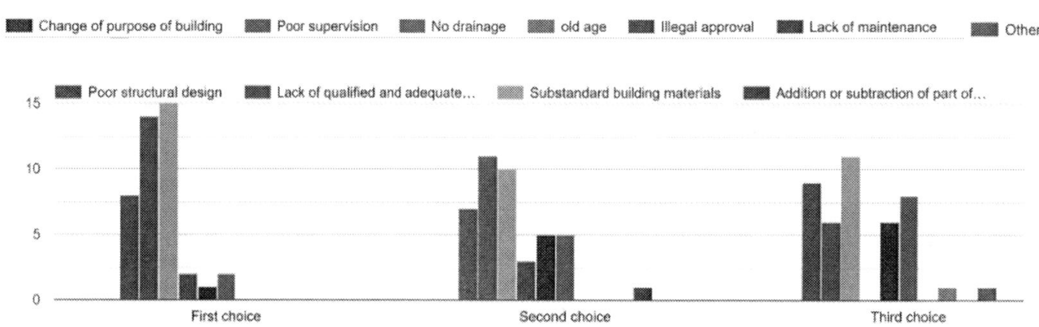

Figure 6: Major man-made causes of building collapse

In addition to these major causes of building collapse, ignorance and greed could also be considered as causes of many of these failures. Religious buildings may also be constructed out of spiritual expediency without factoring actual costs. This, in many cases, could lead to careless designs and compelled supervision out of service to God.

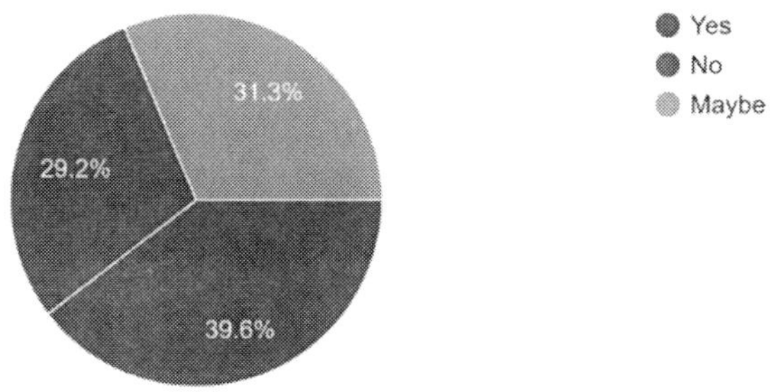

Figure 7: Award of contracts to professionals under registered bodies

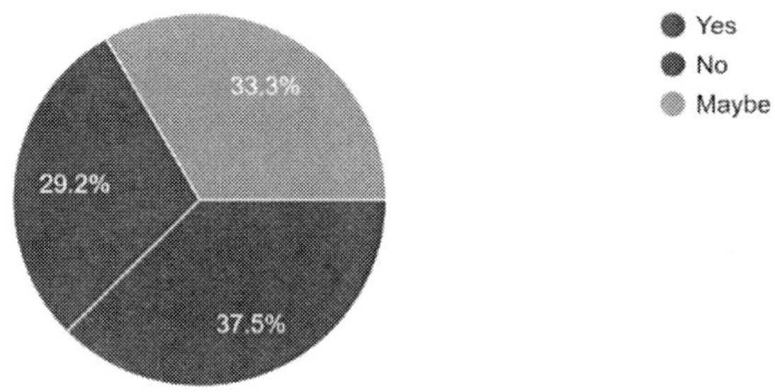

Figure 8: Adherence of professionals to building codes

Although professionals are to abide strictly to building codes, the building codes are also meant to be updated to meet up with the high standard of construction. Figure 9 shows that building codes are not updated regularly in many countries of the world and Nigeria doesn't even have a compiled code for design and practice.

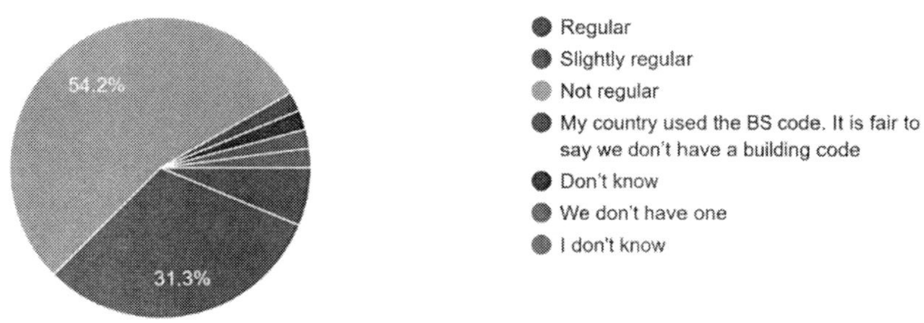

Figure 9: Updating of building codes in countries.

In as much the professionals contributes to the causes of building collapse, the over interference of client on a contract could lead to building collapse. In some cases the clients becomes the contractor while in other cases the contractor listens and obeys the clients instructions forfeiting safety to keep his/her job. In addition, the bribery of law enforcers to achieve construction could result in structural failures.

In providing solutions to the aforementioned causes of building collapse, the use of quality assurance and quality control personnel on a building construction site is highly recommended as shown in figure 10. The quality assurance and quality control personnel should serve as a check on the construction processes, materials, workers and report any fault or foreseen error.

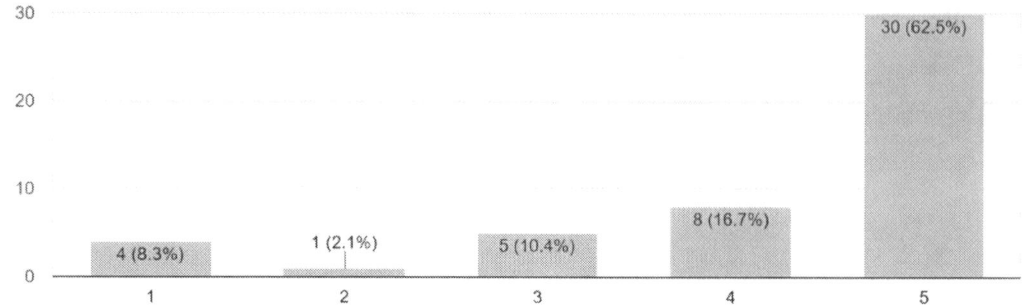

Figure 10: A scale on the importance of quality assurance and quality control personnel on a building construction site

Figure 11, shows that the proper supervision of professionals on site would reduce the structural failures.

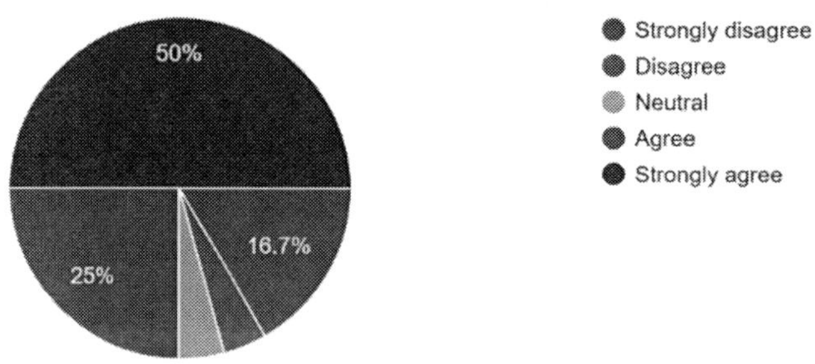

Figure 11: Proper supervion of professionals to reduce stuctural failures

Government agencies also play a significant role in the construction industry, as they are to regulate the codes guiding building construction. The enforcement of penalties on violators should be strongly administered. The design and contractual documents are to be checked thoroughly by government agencies before approval. Lastly, environmental impact analysis and structural analysis should be mandatory.

4. Conclusions
The data from the analysis of the survey carried out show the need for government agencies responsible for construction in Nigeria to enforce quality assurance on every project. Thorough examinations and checks must be carried out during designs and construction by duly certified professionals. Lastly, awareness should be carried out to educate investors and owners of buildings about possible collapse from over interference in the construction process.

Acknowledgements
The authors would like to sincerely acknowledge the Covenant University Centre for Research, Innovation and Discovery (CUCRID), Covenant University, Ota, Nigeria for sponsoring the publication of this article.

References

[1] Ede, A. 2016 Structural Stability in Nigeria and worsening Environmental Disorder: the way forward. *The West Africa Built Environment Research Conference (WABER), Accra, Ghana.*

[2] Akinyemi, A. P., Dare, G. M., Anthony, A. I., & Dabara, D. I. 2016a. Building Collapse in Nigeria : Issues and Challenges Building Collapse in Nigeria : *Issues and Challenges. Conference of the International Journal of Arts & Sciences,* 9(1), 99–108.

[3] Mathebula, A. M., & Smallwood, J. J. 2017. Religious building collapses : The heavy price of short cuts in places of worship and pilgrimage site construction . *Procedia Engineering, 196*(June), 919–929. https://doi.org/10.1016/j.proeng.2017.08.025

[4] Ede, A. 2010 Building Collapse in Nigeria: the Trend of Casualties in the Last Decade (2000 -2010). *International Journal of Civil & Environmental Engineering IJCEE-IJENS IJENS I J E N S, 10* (January 2010), 6–32.

[5] Taiwo, A. A., & Afolami, J. A. 2011. Incessant building collapse: A case of a hotel in Akure, Nigeria. *Journal of Building Appraisal, 6*(3–4), 241–248. https://doi.org/10.1057/jba.2011.1

[6] Clement, O. J. O. I. 2013. Design And Construction Supervision As Structurally Sustainable Tools For Building Failure / Collapse In Nigeria. *International Journal of Computer Science and Information Technology & Security (IJCSITS),* ISSN: 2249-9555 Vol. 3, 3(3), 271–281.

[7] Nwankwo, S. I., Nwankwo, C. V., & Okafor, M. U. 2015. An Investigation of Incessant Building Collapse in Selected Cities of Nigeria. *The International Journal of the Constructed Environment, 6*(2), 17–39. https://doi.org/10.18848/2154-8587/cgp/v06i02/37445

[8] Odeyemi, S. O., Giwa, Z. T., & Abdulwahab, R. 2019. Building Collapse in Nigeria (2009- 2019), Causes and Remedies – A Review. *USEP: Journal of Science and Engineering Production, 1*(1), 123–135.

[9] Ajufoh, M. O., Gumau, W. A., & Inusa, Y. J. 2014. Curbing the Menace of Building Collapse in Nigeria. *International Letters of Natural Sciences,* 168–178. https://doi.org/10.18052/www.scipress.com/ilns.20.168.

[10] Windapo, A. O., & Rotimi, J. O. 2012. Contemporary Issues in Building Collapse and Its Implications for Sustainable Development. *Buildings, 2*(3), 283–299. https://doi.org/10.3390/buildings2030283

[11] Oyegbile, O. B., Tat, T. N., & Olutoge, F. A. 2016. Management of Building Collapse in Nigeria : A Lesson from Earthquake-Triggered Building Collapse in Athens , Greece. 2(6), 36–42.

[12] Michael A, A., Oyewale I, O., & O.A, A. 2018. Assessment of Building Collapse in Lagos Island, Nigeria. *American Journal of Sustainable Cities and Society, 1*(7). https://doi.org/10.26808/rs.aj.i7v1.04

[13] Oloke, O., Ogunde, A., Joshua, O., & Babalola, O. D. 2017. Incessant Building Collapse in Nigeria : A Framework for Post-Development Incessant Building Collapse in Nigeria : *A Framework for Post- Development Management Control.*

[14] Obodoh, D. A. 2019. Analysis Of The Impact Of Building Collapse Risks In The Nigerian Econmy. *Environmental Review ER.* https://www.researchgate.net/journal/0147-2496

Comparative Cost Analysis of Reinforced Concrete and Timber Floor System for Effective Low Housing Scheme

O J Oyedepo*[1] and L M Olanitori[2]
[1&2] Department of Civil Engineering, Federal University of Technology Akure, Nigeria.
*Corresponding author: ojoyedepo@futa.edu.ng

Abstract. In Nigeria, housing is grossly inadequate especially in urban area due to migration from rural area to urban area which results in poor environmental quality. In this paper, the cost analysis of 6 m x 6 m square suspended floor, constructed with reinforced concrete and wooden floor for housing units was carried out. The results show that wooden floor system is 21.6% cheaper than using reinforced concrete in construction. The use of timber in building construction should be encourage, as it will minimize the dead load and reduces the cost of building construction that will be sustainable and affordable for the low income earners.

Key words: Environmental Quality, Suspended Floor, Sustainable, Low Income Earner

1. Introduction

Large dependency on concrete which increases the cost of building construction prevents the low income earner from having access to good shelter in developing countries like Nigeria. This increasing cost of materials for concrete production slowed down the provision of shelter and other infrastructural facilities in developing countries [1]; while [2] aver that majority of people in in urban area lives in slum and some do so at outrageous cost. The increasing needs for infrastructure leads to greater needs for construction materials for housing; however, lack of affordable and good houses is due escalating cost of the traditional processed construction materials such as steel and Portland cement. Hence, people in the construction industry seek to find alternative such as by-products to replace partially or wholly the non-renewable materials. These alternative materials are used as aggregate for concrete production [3]. Natural aggregate from riverbeds constitutes fine aggregate that are used in the production of concrete. The use this non-renewable natural sand is very high and this hindered the supply of infrastructural development. The use of eco-friendly materials by incorporating huge deposit of solid waste to produce light weight concrete and reduces pollution is gaining popularity. Thus, there is need to explore its usage to benefit the environment and maintained the material standard requirements.

Timber is a renewable resource that are used in the construction of infrastructure for various purposes such as beams, columns, trusses, girders, and railway sleepers among others. It is good in both tension and compression which make it suitable as flexural member. According to [4], reinforced concrete is mostly used as building material, but the production process of cement is accompanied by large emission of CO_2, a major factor causing greenhouse effect and climate change aside the depletion of the natural resources (sand and quarry stone). In addition, the incorporation of mass timber for various building elements has offer new opportunities for the building industry. It offers accelerated construction schedules, superior sustainability characteristics, and improved environment for occupants. Mass timber products used in building construction however have a wide application in Europe, Canada and United States of America.

2. Housing Problems and Needs in Nigeria

Many countries in Africa, Asia and Latin America have witness great increase in human populations without a commensurate improvement in infrastructure, which resulted in the shortage of housing. The UN estimates aver that the urban population of the developing world alone will increase from 2.7 billion in the year 2011 to 5.1 billion by 2050 [5]. The implication is that there is need to adequately make provision for shelter to accommodate the growing urban population like Nigeria. However, most low-income households have limited access to affordable housing units in developing nations; thus, focusing on sustainable low-cost housing will serve as a panacea to housing shortages.

housing is also considered as a basic necessities of life after food [6]; On the other hand, millions lived in an overcrowded serviced accommodation, while [7], stressed that housing shortages have been recorded in both rural and urban communities of all African countries but is more critical in urban communities. Hence, housing shortage is a worldwide phenomenon that is mostly evident in developing countries and Nigeria in particular where its delivery is a highly politicized.

The provision of affordable and sustainable housing contributes to the well-being of citizen of every country, which enables her meets the needs of both the present and future generations. [8] in their opinion adduce that sustainable housing provision is the gradual, continual and replicable process of meeting the housing needs of the populace, mostly the poor.

3. Comparative Cost Analysis of Suspended Floor Constructed from Reinforced Concrete and Timber

The cost implication index of the timber structure was compared with reinforced concrete structure using a suspended floor of 6 m by 6 m for the analysis. An imposed load of 1.5 kN/m^2 was used on slab for residential purposes.

3.1 Details of the Reinforced Concrete Suspended Slab

The design parameters for two way reinforced concrete slab, simply supported on all the four edges are:
The characteristic strength of concrete (f_{cu}) = 20 N/mm^2,
The characteristic strength of steel reinforcement (f_y) = 460 N/mm^2,
Thickness of Slab = 150 mm
Reinforcement diameter = 12 mm
Reinforcement spacing: 150 mm centre to centre, both ways.

3.2 Details of the Timber Suspended Floor

The detail of the timber used is as follows:
Type of Timber: Hardwood
The bending strength characteristic ($f_{m,k}$) = 24 N/mm^2.
Safety class 2: $\gamma_d = 0.91$.
The modification factor for load duration class "M: and service class 1 $k_{mod} = 0.8$.
Size factor for depth greater than 150 mm, $k_h = 1.0$
After the design the following dimensions were determined:
Timber floor planks: 1 by 12 by 12 and timber joist of 2 by 6 by 12, spaced at 1.5m centre to centre.

Tables 1 and 2 show that the cross sectional area of timber flooring in square metre is 21.6% cheaper than that of reinforced concrete. For multi-storey buildings, the reduction in self-weight of the flooring system will have reduction effect on the sizes of foundations, columns and beams, which will further reduce, the volume of concrete to be used and the total cost of the building project.

Table 1: Cost implication for construction of concrete slab

S/N	Description	QTY	Unit	Rate
	Amount (₦)			
1.	Concrete slab of mix ratio 1:2:4	5.4	m^3	32,000
2.	172,800			
3.	Y12 bars @ 200 mm c/c	31	mm	2,800
4.	86,800			
5.	Planks for formwork: 1 x 12 planks	40	mm	1,500
6.	60,000			
	Shoring: 2 x 2 @ 1 m c/c	49	mm	500
	24,500			
	Removal of formwork	36	m^2	170
	6,120			
	Rendering	72	m^2	500
	36,000			
	Total			
	386,220			

Table 2: Cost implication for construction of wooden floor

S/N	Description (₦)	QTY	Unit	Rate	Amount
1.	Hardwood: 1 x 12 x 12	42	mm	2,600	109,000
2.	Hardwood: 2 x 6 x 12	13	mm	2,600	33,800
3.	Finishes: 36 + 36 + 21.6	93.6	m^2	800	74,880
	Total				317,680

Embracing the use of wood in our construction industry will bring reduction in the quantity of concrete materials and the volume of concrete, this will reduce the cost of construction for a particular building project. Figures 1 to 4 shows samples of wooden for flooring systems with cost per metre square from United State of America and Nigeria. The usage of wooden flooring systems will expand the market for the wooden industry and create employment opportunities for the teaming youth in the country. It will also reduce the greenhouse effect arising from the production of cement.

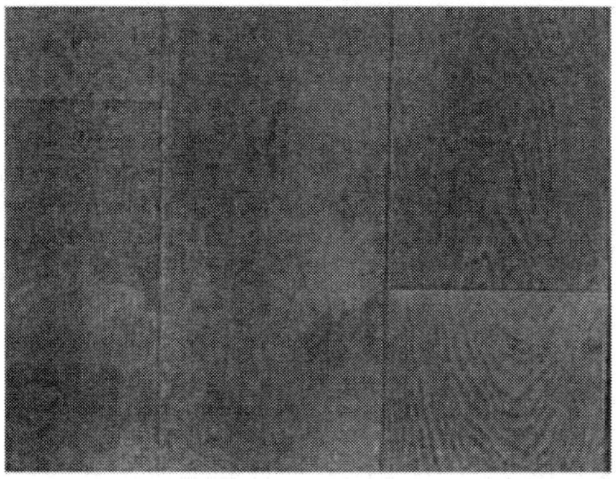

Figure 1: Walnut solid wood flooring US \$9.0 – 25/m^{2}flooring.

Figure 2: White brushed parquet oak solid wood US \$17.0 – 20/m^2

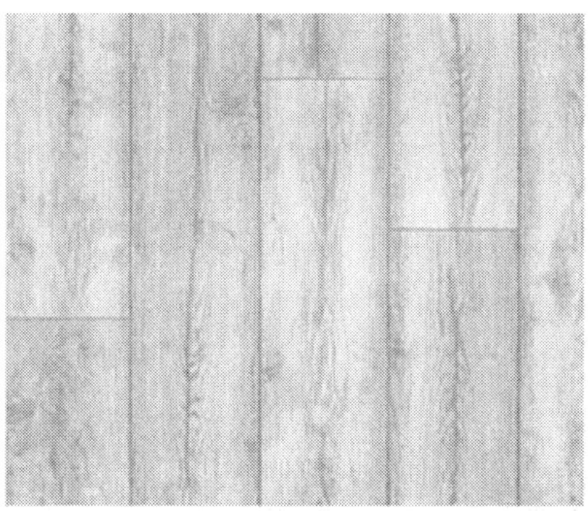

Figure 3: Antique oak vinyl wood flooring ₦6,500/m^2

Figure 4: Sugar oak vinyl wood flooring ₦6,500/m^2

4. Conclusion

The analysis shows using timber in the construction of low housing scheme has advantages such as availability of materials, affordability for the urban poor, reduction to a near zero carbon emission and sustainability among others. Training the end user in the techniques on the usage of wood for building construction purposes will enable low income earners to affordable homes. Also, research should be carried out on factors mitigating the embrace of timber in construction industry.

Reference

[1] Olutoge FA 2010 Investigations on Sawdust and Palm Kernel Shells as Aggregate Replacement, *ARPN Journal of Engineering and Applied Sciences,* **4 (5)**, 7-13.

[2] Olugbenga NT 2001 Housing Finance in Nigeria- Need for Reengineering. Ideal Habitat Cooperative Housing Initiative.

[3] Adewuyi, AP and Adegoke T 2008 Exploratory Study of Periwinkle Shells as Coarse Aggregates in Concrete Works, ARPN *Journal of Engineering and Applied Sciences,* **3 (6)**, 1-5.

[4] Olanitori LM and Tifase TO 2019 Effect of Small Transverse Service Holes on Shear Strength of Reinforced Concrete Slender Beams. *Journal of Applied Sciences and Environmental Management.* **23(3)**:539 – 543.
[5] UN (2012) World Urbanization Prospects. The 2011 Revision. Highlight, NewYork. UN Department of Economics and Social Affairs, Population Division.
[6] Taiwo A and Albert A 2013 Sustainable Housing Supply in Nigeria Through the Use of Indigenous and Composite Building Materials, *Civil and Environmental Research,* **3 (1)**:79-84.
[7] Arayela O and Taiwo AA 2010 Stabilized laterite bricks as an appropriate walling material technology for increasing housing stock in some selected African countries. *Proceedings of XXXVII IAHS,World Housing Congress*, October 26-28, Santacher (Cantabria, Spain) on CD.
[8] Adedeji YMD, Arum C and Ajayi B 2011 Affordable housing initiative in Nigeria: use of composite panels, West Africa Built Environment Research (WABER) Conference 19-21 July 2011 Accra, Ghana Proceedings of the WABER 2011 conference, pp 78-90.

Sustainability Barriers in Nigeria Construction Practice

L O Toriola-Coker[1*], H Alaka[2], W A Bello[3], S Ajayi[4], A Adeniyi[5], S O Olopade[6]
[1] School of Engineering, Yaba College of Technology, Yaba Lagos.
[2] University of Hertfordshire, Hatfield, Hertfordshire, UK.
[3] Department of Quantity Surveying, Faculty of Environmental Sciences, University of Lagos.
[4] School of Built Environment, Engineering and Computing, Leeds Beckett University, UK.
[5] School of Engineering, Yaba College of Technology, Yaba Lagos.
[6] School of Engineering, Yaba College of Technology, Yaba Lagos.
*Corresponding Author: toriolacoker@aol.co.uk

Abstract. The struggles to practise sustainable construction are not gaining the desired traction in Nigeria. This study established the likely barriers to successful application of sustainable construction in the Nigeria construction industry and factors to overcome the possible barriers. A quantitative approach was used for the study and a questionnaire survey was conducted among the professionals and other stakeholders. A descriptive method was used in analysing the collected data. Among the highly ranked sustainability barriers to construction practice are poor sustainability education in academic institutions, lack of incentives for designers to facilitate sustainable design, ignorance of lifecycle cost benefits, sustainable construction regarded as low priority and other issues take priority, and resistance to cultural change in the industry. The research recommends adequate sustainability education in academic institutions to positively impact the required cultural change in the industry. There is call for proper government policies that support implementation of sustainable construction practices. The study also advances the need for construction professionals and stakeholders to embrace the concept of sustainability education through continuing professional development and or postgraduate studies to improve the thinking and practicability of sustainable practice of construction in Nigeria.

Keywords: Barriers, Sustainability barriers, Construction Practice, Professionals, Stakeholders

1. Introduction

Every construction endeavour involves burning of fossil fuels, gas emissions of carbon dioxide, methane and other waste products that pollute the environment including air, water, and noise pollution and destruction of natural habitats. According to [1] construction industry has been found to cause damaging effects to the environment by means of waste generation, energy and water depletion and several other forms of damage to the environment. These damaging effect of construction cannot continue and efforts must be made to curb it by making construction practice sustainable. Recently, nongovernmental bodies, academia and individuals are vigorously propagating sustainability awareness with the Nigeria government in order to promote commercial and professional interest as well as safeguarding the environment [2]. Implementation of sustainable construction practices has become a problem due to desire or courage and strategy which eventually resulted to less of awareness [3] also claimed that present construction practices are unsustainable, and not in agreement with ideal sustainability principles [4] have expressed the deficiency in efforts targeted at sustaining infrastructural development in Nigeria. Therefore, advocating for a more committed government effort and inclusive civil society participation. Sustainable development has been defined decades ago depicted in United Nations document chaired by Brundtland in 1987 as the ability to make development sustainable to ensure that it meets the needs of the present without compromising the ability of future generations to meet their own needs [5], [6]. The concept of sustainable development does imply limits. Sustainability within the construction sector has been a key theme globally and Nigeria is not left out among the developing countries of the world. The construction industry has the largest impact on the economy when compared with other industries [7]. These have shown the importance of sustainability in the construction practice in Nigeria.

Sustainable construction (SC) can be seen as a subdivision of sustainable development applied to the construction industry. It can be viewed as "the creation and liable administration of a healthy built environment based on resource efficient and ecological principles" [8]. The construction sector in Nigeria plays an important role in the development of our nation's economy. The construction industry is said to have contributed about half of the total stock of fixed capital investment in the Nigeria economy [9]. The presence of Asian giant, the Chinese and Indian debar Nigeria construction industry not be the highest employers of labour [10]. The significance contribution of the construction industry in Nigeria cannot be outweighing in terms of socioeconomic development through the proportion of workers and acquisition of skills compared to manufacturing sector. The Nigeria construction industry contribution to the nation GDP has remain abysmally low despite the growth seen in the sector even with 1.4% of its GDP [11].

The influence of the construction industry to sustainability is a worldwide concern. Surprisingly, most organization in Nigeria has a narrow focus because their focus is on profit and not about those that will bring the profit. Development should be seen as the most important formation of any country [12]. However, the problems opposing developing countries are vast and much more complex than those affecting the industrialized world. According to [13] developed countries wanted to reduce the environmental impact of their growth while developing countries were desperate because they were not at similar higher levels of economic growth that industrialized countries had, thus using cheap methods with high environmental impact. Typical problems of developing countries include low income, poor water-supply systems, initial health systems, poor education and health systems, lack of trained human resources, finance and many others. Using healthier and resource-efficient models of construction, renovation, operation, demolition and maintenance will create sustainable building and green construction known as green building [14]. Healthier and eco-friendly environment are encouraged by this process and is undertaken with the greatest possible level of cooperation and coordination of engineers, contractors, design team, and landlords throughout the project in question.

The construction industry is characterized by complex socio-cultural, contextual, structural issues as reflected by its primary resistance to change. A considerable need to move from the native construction system to sustainable construction practice is to eradicate the deficiency in authoritative research in order to understand the current barriers to sustainable construction practice in Nigeria through the lens of stakeholders [15]. However, Nigeria construction process has been faced with challenges and priorities that are completely different from those of a lot of advanced countries. Despite the fact that significant research has revealed different practices and techniques to improve the sustainability of construction, the construction industry seems to be unwilling to apply most of these methods [16].

The quantity of the research focused on this topic, efforts and investments to conduct it, potentials evolving in SC needs to be investigated as to why the owners and developers do not seem to be interested enough to apply the outputs of the research in their projects [17]. This could be attributed to some barriers in the industry that limit the application of the new sustainability guidelines. As an example, cost constraints are the dominant factor while making decisions in the construction industry [18]. The Nigerian construction industry is characterized by construction companies starting from the little and medium enterprises, to the big enterprise and technically competent international construction companies. The government of Nigeria believed more in foreign corporation leaving indigenous contractor with fewer options in terms of execution of construction jobs [19]. This has become a barrier to sustainable construction practice in Nigeria. Government accounts for being the biggest consumer of the trade although, Nigeria recorded magnificent patronage from several private firms within and outside the globe.

[20] proposed that barriers may cause undesirable penalties which are not supportive of organizational aims. He then characterized the barriers into two groups; the barrier with lesser effect and the ones

with greater effect. A suggestion was made for every individual or stakeholders of SC to focus on identifying and acting on the higher effective barrier (root causes) with respective to time. [21] revealed that the major barriers to sustainable construction practices in Nigeria include sustainability knowledge related, regulation and policy related, sustainable materials and technology related, and information and management related. The construction project needs a good management to achieve the satisfied result including efficient satisfaction, appealing satisfaction, accomplishment on time, completion within budget, value for money, and health and safety. Despite the continuous growth in the Nigerian Construction Sector Summary Report 2010 –2012 which has indicated dynamisms of development [22] lack of expertise and professional knowledge; lack of strategy to promote sustainable construction; Lack of demand; Lack of legislation, enforcement and monitoring; and Lack of government incentives are considered to be significant barriers affecting the growth of sustainability in Nigeria [15]. However, projects that are implemented in various developing countries have been linked to poor sustainability in declarations made in recent times, in which the Nigerian construction industry is not an exemption [21]. [23] submitted that the most prevalent barrier was the lack of information that would allow practical implementation of practices, clarify the reasoning behind the need for these practices and the benefits to the trade. This study identifies the barriers to construction practice which would help to understand the barrier and implement the development of strategies to reduce its implications in the country. The results of this research aim at providing insights on the measures to improve and advocate sustainable construction practice in Nigeria.

2. Literature Review

Sustainability has a generally accepted pattern in the industry in that human activities are major cause of the depletion in the quality and standards of infrastructures. The development that presents infrastructures with the ability to serve both the needs of the present and the future was viewed by World Commission on Environmental Development such as [5] and [13]. Nigerian construction industry has faced various disregards and misconception due to its recorded low level of input and has caused contractors in the industry to be reprimanded of constructions that envisage on a large and bulky gauge [24]. This has resulted to a reputation in poor project management due to reasons such as bungling of funds, provision of capital, and abandonment of projects [19]. When challenges are clearly understood, the construction industry decision makers should cooperate to develop practical solutions to overcome these challenges.

Researches have been carried out and have focused on weighing the drivers and barriers of sustainable construction in construction industries globally but with limited factors compared to many available factors identified in this research [25]. It is known that on a global scale, construction industry has a variable impact on the depletion of natural resources, air, water pollutions, deforestation and global warming [26]. Some would argue that the view of sustainability is probably said that it is mostly associated with environmental protection. In as much as it is true, some other factors which are as a result of socio economic reasons cannot be left out as [13] posited that these environmental challenges were intertwined with economic and social conditions. In essence to producing a better impression, the notion of sustainability ought to be reflected as an all-inclusive and integrative approach [27].

In addition, half of the resources used by humans is mostly consumed in construction activities; half of the world's fossil fuel have been used to serve buildings, 37% of the world's total energy is used in construction activities, conversion of about 6500 hectares of rural land to urban of land [28]. The inability of the Nigerian construction industry to astutely practice sustainability of construction at an acceptable standard is linked to the unawareness and ignorance of construction parties such as; Engineers, Contractors, Architects and sometimes the clients [29] as the world realized the importance to create awareness of the need for sustainable development [13].

[30] made a comparable analogy about the drivers and barriers of sustainable construction in Kuwait where it was observed that the main cause is as a result of the lack of awareness of the concept of sustainability by within the country. Nevertheless, it is not the same in a research carried out in the Ethiopian construction industry where awareness of sustainability is ascertained across the industry. This implies that even at its awareness, sustainability needs to be practiced well and accurately. New

methods and tools of checking with the sustainability practiced in construction industries across the world is been used in developed countries [31] but the professionals still perform below expectation. Although, the necessities of moving toward SC has made promising development in the recent years and will still continue to be having impact in the construction industry.

The successes are not sufficient to reverse the negative environmental impacts on the built environment in developing countries [32]. Identifying the barriers to sustainability in construction, specifically from the view of private owners is essential to quicken the pace towards achieving the goals of sustainable construction [33].

The most petrifying barrier to sustainable construction is its inability to actually practice the methods of sustainability. The lack of the ability to perform sustainability of construction further infects the process of maintaining a structure [25]. Consultants and experts would boast of the ability to practice traditional construction, but when asked to address sustainability brings a sudden decrease in their morale. This implies that the understanding of common sustainability process affects the capacity of professionals in developing countries [34].

The challenges affecting sustainable construction must be elucidated if the concept of sustainability is to be enforced. The implications of these challenges are seen across various sectors of the construction industry thus affecting sustainability. In the Palestine, study revealed that the major challenges faced when practicing sustainable construction are; management, skill acquisition, financial constraints and customer's satisfaction [35]. The management skills can further be classified as to skills for coordinating, planning, labelling, motivating and staffing.

All the concept and attributes of a manager can be a deficit to the implementation of sustainable construction. The financial constraint is coined with the clients not allocating funds at the proper time or the extravagance in the management of funds by the project account officer. High investment cost, high capital cost amongst others is a probable reason for challenges facing sustainable construction. [3]

It is paramount we realized that to practice sustainability in construction, a repeated method and approach would be entertained and the construction industry must be ready to embrace new technological methods and innovations that would counter its former methods of practice [36].

There is a long way to go if conventional wisdom is still being used to achieve a sustainable construction industry and this process needs an input from all industry stakeholders; however, it is important to know the current situation at the starting point. Despite the similarity in findings of previous studies worldwide, the situation in each country differs due to its sui generis socio-economic-politic context which requires attention [37]. The perception of sustainability is that it only concerns environmental protection in the current body of knowledge [38].

However, other elements of sustainability, such as economic and social, cannot be ignored and, therefore, the concept of sustainability should be considered as a holistic and integrative approach in order to keep a balance between the three elements. Physical facilities provided by construction industry such as dams, roads, bridges, residential and commercial buildings, factories, recreational facilities amongst others have significant effect on society, environment and economy [39]. Hence, construction industry plays a significant role in the balance between the three elements of sustainability compared to other industries, and it is strongly recommended that rather than traditionally used measures focusing on time, cost and quality the industry's success must be considered based on the triple-bottom-line [40].

For effective implementation of sustainable construction, it is important to identify possible drivers and constraining factors at the beginning, so that frontline industry professionals (government authorities, contractors and project managers) can successively act upon them professionally [41]. It is very crucial to identify the factors as identified in Table 1 (both negative and positive) influencing sustainable construction [42]. This will facilitate the adoption of those that have a positive effect and in the elimination or control of those that have a negative effect. In addition, a robust backing from the literature in order to achieve the aim of this study which is how to improve and advocate sustainable construction practice in Nigeria. This is done because of the gap in the research conducted on the subject in the Nigeria construction context; therefore, an international context is required to identify the barriers, and remove the factors that hinder implementation of sustainable construction.

Table 1: Overviews of the literature on the barriers to sustainable construction practice

Location	Reference	Findings
Canada	[16]	Lack of consideration of sustainability criteria in the evaluation of bids, unavailability of standard methods for procurement, lack of knowledge of local conditions, lack of explicit statutory requirements that cover sustainable procurement.
Malaysia	[43]	Lack of awareness on sustainable building, lack of training and education, the higher cost of sustainable building options, procurement issues, regulatory Lack of professional capabilities/designers, disincentive factors for local material production, lack of case studies/examples.
Chile	[44]	Lack of financial incentives, designers work alone, economic needs of higher priority, environmental costs not included in the cost structure, governmental bureaucracy, lack of knowledge on sustainable technologies, lack of environmental concern, affordability.
USA	[42]	First cost premium of the project, long pay-back periods from sustainable practices, tendency to maintain current practices, and limited knowledge and skills of subcontractors.
Finland	[34]	Steering mechanisms, economics, a lack of client understanding, process (procurement and tendering, timing, cooperation and networking), and underpinning knowledge (knowledge and common language, the availability of methods and tools, innovation).
UK	[45]	Affordability, building regulations, lack of client awareness, lack of business case understanding, lack of client demand, lack of proven alternative technologies, lack of one single labelling/measurement standard, planning policy.

Source: [41]

3. Research methods

Questionnaires can be an effective means of measuring the behaviour, attitudes, preferences, opinions and, intentions of relatively large numbers of subjects more cheaply and quickly than other methods. Local experts with international exposure helped review the variables used in the questionnaire. Their expert listing of variables from practice and literatures led the research to use of questionnaire that was used for data collection. This is to collate a quantitative degree of accuracy based on the opinions of professionals from different fields of professions. Such questionnaire assessments are carried out through a survey approach used in harnessing quantitative data from construction professionals that were selected based on their background knowledge in the management of construction projects.

A total of one hundred and fifty (150) questionnaires were distributed to professional firms using convenient sampling method. This was adopted in order to get timely response by avoiding stringent rules in the selection of respondents, regular cluster locations of professionals that are easily accessible and convenient for the researchers were explored for distribution of questionnaire with minimal cost. The nature of the research as initial probe into awareness of sustainability plan among professionals in the Nigeria construction industry was another reason for adopting convenient sampling technique.

The selected firms have strong presence in the city of Lagos, which is the fastest growing city in the country and has the highest rate of procurement of buildings works. The design was aimed at determining the barriers to SC practices and how it can be improved in Nigeria. This survey was used to get vital information from all drivers on sustainable practice in Nigeria. A total of one hundred (100) questionnaires representing 67% response rate were successfully retrieved and used for the

IOP Conf. Series: Materials Science and Engineering 1036 (2021) 012023 doi:10.1088/1757-899X/1036/1/012023

analysis. A five point Likert scale was adopted, with 5 being strongly agree, and 1 being strongly disagree. The data gathered on the background information of the respondents were analyzed using percentage, while Mean and Standard deviation were used to rank the variable barriers. The population of the participating professionals is substantial compared to similar research in this regard particularly in the engineering.

4. Results and Data Analyses

Background information from the total of 100 questionnaires received, result revealed 10% each of the respondents were Architect and Builder, 36% are Civil/Structural Engineer, 8% each practices as M&E Engineer, Project Manager, Quantity Surveyor, Material Supplier and Site Manager while 2% each practices as subcontractors. Majority of the respondents participated in the research study practices as professional Civil/Structural Engineer. Only 26% of the respondents participated in the study have between ***1-5yrs*** of professional working experience, with the remaining 74% having between 6-25yrs of professional working experience in the industry. All respondents have minimum academic qualification of Higher National Diploma (HND) or Bachelor of Science (B.Sc.) with 18% having Masters Degree and 7% having Doctorate degree. Importantly, all the respondents have professional qualifications which indicate that the respondents are experienced and well informed to provide the required information.

The extents of barriers in preventing the sustainable construction practices in Nigeria are presented in Table 2. According to the ranking, the ten most important barriers to sustainable construction practice are: Poor sustainability education in academic institutions; Lack of incentives for designers to facilitate sustainable design; Ignorance of lifecycle cost benefits; Regarded as low priority and other issues take priority; Resistance to cultural change in the industry; No input tariff for generation of renewable energy; Longer payback period for sustainable resources; Lack of government policies/support; Clients preference for the traditional materials; Insufficient integration and link up within the industry; Lack of commitment from professional bodies (such as NIA, NIOB, etc.); Poor awareness of the benefits of sustainability among the professionals; Limited knowledge of sustainability practices among contractors with the average mean of 3.70, 3.64, 3.58, 3.58, 3.56, 3.56, 3.54, 3.54, 3.52, 3.48, 3.48, 3.48 and 3.48 respectively. However, Sustainable construction does not raise market values of buildings is considered to be the least important barrier to SCP in Nigeria. Therefore, all the barriers are to be considered important as they affect the effective implementation of sustainable construction in Nigeria. There will be need for further research analysis to reassess any underlying causality among these barriers to see likely groupings and trimming down the barriers.

4.1 Discussion

The findings of the research actually depict similar but different outcome in [17], [1] and [36] in Ghana, South Africa, and Zambia respectively with different weightings but similar emphasis on related factors. There are fewer variables in previous researches whereas there are fifty-eight variables identified in this research with more respondents than similar researches conducted in other developing countries. In Ghana [17] discovered that cultural change resistance, lack of government commitment, fear of higher investment costs, lack of professional knowledge, and lack of legislation are the leading barrier to implementation of sustainable construction in the Ghanaian construction industry. All these factors are among the agreed factors in Nigeria according to the outcome of this research however, the leading barrier differs as poor sustainability education in academic institutions ranked first in Nigeria against resistance to cultural change in the industry which was ranked fifth. The research of The two foremost challenges faced by South African construction industry towards the adoption of sustainable construction practices according to [1] are additional cost to building projects followed by limited understanding of the benefits of sustainable construction and of additional cost to building projects. These were revealed in this research as ignorance of the life cycle cost benefits (described as 'lazy view on additional cost to building projects' by [1] and poor sustainability education in academic institutions. Thus, the top two barriers in South African construction industry are among the top three barriers in Nigeria while lack of incentives for designers to facilitate sustainable design emphasised in Nigeria but missing among the listed challenges to sustainable construction in South Africa. The latest in our literature review referenced is the work of [36] which revealed that the major barriers to sustainable construction practices in Zambia Construction Industry are; fear of higher investment costs, no local green certification available, lack of government policies or support, and lack of financial incentives. These four barriers can conveniently be related to the barriers ranked 37th, 35th, 8th, and 2nd (Table 2) in this research with the last two barriers; lack of government policies or support, and lack of financial incentives ranked within the agreed barriers. Professionals in Nigeria considered lack of supporting building regulations fear of higher investment costs, and lack of supporting building regulations (including no local green certification available) at middle course but tend towards agreed barriers that affect the sustainable construction practice similar to challenges being faced in Zambia, another developing country in Africa. Therefore, the challenges facing adoption of sustainable construction practice in developing countries like Nigeria is real and similar to other developing economies of African countries. In this research, none of the fifty-eight listed variable factors is disagreed as not being a barrier to sustainable construction practice thus, as these factors are agreed to be barriers to sustainable construction practice there is need to put additional research efforts to mitigating their impact for sustainable outcome.

The poor sustainability education in academic institutions in Nigeria negatively impacts the resistance to cultural change in the industry. This could exert more negative impact on the ignorance of the life cycle cost benefits or as lazy view on additional cost to building projects. Although, lack of supporting building regulations and fear of higher investment costs also featured among the agreed barriers that affect the sustainable construction practice, their impact will still trail behind more significant barriers that were revealed in this research including adequate education by institutions for the industry and instituting proper government policies and support for regulations that support implementation of sustainable construction practices to change the low sustainable construction (SC) practices adoption in most developing countries, as revealed in this research and corroborated by [36]. Therefore, government intervention in terms of policies is also of key importance since the research revealed that it is lacking because the practitioners are not really seeing sustainability construction practice as a gain owing to the longer payback period for sustainable resources or general poor perception of sustainable construction within the industry.

Table 2: Extent of barriers in preventing the sustainable construction practices

Barriers	Mean	Std. Dev	Rank
Poor sustainability education in academic institutions	3.70	1.10554	1
Lack of incentives for designers to facilitate sustainable design	3.64	1.20202	2
Ignorance of lifecycle cost benefits	3.58	1.06534	3
Regarded as low priority and other issues take priority	3.58	1.27271	4
No input tariff for generation of renewable energy	3.56	.98801	5
Resistance to cultural change in the industry	3.56	1.04756	6
Longer payback period for sustainable resources	3.54	1.10481	7
Lack of government policies/support	3.54	1.15837	8
Clients preference for the traditional materials	3.52	1.24300	9
Insufficient integration and link up within the industry	3.48	1.02966	10
Lack of commitment from professional bodies (such as NIA, NIOB, etc.)	3.48	1.14133	11
Poor awareness of the benefits of sustainability among the professionals	3.48	1.19325	12
Limited knowledge of sustainability practices among contractors	3.48	1.25915	13
Methods and data required for the comparison of drafts in terms of sustainability are missing	3.44	1.10390	14
Affordability	3.44	1.25786	15
Lack of supporting planning policy.	3.42	1.24056	16
Clients preference for the traditional building services	3.42	1.25674	17
Lack of client awareness	3.40	.98473	18
Inadequate innovative and sustainable materials	3.40	1.04447	19
Lack of proven alternative technologies to the prevailing technique	3.40	1.20605	20
Misunderstanding of sustainability as just green (planting of trees and flowers)	3.40	1.22268	21
Lack of sufficient time to address sustainability issues in projects	3.40	1.27128	22
Poor understanding of the benefits of sustainable construction	3.38	1.28535	23
Lackadaisical attitudes towards sustainability	3.38	1.11718	24
Non-consideration of sustainability in the building control (approval) process	3.36	1.21871	25
No adequate information on sustainability performance of materials and products	3.36	1.02020	26
Poor skills among construction craftsmen (bricklayers carpenter, etc.)	3.36	1.20202	27
Lack of general public Interest in Sustainability	3.36	1.26746	28
No professional role for initializing and leading sustainability	3.36	1.44614	29
General perception that sustainability means more expensive buildings	3.34	1.19949	30
Lack of Environmentally Sustainable materials	3.32	1.14486	31
Lack of practioners' awareness of sustainable design and construction practices	3.32	1.22993	32
New solutions are considered too risky to venture into	3.32	1.19663	33
Poor perception of sustainable construction within the industry	3.30	1.27525	34
Lack of supporting building regulations	3.28	1.10170	35
Lack of methods for setting sustainable building requirements	3.28	1.15540	36
Fear of higher investment cost	3.28	1.34900	37
Skill and labour shortage	3.24	1.24819	38

Paucity of exemplar projects	3.24	1.16446	39
Insufficient research and development in local context	3.24	1.16446	40
Fear of long-term warranties for sustainability building's	3.22	1.24381	41
Lack of business case understanding	3.22	1.14221	42
Designers not driving sustainability	3.20	1.20605	43
There is no adequate knowledge in order to consider potential new alternatives by designers	3.20	1.36330	44
Resistance to increment of higher capital cost when needed for facilitating sustainability practice	3.18	1.16671	45
Vagueness of the definition and requirements of sustainable design and construction.	3.16	1.22862	46
No method to verify compliance with sustainable building requirements	3.16	1.30825	47
Lack of assessment method that enables comparison of buildings in terms of sustainability	3.14	1.30283	48
Lack of financial resources	3.12	1.37275	49
Operational and end of life stages are not considered in design	3.06	1.19612	50
Lack of sustainability measurement codes	3.00	1.18918	51
It is not worthwhile to change the construction process due to sustainability consideration	2.98	1.26315	52
Lack of one labeling / measurement standard	2.98	1.32558	53
Buyers / clients do not state any requirement for sustainable buildings	2.96	1.39204	54
There is a lack of resources to supervise the realization of sustainable building requirements	2.96	1.22202	55
Lack of client demand	2.92	1.22003	56
Separation between capital budget and operational budget.	2.80	1.08246	57
Sustainable construction does not raise market values of buildings	2.62	1.15277	58

5. Conclusion and Recommendation

In developing countries like Nigeria, the importance of sustainability of construction can never be over emphasised as highlighted in this research. Developed countries would shrug up in intrepidity that they have been able to properly use the effect of sustainability in construction to cluster their growing economy. This study has focused on the barriers affecting the sustainable practices of construction in Nigeria. The Mean of the opinions of the respondents were analysed as to whether they strongly agree, agree, neutral, disagree or strongly disagree to each of the factor anticipated to be the barriers against the implementation of Sustainable Construction Practice (SCP). The data's gotten from this method of collation was statistically reviewed and analysed using a demographic statistical method using descriptive statistics in identifying the barriers to sustainable construction which include Poor sustainability education in academic institutions, Lack of incentives for designers to facilitate sustainable design, Ignorance of lifecycle cost benefit, Regarded as low priority and other issues take priority, Resistance to cultural change in the industry, No input tariff for generation of renewable energy, Longer payback period for sustainable resources, Lack of government policies/support and Clients preference for the traditional materials. These are the most important barriers that require prompt and timely attention for effective implementation of sustainable construction in Nigeria.

The revelation from the research has described the current construction practice in Nigeria as inadequate in terms of education to sustain the growth of the industry. Academic institutions will need to brace up for sustainability teaching in their curriculum in order to enhance their consciousness and understanding before they graduate so that they can be vanguard of sustainable practice. This view was further echoed by the research participants that participated in the questionnaire. Some of the respondents' views can be summarized as the current knowledge on sustainable construction among practitioners in Nigeria are low and there is less emphasis on it in trainings in schools especially at the managerial level. This will in turn change the current lack of incentives for designers to facilitate sustainable design as revealed in the results. Lifecycle costing is another question raised in the research findings that could make one to ask whether life cycle costing is actually practised at all on projects in Nigeria particularly by Quantity Surveyors. The outlook of regarding sustainability practice as low priority and while other issues take priority over it is another important barrier that has made the practice of sustainability challenging particularly in considering this along with the resistance to cultural change in the industry where many refuse to change the current mode of doing things to the better and sustainable path. Government intervention in terms of policies is also of key importance since the research revealed that it is lacking because the practitioners are not really seeing sustainability construction practice as a gain owing to the longer payback period for sustainable resources and since the clients rarely change their preference for the traditional materials particularly as it aids sustainability. This could also be hinged on client's interest on profit as there is almost a neutral perception of raising market values of buildings just because of sustainable construction practice. Therefore, all the barriers are to be considered important as they affect the effective implementation of sustainable construction in Nigeria. There will be need for further research analysis to reassess any underlying causality among these barriers to see likely groupings and trimming down the barriers.

Therefore, for the industry to accurately tackle these barriers, incentives should be giving to upcoming private sectors to be able to flourish and explore their technological knowledge of construction, appropriate and trained professionals should be assigned to their respective jobs and not vice versa. Higher institutions including Universities and polytechnics will need to brace up for teaching sustainability by reviewing existing curriculum and come up with postgraduate courses and or professional courses in sustainability to checkmate the spread of disbelief in new ideas which is a deterrent to undergraduate and trained professionals. It is also important that professional courses should be observed through continuing professional development by professional bodies and administered to improve the technicality of each stakeholder in his/her field, new ideologies and instruments should be allowed for proper match up of the construction industry to the outside developing world. It can be concluded that adequate government intervention in terms of policy formulation, sustainable building regulations (including green design certification) and implementation coupled with financial grants while embracing an increased level of awareness to clients and construction stake holders would help yield positive results against the barriers affecting sustainable construction practice in Nigeria.

References

[1] Aigbavboa C, Ohioma I, and Zwane T 2017 Sustainable Construction Practices: "A Lazy View" of Construction Professionals in the South Africa Construction Industry *8th International Conference on Applied Energy* DOI: 10.1016/j.egypro.2017.03.743

[2] Ahmed K, Hatira L, and Valva P 2014 The Construction Industry in Ghana, West Africa "How can the construction industry in Ghana become sustainable?". *School of Engineering. Karlskrona, Sweden: Blekinge Institute of Technology.*

[3] Davies OOA and Davies IEE 2017 Barriers to implementation of sustainable construction techniques. *MAYFEB Journal of Environmental Science, 2.*

[4] Olanipekun AO, Aje IO and Awodele OA 2014 Contextualising Sustainable Infrastructure Development in Nigeria *FUTY Journal of the Environment* **8 (1):** 80 – 92.

[5] Brundtland Gro Harlem 1987 Our *Common Future. Report of the World Commission on Environment and Development.* UN Document; 1987

[6] Nwebo OE 2018 The African Union Agenda 2063 and the Imperative of Democratic Governance. *Law and Development Review,* ***11*(2):** 259-276.

[7] Ofori G 2012 New *perspectives on construction in developing countries.* Routledge, 2012.

[8] Kibert CJ 2016 *Sustainable construction: green building design and delivery*. John Wiley & Sons.

[9] Isa RB, Jimoh RA and Achuenu E 2013 An overview of the contribution of construction sector to sustainable development in Nigeria. *Net Journal of Business Management*, ***1*(1)**: 1-6.

[10] Femi OT 2014 Building Construction Technician Training: It's Relevance to Modern Construction Industry in Nigeria. *International Journal of Technology Enhancement and Emerging Engineering Research*, ***2*(3)**: 58-68.

[11] Ikediashi DI, Ogunlana SO, Awodele OA and Okwuashi O 2012 An evaluation of personnel training policies of construction companies in Nigeria. *Journal of Human Ecology*, ***40*(3)**: 229-238.

[12] Aroge, Stephen Talabi 2012 Employee's training and development for optimum productivity: The role of Industrial Training Fund (ITF), Nigeria. *Developing Country Studies* **2(4)**: 50-58.

[13] Anand A and Kumar I 2016 Importance of Brundtland report in the protection of environment: a legal analysis. *South -Asian Journal of Multidisciplinary Studies (SAJMS)* **3 (3)**: 230

[14] Aktas B and Ozorhon B 2015 Green building certification process of existing buildings in developing countries: cases from Turkey. *Journal of Management in Engineering*, ***31*(6)** 05015002.

[15] Daniel EI, Oshineye O and Oshodi O 2018 Barriers to sustainable construction practice in Nigeria. In *Proceeding of the 34th Annual ARCOM Conference, Belfast, UK* (pp. 3-5).

[16] Ruparathna R and Hewage K 2015 Sustainable procurement in the Canadian construction industry: current practices, drivers and opportunities. *Journal of Cleaner Production*, ***109***, 305-314.

[17] Ametepey O, Aigbavboa C and Ansah K 2015 Barriers to successful implementation of sustainable construction in the Ghanaian construction industry. *6th International Conference on Applied Human Factors and Ergonomics (AHFE 2015) and the Affiliated Conferences, AHFE 2015*

[18] Darko A and Chan AP 2017 Review of barriers to green building adoption. *Sustainable Development*, ***25*(3)**: 167-179.

[19] Ugochukwu and Onyekwena 2014 Participation of indigenous contractors in Nigerian public sector construction projects and their challenges in managing working capital. *International Journal of Civil Engineering, Construction and Estate Management;* **1(1)**: 1-21,

[20] McMullen PR and Robert A Strong 1998 Selection of mutual funds using data envelopment analysis. *The Journal of Business and Economic Studies* 4.1

[21] Aghimien D, Aigbavboa C, Ngcobo N and Thwala W 2019 Barriers of Sustainable Construction Practices in Nigeria *Proceedings: 13th Built Environment Conference* 2 - 3 September 2019, Durban, South Africa pg 340 – 348.

[22] Ogunde A, Olaolu O, Afolabi AO, Owolabi J and Ojelabi RA 2017 Challenges confronting construction project management system for sustainable construction in developing countries: Professionals perspectives (a case study of Nigeria). *Journal of Building Performance*, ***8*(1)**: 1-11.

[23] Heidi E S Tomkiewicz 2011 Barriers to implementation of sustainable construction practices in the homebuilding industry: a case study of rochester, NY Unpublished MSc Thesis Presented to the Faculty of The Graduate College at the University of Nebraska, Lincoln, Nebraska

[24] Darko A, Chan APC, Yang Y, Shan M, He BJ and Gou Z 2018 Influences of barriers, drivers, and promotion strategies on green building technologies adoption in developing countries: The Ghanaian case. *Journal of Cleaner Production*, ***200***, 687-703.

[25] Chan APC, Darko A, Olanipekun AO and Ameyaw EE 2018 Critical barriers to green building technologies adoption in developing countries: The case of Ghana. *Journal of cleaner production*, **172**, 1067-1079.

[26] Wang W, Zhang S, Su Y and Deng X 2018. Key factors to green building technologies adoption in developing countries: The perspective of Chinese Designers. *Sustainability*, ***10*(11)**: 4135.

[27] Yin S and Li B 2018 Transferring green building technologies from academic research institutes to building enterprises in the development of urban green building: A stochastic differential game approach. *Sustainable Cities and Society*, ***39***, 631-638.

[28] He B J 2019 Towards the next generation of green building for urban heat island mitigation: Zero UHI impact building. *Sustainable Cities and Society*, ***50***, 101647.

[29] Ayarkwa J, Acheampong A, Wiafe F, and Boateng BE 2017 Factors affecting the implementation of sustainable construction in Ghana: the architect's perspective. In ICIDA 2017-6th International Conference on Infrastructure Development in Africa (pp. 12-14).

[30] AlSanad S 2015 Awareness, drivers, actions, and barriers of sustainable construction in Kuwait. Procedia engineering, **118**, 969-983.

[31] Baron, N., & Donath, D. (2016, March). Learning from Ethiopia–A discussion on sustainable building. In *Proc. of SBE16 Hamburg International Conference on Sustainable Built Environment Strategies–Stakeholders–Success factors, Held from 7th to 11th March in Hamburg, Germany*.

[32] Mahat N, Tah JH and Vidalakis C 2016 Adoption of sustainable construction in the Malaysian residential construction sector: a conceptual framework. In *4th International Conference on Advances in Agricultural, Biological & Ecological Sciences (AABES-16) Dec* (pp. 1-2).

[33] Owusu EK, Chan AP and Ameyaw E 2019 Toward a cleaner project procurement: Evaluation of construction projects' vulnerability to corruption in developing countries. *Journal of cleaner production*, ***216***, 394-407.

[34] Häkkinen T and Belloni K 2011 "Barriers and drivers for sustainable building." *Building Research & Information* **39(3):** 239-255.

[35] Osaily NZr 2010 The key barriers to implementing sustainable construction in West Bank–Palestine." *UK: University of Wales* 63.

[36] Aghimien DO, Adegbembo TF, Aghimien EI and Awodele OA 2018 Challenges of sustainable construction: a study of educational buildings in Nigeria. *International Journal of Built Environment and Sustainability*, ***5*(1).**

[37] Anzagira LF, Duah D and Badu E 2019 A conceptual framework for the uptake of the green building concept in Ghana. *Scientific African*, e00191.

[38] Chan JH and Chan TY 2020 Current Trends of Developing Energy Efficiency Projects in the Building Sector of China. In *Sustainable Energy and Green Finance for a Low-carbon Economy* (pp. 227-244). Springer, Cham.

[39] Salem O, Pirzadeh S, Ghorai S and Abdel-Rahim A 2014 Reducing environmental, economic, and social impacts of work-zones by implementing Lean Construction techniques. In *Annual Conference of the International Group for Lean Construction* **22:** 145-155).

[40] Adegbembo TF, Bamisaye OP and Aghimien DO 2016 Assessment of lean construction practice in the Nigerian construction industry. In *Proceedings of the Joint International Conference (JIC), on 21st Century Habitat: Issues, Sustainability and Development* (pp. 21-24). Akure, Nigeria.

[41] Durdyev S, Zavadskas EK, Thurnell D, Banaitis A and Ihtiyar A 2018 Sustainable construction industry in Cambodia: Awareness, drivers and barriers. *Sustainability*, ***10*(2):** 392.

[42] Ahn YH, Pearce AR, Wang Y and Wang G 2013 Drivers and barriers of sustainable design and construction: The perception of green building experience. *International Journal of Sustainable Building Technology and Urban Development*, ***4*(1):** 35-45.

[43] Shafii F, Arman Ali Z and Othman MZ 2006 Achieving sustainable construction in the developing countries of Southeast Asia.

[44] Serpell A, Kort J, Vera S 2013 Awareness, actions, drivers and barriers of sustainable construction in Chile. *Technol. Econ. Dev. Econ.* ***19*:** 272–288.

[45] Pitt M, Tucker M, Riley M. and Longden J, 2009 Towards sustainable construction: Promotion and best practices. *Construction Innovation: Information, Process, Management,* **9(2):** 201-224.

Strength evaluation of agro waste particle board with melted pure water sachet as the binding agent

O D Atoyebi[1], B O Orogbade[2*], O J Aladegboye[3] and J A Adebiyi[4]
Department of Civil Engineering, Landmark University, Omu-Aran, Kwara State.
*Corresponding Author: orogbade.blessing@lmu.edu.ng

Abstract. Particle board is a construction material that can be used either for construction form works or house furniture, roofing or insulating materials. Due to its usefulness, this research work is aimed at evaluating the mechanical strength and rheological properties of particle board using agro waste namely corncob (CC) and sawdust(SD) with melted pure water sachet as the binding agent. The use of the pure water sachet or low density polyethylene (LDPE) as the binder was also evaluated, by the use of dual purpose kerosene (DPK) to form the homogenous binder that was used to produce the board. The proportion of LDPE to DPK was 50g to 300ml. Three ratio of the LDPE to sawdust and corncob were used; 2:1, 1½: 1, and 1: 1 in which the ratio of CC to SD was interchanged in proportion of 25%:75%, 50%: 50% and 75%: 25% to make 100% aggregate or combination. The production of the particle board was then made with a dimension of 350mm x 350mm x 10mm. The rheological properties of the binder show that it has adequate viscosity, binding power and strength compatible with the sawdust and the corncob. The results from the physical and mechanical property such as density, water absorption (WA), thickness swelling (TS), modulus of rupture (MR) and the modulus of elasticity (ME) concluded that the particle board type 1:1 had favorable physical properties that are recommended for indoor uses in building.

Keywords: Particle board, low density polyethylene (LDPE), dual purpose kerosene (DPK), corncob (CC) and sawdust (SD

1. Introduction

Wood is one of the major construction materials in the world that can be used in so many ways during and after construction. Particle board is the non-specific term for a board produced from lignocellulosic materials (normally wood), fundamentally as discrete pieces or particles, as recognized from strands, consolidated with an engineered resin or other appropriate adhesive[1]. Wasted wood materials may be converted into useful products by producing particle boards, thereby conserving our natural resources[2,3]. Present commercial products of comminuted wood have some very desirable characteristics: availability in large sheets, smooth surfaces, uniformity in properties from sheet to sheet, and freedom from localized defects. The first particle boards were made up from wood related materials which were of wood scraps, shavings, sawdust, and off cuts, all glued together using a specialist resin[4–7]. Before the pieces were glued together, they would be smashed and pounded in order to create small enough pieces that would pass through a specific size wire mesh of 1 mm. In recent times, the thought for a little bit of refinement to improve the properties of the particle board and this was done by making a uniform sized chip, made from processing woods such as birch, larch, pine, and spruce. With the refinement, the particle board produced had better overall strength and required less resin to bind the whole lot together and better appearance of the board was achieved. Today's particle boards are formed by using compression and temperature to make a much harder consistency.

Agricultural waste product like sawdust and corncob has been alternatively used in the production of building material like particle board due to the damaging and continuous exploration of natural resources. The corncob been a lignocelluloses material can be used in the production of particle board which helps in the recycling of agricultural waste and which in return is capable of providing the particle board with high mechanical strength and a pre-established specific weight [3,8]. The application of sawdust as a material in the production of particle board gives rise to it density (compaction ratio), strength and stiffness [2,9,10]. The enormous discharge of this sachet in the street and public places is leaving the street to land pollution and this trigger the use as blinder in this study. This study tries to produce a particle board from corncob and sawdust using melted pure water sachet as binder.

2. Materials and methods

The materials used are sawdust, corncob, pure water sachet, grease and Kerosene or Dual Purpose Kerosene (DPK)

2.1 Materials preparation

The required Corn cob (CC) was room dried for about 3-5 days. The CC was manually crushed into smaller sizes and was treated with water of about 85°C to remove undesirable water soluble chemical substances like phenolic compounds which may interfere with the mixture of the melted pure water sachet used as binder for the board formation. The CC was then milled with the aid of a grinding machine and sieved to obtain particles passing through a British standard sieve of aperture size 3mm and retained in 1.18mm. After the collection of the Sawdust (SD), it was room dried for 3-5 days in other to reduce initial moisture content of 12%. Pure water sachet also known as low density polyethylene (LDPE) was collected within the university community. The LDPE bags were washed thoroughly to remove dirt and debris that may stand as impurities. It was then sundried for 3-5 days in order to remove retained moisture on the sachet. After drying, the sachets were cut into smaller pieces then measured accordingly.

2.2 Production of binder

In order to make a good and effective binder, priority was given to the required viscosity-grade for the anticipated product. This determines the amount of solvent (DPK) that will be used, for complete melting of the solute(LPDE) which was a variable of N-LDPE to Dissolving agent (DPK) proportion- 50g to 300ml was used. The main essence of the DPK is for the achievement of a good binding effect, it is crucial that LDPE solid liquefies having more kinetic energy to translate, rotate and having a desirable molten state and this can only be achieved with the aid of an organic non polar solvent under heat. 300ml of DPK was measured in a volumetric cylinder and 50.0g of the cut tiny bit of the LDPE was added and heated at 101 to 116 °C which was measured with the use of a thermometer for 20 to 40 min until the complete molten state of the mixture was attained. For the board size adopted for this study in terms of length x breadth x thickness of 350mm x 350mm x 10mm, the set value of solute and solvent was determined by the amount of LDPE required for a particular amount of CC and SD for a particular size of particle board to be produced. Thereafter, the mixture of the LPDE and DPK was heated together until a homogenous molten compound was formed.

3. Experimental procedure

3.1 Production of composite

In order to produce the composite (particle board), the calculated volume (according to experimental design in Table 1) of the CC and SD particles required was measured using a graduated bucket before the LPDE was heated. The SD and CC was oven dried for maximum 24 hours due to moisture content rate. The different compositions were then being measured by grams and then kept in black cellophane bags to avoid moisture absorption. After arriving at the appropriate amounts of the materials required, the binder was formed by measuring separately, the amount of LDPE and DPK. In each case the mixture of LDPE

IOP Conf. Series: Materials Science and Engineering 1036 (2021) 012024 doi:10.1088/1757-899X/1036/1/012024

and DPK was heated together until homogenous molten state of the binder was formed. The appropriate amount of the two composite was then added immediately into the binder when it was still under heating and then thoroughly mixed until they are blended together as shown in Plate 1.

Table 1. Experimental design of particle boards

Board Type	% Composition of saw dust(SD)	% Composition of Corncob(CC)	Pure water sachet (LDPE)	Kerosene(DPK) (ml)
$ASD_{25}CC_{75}$	25	75	700	4200
$ASD_{50}CC_{50}$	50	50	700	4200
$ASD_{75}CC_{25}$	75	25	700	4200
$BSD_{25}CC_{75}$	25	75	525	3150
$BSD_{50}CC_{50}$	50	50	525	3150
$BSD_{75}CC_{25}$	75	25	525	3150
$ESD_{25}CC_{75}$	25	75	350	2100
$ESD_{50}CC_{50}$	50	50	350	2100
$ESD_{75}CC_{25}$	75	25	350	2100

SD-Sawdust, CC-Corncob, LDPE to Sawdust Mixing Ratio A- 2:1, Ratio B - $1^1/_2$:1, Ratio E – 1:1

After cleaning the mold with aid of a soft haired brush in order to remove dirt and grease with oil to ensure easy removal of the particle board, the composite mixture was quickly transferred into the metal molds filling it to the top or brim of the panel and evenly spread on it while the top cover was placed as shown in Plate 2.

Constant factors

1. Board thickness-10mm
2. Pressing pressure-$1.230N/mm^2$
3. LDPE to dissolving agent (DPK) proportion-50g to 300ml
4. LDPE to sawdust mixing ratio– Ratio A(1: 1), Ratio B(1½: 1), and Ratio E(2: 1)
5. Sawdust and corncob volume for each level-350g

3.2 Pressing operations

After the top cover plate which has 4mm thickness was placed on the constituent in the molds, still in a semisolid form, the mold was transferred quickly to the hydraulic jack for compression. The pressure was maintained until the mixture becomes fairly cold before stacking took place (the process took place under 10-15min to ensure everything could bedone when it was still in it semi solid form). Thereafter it was demolded and left in air as shown in plate 3.

Plate 1: Composite of sawdust and corncob been mixed

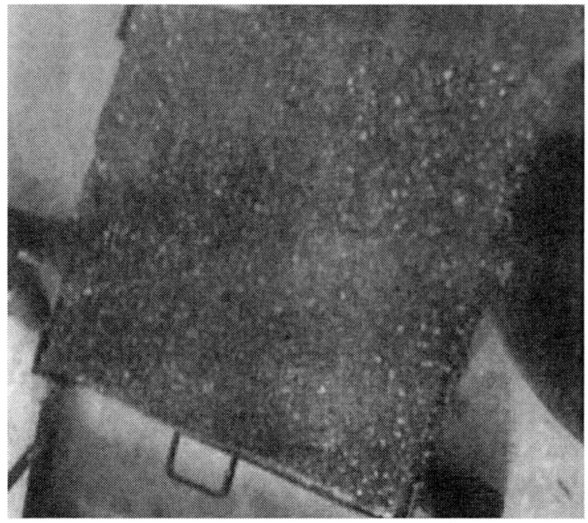

Plate 2:Composite mixture filled in the mold

Plate 3: Particle Board after demolding

4. Testing of particle board properties

The tests for the particle board are divided into two segments which is the physical properties test (Density Test, Water Absorption Test and Thickness Swelling Test) and the mechanical properties test (Static Bending Test (modulus of rupture (MOR) and modulus of elasticity (MOE)).

4.1 Physical Properties Tests

4.1.1 Density Test.

Density test was carried out on the particle board to determine the mean density for each produced board. The mean density values obtained for the 9 boards produced in the three (3) different ratios are shown in Table 2 and Figure 1. The board density obtained ranges from 491kg/$m^3$to 729kg/m^3. For all the three ratios A (2:1), B ($1^1/_2$:1) and C (1:1),boards with higher sawdust content has the highest density. The densities that were obtained are comparable to the particle board densities of 590kg/m^3 and800kg/m^3 of wood product industries. According to the American National standard particle board[11] which categorized board with densities lesser than 640kg/m^3 as low density particle board, three of the boards can be said to be a high density particleboard while the ones below the ANSI standard can be regarded as a low density particleboard.

Table 2. Density result of the particle board produced

Board Type	Density (Kg/m^3)
$ASD_{25}CC_{75}$	501
$ASD_{50}CC_{50}$	537
$ASD_{75}CC_{25}$	577
$BSD_{25}CC_{75}$	659
$BSD_{50}CC_{50}$	491
$BSD_{75}CC_{25}$	688
$ESD_{25}CC_{75}$	502
$ESD_{50}CC_{50}$	605
$ESD_{75}CC_{25}$	729

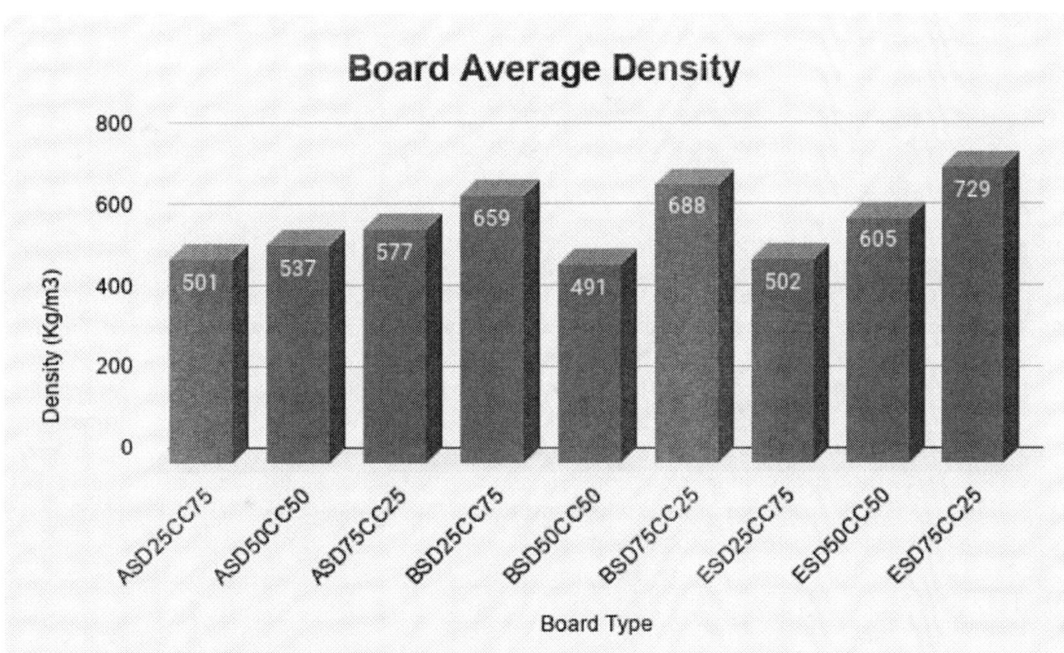

Figure 1: Board Average Density

4.1.2 Water absorption test.

The water absorption (WA) test was done to investigate the amount of moisture that could be retained in the particle board with varying composition of SD and CC and considering the binder that was used. The result of the nine (9) particle boards produced with different ratio after immersion in water for 2 hours and

24 hours at room temperature is presented in figure 2. After 2 hours of immersion, it was observed that ratio 2:1 board sample $ASD_{75}CC_{25}$ has a least value for the WA while ratio 1:1 board sample $ESD_{50}CC_{50}$ has the highest value. After 24 hours of immersion, it was also observed that the ratio 1:1 board sample $ESD_{50}CC_{50}$ still retained the highest value for WA while ratio 2:1 board sample $ASD_{25}CC_{75}$ has the lowest value for WA. It was also observed that there was a wide difference in WA for the boards ratioE (1:1) after the final immersion for 24 hours compared to the initial immersion in water for 2 hours, while the boards ratio A and B has a slight difference in the final and initial immersion.

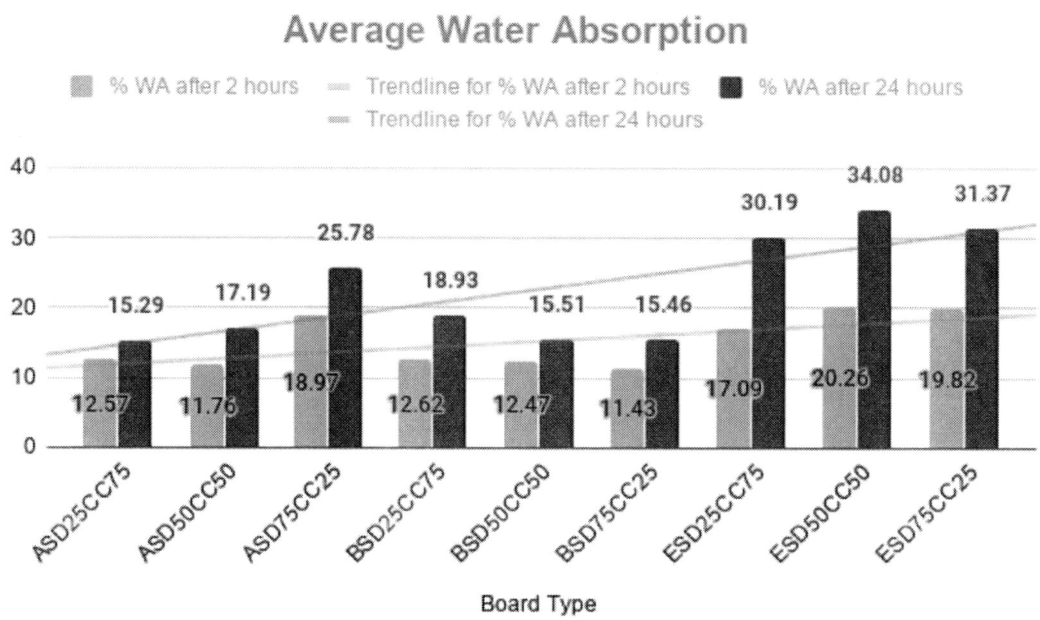

Figure 2. Board Average Water Absorption

4.1.3 Thickness swelling test.

Thickness swelling test (TS) is a dimensional analysis test that was used to determine the change in thickness of the board sample after it has been immersed in water for a given period of time. From the result in Table 3, it was observed that $ASD_{50}CC_{50}$ and $ESD_{50}CC_{50}$ has the lowest value for TS after the immersion in water also bringing to notice that the least board with the TS for each ratios are the ones having 50% CC and 50% SD in it. While $BSD_{25}CC_{75}$ has the highest value for TS. It could be deduced that the quantity of CC is directly proportional to the value of TS for the different ratios. The difference between the TS values for the Initial and final time was not a very high or significant difference while C2 board type did not even change value in thickness at all due to the application of DPK or Kerosene with the pure water sachet which was used as the binder.

Table 3. Result of thickness swelling (TS) of the particle board produced

Board Type	Average Thickness Swelling	
	% T.S. after 2 hours	% T.S. after 24 hours
$ASD_{25}CC_{75}$	7.71	8.59
$ASD_{50}CC_{50}$	1.81	2.1
$ASD_{75}CC_{25}$	4.39	4.78
$BSD_{25}CC_{75}$	8.6	8.9
$BSD_{50}CC_{50}$	3.97	4.2
$BSD_{75}CC_{25}$	4.42	4.44
$ESD_{25}CC_{75}$	4	4.5
$ESD_{50}CC_{50}$	1.81	1.81
$ESD_{75}CC_{25}$	4.21	4.82

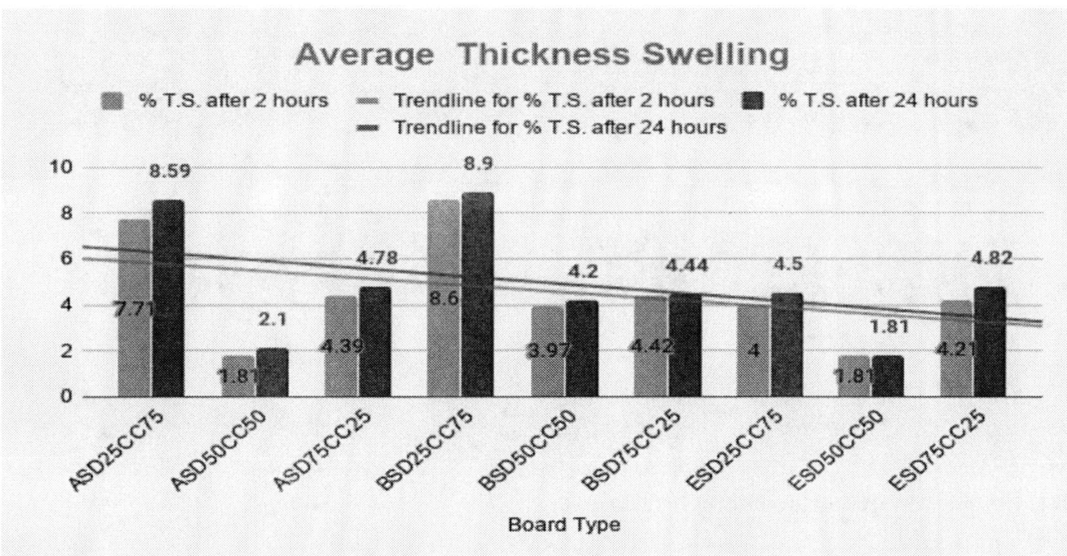

Figure 3: Board Average Thickness Swelling value

4.2 Mechanical properties test

Static bending: This test was carried out in accordance to the British Standard BS EN 310[12].. The MOR and MOE are being measured with the use of a Universal Testing Machine in N/mm^2. From the result, MOR was observed to increase in the board with higher percentage of saw dust i.e75% SD. $ESD_{75}CC_{25}$ has the highest value for MOR while the $BSD_{50}CC_{50}$ has the lowest value. It was also observed that in each ratio, the board with 50% CC and also 50% SD has the least value of MOR and MOE as shown in Table 4. MOE also show a similar value result to that of the MOR in term of board with higher percentage of SD having higher value in MOE, while $ASD_{50}CC_{50}$ possesses the lowest value.

Table 4. Result of MOR and MOE of the particle board produced

Board Type	Average MOR (N/mm^2)	Average MOE (N/mm^2)
$ASD_{25}CC_{75}$	2.39	18.71
$ASD_{50}CC_{50}$	1.52	5.14
$ASD_{75}CC_{25}$	2.93	19.65
$BSD_{25}CC_{75}$	2.64	20.84
$BSD_{50}CC_{50}$	0.86	6.93
$BSD_{75}CC_{25}$	6.35	52.82
$ESD_{25}CC_{75}$	6.02	59.23
$ESD_{50}CC_{50}$	3.81	35.4
$ESD_{75}CC_{25}$	8.19	63.07

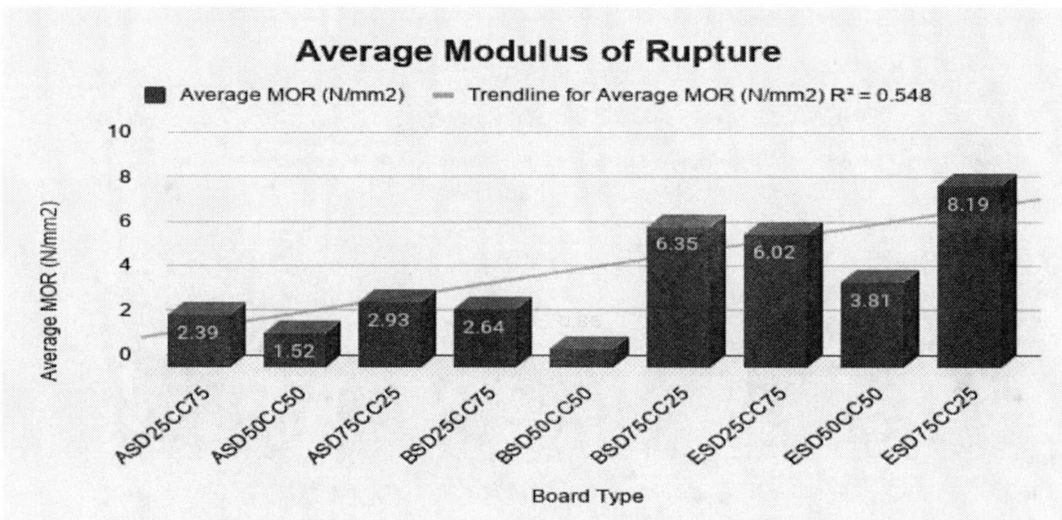

Figure 4: Graphical representation of MOR values

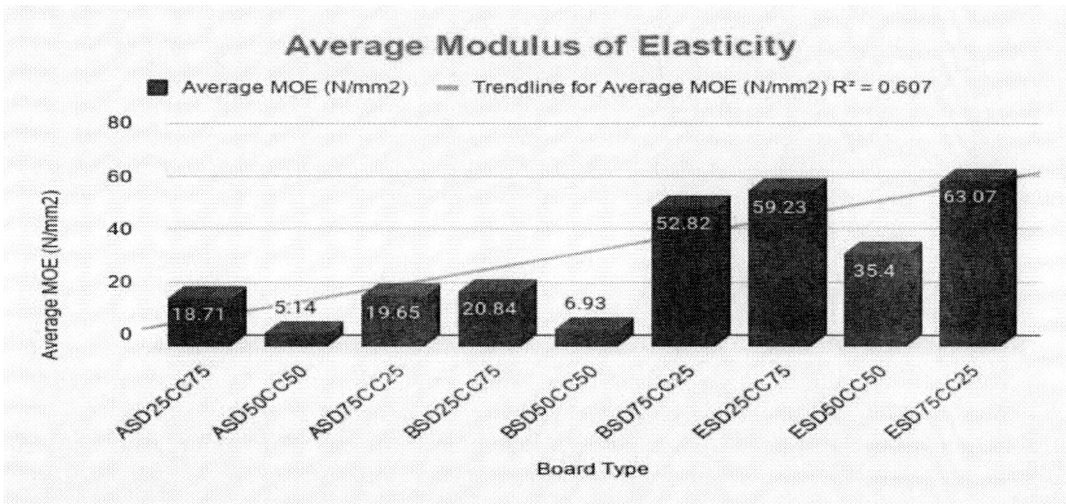

Figure 5. Graphical representation of MOE values

From the MOE and MOR graph in Figure 4 and 5, there was a very high increasein the values of MOR and MOE as the quantity of binder reduces. According to BS EN 13353[13], it was recommended that panel or board of thickness greater than20mm but less than 30mm are expected to have a minimum density of 420kg/m^3. It also required that, boards which are intended for structural purposes which have densities greater than the minimum requirement are required to have their minimum values for MOR and MOE to between a 5N/mm^2 and 400N/mm^2 respectively. From the result acquired, all the boards passed the minimum requirement for board density having it lowest density to be 491kg/m^3.However the requirement for MOR and MOE could not be attained by the board but for the MOR, only three (3) boards where above the 5N/mm^2 while for the MOE, none of the board attained the minimum requirement of 400N/mm^2 and this is in accordance with Sekaluvu, et al. [14]who worked on particle board made out of CC obtained mean values of MOE to range between 5.89 ± 6.00 N/mm^2 to 61.82 ± 10.09 N/mm^2 and also MOR which ranged between 0.32 ± 0.14N/mm^2 to 1.50 ± 0.16 N/mm^2.The low performance of the board in MOE and MOR could be caused by the compaction or pressing and also the curing process and also a slight consideration to the binder used in the production of board. Therefore, the board could be used for ceiling board and wall cladding and not for structural and load bearing because the board did not meet such requirement for work.

5. Conclusion

It is therefore established that particleboard could be produced from the combination of SD and CC with the addition of LDPE as binder which gave result of boards which fell into the two type board according to ANSI A208.1 [11]as three board type passed 640kg/m^3. Also the Mechanical properties of the board such as MOR and MOE were influenced by the amount of agro residue and binder used but they were not influenced by the board density. In conclusion the particle board produced could be used for can be used for ceiling board and wall cladding.

References

[1] Ma D and Kokta B V 1991 Studies on the Preparation and properties of particle boards made from bagasse and PVC : II . *Influence of the Addition of Coupling Agents* **35** 251–261.

[2] Atoyebi O D, Adediran A A and Adisa C O 2018 Physical and mechanical properties evaluation of particle board produced from saw dust and plastic waste *Int. J. Eng. Res. Africa.* **40** 1–8. doi:10.4028/www.scientific.net/JERA.40.1.

[3] Atoyebi O D, Odeyemi S O, Azeez L O and Modupe A E 2019 Physical and mechanical properties evaluation of corncob and sawdust cement bonded ceiling boards *Int. J. Eng. Res. Africa.* **42** 65–75. doi:10.4028/www.scientific.net/JERA.42.65.
[4] Verma D, Gope P C, Shandilya A, Gupta A and Maheshwari M K 2013 Coir fibre reinforcement and application in polymer composites : A Review **4** 263–276.
[5] Baldea B, Furtos G, Antal M, Nagy K, Popescu D andNica L 2013 Push-out bond strength and SEM analysis of two self-adhesive resin cements : An in vitro study *J. Dent. Sci.* **8** 296–305. doi:10.1016/j.jds.2013.01.007.
[6] Monteiro R, Filho D B, Marin L, Monteiro K, Oliveira L. and Roberto V 2011 Hybrid chipboard panels based on sugarcane bagasse, urea formaldehyde and melamine formaldehyde resin *Ind. Crop. Prod.* **33** 369–373. doi:10.1016/j.indcrop.2010.11.007.
[7] Baeza F J, Saval J M, Ivorra S and Maci A 2016 Mechanical properties of boards made in biocomposites reinforced with wood and Posidonia oceanica fibers *Int. J. composit esb* **104** 201-210. doi:10.1016/j.compositesb.2016.08.018.
[8] Atoyebi O D, Osueke C O, Badiru S, Gana A J, Ikpotokin I, Modupe A E and Tegene G A 2019 Evaluation of Particle Board from Sugarcane Bagasse and Corn Cob, *Int. J. Mech. Eng. Technol.* **10** 1193–1200.
[9] Rajaseenivasan T, Srinivasan V, Qadir G S M and Srithar K 2016 An investigation on the performance of sawdust briquette blending with neem powder *Alexandria Eng. J.* **55** 2833–2838. doi:10.1016/j.aej.2016.07.009.
[10] Atoyebi O D, Awolusi T F and Davies I E E 2018 Artificial neural network evaluation of cement-bonded particle board produced from red iron wood (Lophira alata) sawdust and palm kernel shell residues *Case Stud. Constr. Mater.* 00185. doi:10.1016/j.cscm.2018.e00185.
[11] American National Standard Institute., ANSI A208.1-1999 PB.pdf, (n.d.).
[12] BS.EN.310:1993, BS EN 310:1993, "Wood-based panels. Determination of modulus of elasticity in bending and of bending strength, London, UK, 1993.
[13] British, Standard, Institution, BS EN 13353.2003 British Standard Institution Solid wood panel (SWP)-Requirements, 2003.
[14] Sekaluvu L, Tumutegyereize P and Kiggundu N 2014 Investigation of factors affecting the production and properties of maize cob-particleboards *Waste and Biomass Valorization J.* **5** 27–32.

IOP Conf. Series: Materials Science and Engineering 1036 (2021) 012025 doi:10.1088/1757-899X/1036/1/012025

Household energy consumption within a low-income neighbourhood in Akure, Nigeria

S D Olanrewaju[1]* and O B Adegun[1]
[1] Department of Architecture, Federal University of Technology, Akure, Nigeria.
*Corresponding Author: olanrewajso@futa.edu.ng, oyindami@yahoo.com

Abstract. Globally, energy consumption represents a major contributor to climate change and global environmental problems. Identifying the potential factors explaining the variability in household energy consumption in buildings aids the development of appropriate energy conservation policies. The analysis of energy in residential sectors captures energy consumed by households to maintain operational functions of their daily lives. This study examines the household energy consumption in a low-income neighbourhood in Akure. Based on the results from a survey of 80 households, this study presents the energy consumption pattern of households in the study area. The study finds that energy is dominated by fossil fuel sources. Predominant proportion of energy consumed is used for cooking. The study concludes by outlining some practical implications for policy and practice.

1. Introduction

Globally, the building sector consume about 40% of the total energy consumption and accounts for 33% of CO_2 emissions [1]. Building energy demand is constantly increasing due to population growth, high rate of urbanization, technological advancement and lifestyle changes [2]. Energy efficiency is a key strategy effective to minimize energy production cost, utilization of natural resources as well as climate change mitigation [3]. The residential sector is the largest source of energy demand especially in developing economies [4]. For example, residential sector is the second and third largest energy user in china and Malaysia respectively [5, 6]. In Nigeria, about 65% of the annual energy consumption is used by the residential sector [7]. Therefore, residential sector is an important target for improving energy efficiency.

Building energy analysis consists of operational and embodied energy. While embodied energy deals with energy used in the course of building construction, operational energy is the amount of energy required to maintain the operational functions during the lifecycle of a building. The evaluation of energy performance of existing buildings serves as basis for design and non-design strategies for retrofitting or behavioural change to achieve energy-efficient buildings [9]. The potential interventions for energy-efficiency through embodied energy might be quite challenging for the residential sector in Nigeria since the existing housing stock is majorly self-built. In addition, households are made up of individuals with different socio-demographic characteristics, wherein compliance to energy regulation might be difficult [10]. Therefore,

IOP Conf. Series: Materials Science and Engineering 1036 (2021) 012025 doi:10.1088/1757-899X/1036/1/012025

understanding operational energy consumption patterns is imperative and seem a better approach for the development of strategies and policies for energy conservation and energy efficiency.

Studies from the developed countries have explored energy consumption pattern in residential sector [11-13]. Previous studies have also explored the characteristics and factors influencing energy consumption in developing countries such as China [14] and Malaysia [15]. Some studies examined energy consumption for cooking and grid electricity in Nigeria [16-19]. However, there is paucity of research on residential energy consumption that captures operational energy in low-income residential urban setting. Hence this study, seeks to fill this gap by exploring household operational energy consumption pattern in Akure. To explore the existing knowledge base in available literature, the next section presents a review that provides overview on operational energy consumption, residential energy sources and conservation. The subsequent sections discuss the methodological approach for data collection and findings of this study.

2. Literature Review

Residential energy consumption analysis is based on available energy sources utilized by households [20] and the sources can be categorized into that which is renewable and non- renewable. Nigeria is endowed with abundant renewable energy sources such as solar, wind and hydropower. For example, for solar energy there is an average solar radiation of 5.25 kWh/m2/day which, if harnessed, might meet the entire national household energy demand [21]. Despite the abundance of renewable energy sources, Nigerian energy supply is mostly dependent on non-renewable energy sources such as natural gas, petroleum products, and coal. High rate of dependency on fossil fuels can be linked to electricity supply distribution problems. The installation of renewable energy requires technical expertise which is not yet widespread and high installation cost which most homes cannot afford also constitutes an hinderance [22].

Residential energy consumption varies significantly across different geographical region due to climatic variations across cities and household formations. In Nigeria, energy consumed within households is predominately from electricity, LPG, diesel, kerosene. The major end-use activities for residential energy are lighting, cooking and household appliances. Household cooking takes up to 91% of the energy consumed in households [23]. A survey conducted in urban SouthWest Nigeria in 2012 found that the main fuel used for cooking is kerosene (about 90% of the households) while electricity and LPG are used minimally [24]. A more recent survey in Bauchi city (NorthEast Nigeria) revealed something different. Kerosene still was mostly used (41%), followed by wood (35 %), LPG (24 %) and charcoal (20 %) [25]. In Malaysia, LPG and petroleum products are the dominant household energy sources, which accounted for the largest share (27%), followed by wood (25%), electricity and natural gas with a share of 18% and 3%, respectively [15]. In Chinese households, space heating, cooking, and water heating dominate energy-consuming end uses, with 45% of total energy needs from coming district heating supplies [14].

Household energy usage is influenced by individual and household situational factors that affect energy-related practices [26]. Several studies have established the influence of socio-economic status e.g. income [20], home ownership [27], household size [28], dwelling type/size and number of appliances [29] on the variability of energy usage in households. In addition, psychological factors such as occupants' attitude, behavioural control and awareness are also important predictors of energy consumption [27]. For example, primary influencing factors of energy consumption in China include household income and size, presence of children, home ownership, living area, and households' awareness [20]. Corroborating this, single occupant Irish apartment consume less electricity than apartment with two or more occupants.]. It has also been confirmed that occupant's indoor temperature preference, amount of time spend indoor and attitudes are relevant behaviourial variables [31].

IOP Conf. Series: Materials Science and Engineering 1036 (2021) 012025 doi:10.1088/1757-899X/1036/1/012025

Dwelling characteristics have been considered as an interactive factor influencing the energy-related practices of households. . In a survey of American households, it was revealed that dwelling size and number of household appliances mediates the effects of household socio-economic characteristics on energy consumption [12]. Apart from dwelling size, the type of dwelling also have significant impact on energy consumption especially for space conditioning [32]. For instance, [33] observed that households living in detached buildings consume more electricity than those in semi-detached buildings, and these households consume more than those in terrace and apartment buildings. The use of air-conditioners and high-energy consuming appliances in newer apartments contribute to high consumption of electricity [34]. The use of TV in US households have significant effect on domestic electricity use [35].

The influence of some situational factors in Nigeria is evident in the literature accessed. For instance, [17] investigated energy use and emission pattern in south western Nigeria, the study revealed that post-paid low-income earners utilized more energy than prepaid high-income earners. Energy consumption was found to vary across density, as densely populated area consumes more energy than sparsely populated areas [19]. In a survey of energy use in public housing, [36] revealed high energy use intensity based on non-renewable energy sources. The price of petroleum-derived energy sources also affects energy consumption pattern for cooking. According to [16] low-income households switched from kerosene to fuel wood and other inefficient energy sources due to increase in price of fossil fuels. Apart from cost increase, availability of energy sources, increase in household size, dwelling ownership status, seasonal variation, education level and house location influence household cooking energy choice [25].

3. Research methodology

The purpose of this study is to examine energy source and consumption pattern in households within the inner-city area of Akure, Ondo State in the southwestern region of Nigeria. Akure is a medium-sized city with population of around 484,798 people in 2006, which is projected to have grown beyond 588,000 at present [37]. The core area of Akure, which consist of low-income informal housing, was selected as the study area. About one-third of household heads in the area earned less than 10,000 naira (US$30) monthly [38]. A quantitative research approach was adopted, and a survey was conducted. Following the review of literature, a questionnaire (as research instrument) structured into two parts was developed. The first part included questions about demographics such as gender, age, income, educational status, tenure type, household size. The second part consisted of questions about operational energy sources, the pattern of energy used, frequency of usage and amount incurred on energy.

The research population consists of all residential buildings within the study area - the Old Stadium road/Oke Igan neighbourhood in Akure. Satellite image helped to identify all buildings in the area and a sample of 100 households spread across the entire neighbourhood was taken. The questionnaire was administered to every fifth house, where an adult available and willing to respond was engaged with the questionnaire. The research assistants completed the questionnaire by interviewing each respondent. Total of 80 questionnaires administered to the respondents were successfully completed. The data was analyzed using descriptive statistical tool. Based on the data collected, energy consumption of a typical household in the study area was calculated using the following basic equations for primary energy content for electricity and direct fuel consumption [39-41].

$$OE=PEoGE + PEoDFC \quad (1)$$

$$PEoGE = GE x 3.6 x PEF \quad (2)$$

$$PEoDFC= DFC \times LHV \quad (3)$$

IOP Conf. Series: Materials Science and Engineering 1036 (2021) 012025 doi:10.1088/1757-899X/1036/1/012025

Table 1: Description of symbols

Symbol	Description	Value	Unit
OE	operational energy		MJ
PEoGE	primary energy content of grid electricity		
PEoDFC	primary energy content of direct fuel consumption		MJ
GE	grid electricity consumption based on electricity bill and unit price of electricity		kWh
	Conversion factor	3.6	MJ/kWh
DFC	Fuel consumption		Litres
LHV	Lower heating value of fuel		MJ/Kg
	DPK	43.1	
	PMS	44.0	
	LPG	45.3	

4. Findings and Discussion

This section presents findings from the survey. It deals with sources of energy, what activities energy is used for and how much energy is consumed by households in the study area. As shown in table 2, majority of the respondents are males (51.3%). A total of 31.3% are between 18-24years, 23.8% are between 25-34years, 20% are between 35-44years while 15% are between 45-54years. In terms of academic qualifications, 66.3% had secondary education, 21.3% had tertiary education, 10% had primary school education while 2.4% had no form of education. Furthermore, 42.5% earn below #21,000 monthly, 48.7% earn between #22,000 - 60,000 monthly, 7.5% earn between #61,000-120,000 monthly while 1.3% earn above #120,000. These indicate diversity in household composition within the area. As initially stated, the results confirm that the study area is dominated by low-income earners.

Table 2: Household characteristics

Variables	Frequency (N=80)	Percentage (%)
Gender		
Male	41	**51.3**
Female	39	**48.7**
Age		
18-24	25	**31.3**
25-34	19	**23.8**
35-44	16	**20**
45-54	12	**15**
Above 55	8	**10**
Level of Education		
None	2	**2.4**
Primary	8	**10**
Secondary	53	**66.3**
Tertiary	17	**21.3**
Monthly Income		
Below #21,000	34	**42.5**
22,000-60,000	39	**48.7**
61,000- 120,000	16	**7.5**
Above 120,000	1	**1.3**
Employment status		
Self employed	48	**60**

IOP Conf. Series: Materials Science and Engineering 1036 (2021) 012025 doi:10.1088/1757-899X/1036/1/012025

Employed	14	**17.5**
Not employed	18	**22.5**
Household size		
1	5	**6.3**
2-4	30	**37.4**
Above 4	45	**56.3**
Tenure status		
Owner occupied	35	**43.8**
Rented	42	**52.5**
Family House	2	**2.5**
Others	**1**	**1.2**

As shown in table 3, the dominant energy sources are grid electricity, PMS, DPK, LPG. Among the 80 households surveyed, 93.3% use grid electricity, 58.75% use DPK, 67.6% use PMS, and 95%use LPG. This indicates overlapping use of different types of energy as most households use more than a source of energy for an activity e.g. cooking, lighting. The findings align with findings of earlier studies, which posit that LPG, petroleum products, electricity are the predominant energy sources in similar developing economy [14, 24]. Multiple energy use patterns in urban households were identified in Bauchi [25].

Table 3: Major operational energy sources and purpose of usage

Energy source	Usage	Usage ratio
Electricity	Electrical appliances (television, fan, light, hot plate, etc.) water pump	93.3%
DPK	Cooking	58.75%
PMS	Generator	67.6%
LPG	Cooking	95%

The end-use activities for residential energy were cooking, lighting, household electrical appliances and water pumping. In an average household, gas cooker and kerosene stove are the major cooking devices, electric cooker, kettle as well as boiling rings are also available for daily cooking and domestic hot water. Despite the shortage of electricity supply, residents still acquire a quite number of home appliances such as refrigerators, lamps, televisions, computer washing machines. Nearly, 75% of household surveyed have electric fans for space cooling. This supports earlier findings that emphasized the transition from traditional energy sources to modern sources in developing countries [42]. This indicates that the respondents acquire different appliances to improve their quality of life.

Table 4: Electricity connection and expenditure

	Frequency	Percentage
Meter type		
Pre-paid	30	37.5
Postpaid	48	60
Non	2	2.5
Total	80	100
Expenditure		
Below #1500	8	10
#1501-2500	19	23.75
#2501-3500	25	31.25
#3501-4500	12	15
Above #4500	16	20
Total	**80**	**100**

IOP Conf. Series: Materials Science and Engineering 1036 (2021) 012025 doi:10.1088/1757-899X/1036/1/012025

Information about mode of electricity billing and associated monthly expenditure were sought. As shown in table 4, 37.5% of the respondents use prepaid meter, 60% use postpaid meter while 2.5% do not have either of the two. It is evident that post-paid billing system is more prominent in the study area. The findings further revealed that most householders using prepaid meter reported consciousness of the conservative measures for available limited supply. The average expenditure on grid electricity is ₦3000. Considering the electricity tariff of BEDC as of 2019 (#29 per KWh), the average electricity consumption was estimated to be 1236kWh annually per household and the energy content of electricity was estimated to 16,019MJ using equation 1. With an average household size of 4, the energy consumption per capita is 309kWh, this is higher than the per capita estimation of 156kwh in 2012.

Table 5: Weekly direct fuel consumption

	Frequency	Percentage
DPK		
None	33	41.3
1-5litres	16	20
6-10litres	16	20
11-20litres	12	15
21-30litres	2	2.5
Above 30 liters	1	1.3
PMS		
None	26	32.4
1-5litres	5	6.3
6-10litres	15	18.8
11-20litres	12	15
21-30litres	12	15
Above 30litres	10	12.5
LPG 6kg		
None	4	5
Weekly	11	13.75
Every two weeks	11	13.75
Every three weeks	12	15
Every four weeks	42	52.5

The direct fuel consumption was estimated based on the average liquid fuel consumption for each type of fuel. As shown in table 5, average DPK and PMS used per week are 7litres and 15litres respectively. This amounts to 336litres of DPK and 720litres of PMS annually. On the average, the residents use 6kg LPG cylinder monthly. This amount to 72kg annually. Using equation 2, the primary energy content for DPK, PMS and LPG was estimated. The estimated energy contents are 14,482MJ, 31,680MJ and 3,262MJ for DPK, PMS and LPG respectively. Therefore, the annual operational energy is 65,443MJ. This indicates that the energy mix in the study area is fossil fuel driven basically due the use of fossil fuel powered generators as a result of the shortage of electricity supply. Based on the household characteristics obtained, the lower low-income earners use more of DPK for cooking than LPG which is less efficient and have the potential of increasing the greenhouse emission in the area. This finding aligns with previous studies [20, 16, 36] that identified DPK as the most used fuel for cooking. This suggests that the prices of energy source affects choice of energy for cooking. Only one respondent derive energy through solar energy. This confirms that the study area is largely dependent on non-renewable energy sources.

5. Conclusions

This paper presents the operational energy consumption pattern in an Akure inner-city neighbourhood based on data obtained from 80 households. The sources of energy are important aspects of physical development and economic growth. Overall, we find that the major sources of energy are non-renewable which are dominated by fossil fuel resources. The average estimated household energy consumption was 65,443MJ. The postpaid metering system is the mostly available electricity billing system.

Our results are important for policy implications. The diversification of energy basket for the residential sector is important. Hence, the installation of appropriate infrastructure to enhance energy provision using the abundant renewable energy resources present in the country is more than urgent. In addition, initiatives to drive energy efficiency such as use of incandescent light bulbs, energy performance labeling of electrical appliances should be enhanced. In addition, the full implementation of prepaid metering system is necessary as it is in consonance with global best practices. Promotion of energy-efficiency among households through relevant information, awareness and rewards are necessary.

References

[1] Baek C, Park S, Suzuki M and Lee S 2013 Life cycle carbon dioxide assessment tool for buildings in the schematic design phase. *Energy Build* **61** 275-287.

[2] Pérez-Lombard L, Ortiz J and Pout C 2008 A review on buildings energy consumption information. *Energy Build* **40(3)** 394-398.

[3] Oyedepo S 2014 Towards achieving energy for sustainable development in Nigeria. *Renewable and Sust Energy Reviews* **34** 255-272.

[4] Nejat P, Jomehzadeh F, Taheri M, Gohari M and Majid M 2015 A global review of energy consumption, CO2 emissions and policy in the residential sector *Renewable and Sust Energy Reviews* **43** 843–862.

[5] Yuan X, Wang X and Zuo J 2013 Renewable energy in buildings in China—a review. *Renewable and Sust Energy Reviews* **24** 1–8

[6] Malaysia Energy Commission 2014 National Energy Balance. Putrajaya.

[7] Oyedepo S, Adekeye T, Leramo R, Kilanko O, Babalola P, Balogun A, and Akhibi V 2016 Assessment of energy saving potentials in covenant university, Nigeria, *Energy Eng* **113(3)** 7–26.

[8] Giordano R, Giovanardi M, Guglielmo G and Micono C 2017 Embodied energy and operational energy evaluation in tall buildings according to different typologies of façade. *Energy Procedia* **134** 224–233.

[9] Olanrewaju S, Adetunji O and Ogundepo T 2019 Achieving Energy Efficient Building through Energy Performance Analysis of Building Envelope in Student Housing. *Journal of Physics: Conference Series*, doi:10.1088/1742-6596/1378/4/042023.

[10] Hori S, Kondo K, Nogata D and Ben H, 2013 The Determinants of Household Energy-saving behavior: Survey and Comparison in Five Major Asian Cities. *Energy Policy* **52** 354–362.

[11] Fertel C, Bahn O, Vaillancourt K and Waaub J 2013 Canadian energy and climate policies: a SWOT analysis in search of federal/provincial coherence. *Energy Policy* **63** 1139–50.

[12] Karatasou S and Santamouris M 2019 Socio-economic status and residential energy consumption: A latent variable approach. *Energy Build* **198** 100–105.

[13] Heinonen J and Junnila S 2014 Residential energy consumption patterns and the overall housing energy requirements of urban and rural households in Finland. *Energy Build* **76** 295–303.

[14] Zheng X, Wei C, Qin P, Guo J, Yu Y, Song F and Chen Z 2014 Characteristics of residential energy consumption in China: Findings from a household survey. *Energy Policy* **75** 126–135

[15] Azlina A, Siti E, Abdullah E, Kamaludi M and Radam M, 2015 Energy conservation of residential sector in Malaysia. *Journal of Business and Social Development* **3(2)** 51-62.

[16] Adelekan I, and Jerome A, 2006 Dynamic of household energy consumption in a traditional African city, Ibadan. *Environmentalist* **26** 99–110.

[17] Abimbola O, Amori A, Omotosho O, Igbode I, Omoyeni D and Ajayi-Banji A 2015 Investigation of energy use pattern and emission discharge in Nigeria: a case study of South-west zone. *International Journal of Engineering and Technology Innovation* **5(1)** 56-65.
[18] Ubani O, Umeh L and Ugwu L 2013 Analysis of the electricity consumption in the south - east geopolitical region of Nigeria. *Journal of Energy Technologies and Policy* **3(1)** 2224-3232.
[19] Akinola A, Oginni O, Rominiyi O and Eiche J, 2017 Comparative study of residential household energy consumption in Ekiti state-Nigeria. *British Journal of Applied Science & Tech* **21(2)** 1-10
[20] Zhang J, 2010 Operational energy consumption and GHG emissions in residential sector in urban China: An empirical study in Jinan. Unpublished Master's thesis, Massachusetts Institute of Technology.
[21] Abam F, Nwankwojike B, Ohunakin O and Ojomu S, 2014. Energy resource structure and on-going sustainable development policy in Nigeria: A review. *Int Journal of Energy and Env Engr* **5(2-3)**102.
[22] Ogunleye I and Awogbemi O 2011 Constraints on the use of solar photovoltaic as a sustainable power source in Nigeria, *American Journal of Scientific and Industrial Research* **2(1)** 11-16.
[23] Oyedepo O 2012 Energy and sustainable development in Nigeria: the way forward. *Energy Sustain Soc* **2**15.
[24] Ajayi P 2018 Urban household energy demand in southwest Nigeria. *African Development Review* **30(4)** 410-422.
[25] Bisu D, Kuhe A and Iortyer H, 2016. Urban household cooking energy choice: an example of Bauchi metropolis, Nigeria. *Energy, Sustainability and Society* 6(1) 15-27.
[26] Frederiks E, Stenner K and Hobman E 2015 The socio demographic and psychological predictors of residential energy consumption: a comprehensive review. *Energies*, *8*, 573-609.
[27] Rehdanz K, 2007 Determinants of residential space heating expenditures in Germany. *Energy Econ.* **29** 167–182.
[28] Bedir M, Hasselaar E and Itard L 2013 Determinants of electricity consumption in Dutch dwellings. *Energy Build* **58** 194–207.
[29] Kavousian A, Rajagopal R and Fischer M 2013 Determinants of residential electricity consumption: using smart meter data to examine the effect of climate, building characteristics, appliance stock, and occupants' behavior *Energy* **55** 184–94.
[30] Leahy E and Lyons S 2010 Energy use and appliance ownership in Ireland. *Energy Policy* **38(8)** 4265–79.
[31] Nair G, Gustavsson L and Mahapatra K, 2010 Factors influencing energy efficiency investments in existing Swedish residential buildings. *Energy Policy* 38, 2956–2963.
[32] Kaza N 2010 Understanding the spectrum of residential energy consumption: A quantile regression approach. *Energy Policy* **38** 6574- 6585.
[33] Chong H, 2012 Building vintage and electricity use: old homes use less electricity in hot weather. *European Economic Review* **56(5)** 906-930.
[34] Carlson D, Mattews H and Bergés M 2013 One size does not fit all: averaged data on household electricity is inadequate for residential energy policy and decisions. *Energy Build* **64** 132-144
[35] Sanquist T, Orr H, Shui B and Bittner A 2012 Lifestyle factors in U.S. residential electricity consumption *Energy Policy* **42** 354-64.
[36] Ezema I, Olotuah A and Fagbenle O, 2016 Evaluation of energy use in public housing in Lagos, Nigeria prospects for renewable energy sources. *Int Journal of Renewable Energy Dev* **5** 15-24
[37] National Population Commission, 2006. Federal Republic of Nigeria Official Gazette Abuja.
[38] Ayoola H, Fakere A and Olusoga O, 2019. Place Attachment in Poor Residential Neighbourhoods of Akure Nigeria. *Covenant Journal of Research in the Built Environment* **7(2)** 20-36.
[39] GREET 2010 The greenhouse gases, regulated emissions and energy use in transportation model, GREET1.8d.1. http://greet.es.anl.gov/ accessed 08/05/2020.
[40] Hofstrand D 2008 Energy measurements and conversions, file C6-86, University Extension, Iowa State University, www.extension.isate.edu/agdm.

[41] IINAS 2015 Development of the primary energy factor of electricity generation in the EU-28 from 2010-2013, International Institute for Sustainability Analysis and Strategy, Darmstadt, Germany.
[42] Alam M, Sathaye J and Barnes D 1998 Urban Household Energy Use in India: Efficiency and Policy Implications. *Energy Policy* **26(11)** 885-89.

Assessment of construction sites' accident: strategies for sustainable construction

I A Ijaola[1], A O Akerele[1] and O H Omolayo[1]
[1]*Department of Building Technology, School of Environmental Studies, Yaba College of Technology, Yaba, Lagos, Nigeria*
Corresponding author email: iredbuilder@gmail.com

Abstract. Diverse causes of accidents abound on construction sites and these have negative impacts on the construction process. These negative impacts are enormous, leading to non-sustainable construction. However, achieving sustainable construction requires reduction in the occurrence rate of accidents. The aim of this study is to investigate the occurrence rate of accident types on construction site. Accident types were majorly categorised into four namely, earth moving machinery, electrocution, construction fall, and others. From the positivist research paradigm and adopting cross-sectional survey research design, questionnaires were used to elicit information from personnel on construction sites. The mean of the occurrence rate of accidents was calculated. Findings show that construction site fall has high occurrence rate while earth moving machinery, electrocution and others have moderate rate of occurrence. The study concludes that the most frequently occurring type of accidents on construction site is construction fall.

Keywords: Accident, construction site, sustainable construction

1. Introduction

A sustainable construction is one in which construction activities, which have negative impacts, are minimised and positive impacts maximised for achieving balance between environmental, economic and social performance [1]. The idea of sustainable construction is to minimise negative impacts, hence sustainable construction is one in which the occurrence rate of accidents is minimised to achieve environmental, economic and social performance. Accident is an undesirable event yet it occurs on construction sites with consequences that can be described as negative impacts. Such consequences are fatal, major injury, minor injury, severe injury, damage, disabilities, deaths, increase in construction projects cost and time [2, 3, 4]. Further consequences include negative impacts on the economy, employers, employees and families [5]. Other negative impacts include demotivation of workers or reduced morale, conflict with owners and loss of productivity [6].

The diverse negative impacts of accidents negate the principle of sustainable construction; it is thus pertinent to examine the occurrence rate of accidents on construction site for proper understanding of this menace that impede sustainable construction. Previous studies on accidents focused on diverse areas. Wang [7] studied construction statistics with focus on types of accident, kind of work and location of accidents without considering the occurrence rate of accident based on it types. Jo [5] investigated the types of accident and the occurrence of accident but did not consider the occurrence rate based on types of accident. Gonzalez [8] studied the type of injury, part of body affected, agent in accident, accident mechanism and causes of accidents. Again, the occurrence rate of accident types was not investigated. Similarly, Yimaz [9] investigated type of accidents, affected body part and dangerous behaviours of workers that causes accident, but the occurrence rate of accident types was not considered.

There seems to be a gap in research on the categorisation of the types of accident and occurrence rate of accident types in literature. This study aims at minimizing accidents on construction sites by investigating the occurrence rate of accident types. The objectives of the study are; to identify the types of accident on construction sites and to investigate the frequency of occurrence of the accident types on construction site.

Achieving sustainable construction through minimizing negative impacts of accidents is important in construction industry. Knowledge of the occurrence rate of accident types will provide information for establishing strategies to minimise accident rate on construction sites. In addition, knowledge of the occurrence rate of accidents will also create awareness among construction workers and aid policy makers in formulating policy in construction industry; hence avoidance of such acts and behaviours that lead to accidents which in turn leads to sustainable construction.

2. Literature Review

2.1. Sustainable Construction

Sustainable construction coined from the two words 'sustainable' and 'construction' is a complex concept with different scope and meaning [10]. A clear definition of sustainable construction is important so as to adopt it into diverse areas of construction especially construction accident. Kamar [1] defined sustainable construction as 'construction activities whose negative impacts are minimised and positive impacts maximised so as to achieve a balance in terms of environmental, economic and social performance'. This definition explains the fact that the occurrence of accidents during construction activities are negative impacts, thus achieving balance in terms of environmental, economic and social performance emphasises reduction in negative impacts such as accident and increase in positive impacts.

Sustainable construction aims at achieving a balance between social, economic and environmental dimensions of construction [11]. The social dimension focuses on issues relating to health and safety, involvement of stakeholders, equality and diversity in workplace and creating employment opportunity [11]. Sustainable construction also considers safety, security, environment and health in construction [12]. Achieving sustainable construction from social dimension implies minimising the occurrence rate of accidents on construction site.

2.2 Types of Accident

An accident is described as unpleasant events that result in fatality, injury of the person involved and damage to property [2]. Different authors identified different types of accident on construction site. Such accident types include fall from height, fall from same level, fall from roofs, ladders and scaffold [5, 7, 13], struck by objects or falling object, hit by object [4, 7, 14, 15], electrocution or electric shock [5, 15, 16]; cut or caught in between [4, 5, 7], fire explosion [17] and others [7, 13].

Al-Tabtabal [17] discovered that the three major types of accident are falling from height, falling objects and tools while Adekunle [13], concluded that falls are the most common home building fatality. In addition, the four major types of accident are falls, contact with objects and equipment, exposure to harmful substance or environment and transportation accidents. Lee [15] identified falls, hit by objects and machinery and equipment error as the most common types of accident on construction site. In contrast, Bhole [14] found falls from height, slips, trips or falls on same level and handling, lifting or carrying as the three major types of accidents.

Previous authors [13, 14, 15, 16, 17] identified the different types of accident; however the types of accident were not categorised into major groups for easy identification and understanding. Based on the accident types identified in previous literature, this study categorises the types of accidents into four, namely earth moving machinery, electrocution, construction site fall and others.

3. Research Method

A cross-sectional survey design which investigates a phenomenon at a specific time from more than one respondent was adopted [18]. This is because it was required that data should be collected at a point in time and a representative sample taken from the population be used to generalise the findings.

Furthermore a cross-sectional survey research design was adopted because there is a need to determine the prevalence of accidents on construction sites [18]. Questionnaires were used in eliciting information from the respondents. The respondents of this study are construction professionals or workers namely, architects, builders, quantity surveyors, civil engineers, mechanical engineers and electrical engineers who have on one or more occasion be involved in an accident or witness an accident on construction sites. The study adopts a purposive sampling technique. A purposive sampling technique is a nonprobability sampling technique in which the researcher chooses the samples of the study based on their knowledge of the information and their willingness to share the information [19]. Purposive sampling technique was adopted in the selection of the samples because there is no list of professionals working on construction site in the study area [Lagos state] and because the study requires specific group of persons who have knowledge of construction site accidents to give the required information. One hundred and twenty questionnaires were distributed and one hundred returned questionnaires were valid and used for analysis representing 95% response rate.

3.1 *Measures of variables*

Types of accident was conceptualised as four latent variables namely; earth moving machinery, electrocution, construction site fall and others. All the observable variables that were used to measure all the four latent variables are shown in Table 1. The first latent variable 'earth moving machinery' was measured with 8 observable variables such as fall from machine, starting of vehicle when in gear. The second latent variable 'electrocution' was measured with 3 observable variables. Examples are, stepping on expose electric, electric shock, etc. The third variable addresses construction fall and it was measured with 5 observable variables. For example, fall from roof, fall from scaffold, etc. The fourth variable classified as 'others' was measured with 7 observable variables such as welding accident, fire and explosions, etc.

To identify the types of accident on construction sites, the types of accident were categorised into four. Respondents were asked to choose the types of accident on construction site. A two-point rating scale was used; Yes [1], No [2] respectively. Percentage of the respondents that answered "yes" was thus calculated.

To determine the rate of occurrence of accident types, respondents were asked to rate the frequency of occurrence of each accident type using 1.00-1.49 for very low level; 1.50-2.49 for low level; 2.50-3.49 for moderate level; 3.50-4.49 for High level and 4.50-5.00 for very high level. Mean score was then used to calculate the rate of occurrence of accidents on construction site.

3.2 *Demographic details of respondents*

43% of the respondents are builders by profession, 32% are civil engineers, 17% are architects, 15% are project managers while 2% and 1% are electrical and mechanical engineer respectively. This shows that respondents are professionally qualified to answer questions correctly. Furthermore, 35% are site managers, 28% are site supervisors, 16% occupy other positions such as human resource manager, 15% are project managers while 6% are safety manager. The diverse range of professionals' positions on construction sites indicates that they are aware of construction site accidents. In terms of site experience, 44% of the respondents have experiences of 5years and below while 42% have experiences between 6-10 years. 9% of the respondents have experience between 11-15 years, 2% have between 10-20 years and 1% has experiences above 20 years. A large percentage of the respondents have experience of 10 years and below, this suggests that most professionals working on construction sites are middle management staff, hence the reason for having large proportion of the respondents with experiences of 10 years and below. In addition, the diverse range of respondents' site experiences shows that they have vast knowledge of construction site accidents.

4. Result and Discussion of Findings

4.1 Types of accident

The result in Table 1 shows that on average 90% of the respondents identified 'other types' as construction site accident, 89% identified construction site fall as construction site accident, 87% of the respondents identified electrocution as construction site accident, while 79% identified earth-moving machinery as construction site accident. Overall, large percentage of the respondents identified all the listed types of accident in Table 1 as construction sites accident. Previous studies [13, 17] also identified fall from height, fall from roof, and electrocution as type of construction sites accident. The result of this study differs slightly from [13, 16, and 17] in that the types of construction accident were categorised into groups for easy identification and understanding.

Table 1: Types of Construction Site Accident

S/N	Types of Accident on Construction Site	%	S/N	Types of Accident on Construction Site	%
	Earth moving machinery	**79**		**Construction site fall**	**89**
1	Person fall from machine	70	12	Fall from roof	88
2	Vehicle started when in gear	78	13	Fall from crane	78
3	Person hit by machine which was moving in a forward direction	79	14	Fall from scaffold	95
5	Person injured when conducting maintenance on the machine	85	15	Slips and fall	88
6	Struck by backhoes	85	16	Falling debris, materials or object	96
7	Machine cutting worker	85		**Other types**	**90**
8	Reversing excavators	71	17	Building Collapse	96
	Electrocution	**87**	18	Wielding accident	89
9	Stepping on expose electric wire	90	19	Getting caught in-between objects or material	90
10	Electric shock	86	20	Fires and explosions	90
11	Power line contact	84	21	Trench collapses	92
			22	Gas explosion	88

4.2 Rate of occurrence of construction sites accident

The result in Table 2 shows that the frequency of occurrence of construction site accident is moderate except for construction site fall, which is high. This suggests that the rate at which falling from scaffold, falling of debris, materials or object on workers, fall from roof, fall from crane is high. The implication is that proper inspection of scaffold is not carried out before use. Furthermore, training of construction site workers on safety practices especially when working on height is not properly carried out on construction site. The result correlates with that of Jo [5] in which fall was identified as the dominant type of accident. In addition, the result differs from Lee [15] in which the three most occurring types of accident are fall, hit by objects, machinery and equipment. The difference lies in that fact that this study categorises accident types into major groups while Lee [15] did not categorise accident types. Individually, fall debris, materials or objects, slips and fall, and fall from scaffold were the three most occurring types of accidents. This differs from Adekunle [13] in that fall from roof, fall from ladder and fall from scaffold were identified as the types of accident with highest number of fatalities. The difference could be due to the scope of the study, while Adekunle's [13] finding is based on accidents in home building; this study generally focuses on construction projects.

The second most occurring group of accident types is 'others'. Others imply types of accident not mentioned. Previous studies also identified others as types of accident [5, 7, and 16]. Due to the fact that different types of accident occur, it may be difficult to classify them; as such they are categorised as others. In contrast to Jo [5] in which other types of accident was the least dominant type of accident, this study discovered that other types was the second most occurring types of accident. Furthermore, the study identified building collapse and trench collapse as the most occurring types of accident under the main category of 'others'. Although, Jo [5] did not consider the subtypes of other types of accidents, collapse was found to be the 6th dominant types of accident. The difference could be as a result of the study area; the implication is that building collapse occurs more in Lagos state compare to Korea.

Table 2: Rate of Occurrence of Accidents on Construction Site

S/N	Occurrence Level of Accident	1	2	3	4	5	TS	MS	SD	Remark
A	**EARTH MOVING MACHINERY**	**8**	**18**	**43**	**24**	**7**	**100**	**3.04**	**0.59**	**Moderate**
1	Person fall from machine	19	16	39	16	10	100	2.82	1.21	
2	Vehicle started when in gear	6	14	56	18	6	100	3.04	0.90	
3	Person hit by machine which	5	14	57	21	3	100	3.03	0.82	
4	Person injured when conducting	0	8	47	30	15	100	3.52	0.85	
5	Struck by back hoes	4	26	39	28	3	100	3.00	0.91	
6	Machine cutting worker	1	11	39	39	10	100	3.46	0.86	
7	Reversing excavators	21	39	24	13	3	100	2.38	1.05	
B	**ELECTROCUTION**	**6**	**17**	**37**	**27**	**14**	**100**	**3.27**	**0.88**	**Moderate**
8	Stepping on expose electric wire	3	10	39	29	19	100	3.51	1.01	
9	Electric shock	4	18	35	28	15	100	3.32	1.06	
10	Power line contact	10	22	36	24	8	100	2.98	1.09	
C	**CONSTRUCTION SITE FALL**	**2**	**10**	**24**	**45**	**19**	**100**	**3.70**	**0.73**	**High**
11	Fall from roof	3	11	37	38	11	100	3.43	0.94	
12	Fall from crane	1	12	24	51	12	100	3.61	0.89	
13	Fall from scaffold	2	10	24	46	18	100	3.68	0.95	
14	Slips and fall	4	6	22	44	24	100	3.78	1.01	
15	Falling debris, materials or object	0	10	13	45	32	100	3.99	0.93	
D	**OTHERS**	**2**	**16**	**36**	**33**	**14**	**100**	**3.43**	**0.63**	**Moderate**
16	Building Collapse	1	7	18	44	30	100	4	1	
17	Wielding accident	2	26	37	23	12	100	3	1	
18	Getting caught in-between objects	3	11	35	41	10	100	3	1	
19	Fires and explosions	2	17	41	28	12	100	3	1	
20	Trench collapses	1	9	29	43	18	100	4	1	
21	Gas explosion	1	23	44	22	10	100	3	1	
22	Cutting Accident	1	16	45	29	9	100	3	1	

[a] MS= Mean Score, TS= Total Score
[b]1.00-1.49 for *very low level*; 1.50-2.49 for *low level*; 2.50-3.49 for *moderate level*; 3.50-4.49 for *High level* and 4.50-5.00 for *very high level*

5. Conclusion

Achieving sustainable construction through determining the occurrence rate of accidents on construction sites is important. The study examined the occurrence rate of accidents on construction sites. The result indicates that four major types of accident namely earth moving machinery, electrocution, construction site fall and other types of construction accident occur on construction sites. Thus, the study concludes that accidents can be majorly classified into four types on construction site. In addition, the study discovered that the occurrence rate of construction site fall is high while the occurrence rate of earth moving machinery, electrocution, and other types are moderate. The study therefore, concludes that the frequency of; fall from roof, fall from crane, fall from scaffold, slips and fall, falling debris, materials or object is high on construction sites. Construction workers should therefore take cognise of these types of accident when working on construction site so as to prevent the frequent occurrence. This can be achieved by placing the accident types on signboard on construction sites to serve as a means of awareness and training of workers on sites.

References

[1] Kamar KM and Abdul Hamid Z 2009 Collaboration initiative on green construction and sustainability through industrialized buildings systems [IBS] in the Malaysian construction industry *Proc. of ICON-BSE 09 UTHM [Johor Baharu]*

[2] Health and Safety Executive [HSE] 2004 Investigating accidents and incidents: a workbook for employers, unions, safety representatives and safety professionals.

[3] Asanka WA and M Ranasinghe 2015 Study on the impact of accidents on construction projects In *6th Int. Conf. on Structural Engineering and Construction Management*, [Kandy, Sri Lanka] pp. 58-67.

[4] Pillay K and Haupt T C 2008. The cost of construction accidents: an exploratory study *Proc. of the CIB W99 14th Int. Conf. on Evolution of and Directions in Construction Safety and Health*, pp. 456-64.

[5] Jo BW, Lee YS, Kim JH and Khan RM 2017 Trend analysis of construction industrial accidents in Korea from 2011 to 2015 *Sustainability* **9** 1297 1-12.

[6] Ali AS, Kamaruzzaman SN and Sing GC 2010 A Study on causes of accident and prevention in Malaysian construction industry. *J. of Design+ Built*, **3** 1 95-104.

[7] Wang YS, Zhang Y, Poon SW, Huang H 2002 A study of construction site accident statistics *In Proc. of CIB W099 Implementation of Safety and Health on Construction Sites*. 223-7.

[8] González A, Bonilla J, Reyes C and Chavarro A 2016 Analysis of the causes and consequences of accidents occurring in two constructions projects *Revista Ingeniería de Construcción*, **31**1 5-16.

[9] Yilmaz F and Çelebi UB 2015 The importance of safety in construction sector: Costs of occupational accidents in construction sites *Bus. and Econ. Res. J.* **6** 2 25-37.

[10] Du Plessis C [2007] A strategic framework for sustainable construction in developing countries. *Construction Management and Economics* **25** 67-76.

[11] Sourani A and Sohail M 2011 Barriers to addressing sustainable construction in public procurement strategies *In Proceedings of the Institution of Civil Engineers-Engineering Sustainability* 2011 Dec [Vol. 164, No. 4, pp. 229-237]. Thomas Telford Ltd.

[12] Abdul Hamid ZA and Kamar KA [2012] Aspects of off-site manufacturing application towards sustainable construction in Malaysia *Construction Innovation* **12** 1 4-10.

[13] Adekunle A, Umanah II, Adewale AK and Egege CC 2018 Analytical study of casualties in the construction industry in Nigeria with a view to provide remedial measures: case study of Lagos State. *Int. J. Eng. Res. Adv. Tech.* **4** 8 6-14.

[14] Bhole MS 2016 Safety problems and injuries on construction site: a review *Int. J. of Eng. Tech.* **2** 4 24-35.

[15] Lee BH, Chen JC and Fo KW 2018 Accidents in construction sites: a study on the causes and preventive approaches to mitigate accident rate *INTI. J.***1** 3 1-12.

[16] Betsis S, Kalogirou M, Aretoulis G, Pertzinidou M. 2019 Work accidents correlation analysis for construction projects in Northern Greece 2003–2007: a retrospective study *Safety* **5** 33 1-12.
[17] Al-Tabtabai HM 2002 Analyzing construction site accidents in Kuwait *Kuwait J. Sci. Eng.* **29** 2 213-238.
[18] Greener S 2008 *Business Research Methods* [Dr. Sue and Ventus Publishing Aps] p 36
[19] Kumar R 2011 Research Methodology: a step-by-step guide for beginners 3rd [London: SAGE]

Reviewing sustainability, preservation and semiotics of potsherd pavements in southwest Nigerian cultural spaces and built-forms

A A Adeyemo[1,2*], I C Ezema[2], E A Adeyemi[2] and C A Anakor[3]

[1]Department of Architecture Technology Lagos State Polytechnic, Ikorodu, Lagos State Nigeria.
[2]Department of Architecture Covenant University, Ota, Ogun State, Nigeria.
[3]Department of Quantity Surveying Lagos State Polytechnic, Ikorodu, Lagos State Nigeria.

*Corresponding author's email: ajibadeadeyemo@gmail.com

Abstract. Potsherds as used in southwest Nigeria Yoruba ancient civilization to pave walkways are continuously discovered along major migratory route from Ile-Ife to Notche in Togo. Its durability was compared to early terracotta pioneer road materials of the Roman and pre-modern Europe. Popularly called Potsherd pavement; the differently beautifully patterned ceramic materials were original to Queen Luuwo; a past and only female *Ooni* of, the Yoruba historical capital Ile-Ife. Despite not being preserved they remain sustainable indigenous material art and technology used around cultural spaces and built-forms since the 9th-12th century. Reviewed by visitations were some sites of the waste-to-wealth sustainable ceramic paving materials on which much work had not been done by built environment experts. Apart from its original denotative function as a walkway material, it also had ritual, storage, community and presently historic semiotic implications adducible to it. The research used the theories of pioneer semioticians; Saussure and Pierce's as well as denotation and connotation by Eco. "Archaeological" method to compare the cultural material use of spaces, the general semiotic and Elleh's African triple heritage concept. The study discovered that the cultural and tourism imperative of pavements are not appreciated and so they are not preserved. The use of physical observation, interview and systematic review of past low traffic-making and contemporary engineering applications for the qualitative research helped the discovery. The study concluded that proper preservation of potsherd pavements, drawing out their symbolic meanings and continued application would contribute to build environment sustainability in the study area.

Keywords: Paving, Potsherd Pavement, Material, Southwest, sustainable, Global Semantic, Semiotics.

1. Introduction

The greatly extolled Ife-Yoruba artistic golden age culture was alluded, to the Greek civilizations for over a century by explorers but is now being continuously debunked according to [1]. Closely related to this is description of the early 18th century London-Norfolk, four days' journey through a sparsely lit, poorly maintained roads that dissolve into mud in wet weather [2]. Interestingly, oral tradition also spoke of splattering of mud on Ife's queen *Luuwo,* which made her order the construction of potsherd pavements, around the all public and religious spaces (obviously her path). Queen *Luuwo,* was the only female monarch ever in Ife, all because she turned the pavement art into a sort of compulsory enlistment issue, even women while the males were at war. Potsherd pavements were alloy road materials or early ingenious efforts highway engineering about the time of Roman terracotta road developments. Potsherd pavement was regarded by an archeologist as a regional architectural phenomenon, but may be at best seen as a landscape architectural element in modern architectural interpretation. However, the resilience of the indigenous floor (walkway) paving art for over ten centuries in some cases is such that its sustainability quality cannot be downplayed. It also has various implicit historic values ranging from its use to classify the history of Ile-Ife into pre and post pavement respectively and for migrations borders and identities, [3;4;5;6].

Architecture is variously represented in the form of building, interior and landscape designs, in this context are man-made material ordering of our environment. This include material and constructed spaces, like curbs and skate ramps to high-rise hotels, retail parks and urban landscapes [7;8]. Architecture and urbanism are related disciplines of built-forms and cities design respectively with the city being a socio-physical network of subjects and objects [9]. Built- forms of the study area are within indigenous, Islamic and western classifications of the African architecture triple heritage [10]. This was used earlier for better understanding African history by [11], to relativelyfit in as a 3-era balanced holistic sequential dateline approach. Indigenous heritage consists of things before or without Islamic or western culture influences such as; the Potsherd pavement, moats cum city walls like the *Ijebu-Ode's Sugbon Eredo* and the *Olumo* rock war time refuge rooms.

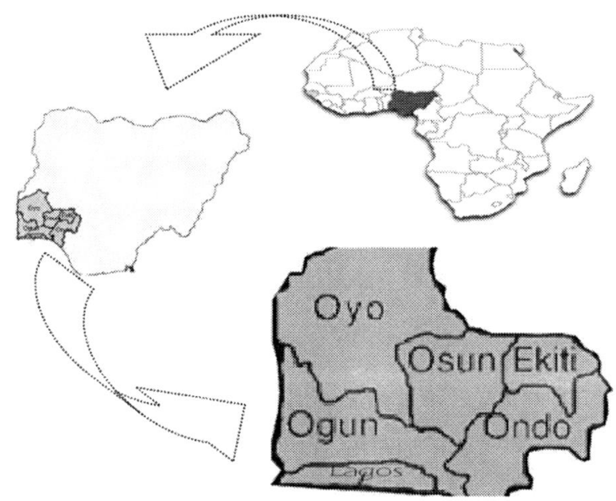

Figure 1: Nigeria's southwest geopolitical

For this study, the strictly utopian indigenous time can modestly be dated to be about the 7th century and the 12th century with some faint traces till date, as the traditional built-forms typology [12]. The Islamic heritage post 12th century had built-forms with separated harems for women and mosques. The western heritage period includes the Brazilian built-form styles that led to tropical modern buildings (international 'style') and the post-modern style, from the 17th century of the colonial, missionaries, to about the time of Nigeria's independence.

2. Sustainable Environment and Architecture

There is a necessity for ingenuity to be able to ensure a development that can match the rate of urban growth and the ever increasing population of the study area without compromising quality of life [14]. Environmental sustainability came from response to the industrial revolution and the subsequent global challenges of the eco system, resource depletion and pollution. Modern sustainable development concept became clear, as a new development paradigm through the World Commission on Environment and Development (WCED). From the 1970 first earth day by the U.S.A and the Stockholm, 1972 first UN conference on human environment, also the 1980 world conservation meeting where the word 'sustainable development' appeared first used.

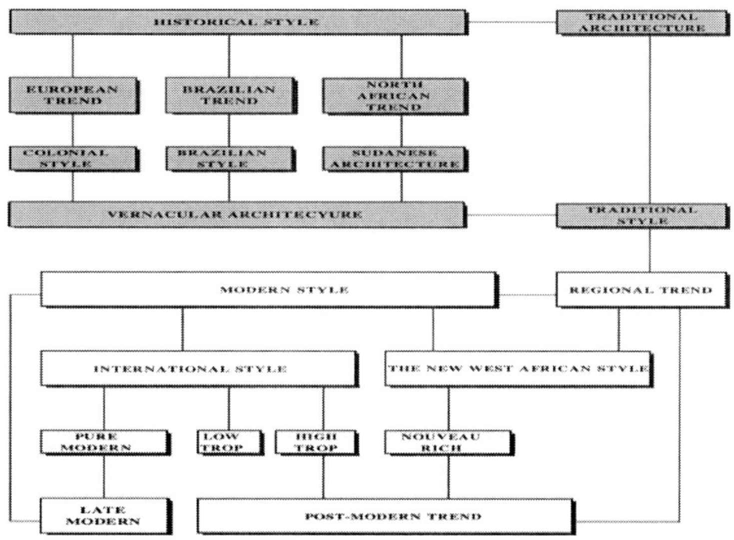

Figure. 2: Architectural classifications of architecture in Nigeria along the line of styles
Source: [13].

Therefore, sustainable development was defined as what is done to meet present generation needs, without compromising the ability of the future meeting of needs and aspirations (1987 WCED). Important to sustainability discourses, was ensuring appropriate relationship between two complementary but sometimes conflicting concepts of environment and development, while maintaining equitable balance between them for the present and the future. As human-induced development is largely responsible for environmental problems, sustaining the environment must ensure developments with minimum negative impact on the environment. A sustainable environment is one, where resource utilization and associated emissions, do not impair the self-regenerating and self-regulating capacity of the natural environment. The construction sector of the built environment plays a major role in the pursuit of a sustainable environment in terms of energy, resource consumption and GHG emissions [15]. The built environment performs a positive role through the provision of shelter, infrastructure and employment; however, its negative role is of particular importance to sustainability. Therefore, making the entire building stock sustainable to protect occupants' health and well-being is a goal to be pursued by all stakeholders in the built environment. Buildings are Sustainable, if designed, built, renovated, operated or reused resources efficiently, (including water, energy and other resources) to reduce negative environmental impacts. The observed reduction in use of local materials; a key to energy efficiency of buildings and high imported materials in Nigeria, must stop. Sustainable architecture produces healthy comfortable, energy efficient, flexible and long lasting built-forms to show the wide implication of sustainable development on the built environment. Sustainable or green architecture entail environmental conscious, earth friendly design that encourages recycling existing resources and reducing harmful substances, emission to the environment [13, 14]. This will entail built-forms using the six logic and methods of eco-technic, eco-centric, eco-aesthetic, eco-medical, eco-social and eco-cultural as well and the six basic sustainable activity principles which implies procuring built-form that is environmentally responsive, energy efficient, socially inclusive and economically affordable [14].

Table 1: Basic principles of sustainable activities

Sustainability Principle	Remark
Conserving energy	A building should be constructed so as to minimise the need fossil fuel to run it
Working with climate	A building should be designed to work with climate and natural energy sources
Minimise new resources	A building should be designed to minimise the use of new resources and at the end of its useful life to form the resources for other architecture
Respect for users	A sustainable architecture recognises the importance of all the people
Respect for site	A building design for sustainability will touch the earth lightly
Holism	Principles of green architecture need to be embodied in a holistic approach to the built environment

Source: [16]

2.1 Sustainability, Preservation, Semiotics, Meaning and Architecture

The basis of Sustainable development is triple, thus; environmental protection, social equity and economic growth, with culture as a fourth or self- standing additional bottom-line, as a tree on which the identity people is rooted. Economic and social system entails enhancing peoples' living standards by balanced preservation, conservation and maintenance oriented resources exploitation methods [17]. The cultural aspect is where semiotics, having been used to variously interpret meaning in architecture and meaningful urban development come relevant. There are elements of sustainability in [18] overview of visual arts meanings in the semiotic analysis of the art and architecture of ancient Greece and Rome where, five different levels of meanings were given. They are; factual meaning, conceptual meaning, explicit historical meaning of images, implicit historical meaning of images, and actualized meaning of images. Modern methods for meaning in ancient arts and interpretive architecture were further discussed to be Hermeneutics, phenomenology, grounded theory and visual agency.

Semiotic was a visual language tool in Egyptian Mamluk architecture by [19], for analysis using architectural linguistic properties, elements and tools such as, signs, codes, denotations, connotations and symbols. These were described as primary cum secondary, metaphors cum metonymy, articulations cum semiotic reading of texts in the interior and exterior of the funerary. Symbolic variable attributes were differentiated from the usually fixed sign attribute that sees the minaret as a sign to imply function of call for prayer, and symbolic to remind someone of Islam. Others could be structures like the spiral minaret of Samarra placed on postage stamps identify some countries. Objects as signs have (literal) functional denotative physical interpretation; and figurative connotative or deeper symbolic meanings at the spiritual, conceptual or idea level.

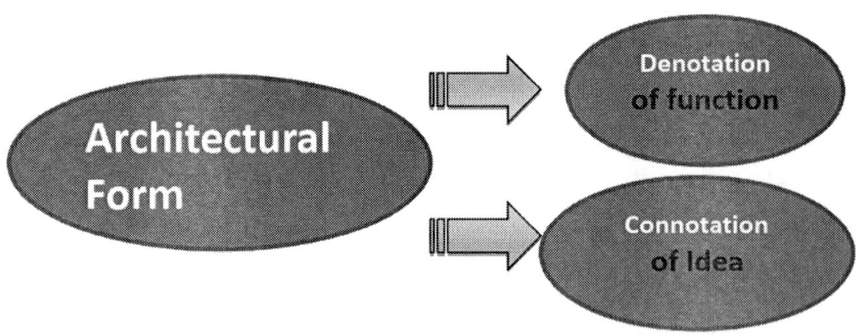

Figure 3: Eco's theory of the Denotation and Connotation
Source: Adopted from [20]

A socio-semiotic investigation on meaning in the change of urban heritage of historic areas in Yogyakarta, Indonesia was carried out by [21]. Relevant here also is the examination by [22] of the semiotics of social memory and cultural values in the urban space of Volgograd (Stalingrad) [23]. Semiotic case study of Tashkent pre and post-soviet city centre emphasized the use of the built environment for the assertion of cultural values. The connection of collective memory with the strategies of perpetuating memory of important events was explained, making the city to be seen as a commemorative space that is viewable as a sign and a text. This was the case in the study of the street layout of an Omani walled city.

Semiotics in urban studies can be viewed in two ways; the structural analysis of sign systems which focuses on the interrelationships in the semiosphere of the city and the phenomenological analysis of emphasized sign processes. A sign (for example, the word "ship") may be recognized by the presence of its constituent parts. This in semiotic theories is based on Saussure's concept, at least, here the signifier (the container, or the sign's perceptible form: the word; ship) and the signified (the meaning or content). All life form engages in the use of signs, however only humans are aware of their existence. Only humans inquire into semiotics, or sign activity which without an exception has its core in the realization that, every human experience is an interpretive structure mediated and sustained by signs. There was a Zimbabwean study on heritage, semiotics and innovations: architectural space, object-designs, meanings and implications in sustainable development on the post-colonial changes in land use policies, slum improvements and urgent built-form development. The development was such that tilted more towards global forms that down played the rich indigenous heritage. It is such that areas that are disregarded are intentionally visually articulated [24].

3. Methodology

The paper, reviewed the sustainability and preservation status of potsherd pavement as a cultural urban landscape architecture material found in course of studying the cultural spaces and built-form of some selected cities of southwest Nigeria adopting a qualitative research method. The historical research was out to prove architecture as part of a past transcending Yoruba civilization seeing potsherd as a cultural generative concept that came from within the architect's *(Akole-Inu)*. For a comprehensive appreciation of the case studies beyond visual quality, the archaeological method [24] was adopted. An interpretive research method was used in natural settings with the goal of making sense and interpreting the phenomena with emphasis on meanings attached to cultural built-forms and materials by people [25]. Data generated from observations and interview of people, individually and collectively was analysed using Eco's semiotic theory of denotation and connotation as a tool. Semiotics was used for the analysis of potsherd pavement art and architecture is applied to bring out, whichever is applicable of the five levels of meaning by [18]. Insights from the 'African triple heritage classifications of built-forms selected historical cities in the six states of southwest Nigeria geopolitical zone's cultural built-form, spaces and practises especially the indigenous are used.

4. Findings and Discussion

4.1. *Ile-Ife potsherd pavements sites*

4.1.1. *Ita Yemoo (potsherd square)*

The renowned National Commission for Museums and Monuments (NCMM) archaeological site under the Federal Ministry of Culture and Tourism is located on Moore Street in Ife East LGA, Ile-Ife. It used to be an originally fenced-off and guarded 3-acre piece, with a pottery, bronze sculpture and artefacts museum and mass herringbone potsherd pavements. Despite being so known, a 2018 visit, showed a multi-purpose hall built on most of the pavement, without due recourse to the potsherd pavements' continued preservation. This shows the careless stance of stakeholders even when potsherd retain its naturally transcendental cultural sustainable quality.

Figure 4: potsherd pavement in Ita Yemoo and virtually all over Ife vicinity
Source: Author's field survey in 2018.

4.1.2. *Yemoo Grove*

It is one among the few cultural spaces found preserved in the *Aafin* (palace), Ile-Ife whose past legendary queen Yemoo became a well and is deified. There is the well, green landscape element on the site such as, an old tree, shrubs, with the potsherd pavement on the walk way in figure 5a and b, as this research's focus. The grove remains a great symbol of indigenous built-form and space, though when it was built is not known. The potsherd is primarily a paved walkway from the fenced grove entrance to the well at its centre, near the other end of the fence.

Figure 5 (b) shows a beautiful sight, rhythmically tightly laid in straight short rows piece interspersed with green growths by design. It is well maintained and remains one of the few purpose designed potsherd pavement site still preserved. It is fenced off and gated with, beautiful cobble stone finished arched entrance with metal frontage part of which is the entrance door.

Figure 5a. The entrance Yemoo grove entrance view in Ile-Ife **Figure 5b**. Inside view of the *Yemoo* Grove
Source: Author's field survey in 2018.

4.2 Oshogbo potsherd pavements sites

The *Osun Osogbo* grove curating team, led as guides to a Potsherd Pavement site in the town, found years back, yet to be studied or recognised, it was met being used as a mechanic yard. Movement of cars daily remove the heritage architectural element. The site users hinted readiness to vacate if compensated. However without a drastic action, the site's intricately laid near herringbone patterned potsherd pavement, might be gone within a year. The slightly steep, uneven site; exposed to the typical cut and fills activities, of the road facing site, from which there is no designed manner of approach for vehicles.

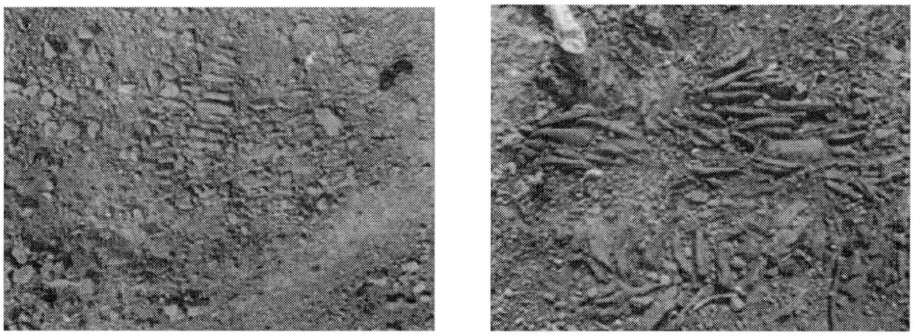

Figures 6a and b: Potsherd Pavements in Osogbo
Source: Author's field survey in 2018.

Continuous driving friction caused "dry erosion", (questioning potsherds' pressure resistance ability) shown in Figure 4: the little left of the continued destruction of the large paved surface of this great past work of art to learn from.

4.3. Ibadan potsherd pavements site

This site was studied with information from [26]. *Ope-Odu* was found to be part of a high spate of non-strictly controlled urban development given the expansion and movement from the city core as such all cultural built-forms were gone. The potsherd pavement site was located at the centre of the *Elewuro* earth road construction in *Ope-Odu*, which from hindsight, in *Ile-Ife* and *Oshogbo* was still recognisable even though sadly the potsherd were gone. Figures 7 below shows the details of the earthen road without a sign of potsherd pavement having ever been there before.

A nearby resident gave the opinion that the group of archaeologists from the University of Ibadan went away with samples found in their earlier study. Potsherds in the Ibadan *Ope-odu* sub-urban area were gone with the earth road construction apart from locating the spot where they were found. This trend of potsherd removal seemed normal to archaeologists in literature, who including Dr. Oriejeme, one of the earlier researchers, when interviewed supplied some pictures, taken previously as the only evidence of potsherd being ever on the site. The pictures shown in figure 8a-e, reveal the design of the potsherd pavement to be of herringbone pattern. This is a sad pointer to increasing spate of losing our built environment heritages, because of the way stakeholders handle them with impunity. Figure 7 and 8 show sites of the potsherd pavement when exhumed and as it was with associated pots during *Ope-Odu* earthen road construction

Figure 7: site with potsherd gone

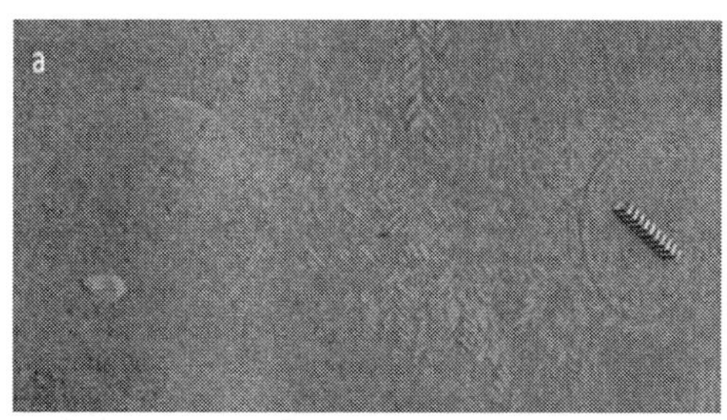

Figure 8a potsherd pavements with two buried pots

IOP Conf. Series: Materials Science and Engineering 1036 (2021) 012027 doi:10.1088/1757-899X/1036/1/012027

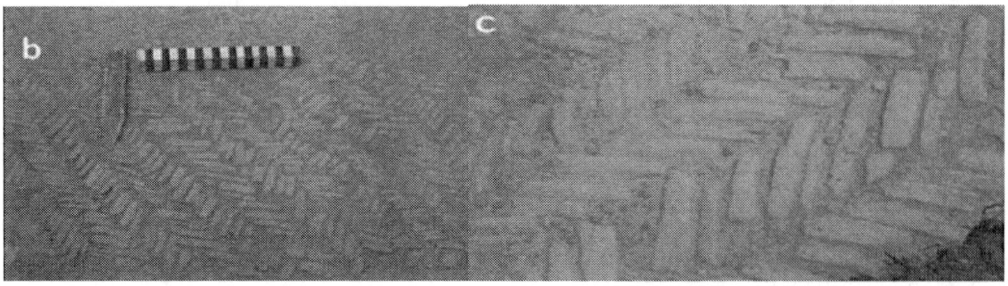

8b) herringbone patterned potsherd tiles 8c. pavement showing laterites in the soil matrix

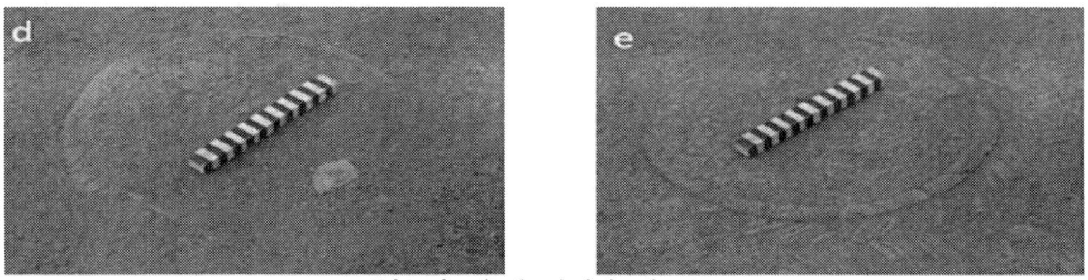

8d-e. in-situ buried ritual pots.
Source: [26] and author's field survey in 2018.

4.4 Discussion

Potsherd pavements were found as part of other cultural spaces and built-form being studied in the southwest geo-political zone in Nigeria. They were only found in five sites three of which were in Ile-Ife and the others in Osogbo and Ope-Odu Ibadan respectively. Out of these sites it was only the ones in the less traversed grove and shrine that were met preserved and have the opportunity semiotic interpretations for meaning. Where found the pavements functioned historically in denotation as walkways, for water collection in impluvium courtyards and to keep medicinal or ritual pots as evidenced in Figures 8d-e. Potsherd pavements sites now serve to symbolise old Yoruba civilization and as such as connotation for tourism. Potsherd pavement in line with literature belongs to the indigenous class of the three heritages of African architecture because in the few places where they were still found they remained intact without Islamic or western influences. It is now more like a relic and it can only be regarded as a regional phenomenon as far as it is being found more in literature widely across the west-African region in archaeological literature. An architect and landscape architect in the focus group in authenticating data saw potsherd pavement rather as a landscape element and not as a major architectural phenomenon. It was the original documentation and subsequent telephone interview with the archaeologists that remained as the only authentic proof there ever having been potsherd pavement at Ope-Odu in Ibadan. This lends further credence to the importance of at least documentary records where the built-form has to give way. Comparatively archaeologists rather see the good in planned or accidental exhumation of pavements since that the easiest way of knowing it exists in a place before plans can be made for preservation. Despite the historical importance of potsherd pavement having been listed for preservation as a museum artefact, there has not been holistic implementation of preservation plans. It is only archaeologists who have in their best possible ways, which in grossly inadequate as Osogbo case of non-response over a long time has shown. It is however evident from this review that the best approach to harness the natural sustainable and symbolic characteristics to ensure adequate preservation is a multi-disciplinary all stake holder to handling potsherd pavement issues.

5. Conclusions

The paper reviewed poor state of preservation of the historic and cultural west-African potsherd pavements revealing their great semiotic and sustainability characteristics. The use of "archaeological"

method, in course of the study brought a new vista of appreciating the cultural built-forms and spaces of the study area and thus a multi-disciplinary approach. From the review it can be concluded that:
Potsherd pavements if preserved has sustainable qualities of conservation, minimal maintenance, waste to wealth material and art practise that allowed minimal use of new resources found to have remained for over one thousand years where used around sacred places.

The process of rhythmical laying in straight short pieces of rows or herringbone pattern and either on hedge or flat surface, baked in waste palm oil water, presented a strong decorative yet durable surface despite the fact of being broken pots wastes.
The tight fixing together of the shards in a way that growth of green intermingling with hard landscape elements as ideally allowed by design did not destroy a holism and green architecture sustainability trait but has transcended globally into city greening.

Traditional potsherd remain stronger showing the better quality of pottery in the past as seen in strength of materials continues to survive effects of dry and wet erosion; their being washed away is rather taken as the advantage of having it exposed.
It is however sad that upon exposure their multi-disciplinary cultural symbolic significance is never appreciated especially complimentarily in architecture, semiotics and sustainable development areas. It therefore seems that they only remained naturally preserved and not necessarily appreciated except in traditional religious and historic places like the *Ogun Laadin* shrine (picture taking was disallowed) and *Yemoo* grove, in Figures 5a and b at Ife palace.
Potsherd pavement is versatile as it also serves religious and health storage symbolic purpose, fulfilling the respect for people sustainability principle, *Ife* palace sites and the *Ope-odu* sites as well as earlier mentions in literature of use of the central pots for *ero* traditional first aid storage.
It still possesses historic tourist attraction quality with original meaning and use by devotees for rites, as was observed during field observation (white basins in Figure 5b).
The local traditional material like earth, the potsherd pavement and other methods local content and technology are appropriate in fulfilling this sustainability indexes and are not far-fetched to mitigate negative effects of importation.

6. Recommendations
All stakeholders especially the government must come to the realization of the importance of existing potsherd pavements sites as great historic relics to be properly preserved. Lessons from the traditional method of construction should be adapted and improved upon to resuscitate the art, for its benefits such as; cultural, tourism, religious, health, storage, economic and environmental sustainability.

References
[1] Osasona CO and Ewemade FO (2011). Ile timi: the interface between traditional and vernacular architecture in Ile-Ife. *WIT transactions on the built environment,* **118**, 99-114.

[2] Hil J (2012). A Hellish Cloud and a Very Clear Air: Industry, Nature and Weather in Early Eighteenth- Century England. In A. Sharr, *Reading Architecture and Culture; Researching Buildings Spaces and Documents* (pp. 74-91). New Year: Routledge; Taylor & Francis Group.

[3] Aregbesola RA (2020). *Infrastructural deficit and technologicaldevelopment in Nigeria: The role of technical education.* Convocation lecture, Lagos State Polytechnic Ikorodu, Lagos, Nigeria.

[4] Adeyemo A (2014). Historical Architectural Maintainable Materials: The Transcending Art and Technology of African Potsherd Pavement. *Int J. of Nov. Res. in Eng and App. Scs (IJNREAS)*, 32-53.

[5] Adewumi I (2001). Indigenous 16th ceentury wearing course for low-traffic walkways in ile-ife: contemporary engineering application. Obafemi Awolowo University Institute of Cultural Studies Weekly Seminar Series.

[6] Usman A and Falola T (2019) The Growth of Complex Societies. *In The Yoruba from Prehistory to the Present.* (Pp 58-86) Cambridge: Cambridge University Press. (Online DOI:https://doi.org//10.1017/97811075).

[7] Hendrix J (2012) Part Three: A Time of Aspirations: Cultural Understanding of the Roles of Architecture. In P. Emmons, P. Emmons, J. Hendrix, J. Lomholt, J. Hendrix, & J. Lomholt (Eds.), *The Cultural Role of Architecture Comtemporary and Historical Perspectives* (pp. 151-154). New York, London: Routledge; Taylor and Francis Group London.

[8] Karrholm M (2012). *Retailing Space: Architecture, Retail Territorialization of Public Space.* New York, USA: Routledge; Taylor and Francis Group London USA.

[9] Muntañola J and Muntanyola D (2012). Universidade Da Coruña (España / Spain), World: Distributive Cognitive Paradigm. *the 10th World Congress of the International Association for Semiotic Studies (Iass/Ais)* (pp. 905-914). Barcelona, (España /Spain): Technical University of Catalonia / Autonomous University of Barcelona (Spain).

[10] Elleh N (1996). *African Architure Evolution and Transformation.* New York USA: Mcgraw-Hill USA.

[11] Munro C (1987). Semiotics, Aesthetics and Architecture. *British Journal of Aesthetics,* **27**(2), 115-128.

[12] Majekodunmi O (2017). The Cradle of Civilisation. *Nigerian Institute of Architects, 3rd Distinguished Architect's Lecture.*

[13] Ogunsote P (2001). *Classification of Nigerian Architecture.* Ile-Ife, Osun State: Association of Architectural Educators in Nigeria.

[14] Adebamowo M (2017). *Sustainable Architecture: A Return To Nature through Biomimicry, Inaugural Lecture Series 2017.* Akoka: University of Lagos, Nigeria: Press and Bookshop Limited.

[15] Ezema I (2015). *Profiling The Environmental Sustanability of Residential Buildings in Lagos Nigeria using Life Cycle Assessment.* Ota, Ogun State, Nigeria: Un-published thesis Submitted for the degree of doctor of philosophy in the department of Architecture, school of postgraduate studies, Covenant University.

[16] Adeyemo A (2014). Historical Architectural Maintainable Materials: The Transcending Art and Technology of African Potsherd Pavement. *International Journal of Novel Research in Engineering and Applied Sciences,* **1**(3).

[17]Dare-Abels OA, Obaleye JO, Oyesode S, and Adeyemo A (2019). Art, Culture, Architecture, and urbanization convergence:Case of two centres. *Caleb International Journal of Development Studies,* **2**(1),

[18] Holscher T (2014). *Semiotics to Agency. In C. Marconi, The oxford Hohandbook of greeks and Roman arts and Architecture.* Oxford: Oxford University Press England.

[19] Ramzy NS (2013). Visual language in Mamluk architecture: A semiotic analysis of the Funerary Complex of Sultan Qaitbay in Cairo. *Frontiers of Architectural Research,* **2**(1), 338–353.

[20] Sinha KK (2010) Semiotics in Architecture: Architectural Representation and Meaning in Legislative Buildings: https://viewer.pdfrock.com/view.php?hash=72ece7d474a8cb07f5c6f9d99048be0d&title=%5BPDF%5D+Kapil+Sinha+REPORT&source=dl

[21] Siregar (2018) The meaning change of urban heritage: a socio-semiotic investigation of historic areas in Yogyakata, Indonesia. Ph.D thesis QUT.

[22] Yanushkevich I (2014). Semiotics of Semiotics of Semiotics of Social Memory in Urban Space: The Case of Volgograd. *Int. J. Cognitive Res. in Sc., Eng and Edu.,* **2**, 2-10.

[23] Demytrie D (2000). *A semiological analysis of urban space in transitional cultures: A case study of Tashkent's city centre (Master's Thesis, University of Manitoba, Winnipeg, Manitoba, Canada).* Retrieved from http://hdl.handle.net/1993/2415

[24] Saidi U (2019): *Heritage, semiotics and innovations: architectural space, object-designs, meanings and implications in sustainable development, Social Semiotics*, **29**(4), 448-462

[25] Adejumo T (2002). *Place Making: A Study of 'Oja – Oba' Open Space Design in Yoruba Urbanism.* Akoka, Lagos, Nigeria: Unpublished Ph.D Thesis Submitted to Department of Architecture, University of Lagos, Akoka.

[26] Orijieme E and Ogiogwa J (2016). Potsherd Pavements in ope-odu, ibadan: autochthonous or migrant phenomenon. *Nyame Akuma* **(85),** 104-113.

Building structural health monitoring: a tool for building collapse mitigation

S O Ongbali[1,*], S A Afolalu[1], S Oladipupo[1], S Akra[2] and K A Bello[3]
[1]Mechanical Engineering Department, Covenant University, Nigeria.
[2]Chevron Nigeria Limited, Onshore Construction, Nigeria
[3]Department of Mechanical Engineering, Afe Babalola University, Nigeria
[*1]samson.ongbali@covenantuniversity.edu.ng

Abstract. Building collapse occurs recurrently worldwide on account of a myriad of variables associated with man or nature whose consequences are loss of life and properties that are sometimes incalculable. This paper aims to appraise the current advances in the topic area to identify and provide a holistic document on building structural health monitoring approaches that are fragmented in the literature to enable practitioners, stakeholders, and inspectors gain insight into the modern and cost-effective methods of building structural health monitoring to alleviate the incessant problem. It appeared that most of the approaches are interwoven and dependent on one another to obtain reliable information on building structural integrity. Apart from the traditional methods of visual inspection and non-destructive assessment of building health monitoring, it appeared the predominant and recent approaches to building health monitoring are analytics or statistics-based, statistics-sensor based, fiber optic framework, fiber optic-sensor network, and sensor-based. Arguments in the literature suggest that the analytics or statistics approach may not provide accurate information on account of the outlier and data corruption. The fiber optic approach is expensive and time consuming compared with the fiber-sensor network method. Finally, the sensor-based building health monitoring is less expensive over other approaches and provides reliable and accurate information on building integrity. Hence, we suggest future research direction should focus on the development and integration of fiber optic-sensor network and sensor-based methods of building health monitoring to improve the accuracy of the methods

Keywords: *Building, structure, collapse, health monitoring, mitigation*

1. Introduction

Building collapse occurs frequently worldwide on account of a myriad of variables associated with man or nature whose consequences are loss of life and properties that are sometimes incalculable depending on the response time. Hence, there is a necessity for periodical building health monitoring to determine and remedy any potential risk on the integrity of the building. Building structural health monitoring involves regular evaluation of structures to detect any damage for maintenance to sustain the build against collapse. Existing building structures deteriorate with time on account of many variables such as environmental hazards, excessive loading, and manmade problems among others. Building damage detections are normally based on changes in dynamic response to load and environmental conditions of the structure. However, if the responses are not monitored, the building may degrade and eventually collapse. The traditional building health monitoring frameworks that relies on visual inspection and non-destructive assessment to evaluate and provide information on the integrity and functionality of the structure is inadequate. Building health monitoring identifies variations in the properties of a building concerning its functionality over some time. The health monitoring procedure includes checking the structure over a while through sampled response measurement via sensors, retrieval of damage-sensitive features from the measurements followed by analysis of the features to ascertain the status of the building and the result of the process can be occasionally updated concerning the capability of the building to perform its intended function regardless of aging and operational environment degradation.

Building health monitoring is thematic and has gained the attention of researchers over the past decades because of its relevance to building collapse mitigation. Researchers have proposed diverse methods for building structural health monitoring that are fragmented in literature. Consequently, this paper seeks to evaluate the current efforts to identify and provide a holistic document on building structural health monitoring approaches to enable practitioners, stakeholders, and policymakers gain insight into the modern and cost-effective methods of building health monitoring schemes to alleviate the incessant problem of building collapse to save lives and properties. According to [1], condition assessment (CA) is one of the many approaches for building health monitoring systems and stressed that building health monitoring is aimed at enhancing the understanding of loading and response mechanisms of earthquakes, wind loads, and other natural sources.

[1] noted that a building health monitoring scheme time-to-time screen building's ability to perform its function as it is aging in the presence of damages arising from operational environment and unexpected situations like earthquake and blast loading to provide information about the integrity of the building [2] defined building structural health monitoring as a scheme applied to structures to determine their fitness for use under changes occasioned by load and response mechanism of the structures such as the dam, cable-supported bridges, and offshore production platforms among other forms of structures. [3] Described building health monitoring as a procedure for implementing a damage detection system for a structure. Similarly, [3] defined building health monitoring as a combination of sensing and intelligence system that records changes in structure, analyzed, localized, and predict the integrity of the building in conjunction with the nondestructive testing approach. [4] Stated that building health monitoring as a scheme deployed to identify damage in buildings. The damage detection system is capable of indicating (i) the presence of damage, (ii) position of damage, (iii) severity of the damage.

Additionally, [5] stressed that building health monitoring is to show the existence of damages in building based on the measured dynamic or static response of the building. [6] argued that aging buildings deteriorate further on account of incessant loading and tough environmental condition and noted that the health monitoring frameworks consist of sensors such as accelerometers, temperature sensors, and displacement transducers which are mounted in building to obtain response measurements data caused by environmental or internal factors that are further analyzed to ascertain the state of a structure. Also, [7] claimed that structural health monitoring assesses the level of deterioration and residual service life of a civil infrastructure. [8] Argued that infrastructure health monitoring entails detecting the position and severity of impairment in a building, dam, or bridge using measured parameters as they occur. The core purpose of building structural health monitoring is to detect and identify the onset of structural deterioration and damage to protect lives and properties. The monitoring scheme is used to track the characteristics of building during mandatory vibration investigation or natural excitation [9] asserted. Similarly, [10] argued that building health monitoring provides information on 'fitness for use' of the structures under changes in their state occasioned by load and response mechanisms.

Besides, [11] described building health monitoring as the key components of building safety. Usually, early warning systems comprising of sensors networks, signal processing devices, and communication amenities are deployed to provide real-time information that are used for planning maintenance and repair works on buildings to minimize the impact of building degradation [12] asserted. The system combines a wide range of sensors and evaluation methods to provide accurate forecasting and enables users to improve the level of emergency preparedness. Usually, the traditional approach to building health monitoring is based on visual inspection for damage propagation. Buildings are expected to meet standard safety requirements, maintainability, and sustainability during their operational life cycle [13] argued. The professionals in the building and construction industry therefore must ensure that these properties are incorporated in building design from the onset. [14] claimed that using fiber optic as a sensing device in the building provides a more precise and dependable results than other sensors such as strain gauge. Advances in the building health monitoring schemes aid researchers in building characteristic identification, exposure of harms to the building, model improvement, security, and sustainability evaluation by using observed response data obtained from the building health monitoring system. Notwithstanding the advances in building health monitoring systems [15] stressed that there are still challenges in to be addressed such as accuracy of sensory devices, frequency and exact data sampling, and data mining among others for decision making on maintenance and management of buildings. [16] Asserted that proper management of civil infrastructure requires condition, serviceability, and reliability assessment for which SHM provides information on the present status of the building by measuring the vibration responses and other physical disorders.

Section one provides background to building structural health monitoring, section two itemizes approaches for building structural health monitoring and section three present conclusions to the appraisal.

2. Methods for building health monitoring (BHM)

Diverse approaches for building health monitoring have been proposed by researchers over the years. For example, [17] presented a vibration monitoring system and identification method of building structural health monitoring aimed at increasing the security and consistency of buildings by identifying impairment in buildings before it reaches a dangerous stage. [18] Projected ARMA-eigenvalue analysis for detecting building damages under wind loads. Similarly, [19] proposed a local information analysis method for online building health monitoring via wireless sensor frameworks where each sensor estimates the coefficient of damage statistically by measured acceleration data for every monitoring exercise. Subsequently, a nonlinear programming analytic was used to identify the existence of impairment, the local impairment location, and quantify the impairment severity from the damage-sensitive coefficients in the integrated sensing module.

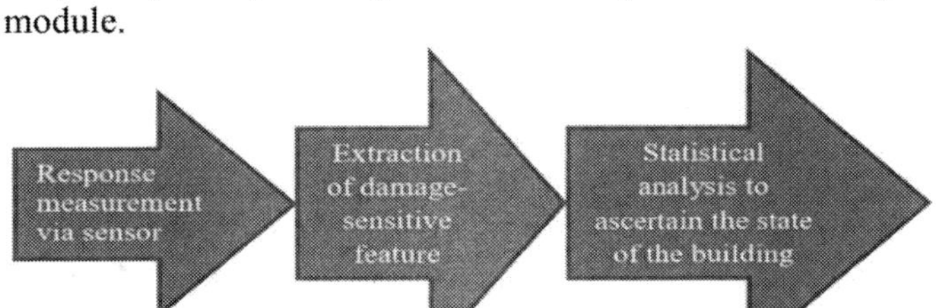

Figure 1 Statistics-sensor-based scheme for building health monitoring

Figure 1 illustrates the statistics-sensor-based approach of building health monitoring where the characteristics response measurement data of a structure is obtained by the sensor device. Subsequently, the data containing damage-sensitive features are retrieved and statistically analyzed to ascertain the integrity and state of the building.

[20] Proposed statistical process control for building health monitoring contextualized in pattern recognition using statistical control chat to diagnose and identify building impairment by vibrational method. Also, [20] stressed that autoregressive coefficients of the models fitted thereafter to a new set of

data were supervised relative to the control chat and the statistically noteworthy numbers of characteristics outside the control limits suggest that the building is in transition from a healthy status to a damaged status. Likewise, [21] projected the ARMA model for building health monitoring with the ARMA coefficient been fed to a classifier capable of learning and establishing new classes when the building response show variations conforming to diverse structural status. Also, [21] employed a statistical pattern recognition approach in building health monitoring. Similarly, [22] presented a Bayesian probabilistic method for building health monitoring.

[23] Advocated wireless sensor networks as an emerging substitute for traditional structural health monitoring device whose benefits include no wiring cost between sensors and the data retrieval systems on like the traditional approach. Besides, they argued that wireless sensors play a vital role in processing structural dynamic response data used to screen data for symptoms of structural damage. However, [23] further argued that wireless sensors have limitations that need innovative designs and manners of operation. Furthermore, [23] stressed that bridges, dams, pipelines, aircraft, ships, and buildings are complex and multifaceted engineering structures that enhance the economic prosperity of a nation. However, these structures are usually endangered with excessive loading and harsh operational environment. Consequently, there is a justification for the health monitoring of structural performances to achieve the desired goal of constructing the structure in the first instance.

Also, [24] emphasized that it is mandatory for some buildings located within the region of high seismic activities to incorporate structural health monitoring devices as a requirement that is responsible for acquiring measurement and storing outputs from sensors with data center repository, an operation that can be expensive and labor-intensive, unlike the wireless. The size of a structural health monitoring system is determined by the overall number of sensors which in turn determine the cost. Hence, the cost associated with installing and maintaining wires is always high. For instance, the cost of 350 sensing channels on the Tsing Ma bridge exceeds US$8million [24] with similar cases reported in monitoring systems installed within aircraft, ships, and other structural systems.

Figure 2 Distributed sensor frameworks for instrumented building health monitoring system

Figure 2 depicts sensor networks for a building health monitoring scheme where the sensors obtain the building dynamic response measurement data, retrieves and analyze the data, and subsequently detects damage therein for actionable plan by the operator. According to [24] building damage detection approaches are usually grouped into local-base or global-based. The local-based approach identifies building impairment by screening buildings at their component or subparts and non-destructive evaluation (NDE) methods such as ultrasonic inspection to detect cracks and yielding of the steel moments frame connections used by experts, an expensive approach. The global-based impairment identification approach involves statistical methods that take into account global vibration behavior such as the natural frequencies of the building to detect building impairment. [25] Asserted that wireless sensors networks are capable of autonomous operation such as the collection, storage, analysis of data, and decides what to transmit to other sensors within the network with the aid of the embedded software in the network. Also, [25] Presented impedance-based structural health monitoring by applying excitations via the surface-bonded piezoelectric transducers that records the impedance of building by checking the current and voltage applied to the transducers. Furthermore, [25] noted that variations in impedance show variations in the structure that indicate that damage has occurred in the building.

IOP Conf. Series: Materials Science and Engineering 1036 (2021) 012028 doi:10.1088/1757-899X/1036/1/012028

[24] and [25] pointed out that building damage detection approaches include damage identification and it is conducted in conjunction with five related areas that include (i) SHM, (ii) Condition monitoring (CM). (iii) Non-destructive evaluation (NDE), (iv) Statistical process control (SPC), and (v) Damage prognosis (DP). However, SHM is associated with online–global damage detection in structures. [26] Argued that modern building monitoring systems should be able to identify and determine impairments, locate impairment, the severity of the impairment, consequences of impairment, and self-diagnose to restore the damage. In building a health monitoring system the sensors do not measure damages rather it requires extraction of features via signal processing and statistical identifier to convert the sensor data into damage information for actionable plan. [28] Noted that structures are designed to operate within specified limits of the environment that they are intended to be used and that is why it is justifiable to ensure continuous health monitoring to detect the onset of damages. Structure loading is approximated during design and required materials are specified, however, engineers depend on modeling techniques to gain knowledge of the behavior of the structure in the operating environment based on approximations and assumptions. Modern structural health monitoring schemes employs and implement implements sensor networks typically optical fibers, electrical resistance strain gauge or acoustic systems to monitors the behavior of the structure online.

Table 1 Reference of building structural health monitoring techniques

No	Building Structural Health Monitoring Technique	Ref.	No.	Building Structural Health Monitoring Technique	Ref.
1	ARMA-eigenvalue analysis	[1,10]	13	Vibration-based monitoring	[1,4,7,19]
2	Ultrasonic testing	[1]	14	Fiber optic-sensor networks	[1,2,22,24
3	Eddy current testing	[1]	15	Optical inspection	[1,7]
4	Magnetic particle inspection	[1]	16	Optoelectronic scanning	[1]
5	Liquid penetrant testing	[1]	17	Laser scanning systems	[1]
6	Visual inspection	[1]	18	IoT-based system	[1]
7	Early warning systems	[1,6]	19	Stiffness reduction system	[2,7]
8	Multivariate ARMA model	[7]	20	Inter-story drift measurement	[2]
9	Signal processing method	[5]	21	Modal-based algorithm	[3]
10	Impedance-based method	[6]	22	Bayesian probabilistic	[8]
11	Statistical process control	[7]	23	Piezoelectric transducers	[7,9,12,17,24,28]
12	Statistical pattern recognition	[7]	24	Wireless sensor networks	[2,3,7,8,13,14,18,20,21,27]

Table 1 depicts various approaches used for building health monitoring to detect the onset of deterioration and damage. The methods appeared to be interwoven and dependent on one another.

A vibration-based method in conjunction with signal processing for beams can be used in building health monitoring. Also, smart sensors, with onboard computational and communication capabilities, offer new opportunities for building health monitoring without the need for power, communication cables, and installation cost. Thus, smart sensors can aid in building monitoring with an array of sensors economically. Similarly, a piezoresistive strain sensor from carbon nanotube-polymer material for building structural health monitoring that is used to detect large strains, cracks, and reduces the number of data acquisition required for health monitoring and claimed that it is cost-effective, lightweight and easy to install. Similarly, dynamic response measurements have been used over the years in which outliers may exist in the measurements that lead to unreliable results. Hence, an outlier-resistant extended Kalman filter was proposed to detect outlier in an online structural parameter identification using dynamic response data that may be corrupted.

[29] Presented damage prognosis that forecasts building performance by measuring the existing damage, assessing the future environmental loading, and predicting the remaining useful life of the

building by simulation and experience. However, the capability of the approach requires the design and integration of technology that include measurement, processing, telemetry hardware couple with deterministic and probabilistic modeling capacities to quantify the uncertainties therein the predictions.
The application of piezoelectronic transducer patches is used in health monitoring of civil infrastructure by attaching the patches to the surface of the structure that is electrically excited, and the electrical conductance part was extracted as a function of frequency. Subsequently, an impedance analyzer was used to scan the patches over some frequency to obtain the signature and was discovered that the deviation from the obtained record for a healthy state is an indication of the new state of the health of the structure.

[30] Emphasized that building structural health monitoring (BSHM) involves four basic steps including, (i) acquisition of data, (ii) validation of data, (iii) analysis of data, and (iv) prognosis and management. Hence, structural health monitoring (SHM) scheme integrally comprise five features namely (i) sensors and sensing technology, (ii) diagnostic signal generation, (iii) signal transmission and processing, (iv) event identification and interpretation, and (v) integration into an actionable plan for management of the structure. Furthermore, [30] stressed that though some non-destructive evaluation (NDE) methods are integrated into SHM techniques, there is a clear difference between SHM and NDE. While NDE measures specific characteristics and assesses the state of a single point without determining the level of deterioration, the SHM scheme entails the diagnosis, prognosis of damages, and the order of damages based on the parameters and as well determines the residual service-life of the structure.

3. Conclusion

The various approaches presently available for building health monitoring converges in the materials reviewed and thus, bring the appraisal to a compendium as outlined below. The two categories of solutions to building structural health monitoring are local and global solutions. The non-destructive evaluation (NDE) presents local solutions while the structural health monitoring (SHM) approach offers a global solution. The NDE measures specific characteristics and assesses the state of a single point in a structure without determining the level of deterioration whereas, the SHM scheme entails diagnosis, the prognosis of damages and the order of the damages based on the parameters as well as the determination of the residual service life of the structure.

It is evident from the foregoing that the approaches used in building structural health monitoring are interwoven and dependent on one another to obtain real-time reliable information on building structural integrity. The common methods include wireless sensors networks, fiber optic networks, fiber optic-sensor networks, and statistics or analytics methods. It appeared that the sensor network is increasingly gaining attention than any other method of building a structural health monitoring scheme. Apart from the cost-effectiveness of the approach, it is easy to install, easy to operate, and provides real-time and reliable information than other methods presently available it is capable of obtaining and analyzing data autonomously for actionable plan unlike the statistic or analytic methods that generate data with the possibility of outliers and likelihood that the data may be corrupted before analysis thereby resulting in providing wrong information.

4. Recommendation

Despite the advances in building structural health monitoring, there are still challenges to be addressed such as accuracy of the sensory devices, frequency, and exact data sampling among other issues for decision making on maintenance and management of buildings. However, we recommend that the future research direction should explore more of the sensor network and sensor-fiber optic networks techniques because of the potential therein particularly the generation of real-time and reliable information about the status and integrity of a building as well as determining the remaining service life of a structure.

Acknowledgment
We sincerely acknowledge the support offered by Covenant University to facilitate publication of this manuscript.

References

[1] Tudor M 2012 Structural Health Monitoring of high-rise building structures B.Sc. *Thesis at Worcester Polytechnic Institute,*
[2] Jun W, Yongfeng F and Xiaokang Y 2017 An integrated system for building structural health monitoring and early warning based on an Internet of things approach, International Journal of Distributed Sensor *Networks, 13(1),*
[3] Yoshihiro N and Akira N 2004 Structural Health Monitoring Methodology Consisting of Two Stages with Different Purposes, *13th World Conference on Earthquake Engineering Vancouver, Canada, No. 1168*
[4] Takuji H. 2003 Experimental verification for damage detection strategies of multistory buildings based on vibration monitoring, *Proc. SPIE 5057, Smart Structures and Materials: Smart Systems and Nondestructive Evaluation for Civil Infrastructures,*
[5] Kondo I and Hamamoto T 1993 Identification of Damage Location of Existing Structures Using a Multivariate ARMA Model, *Journal of Structural Engineering*, **39B**, pp.87-94
[6] Skolnik DA, Kaiser WJ and Wallace JW 2008 Instrumentation for Structural Health Monitoring: Measuring Inter-story Drift, T*he 14th World Conference on Earthquake Engineering, Beijing, China*
[7] Zengshun CC, Xiao ZZ, et al. 2017 Deployment of a Smart Structural Health Monitoring System *for Long-Span Arch Bridges: A Review and a Case Study, Sensors,*
[8] Lynch JP 2006 A Summary Review of Wireless Sensors and Sensor Networks for Structural Health Monitoring. *The Shock and Vibration Digest*, ***38*(2):** 91–128.
[9] Alamos L. 2000 *S h m u s p c*. (May 2011).
[10] Ore DIN. 2016a *ORE Open Research Exeter.*
[11] Park G and Inman DJ 2007 Structural health monitoring using piezoelectric impedance measurements. *Philosophical Transactions of the Royal Society A: Mathematical, Physical and Engineering Sciences*, ***365*(1851)**: 373–392. https://doi.org/10.1098/rsta.2006.1934
[12] Ore, DIN 2016b *ORE Open Research Exeter*
[13] Farrar CR and Worden K 2007 An introduction to structural health monitoring. *Philosophical Transactions of the Royal Society A: Mathematical, Physical and Engineering Sciences,* ***365*(1851),** 303–315. https://doi.org/10.1098/rsta.2006.1928
[14] Loupos K and Amditis A.2017 Structural health monitoring fiber optic sensors. *Smart Sensors, Measurement and Instrumentation*, *21*(4), 185–206. https://doi.org/10.1007/978-3-319-42625-9_9
[15] Worden K, Farrar CR, Manson G and Park G 2007 The fundamental axioms of structural health monitoring. *Proceedings of the Royal Society A: Mathematical, Physical and Engineering Sciences*, ***463*(2082)**, 1639–1664. https://doi.org/10.1098/rspa.2007.1834
[16] Worden K and Dulieu-Barton, JM 2004 An Overview of Intelligent Fault Detection in Systems and Structures. *Structural Health Monitoring*, *3*(1), 85–98. https://doi.org/10.1177/1475921704041866
[17] Sohn H, Farrar CR, Hunter NF, and Worden K 2001 Structural health monitoring using statistical pattern recognition techniques. *Journal of Dynamic Systems, Measurement and Control, Transactions of the ASME*, ***123*(4)**, 706–711. https://doi.org/10.1115/1.1410933
[18] Segundo E 2000 *B p a s h m*. (July)

[19] Fan W and Qiao P 2011 Vibration-based damage identification methods: A review and comparative study. *Structural Health Monitoring*, *10*(1), 83–111. https://doi.org/10.1177/1475921710365419

[20] Nagayama T, Spencer JF and Fujino Y 2008 Structural health monitoring using smart sensors. *EASEC-11 - Eleventh East Asia-Pacific Conference on Structural Engineering and Construction*, (November). https://doi.org/10.1201/9781439828441.ch129

[21] Kang I, Schulz MJ, Kim JH, Shanov V and Shi D 2006 A carbon nanotube strain sensor for structural health monitoring. *Smart Materials and Structures*, ***15*(3):** 737–748. https://doi.org/10.1088/0964-1726/15/3/009

[22] Mu HQ and Yuen KV 2015 Novel outlier-resistant extended Kalman filter for robust online structural identification. *Journal of Engineering Mechanics*, ***141*(1):** 1–9. https://doi.org/10.1061/(ASCE)EM.1943-7889.0000810

[23] Sohn H 2007 Effects of environmental and operational variability on structural health monitoring. *Philosophical Transactions of the Royal Society A: Mathematical, Physical and Engineering Sciences*, ***365*(1851)**, 539–560. https://doi.org/10.1098/rsta.2006.1935

[24] Farrar CR and Lieven NAJ 2007 Damage prognosis: The future of structural health monitoring. *Philosophical Transactions of the Royal Society A: Mathematical, Physical and Engineering Sciences*, ***365*(1851),** 623–632. https://doi.org/10.1098/rsta.2006.1927

[25] Ling Q, Tian Z, Yin Y, and Li Y 2009 Localized Structural Health Monitoring Using Energy-Efficient Wireless Sensor Networks. *IEEE Sensors Journal*, ***9*(11):** 1596–1604. https://doi.org/10.1109/JSEN.2009.2019318

[26] Karbhari VM and Ansari F 2009 Structural health monitoring of civil infrastructure systems. In *Structural Health Monitoring of Civil Infrastructure Systems*. https://doi.org/10.1533/9781845696825

[27] Oliver J 2019 済無No Title. *Hilos Tensados*, *1*, 1–476. https://doi.org/10.1017/CBO9781107415324.004

[28] Kim S, Pakzad S, Culler D, Demmel J, Fenves G, Glaser S, and Turon M 2007 Health monitoring of civil infrastructures using wireless sensor networks. *IPSN 2007: Proceedings of the Sixth International Symposium on Information Processing in Sensor Networks*, 254–263. https://doi.org/10.1145/1236360.1236395

[29] Chang PC, Flatau A and Liu SC 2003 Review paper: Health monitoring of civil infrastructure. *Structural Health Monitoring*, *2*(3), 257–267. https://doi.org/10.1177/1475921703036169

[30] Siddiqui M F, and Darade M M 2017 Failures in Construction; Types and Causes and Its Assessment, Imperial Journal of Interdisciplinary Research, **3(4)**

Evaluation of compacted lateritic soil –*bacillus coagulans* induced calcite precipitate for landfill application using bacterial foraging optimization algorithm

P Yohanna[1], K J Osinubi[2], A O Eberemu[2*], T S Ijimdiya[2] and A T Salawudeen[3]

[1]Department of Civil Engineering, University of Jos, Plateau State, Nigeria.
[2]Department of Civil Engineering, Ahmadu Bello University Zaria, Nigeria.
[2*]Department of Civil Engineering and African Centre of Excellence on New Pedagogies in Engineering Education (ACENPEE), Ahmadu Bello University Zaria, Nigeria.
[3]Department of Electrical and Electronics Engineering, University of Jos, Plateau State.

Corresponding Author: paulyohanna45@yahoo.co.uk

Abstract: Evaluation of compacted lateritic soil –*Bacillus coagulans (B.coagulans)* induced calcite precipitate for landfill application was carried out using Bacterial Foraging Optimization Algorithm (BFOA). Experimental results were obtained through laboratory tests. Measured soil factors include; Coefficient of permeability (hydraulic conductivity) as a dependent factor and void ratio, bulk density, *B. coagulans* suspension, pH, Compactive effort, water content relative to optimum, plasticity index, viscosity of microbes and liquid limit as self-determining (independent) factors. GeneXproTools 5.0 software was used to develop the fitness (objective) function used for Bacterial Foraging Optimization (BFO) to predict hydraulic conductivity of treated lateritic soil with *B. coagulans* suspension density. The fitness function was then integrated using BFO codes in Matlab 2016 version to predict the minimum hydraulic conductivity for a given set of self-determining factors. The result shows a strong association between the predicted hydraulic conductivity values and the measured hydraulic conductivity values in the laboratory using a regressed line with coefficient of determination R=0.85. It is clear from the optimization results that the hydraulic conductivity values and the self-determining variables have strong correlation. The least hydraulic conductivity value of 2.38×10^{-11}m/s which falls below the permissible regulatory value of 1.0×10^{-9} m/s for landfill application was achieved at 100 iterations. Based on all the optimization results, it is recommended that lateritic soil-*B.coagulans* mixtures be compacted with 1.20×10^{-9} - 2.40×10^{-9} cells/ml *B.coagulans* suspension density compacted using Britsh standard heavy (BSH) energy at 15.3 % moisture content for landfill application
Keywords: Bacterial foraging optimization, Bacillus coagulans, Iterations, Lateritic soil, Objective function.

IOP Conf. Series: Materials Science and Engineering 1036 (2021) 012029 doi:10.1088/1757-899X/1036/1/012029

1. Introduction

Soil improvement using soil microbes is a new practice which uses bacterial activity in the improvement of properties of soil. This method applies biochemical mechanism to increase soil engineering performance (i.e. strength and hydraulic properties). Bio-clogging is a process that involves the reduction in hydraulic conductivity of soil and porous rocks by pore filling materials (calcite).Calcite is produced as a result of microbial activity [1]. Microbial activity involves the production of calcite via microbial means, this helps to reduce the soil porosity, hydraulic conductivity, blocks soil pore throats, subsequent establishment of aggregations and as well as biofilms to restrict movement of fluid within the soil matrix [1-2]. Optimization or optimality is an act of minimizing or maximizing an equation/function done by choosing the values of integer variables analytically from inside a tolerable interval [3]. It involves the selection of a best option or amount in several options using mathematical and computer programming approach. The best option is referred to as the optimal value in terms of minimum or maximum among set of outcomes. [3] reported that optimization problems are categorised into two major parts namely, the objective or fitness function, and the set of constraints. Objective or fitness function defines the routine principles for the structure. Constraints define the margins and limits beneath which the structure is investigated.
However, the constraints comprise physical properties of the system to be optimized. The best solutions also referred to as the optimal values are the established values of the choice parameter that fulfil the requirement for the constraints and give a best value of the objective or fitness function [3]. The procedure for optimization is as follows:

Identify decision variables: These are indeterminate factors linked with the problem statement that is requisite for solving the problem. Coming up with specific definition for decision variables is a vital phase in coming up with an optimization model or equation. Also, the outcome of the results and the level of relevance to the subject under study are linked to and heavily rely on the decision variables.

Determine the objective/fitness function: Fitness function demonstrates the decision-maker's main aim. It defines the purpose for the optimization via the definition of the factors in the equations. Also the subject to minimise or optimised is stated. It is the most important component of the entire estimation processes.

Determine the constraints: Constraints describe the necessities that the preferred problem shall encounter. Constraints are expressed as equalities and in some situations as inequalities. It is a vital component for the estimation of the desire property. Constraints defined the boundaries of the results so as to minimise errors and achieve a feasible results.

There are different forms of optimization techniques which include neural network, Swarm Optimization, Random Search Method, Bacterial foraging optimization, Particle swarm optimization, Ant colony optimization etc [4].

2. Bacterial foraging optimization Theory

The idea of bacteria foraging optimization was first suggested by Prof. K.M. Passino in 2002 [5]. Passino developed a novel nature stimulated calculation technique (Algorithm) after simulating the food scavenging, evolutionary replica and ecological elimination-dispersal activities of an Escherichia Coli (E.Coli) bacteria and called it Bacteria Foraging Optimization Algorithm (BFOA). BFOA is stimulated by the social foraging behaviour of Escherichia coli bacteria in hunt of food [4, 6]. BFOA provides a universal outline and established rules for generating explanation to a problem than giving the comprehensive arrangement and analysis [7]. Numerical bio-inspired optimization system is a novel associate in the swarm intelligence optimization paradigm [8]. BFOA has been broadly recognized as a universal optimization algorithm of present concern for optimization and regulations in many engineering applications.

Foraging theory is established on the theory that animals quest for and get food for leaving in a manner that make best use of their energy intake E per unit time T use in searching. Mathematically represented as $\frac{E}{T}$ .Optimal expansion of such function makes available nutrient bases to live and extra period for supplementary vital actions (such as mating, reproducing, fighting, sleeping). Ideal foraging concept forms the foraging problem as an optimization problem through the use of computational and also analytical approaches [5]. [5] proposed that generally, animals have different methods of search for food in their surroundings by either cruise or ambush approach. In the case of cruise method to searching of food or nutrients, the forager changes position constantly over the environment, continuously hunting for prey. In the case of ambush search, the forager (such as a rattlesnake) rests motionless and pauses waiting for prey to pass cross its range of striking [5].

The techniques of Bacterial foraging optimization (BFO) has been successfully applied in many engineering fields [4-11].

2.1. Steps for Bacterial Foraging Optimization Algorithm

2.1.1 Chemotaxis

Chemotaxis processes mimic the behaviour of an E.coli cell via swimming and tumbling using flagella [4,6]. Assume θi (j, k, l) signifies the ith bacterium at jth chemotactic, kth reproductive, and ith elimination–dispersal step. C (i) is said to be a scalar and designates the size of the phase occupied in the arbitrary direction stated by the tumble (run length unit). Formerly, in computational chemotaxis, bacterium movement could be signified by

$$\theta^i(j+1,k,l) = \theta^i(j,k,l) + C(i)\frac{\Delta(\Delta i)}{\sqrt{\Delta^T(i)\Delta(i)}} \quad (1)$$

where Δ designates a unit length vector in the random direction [4].

2.1.2 Swarming

Motivating group performance has been described for numerous species of bacteria, where swarms are made in semisolid nutrient medium [4]. Swarming involves the arranging of an assembly of E.coli cells moving in a ring toward a nutrient slope when positioned within a semisolid medium with its own nutrient chemo-effecter [4]. The cells when motivated by a high level of succinate release an attractant aspartate, which aids them to conglomerate into clusters and transfer as concentric arrangements of swarms with high bacteriological concentration [12].

2.1.3 Reproduction

In this step, the smallest healthy bacteria eventually die whereas the ones in good health asexually fragment into two bacteria, which are at that moment positioned in the same location. This retains the swarm size constant [4].

2.1.4 Elimination and Dispersal

In other to estimate elimination and dispersal phenomenon in BFOA, [5] suggested that selected bacteria occupying best points are reserved and the remaining population of the bacterial are killed. The bacteria with best locations are eventually moved to a new location within the surroundings [13].

The aim of the study is to evaluate the potentials of compacted lateritic soil –*Bacillus coagulans (B.coagulans)* induced calcite precipitate for use as liner and cover material in waste containment system with the aid of BFOA. Also, to determine the optimal combination of compaction energy, water content and *B.coagulans* needed to achieve the optimal hydraulic properties for the intended application.

3. Material and Methods

3.1 Bacteria Foraging Optimization Procedure

Experimental results were obtained through laboratory tests. Measured soil factors include; Coefficient of permeability (hydraulic conductivity) as dependent factor and void ratio (d_1), bulk density (d_2), *B. coagulans* suspension density (d_3), pH (d_4), Compactive effort (d_5),water content relative to optimum (d_6), plasticity index (d_7),viscosity of microbes (d_8) and liquid limit (d_9) as self-determining (independent) factors. GeneXproTools 5.0 was used to develop the fitness (objective) function used for Bacterial Foraging Optimization to predict hydraulic conductivity of treated lateritic soil with *B. coagulans* suspension density. The fitness function was integrated using codes in Matlab 2016 version to predict the minimum hydraulic conductivity for a given set of self-determining factors. The fitness function used is shown in equation 2.

3.2 Fitness/Objective function

$$y = d_1 - \frac{5.25}{d_9} + 2.52 + d_6 - d_7 + \mathrm{Cos}(\mathrm{Cos}d_1\, d_6) - 8.14\left[\frac{0.18}{\mathrm{Cos}\left[\mathrm{Sin}\left(\mathrm{Tanh}\left(\frac{d_4 - d_5}{2}\right)\right)\right]}\right]d_7 + \mathrm{Cos}\left[\frac{d_6}{\mathrm{Cos}(d_8 + d_6) + \left(\frac{d_7 + d_2}{d_2}\right)}\right] - d_2 + \mathrm{Tanh}[\mathrm{Sin}\, d_5(\mathrm{Cos}(-5.95d_6) - 3.82d_3)] + d_7 \quad (2)$$

$R^2 = 85.5\%$

Where; y= Coefficient of permeability (hydraulic conductivity), d_1 = Void ratio, d_2 = Bulk density, d_3 =*B. coagulans* suspension density, d_4 = pH, d_5 = Compactive effort, d_6 = water content relative to optimum, d_7 = Plasticity index, d_8 =Viscosity of microbes, d_9 = Liquid limit

3.3 Constraints
The constraint considered include *B. coagulans* suspension density (d_3), Compactive effort(d_5) and water content relative to optimum(d_6) .

$$0 \leq d_3 \leq 2.4\mathrm{E}-09$$

$$-1 \leq d_5 \leq 1$$

$$13.8 \leq d_6 \leq 15.3$$

4 Results and Discussion

4.1 Bacterial Foraging Optimization

BFO method was applied to determine the minimum hydraulic conductivity based on laboratory results using the possible combination of the self-determining variables (Void ratio (d_1), bulk density (d_2), *B. coagulans* suspension density(d_3), pH(d_4), Compactive effort(d_5), water content relative to optimum(d_6), plasticity index(d_7),viscosity of microbes(d_8) and liquid limit(d_9)) measured in the laboratory that give rise to the minimum hydraulic conductivity value. GeneXproTools 5.0 was used to develop the fitness (objective) function for optimizing hydraulic conductivity. The various output for the model developed are contained in Table 1.

Table 1. Variables measured for Fitness function generated by GeneXproTools 5.0

S/No	Variable	Quantity
1	Fitness	856.11
2	MSE	2.82E-02
3	RMSE	0.168
4	MAE	0.136
5	RSE	0.146
8	Correlation Coefficient	0.924
9	R-square	0.854
15	Number of gens	4

MSE = Mean Square Error; RMSE = Root Mean Square Error; MAE = Mean Absolute Error; RSE= Residual Standard Error.

The relationship between the predicted hydraulic conductivity (Target) against the measured values from the laboratory used for the model is shown in figures 1 and 2. The result shows a strong association between the predicted values and the measured values in the laboratory using a regressed line with coefficient of determination R=0.85. This indicate that the self determinning variables used to develop the model has strong effect on the hydraulic conductivity of lateritic soil-*B.coagulans* mixtures and should be carefully studied and considered during the design and construction phase of a waste containment system.

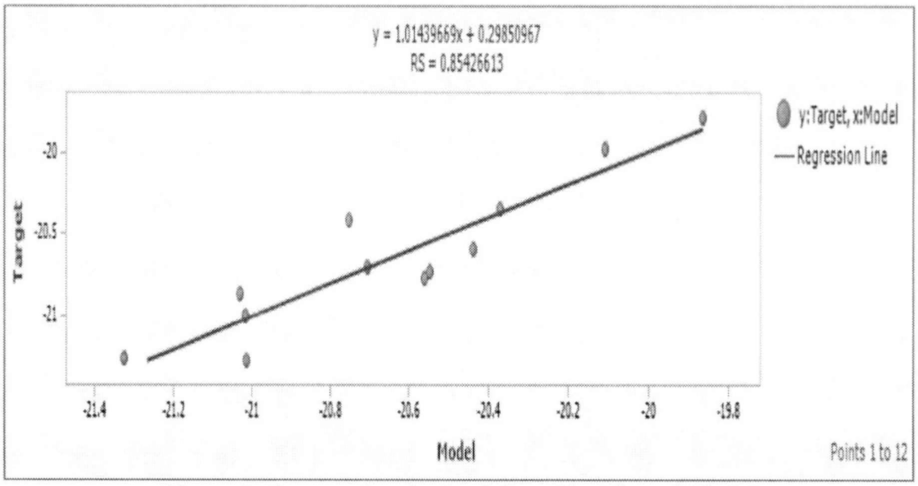

Figure 1. Scatter plot obtained from GenXproool Model

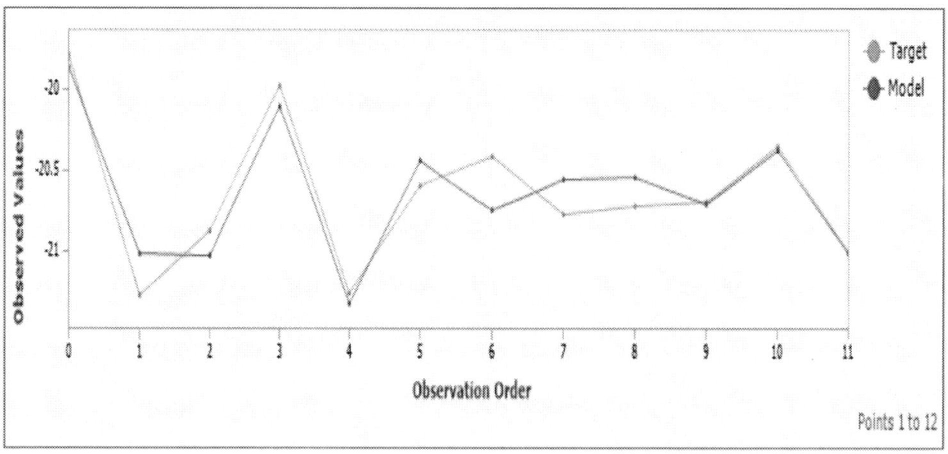

Figure 2. Plot of observed value against observation from GenXproool Model

Results of measured hydraulic conductivity and predicted hydraulic conductivity values from Fitness function generated by GeneXpro Tools 5 is presented in table 2 with absolute percentage error of 0.68-30.12%. The graphical plots of hydraulic conductivity against iteration number for (A) 50 (B) 100 (C) 500 and (D) 1000 iterations are shown in figure 3. Table 3 presents summary of the optimization results for 50,100, 500,750, 1000, 1500 and 2000 iterations. It is clear from the optimization results that the hydraulic conductivity values and the self-determining variables are close to each other irrespective of the number of iterations. However, the least hydraulic conductivity value of 2.80×10^{-11}m/s was recorded from 750 iterations and above, this suggest that the minimum iteration required to minimize hydraulic conductivity of lateritic soil-*B.coagulans* mixtures is 750 iterations. This results falls within the range of optimal values from overall acceptable zone generated from laboratory result. Therefore, the compactive effort, moisture content and *B.coagulans* suspension density should be considered primarily during design and construction of the compacted clay liner in the field or in any related geotechnical engineering applications. Also, [14] suggested that hydraulic head difference should be strictly controlled in the course of construction of landfill to mitigate the movement of contaminant into the ground water. Higher hydraulic gradient can significantly encourage the migration of contaminant. The application of BFOA has proven be a good mechanism for this application. However, the development of a proper objective function with the required variables that have direct impact on the hydraulic performance of a containment system is key to achieving the expected result in the field. Hence care should be taken in developing the objective function and setting of the appropriate equation constraints in order to arrive at expected output in the field.

Table 2. Results of Measured and predicted hydraulic conductivity values from Fitness function generated by GeneXproTools 5.

S/No	Measured hydraulic conductivity(m/s)	Predicted hydraulic conductivity(m/s)	Absolute error	% Error
1	5.75E-10	7.48E-10	1.7E-10	30.12
2	5.85E-10	5.47E-10	3.8E-11	6.51
3	7.52E-10	7.47E-10	5.1E-12	0.68
4	8.64E-10	7.37E-10	1.3E-10	14.71
5	9.56E-10	1.18E-09	2.2E-10	23.13
6	9.99E-10	1.19E-09	2.0E-10	19.65
7	1.03E-09	1.02E-09	8.1E-12	0.79
8	1.14E-09	1.33E-09	1.9E-10	16.96
9	1.36E-09	9.72E-10	3.9E-10	28.62
10	1.46E-09	1.42E-09	3.8E-11	2.58
11	2.12E-09	1.85E-09	2.7E-10	12.72
12	2.56E-09	2.36E-09	2.0E-10	7.92

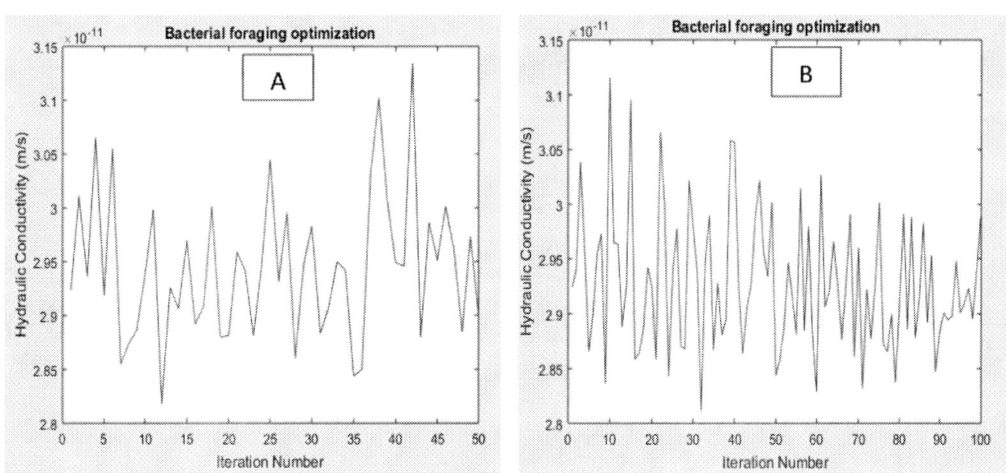

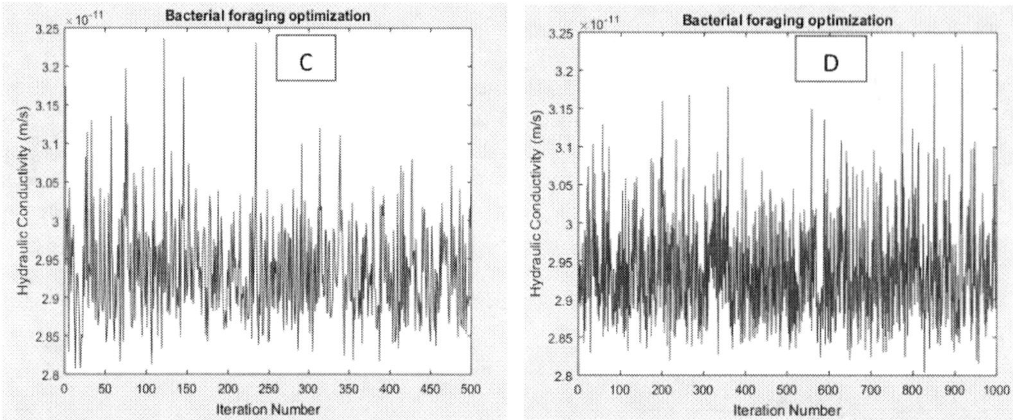

Figure 3. Plots of Hydraulic conductivity against Iteration number using Bacterial foraging optimization algorithm for (A) 50 (B) 100 (C) 500 and (D) 1000 iterations.

Table 3. Bacterial Foraging Optimization Results

No of Iteration	y	d_1	d_2	d_3	d_4	d_5	d_6	d_7	d_8	d_9
10	2.84e-11	0.415	2.193	6.708	7.301	0.969	15.300	26.090	33.991	37.996
50	2.82e-11	0.406	2.191	3.095	7.250	0.730	15.300	26.100	33.991	37.983
100	2.81e-11	0.404	2.191	6.423	7.293	0.960	15.299	26.100	34.000	38.000
250	2.82e-11	0.400	2.193	6.383	7.300	0.850	15.299	26.099	34.000	37.998
500	2.81e-11	0.408	2.193	6.620	7.305	0.851	15.299	26.100	33.990	37.996
750	2.80e-11	0.406	2.193	5.955	7.301	0.866	15.300	26.100	33.984	37.984
1000	2.80e-11	0.405	2.193	3.361	7.264	0.874	15.300	26.095	33.986	38.000
1500	2.80e-11	0.409	2.193	4.048	7.240	0.967	15.300	26.087	33.967	37.985
2000	2.80e-11	0.408	2.193	6.375	7.296	0.968	15.300	26.090	33.976	37.987

Where y = Least Hydraulic conductivity (m/s), d_1 = Void ratio, d_2 = Bulk density, d_3 =*B. coagulans* suspension density, d_4 = pH, d_5 = Compactive effort, d_6 = Water content relative to optimum, d_7 = Plasticity index, d_8 =Viscosity of microbes, d_9 = Liquid limit

5. Conclusion

Lateritic soil –*Bacillus coagulans (B.coagulans)* induced calcite precipitate was analysed for hydraulic behaviour using Bacterial Foraging Optimization Algorithm (BFOA). Laboratory experiments were carried out to generate results used for BFOA. GeneXproTools 5.0 software was used to develop the fitness (objective) function used for BFO to predict hydraulic conductivity of treated lateritic soil with *B. coagulans* suspension density for use as a hydraulic barrier material. The fitness function was then integrated using BFO codes in Matlab 2016 version to predict the minimum hydraulic conductivity for a given set of self-determining factors. The results showed a strong association between the predicted and the measured hydraulic conductivity values in the laboratory using a regressed line with coefficient of determination R=0.85. It is clear from the optimization results that the hydraulic conductivity values and the self-determining variables have strong correlation irrespective of the number of iterations. The least hydraulic conductivity value of 2.80×10^{-11}m/s was recorded after 750 iterations and above which fell below the permissible regulatory value of 1.0×10^{-9} m/s for waste containment barrier application. Based on all the optimization results, it is recommended that lateritic soil-*B.coagulans* mixtures be compacted with 1.20×10^{-9} - 2.40×10^{-9} cells/ml *B.coagulans* suspension density using Britsh standard heavy (BSH) at 15.3% moisture content for waste containment barrier application. Therefore, the compactive effort, moisture content and *B.coagulans* suspension density should be considered primarily during design and construction of the compacted clay liner in the field or in any related geotechnical engineering applications.

References

[1] Muthukkumaran Kasinathan, Bettadapura. Subramanyam S. 2016. Durability of microbial Induced Calcite Precipitation (MICP) treated cohesionless soil. *Japanese geotechnical society,* **2**(56), 1946-1949. doi:10.3208/jgssp.IND-23.

[2] Mitchell, J.K. and Santamarina, J.C.2005.Biological considerations in geotechnical engineering. *Journal of Geotechnical and Geoenvironmental Engineering.* **131** (10), 1222–1233.

[3] Ehsan, G., Mina, Z. and Edward, Z. H. 2014. *Introduction to Optimization Analysis in Hydrosystem Engineering*, Topics in Safety, Risk, Reliability and Quality 25, DOI: 10.1007/978-3-319-04400-2_2, pringer International Publishing Switzerland.

[4] Nitin, K J. and Dipak, A. 2014.Comparative Analysis of Bacterial Foraging Optimization Algorithm with Simulated Annealing. *International Journal of Science and Research (IJSR)* **3** (3), 10-13.

[5] Passino, K. M. 2002. Biomimicry of Bacterial Foraging for distributed optimization and control, *IEEE Control System Magazine*, 22(2), 52-67.

[6] Dharminder, K and Vinay, C. 2012. Image Quantization using HSI based on Bacteria Foraging Optimization. *International Journal of Information Technology and Knowledge Management.* **5**, (2) 335-343

[7] Gautam, M and Soumya, B. 2013. A Study of Bacterial Foraging Optimization Algorithm and its Applications to Solve Simultaneous Equations. *International Journal of Computer Applications,* **72**(5)1-7.

[8] Badamchizadeh, M.A., Nikdel,A., Kouzehgar, M. 2010. Comparison of Genetic algorithm and particle swarm optimization for data fusion method based on kalman filter' *International Journal of Artificial Intelligence*, **5**(10), 67-78.

[9] Jayaram, M.A.,Nataraja, M. C and Ravikumar, C. N.2009.Particle Swarm Optimization: Design of High Performance Concrete Mixes. Conference: *Proceedings of the 4th Indian International Conference on Artificial Intelligence*, IICAI 2009,Tumkur,Karnataka, India, December 16-18,1-10.

[10] Wang, X. S. Cheng, Y. H. and Hao, M. L. 2010. Estimation of distribution algorithm based on bacterial foraging and its application in predictive control," *Acta Electronica Sinica,***38**(2) 333-339.

[11] Yang, D., Li, X. and Jiang, L. 2012. Improved algorithm of bacterium foraging and its application

Computer Engineering and Applications, **48**(13),31-34

[12] Lowe, G., Meister, M. and Berg, H. 1987. Rapid rotation of flagellar bundles in swimming bacteria, *Nature,* **325**, 637-640.

[13] Hanning Chen, Yunlong Zhu, and Kunyuan Hu (2009) Cooperative Bacterial Foraging Optimization", Hindawi Publishing Corporation, Discrete Dynamics in Nature and Society, *Article ID 815247*, DOI:10.1155/2009/ 815247.

[14] Xie, H., Wang, S. and Qiu, Z. 2017. An analytical model for contaminant transport in a vertical cut-off wall and an aquifer system. *Proceedings of the 2nd Symposium on Coupled Phenomena in Environmental Geotechnics (CPEG2)*, Leeds, United Kingdom, 6 – 8 September. Session: Surface containment , Paper #92, Pp. 1 – 6.

Evaluation of the Free Swell and Physio-chemical Properties of Black Cotton Soil Treated with Bacillus *Coagulans*

A O Eberemu[1], A B Bassey[1*] and K J Osinubi[1]

[1] Department of Civil Engineering, Ahmadu Bello University, Zaria, Kaduna state, Nigeria.
[*] Corresponding Author: basseyaniekpeno@gmail.com

Abstract. The effect of Bacillus *coagulans* (B. *coagulans*) on the free swell and physio-chemical properties of black cotton soil (or expansive tropical black clay) was investigated using the Microbial Induced Calcite Precipitation (MICP) technique. Black cotton soil was treated with different mix ratios of B. *coagulans* suspension density (B) and cementation reagent (C) for the volume obtained as product of the liquid limit and a specific weight of soil sample. The mix ratios considered were 25B-75C, 50B-50C and 75B-25C for the treated specimens, while the control specimen was mixed with 100C only. The results obtained show that for treatment of soil with 25B-75C mix ratio, free swell value decreased with increased B. *coagulans* suspension density from 70.0 % and 50.0 % for the natural and control specimens, respectively, to minimum value of 40.0 %, at B. *coagulans* suspension density of 2.4×10^9 cells/ml. The cation exchange capacity (CEC) value for the natural and control specimens were 49.7 and 29.6 Cmol/Kg, respectively, while for the treated specimens the values generally decreased a minimum value of 29.0 Cmol/Kg at B. *coagulans* suspension density of 2.4×10^9 cells/ml. The pH test results initially decreased from 7.4 for the control specimen to a minimum at of 7.37 at B. *coagulans* treatment suspension density of 1.5×10^8 cells/ml and thereafter gradually increased to 7.51 at peak B. *coagulans* suspension density of 2.4×10^9 cells/ml. Tests results indicate that treatment of black cotton soil with 25B-75C mix ratio at 2.4×10^9 cells/ml B. *coagulans* suspension density significantly reduced the free swell value and makes it useable for engineering application.
Keywords: Bacillus coagulans, Black cotton soil, Free swell, physio-chemical properties, cation exchange capacity, pH.

1. Introduction

Black cotton soils (BCS) are problem tropical clay deposits with high swelling and shrinkage potentials. Hence, they are generally unsuitable for use as construction materials. They are classified under the smectite group of clay minerals which are known to contain the highly expansive clay mineral, montmorillonite [1]. BCS are formed mainly in semi-arid tropics with pronounced varying dry and wet seasons [2]. They can be found, especially in the North Eastern region of Nigeria, and in other countries around the world like Australia, India, and Asia [3].
In Nigeria, BCS are formed through the disintegration of basaltic rocks and clayey and shaley sediments. The soil has excessive montmorillonite content with subsequent exhibition of expansive tendencies. In the North Eastern region of Nigeria where the soil is predominantly found, it spans an area of about 104,000 km^2. Major roads connecting this region with bordering West African countries such as Chad Republic, Cameroun and Niger traverse through this soil [4].
The chemical composition of BCS varies depending on factors such as parent rock, generic characteristics of soil (transported or residual), degree of weathering, etc. However, BCS are generally

rich in silica, lime, iron, magnesia and alumina. In addition, it contains titanium oxide in small concentrations, which is responsible for the characteristic black color to the soil. The organic matter contents of black cotton soils are low [5].

Generally, BCS exhibits properties that make it a candidate for improvement before use in any engineering construction. Stabilization of soil becomes necessary when the soil does not have sufficient strength to support the designed structural load. Also, stabilization can also be used to improve or regulate the shrink-swell characteristics of the soil as well as decrease compressibility and permeability of a mass of soil in earth structures [6].

The use of chemical admixtures to enhance the properties of BCS is widespread however, the method is neither eco-friendly nor economical [7]. This prompted the search for novel soil improvement methods that resulted in the use of the Microbial Induced Calcite Precipitation (MICP) technique that entails biological processes, nutrients, microorganisms that are environmentally sustainable solution to soil improvement [8-10]. This research was conducted to evaluate the effect of B. *coagulans* - induced calcite precipitate on the free swell and physio-chemical properties of black cotton soil using MICP technique. The specific objective was to ascertain the changes in the free swell and physio-chemical properties of BCS when treated with stepped B. *coagulans* suspension density.

2. Materials and methods

2.1 Materials

2.1.1 Black cotton soil. The black cotton soil samples utilised in this work was acquired along Gombe-Biu road in Yamaltu Deba Local Government Area (latitude 10° 190N and longitude 11° 300E) in Gombe state, Nigeria.

2.1.2 Microorganism. The species of bacteria used as the soil improvement agent is B. *coagulans*. It is classified as ATCC 7050 in the American Type Culture Collection (ATCC). It is Gram-positive rod-shaped bacteria. Five different bacteria suspension densities on the McFarland turbidity scale, MFS (i.e., 0.5, 2.0, 4.0, 6.0 and 8.0 with equivalents 1.5×10^8 cells/ml, 6×10^8 cells/ml, 1.2×10^9 cells/ml, 1.8×10^9 cells/ml and 2.4×10^9 cells/ml, respectively) were used in this study. The control soil was referenced as 0 cells/ml (0 MFS).

2.1.3 Cementation reagent. Cementation reagent was used to activate the urea hydrolysis process. The constituents of the cementation reagent used in this study comprised of 2.8 g of calcium chloride ($CaCl_2$), 3g nutrient broth, 10 g ammonium chloride (NH_4Cl), 20 g of urea ($CO\ (NH_2)_2$) and 2.12 g sodium bicarbonate ($NaHCO_3$) per litre of de-ionized water [11-14].

2.2 Methods

2.2.1 Free swell. The test was conducted according to the methods outlined in the United States Bureau of Reclamation (USBR) [15]. About 10 g of soil passing BS No.40 sieve (425 μm aperture) was oven-dried and allowed to cool in a desiccator. The sample was slowly poured into a 100 cm^3 measuring cylinder and the cylinder was then filled with water. Before the free swell; the initial volume of soil in the measuring cylinder was recorded. The cylinder was then agitated in order to obtain a homogenous mixture of water and soil after which it was left for 24 hours to settle, Free swell was calculated using:

$$F(\%) = \frac{V1-V2}{V1}\ x\ 100 \qquad (1)$$

Where; F = Free Swell (%)
V_1 = Initial Volume (cm^3)
V_2 = Final Volume (cm^3)

This procedure was repeated for the various bacteria suspension densities.

2.2.2 Cation exchange capacity.

This test was conducted in accordance to the procedures outlined by ISRIC [16]. 10 g of soil sample passing through BS Sieve No.10 (2.00 mm aperture) was poured inside a 100 cm^3 beaker. 40ml of Ammonium acetate (1N pH7.0) was poured into the beaker. With the aid of a glass rod, the mixture was stirred and left for 24 hours. The soil was filtered with a light suction using a 55 mm Bucher funnel, the soil was leached so that it could fit in a funnel with Ammonium acetate of a volume of 250 cm^3. The filtered product was tested to ascertain the absence of calcium. White turbidity or precipitate indicated calcium presence. With 150-200 ml of isopropyl alcohol, the electrolyte was washed out.
With (0.1N $AgNO_3$), chloride content was tested in the leachate until the leachate became negligible. A volume of 250ml of the leached soil was acidified, after the soil was verified to drain thoroughly. 50ml of boric acid and mixed indicator, a few drops of it, was poured into a 250 ml conical flask. The acidified soil was turned inside a 500 ml flask, with the flask still connected to the steel, anti-bump and 10 ml of 1N NaOH was poured inside the flask. The content in the flask was then distilled over the boric acid and 150ml distilled sample was collected. A standard acid of 0.1N HCl was used to titrate the NH_4-borate. The CEC was calculated using:

$$\text{CEC (Cmol/kg)} = \frac{(\text{Titre - B}) \text{ x NA}}{\text{Weight of soil}} \times 100 \quad (2)$$

Where:
B = Blank
NA = Normality of acid
This process was repeated for both the untreated and treated samples.

2.2.3 pH.

This test was conducted according to the procedures stated in BS 1377-3: 1990 [17] using electrometric method with a pH meter (PHS-25; Techmel & Techmel, USA). A soil to water ratio of 1:2 (10 g of soil to 20 ml of distilled water) was adopted for this test. 10 g of air-dried soil sieved through BS Sieve No.10 (2 mm aperture) was poured into a 50 ml plastic beaker and distilled water of 20 ml volume was poured inside the beaker. The mix was stimulated repeatedly for about 30 minutes and was left undisturbed for another 30 minutes. Using pH buffer of 4, 7, and 9, the pH meter was regulated. An electrode was inserted into the soil but was not allowed to touch the bottom of the beaker and the pH read after 30 seconds. These steps were repeated using 0.01M $CaCl_2$ and 1N KCl solutions.

2.2.4 Soil sample preparation and treatment for test procedures.

400 g of Soil sample of passing through BS Sieve No.40 (425 μm aperture) was utilised for this process. For the control specimen, only cementation reagent which volume was obtained from the total volume stated in equation (3) was utilized for the mixing process of the soil sample. For the treated specimens, the soil was mixed with each B. *coagulans* suspension density and cementation reagent in a bowl. Three trial mixes were adopted for the calculation of the volume of each of the bacterial suspension density (1.5×10^8, 6×10^8, 1.2×10^9, 1.8×10^9 and 2.4×10^9 cells/ml) and cementation reagent added to the soil. The mixes are:

i. 25B-75C: In this mix, the soil was mixed with 25% bacterial suspension (25B) and 75% cementation reagent (75C) of the total volume of soil sample used which was obtained from the equations;

$$Total\ Volume = \frac{Liquid\ Limit\ of\ Natural\ Sample}{100} x\ Weight\ of\ Soil\ Sample \quad (3)$$

$$Volume\ of\ Bacteria = \frac{25}{100} x\ Total\ Volume \tag{4}$$

$$Volume\ of\ Cementation\ Reagent = \frac{75}{100} x\ Total\ Volume \tag{5}$$

ii. 50B-50C: In this mix, the soil was mixed with 50% bacterial suspension (50B) and 50% cementation reagent (50C) of the total volume which was obtained from the equations;

$$Volume\ of\ Bacteria = \frac{50}{100} x\ Total\ Volume \tag{6}$$

$$Volume\ of\ Cementation\ Reagent = \frac{50}{100} x\ Total\ Volume \tag{7}$$

iii. 75B-25C: In this mix, the soil was mixed with 75% bacterial suspension (75B) and 25% cementation reagent (25C) of the total volume which was obtained from the equations;

$$Volume\ of\ Bacteria = \frac{75}{100} x\ Total\ Volume \tag{8}$$

$$Volume\ of\ Cementation\ Reagent = \frac{25}{100} x\ Total\ Volume \tag{9}$$

After mixing the soil sample with the three trial mixes for the various bacterial suspensions, the treated samples were left to air-dry before being pulverized and sieved with BS Sieve No. 40 (425 μm aperture) for free swell, CEC and pH test procedures. However, for the CEC and pH test procedures, only the most suitable mix ratio obtained from the free swell test for the various mix ratios was utilized.

3. Results and discussion

3.1 Natural soil

The natural black cotton soil had a high natural moisture content of 15.1 %. Its high moisture content was probably influenced by the period in which the samples were collected. The index properties of the natural soil are shown in Table 1, while its oxide composition is presented in Table 2. The sample is greyish-black in colour and has a free swell value of 70.0 %. Its liquid limit, plastic limit and plasticity index values are 53%, 26.7% and 26.3 % respectively. It has 78.7 % fines passing BS sieve No.200. The soil was classified as A-7-6(23) soil group according to American Association of State Highway and Transportation Officials [18] as well as medium and high plasticity clay (CH) according to the Unified Soil Classification System, USCS [19].

Table 1. Basic properties of the black cotton soil used in the study.

Properties	Quantity
Percentage passing 0.075mm sieve	78.7
Natural moisture content, %	15.1
Free Swell, %	70.0
Specific gravity	2.42
Liquid limit, %	53.0
Plastic limit %	26.7
Plasticity index, %	26.3
Linear shrinkage, %	11.5
AASHTO classification	A-7-6(23)
USCS	CH
Maximum dry density, Mg/m^3	1.65
Optimum moisture content, %	17.5
Colour	Greyish - Black

Table 2. Oxide compositions of black cotton soil.

Oxide	Concentration (%)
Sodium Oxide (Na_2O)	0.064
Magnesium Oxide (MgO)	0.572
Alumina (Al_2O_3)	20.601
Silica (SiO_2)	63.208
Potassium oxide (P_2O_5)	0.141
Sulphur oxide (SO_3)	0.337
Chloride (Cl)	0.026
Potassium Oxide (K_2O)	2.63
Lime (CaO)	2.017
Titanium Dioxide (TiO_2)	1.736
Chromium (III) Oxide (Cr_2O_3)	0.011
Manganese Oxide (Mn_2O_3)	0.091
Iron oxide (Fe_2O_3)	8.522
Zinc oxide (ZnO)	0.009
Strontium Oxide (SrO)	0.034
Loss on Ignition (LOI)	0.001

3.2 Mechanism of microbial induced calcite precipitate (MICP) process

The concept of utilizing microbes in bio-geotechnical engineering purposes was first initiated in 1992 [20]. The concept involves the catalysis of microbes to generate deposits that can cement soil particles. This new technique, where calcite precipitate is induced is termed Microbial induced calcite precipitation (MICP).

MICP is a soil stabilization method, which involves hydrolyzing urea by using bacteria, to generate carbonate ions, that react with a solution rich in calcium (i.e., calcium chloride); to create calcium carbonate (calcite) which results in an improved soil strength and stiffness by increasing the bond between the particles of the soil [21,22]. The technique can be applied in diverse means such as permeability reduction, soil strength improvement, surface erosion control, and seismic remediation [11,8,23,24].

3.3 Influence of B. coagulans induced calcite precipitate on free swell and physio-chemical properties of black cotton soil

3.3.1 Free swell. The variation of free swell of black cotton soil with B. *coagulans* suspension density is presented in Figure 1.

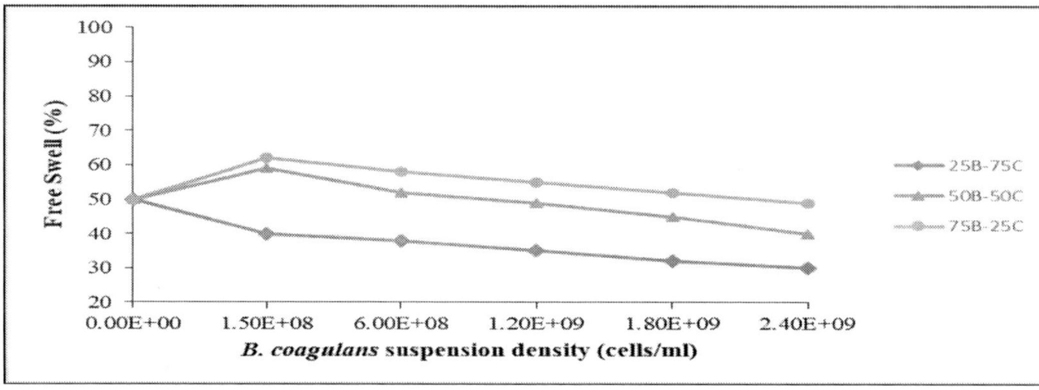

Figure 1. Variation of free swell of black cotton soil with B. *coagulans* suspension density.

Generally, free swell (FS) decreased from 70 % for the natural BCS with increase in B. *coagulans* suspension density for the mixes considered, while the free swell value obtained for the control specimen was 50.0 %. The lowest FS values of 30.0 %, 40.0 % and 49.0 % were recorded for 25B – 75C, 50B – 50C and 75B – 25C mix ratios, respectively, at *B. coagulans* suspension density of 2.4 x 10^9 cells/ml. The observed decrease in FS with increase in B. *coagulans* suspension density can be attributed to the reduction in the clay sized particles and increase in coarser fraction because of the precipitation of calcite from the reaction between B. *coagulans* and cementation reagent [25,26]. The decrease in the fines fraction was probably due to flocculation and agglomeration of the particles in the treated soil samples which resulted in the formation of larger-sized particles from the clay fractions [27]. In addition, it was observed from Figure 1 that the lowest FS values were obtained for 25B-75C mix when compared with 50B-50C and 75B-25C mixes. This could probably be due to the volume of cementation reagent being insufficient to activate the desired reaction with the microbe colony in each suspension density. The amount of cementation reagent and B. *coagulans* suspension density are two of the numerous factors that influence the rate of calcite precipitation [25,26].

3.3.2 Cation exchange capacity (CEC)

Cation exchange capacity (CEC) can be utilized to ascertain a soil specimen's mineral composition. A high CEC value is often indicative of a high expansiveness rate [28].Some clay experience replacement of cation of one kind by another while retaining the same crystal structure in a process known as isomorphous substitution. This substitution, which occurs along with the dissociation of hydroxyl ions, often leads to a residual negative charge on the clay mineral's particle surface. As a result, positively charged ions, known as cations, are adsorbed on the clay mineral's surface. The bonds between the ions are weak and can be replaced by other ions present in water. This phenomenon is termed as cation exchange. The amount of exchangeable cation available in a soil is designated as the exchange capacity [29,30].

The variation of the CEC of black cotton soil with B. *coagulans* suspension density is presented in Figure 2. A general decrease of CEC with increase in B. *coagulans* suspension density was recorded.

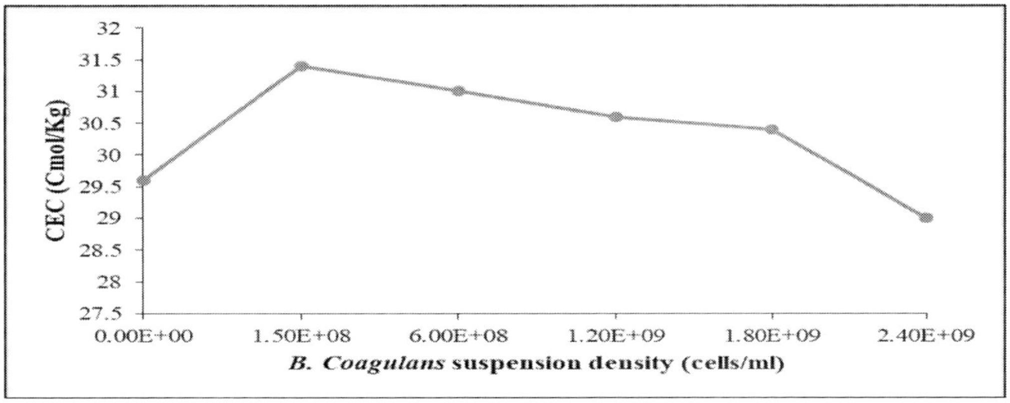

Figure 2. Variation of cation exchange capacity of black cotton soil with B. *coagulans* suspension density.

The CEC value for the natural and control sample were 49.7 and 29.6 Cmol/Kg. For the treated samples, the CEC values initially increased from the value obtained for the control sample to a peak value of 31.4 Cmol/Kg at B. *coagulans* suspension density of 1.5 x 10^8 cells/ml and thereafter gradually decreased to a value of 29.0 Cmol/Kg at B. *coagulans* suspension density of 2.4 x 10^9 cells/ml. This decrease could be attributed to the changes in the soil's granular matrix and mineralogical composition due to the flocculation of the treated soil's particles resulting in decreased specific surface area and the clay sized fractions within the soil [31,27]. The changes in the soil's granular matrix was due to the biogeochemical reaction between the carbonate ions from the urea hydrolysis process and the calcium ions present in the cementation reagent, which causes the formation of calcite within the soil mass [25].

3.3.3 pH.

Under proper conditions, urease positive bacteria are capable of hydrolyzing urea into carbonate ions and ammonium ions, forming an alkaline micro-environment around the microorganism cells and increasing the pH of the environment in the process. The carbonate ions in the presence of calcium ions react to produce calcium carbonate. The precipitated calcium carbonate bound soil grains together to enhance the strength and reduce the expansive characteristics of clayey soils [26,28].
The variation of pH of black cotton soil with B. *coagulans* suspension density is presented in Figure 3. The pH value for the control specimens initially decreased to a minimum before increasing with increase in B. *coagulans* suspension density.

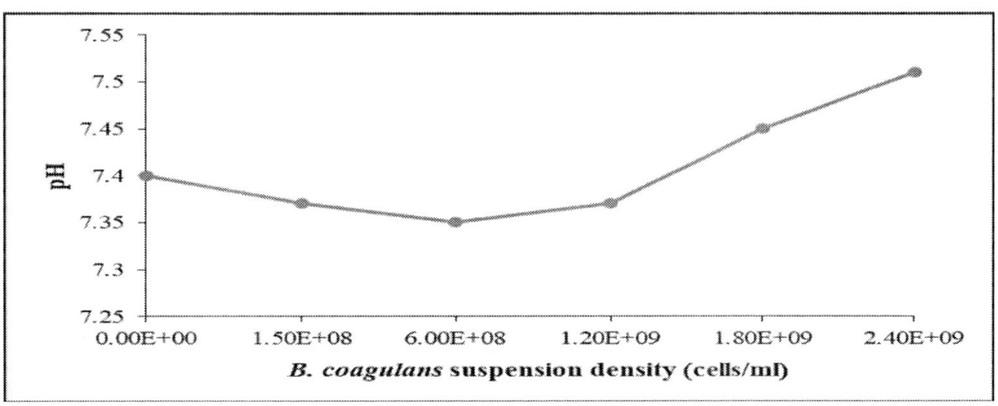

Figure 3. Variation of pH of black cotton soil with B. *coagulans* suspension density.

The pH value for the control specimens of 7.4 decreased to a minimum value of 7.35 at 6.0 x 10^8 cells/ml and then gradually increased to a peak value of 7.51 at B. *coagulans* suspension density of 2.4 x 10^9 cells/ml. This trend correlates with the results obtained for the free swell values for the control and treated samples. The increase in pH was probably due to the breakdown of urea by the urease enzyme that formed ammonia as a by-product and eventually resulting in the precipitation of calcite [32]. Therefore, the gradual increase in pH is an indication that more urease activities occurred at higher bacteria suspensions leading to a larger amount of precipitated calcite at 2.4 x 10^9 cells/ml bacteria suspension density.

4. Conclusion

The black cotton soil classified as A–7-6 (23) and medium to high plasticity clay (CH) in AASHTO and USCS, respectively, was treated with stepped B. *coagulans* suspension density (B) of 0, 1.5 x 10^8, 6.0 x 10^8, 1.2 x 10^9, 1.8 x 10^9 and 2.4 x 10^9 cells/ml and cementation reagent (C). Three mix ratios 25B-75C, 50B-50C and 75B-25C were adopted for the study, while treatment with cementation reagent only was used as control specimen. Results indicated that the free swell improved with higher B. *coagulans* suspension density with the best results recorded for the 25B-75C mix ratio. Also, the pH and CEC test results indicate increase in urease activity and calcite formation with increase in B. *coagulans* suspension density. From the results obtained it can be concluded that B. *coagulans* can be effectively used to reduce the shrink-swell properties of black cotton soil in engineering construction.

5. References

[1] Osinubi K J, Soni E J and Ijimdiya T S 2010 Lime and slag admixture improvement of tropical black clay road foundation *In: Transportation Research Board (TRB) 89th Annual Meeting CD-ROM Subject: Recycled Materials in Transportation Infrastructure, Session 243: AFP40 – Physico-chemical and Biological Processes in Soils Committee* Paper # 10–0585 (Washington DC, USA) pp. 1–18

[2] Madedor A O and Lal N B 1985 Engineering Classification of Nigerian Black Cotton Soils for Pavement Design and Construction, Geotechnical Practice in Nigeria *International Society of Soil Mechanics and Foundation Engineering*, Nigerian Geotechnical Association, Nigerian Commemorative Edition, Nigeria. Pp 49-67.

[3] Oluyemi-Ayibiowu B D and Ola S A 2015 Stabilization of Black Cotton Soils from North-Eastern Nigeria with Sodium Silicate *International Journal of Scientific Research and Innovative Technology* **2(6)** 189-203

[4] Ola S A 1983 The Geotechnical Properties of the Black cotton Soils of North – Eastern Nigeria *Tropical Soils of Nigeria in Engineering Practice* A A Balkema (Rotterdam: Netherlands) pp 85–101

[5] Gidigasu S S R and Gawu S K Y 2013 The Mode of Formation, Nature and Geotechnical Characteristics of Black Cotton Soils - A Review *Standard Scientific Research and Essays* **1**(14) 377-90

[6] Afrin, H 2017 A Review on Different Types Soil Stabilization Techniques *International Journal of Transportation Engineering and Technology* **3**(2) 19-24

[7] Tingle J and Santoni R 2003 Stabilization of Clay Soils with Non-traditional Additives *Journal of the Transportation Research Board* **2** 72–84

[8] Dejong J T, Mortensen B M, Martinez B C and Nelson D C 2010 Bio-mediated Soil Improvement *Ecological Engineering* **36**(2) 197 – 210

[9] Wath R B and Pusadkar S S 2016 Soil Improvement Using Microbial: A Review *Indian Geotechnical Conference IGC2016* Vol 2, ed T Thyagaraj (Chennai, India: Indian Institute of Technology, Madras) pp 329-36

[10] Osinubi K J, Eberemu A O, Ijimdiya T S, Yakubu S E, Gadzama E W, Sani J E, Yohanna P 2020 Review of the Use of Microorganisms in Geotechnical Engineering *SN Applied Sciences Springer Nature Journal* **2** 207
[11] Stocks-Fischer S, Galinat J K and Bang S S 1999 Microbiological precipitation of $CaCO_3$ *Soil Biology and Biochemistry* **31**(11) 1563–71.
[12] Stoner, D L, Watson S M, Stedtfeld R D, Meakin P, Griffel L K, Tyler T L, Pegram L M, Barnes J M and Deason V A 2005 Application of Stereo Lithographic Custom Models for Studying the Impact of Biofilms and Mineral Precipitation on Fluid Flow *Applied and Environmental Microbiology* **71**(12) 8721–28
[13] DeJong J T, Fritzges M B and Nüsslein K 2006 Microbial Induced Cementation to Control Sand Response to Undrained Shear *Journal of Geotechnical Geoenvironmental Engineering* **132**(11) 1381-92
[14] Al Qabany A, Mortensen B, Martinez B, Soga K and DeJong J 2011 Microbial Carbonate Precipitation: Correlation of S-Wave Velocity with Calcite Precipitation *Proc. ASCE Geo-Frontiers 2011 Conf.* (Reston VA, USA: ASCE) 3993-4001
[15] Holtz W G and Gibbs H J 1956 Engineering Properties of Expansive Soils *Transactions of ASCE* **121** 641-79
[16] ISRIC 1998 *World Soil Information* International Soil and Reference Information Center.
[17] BS 1377-3: 1990 *Method of Testing Soils for Civil Engineering Purpose* (London: British Standard Institute BSI)
[18] AASHTO 1986 *Standard Specifications for Transport Materials and Methods of Sampling and Testing* 14th Ed (Washington DC: American Association of State Highway and Transport Officials AASHTO)
[19] American Society for Testing and Materials ASTM 1992 *Annual Book of Standards* Vol. 04.08 (Philadelphia, USA: ASTM)
[20] Ferris F G and Stehmeier L G 1992 Bacteriogenic Mineral Plugging *Journal of Canadian Petroleum Technology* **36**(9)
[21] Cheng L, Shahin M A, Cord-Ruwisch R, Addis M, Hartanto T and Elms C 2014 Soil Stabilisation by Microbial-Induced Calcite Precipitation (MICP): Investigation into Some Physical and Environmental Aspects *7th International Congress on Environmental Geotechnics Conf.* (Melbourne, Australia: Engineers Australia) pp 1105-12
[22] Osinubi K J, Eberemu A O, Ijimdiya S T, Yakubu S E and Sani J E 2017 Potential use of B. *Pumilus* in microbial induced calcite precipitation improvement of lateritic soil *Proceedings of the 2nd Symposium on Coupled Phenomena in Environmental Geotechnics CPEG2* Paper #64 (Leeds, United Kingdom) pp 1– 6
[23] Kim G and Youn H 2016 Microbially Induced Calcite Precipitation Employing Environmental Isolates *Materials* **9**(6) 468
[24] Zhang J, Zhou A, Liu Y, Zhao B, Luan Y, Wang S, Yiu X and Li Y 2017 Microbial Network of the Carbonate Precipitation Process Induced by Microbial Consortia and the Potential Application to Crack Healing in Concrete *Science Reports* **7** 1-10
[25] Ng W, Lee M and Hii S 2012 An Overview of the Factors Affecting Microbial-Induced Calcite Precipitation and its Potential Application in Soil Improvement *World Academy of Science, Engineering and Technology International Journal of Civil and Environmental Engineering* **6**(2) 188-193
[26] Anbu P, Kang C, Shin Y and So J 2016 Formations of Calcium Carbonate Minerals by Bacteria and its Multiple Applications *Springerplus Springer International Publishing* **5**(1) 250
[27] Yohanna P 2015 *The Use of Iron Ore Tailing as Admixture in Cement Modification of Black Cotton Soil* (MSc Thesis, Ahmadu Bello University, Zaria, Nigeria)
[28] Neupane S 2016 *Evaluating the Suitability of Microbial Induced Calcite Precipitation Technique for Stabilizing Expansive Soils* (MSc Thesis, Department of Civil Engineering, Boise State University, Idaho)

[29] Brady N C and Weil R 2002 *The Nature and Properties of Soils* 13th Ed. (New Jersey, USA: Prentice Hall Inc.) p 960
[30] Ovuarume U B 2011 *Effect of Locust Bean Waste Ash on Lime Modified Black Cotton Soil* (MSc Thesis, Ahmadu Bello University, Zaria, Nigeria)
[31] Salahedin M 2013 Effects of CEC on Atterberg limits and Plastic Index in Different Soil Textures *International Journal of Agronomy and Plant Production* **4**(9) 2111-18.
[32] Kirk G J D, Versteegen A, Ritz K and Milodowski A E 2015 A Simple Reactive-Transport Model of Calcite Precipitation in Soils and other Porous Media *GeochimicaetCosmochimicaActa* **165** 108–122

Microbial-Induced Calcite Precipitation Study on the Plasticity and Compaction Characteristics of Lateritic Soil Treated with Bacillus *Megaterium* in Urea-$CaCl_2$ Culture Medium

K J Osinubi[1], A O Eberemu[1,2], T S Ijimdiya[1] and R K Etim[3*]

[1]Department of Civil Engineering, Ahmadu Bello University, Zaria, Kaduna State, Nigeria.
[2]Department of Civil Engineering and Africa Center of Excellence on New Pedagogies in Engineering Education (ACENPEE), Ahmadu Bello University, Zaria, Kaduna State, Nigeria.
[3]Department of Civil Engineering, Akwa Ibom State University, Ikot Akpaden, Nigeria.

*Corresponding Author: rolandkufre24@gmail.com

Abstract. An ecofriendly method of soil improvement known as Microbial Induced Calcite Precipitation (MICP) has received significant recognition in the past decade. This study presents a report on the capability of MICP in modifying the plasticity and compaction properties of lateritic soil bio-treated with different suspension densities of a ureolytic microorganisms - Bacillus *megaterium* (B. *megaterium*). Specimens were prepared with liquid limit moisture of the natural soil in three mix ratios of B. *megaterium* (B) and cementation reagent (C) (urea medium): 25 % B: 75 % C, 50 % B: 50 % C and 75 % B: 25 % C. Another set of specimens were prepared at optimum moisture content (OMC) equivalent covering 1/3 pore volume of B. *megaterium* and 2/3 pore volume of urea medium introduced in three sequences at 6-hour interval to steadily initiate MICP mechanisms. The results of the various mix ratios indicated improvement in plasticity index (PI) with higher B. *megaterium* suspension density. The best improved PI value was obtained for specimen treated with 75 % B: 25 % C mix ratio at B. *megaterium* suspension density of 2.40×10^9 cells/ml with an equivalent maximum 5.3 % $CaCO_3$ content. The maximum dry density (MDD) and OMC values marginally increased with higher B. *megaterium* suspension density as well as calcite content. The micrographs obtained from scan electron microscope (SEM) showed changes from untreated to bio-treated state, which resulted in non-uniform precipitation of calcite in the soil. The variation in elemental quantifications displayed by the X-ray energy dispersive spectroscopy (EDS) validate the formation of calcite within voids and on surface of the soil particles.

1. Introduction

Concerns over use of chemical stabilizers for soil strengthening and indeed improvement have been raised, especially as most chemical additives have some debilitating effects on the environment [1]. The increasing awareness on environmental issues caused by the use of chemical stabilizers has brought about an innovative approach which adopts green and eco-friendly technologies. A unique soil improvement practice involving biologic alteration in properties of soil can be achieved by means of microbiologically induced calcite precipitation (MICP). MICP is a relatively novel technique that utilizes microorganism, nutrients and biological processes to strengthen bearing strength, mitigate liquefaction, reduce permeability and shrinkage cracks in soils [1-8].

The eco-friendly features of ureolytic-microorganisms are most often than not the basis for its use. MICP has presented tremendous ground-breaking researches ranging from laboratory scale investigation to potential full-scale field implementation [9-14]. Although not without challenges

consistent with conditions that favour precipitation as well as deposition of calcite, a case study approach to upscale MICP into field or large scale application based on specific applications/objectives was used and reported by [15]. Studies have shown that MICP techniques have been used in bio-cementation of leaky media, environmental remediation of contaminated sites, increased soil strength, reduction of desiccation cracks, improved plasticity properties, reduction in permeability of fine-grained soil, etc. [1, 7-8, 15-23].
The reaction mechanisms for the formation and spread of calcites products in MICP have been reported in several studies [1, 24]. Detailed reactions are given in eqs. (1) - (7). Urease enzymes released by the microorganism is used in the spontaneous hydrolysis of urea - calcium source. The bacteria cell offers itself as nucleation site from where calcium ions (Ca^{2+}) concentration and calcite in the form of Ca^{2+} and $CO_3{}^{2-}$ in solution (see eqs. (6) and (7)) is formed.

$$CO(NH_2)_2 + H_2O \xrightarrow{Microbial\,urease} NH_2COOH + NH_3 \quad (1)$$
$$NH_2COOH + H_2O \xrightarrow{Spontaneous} NH_3 + H_2CO_3 \quad (2)$$
$$H_2CO_3 \leftrightarrow HCO_3 + H^+ \quad (3)$$
$$NH_3 + 2H_2O \leftrightarrow 2NH_4^+ + 2OH^- \quad (4)$$
$$HCO_3 + H^+ + 2OH \leftrightarrow CO_3^{2-} + 2H_2O \quad (5)$$
$$Ca^{2+} + Bacteria\ cell \rightarrow Cell - Ca^{2+} \quad (6)$$
$$Cell - Ca^{2+}CO_3^{2+} \rightarrow Cell - CaCO_3 \quad (7)$$

Most studies in the field of MICP have mainly focused on engineering properties of coarse grained soils. However, few systematic studies have been carried out on the effect of MICP process on the plasticity behaviour of fine-grained soils [6, 18-19, 25]. Furthermore, limited literature exists which compare evaluation from different methods of sample preparation prior to consistency limit test. This study focused on the growing area of research on microbiological interactions in biogeotechnics exclusive to soil plasticity in MICP mechanisms. The potential of B. *megaterium* in the swamp of spore forming and urease producing enzymatic bacteria is explored in studying the plasticity and compaction behaviour of fine-grained soil. The specific objectives include, but are not limited to the determination of liquid limit, plasticity index and calcite content of lateritic soil bio-treated with B. *megaterium* in urea-calcium medium as well as the most effective bacteria - cementation reagent mix ratio.

2. Materials and Methods

2.1. Materials

2.1.1. Soil

Lateritic soil sample was collected from Abagana, Anambra state, Nigeria, using the method of disturbed sampling.

2.1.2. Bacteria

The microorganism Bacillus species (B. *megaterium*) bacteria was used. The microorganism was cultured and grown from the soil sample used in this study. B. *megaterium* is characteristically a gram-positive, rod-like, endospore forming and urease positive.

2.1.3. Cementation reagent

The cementation reagent is composed of 20 g$(NH_2)_2CO$ (urea), 10 g NH_4Cl (Ammonium chloride), 3 g nutrient broth, 2.8 g $CaCl_2$ (Calcium chloride) and 2.12 g $NaHCO_3$ (Sodium hydrogen trioxocarbonate (iv)) per 1 dm^3 (de-ionized water) as reported by researchers [25-29]. After autoclaving the other solutions in the cementation reagent, urea was aseptically added to avoid the decomposition of urea by heat [7, 8, 20, 21, 25, 30].

2.2. Methods
2.2.1. Index properties

Tests on the natural and treated soil were performed based on procedures outlined in [31] and [32], respectively.

2.2.2. Sample preparation
Two methods were adopted for the preparation of samples for Atterberg limits tests. The first method adopted is the procedure reported by [18, 19, 33]. The method involved the introduction of 1/3 and 2/3 pore volume of bacterial and urea medium, respectively, to activate MICP process. In this study, air-dried sample of lateritic soil was pulverized and sieved through 4.76 mm openings. The pore volume was obtained using the OMC of natural soil compacted with British Standard light (or standard Proctor) energy. The soil - bacteria mixture for each test was obtained by mixing each 250 g of the soil with 1/3 pore volume of B. *megaterium* at suspension densities of 0, 1.5 x 10^8, 6.0 x 10^8, 1.2 x 10^9, 1.8 x 10^9 and 2.4 x 10^9 cells/ml, respectively. The MICP process was initiated by the introduction of 2/3 pore volume of urea medium (cementation reagent) into the soil-bacteria mixtures in three cycles at 6 hours' intervals. After the third cycle, the lateritic soil - B. *megaterium* - cementation reagent mixtures were sealed in polythene bags for 24 hours to enhance saturation and hydration before drying under laboratory condition.
The second method of sample preparation adopted is procedure reported by Osinubi *et al*. (2019e). Three specimens were prepared for different mix ratios of B. *megaterium* + urea medium. The composition ratio was considered based on the liquid limit moisture of the natural soil. Each 250 g of soil that was passed through sieve with 425 μm openings was mixed with the pre-determined liquid limit of the natural soil of 42.0 % prepared in three ratios of 14 %: 28 % (i.e., 25 % B: 75 % C), 21 %: 21 % (i.e., 50 % B: 50 % C) and 28 %: 14 % (i.e., 75 % B: 25 % C). The treated specimens were dried under laboratory condition to enable MICP modification of the soil structure. The stepped B. *megaterium* suspension density of, 0, 1.50 × 10^8, 6.0 × 10^8, 1.20 × 10^9, 1.80 × 10^9 and 2.40 × 10^9 cells/ml (i.e., equivalent of 0, 0.5, 2.0, 4.0, 6.0 and 8.0 McFarland standards) used in this study is based on McFarland standards [34].

2.2.3. Atterberg limits

The treated air-dried lateritic soil - B. *megaterium* - cementation reagent specimens were pulverized and passed through sieve with 425 μm aperture before testing using the procedure outlined in [31]. The average of three consistent results was used in the computations.

2.2.4 Calcite content

The calcite content was determined using the washing method reported by [35]. 5 g each of natural soil and treated soil were dissolved in 20 ml of 1-M HCl (gravimetric acid washing) to remove $CaCl_2$ residue. The solvent was then filtered using filter paper with openings corresponding to No. 200 sieve. The filtrate on the filter paper was then rinsed and washed for about 10 minutes with distilled water to completely remove dissolved calcium from the soil. Thereafter, particles of solid which were left on the filter paper were oven-dried at 105°C for 24 hours and weighed. The change in mass between the original sample (B_1) and washed sample (B_2) was determined as the mass of $CaCO_3$. The test was repeated in triplicate and the average taken for the mixtures considered. The calcite content (CC) was determined using eq. (8):

$$CC\ (\%) = 100 - \frac{B_1}{B_2} \times 100 \quad (8)$$

2.2.5 Compaction

The procedure outlined in [31] for BSL compaction was used.

2.2.6 Microanalysis

Scanning electron microscopy, SEM (Phenom-world analyser) was used to assess the morphological variations due to the precipitation and spread of calcite imprints that bonded the particle contact points of the bio-treated specimens. The results were complimented with x-ray Energy dispersive spectroscopy (EDS) and automated statistical analyser that aided the display of semi quantitative elemental analysis of the basic component of untreated and bio-treated soil.

3. Results and Discussion

3.1. Index property

The tested soil was classified as A-4(3) soil using AASHTO [36] or SC soil using USCS [37]. The particle size distribution and summary of the properties of the natural lateritic soil are provided in Figure 1 and Table 1, respectively.

Table 1. Properties of natural lateritic soil

Property	Quantity
Percentage passing 0.075 mm sieve	35.3
Natural moisture content (%)	11.6
Liquid limit (%)	41.5
Plastic limit (%)	15.9
Plasticity index (%)	25.6
Specific gravity	2.65
AASHTO classification	A-4(3)
USCS classification	SC
Colour	Reddish-brown
Dominant clay mineral Kaolinite	Kaolinite

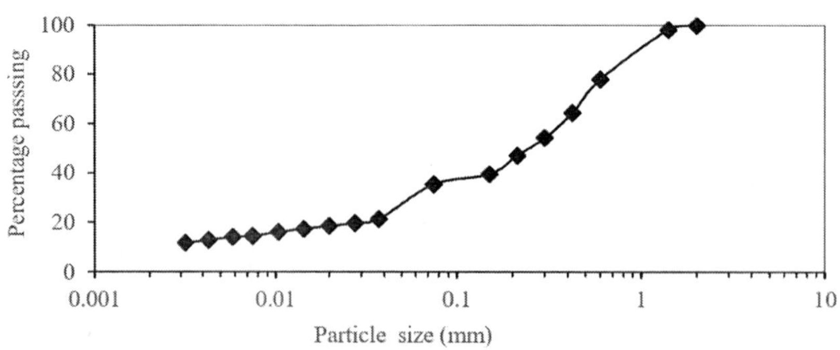

Figure 1. Particle size distribution curve for natural soil

3.2. Effect of B. megaterium on liquid limit of lateritic soil

The effect of the B. *megaterium* suspension density on liquid limit (LL) of lateritic soil for the two methods of treatment considered is presented in Fig. 2. The LL of specimen bio-treated with 1/3 pore volume bacteria (PVB) and 2/3 pore volume urea medium (cementation reagent) (PVC) i.e., 13 PVB: 2/3 PVC, increased from 41.5 % for the untreated to a peak value of 45.7 % at B. *megaterium* suspension density of 2.4 × 10^9 cells/ml. The steady increase in liquid limit of 1/3PVB: 2/3VPVC specimen could probably be due to inadequate calcium (Ca^{2+}) from urea-$CaCl_2$ medium/cementation reagent required to precipitate the CO_3^{2-} into $CaCO_3$ or inadequate nucleation site (from bacteria); required to absorbed/attract mobile positive end of ions of $CaCO_3$ ions and free divalent Ca^{2+} onto its surface [1].

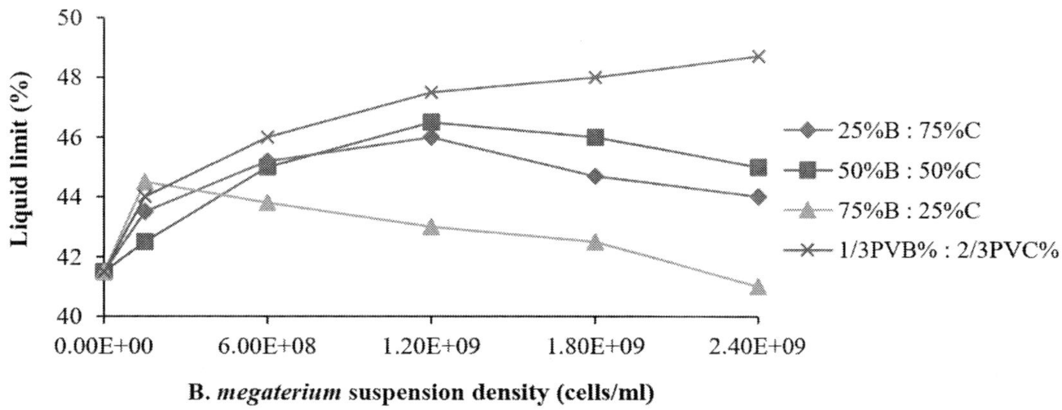

Figure 2. Variation of liquid limit of lateritic soil with B. *megaterium* suspension density

For the second method of treatment, the liquid limit values of specimens treated with 25 % B: 75 % C and 50 % B: 50 % C mix ratios initially increased from 41.5 % for the untreated soil to peak values of 46 % and 46.5 %, respectively, at B. *megaterium* suspension density of 12 × 10^9 cell/ml. Thereafter, LL values decreased marginally to 44 % and 45 %, respectively, with increase in B. *megaterium* suspension density up to 2.4 × 10^9 cell/ml. Also, the LL of specimen treated with 75 % B: 25 % C mix proportion marginally increased from 41.5 % for the untreated to 42.5% at B. *megaterium* suspension density of 1.5× 10^8cells/ml and thereafter decreased to a minimum value of 41.0 % at 2.4 × 10^9 cell/ml. The initial increase in LL value could be attributed to increase in water content required for the continuous hydrolysis of urea. The biochemical reaction produced urease positive enzymes required to actively hydrolyse urea into ammonium ions (NH_4^+) and carbonate ions (CO_3^{2-}) in solution which increased the pH of the mixtures [25, 38]. The process probably increased the affinity of the soil clay mineral for water that increased its diffuse double layer thereby increased the liquid limit of the mixtures [39, 40] The recorded decrease in LL value with higher B. *megaterium* suspension density could be due to increasing cell forming unit, which provided greater surface sites for nucleation of $CaCO_3$ and free Ca^{2+} [23, 41]. The non-plastic calcium carbonate that was formed in the pore of the soil contributed to the decrease in LL values of the treated soil specimens [20, 42]. Furthermore, the decrease in LL value could also be due to the presence of multivalent cations (Ca^{2+}, Al^{3+}, Fe^{3+}, etc.); present in the soil which may have led to cation exchange between the soil clay minerals and precipitated calcite. This stimulates flocculation of clay minerals as reported in [25, 43-45]. The results of this study show that, higher B. *megaterium* population in colony forming unit/ml (cfu/m) or suspension density/ml is required to drastically reduce the LL value of lateritic soil [25].

3.3. Effect of B. megaterium suspension density on plastics limits

The plastic limit (PL) generally increased with increasing B. *megaterium* suspension density to peak values at 2.4×10^9 cell/ml for the two methods of treatment considered (see Fig. 3). A probable explanation for the trend could be due to insufficient Ca^{2+} in solution thus leading to the possible formation of vaterite (anhydrous calcite $CaCO_3.H_2O$) [17, 46]. Adequate amount of Ca^{2+} was not available and neither adsorbed on surface of the calcite causing a reduced deposition of ion pairs ($CaCO_3$) and conversion of vaterite to calcite which decreased the water content (i.e., plastic limit) [17, 47]. It implies that the crystallized water in vaterite could have initiated an intrinsic rise in moisture of the mixture that resulted in the increasing PL values that increased with higher B. *megaterium* suspension density. Similar findings were reported by [6, 25, 42].

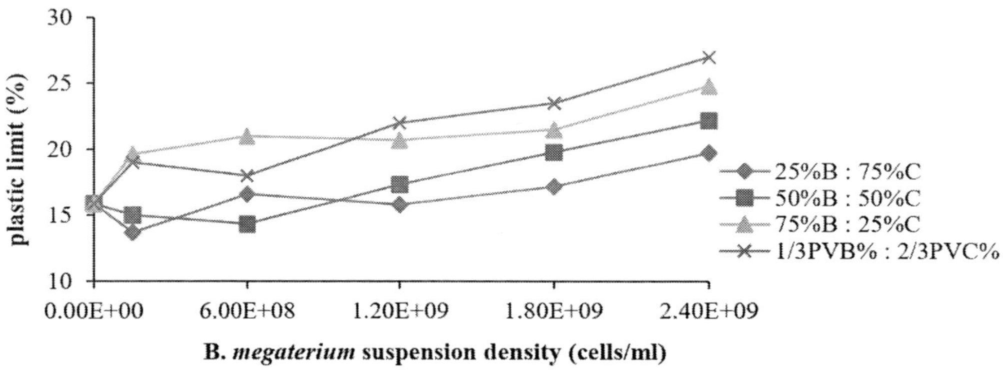

Figure 3. Variation of plastic limit of lateritic soil with B. *megaterium* suspension density

3.4. Effect of B. megaterium suspension density on plasticity index

The changes observed in plasticity index (PI) with increasing B. *megaterium* suspension density recorded for the two methods of bio-treatment is shown in Fig. 4. The trends were affected by the LL and PL values because PI is the mathematical difference between the former and latter. In the method of treatment of soil with 1/3 PVB: 2/3 PVC mix ratio, the PI value initially increased to a peak value of 28.0 % at 6.0×10^8 cell/ml and thereafter decreased to a minimum value of 21.7 % at B. *megaterium* suspension density of 2.4×10^9 cell/ml. On the other hand, for the second method, specimen treated with 25 % B: 75 % C mix ratio, the PI value increased from 25.6 % (untreated soil) to 30.19 % at B. *megaterium* suspension density of 1.2×10^9 cell/ml and thereafter decreased to a minimum value of 24.3 % at 2.4×10^9 cell/ml. For specimen treated with 50 % B: 50 % C mix ratio, PI value initially increased from 25.6 % (untreated soil) to a peak value of 30.7 % at B. *megaterium* suspension density of 6.0×10^8 cell/ml and thereafter decreased to a minimum value of 22.8 % at 2.4×10^9 cell/ml. The PI of specimen treated with 75 % B: 50 % C mix ratio progressively decreased with increased B. *megaterium* suspension density from 25.6 % (untreated soil) to 16.2 % at 2.4×10^9 cell/ml.

Overall, the treatment of lateritic soil using the two methods considered indicated improvement in plasticity of the natural soil, which can be attributed to the formation of calcite during the MICP process. It is pertinent to reiterate that, the methods adopted in this study are based on the findings reported by Osinubi *et al.* [18, 19, 25]. Since plasticity index is a soil improvement criterion, the performance of the methods of treatment considered can be ranked as 75 % B: 25 % C > 1/3 PVB%: 2/3 PVC% > 50 % B: 50 % C > 25 % B: 75 % C. The results recorded in this study show that compositional variables (Bacteria: Cementation reagent mix ratio) could affect the plasticity characteristics of fine-grained soils during the MICP process in agreement with Osinubi *et al.* [25].

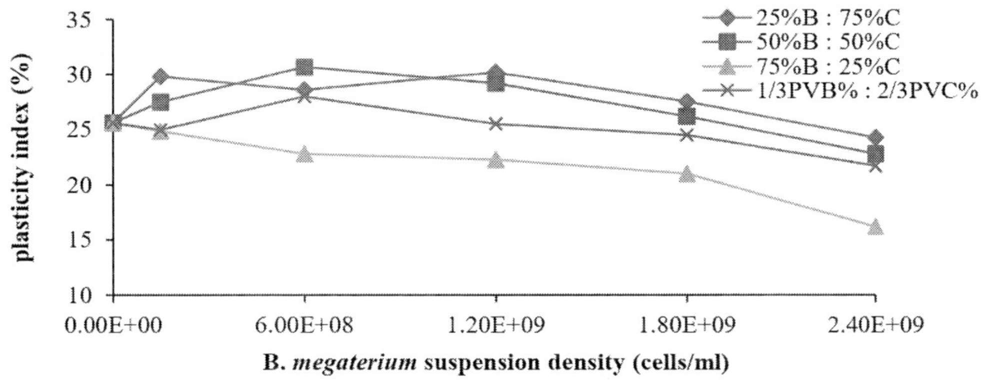

Figure 4. Variation of plasticity index of lateritic soil with B. *megaterium* suspension density

3.5. Calcite content

The calcite content increased as B. *megaterium* suspension density increased for the two methods of treatment considered as shown in Fig 5. The highest calcite content of 5.3 % was recorded for the treatment of soil with 75 % B: 25 % C mix ratio that produced the best PI value of 16.2 % at B. *megaterium* suspension density of 2.4×10^9 cells/ml. The result is not unconnected with the increased microorganism due to higher bacteria cell concentration which produced more urease enzyme to commence the hydrolysis of urea present in the cementation reagent [48] as well as offering more nucleation site which facilitated precipitation of calcite [23, 26].

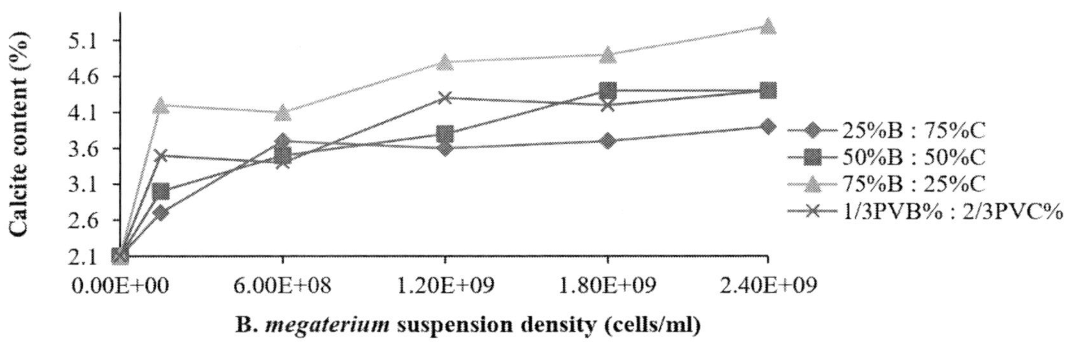

Figure 5. Variation of calcite content of lateritic soil with B. *megaterium* suspension density

3.6. Compaction characteristics

The maximum dry density (MDD) of lateritic soil marginally increased from 1.72 Mg/m^3 for the untreated soil to 1.79 Mg/m^3 at 24×10^8 cells/ml B. *megaterium* suspension density (Fig. 6). The observed increase in MDD was probably due to the 65.0 % sand fraction (i.e., 0.075 - 4.73 mm) which is higher than 35.0% silt content (i.e., passing 0.075 mm) recorded for the natural soil that enhanced the geometric compatibility and pore throat requirement relative to the bacteria size that is essential for effective MICP process. Another reason for the increased MDD could be due to supposed higher specific gravity of the precipitated calcite that filled the pores of the soil with lower specific value of 2.65. Findings reported by researchers [18, 19, 49-51] suggest that calcites precipitated are aggregated as one in both weight and specific density.

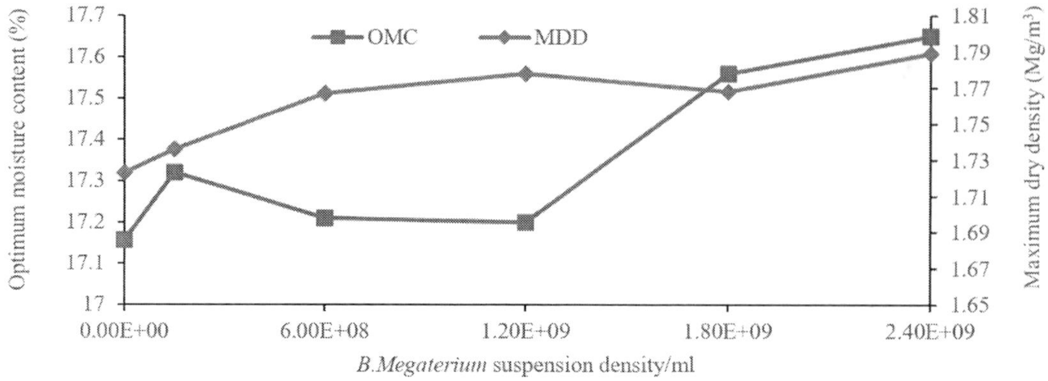

Figure 6. The variation of compaction characteristics of lateritic soil with B. *megaterium* suspension density

The optimum moisture content (OMC) of lateritic soil marginally increased from 17.2 % for the untreated soil to 17.7 % at B. *megaterium* suspension density of 24 × 10^8 cells/ml. The increase could be due to insufficient supply of Ca^{2+} from urea-$CaCl_2$ medium with increased B. *megaterium* cells/ml within the suspension that possibly facilitated the deposition of both calcite and vaterite (anhydrous calcite $CaCO_3.H_2O$) [17, 46, 47]. The crystallized water in the vaterite went into solution during compaction thereby increasing the moisture content of the soil. This reaction suggests that greater quantity of water would be required to obtain the optimum value.

3.7. Scanning electron microscopy

The scanning electron microscope (SEM) micrographs of untreated soil (0 cell/ml) and soil treated with 75 % B: 25 % C mix ratio that produced the best PI value at B. *megaterium* suspension density of 24 × 10^8 cell/ml is shown on Plate I. The micrographs of untreated soil (see Plate Ia) depict larger and smaller particle sizes with characteristically smooth-to-rough surface structure. The central thick lump particle could be due to the cohesive structure of the soil due to addition and mixing with water. This possibly suggests that a thick diffuse double layer due to reaction of clay minerals and multivalent cations was still holding or still held a dominant influence on the soil-water mixture. The microstructure of the treated soil (see Plate Ib) show imprints of calcites precipitated and bonded to the inter particle surface that bridged contact points. However, the precipitated calcite was not uniformly distributed. The additional irregular calcite deposition configuration might be attributed to the non-uniform circulation/percolation of the cementation reagent relative to the B. *megaterium* cells/ml [49]. The magnifications reported in this study was based on the most probable clarity offered during testing.

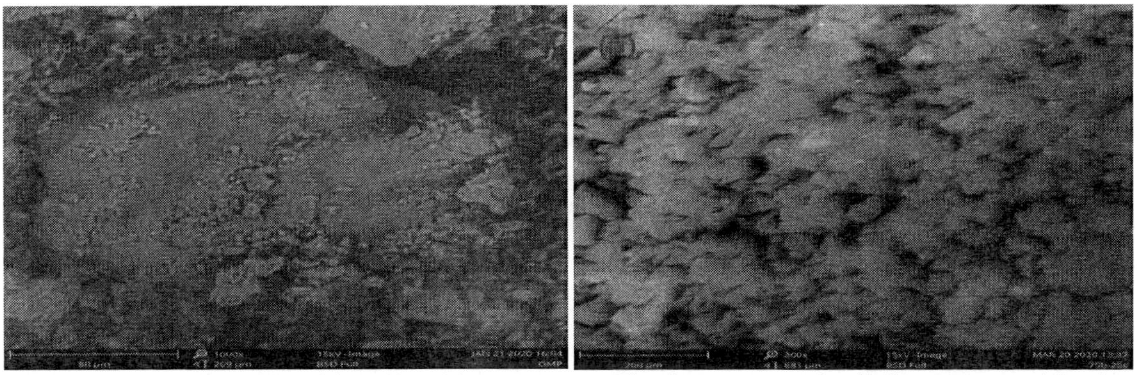

Plate I. Micrographs of lateritic soil at (a) x1000 magnifications untreated soil (b) x300 magnifications soil treated with 75 % B: 25 % C mix proportion at B. *megaterium* suspension density of 24 × 10^8 cells/ml

The results of energy dispersive spectroscopy (EDS) obtained from the SEM of the untreated and bio-treated specimen are shown in Fig 7a and Fig 7b, respectively. The presence of Si, Al and Fe in the soil indicated the predominant presence in the soil. However, there were some observed slight variation in other elements. Basically, the higher Ca in treated soil compared to untreated could be associated with formation of calcite and free Ca^{2+} ions in solution available at the nucleation site of bacterial cells [26. 27, 52]

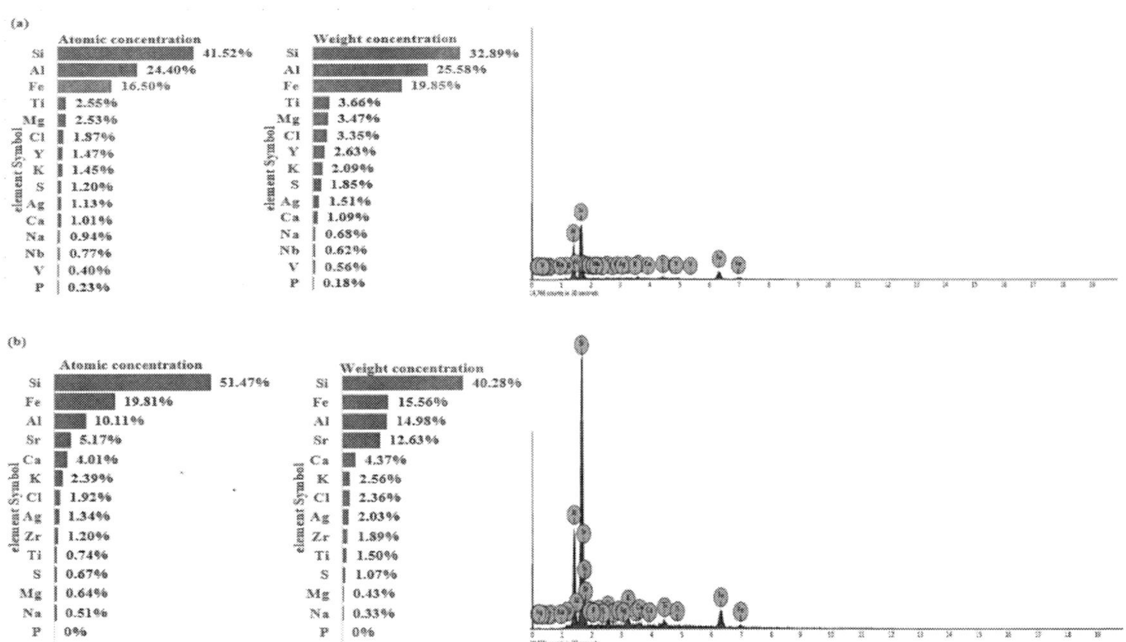

Figure 7. EDS with quantitative atomic and weight concentration; (a) Untreated soil (b) Soil treated with 75 % B: 25 % C mix proportion at B. *megaterium* suspension density of 24 × 10^8 cells/ml

4. Conclusion

The study investigated the plasticity and compaction characteristics as well as calcite content of of bio-treated lateritic soil using different compositional variables (i.e., B. *megaterium* -cementation reagent mix ratios). Two methods of specimen preparations were adopted for the evaluation of Atterberg limits, compaction and calcite content. SEM complimented with EDS was used to study the variation in microstructural morphology and elemental quantification analysis of the untreated and bio-treated specimen. Treatment of lateritic soil with 75% B. *megaterium*: 25% urea medium at B. *megaterium* suspension density of 2.40×10^9 cells/ml can be effectively utilized to reduce the plasticity index of A-4(3) or SC to a more desirable value using the MICP technique. The compaction characteristics of the treated soil significantly improved. The microanalysis carried out using SEM and EDS support the improved properties of the treated soil. In order to obtain significantly higher values for the properties considered in the study, it is recommended that higher B. *megaterium* suspension density be used.

References

[1] DeJong JT, Mortensen M, Martinez BC and Nelson DC 2010 Biomediated soil improvement. *Ecol Eng* 36:197–210. https://doi.org/10.1016/j.ecoleng.2008.12.029
[2] Rebata-Landa V and Santamarina JC 2006 Mechanical limits to microbial activity in deep sediments. *Geochem. Geophys. Geosyst.*, **7**(11), Q11006.
[3] Handley-Sidhu S, Sham E, Cuthbert MO, Nougarol S, Mantle M, Johns ML, Macaskie LE and Renshaw JC 2013 Kinetics of urease-mediated calcite precipitation and permeability reduction of porous media evidenced by magnetic resonance imaging. *Int J. Environ. Sci. Technol.* DOI: 10 881–890. 10.1007/s13762-013-0241-0.
[4] Park SS, Choi SG, Kim WJ and Lee JC 2014 Effect of Microbially Induced Calcite Precipitation on the Strength of Cemented Sand. In *New Frontiers in Geotechnical Engineering* (pp. 47-56).
[5] Umar M and Kassim KA 2017 Effect of cementation reagents concentrations on microbial calcite precipitation in residual soil. *Malaysian Journal of Civil Engineering 29 Special Issue (1):79-90*
[6] Osinubi KJ, Eberemu AO, Ijimdiya TS, Yakubu SE. and Sani J. E. 2017 Potential use of B. pumilus in Microbial-Induced Calcite Precipitation Improvement of Lateritic Soil. *Proc. of the 2nd Symposium on Coupled Phenomena in Environmental Geotechnics* (CPEG2), Leeds, UK 2017
[7] Osinubi KJ, Yohanna P, Eberemu AO, Ijimdiya TS 2019a Evaluation of hydraulic conductivity of lateritic soil treated with Bacillus *coagulans* for use in waste containment applications. In: Zhan L, Chen Y, Bouazza A (eds) *Proc. of the 8th Int. congress on environmental geotechnics* (ICEG 2018). Towards a sustainable geoenvironment, vol 3. Springer, Hangzhou, pp 401–409. doi. 10.1007/978-981-13-2227-3_50
[8] Osinubi KJ, Sani JE, Eberemu AO, Ijimdiya TS and Yakubu SE 2019b Unconfined compressive strength of Bacillus Pumilus treated lateritic soil. In: Zhan L, Chen Y, Bouazza A (eds) *Proc. of the 8th Int. congress on environmental geotechnics* (ICEG 2018). Towards a sustainable geoenvironment, vol 3. Springer, Hangzhou, pp 401–409. doi. 10.1007/978-981-13-2227-3_51
[9] Cuthbert MO, McMillan LA, Handley-Sidhu S, Riley MS, Tobler DJ, and Phoenix VR 2013 A field and modeling study of fractured rock permeability reduction using microbially induced calcite precipitation. *Environ. Sci. Technol.*, **47**(23):13637-13643.
[10] Phillips AJ, Cunningham AB, Gerlach R, et al. 2016 Fracture sealing with microbially induced calcium carbonate precipitation: A field study. *Environ. Sci. Technol.*, **50**(7):41114117.
[11] Hodges TM and Lingwall BN 2020 Case Histories of Full-Scale Microbial Bio-Cement Application for Surface Erosion Control. *In proceedings of Geo-Congress 2020. American Society of Civil Engineers, GSP* **320** pp. 9 – 19
[12] Terzis D, Laloui L, Dornberger S and Harran R 2020 A Full-Scale Application of Slope Stabilization via Calcite Bio-Mineralization Followed by Long-Term GIS Surveillance. *In proceedings of Geo-Congress 2020. American Society of Civil Engineers, GSP* **320** pp. 65 – 73
[13] Bhaskar CS, Chittoori BCS, Pathak A, Burbank M and Islam MT 2020 Application of Bio-Stimulated Calcite Precipitation to Stabilize Expansive Soils: Field Trials. *In proceedings of Geo-Congress 2020. American Society of Civil Engineers, GSP* **320** pp. 111 – 120
[14] Wang Z, Zhang N, Cai G, Jin Y, Ding N and Shen D 2017 Review of ground improvement using microbial induced carbonate precipitation (MICP), *Marine Georesources & Geotechnology*, http://dx.doi.org/10.1080/1064119X.2017.1297877

[15] van Paassen, LA 2011 Bio-mediated ground improvement: from laboratory experiment to pilot applications. In: *Proceedings of geofrontiers in geotechnical engineering 2011: technical papers, ASCE*, pp 4099–4108
[16] Fujita Y, Taylor J, Wendt L, Reed D and Smith R 2010 Evaluating the potential of native ureolytic microbes to remediate 90Sr contaminated environment. *Environ Sci Technol* **44**, 7652–7658
[17] Li W, Cheng W, Zhou P, Zhu S and Yu L 2013 Influence of initial calcium ion concentration on the precipitate and crystal morphology of calcium carbonate induced by bacteria carbonic anhydrase. *Chem Engrg J.* **218**, 65-72. http://dx.doi.org/10.1016/j.cej.2012.12.034
[18] Osinubi KJ, Eberemu AO, Gadzama EW, Ijimdiya TS, Jonathan OJ and Al-Mustapha I 2018a Comparative Effect of Sporosarcina Pasteurii on the Index Properties and Compaction Characteristics of Black Cotton and Lateritic Soils. Proceedings of NBRRI International Conf. on Sustainable Development Goals & Nigerian Construction Industry, 12-14th June 2018. Pp. 351-361.
[19] Osinubi KJ, Eberemu AO, Yohanna P, Ijimdiya TS, Adjei A and Sada YY 2018b Effect of Bacillus Coagulans-Induced Calcite Precipitate on the Index and Compaction Properties of Lateritic and Black Cotton Soils. Proceedings of NBRRI International Conf. on Sustainable Development Goals & Nigerian Construction Industry, 12-14th June 2018. Pp. 362-372
[20] Osinubi KJ, Gadzama EW, Eberemu AO, Ijimdiya TS and Yakubu SE 2019c Evaluation of the strength of compacted lateritic soil treated with Sporosarcina pasteurii. In: Zhan L, Chen Y, Bouazza A (eds) Proceedings of the 8th international congress on environmental geotechnics (ICEG 2018). Towards a sustainable geoenvironment, 3. Springer, Hangzhou, pp 419–428. https://doi. org/10.1007/978-981-13-2227-3_52
[21] Osinubi KJ, Yohanna P, Eberemu AO, Ijimdiya TS 2019d Unconfined Compressive Strength of Lateritic Soil Treated with Bacillus Coagulans for use as Liner and Cover Material in Waste Containment System. 1st International Conference on Sustainable Infrastructural Development. IOP Conf. Series: Materials Science and Engineering **640** (2019) 012081 doi:10.1088/1757-899X/640/1/012081
[22] Osinubi KJ, Eberemu AO, Ijimdiya TS and Yohanna P 2020a Interaction of landfill leachate with compacted lateritic soil treated with bacillus coagulans using microbial-induced calcite precipitation approach. *J Hazard Toxic Radioact Waste*, **24**(1), 040190249. https://doi.org/10.1061/(asce)hz.21535515.0000465
[23] Osinubi KJ, Eberemu AO, Ijimdiya TS, Yakubu SE, Gadzama EW, Sani JE and Yohanna P 2020b Review of the use of microorganisms in geotechnical engineering applications. **2**, 207. https://doi.org/10.1007/s42452-020-1974-2
[24] Castro-Alonso MJ, Montañez-Hernandez LE, Sanchez-Muñoz MA, Franco M, Rubi M, Narayanasamy R and Balagurusamy N 2019 Microbially induced calcium carbonate precipitation (MICP) and its potential in bioconcrete: microbiological and molecular concepts. *Front Mater* 6:126. https ://doi.org/10.3389/fmats .2019.00126
[25] Osinubi KJ, Eberemu AO, Gadzama EW, Ijimdiya TS 2019e Plasticity characteristics of lateritic soil treated with Sporosarcina pasteurii in microbial-induced calcite precipitation application. *SN Applied Sciences* **1**,829. https://doi.org/10.1007/s42452-019-0868-7
[26] Stocks-Fischer S, Galinat JK and Bang SS 1999 Microbiological precipitation of $CaCO_3$. *Soil Biology and Biochemistry* **31**(11): 1563–1571.
[27] Dejong JT, Fritzges MB and Nusslein K 2006 Microbially induced cementation to control sand response to undrain shear. ASCE *J Geotech Geoenvironmental Eng* **132**(11):1381–1392. https://doi.org/10.1061/(ASCE)1090-0241(2006)132:11(1381)
[28] Kim G and Youn H 2016 Microbially induced calcite precipitation employing environmental isolates. *Materials* **9**(468):1–10. https://doi.org/10.3390/ma906 0468
[29] Tirkolaei HK and Bilsel H 2017 Estimation on ureolysis-based microbially induced calcium carbonate precipitation progress for geotechnical applications. *Mar Georesour Geotechnol* **35**(1):34–41. https://doi.org/10.1080/10641 19X.2015.10990 62

[30] Pakbaz MS, Behzadipour H and Ghezelbash GR 2018 Evaluation of shear strength parameters of sandy soils upon microbial treatment. *Geomicrobiol J.* https://doi.org/10.1080/01490 451.2018.14557 66
[31] BS 1377 (1990). *Method of Testing Soils for Civil Engineering Purpose*. British Standard Institute, BSI, London.
[32] BS 1924 (1990). *Method of Test for Stabilized Soils*. British Standard Institute, BSI, London.
[33] Rowshanbakhta K, Khamehchiyana M, Sajedib RH and Nikudela MR 2016 Effect of Injected Bacterial Suspension Volume and Relative Density on Carbonate Precipitation Resulting from Microbial Treatment. *J. Ecol Engrg,* **89**,49-55. https://doi.org/10.1016/j.ecoleng.2016.01.010
[34] Chapin KC and Lauderdale T 2003 Reagents, stains, and media: bacteriology. In: Murray, P. R., Baron, E. J., Jorgensen, J. H., Faller, M. A. P., Yolken, R. H. (eds) Manual of clinical microbiology, 8th edn. ASM Press, Washington, DC, p 358
[35] Choi, S. G., Park, S. S., Wu, S., Chu, J. (2017) Methods for Calcium Carbonate Content Measurement of Biocemented Soils. Journal of Materials in Civil Engineering, Technical note **29**(11): 06017015.
[36] AASHTO 1986 Standard Specification for Transportation, Material and Methods of Sampling and Testing. 14th Edition. Amsterdam Association of State Highway and transportation officia Washington D.C.
[37] ASTM 1992 Annual Book of Standards. Vol. 04.08, American Society for Testing and Materials. Philadelphia.
[38] Wang L, van Paassen, LA and Edward Kavazanjian Jr., E 2020 Feasibility Study on Liquefaction Mitigation of Fraser River Sediments by Microbial Induced Desaturation and Precipitation (MIDP). *In proceedings of Geo-Congress 2020. American Society of Civil Engineers, GSP* **320**, pp. 121 – 131
[39] Neumann AW, David R and Zuo Y 2011 *Applied surface thermodynamics*. Focus Surf. 2011, **6**.
[40] Sani JE, Etim RK and Joseph A 2019 Compaction Behaviour of Lateritic Soil–Calcium Chloride Mixtures. *Geotechnical and Geological Engineering*. Springer Nature Switzerland. **37**,2343-2362. https://doi.org/10.1007/s10706-018-00760-6
[41] Muynck WD, Belie ND and Verstraete W 2020 Microbial carbonate precipitation in construction materials: a review. *Ecological Engineering* **36** (2):118–36.
[42] Neupane S 2016 *Evaluating the suitability of microbial induced calcite precipitation technique for stabilizing expansive soils*. An unpublished thesis submitted to Boise State University
[43] Coka E 2001 Use of class C fly ashes for the stabilization of an expansive soil. *ASCE J Geotech Geoenvironmental Eng* **127**(7):568–573
[44] Etim RK, Attah IC, Eberemu AO, Yohanna P 2019 Compaction behaviour of periwinkle shell ash treated lateritic soil for use as road sub-base construction material. *Journal of GeoEngineering* **14** (3), pp. 179-190, http://dx.doi.org/10.6310/jog.201909_14(3).7
[45] Etim RK, Attah IC, Yohanna P 2020 Experimental study on potential of oyster shell ash in structural strength improvement of lateritic soil for road construction. *International Journal of Pavement Research Technology* https://doi.org/10.1007/s42947-020-0290-y
[46] Dhami NK, Mukherjee A and Reddy MS 2016 Micrographical, minerological and nano-mechanical characterization of microbial carbonates from urease and carbonic anhydrase producing bacteria. *Ecol Engrg* **94**, 443–454. http://dx.doi.org/10.1016/j.ecoleng.2016.06.013
[47] Dhami NK, Reddy, MS and Mukherjee A 2013 Biomineralization of calcium carbonate polymorphs by the bacterial strains isolated from calcareous sites. *Journal of Microbiology and Biotechnogy*, l 23:707–714
[48] Okwadha GD and Li J 2010 Optimum conditions for microbial carbonate precipitation. *Chemosphere*, **81**(9), 1143-1148.
[49] Al Qabany A, Soga K and Santamarina C 2012 Factors Affecting Efficiency of Microbially Induced Calcite Precipitation. *Journal of Geotech and Geoenv. Engrg,* **138**(8), pp. 992-1001. https://doi.org/10.1061/(ASCE)GT.1943-5606.0000666

[50] Pan X, Chu J, Yang Y and Cheng L 2020 A new biogrouting method for fine to coarse sand. *Acta Geotechnica* 15:1–16. https://doi.org/10.1007/s11440-019-00872-0
[51] Wen K, Li Y, Amini F and Li L 2020 Impact of bacteria and urease concentration on precipitation kinetics and crystal morphology of calcium carbonate. *Acta Geotechnica* **15**,17–27 https://doi.org/10.1007/s11440-019-00899-3
[52] Rebata-Landa V 2007 Microbial activity in sediments: Effects on soil behaviour." Doctoral dissertation, Georgia Institute of Technology, Atlanta, GA.

Effect of curing method on unconfined compressive strength of silty sand treated with Bacillus *Megaterium*

A O Eberemu[1], M Onah[1*] and K J Osinubi[1]

[1] Department of Civil Engineering, Ahmadu Bello University, Zaria, Kaduna state, Nigeria.
[*]Corresponding Author: maryonah2664@gmail.com

Abstract. Laboratory tests were carried out to evaluate the effect of curing method on the unconfined compressive strength (UCS) of silty sand treated with Bacillus *megaterium.* (B. *megaterium*) using the Microbial Induced Calcite Precipitation (MICP) technique. Compaction test was conducted using the British Standard light (BSL) energy. UCS tests were performed on two sets of specimens treated with varying B. *megaterium* suspension densities and cementation reagent using trial mix ratio of 50% B. *megaterium* (B) and 50 % cementation reagent (C) (i.e., 50B-50C) for the treated specimen and 100 % cementation reagent (C) (i.e., 100C) for the control specimen. One set of specimens extruded from compaction mould was air-cured at room temperature for 24 hours, while the second set of specimens were placed in sealed polythene bags immediately after extrusion from the moulds. The results obtained show that the peak UCS values of the air-cured specimens and the sealed specimens are 448.8 kN/m^2 and 19.8 kN/m^2, respectively, for treatment with B. *megaterium* suspension density of 6.0 x 10^8 cells/ml. The study shows that the UCS of silty sand treated with B. *megaterium* is sensitive to curing method.
Keywords: Bacillus megaterium, Cementation reagent, Silty sand, Unconfined compressive strength

1. Introduction

The dynamics of the world we live in change day in, day out with the constant need to make improvements on the way we do things as well as the preservation of the environment becomes paramount. Conventional soil improvement technology causes damage to the environment and as such, greater assiduity/attention is being given to environmentally friendly soil modification techniques [1].
The concept of employing microorganisms in bio-geotechnical engineering field was first pioneered in 1992 [2]. Thereafter, more studies have been carried out with respect to the utilization of microbes as alternative to the various eco-friendly soil improvement technologies considered in the engineering field [3-6]. The motivation for this technique is basically the ability of the microbes to catalyse chemical reactions that generates precipitates that has the ability to coat or/bind soil grains. This method of inducing calcite precipitate is identified as Microbial induced calcite precipitation (MICP). The MICP technique generally increases soil strength, improve surface erosion control, reduces permeability and seismic remediation [6,7]. Microbial-induced calcite precipitation (MICP) method is regarded as a current and an innovative technique of soil improvement [8-13]. Current investigations have suggested that employing special bio-mediated soil

improvement techniques have justified the effective performance and environmental sustainability that can prevent collapse of loose soil sand that results in distortion or failure in foundation on the one hand, and liquefaction on the other [8, 14 - 18].
In a study using Bacillus *pasteurii* and sandy soil, Soon [19] reported that the treated soil specimen showed a behaviour of low failure of strain softening shear with improvement in shear stiffness and maximum shear capacity as compared to a soil specimen that hasn't undergone treatment. DeJong [8], reported that MICP treated sand is projected to have a life span of 50 years as long as alkaline conditions remain, which is in alignment with expected life span for most engineering structures. The longevity can be extended through occasional retreatment. Soon [20] also stated that there was sufficient bettering in the shear strength and hydraulic conductivity of sandy and residual soil samples, however, the residual soil samples recorded higher shear strength values than the sandy soil, while the reverse was the case in the hydraulic conductivity results. He further stated that MICP treatment was dependent on density of soil, condition of treatment and soil type. Esringu [21] utilized Bacillus *megaterium* on contaminated sandy soil but the study evolved around the remediation of Boron, Cadmium and Lead, polluted soils. Muthukkumaran [22] studied the effect of MICP on the strength of cohesionless soils under dissimilar environmental circumstances. The author reported an exponential rate of strength gain in the treated samples and stated that the level of improvement was dependent on the calcium carbonate precipitated, which was in turn dependent on a favourable environment for urease production from the microbes and sufficient reaction time for the reagents. However, most investigations on MICP have focused on the utilization of sand and very little work has been done on silty soils, which underscores the need for this study.
Silty soils are usually found in semi-arid environments and are referred to as a collection of macroscopic particles of a size between sand and clay, which originated from the minerals quartz and feldspar [23]. Silty soils act differently from clay and sand. Silt may exist as a soil (frequently combined with clay or sand) or as sediment combined in suspension with water (also called suspended load) and in a water body such as a river. In addition it may exist as soil located at the base of water bodies, like mudflows from landslides. Silty soil is usually associated with a floury feel when free from moisture, and a slippery feel when moist. During shear, Silty Sand due to increasing strains has a propensity to dilate and decrease in pore pressure [24, 25]. Silty sand soils usually do not have cohesion amidst their particles and most of their shear strength values are based on their internal friction.
This research was aimed at the comparative assessment of the effect of curing method on the unconfined compressive strength (UCS) of a B. *megaterium* - induced calcite precipitate treated silty sand soil using MICP technique. The specific objective was to ascertain the alterations in the UCS properties of silty sand soil when treated with stepped B. *megaterium* suspension density.

2. Materials and Methods

2.1 Materials

2.1.1 Silty sand soil. The silty sand sample used for this research were obtained from Wudil Local Government Area (latitude 11° 786N and longitude 8° 840E) in Kano State, Nigeria.

2.1.2 Microorganism. The bacterium used as the soil improvement agent in this investigation is Bacillus *megaterium*. It is Gram-positive rod-shaped bacteria classified as ATCC 14581 in the American Type Culture Collection (ATCC). Five separate bacteria suspension density of distinct turbidities on the McFarland turbidity scale, MFS (i.e., 0.5, 2.0, 4.0, 6.0 and 8.0 with equivalents 1.5×10^8 cells/ml, 6×10^8 cells/ml, 1.2×10^9 cells/ml, 1.8×10^9 cells/ml and 2.4×10^9 cells/ml, respectively) were used in this study. The control soil was referenced as 0 cells/ml (0 MFS).

IOP Conf. Series: Materials Science and Engineering 1036 (2021) 012032 doi:10.1088/1757-899X/1036/1/012032

2.1.3 Cementation Solution. Cementation solution was prepared using a mixture of chemicals at varying proportions. Chemicals that were used for the cementation solution include the following per litre of water purified by distillation: 3 g of nutrient broth (Bacto); 20 g of urea; 10 g of NH_4Cl; and 2.12 g (the same as 25.2 mM) of $NaHCO_3$. The pH of the medium was adjusted to 6.0 with 6 M HCL preceding in time to autoclaving. 10 ml of filter, completely cleaned or uncontaminated solution containing 2.80 g $CaCl_2$ were added afterward as reported by Stocks-Fischer *et al.* [26].

2.2 Methods

2.2.1 Unconfined compressive strength. The test for the unconfined compressive strength (UCS) of the soil samples were performed in agreement with specification outlined in BS 1377-7: 1990 [28] as well as [29], utilizing the British Standard light (BSL) compactive effort. Compaction of the soil samples was done in a 1000 cm^3 mould at each of their OMC. With the use of an extruder, each of the compacted soil specimen was then gently removed from the mould. Thereafter, three cylindrical specimens with 38.1 mm diameter and 76.2 mm length each were cored from the extruded specimen. The cored specimens were cured for 7 days and each specimen was placed on the bottom platen of a compression machine before applying a compressive force at a strain controlled rate of 0.10 %/min. Results of the axial deformation and force were recorded simultaneously at fixed intervals until the specimen failed. The compressive stress was computed using equation (1):

$$\sigma\ (kN/m^2) = \frac{[R\ x\ Cr\ x\ (100 - E\%)\ x\ 100]}{(100\ x\ Ao)} \quad (1)$$

Where:
E % = l / lo
E % = Strain Percent
l = Amount of compression at any stage
R =Load ring reading at Strain E
Cr =Mean calibration of load ring
lo =Initial Length of the specimen.
Ao = Initial cross sectional area
σ (kN/m^2) =Compressive stress at strain E.

2.2.2 Soil sample preparation and treatment. Two sets of specimen were utilized. 3000 g of soil sample passing BS No. 4 sieve (4.76 mm aperture) was used for each set. For the control specimen, only cementation reagent which volume was obtained from the total volume determined from equation (2) was mixed with the soil sample. For the treated specimens, the soil was mixed with each bacterial suspension density and cementation reagent in a pan. A trial mix ratio of 50% B. *megaterium* suspension density (B) and 50% cementation reagent (C) (i.e., 50B-50C) was adopted for the two sets of specimens. The calculation of the volume of each of the bacterial suspension density (1.5×10^8, 6×10^8, 1.2×10^9, 1.8×10^9 and 2.4×10^9 cells/ml) and cementation reagent added to the soil were determined using equations (2) and (3), respectively:

1) For 50B-50C mix ratio: In this mix, the soil was mixed with 50% bacterial suspension (50B) and 50% cementation reagent (50C) of the total volume which was obtained from equations (2) – (4):

$$Total\ Volume = \frac{\text{Optimum Moisture Content}}{100} \times Weight\ of\ soil\ sample \quad (2)$$

$$Volume\ of\ Bacteria = \frac{50}{100} \times Total\ Volume \tag{3}$$

$$Volume\ of\ cementation\ reagent = \frac{50}{100} \times Total\ Volume \tag{4}$$

After treating the soil sample with the various mix ratios, the samples were sealed with polythene sheet for 24 hours for proper hydration, before being subjected to the UCS test procedure. However, the curing process was different for each of the two sets of specimen with one set air-cured (i.e, Specimen 1) at room temperature for 24 hours, while the second set was sealed immediately after extrusion from the moulds (i.e, Specimen 2).

3. Results and Discussion

3.1 Natural soil

The natural moisture content of the soil was 20 %. The high moisture content was as a result of the period of the year in which the soil samples were collected. The index properties of the natural soil are shown in Table 1. The sample is brown in colour and has 6.65 % fines passing No. 200 sieve. The soil was classified as A-3(0) in agreement with American Association of State Highway and Transportation Officials [30] and as a poorly graded sand with silt (SP-SM) according to the Unified Soil Classification System [31].

Table 1. Basic Properties of the Silty Sand Used in the Study.

Properties	Quantity
Percentage passing 0.075 mm sieve	6.7
Natural moisture content, %	20.0
Specific gravity	2.63
Liquid limit, %	-
Plastic limit %	Non plastic
Plasticity index, %	Non Plastic
Linear shrinkage, %	-
AASHTO classification	A-3(0)
USCS classification	SP-SM
Maximum dry density, Mg/m^3	1.67
Optimum moisture content, %	8.5
Colour	Brown

3.2 Mechanism of Microbial Induced Calcite Precipitation (MICP) Process

In order to stimulate calcite mineralization, which is an after-effect of microbial metabolic activity [32], an aerobic urease producing bacteria is used to induce the calcite precipitate. This MICP process is known as Urea Hydrolysis. Although there are other processes (aerobic oxidation, sulphate reduction, denitrification, etc.), Van Paassen [33] suggested the adoption of Urea Hydrolysis for MICP process because of their high calcite conversion rate in contrast with other processes [26].

3.3 Influence of B. megaterium Induced Calcite Precipitate on Unconfined Compressive Strength of Silty Sand Soil

3.3.1 Unconfined compressive strength.

The variation of the unconfined compressive strength of silty sand with B. *megaterium* suspension density is shown in Figure 1. The UCS values for the control specimen

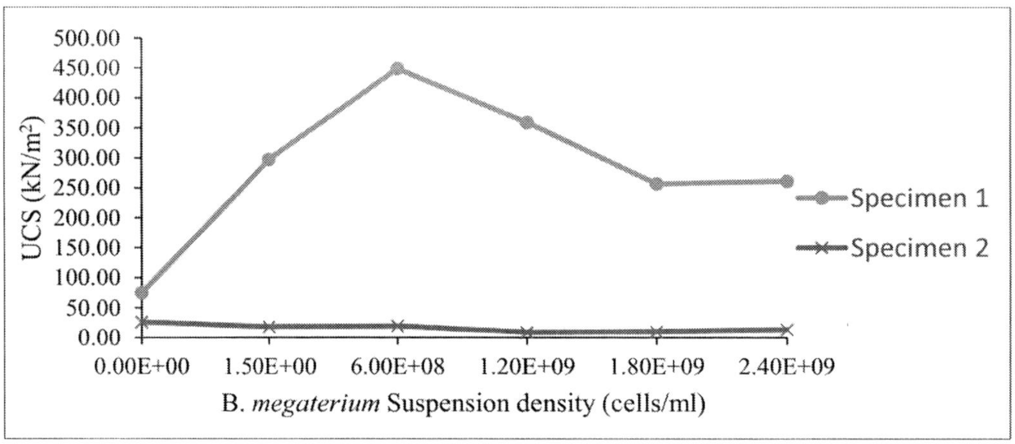

Figure 1. Variation of unconfined compressive strength of silty sand with B. *megaterium* suspension density

(i.e., treatment with cementation reagent only) are 75.10 and 26.08 kN/m^2 for Specimen 1 and Specimen 2 respectively. For Specimen 1, the UCS initially increased from the value recorded for the control specimen to a peak value of 448.84 kN/m^2 at B. *megaterium* suspension density of 6.0 x 10^8 cells/ml and thereafter decreased to 261.56 kN/m^2 at B. *megaterium* suspension density of 2.4 x 10^9 cells/ml. For Specimen 2, the value recorded for the control specimen generally decreased to 13.35 kN/m^2 at B. *megaterium* suspension density of 2.4 x 10^9 cells/ml.

The higher UCS value obtained for the Specimen 1 in comparison with Specimen 2 is due to a higher calcite precipitated within air-dried Specimen 1 mainly as a result of greater oxygen available for reaction. B. *megaterium* is an aerobic microbe which thrives better in the presence of air to promote the growth of the microbes and increase the conversion rate that resulted in the production of more calcium carbonates than in sealed Specimen 2 which was not exposed [34]. Bacteria cell density has been reported as one of the factors that influence the calcite formation, as urease hydrolysis has a direct relationship with density of microbes in a given cell, since microbes act as nucleation sites for the precipitation of calcite during the MICP metnod. Hence, a higher concentration of microbes which means more available nucleation sites, often results in higher urease activity and consequent increase in calcite precipitated [19,26]. This resulted in the deposition of a greater amount of $CaCO_3$ within the pore spaces of Specimen 1, than Specimen 2. Similar findings were reported by Lian [35] in their study of Bacillus *megaterium* induced carbonate mineralization, that the process of crystallization was affected by the existence of bacterial cell surfaces and metabolic products.

The decrease in the UCS values recorded in the two specimens could be ascribed to the low volume of cementation reagent when compared to the bacteria cell population in those suspension densities. As reported by Anbu [36], urea hydrolysis is influenced greatly by the volume of cementation reagent and concentration of cells in a given bacteria suspension density. Hence, the volume of cementation reagent supplied to those suspension densities with higher cell concentrations was not sufficient to generate sufficient calcite that would enhance the bond and strength of the treated soil mass [19, 37 and 38].

4. Conclusion

The silty sand classified as A–3 (0) and a poorly graded sand with silt (SP-SM) in AASHTO and USCS, respectively, was treated with five *B. megaterium* suspension density (B) of 0, 1.5 x 10^8, 6.0 x 10^8, 1.2 x 10^9, 1.8 x 10^9 and 2.4 x 10^9 cells/ml and cementation reagent (C). A mix ratio of 50B-50C was adopted for the treatment of samples, while the control specimen was treated with cementation reagent only. Results obtained indicate that the UCS values recorded for Specimen 1 (i.e., air-cured) were higher than those of Specimen 2 (i.e., sealed) due to a higher urease activity and calcite formation resulting from the exposure to oxygen. It can be deduced that the curing method influences the strength gain of B. *megaterium* treated silty sand.

References

[1] Dejong J T, Mortensen B M, Martinez B C and Nelson D C 2010 Bio-mediated Soil Improvement *Ecological Engineering* **36**(2) 197 – 210
[2] Ferris F G and Stehmeier L G 1992 Bacteriogenic Mineral Plugging *Journal of Canadian Petroleum Technology* **36**(9) Doi: 10.2118/97-09-07.
[3] Mitchell J K and Santamarina J C 2005 Biological considerations in geotechnical engineering *Journal of Geotechnical and Geoenvironmental Engineering* **131**(10) 1222–33.
[4] DeJong J T, Michael B Fritzges, and Klaus N 2006 Microbially induced cementation to control sand response to undrained shear. *Journal of Geotechnical and Geoenvironmental Engineering*, **132**(11): 1381-1392.
[5] Chu J, Ivanov V, Naeimi M, Stabnikov V, Liu H L 2014 Optimization Of Calcium-Based Bioclogging and Biocementation of Sand *Acta Geotech* **9** 277-285
[6] Kim G and Youn H 2016 Microbially Induced Calcite Precipitation Employing Environmental Isolates *Materials* **9(6)** 468; Doi:10.3390/ma9060468.
[7] Osinubi K J, Eberemu A O, Ijimdiya T S, Yakubu S E, Gadzama E W, Sani J E and Yohanna P 2020 Review of the use of Microorganisms in Geotechnical Engineering Applications. *SN Applied Sciences Journal. Article* **2** 207
[8] DeJong J T, Soga K S, Kavazanjian E, Burns S, Van Paassen L A, Al Qabany A, Aydilek A 2013 Biogeochemical Processes and Geotechnical Applications: Progress, Opportunities and Challenges *Geotechnique* **63**(4).
[9] Gat D, Tsesarsky M, Shamir D and Ronen Z 2014 Accelerated Microbial-Induced $CaCO_3$ Precipitation in a Defined Coculture of Ureolytic and Non-ureolytic Bacteria *Biogeosciences* **11** 2561–69, doi:10.5194/bg-11-2561-2014
[10] Wei S, Cui H, Jiang Z, Liu H, He H, Fang N 2015 Biomineralization Processes of Calcite Induced by Bacteria Isolated from Marine Sediments *Brazilian Journal of Microbiology* **46**(2) 455–64
[11] Murtala U, Khairul A K, Kenny T P C 2016 Biological Process of Soil Improvement in Civil Engineering: A Review *Journal of Rock Mechanics and Geotechnical Engineering* **8** 767-774.
[12]Dhami N K, Mukherjee A and Watkin E L J 2018 Microbial Diversity and Mineralogical-Mechanical Properties of Calcitic Cave Speleothems in Natural and in Vitro Biomineralization Conditions *Frontiers in Microbiology* **9**(40) 1-22 doi: 10.3389/fmicb.2018.00040.

[13] Delgado-Baquerizo M, Oliverio A M, Brewer T E, González A B, Eldridge D J, Bardgett R D, Maestre F T, Singh B K, Noah F 2018 A global Atlas of the Dominant Bacteria Found in Soil. *Science* **359** 320–325 DOI: 10.1126/science. aap9516.

[14] Alvarado, D 2009 *Bio- Mediated Soil Improvement: Cementation of Unsaturated Sand Samples* (Ph.D Thesis, Arizona State University, USA).

[15] Van Paassen, L A, Harkes, M P, Van Zwieten, G A, Van der Zon, W H, Van der Star, W R L, Van Loosdrecht, M C M 2009 Scale up of BioGrout: A Biological Ground Reinforcement Method *In: Proceedings of the 17th International Conference on Soil Mechanics and Geotechnical Engineering* (Alexandria: Egypt) pp 2328–33.

[16] Montoya B, DeJong J T and Boulanger R 2013 Dynamic response of liquefiable sand improved by microbial-induced calcite precipitation *Géotechnique* **63**(4) 302–312 http://doi.org/10.1680/geot.SIP13.P.019.

[17] Zhao Q, Li L, Li C, Li M, Amini F and Zhang H 2014 Factors Affecting Improvement of Engineering Properties of MICP-Treated Soil Catalyzed by Bacteria and Urease. *Journal of Materials in Civil Engineering* **26**(12) 4014-94 http://doi.org/10.1061/(ASCE)MT.1943-5533.0001013.

[18] Jiang N J, Soga K and Kuo M 2017 Microbially Induced Carbonate Precipitation for Seepage-Induced Internal Erosion Control in Sand–Clay Mixtures *Journal of Geotechnical and Geoenvironmental Engineering* **143**(3) 4016-100, http://doi.org/10.1061/(ASCE)GT.1943-5606.0001559.

[19]Soon N W, Lee L and Hii S 2012 An overview of the factors affecting microbial-induced calcite precipitation and its potential application in soil improvement *World Academy of Science, Engineering and Technology* **62** 723-729

[20]Soon N W, Lee M, Tan C K and Hii S 2013 Improvements in Engineering Properties of Soils through Microbial-Induced Calcite Precipitation *KSCE Journal of Civil Engineering* **17** 718-728

[21]Esringü A, Turan M, Güneş A, Karaman R 2014 Roles of Bacillus megaterium in Remediation of Boron, Lead, and Cadmium from Contaminated Soil *Communications in Soil Science and Plant Analysis* **45** 1741-59

[22]Muthukkumaran K, Shashank B S 2016 Durability of Microbially Induced Calcite Precipitation (MICP) treated Cohesionless Soils *The 15th Asian Regional Conference on Soil Mechanics and Geotechnical Engineering* Vol. 2 (Japan: Japaneses Geotechnical Society Special Publication) pp 1946-49

[23]Assallay A 1998 Silt: 2-6 μm 9-4ϕ *Earth-Science Reviews* **45**(1-2) 61-88

[24] Brandon T L, Rose A T and Duncan J M 2006 Drained and Undrained Strength Interpretation for Low-Plasticity Silts *Journal of Geotechnical Engineering and Geoenvironmental Engineering ASCE* **132**(2) 250-7

[25] Mohammad N, Ineisane B and Harer N 2013 The Strength Characteristics of Silty Soil Stabilized Using Nano-Clay *Paper presentation at the 7th SAStech conference on geotechnical surveys* (Bandar-Abbas, Iran: SAStech)

[26] Stocks-Fischer S, Galinat J K and Bang S S 1999 Microbiological precipitation of $CaCO_3$ *Soil Biology and Biochemistry* **31**(11) 1563–71.

[27] Sharma A and Ramkrishnan R 2016 Study on Effect of Microbial Induced Calcite Precipitates on Strength of Fine Grained Soils *Perspectives in Science* **8** 198-202.

[28] BS 1377-7: 1990 *Method of Testing Soils for Civil Engineering Purpose* (London: British Standard Institute BSI)

[29] Head, K.H 1992 *Manual of Soil Laboratory Testing* Vol. 2 (Plymouth, London: Pentech Press)

[30] AASHTO 1986 *Standard Specifications for Transport Materials and Methods of Sampling and Testing* 14th Ed. (Washington DC: American Association of State Highway and Transport Officials AASHTO)
[31]American Society for Testing and Materials ASTM 1992 *Annual Book of Standards* Vol. 04.08 (Philadelphia, USA: ASTM)
[32] Neupane S 2016 *Evaluating the Suitability of Microbial Induced Calcite Precipitation Technique for Stabilizing Expansive Soils* (MSc Thesis, Department of Civil Engineering, Boise State University, Idaho)
[33] Van Paassen L A, Ghose R, van der Linden T J M, van der Star W R L and van Loosdrecht M C M 2010 Quantifying Biomediated Ground Improvement by Ureolysis: Large-Scale Biogrout Experiment *Journal of Geotechnical and Geoenvironmental Engineering* **136**(12) pp 1721-28
[34] Dhami N K, Reddy M S, Mukherjee A 2013 Bacillus Megaterium Mediated Mineralization Of Calcium Carbonate As Biogenic Surface Treatment Of Green Building Materials *World Journal of Microbiology and Biotechnology* **29** (12) 2397-406. DOI: 10.1007/s11274-013-1408-z
[35] Lian B, Hu Q, Chen J, Ji J and Teng H 2006 Carbonate Biomineralization Induced by Soil Bacterium Bacillus Megaterium Geochimica et Cosmochimica Acta **70** 5522–5535
[36] Anbu P, Kang C, Shin Y and So J 2016 Formations of Calcium Carbonate Minerals by Bacteria and its Multiple Applications *Springerplus Springer International Publishing* **5**(1) 250
[37] De Muynck W, Verbeken K, De Belie N and Verstraete W 2010 Influence Of Urea and Calcium Dosage on the Effectiveness of Bacterially Induced Carbonate Precipitation On Limestone *Ecological Engineering* **36**(2) 99-111. DOI: http://dx.doi.org/10.1016/j.ecoleng.2009.03.025
[38] Osinubi K J, Eberemu, A O, Gadzama, E W, Ijimdiya, T S 2019 Plasticity characteristics of lateritic soil treated with Sporosarcina Pasteurii in Microbial-Induced Calcite Precipitation Application. *SN Applied Sciences Journal* **1** 829.

Potentials of Cement Kiln Dust-Periwinkle Shell Ash Blends on Plasticity Properties of Two Selected Tropical Soils for use as Sustainable Construction Materials

D U Ekpo[1], A B Fajobi[2], A L Ayodele[2] and R K Etim[1]

[1]Department of Civil Engineering, Akwa Ibom State University, Ikot Akpaden, Mkpat Enin, Nigeria.
[2]Department of Civil Engineering, Obafemi Awolowo University, Ile-Ife, Nigeria.

Corresponding Author: davidekpo@aksu.edu.ng

Abstract: This paper investigated the effect of cement kiln dust (CKD) and periwinkle shell ash (PSA) on the plasticity characteristics of two lateritic soil. Two Soils (termed soil A and B) were collected from two selected locations within Osun State, Nigeria. Air-dried soil was treated with CKD and PSA blend in order of 0, 5, 10, 15, and 20% for CKD and 0, 2, 4, 6, 8 and 10% for PSA by dry weight of soil. The specific gravity of both soils increased with increase in CKD and PSA fillings. The Index properties such as Atterberg's limit (liquid limit, LL, plastic limit, PL, and plasticity index, PI), percentage fines content of the soils was determined. The index properties results showed that the two soils are clayey soil of low plasticity. There was a general decrease in the Atterberg's limits (liquid limit, LL and plastic limit, PL). The scan electron microscopy (SEM) and electron diffraction spectroscopy of treated soil revealed the formation of crystalline products of calcium silicate hydrate (CSH) and calcium aluminate hydrate (CAH) which was responsible for strength gain. Based on preliminary investigations, the study concluded that the combination of CKD-PSA blends is a viable stabiliser for the two tropical soils under consideration to be used as sub-base construction.
Keywords: Lateritic soil, Periwinkle shell ash, Cement kiln dust, Atterberg's limits, and Scanning electron microscope.

1. Introduction

The increase in urbanization has led to the development of more facilities like roads, railways, underground tunnels etc. Generally, in developing and advanced countries, laterite and lateritic soils form the components of roads which in turn are composed of sub-base and subgrade materials. Due to the poor engineering behaviour inherent in most of these soils (such as high shrinkage and swell potential, low plasticity and a reduced bearing capacity), these renders the soil unsuitable for construction, hence stabilization becomes an alternative. In road stabilization, recycling those materials considered as waste (agricultural and industrial wastes) has gradually become more attractive rather than discarding them off. This has prompted researchers into harnessing the possibilities of reusing and recycling those wastes instead of disposing them off. Several researchers have utilized the combination of both agricultural and industrial wastes for instance rice husk ash [13]; bagasse ash [24, 25]; steel slag [4]; sludge ash [20, 9]; cement kiln dust [21, 32]; periwinkle shell ash [15, 27] to improve soils with poor engineering properties. Results showed that a great improvement was recorded in the properties of these soils.

The by-products left during the production of cement is called cement kiln dust (CKD). They are usually obtained in the kiln during the production of cement clinker. Approximately 30 million tons of CKD worldwide is generated every year [13], out of which 4 million tons per year are produced in the US. CKD has found applications in various fields ranging from masonry and concrete blocks; soil stabilization; waste stabilization; agricultural amendments etc. [31, 32, 19, 18, 39].

Periwinkles on the other hand are marine shells predominantly dominant in the coastal regions. Their shells are usually deposited in open spaces which poses an environmental risk and as such, managing it have become a challenge. Ekop *et al.* [14] reported that researchers have recently employed periwinkle shells for construction purposes. Several studies have employed periwinkle shells in replacing coarse aggregates in lightweight concrete [3,11]. Periwinkle shell ash (PSA) is a waste that comes from the calcination of periwinkle shells. Recent researches have shown that the geotechnical properties of soils have been improved on upon the addition of PSA and as such, it is considered a potential soil stabilizer [15, 27].

Studies conducted by Umoh and Olusola [37], Attah et al. [8] shows that PSA is rich in Calcium oxide (CaO) similar to CKD [38, 32]. The richness of these two additives in calcium has the potential of reducing diffuse double layer (DDL) through cation exchange reaction, flocculation and agglomeration, reduction of adsorbed water and formation of cementitious compounds which has the potential of improving the properties of soil significantly.

However, there are no open literatures that has explored the possibilities of combining both additives to serve as an alternative to the traditional soil stabilizers in order to render it fit as a paving layer material. Hence, authoritative data on the combination of the blends is clearly needed. The specific objective of this study is to evaluate the effect of CKD and PSA on the Atterberg's limits and microstructure of two selected tropical lateritic soils. On the basis of this, it is presumed that the combination of these blends will serve as an alternative to cement, lime and bitumen; so that both soil improvement and safe disposal will be realized.

2. Materials and Methods

2.1 Materials

2.1.1. Laterite soil

The two soils were sourced from an existing borrow pit at Ilesha West LGA and along Ibadan-Ile-Ife expressway, both in Osun State, South West Nigeria. The first soil was designated as Soil A (Latitude 7° 37' 1652.25"N and Longitude 4° 44' 1782.08"N) while the other soil was designated as Soil B (Latitude 7°30' 1813.27" and Longitude 4° 27' 1667.09"N).

2.1.2. Periwinkle shell ash

The shells of the periwinkle were sourced from a dumpsite in Nwaniba road, Uyo, Akwa Ibom State, Nigeria. It was further calcined and allowed to cool inside the kiln. The calcined ash was brought out of the kiln and further pulverized to powdered form. The ash was then sieved through 75μm and thereafter stored in a sealed polythene bag before it was mixed with requisite percentages by dry weight of the soil.

2.1.3. Cement Kiln Dust. The CKD was sourced from a local cement industry which was herein after sealed in tight sacks in order to prevent moisture from coming in contact with it which might result in hardening and loss of properties before being transported to the laboratory.

2.1.4. Physicochemical analysis of soil samples, CKD and PSA

The pH of the soil samples, CKD and PSA in its natural states was determined using standardized pH meter while the oxide composition of the specimens (soils, PSA and CKD) were investigated by X-ray Fluorescence (XRF) technique using the Thermo Fisher Model ARL 9990.

2.2 Methods

2.2.1. Index tests

The liquid limit, LL and plastic limit, PL otherwise referred to as the Atterberg's limits was done on soil specimen passing through sieve No. 425μm, the particle size distribution (PSD) were both investigated according to ASTM D 4318 and ASTM D 422. The plasticity index, PI of the samples, was also determined. The soil samples were classified according to [1] based on Atterberg's limits and PSD.

2.2.2. *Microanalysis*
The structural and morphology properties of the treated and typically treated soil was determined using the scanning electron microscope (SEM). The PHENOM WORLD scanning electron microscope together with the energy dispersive X-ray spectrometers (EDS) operated at 5kv and 15kv at a resolution of 50μm, 80μm, and 100μm was employed for micro analysis. The morphology and structural orientation of the soil structure is predicted with a high degree of accuracy and precision using the SEM.

3. Results and Discussion

3.1. *Index Properties*
Presented in Table 1 are the summary of the properties of the untreated soils in their natural states. The results show a higher natural moisture content for soil A in comparison with soil B. The shows that both soil samples contain some appreciable amount of moisture. The pH shows that both soils is acidic probably as a result of intensive weathering caused by high temperature and rainfall which eventually results into the leaching of alkaline and alkaline earth basic metals from the soil [16, 40]. The two soils (A and B) according to AASHTO [1], are classified as A-7-5(3) and A-7-6 (7), respectively. The USCS [41] classified the two soils as CL, which implies a low plasticity clay. The chemical composition of both soils, CKD and PSA are shown in Table 2, in the form of their oxide composition.

Table 1. Properties of the natural soils used in this study

Property	Soil A	Soil B
Natural Moisture Content (%)	19.4	11.6
Specific Gravity (Gs)	2.55	2.51
Colour	Reddish brown	Yellowish brown
pH	6.2	6.5
LL (%)	45.0	43.5
PL (%)	28.5	22.5
PI (%)	16.5	21.0
Percentage passing sieve No. 200	45.46	40.94
Percentage passing sieve No. 40	81.34	57.36
AASHTO classification	A-7-5 (3)	A-7-6 (7)
USCS	CL	CL
Maximum Dry Density, MDD (Mg/m^3)	1.600	1.770
Optimum Moisture Content, OMC (%)	20.50	17.00
Group Index	3	7

Table 2. Oxides composition of natural lateritic soils, CKD and PSA

Chemical (% Oxide Concentration)	Chemical compositions by weight (%)			
	Soil A	Soil B	CKD	PSA
Silica (SiO_2)	57.02	56.42	9.87	33.85
Alumina (Al_2O_3)	21.32	19.17	3.80	9.24
Iron oxide (Fe_2O_3)	6.01	10.49	2.41	7.20
Lime (CaO)	0.46	0.13	45.08	53.87
Magnesium oxide (MgO)	0.21	0.15	1.08	0.06
Sulphur oxide (SO_3)	0.00	0.00	0.22	0.16
Potassium oxide (K_2O)	2.29	1.46	0.17	0.06
Dinitrogen pentoxide (N_2O_5)	0.11	0.13	0.07	1.40
Phosphorous pentoxide (P_2O_5)	0.09	0.12	0.60	0.37
Tin oxide (TiO_2)	0.76	1.53	0.24	0.09
Loss on Ignition	9.73	8.45	35.66	38.68

3.2. Specific Gravity

Specific gravity is a vital index which is used in determining the void ratio and particle size of any soil particle. The variation of specific gravity of soil-CKD and PSA contents are presented in Fig.1. The specific gravity of both soils increases with an increase in CKD and PSA contents as shown in the results. The specific gravity of soil A shows a general increase of about 9.41% while that of soil B shows an increment of about 9.96%. Peak value of 2.79 was obtained at 20% CKD/10% PSA for soil A while 2.76 was obtained at 20% CKD/ 8% PSA for soil B. The general increasing trends in specific gravity for both soils may be attributed to the higher specific gravity of the additives (CKD, 2.93 and PSA, 2.63) replacing the soils A and B with specific gravity of 2.55 and 2.51 respectively. A similar trend of increase in specific gravity was reported by Etim [43] and Yohanna [44].

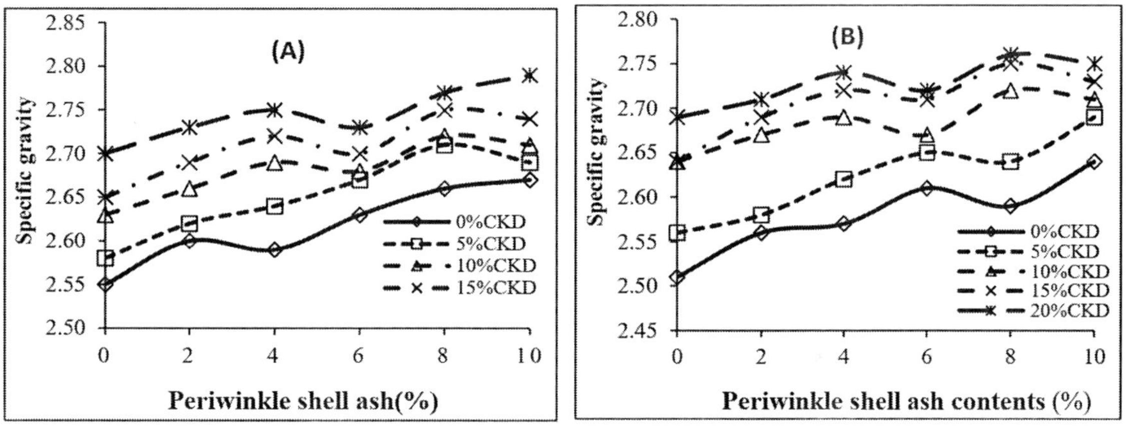

Figure 1. Variation of specific gravity of CKD and PSA treated soils for; (a) soil A and (b) soil B

3.3. Atterberg's Limits

3.3.1. Liquid limit

The variation of liquid limit, LL of soil-CKD with PSA for both soils are presented in Figs.2. The LL values of both soils decreases with an increase in CKD and PSA contents as presented in the results. The LL of soil A shows a reduction from 45% to a minimum value of 32.2 % at 15% CKD/6% PSA while soil B reduced to a minimum value of 31.7% at 20% CKD/10% PSA. Both soils show a general decrease with as high as 28% decrement. The formation of flocs coupled with the clay particles clustering together due to exchange of cations might have attributed to the overall reduction in LL of both soils as well as reduction in diffuse double layer (DDL) of clay minerals. Sharma and Lewis [35], Abood *et al.* [2], Al-Zoubi [6], Ramesh *et al.* [30], Ayodele *et al.* [42] reported similar trends of decrease in liquid limit.

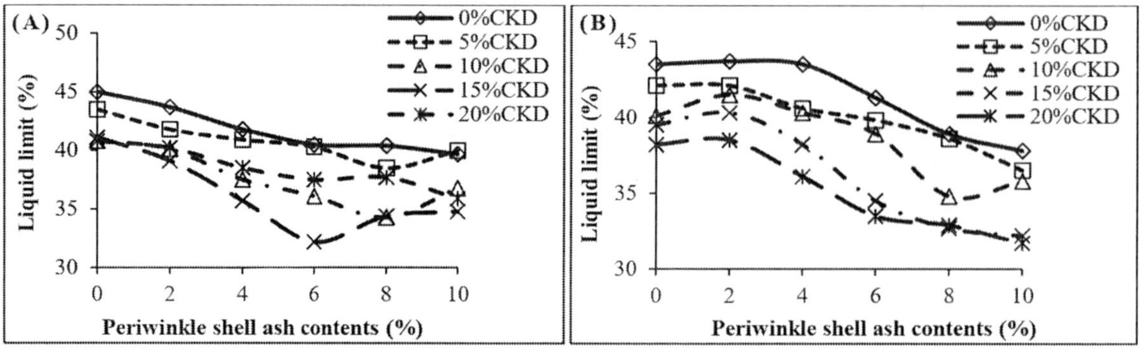

Figure 2. Variation of liquid limits of CKD and PSA treated soils for; (a) soil A and (b) soil B

3.3.2. Plastic limit

The effects of CKD and PSA on plastic limit, PL of soil-CKD-PSA for both soils are shown in Figs.3. A general decreasing trend with increase in CKD and PSA admixtures was observed for soil A while the reverse was the case for soil B. The decrement in soil A was about 19.3% decrease and an increment of about 13.7% increase for soil B. An increase or decrease in PL is inconsequential as long as the PI decreases which actually typifies an improvement in the properties of the soil. The presence of divalent cations in the admixtures such as Ca^{2+} might have replaced those weakly bonded monovalent cations of Na^+ and K^+ in the clay structures; thus, resulting in the formation of flocs and escape of water that has been bonded at the outer layers. The reduction in the trend is similar to the findings of Osinubi [28], Amadi [7], Ramesh *et al.* [30], and Ayodele *et al.* [42]. Furthermore, it could also be attributed to the presence of silica (SiO_2) and alumina (Al_2O_3) offered by the PSA that further facilitated the exchange of cations between clay particles of the soil and Ca^{2+} from the PSA.

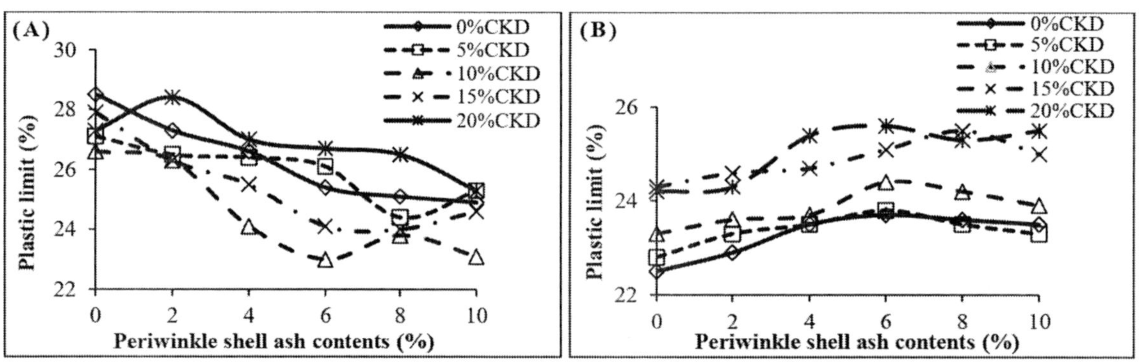

Figure 3. Variation of plastic limits of CKD and PSA treated soil for; (a) soil A and (b) soil B

3.3.3. Plasticity index

The general decrease in LL and increase in PL resulted to a decrease in PI of the soil. The graph of PI of soil-CKD mixture with PSA for both soils is shown in Figs. 4. The PI values decreases generally with increase in CKD and PSA contents. The PI of untreated soil A at 16.5% reduced to a minimum value of 8.1% at 15% CKD/6% PSA. A minimum value of 10.6% was obtained for soil B at 10% CKD/8% PSA. The general decrement in PI of both soils is a pointer that the natural soil has actually improved. The CKD which is rich in lime (CaO) undergoes cation exchange reaction with water in the stabilized; further dissociation might be accountable for the decrease in PI of the soil on addition of different percentages of PSA blends. The PSA with a divalent electron replaces the monovalent electrons in the soil thereby leading to a reduction in DDL i.e. diffuse double layer so that the soil becomes more workable with field equipment. Peter [29] and Abood et. al [2] reported similar studies of decrease in plasticity index.

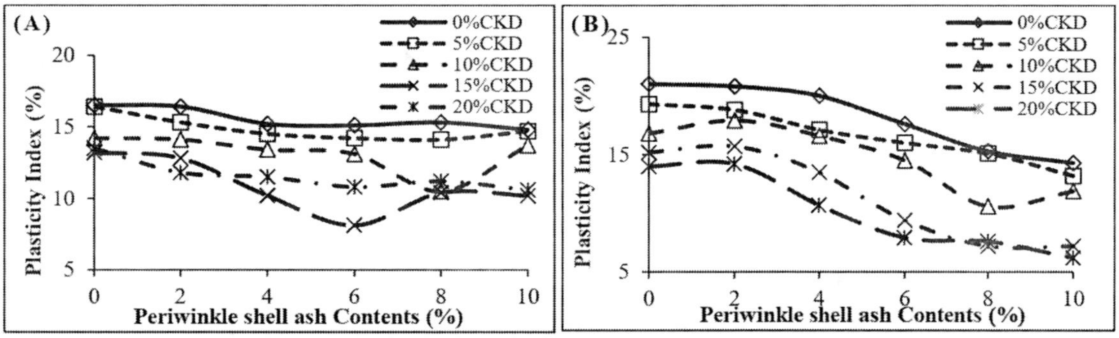

Figure 4. Variation of plasticity index of CKD and PSA treated soils for; (a) soil A and (b) soil B

3.4. Micrograph of Specimens

3.4.1. Scanning Electron Microscope

Essentially, microstructural studies are employed with the use of aggregation scale to analyze the arrangement of particles, particle assemblies and pores and their contacts and connectivity in different soils [5, 22]. The micrographs of untreated and optimally stabilized soils with 10% CKD/8 % PSA are presented Figs. 5 & 6. The micrograph of untreated soil A is composed mainly of particles that appear as that a blocky arrangement of closely parked particles while that of soil B is mainly of smooth surface with lots of shrinkage cracks at the background which is suggestive of clay contents. Clay particles possesses some polar properties that allows them to absorb water easily. The presence of whitish particles at the background is perhaps more of SiO_2 probably due to its size in relation with the background. The micrographs of the treated soils show that the clay particles and the flocs were grown within the pore. The morphology of the optimally treated soil consists of scaly and smooth surface. The surface being smooth might be due to the clay particles hovering around the surface of the silt particles during specimen treatment. While the surface being scaly on the other hand could be due to the clay minerals as undergoing both dissolution and precipitation combined as reported by Simms and Yanful [36], Monroy et al. [23] and Sani *et al.* [33]. Generally, for the optimally treated soils, pozzolanic reaction between the soil constituents and the additives between causes the formation of new minerals that are cementitious in nature. This is in agreement with the findings of Sathyapriya and Arumairaj [34], Jaritngam et al. [17], Negi et al. [26], Bhuvaneshwari et al. [10].

3.4.2 Energy-Dispersive X-Ray Spectroscopy

The energy dispersive x-ray is incorporated in to the scanning electron microscope to make it more robust. The EDS of natural soil sample A and soil optimally stabilized with 10% CKD/8 % PSA cured for 28 days is presented in Figs. 7 & 8.

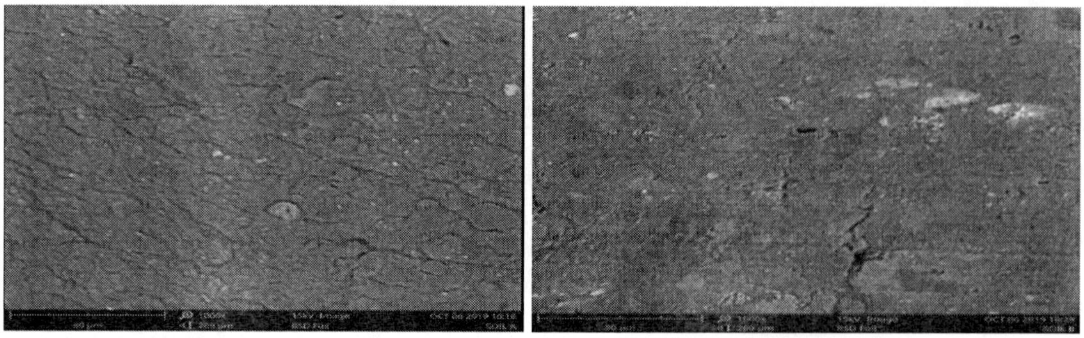

Soil A Soil B

Figure 5. Micrographs of natural soil A and B after 28days curing at 80μm magnification

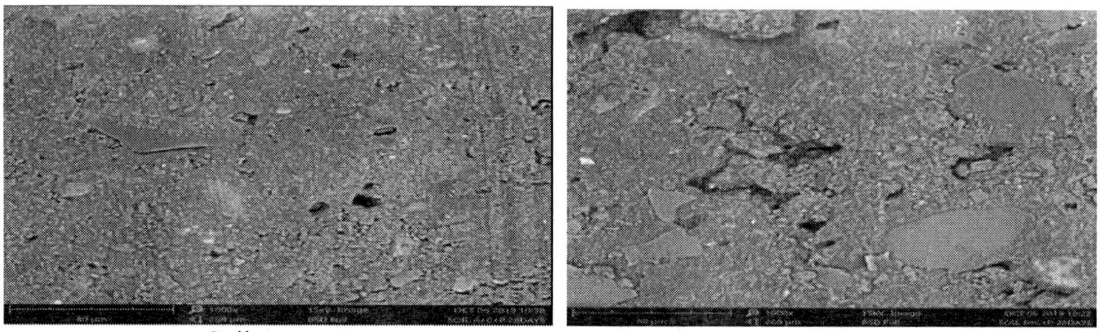

Soil A Soil B

Figure 6. Micrographs of soil + Optimal blend of 10% CKD/8%PSA viewed at 80μm magnification.

It was observed that the natural soil had a higher silicon and aluminum content of 27.46 and 24.85 % respectively than the optimally stabilized soil with a silicon and aluminium value of 21.14 and 19.82 %. The pozzolanic reaction that causes CSH and CAH to be formed which eventually leads to improved plsticity might have resulted in the reduction in aluminium and silicon content. Similar observations were made for both the natural and typically treated soil B as shown in Figs. 9 & 10.

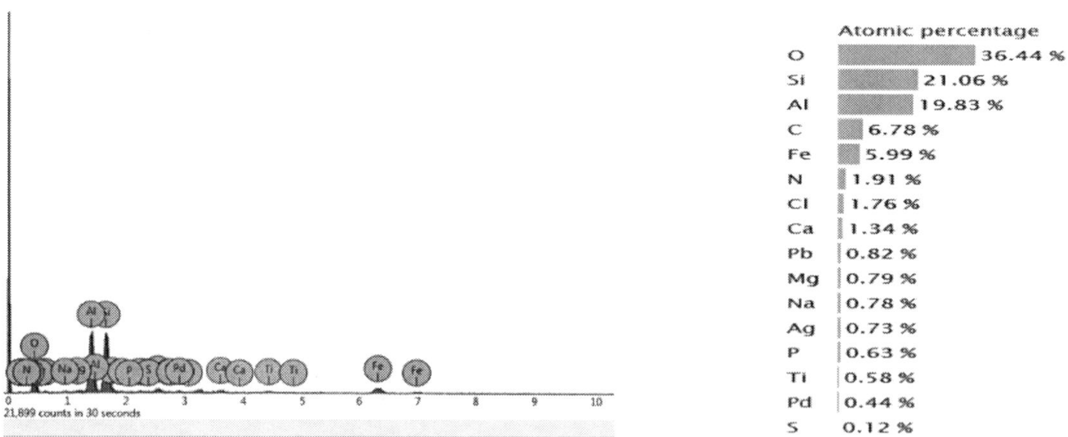

Figure 7. Spectroscopy of untreated soil A after 28 days curing

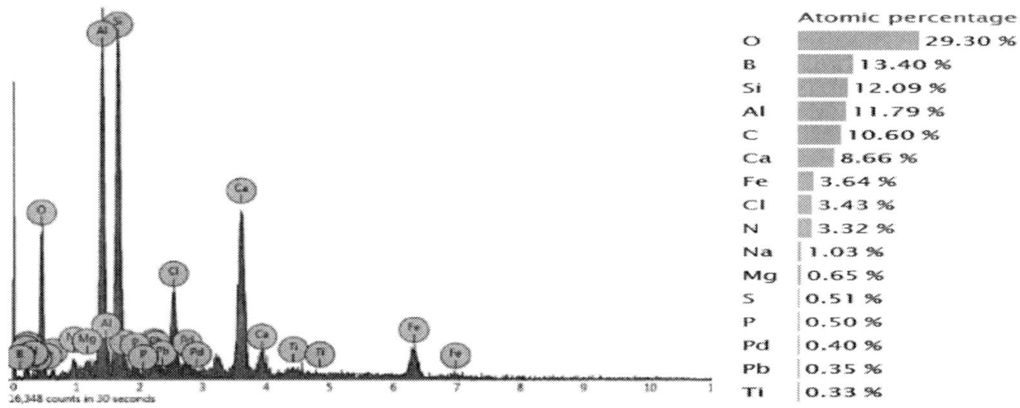

Figure 8. Spectroscopy of optimally treated soil A (soil +10% CKD/8% PSA) after 28 days curing

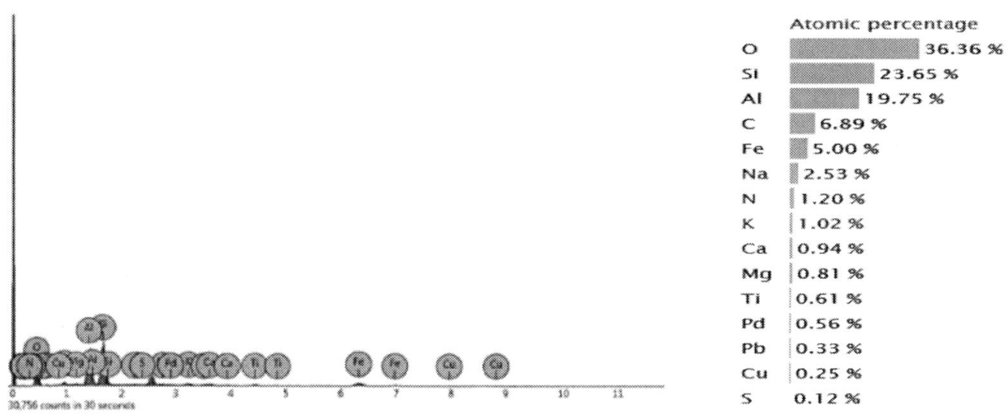

Figure 9. Spectroscopy of untreated soil B after 28 days curing

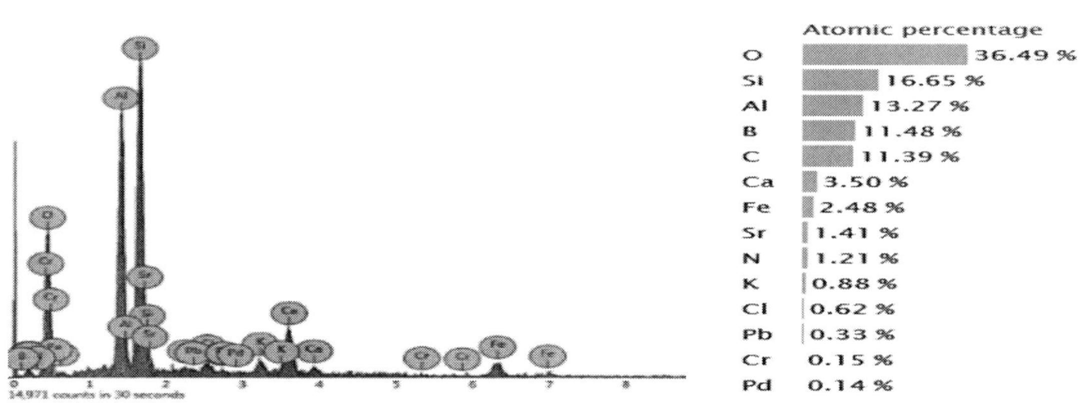

Figure 10. Electron diffraction spectroscopy of typically treated soil B (soil +10% CKD/8% PSA) after 28 days curing

4. Conclusions

The two soils under consideration based on AASHTO classification system were classified as A-7-5 (3) for soil A, and A-7-6 (7) for soil B. The USCS classification system classified both soils as CL, poorly graded with low plasticity. The Atterberg's limits and plasticity characteristics of the two lateritic soils indicated that there was need for soil improvement. The Atterberg's limits (LL, PL and PI) greatly reduced on addition of up to 15% CKD-6% PSA. The plasticity indices for both lateritic soils after stabilization was less than 12% which represents the benchmark for a material to be considered as subgrade material as specified by the Federal Ministry of Works and Housing, FMWH. Microanalysis conducted via scanning electron microscope showed that the treated soils were aggregated coupled with changes in the soil's fabric orientation. The SEM micrographs also revealed the presence of crystalline hydration products of CSH and CAH which were presumed to be the major factors that reduce the plasticity indices of the soils. It can therefore be concluded that CKD and PSA proves to be a viable material for improving the plasticity indices of clay soils as sustainable construction materials.

5. References

[1] AASHTO 1986 Standard Specifications for Transport Materials and Methods of Sampling and Testing. 14th Edition. American Association of State Highway and Transport Officials, Washington, D.C., U.S.A.

[2] Abood T T, Kasa A B and Chik Z B 2007 Stabilisation of silty clay soil using chloride compounds. *J Eng Sci Technol* **2**(1), 102–110.

[3] Agbede O I and Manasseh J 2009 Suitability of periwinkle shell as partial replacement for river gravel in concrete. *Leonardo Electronic Journal of Practices and Technologies*, **8**(15), 59-66.

[4] Akinwumi I I 2014 Soil modification by the application of steel slag. *Periodica Polytechnica Civil Engineering,* **58**(4), 371-377. doi: 10.3311/PPci.7239

[5] Al-Rawas A A 2002 Microfabric and mineralogical studies on the stabilization of an expansive soil using cement by-past dust and some types of slags. *Can. Geotech. J.* **39**(5), 1150 – 1167. https://doi.org/10.1139/t02-046.

[6] Al-Zoubi M S 2008 Undrained Shear Strength and Swelling Characteristics of Cement Treated Soil. *Jordan Journal of Civil Engineering,* **2**(1), 53-62.

[7] Amadi A A 2010 Evaluation of Changes in Index Properties of Lateritic Soil Stabilized with Fly Ash. Leonardo *Electronic Journal of Practices and Technologies.* **17**, 69-78.

[8] Attah I C, Etim R K and Ekpo D U 2018 Behaviour of periwinkle shell ask blended cement concrete in sulphuric acid environment. *Nigerian Journal of Technology (NIJOTECH),* **37**(2), 315-321. http://dx.doi.org/10.4314/nit.v37i2.5

[9] Ayodele A L, Adebisi A O and Kareem M A 2016 Use of sludge ash in stabilizing two tropical laterite. *International Journal of Scientific and Engineering Research,* **7**(8), 104-108. http://dx.doi.org/10.1016/j.jrmge.2016.04.002

[10] Bhuvaneshwari S Robinson R and Gandhi S 2010 Micro-fabric and mineralogical studies on the stabilization of an expansive soil using inorganic additives. *International Journal of Geotechnical Engineering*, **4**(3), 395-405.

[11] Dahunsi B I 2002 Properties of Periwinkle-Granite Concrete. *Journal of Civil Engineering, JKUAT,* **8**(1), 27-35.

[12] Dyer T D, Halliday J E and Dhir R K 1999 An investigation of the hydration chemistry of ternary blends containing cement kiln dust. *J Mater Sci.,* **34**(20), 4975–4983.

[13] Eberemu A O and Sada H 2013 Compressibility characteristics of compacted black cotton soil treated with rice husk ash. *Nigerian Journal of Technology (NIJOTECH)*. **32**(3), 507- 521. Faculty of Engineering, University of Nigeria, Nsukka, ISSN 1115-8443.

[14] Ekop I E, Adenuga O A and Umoh A A 2013 Strength characteristics of granite-pachimalania aurita shell concrete. *Nigerian journal of Agriculture, Food and Environment,* **9**(2), 9-14.

[15] Etim R K, Attah I C, Eberemu A O and Yohanna P 2019 Compaction behaviour of periwinkle shell ash treated lateritic soil for use as road sub-base construction material. *Journal of GeoEngineering*, **14**(3),191-202. http://dx.doi.org/10.6310/jog.201909_14(3).6
[16] Gidigasu M D 1976 Laterite Soil Engineering: Pedogenesis and Engineering Principles. Elsevier Scientific Publication Company, Amsterdam.
[17] Jaritngam S, Somchainuek O and Taneerananon, P 2014 Feasibility of lateritic cement mixture as pavement base coarse aggregate. *Transactions of Civil Engineering, IJST*, **38**(1), 275-284.
[18] Kanare H M 1999. Comparison of Trace Metal Concentrations in Cement Kiln Dust, Agricultural Limestone, Sewage Sludge, and Soil, SN2080, Portland Cement Association, Skokie, Illinois, USA.
[19] Kigel M Y, Shultis J F, Goldman E S and Demytrk M K 1994 Method of Detoxification and Stabilization of Soils Contaminated with Chromium Ore Waste, Patent and Trademark Office, United States Department of Commerce, Washington, DC, USA, Patent No.: 5, 304, 710.
[20] Li Chen and Deng-Fong Lin 2008 Stabilization treatment of soft subgrade soil by sewage sludge ash and cement. *Journal of Hazardous Materials*, **162**, 321-327. https://doi.org/10.1016/j.jhazmat.2008.05.060
[21] Miller A and Zaman R 2000. Utilization of cement kiln dust in cement mortar and concrete. *Resour Conserv Recycl*, **48**(4), 315–33.
[22] Mitchell J K and Soga K 2005 Fundamentals of soil behaviour, 3rd edition. John Wiley, Sons, Inc, New Jersey.
[23] Monroy R, Zdravkovic L and Ridley, A 2010 Evolution of microstructure in compacted London clay during wetting and loading. *Geotechnique*, **60**, 105–119. https://doi.org/10.1680/geot.8.P.125
[24] Moses G 2006. Stabilization of Black Cotton Soil with Cement and Bagasse Ash. Unpublished M.Sc. Thesis Department of Civil Engineering, Ahmadu Bello University, Zaria.
[25] Neeraj K S, Swain S K and Sahoo U C 2012 Stabilisation of a Clayed Soil with Fly ash and Lime: A Micro Level Investigation. *Geotech Geol Eng*, **30**, 1197-1205. https://doi.org/10.1007/s10706-012-9532
[26] Negi A S, Faizan M, Siddharth D P and Singh R 2013 Soil Stabilization Using Lime. *International Journal of Innovative Research in Science, Engineering and Technology*, **2**(2), 448- 453.
[27] Nnochiri E S and Aderinlewo O O 2016 Geotechnical Properties of Lateritic Soil Stabilized with Periwinkle Shell Ash in Road Construction. *International Journal of Advanced Engineering, Management and Science*, **2**(5), 484-487.
[28] Osinubi K J 1999 Evaluation of admixture stabilization of Nigerian black cotton soil. Nigerian Society of Engineers Technical Transactions, **34**(3), 88-96.
[29] Peter G M 1993 Fly ash stabilization of tropical hawai soils. In: Sharp KD (ed) Fly ash for soil improvement. Geotechnical Engineering Division of ASCE, Geotechnical Special Publications, **36**, 15–20.
[30] Ramesh H N, Krishnaiah A J and Supriya M D 2013 Role of Moulding Water Content on the Strength Properties of Red Earth treated with Mine tailings. *International Journal of Scientific & Engineering Research*, **4**(5), 47-50.
[31] Ravindrarajah R S 1982 Usage of Cement Kiln Dust in Concrete. *International Journal of Cement Composites and Lightweight Concrete*, Harlow, U. K, **4** (2). https://doi.org/10.1016/0262-5075(82)90013-6.
[32] Salahudeen A B, Eberemu O A and Osinubi K J 2014 Assessment of cement kiln dust treated expansive soil for the construction of flexible pavements. *Geotechnical and Geological Engineering, Springer*, **32**(4), 923-931. https://doi.org/10.1007/s10706-014-9769-0.
[33] Sani J E, Etim R K and Joseph A 2019 Compaction Behaviour of Lateritic Soil–Calcium Chloride Mixtures. *Geotechnical and Geological Engineering*. Springer Nature Switzerland. **37**, 2343-2362. https://doi.org/10.1007/s10706-018-00760-6.

[34] Sathyapriya S and Arumairaj P D 2016 Micro fabric and mineralogical studies on the stabilization of expansive soil using cement industry wastes. *Indian Journal of Geo-Marine Sciences*, **45** (6), 807-815.

[35] Sharma H D and Lewis S P 1994 Waste containment systems, waste stabilization, and landfills: design and evaluation. Wiley, Hoboken, p 588

[36] Simms P and Yanful E 2001 Measurement and Estimation of Pore Shrinkage and Pore Distribution in a Clayed Fill during Soil Water Characteristic Curve Test. *Canadian Geotechnical Journal*, **38**, 741-754. https://doi.org/10.1139/t01-014.

[37] Umoh A A and Olusola K O 2012 Compressive strength and static modulus of elasticity of periwinkle shell ash blended cement concrete. *International Journal of Sustainable Construction Engineering & Technology*, **3**(2), 45-55.

[38] Wayne S A and Donald H T 2008 Beneficial uses of cement kiln dust. In: 50th cement industry technical conference, paper no. 1627, Miami, USA. https://doi.org/10.1109/CITCON.2008.24.

[39] Womack G, Holley C and Greskovich E 2001 *Soil Conditioning Agglomerates Containing Calcium*, Patent and Trademark Office, United States Department of Commerce, Washington, DC, USA, Patent No.: 6, 325, 826.

[40] Eisazadeh A, Kassim K and Nur H 2011 Characterization of Phosphoric Acid and Lime Stabilized Tropical Lateritic Clay. *Environmental Earth Sciences,* **63**, 1057-1066. https://doi.org/10.1007/s12665-010-0781-2.

[41] ASTM 1992 Annual Book of Standards, Vol. 04.08, American Society for Testing and Materials, Philadelphia.

[42] Ayodele A L, Pamukcu Sibel and Agbede O A 2020 Plasticity modification of a tropical laterite by electrochemical stabilization. *Journal of Electrochemica Acta,* **341**, 1-9. https://doi.org/10.1016/j.electacta.2020.136047.

[43] Etim R K 2015 Stabilization of black cotton soil with lime-Iron Ore Tailing blends. *Unpublished M.Sc. thesis,* Department of Civil Engineering, Ahmadu Bello University, Zaria.

[44] Yohanna P 2015 The use of iron ore tailing as admixture in cement modification of black cotton soil. *Unpublished M.Sc. thesis,* Department of Civil Engineering, Ahmadu Bello University, Zaria.

Performance evaluation of asphalt concrete mixture incorporating calcite powder as filler substitute for sustainable pavement construction

A E Modupe[1,3*], O J Aladegboye[1], E A Alejolowo[2], B O Orogbade[1], G O Akade[1], and O P Akinmolayan[1]
[1]Department of Civil Engineering, Landmark University, Omu-Aran, Nigeria
[2]Department of Geophysics, Landmark University, Omu-Aran, Nigeria
[3]Department of Civil Engineering, University of KwaZulu-Natal, Durban, South Africa
*Corresponding Author

Abstract: The increasing cost and paucity of conventional pavement construction materials has in recent times become very worrisome and a serious challenge in the road construction industry. This has spurred the search for a less expensive and environmentally friendly material, readily available and suitable for the purpose of constructing sustainable pavements both in the rural and urban areas of developing countries. In this research, the suitability of calcite powder in asphalt as filler in substituted proportions was investigated through tests on the modified samples for enhanced engineering properties. This was achieved by partially replacing conventional filler with calcite in proportion of "10%, 20%, 30%, 40% and 50%" respectively. After the production of the modified bituminous concrete at the various proportions, the samples produced were placed in a water bath for 30 minutes at a temperature of 60°C before conducting the Marshall stability test and flow. Marshall stability values of 37.7 kN, 37.9 kN, 34.0 kN, 35.0 kN, 28.9 kN and corresponding flow values of 12.0 mm, 11.3 mm, 10.5 mm, 9.60 mm, 10.3 mm, were obtained for the modification proportions respectively. The stability values were observed to be higher than the control mix and likewise satisfy the minimum requirement of 3.33 kN or 340 kg according to the Marshal mix design specification for wearing course surfacing. Similarly, the flow values obtained where found to be within the specified range of 8 – 17 mm at 0.25mm units. Percentage air voids (Vv) values of 3.7%, 5.2%, 4.7%, 3.0%, and 5.9% obtained where also found to satisfy the required range of 3-5% according to the specification. Voids filled with bitumen (VFB) values of 78.58%, 69.56%, 69.73%, 76.63%, and 58.93% derived where also within specified range 75 – 85 % of the specification. The X-Ray diffraction (XRD) micrograph showed the distinct presence of calcium and oxygen in the hot mix asphalt modified with calcite, while the scanning electron microscope (SEM) provided an in-depth perspective of the concrete grains in the hot mix matrix and the improved bonding of the aggregates, bitumen and calcite due the presence of calcium. The results have shown the possibility of calcite utilization in bituminous mixes, towards reducing cost of construction and best suitable for medium trafficked roads, parking lots and walkways.
Keywords: Pavement Engineering, Calcite Powder, Marshall Stability, Flow, Optimum Bitumen Content, Sustainable Transportation

1. Introduction

Recently, there has been an increase in the demand for the construction of more sustainable roads in developing countries taking advantage of innovative and substitute materials [1]. This is as a result of the increasing and worrisome high cost of conventional pavement construction materials [2]. This has motivated a lot of researchers to begin to look into this challenge, hence their search for a sustainable and alternative material with the similitude of the properties of the conventional mineral aggregates especially fillers in bituminous pavement mixes [3]. The wearing surfaces of flexible pavements comprise of coarse aggregates, fine aggregates, mineral filler and binder (bitumen), forming a complete compact composite [4]. The filler material in particular performs an important function in the mix as it forms a mastic with bitumen, filling all the cavities of the mixture and technically seals all the voids, hence resulting into a dense and stiff mix, resistant to the impacts of heavy loads [5]. Several investigations have been made on the substitution of conventional fillers with alternative materials from different sources for sustainable pavement works. For instance materials such as carbon powder [6], periwinkle shells [7], fly ash [8] , rice husk ash [9], waste plastics [10], polymer[11], milled egg shells [12], brake pad waste [13], steel slag [14], date seed ash [15] coal gangue [16], recycled brick powder [17], oil shale [18], oyster shell powder [19], and cow-bone ash [20] have been investigated. A promising material with similar properties with the conventional is calcite, which is the focus of this paper. The use of calcite in asphalt as filler in substituted proportions poses a very economical move as this is a product that is abundant. Calcite a is rock forming crystalline mineral with a chemical formula of $CaCO_3$ [21]. It is usually found in sedimentary, igneous and metamorphic rocks. It is a mineral constituent of Limestone and Marble and is extremely common [22]. Calcite has more uses than any other mineral. This material is relatively cheap and can be found in most developing countries and if put to use, will make the production of asphalt cheaper [23]. Research has shown that the fatigue cracking resistance was enhanced by the addition of CaCO3 in asphalt mixtures [24]. Such modifications will therefore increase the performance characteristics of the hot mix asphalt such as fatigue life, resistance to rutting and thermal cracking. Furthermore and according to [25] it was found that the complex shear modulus of asphalt was improved as a result of using sludge waste containing calcite in asphaltic concrete. Ameliorating the mechanical properties of asphaltic concrete has shown to improve the road service performance in the light of constant usage [22]. As a result of the search for cheaper and more environmentally friendly materials in road construction, the use of calcite as a partial replacement for filler in asphaltic concrete was conceived.

2. Materials used and Methodology

2.1 Materials

2.1.1 Aggregates

The aggregates used in this work are conventional and were sourced from an on-going construction site in Landmark University, Omu Aran, Nigeria. Coarse aggregates (granite chippings) with sizes larger than 2.36 mm were used, while aggregates whose particle sizes were less than 2.36 mm were used as fine aggregates (river bed sand). Prior to the laboratory experiments and tests, they were washed and subsequently sun dried.

2.1.2 Bitumen

Bitumen of 60/70 Penetration Grade (PG) [26] was used in the production the samples to be tested, both the calcite powder modified and control. It was sourced from Landmark University Omu-Aran, Nigeria.

2.1.3 Filler

Sieved particles of the calcite powder retained on the number 200 mesh were utilized in this investigation. Likewise, particles of the virgin or base mineral filler (quarry dust), passing sieve number 200 and retained on the pan were used. Calcite rocks for the study were sourced from the geophysics laboratory, Geophysics Department, Landmark University, Omu Aran, Nigeria. They were grinded to a powdery form and filtered with the 75 μm sieve. The base filler was blended with other aggregates and bitumen to obtain the control samples while the calcite powder for the samples [20].

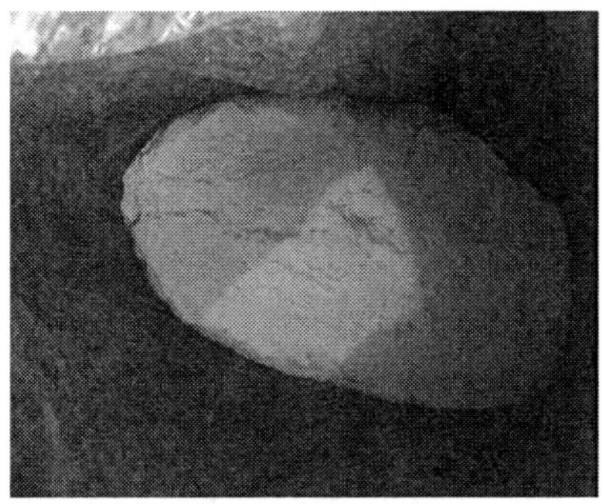

Figure 1. Calcite Powder

2.2 Method

In order to determine the appropriateness of materials sourced for this research, they were subjected to various tests in the laboratory. The suitability of these materials in terms of expected desirable properties to produce the control samples, and in accordance with the standard was verified by the tests.

2.2.1 Tests on Mineral Aggregates

Physical tests such as specific gravity [32], voids percentage [40], test for water retention [33], test on Abrasion [34], bulk density test [32], flakiness index test [36], aggregate impact test[37], test on elongation [36] and moisture content [38] tests were carried out on materials.

2.2.2 Test for Water Content

Empty cans were branded and their weights were determined. Representative wet soil samples were placed in the can, weighed transferred into the oven at 105°C for 20 to 24hours. Water content for the experiment was thereafter calculated using equation 1

$$\text{Water Content} = \frac{(W2 - W3)}{(W3 - W1)} \times 100 \quad (1)$$

Where; W_1 = can weight
W_2 = can + wet soil weight
W_3 = can + dry soil weight

2.3 Tests on Bitumen

2.3.1 Ductility test

The 60/70 Pen bitumen was softened to temperature of 75°C to 100°C greater than the nearest softening point until it was almost liquid. It was strained with the aid of IS sieve 30, transferred into a mold assembly and placed on a brass plate and afterwards a glycerin solution was applied over all the surfaces of the mold exposed to bitumen. Later the sample was transferred into the molds, the plate assembly along with the sample were placed in a water bath and kept at 27°C for 30 minutes. The sample and mold assembly were then removed from the water bath and excess bitumen material was cut off through leveling the surface with a hot knife. After trimming the specimen, the mold assembly in which the sample was poured was replaced in water bath and maintained at 27°C for 85 to 95 minutes. The slides of the mold were then removed and the clips were carefully hooked on the machine without causing any initial strain. The pointer was set to read zero. The distance at which the bitumen thread broke was recorded (in cm) and reported as the ductility value.

2.3.2 Flash and Fire Point Test

Using the Pensky-Martens Closed Tester method [28], the flash and fire point tests were carried out. The require temperature at which it would be safe to heat the specimens was determined.

2.4 Blending of aggregates

For the purpose of achieving a mix that would meet the requirements for gradation of bituminous mixtures, coupled with aid of a trial method, an integration of the mineral aggregates was done. Calculation of the various aggregate size percentages was carried out and benchmarked with the expected requirement. Repeated trials were done until the proportions were within acceptable limits.

2.5 Optimum Bitumen Content (OBC) Determination

To determine the OBC, mean values of three of bitumen contents at the highest values of stability and bulk density, and the bitumen content corresponding to the median value of air voids [27], were determined. The three values were determined through Marshall plots.

$$\text{Optimum Bitumen Content} = \frac{(\%)\ \text{Stability} + (\%)\ \text{Bulk Density} + (\%)\ \text{Vvoids}}{3} \quad (2)$$

Volumetric properties for asphalt mixes usually obtained from OBC such as VMA, VFB, stability, flow, volume of voids (Vv) and bulk density were determined and varied with the design requirements of medium traversed roads shown in Table 1.

2.6 Marshall test

In this research, it became necessary to calculate the optimum bitumen content required to be mixed with the mineral aggregates through the Marshall mix design method. In compliance with standard (ASTM D 1559-89), ten samples weighing 1200gm were produced. They were given 75-blows and maintained at varied bitumen contents (i.e between 4.45 – 6.45% with 0.5% augmentation). Average values each of Marshall flow and bulk density were obtained for one out of two samples prepared separately. OBC was determined for each of the Marshall properties determined viz stability, flow Voids in Total Mix (VTM), Voids Filled with Bitumen (VFB), air voids (Vv), and density (Gm). The standard plots of each of these properties against bitumen content were done to obtain the optimum bitumen content which was benchmarked with the standard Marshall requirement for wearing courses as shown in Table 1.

Table 1. Standard Marshall Requirements for Wearing Courses

Properties	Type of Mix	Standard Values
Air voids (%)	DG wearing course	3 – 5
Voids filled with bitumen VFB (%)	DG wearing course	75 – 85
Stability (KN)	DG wearing course	Not less than 7.5
Flow (mm)	DG wearing course	8 – 17
Optimum bitumen content (%)	DG wearing course	5.0 – 8.0

Source: AASHTO T245

2.7 Preparation of Calcite Modified Asphalt

Cylindrical samples containing calcite powder as shown in Figure 2, were prepared for testing to determine their Marshall and volumetric properties. A total of five samples weighing 1200gm each were prepared to assess the influence of mixing calcite powder to the asphalt mixture samples using five proportions of the calcite powder (10%, 20%, 30%, 40% and 50%) by the weight of OBC. Calcite rock was crushed to powdery form and sieved to size 0.075mm. The resultant powder was mixed with the base filler at the specified proportions and heated to about 150-170°C, as well as the mineral aggregates. Bitumen was also heated to about 150-165°C and thereafter mixed with the calcite powder – quarry dust mixture and with both the fine and coarse aggregates OBC. At 165°C, after a vigorous mixing of all the materials, a homogeneous mix was achieved. Mixes were put in cylindrical molds and given 75blows each according to ASTM D1559[30] for adequate compaction.

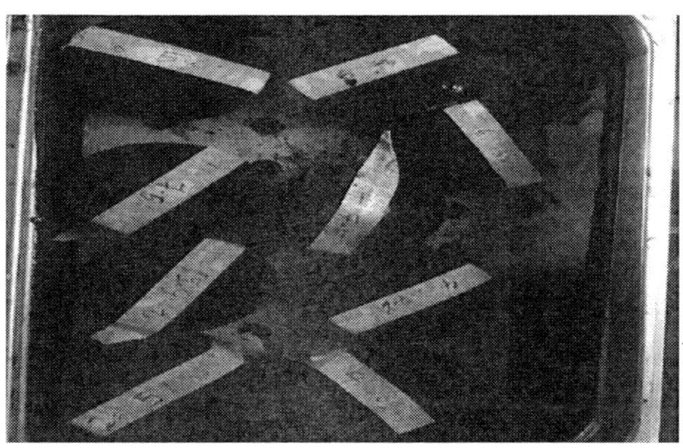

Figure 2. Calcite Modified Samples

3. Results and Discussion

Control sample test results, and of the samples partially modified are as presented below. Others such as Marshall stability test, SEM and XRF are also presented. All tests except SEM and XRF were conducted at Landmark University. The SEM test was carried out at the college of agriculture, University of Ibadan,

Oyo State while X-RAY Florescence test and analysis of materials was conducted at the Engineering Materials Development Institute (EMDI), km 4, Ondo road, Akure, Ondo State, Nigeria.

3.1 Combined Particle Size Distribution Graph

The combined graph of sieve analysis for the mineral aggregates and calcite powder used in the preparation of the asphalt is as shown in Figure 3.

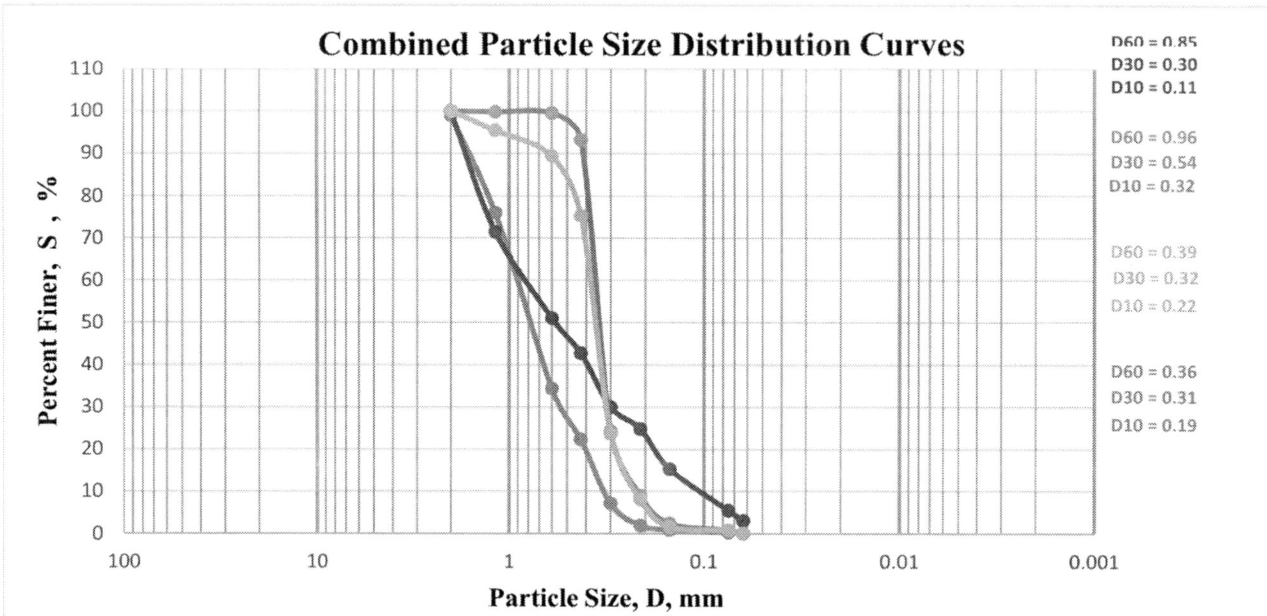

Figure 3. Combined Particle Size Distribution Curves (PSDCs) for coarse, fine, mineral filler aggregates and Calcite Powder

From the combined particle size distribution graph of the aggregates above, the values of D_{10}, D_{30} and D_{60} on the fine aggregate curve in orange were obtained as 0.32, 0.54 and 0.96 respectively. The values of Cu and Cc were calculated using equations 3 and 4 as follow:

$$Cu = \frac{D_{60}}{D_{10}} = \frac{0.96}{0.32} = 3.0 \tag{3}$$

$$Cc = \frac{{D_{30}}^2}{D_{60}\ X\ D_{10}} = \frac{0.54^2}{0.96\ X\ 0.32} = 0.95 \tag{4}$$

From ASTMD-2487 soil classification it can be said that the fine aggregate soil is a fairly graded soil sample because Cu < 6. Similarly, the values of D_{10}, D_{30} and D_{60} on the coarse aggregate curve in blue were 0.11, 0.30 and 0.85 respectively. The value of Cu and Cc were using equations 5 and 6 as follow:

$$Cu = \frac{D_{60}}{D_{10}} = \frac{0.85}{0.11} = 7.72 \tag{5}$$

$$Cc = \frac{D_{30}{}^2}{D_{60} X D_{10}} = \frac{0.30^2}{0.85 X 0.11} = 0.96 \quad (6)$$

From ASTMD-2487 soil classification, it can be said that the particles of the coarse aggregates are well graded as $1 \leq C_c \leq 3$. The results in Table 2 are indicators of the suitability of the materials for use in construction having met the required standard and specification.

Table 2. Coarse Aggregates Mechanical Test Results

Properties	Results	Standard
Aggregate Impact Test	19.20%	30% Max
Aggregate Crushing Test	42.40%	45% Max
Los Angeles Abrasion Test	48.92	60% Max
Flakiness Index	27.73	30% Max
Elongation Index	28.42	30% Max
Density	1500.20kg/m^3	
Specific Gravity	3	3

Source: ASTM D2419 [29] – for Aggregates' Soundness

3.2 Bitumen

The summary of the properties of bitumen investigated are shown in Table 3. These properties were compared with the values stipulated in the specification.

Table 3. Standard Engineering Test Results on Bitumen

Parameters Tested	Test Values	Expected Standard Values (in ranges if possible)	Standard
Penetration (mm)	72	60-70 IS Standard	ASTM D-5
Softening (C)	58	45-60 IS Standard	ASTM D-36
Ductility (cm)	82	75 Minimum IS Standard	ASTM D-113
Viscosity (secs)	77	Nil	ASTM D-3381
Flash Point (C)	275.7	175 Minimum BIS Standard	ASTM D-92
Fire Point (C)	307.7	Nil	ASTM D-92
Specific gravity	0.96	0.97-1.02	ASTM D-70
Loss on heating	0.86	1.00% max	ASTM D-6
Moisture Content	0.0022	Nil	ASTM D-41

3.3 Results for Marshall Stability of Control Samples

Table 4 shows the result from the carried-out Marshall test on the control sample. It also shows the uncorrected and corrected Marshall stability.

Table 4. Results of Marshall Properties for the Control Samples

Test No	% BC	Weight in air	Weight in H_2O	Gt	Gm	Vol. of void (vv)	% Vol of bitumen (vb)	Vma (%)	Vfb	Uncorrected stability (kN)	Cf	Corrected stability (kN)	Flow (mm)
1	5.5	1017	578	2.404	2.317	3.619	13.274	16.893	78.577	37.7	1.000	37.7	12
2	5	1025	579	2.425	2.298	5.237	11.969	17.206	69.563	37.9	1.000	37.9	11.3
3	4.5	1002	572	2.446	2.330	4.742	10.922	15.664	69.727	34	1.000	34	10.5
4	4	897	522	2.467	2.392	3.040	9.967	13.007	76.628	35	1.000	35	9.6
5	3.5	995	570	2.489	2.341	5.946	8.535	14.481	58.939	28.9	1.000	28.9	10.3

3.4 Optimum Bitumen Content (OBC)

It is necessary to obtain the Optimum Bitumen Content (OBC) in order to produced asphalt concrete modifier with an alternative filler material. The OBC was determined based on Marshall mix design and analysis technique which ranges from 4.5% to 6.5% of bitumen by standard. A total of 5 samples were tested representing the control mix in this research. As shown in figure 5, The plots each of density, stability, flow, stiffness, void fill with bitumen and void in total mix against % of bitumen were used to determine the OBC.

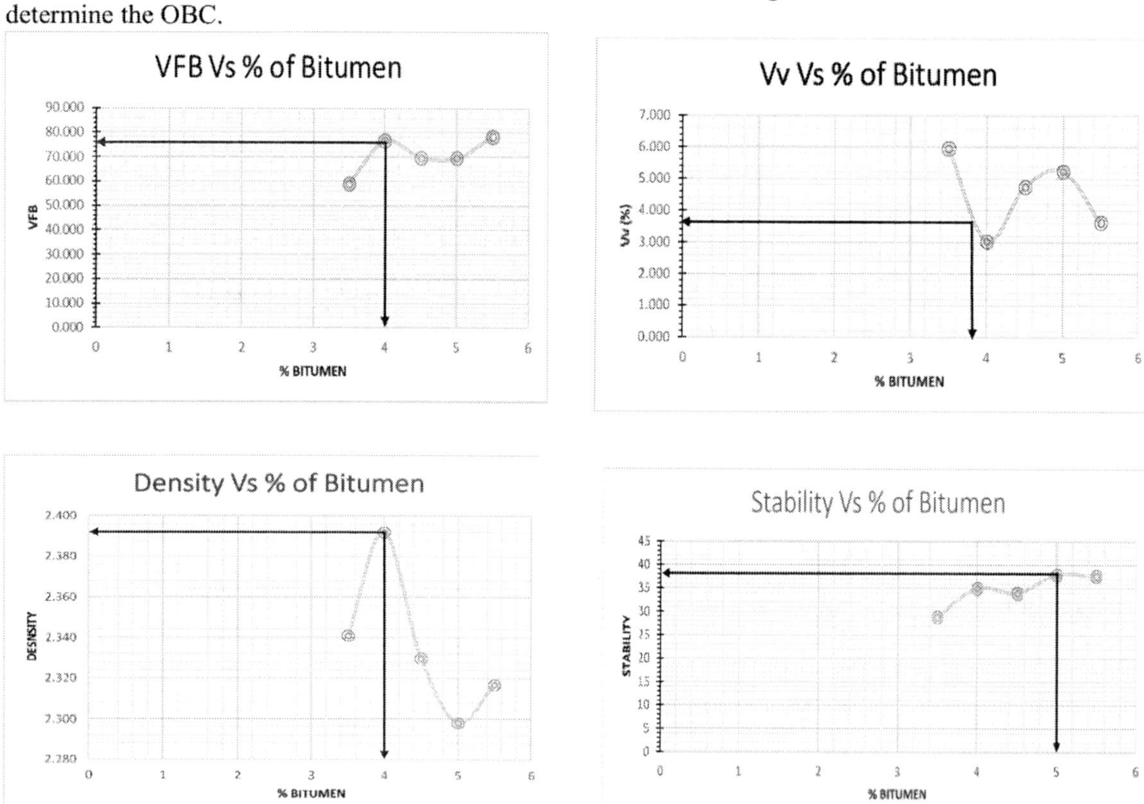

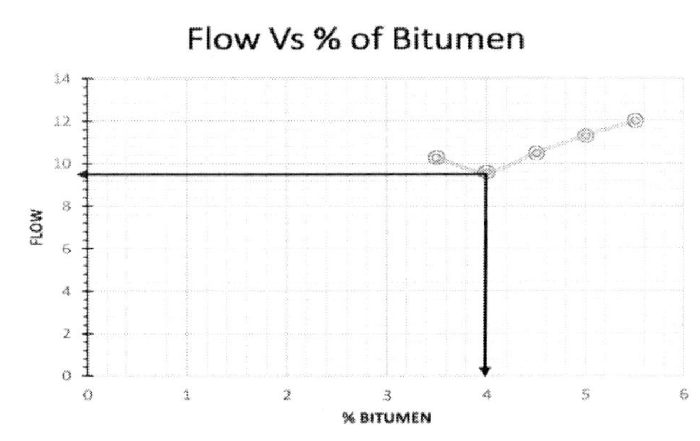

Figure 4. Marshall Plots for the determination of OBC

3.5 Marshall Properties Results of the Calcite Powder Modification as filler

The Marshall properties of the calcite powder modified asphalt mixture were compared with the control mix and checked with standard expectations for sustainable highway construction purposes. the Marshall experiment on the samples modified using the OBC was conducted and the resulting parameters of the asphalt concrete mix with calcite powder and without calcite powder were compared. The required Marshall parameters of the calcite modified mix were same as the control mix. Table 5 shows the summary of the Marshall properties of the calcite modification.

Table 5. Marshall properties of the calcite modification

CALCITE POWDER	VB	STABILITY (kN)	Vv (%)	VMA	VFB	OBC	FLOW	GM
0%	10.93	34.7	4.517	15.45	70.7	4.27	10.74	2.36
10%	10.36	17.91	8.07	12.76	100.61	4.03	16.8	2.56
20%	9.45	16.45	6.82	16.28	57.84	3.15	14.74	2.34
30%	9.27	16.77	6.57	15.84	59.06	3.3	19.83	2.35
40%	9.41	30.64	3.68	12.87	78.34	3.68	12.85	2.43
50%	44.74	19.77	3.89	41.79	46.63	3.73	18.63	13.12

4. Scanning Electron Microscopy (SEM) and X-Ray Fluorescence (XRF) Test Results

4.1 SEM Analysis Test Results

Figure 5 shows the SEM micrograph of the control and calcite powder modified samples using SEM machine at different magnifications. The micrograph of the modified samples at the different magnifications (b), (c), and (d) show proper interaction between the particles of the conventional aggregates and the calcite powder compared to the voids observed in the SEM micrographs of the control sample.

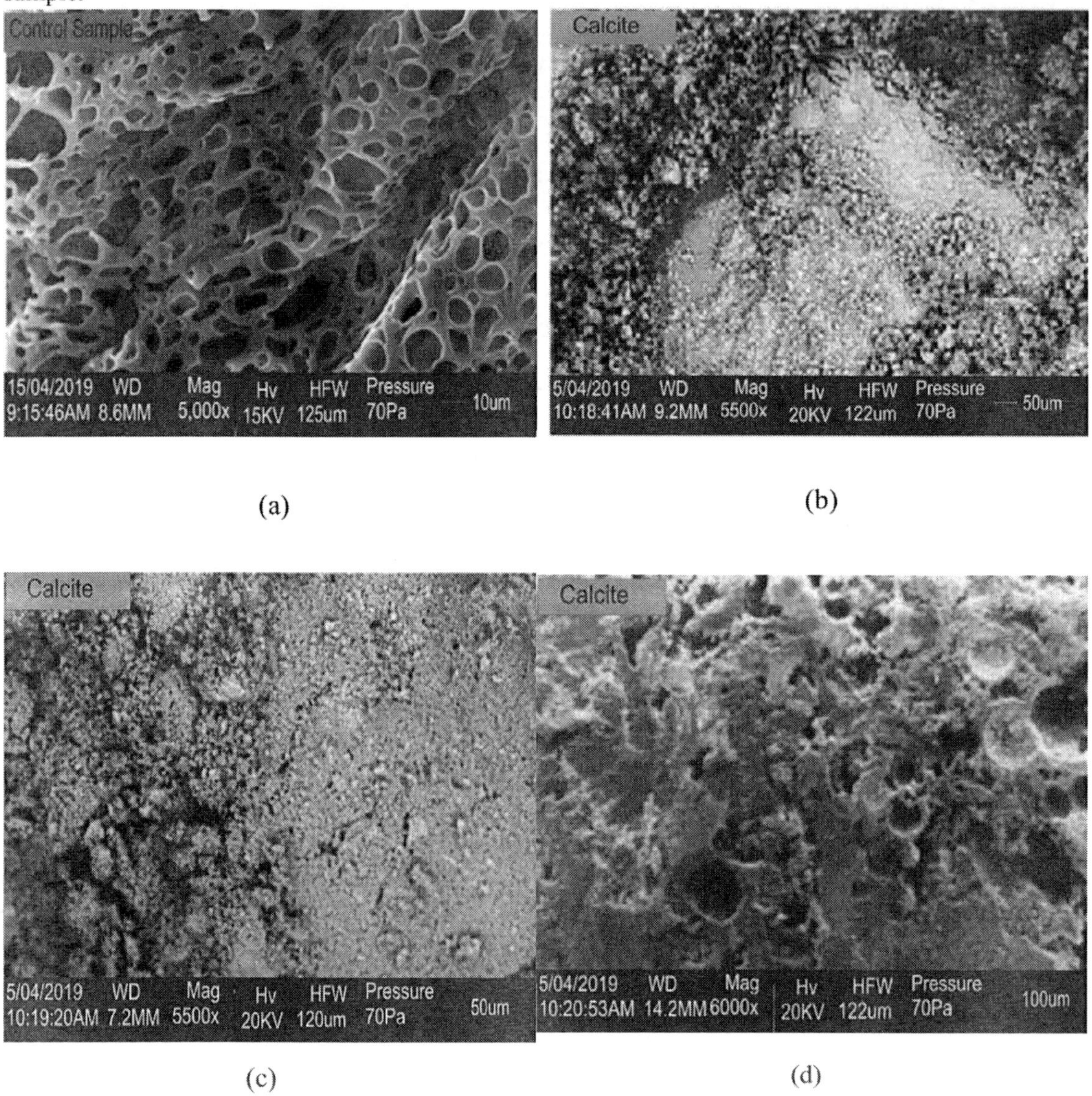

Figure 5: SEM Micrographs of the Control and Calcite Powder Modified Samples at (a) 5000x mag, (b) 5500x mag (9.2mmWD), (c) 5500x mag (7.2mmWD), and (d) 6000x mag.

Figure 6 shows the X-Ray Florescence results, displaying the level and intensity of the elements present in calcite. It was observed to contain Calcium in high proportion which contributes to adhesion in bituminous mixtures.

2	Sample Name	CALCITE	Test Time(s)	100	[redacted]
3	Suppliers		Work Curve	ORE	
4	Voltage(KV)	40.0	Operator	001	
5	Current(μA)	350	Date	18/03/2019	
6	Element	Intensity	Content		
7	Mg	0.0000	0.0127		
8	Al	0.0019	0.5866		
9	Si	0.0065	0.5649		
10	P	0.0056	0.2656		
11	S	0.0071	0.5261		
12	K	0.0000	0.0000		
13	Ca	0.6874	67.4786		
14	Ti	0.0000	0.0000		
15	V	0.0000	0.0010		
16	Cr	0.0000	0.0000		
17	Mn	0.0000	0.0000		
18	Co	0.0000	0.0000		
19	Fe	0.0012	0.1904		
20	Ni	0.0003	0.0104		
21	Cu	0.0005	0.0098		
22	Zn	0.0007	0.0205		
23	As	0.0001	0.0000		
24	Pb	0.0001	0.0006		
25	W	0.0001	0.0000		
26	Au	0.0001	0.0351		
27	Ag	0.0000	0.0000		
28	Rb	0.0000	0.0000		

29

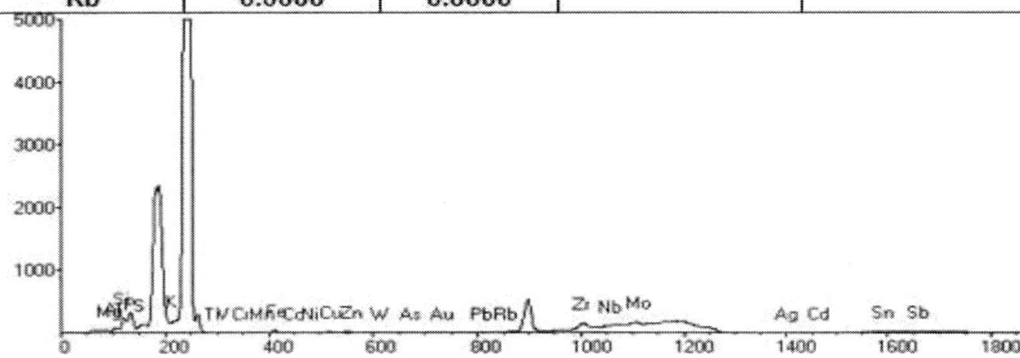

30	Nb	0.0002	0.0006		
31	Mo	0.0014	0.1731		
32	Cd	0.0000	0.0000		
33	Sn	0.0017	0.2998		
34	Sb	0.0026	0.3157		

Figure 6. X-Ray Florescence

4.2 Energy-Dispersive Spectroscopy (EDS)

Figure 7 shows the level and intensity of elements present in calcite powder. It can be observed that the calcite powder is filled with high intensity of calcium, which is responsible for the pozzolanic characteristic of the material.

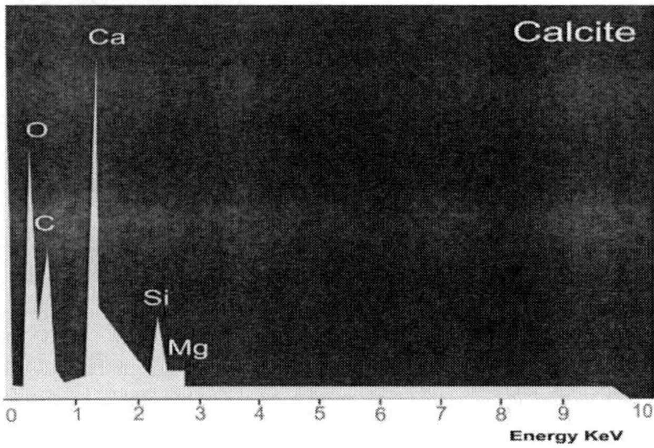

Element	Wt (%)
O	20.92
C	9.17
Ca	65.13
Si	3.20
Mg	1.58

Figure 7. Energy-Dispersive Spectroscopy (EDS)

From the above readings, a high proportion of Calcium (65.13%) by weight of other elements present in the calcite powder was observed. This further strengthens the bond between the constituent materials in the mix.

5. CONCLUSION

Findings from this study have shown that calcite powder can be incorporated into asphaltic concrete mixtures with the intention of improving its Marshall properties due to its similarities in terms of aggregate shape, size and texture with the conventional filler. Calcite is readily available and can be explored for sustainable pavement construction or rehabilitation. According to the results from the calcite powder modified proportions viz 10%, 20%, 30%, 40%, 50% as substitute for the conventional filler, it was found suitable for the construction of medium trafficked roads at 10%, 20%, 30%, 40% and 50%, which corresponds to stability values 17.91 kN, 16.45 kN, 16.77 kN, 30.64 kN and 19.77 kN respectively, having satisfied the minimum standard requirement for stability (not less than 7.5 kN). Furthermore, the XRF, SEM, stability and the flow results obtained from the experimental work show that the physical and chemical properties of the asphaltic concrete had a significant improvement to them.

References

[1] Balaguera A, Carvajal GI, Albertí J, and Fullana-i-Palmer P 2017 Life cycle assessment of road construction alternative materials: A literature review *Resour. Conserv. Recycl.* **132** pp 37–48.

[2] Wu S, and Tahri O, 2019 State-of-art carbon and graphene family nanomaterials for asphalt modification *Road Materials Pavement Design.*, **0** pp 1–22.

[3] Thom N, and Dawson A 2019 *Sustainable road design* Promoting recycling and non-conventional materials *Sustain.* **11** no 21.

[4] Oluwasola EA, Hainin MR, Idham MK, and Modupe AE 2018 Workability and rheological properties of eva-modified bitumen compared with PG 76 binder," *J. Teknol.* **80**.

[5] Zaumanis M, Poulikakos LD, and Partl MN 2018 Performance-based design of asphalt mixtures and review of key parameters. *Materials & Design* **141** pp 185-201.

[6] Ziari H, Divandari H, Hajiloo M, and Amini A 2019 Investigating the effect of amorphous carbon powder on the moisture sensitivity, fatigue performance and rutting resistance of rubberized asphalt concrete mixtures *Constr. Build. Mater.* **217** pp. 62–72.

[7] Nwaobakata C, and Cagwunwamba J 2014 Influence of Periwinkle Shells Ash as Filler in Hot Mix Asphalt **3**, no. 7 pp. 2369–2373.

[8] Woszuk A, Bandura L, and Franus W 2019 Fly ash as low cost and environmentally friendly filler and its effect on the properties of mix asphalt *J. Clean. Prod.*, **235**, pp. 493–502

[9] Shafabakhsh G, Divandari H, and Reza Sajadi S 2018 Evaluation of Optimum Content of Rice Husk Ash to Improve the Hot Asphalt Concrete Performance *Int. J. Eng. Technol* **7** no. 4.20 p. 338

[10] Badejo AA 2017 Plastic waste as strength modifiers in asphalt for a sustainable environment *African J. Sci. Technol. Innov. Dev.*, vol. **00** no. 0 pp. 1–5

[11] Modupe AE, Olumoyewa AD, Adebayo BO, Gana AJ, John RAL, and Raphael OD 2019 Development and performance evaluation of crumb rubber – Bio-oil modified hot mix asphalt for sustainable highway pavements *Int. J. Mech. Eng. Technol.* **10**, no. 2 pp. 273–287.

[12] Oluwatuyi OE 2018 Ameliorating effect of milled eggshell on cement stabilized lateritic soil for highway construction *Case Stud. Constr. Mater.* **9**

[13] Hu X, Wang N, Pan P, and Bai T, 2017 Performance evaluation of asphalt mixture using brake pad waste as mineral filler *Constr. Build. Mater.* **138** pp. 410–417

[14] Hasita S, Suddeepong A, Horpibulsuk S, Samingthong W, Arulrajah A, and Chinkulkijniwat A 2020) Properties of Asphalt Concrete Using Aggregates Composed of Limestone and Steel Slag Blends *J. Mater. Civ. Eng.* **32** no. 7 pp. 1–7

[15] Foroutan SA, Mirhosseini MM, Khabiri AK, and Jalal Kamali MH 2016 Applying surface free energy method for evaluation of moisture damage in asphalt mixtures containing date seed ash *Constr. Build. Mater.* **125** pp 408–416.

[16] Bao Hong R, Rong Wu J, and Bing Cai H 2019 low-temperature crack resistance of coal gangue powder and polyester fibre asphalt mixture *Constr. Build. Mater.* **238** pp. 1176-1178.

[17] Kathem Taeh Alnealy DS 2015 Effect of Using Waste Material as Filler in Bituminous Mix Design *Am. J. Civ. Eng.* **3** no. 3 pp. 88

[18] Wang W, Cheng Y, Tan G, and Shi C 2020 Pavement performance evaluation of asphalt mixtures containing oil shale waste *Road Mater. Pavement Des.* **21** no. 1 pp. 179–200

[19] Nciri N, Shin T, Lee H, and Cho N 2018 Potential of waste oyster shells as a novel biofiller for hot-mix asphalt *Appl. Sci.* **8** no. 3

[20] Modupe AE, Adeyanju TMA, Atoyebi OD, Aladegboye OJ, Aderemi PA and Modupe OC 2019 Performance evaluation of hot mix asphaltic concrete incorporating cow bone ash (CBA) as partial

replacement for filler in *IOP Conference Series: Materials Science and Engineering* **640** no. 1.

[21] Avaro J, Moon EM, Rose J, and Rose A L, 2019 Calcium coordination environment in precursor species to calcium carbonate mineral formation *Geochim. Cosmochim. Acta* **259** pp. 344–357.

[22] Sangiorgi CP, Tataranni F, Mazzotta A, Simone VV, and Lantieri C 2017 Alternative fillers for the production of bituminous mixtures: A screening investigation on waste powders *Coatings* **7**, no. 6.

[23] Scrivener KL, John VM, and Gartner EM 2018 Eco-efficient cements: Potential economically viable solutions for a low-CO2 cement-based materials industry *Cem. Concr. Res.* **114** pp. 2–26.

[24] Mohd Hasan MR, You Z, Mohd Satar MKI, Mohd Warid MN, Mohd Kamaruddin NH, and Poovaneshvaran S, 2019 Rheological and Mechanical Performance of Asphalt Binders and Mixtures Incorporating CaCO3 and LDPE *Org. Polym. Mater. Res.* **1** no. 1 pp 10–21.

[25] Haouzi F, and Esnault-filet A 2019 *Advancements on Sustainable Civil Infrastructures*, vol. **1** no. 1. Springer International Publishing.

[26] Modupe AE, Atoyebi OD, Oluwatuyi OE, Aladegboye OJ, Busari AA, and Basorun AO 2018 Dataset of mechanical, marshall and rheological properties of crumb rubber – Bio-oil modified hot mix asphalt for sustainable pavement works *Data in Brief.* **21**.

[27] Choudhary J, Kumar B, and Gupta A 2019 Feasible utilization of waste limestone sludge as filler in bituminous concrete *Constr. Build. Mater.*, **239** pp 1177-81.

[28] ASTM International. (2012). ASTM D93-12: Standard Test Methods for Flash Point by Pensky-Martens Closed Cup Tester. ASTM.

[29] ASTM D2419. 2014. ASTM D2419-14: Standard test method for sand equivalent value of soils and fine aggregate.

[30] ASTM, D. 2003. D1559-89. Standard test method for resistance to plastic flow of bituminous mixtures using Marshall Apparatus. Annual Book of ASTM Standards USA.

[31] Soils, B. S. ASTM D 2487 soil classification groups GW. GP, GM, SW, SP, and SM or a combination of these groups.

[32] ASTM, S. 2012. Standard test method for density, relative density (specific gravity), and absorption of coarse aggregate. C127-12.

[33] ASTM International. 2010. ASTM D570-98 (2010) e1-Standard Test Method for Water Absorption of Plastics.

[34] ASTM, ASA 2008 C131/C131M: Standard Test Method for Resistance to Degradation of Small-Size Coarse Aggregate by Abrasion and Impact in the Los Angeles Machine.

[35] ASTM C 2000 127 Standard Test Method for Specific Gravity and Adsorption of Coarse Aggregate. Annual Book of ASTM Standards, Road and Paving Materials, 4.

[36] ASTM SD 2010 Standard test method for flat particles, elongated particles, or flat and elongated particles in coarse aggregate. D4791-10.

[37] ASTM D. 5874 2007 Standard Test Method for Determination of the Impact Value (IV) of a Soil." American Standard Testing Methods, West Conshohocken, Philadelphia, 1-9.

[38] ASTM D 2010 4791 Standard Test Method for Flat Particles, Elongated Particles, or Flat and Elongated Particles in Coarse Aggregate

[39] ASTM D 2010 Standard test methods for laboratory determination of water (moisture) content of soil and rock by mass. Annual Book of ASTM Standards.

[40] Standard ASTM 007. C128-07a: Standard Test Method for Density, Relative Density (Specific Gravity), and Absorption. ASTM International.

Investigation of geotechnical properties of soil samples in Isu-Njaba Imo State

A C Ekeleme[1*], O H Ibearugbulem[2], E I Ugwu[3], C E Njoku[4], E C Amanamba[1], O P Ibe[2], E W Igwegbe[2] and P C Okpara[2]

[1]Department of Civil Engineering, Abia State University Uturu, P.M.B 2000 Abia State, Nigeria.
[2]Department of Civil Engineering, Federal, Polytechnic Nekede, P.M.B 1036 Imo State, Nigeria
[3]Department of Civil Engineering, Michael Okpara University of Agriculture Umudike, P.M.B. 7267 Abia State, Nigeria
[4]Department of Materials and Metallurgical Engineering, Federal University of Technology, Owerri, PMB 1526, Imo State Nigeria.
[5]Department of Civil Engineering, Imo State University Owerri, P.M.B 2000 Imo State Nigeria.
*Corresponding author: chytony4real09@yahoo.com

Abstract: The properties of the geotechnical characteristics of soil samples from Isu-Njaba borrow pit was carried out in this study. Ten soil samples from the depth of 3m were taken from the pit at day1, day7 and day14 respectively. The tests carried out on the soil samples include sieve analysis, natural moisture content, Atterberg limit, compaction and California bearing ratio tests in line with standards. From the results, the fine soil content is 4.19%, and the soil samples contain more of sand and gravel and are uniformly graded. Thus the soil is classified as A-2-4 soil in line with AASHTO standard of soil classification. The obtained liquid limit is 38.02%, plastic limit of 27.53% and plasticity index of 10.49%. The results from the test on compaction showed the maximum dry density to be 1.56 Mg/m and optimum moisture content as 14.30%. The values of the California bearing ratio for the non-soaked soil is 20.64%, and 12.63% for the soaked soil. The study showed that the soil samples from Isu-Njaba borrow-pit could be applied as materials for sub grade, sub base and base purposes, as most of the properties meet the specifications of the Federal Ministry of Works and Housing guidelines for road construction.
Keywords: Geotechnical characteristics, Soil, Atterberg limit test, Isu-Njaba

1. Introduction

In the course of construction expansion in structural engineering, some of the frequently faced problems include cracks, pavement break-ups, lines for reservoir, water and sewers, foundations of buildings, roadways, highways, and irrigation systems amongst others. These problems occur because of not understanding the behavior of the soil or lack of information as it concerns the soils' engineering properties [1]. Investigation on the properties of the geotechnical attributes of the soil is highly needed, because these data are very important in the preliminary designs and in the designs involving building foundations, pavement, and retaining structures for construction projects. Investigation of engineering properties of soil should be done for economical design of sub-structural elements and almost everything we build such as buildings, bridges, dams and others that are in, on or within soils [1]. Sufficient information is necessary for studies on economic and applicability of the anticipated project. Not carrying

IOP Conf. Series: Materials Science and Engineering 1036 (2021) 012035 doi:10.1088/1757-899X/1036/1/012035

out enough geotechnical analysis, interpretation and representation of results wrongly may contribute to inappropriate design and construction delays. This also may cause expensive modifications on constructions, use of inferior materials, environmental pollution, repair and failure of construction works, succeeding legal actions amongst others. Thus in order to get more information on the kind of soil, its attributes, and how it is distributed, investigations are made on the geotechnical qualities of the main soil or neighbouring soils to the site of the proposed structure.

The soil of Adama town is categorized as silt and silty sand in line with Unified Soil Classification System. Almost all the samples of the soil have free swell value of less than 50%. This shows that the soil in the Adama town is non-expansive [2].

The soils of Ambo town are highly plastic. Black and gray soils have higher plasticity index than reddish brown and brown soils. Grain size analysis shows that the predominant proportion of the soils is clay size fraction. The soils in Ambo town are categorized under A-7-5 soils in keeping with AASHTO system of soil classification. It implies that the quality of the soil is poor and would not be suitable for use as grade materials. However during dry season, the consistency of Ambo soils changes from rigid to hard because the unconfined compressive strength (UCS) varies from 112 KPa to 545 KPa [3].

The grain size distribution of Sebeta town indicates majority samples have clay material more than 50%. In line with AASHTO system of soil classification, the soils are classed under group A-7 and A-6, and thus cannot be used as sub grade materials due to the quality is poor [4].

There are both black clay soils and red soils in the town of Adet. The black clay soils are found in the relatively lower elevation and flat lying areas of the town. However, most of the areas in the town are covered with red soils. The free swell values of black clay soils, ranges from 67-100%, which shows that the soils are fairly expansive. Most of the red soils' free swell values are below fifty (50) percent whereas the whole red soil's free swell values ranges from 27-67.5%. The consistency of Adet soils showed medium rigid to hard for the UCS varies from 84 to 442 KPa at natural moisture content of 28-41% [5]. Thus this study is aimed at investigating the engineering attributes (properties) of samples of soils from Isu-Njaba area of Imo State.

2. Materials and methods

Ten soil samples were taken from the depth of 3m. Three soil samples at day1 and day7, while four samples were taken at day 14. The collected soil samples were positioned in weighed containers and properly kept for the needed tests. The following tests were performed on the samples of the soils; sieve analysis, moisture content, Atterberg limit, compaction and California bearing ratio tests.

2.1 Sieve analysis

To understand the distribution of the sizes of the soil particles, sieve analysis was done, in line with BS 1377 [6] method. 700g of sample that has been dried in the oven was discharged into a set of sieves and placed on the sieve shaker. Later on, the set of sieves was taken out and the weights of the quantities remaining on the different sieves were taken. The weight of soil retained in percentage and weight of soil that passed in percentage were calculated using equations (1) and (2) [7]:

$$\%\,weight\ of\ soil\ retained = \frac{weight\ retained}{Initial\ weight}\ x\ 100\% \tag{1}$$

$$\%\,of\ soil\ passing = 100 - \%\,weight\ retained \tag{2}$$

The results are plotted in a semi-logarithmic graph identified as the particle size distribution curve.

2.2 Test for moisture content

The natural moisture content analysis was done in line with BS 1377 [6] method. A tiny quantity of the sample of the soil was placed on a weighed container and then kept for drying in the oven for a temperature of 105°C for a period of 24 hrs. Then the sample that was dried in the oven was brought out to cool. Thereafter, moisture content was calculated using equation (3) [7]:

$$Moisture\ content = \frac{M_2-M_3}{M_3-M_1} \times 100\% \quad (3)$$

where M_1 gives the weight of the container, M_2 gives the weight of the soil before oven drying, M_3 gives the weight of the soil after oven drying.

2.3 Test for Atterberg limit

The liquid limit was determined using the ASTM D423 [8] method. To carry out the test on liquid limit, 300g of the sample of the soil that went through sieve No. 40 was carefully blended with water, in a clay dish. Water is repeatedly added to the soil mixture, and steadily stirred, kneaded and chopped using a spatula. When a homogenous mixture was produced, a small quantity of the mixture was forced down into a cup and cautiously positioned with a spatula while avoiding air entrapment. Later the moisture content was plotted in respect to the number of blows and the moisture content at 25 blows gives the liquid limit. Also the plastic limit and the shrinkage limit tests were performed using the BS 1377 [6] standard procedures.

2.4 Test for compaction

The test for compaction was conducted using BS 1377 [6] method. The maximum dry density was carried out by varying the content of moisture and about 3 kg of the soil sample was used. Three equal layers of compacted soil samples and each of the layers was subjected to 27 blows, which were shared evenly on top of every layer. Then the collar was taken away and the sample that was compacted was stabilized using a straight rim. The mould that contains the sample that was stabilized was brought to a weight of 1g. A little sample was removed from the soil that was compacted with the aim of finding out the moisture content. In the bid to find out the moisture content, similar practice was undertaken again. Later, a graph of dry densities was plotted against the individual moisture contents, and thus the maximum dry density (MDD) and the optimum moisture content (OMC) were deduced from the plot.

2.5 Test for California bearing ratio

The BS 1377 [6] method was followed in performing the test for California bearing ratio (CBR). The sample of soil about 6 kg was carefully blended with moisture at their particular optimum moisture contents. The compaction of the soil sample was done and placed in a mould with dimensions (10 x 10 x 10 cm), in five layers. Every layer was compacted by 62 blows using a 4.5 kg rammer. From the CBR test, an indirect measurement of the resistance of the soil to shear under monitored conditions of density and moisture content would be achieved. The ratio of the amount of load needed to penetrate to a particular depth in a compacted sample at given density and moisture content, to the standard load needed to obtain similar depth penetration on a standard crushed stone sample, gives the CBR. This can be mathematically expressed using equation (4):

$$CBR = \frac{Test\ load\ on\ the\ sample}{standard\ load\ on\ the\ crushed\ stone} \times 100 \quad (4)$$

3. Results and discussions

The sieve analysis was carried out to find out the distribution of the grain size of the soil. The result from the sieve analysis is shown in Figure 1 and the clay content of the soil sample is 4.19%. This fits into the

Federal Ministry of Works and Housing (FMWH) [9] specification, which directs that the content of clay for materials used for sub-grade, sub-base and base purposes going through No. 200 BS Test sieve or 0.075 mm should be less than or equal to 35%.This is to ensure that the amount of clay in the small is not too much so as to enhance the mechanical stability of the soil [10]. The soil sample studied belongs to the class of A-2-4 and SW soils in accordance with AASHTO and USCS standards for classifications of soils respectively, which implies that the soil contains more of sand and gravel. The coefficient of uniformity (C_u) is 2.61, and the coefficient of curvature (C_c) obtained is 1.01. This classifies the soil as a uniformly (or poorly) graded sandy soil [11].

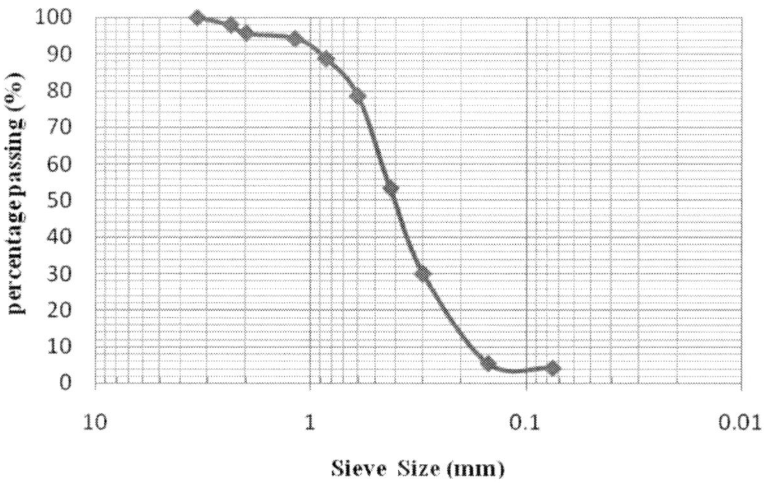

Figure 1. Distribution of the particle size of the soil samples

The properties of the geotechnical characteristics of the soil sample are presented in Table 1. From Table 1, the obtained natural moisture content is 3.45%. The reduced moisture content of the soil sample would enhance the shear strength of the soil and thus makes it fitting for use in construction of roads. There is an implication also that there would be no interaction of moisture between the sub grade, sub base and base rocks, thus keeping the soil intact [12].

The results of the Atterberg test carried out as presented in Table 1, showed that the soil samples studied have average liquid limit of 38.02%, average plastic limit of 27.53% and average plasticity index of 10.49%. According to the plasticity chart of Casagrande [13], soils that have liquid limits less than 30% are said to depict low plasticity and compressibility, those that have liquid limits amid 30% and 50% show medium plasticity, while as the soils having liquid limits that are above 50% are said to display high plasticity and compressibility. Thus this soil sample displays medium plasticity. Moreover, the Federal Ministry of Works and Housing [9] has specified that the liquid limit for sub grade materials should be maximum of 40%, maximum of 35% for sub base, and maximum of 30% for base materials. Also that the plasticity index should be maximum of 20% for sub grade, maximum of 16% for sub base and maximum of 13% for base materials [7]. Comparing the results obtained from this study to the FMWH [9] guidelines, the liquid limit obtained showed that the soil may not have stretchy clayey materials and would be good for sub grade since it is below 40%, but would be poorly suitable or sub base and base materials [7]. However, obtained plasticity index of 10.49% from the study meets the specification of FMWH [9] for application as materials for sub grade, sub base and base purposes. The increase in the plasticity index more than the required value may lead to decreased permeability and increase in soil's compressibility, which can in turn cause the soil to be easily flooded and water logged [14].

The result of the test on linear shrinkage test is presented in Table 1. The linear shrinkage shows how the soil expands or swells during wet or dry seasons [15]. For this study, the linear shrinkage is 8.10% which fairly satisfies the recommendation by FMWH [9] that the linear shrinkage of soils for sub grade materials should be at maximum of 8%. Thus the soil studied would be fairly fit to be used as material for sub grade purposes, for its expansion would be as required.

From the test on compaction, the maximum dry density (MDD) and optimum moisture content (OMC) were obtained as 1.56 Mg/m^3 and 14.30% respectively as presented in Table 1. It has been reported that soils with increased maximum dry density and reduced optimum moisture content values are most appropriate for sub grade and sub base purposes [16, 17]. FMWH [9], specified that for sub grade and sub base materials that the MDD should be $\leq$ 1.7 Mg/m^3 and OMC should be < 18%. The soil sample in this study met this specification and would be fitting for application as sub grade and sub base materials.

To determine the capacity of the soil to bear loads, the California bearing ratio (CBR) was investigated on the samples of soils for unsoaked and soaked conditions and the results are included in Table 1. The FMWH [9] specifications for unsoaked soil is $\leq$ 10% for sub grade, $\leq$ 30% for sub base, and $\leq$ 80% for base materials. It could be seen that the value of the CBR of the investigated unsoaked sample of soil (20.64%) showed that the soil would be best fit to be applied as sub grade and sub base materials, and also can be utilized as base materials. However the CBR of the soil sample reduced after soaking for 24 hours, which implied that if the soil becomes water logged, its capacity to bear loads or its strength would be reduced.

Table 1. Properties of the geotechnical characteristics of the soil.

Parameters	Quantities
Percentage passing BS Sieve No 200 (%)	4.19
Average moisture content (%)	3.45
Liquid limit (%)	38.02
Plastic limit (%)	27.53
Plasticity index (%)	10.49
Linear shrinkage (%)	8.10
AASHTO Class	A-2-4
USCS	SW
Maximum Dry Density (Mg/m^3)	1.56
Optimum Moisture Content (%)	14.30
California Bearing Ratio (unsoaked) (%)	20.64
California Bearing Ratio (after 24 hour soaking) (%)	12.63

4. Conclusion

The properties of the geotechnical characteristics of soils from Isu-Njaba borrow pit was investigated in line with established standards. The results indicated that the sample of soil investigated is a uniformly graded sandy soil having the percentage passing less than 35%. Low values of natural moisture content, maximum dry density and optimum moisture content were obtained from the results. Also, the values of the plasticity results and CBR results for the non-soaked soil samples showed that the Isu-Njaba borrow-pit soils would be suitable for application as sub grade, sub base and base materials for the construction of roads. In addition, the results from this study can offer preliminary information needed for the planning and infrastructural development of Isu-Njaba Imo State in future.

This study has shown that the soil in the Isu-Njaba area is underlain predominantly by sandy soil (fine and medium sands).The soil is a uniformly graded sandy soil, and has desired properties needed as a construction material without or less improvement in its properties. Therefore, the soil would be apt for use as materials for sub-grade material in the construction of roads. Also, results of this study gives preliminary information required for future planning and infrastructural development of Isu-Njaba in Imo State.

References

[1] Zekrab N 2016 *Investigation on Some of the Engineering Properties of Soils Found in Haromaya Town, Ethiopia (*M.Sc Thesis Addis Ababa University)

[2] Dagnachew D 2011 *Investigation on some of the Characteristics of Soils in Adama Town, Ethiopia* (M.Sc Thesis Addis Ababa University)

[3] Behaylu H 2014 *Investigation on some of Engineering Properties of soils found in Ambo Town, Ethiopia* (M.Sc Thesis Addis Ababa University)

[4] Selamawit M 2015 *Investigation on some of Engineering Properties of soils found in Sebeta Town, Ethiopia* (M.Sc Thesis Addis Ababa University)

[5] Tadesse A, 2015 *Investigation on some of Engineering Properties of soils in Adet Town, Ethiopia* (M.Sc Thesis Addis Ababa University)

[6] BS 1377 1990 *Methods of Tests for Soils for Civil Engineering Purposes* (BSI Publications)

[7] Nwadike H O and Nweke O M 2016 *Eng. Appl. Sci.* 1 **4** 74-86

[8] ASTM International 2006 *Standard Practice for Classification of Soils for Engineering Purposes (Unified Soil Classification System)* p 12

[9] Federal Ministry of Works and Housing (FMWH) 1997 *Nigerian General Specification for Roads and Bridge* Revised Edition, **2** 37–275

[10] Ademilua O 2018 *Proc. of 1st and 2nd Intern. Conf. of the Nigerian Association of Engineering Geology and the Environment* (Lagos Nigeria) **1** 42–8

[11] NYSDOT Geotechnical design manual 2013 *Engineering properties of soil and rock* chapter 6 pp 1-85

[12] Oyelami A C and Alimi S A 2015 *Ife J. Sci.* 17 **1**, 87-95

[13] Casagrande A 1948 *Trans. Am. Soc. Civil Engrs.* 13 901-30

[14] Onuoha D C, Onwuka S U and Okoye C O 2014 *J. Env. Earth Sci.* 4 **16** 125-34

[15] Olofinyo O O, Olabode O F and Fatoyinbo I O 2019 *SN Appl. Sci.* 1 **507** 1-10

[16] Bello A A, Ige J A and Tajudeen S 2007 *LAUTECH J. Engr. Technol.* 4 **2** 34 – 8

[17] Jegede O G 2004 *Ife J. Sci.* 6 **1** 41–4

Classification of surface condition of flexible road pavement using Naïve Bayes theorem

A T Olowosulu[1], J M Kaura[1], A A Murana[1] and P T Adeke[2]*
[1]*Department of Civil Engineering, Faculty of Engineering, Ahmadu Bello University Zaria, Kaduna State, Nigeria*
[2]*Department of Civil Engineering, College of Engineering, Federal University of Agriculture Makurdi, Benue State, Nigeria*
*Corresponding Author's email: adeke.pt@outlook.com

Abstract. Data collection, availability and accuracy on road pavement surface condition have been identified as the major setback for pavement management system in low and medium income countries. This situation necessitates the adoption of historic dataset which often is characterised by missing and noisy data elements. The Naïve Bayes theorem which depends on conditional probabilities of prior events or attributes for condition classification and prediction was used to investigate the authenticity of surface condition classification of flexible road pavement. Some selected links of Federal Highways in Nigeria based on the availability of historic dataset were considered. The Naïve Bayes theorem as a data mining approach was implemented using Waikato Environment for Knowledge Analysis (WEKA) software which performed satisfactorily in surface condition classification with minimal margin of errors due to its ability to handle challenges of missing and noisy dataset. Results of classifications indicated that; links 716, 5, 8, 22 and 136 in Borno, Kwara, Lagos, Oyo and Plateau states respectively had relatively high level of unworthiness as at the time the data was collected, hence call for immediate maintenance and rehabilitation actions, while links 89, 375, 370, 130, 17, 332, 138, 144 and 255 from Anambra, Bendel, Imo, Kaduna, Ogun, Plateau, Rivers and Sokoto states respectively were worthy and had no cause for immediate maintenance.
Keywords: Pavement Surface condition, Data mining, Naïve Bayes theorem, WEKA Software.

1. Introduction

The choice of flexible road pavement for highway construction in low and medium income countries is due to its relatively cheap cost of construction, maintenance and rehabilitation [1][2]. The management of road pavement system requires periodic surface condition assessment and classification for optimum maintenance strategy [3][4]. In most cases, the act employs manual approaches which are associated with subjective measurements by trained personnel whose reportage are prone to significant errors due to physical limitations and human intuitive reasoning. The aim of developing road pavement management system is to build and maintain road facilities that can carter for the fast growing human population and corresponding travel demand over time [5]. Since road infrastructures deteriorate over time, hence the need for effective planning and efficient management system arises [6][7][8]. The culture of investigating and classifying road pavement condition periodically for timely scheduling of repairs is essential for achieving reliable road network that could connect people and places, based on proper planning, timely maintenance and rehabilitation actions, budgetary allocation and execution to aid socio-cultural, administrative and economic activities in the country [9][10][11].

The major disadvantages of flexible pavement are its relatively short life cycle and rapid deterioration over time which leads to reduction in performance index due to the impact of damaging factors [12][13][14]. The rate of deterioration of flexible road pavement which shortens its life cycle is a function of several factors such as; traffic load, climatic condition (rain, temperature, etc.), material properties, drainage condition, maintenance policy, design of pavement structure and workmanship, etc. [8] [10][15] [16][17]. Because huge funds are required for highway construction and maintenance

practices, periodic and timely maintenance strategies are essential for efficient management system [13][18][19][20][21][22][23][24][25] [26].
The assessment of pavement performance usually employs different engineering techniques that require field data generated on-site using manual or automated methods, or obtained from database of pavement management agencies, or from experienced engineers working on pavement distresses [18][27]. According to [3] and [28], the investigation of pavement behaviour using insufficient or incomplete dataset necessitate basic assumptions and engineering judgement which may be misleading so often.
The performance of road pavement could be classified on a scale of good to worse. The classification is a function of the severity level, quantity and frequency of surface defects caused by deterioration of the pavement materials and its drainage condition [3][29][30]. The periodic deterioration and consequent failure of road pavement has become a serious concern to road management agencies, since it necessitate initial condition classification for accurate performance prediction over time and planning for repairs [29][31]. This therefore requires accurate condition classification and performance prediction for effective road management decision [12][13] [14] [25][32][33].
Following the rapid technological advancement in recent times, the use of intelligent algorithms known as Artificial Intelligence (AI) techniques for developing pavement management tools has become a common research methodology [4][17][18][25][34][35]. This intelligent approach employs methods of machine learning or data mining (knowledge discovery) for analysing system behaviour [31][36][37][38][39][40][41]. Data mining is a mathematical technique used for examining hidden relationship between variables in a dataset [18][42]. It is a machine learning technique which has gained acceptance from many researchers of different fields [43]. Examples of data mining methods include; Rough Set theory, Artificial Neural Network (ANN), Decision Tree, Support Vector Machine, Bayesian models (Naïve Bayes), etc. [18][44][45].
Data mining techniques are capable of analysing problems characterised by imprecise, uncertain or incomplete dataset and knowledge based elements with associated attributes to discover patterns and relationship between elements [18][41][44][46][47][48]. Other functions of the algorithms include to identify partial or total dependencies in a given dataset, eliminate redundant data, give approach to null values, missing data, dynamic data, etc. They are suitable for sourcing core information in a relatively poor dataset generated from system behaviour such as pavement condition for the purpose of performance prediction.
Before the adoption of any road pavement maintenance model or performance predictions are made, it is required that the initial surface condition be known [17][49][50]. Surface condition classifications are usually based on pavement surface distress indices such as; Condition Rating Survey (CRS), Pavement Condition Index (PCI) [25][30], International Roughness Index (IRI) [51], Condition Survey Rating Scores (CSRS) [3], etc. The measurement of these parameters requires the assessment of pavement surface distress attributes such as; cracks, potholes, rutting, roughness, drainage condition, etc. over time. In most instances, the major challenge associated with pavement management systems include data collection, availability and accuracy which necessitate adoption of historic or existing dataset which defines its behavioural attributes that were generated by individuals whose degree of accuracy of subjective measurements may be unknown or questionable [5][29]. In addition to challenges associated with data gathering and processing in developing countries like Nigeria, such data are prone to significant errors (noise and outliers) that may be hiding and misleading in the interpretations of performance attributes, hence requires processing using advanced analytical techniques for cleansing to help reduce margin of error from the subjective estimations on the field [5]. This therefore necessitate the use of intelligent algorithms of data mining techniques such as the Naïve Bayes classifier. It is an intelligent algorithms which is capable of handling the imprecise (vagueness) or inaccurate dataset on road pavement surface condition for better initial condition classification to pave the way for accurate prediction of future condition.
Though several data mining techniques such as Decision Trees, Support Vector Machine, Neural Networks and Meta-algorithms were employed for inverse analysis of pavement response to load effect in order to overcome limitations of the conventional backcalculation of analysing pavement moduli in

[44], techniques like the Random SubSpace, Rough Set theory, Pace Regression, etc. are also recommended for pavement analysis [18]. The choice of Naïve Bayes algorithms for this study was based on it relatively high sensitivity to missing data which affects estimation of prior probabilities during classification runs, hence results to accurate and reliable classification outputs [5].

1.1. Naïve Bayes Theorem

This algorithm is based on the Bayesian probability theory which operates on conditional rules [5][31]. It is an intuitive approach which uses conditional probabilities of prior events classified by attributes of a class label for making predictions or classifications of the targeted attributes. The Naïve Bayes algorithm is mostly used for predictive modelling using discrete events of a database without missing data to predict class labels [52]. The discretization of continoues dataset is required in order to suite the Naïve Bayes algorithm or classifer [53]. The model computes or predictes probability of an attribute belonging to a given class based on the assumption that the effect of the attribute is independent of the other attributes in the same class, hence is known as class conditional independence model [54]. According to [52] the Bayes theorem calculates the probability of a targeted attribute or classification by counting the frequency and combinations of values in the dataset, then estimating the parameter using method of maximum likelihood. The algorithm suites most complex real life problems. Also, its ability to estimate the parameter using small amount of training data is considered the major advantage of the Naïve Bayes algorithm. But its robustness is affected by noisy element in the dataset. The formate of dataset used for prediction and classification problems assums the formate shown in equation (1);

$$y = f(x_1,\ x_2,\ x_i, \dots) \qquad (1)$$

where y is the target or classification and x is an array of independnet attributes used for condition prediction or classifiaction. According to [55], the Naïve Bayes model is expressed as shown in equation (2);

$$P(y|x) = \frac{P(x|y).P(y)}{P(x)} \qquad (2)$$

where $P(y|x)$ is the posterior probability of class y (target) given the predictor x (attribute), $P(x)$ is the prior probability of predictor, $P(y)$ is the prior probability of class and $P(x|y)$ is the measure of maximum likelihood or probability of predicting a given class label.

Generally, previous studies on the use of probability theorems for pavement management system have recorded successes [56][57][58][59]. The use of Naïve Bayes algorithm as resported in [31], and [5] also confirm the assertions. Though its inability to handle missing and noisy data elements yields results with relatively significant margin of error [29][52][55][60], estimations of the Naïve Bayes approach are relatively more reliable.

Data classification using the Naïve Bayes theorem could be implemented in the Waikato Environment for Knowledge Analysis (WEKA) software [61][62]. It is a data mining toolkit developed at the Waikato University, New Zealand. It is an open source software written in Java programming language, used for research and project works [63]. The software is capable of carrying out data mining in the form of data pre-processing, classification, visualisation, association, clustering and filtering [64][65]. Algorithmic models used by WEKA for behavioural classification and predictions of systems behaviour include the Random Forest, Naïve Bayesian, Decision Tree (J48), etc.

The aim of this study therefore is to investigate surface condition classification of flexible road pavement using Naïve Bayes theorem. The objective of the study is to adopt historic dataset and apply the Naïve Bayes classifier for surface condition classifications based of surface distresses implemented using the WEKA software.

2. Methodology

2.1. Study Area

Nigeria is among the most populous countries on the Africa continent with an estimated human population of 180 million. Its land mass is estimated at 923.7768 square kilometres area [66], upon which her roads are constructed. Properties of some Federal Highway links in Nigeria considered by this study are as presented in Table 1.

Table 1. Details of Some Highway Links in Nigeria [3]

S/N	State	Link ID	Length (km)	No. of Sections	Age (Years)	Surface Finishing
1.	Anambra	89	43.7	106	6	Asphalt concrete
2.	Bauchi	285	51.3	77	8	Asphalt concrete
3.	Bendel	375	69.0	142	6	Asphalt concrete
4.	Borno	716	50.7	173	9	Surfaced dressed
5.	Imo	370	45.6	47	2	Asphalt concrete
6.	Kaduna	130	46.8	123	5	Asphalt concrete
7.	Kwara	5	22.8	74	9	Asphalt concrete
8.	Lagos	8	8.1	27	6	Asphalt concrete
9.	Ogun	17	18.3	61	12	Asphalt concrete
10.	Ogun	332	30.9	105	10	Asphalt concrete
11.	Oyo	22	23.4	56	10	Asphalt concrete
12.	Plateau	136	10.0	24	9	Asphalt concrete
13.	Plateau	138	23.2	59	5	Asphalt concrete
14.	Rivers	144	18.0	27	8	Asphalt concrete
15.	Sokoto	255	57.3	200	8	Asphalt concrete

2.2. Data Collection and Description

This study considered secondary sources of data on road pavement surface condition of the selected sites across Nigeria obtained from the database of pavement condition evaluation unit of the present Federal Ministry of Power, Works and Housing Nigeria [3].

The dataset on classified pavement surface condition using the Condition Survey Rating Scores (CSRS) were based on the Average Rut Depth (cm), percentage area of Alligator or Fatigue Cracking per segment and the Drainage condition.

Average Rut Depth: it was measured along the wheel path as the depth required to level up the existing pavement surface. It was measured every 20 m when exceeded 3 cm. One measurement for each 100 m was enough when the rutting was not severe (< 3cm). The measurement of rut depth was done using a long straight edge laid across the wheel path. Figure 1 presents pavement rutting and how to measure severity levels.

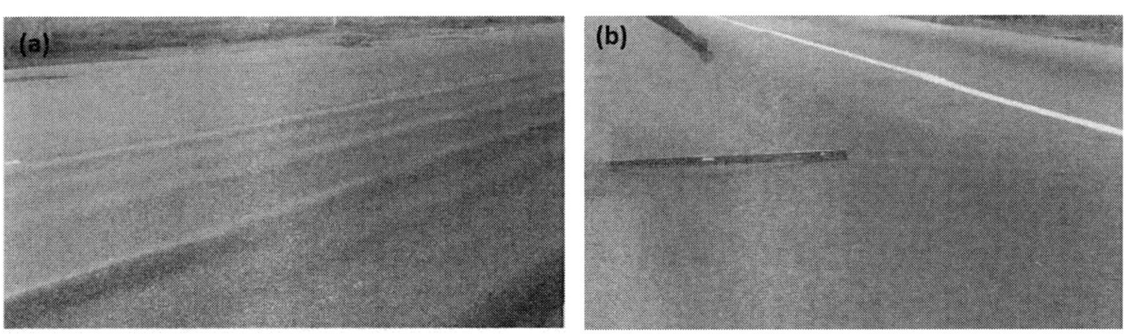

Figure 1. Pavement Rutting (a) Sample of Rutting (b) Measurement of Severity Level.

The average of both wheel path were determined for the same pavement length. The level of severity of measurements was determined using Table 2.

Table 2. Level of Severity of Rut Depth Classification

Level of Severity	Rut Depth (cm)
None	0 – 1
Slight	1 – 2
Moderate	2 – 3
Severe	> 3

Alligator or Fatigue Cracking: it was measured and recorded using the AASHO T-274-82 Road Test procedure. The cracking was divided into Class I, Class II and Class III. The AASHO Class I cracking was defined as longitudinal cracks in the wheel path which were generally wavy and in places they crossed and crisscrossed. They were usually numerous separate cracks of 2 – 3m length rather than one single long crack. The AASHO Class II cracking were those that had progressed and interconnected to form chicken wire or alligator skin pattern with numerous small paragon and octagon shapes pieces. At this stage, the pieces were still together and disintegration had not begun. The AASHO Class III cracking were the progression of Class II where cracks were spalled more separately at the edges, the small pieces lost integrity between the blocks, and the segments of the pavement surface loosen, moved or rock under traffic. Some pieces may have moved apart or miss out. Figure 2 presents a typical view of alligator cracks.

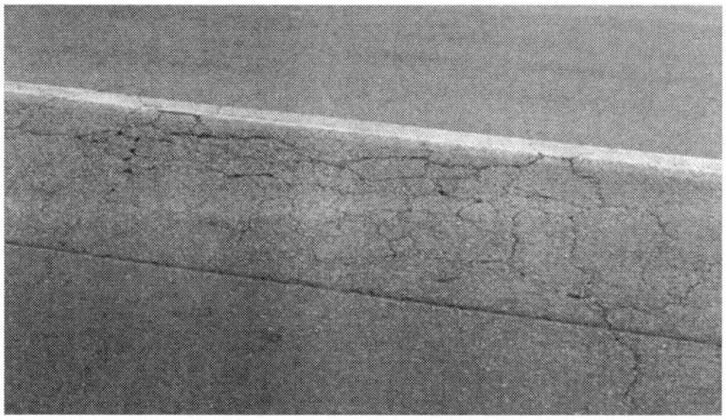

Figure 2. Alligator Cracks

The percentage area of total cracking for each class per road segment was then computed using Equation (3);

$$Density\ of\ Section\ (\%) = \frac{Total\ Affected\ Area\ Per\ Section}{Total\ Area\ of\ Road\ Section} \times 100 \qquad - \quad - \quad - \qquad (3)$$

The severity levels were classified using the scale shown in Table 3;

Table 3. Level of Severity of AASHO Classification of Cracks [3]

AASHO Classes	Level of Severity	Density of Section Area (%)
None	None	1 – 10
Class I	Slight	11 – 26
Class II	Moderate	26 – 50
Class III	Severe	> 50

Drainage condition: this was classified based on the following descriptions;
Very Good: a drainage section characterised with well drained side slopes, high embankments and good runoff ditches.

Good: a drainage section characterised with moderate side slopes and ditches, and water probably runs off reasonably well to the surrounding streams and rivers.
Adequate: this drainage condition is characterised with some side slope and ditches of moderate size, however, there are evidence that the rate of water runoff is less than desired and some water stand along the roadway. The roadway section may be close to level with the surrounding terrain.
Poor: this is when the drainage is characterised with very little side slope and poor ditches. It may be slightly lower than the surrounding terrain; water probably stands and the rate of runoff is very slow.
Very Poor: this is a bathtub type of section which is clearly lower than the surrounding terrain. Some side slopes and ditches may be present, but the water is not expected to be carried away and most likely collects and slowly percolates into the sub-base and subgrade surrounding soils of the section.

The classification of CSRS employed the scale presented in Table 4;

Table 4. Classification of Pavement Surface Condition [3]

Classification of Pavement Condition	Limits of CSRS
Excellent	> 81
Good	66 - 81
Fair	46 – 65
Poor	25 – 45
Very poor	< 25

2.3. Implementation of Naïve Bayes Theorem using WEKA Software

This involved using Naïve Bayes classification algorithms or classifier of WEKA software. It required data preparation in Microsoft Excel worksheet in form of column-separated value (.csv) file which was imported into WEKA explorer as an Attribute-Relation File Format (ARFF). The test or observed data file saved in column-separated value (.csv) file format contained values for the independent attributes responsible for pavement condition which included percentage creaking, rut depth and drainage condition entered in the spreadsheet with rows representing instances, columns for attributes per segment for the classification of initial pavement condition as the dependent (target) attribute on a Condition Survey Rating Scores (CSRS) scale translated into Excellent, Good, Fair, Poor and Very Poor according to thresholds proposed by the study. The output of analysis using WEKA software presents confusion matrices with perfect diagonal to fit the targeted attributes into classes, while erroneous classifications were caused by incorrect classification which were estimated using statistical errors.

3. Results and Discussion

The distribution of classified flexible road pavement surface condition of the investigated Federal Highway links in Nigeria using Naïve Bayes classifier is as presented in Figure 3;

IOP Conf. Series: Materials Science and Engineering 1036 (2021) 012036 doi:10.1088/1757-899X/1036/1/012036

```
=== Confusion Matrix ===

  a  b  c   <-- classified as
 10  2  0 |  a = Good
  6 87  0 |  b = Excellent
  0  0  1 |  c = Fair
```

(a) Link 89 in Anambra State

```
=== Confusion Matrix ===

 a  b   <-- classified as
 7  0 |  a = Good
 2 68 |  b = Excellent
```

(b) Link 285 in Bauchi State

```
=== Confusion Matrix ===

  a   b   <-- classified as
 23   0 |  a = Good
  1 118 |  b = Excellent
```

(c) Link 375 in Bendel State

```
=== Confusion Matrix ===

  a  b  c  d  e   <-- classified as
 43  2  0  6  0 |  a = Fair
  2 27  0  0  6 |  b = Poor
  0  0 34  1  0 |  c = Excellent
  0  0  1 41  0 |  d = Good
  0  1  0  0  9 |  e = Very Poor
```

(d) Link 716 in Borno State

```
=== Confusion Matrix ===

  a  b  c   <-- classified as
 21  0  0 |  a = Excellent
  0 16  0 |  b = Good
  0  2  8 |  c = Fair
```

(e) Link 370 in Imo State

```
=== Confusion Matrix ===

  a   b   c   <-- classified as
  1   0   0 |  a = Fair
  0  12   0 |  b = Good
  0   4 106 |  c = Excellent
```

(f) Link 130 in Kaduna State

```
=== Confusion Matrix ===

  a  b   <-- classified as
 68  1 |  a = Excellent
  0  5 |  b = Good
```

(g) Link 5 in Kwara State

```
== Confusion Matrix ===

  a  b  c  d   <-- classified as
 12  0  0  0 |  a = Excellent
  0  4  0  0 |  b = Good
  0  0 10  0 |  c = Fair
  0  0  0  1 |  d = Poor
```

(h) Link 8 in Lagos State

```
=== Confusion Matrix ===

  a  b  c  d   <-- classified as
 25  4  4  0 |  a = Good
  2 14  0  0 |  b = Fair
  0  0 10  0 |  c = Excellent
  0  0  0  2 |  d = Poor
```

(i) Link 17 in Ogun State

```
=== Confusion Matrix ===

 a  b   <-- classified as
 6  0 |  a = Good
 1 98 |  b = Excellent
```

(j) Link 332 in Ogun State

```
=== Confusion Matrix ===

  a  b  c   <-- classified as
 27  2  0 |  a = Good
  1 25  0 |  b = Excellent
  0  0  1 |  c = Fair
```

(k) Link 22 in Oyo State

```
=== Confusion Matrix ===

a b c   <-- classified as
7 1 1 | a = Good
0 6 0 | b = Fair
1 0 9 | c = Excellent
```

(l) Link 136 in Plateau State

```
=== Confusion Matrix ===

  a  b   <-- classified as
 53  4 |  a = Excellent
  0  3 |  b = Good
```

(m)Link 138 in Plateau State

```
=== Confusion Matrix ===

  a  b  c   <-- classified as
 20  0  0 |  a = Good
  0  4  0 |  b = Excellent
  0  0  3 |  c = Fair
```

(n) Link 144 in Rivers State

```
=== Confusion Matrix ===

   a   b   <-- classified as
 184   2 |   a = Excellent
   1  13 |   b = Good
```

(o) Link 255 in Sokoto State

Figure 3. Confusion Matrices of Pavement Surface Condition Classifications

Figure 3 presents confusion matrices of classifications based on attributes into Excellent, Good, Fair, Poor and Very Poor as the case may be. The matrices showed the certainty or accuracy of classifications based on the diagonal entries, where a perfect diagonal matrix as shown in Figures 3 (h) and (n) depicted perfect road pavement surface condition classification, while classifications in Figures 3 (a), (b), (c), (d), (e), (f), (g), (i), (j), (k), (l), (m) and (o) showed erroneous classifications indicated by the dispersed entries outside the diagonal forming the upper and lower triangles of the matrices. The imperfect classifications were attributed to missing or incomplete data and noisy or outlier entries in the dataset [5][43]. The numerical performance of the model was also evaluated using Figures 4;

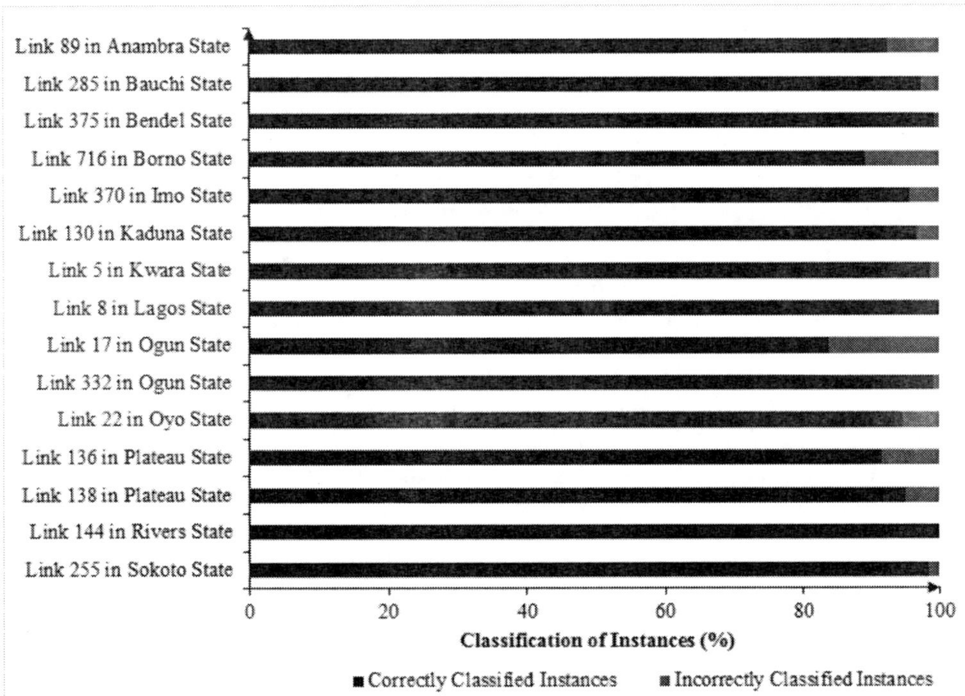

Figure 4. Correctness of Classifications

Figure 4 revealed the correctness of classifications by the Naïve Bayes model, where each classified link had over 90% correctly classified instances except links 89, 716, 17 and 138 in Anambra, Borno, Ogun and Plateau states respectively. This was attributed to the significant deviation of pattern of attributes caused by missing and noisy elements in the dataset [5][52]. The classification was further verified using estimated errors by the model as shown in Figure 5;

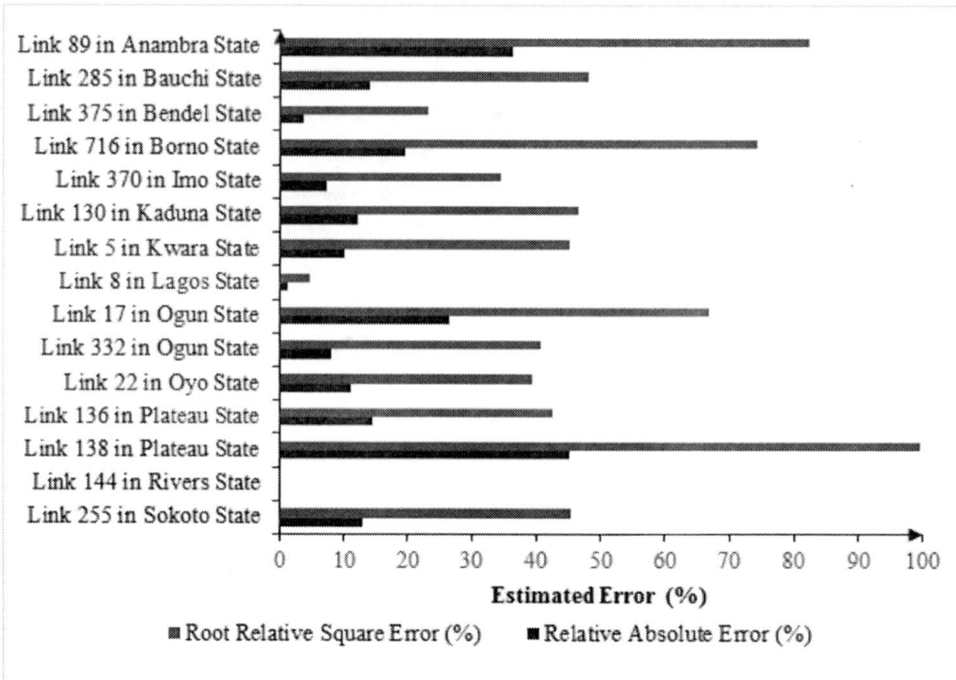

Figure 5. Estimated Errors of Classifications

Figure 5 revealed that the computed Root Relative Square Error (RRSE) for links 89, 716, 17 and 138 in Anambra, Borno, Ogun and Plateau states respectively were relatively high compared to the other sampled links, with corresponding increase in the estimated Relative Absolute Error (RAE). The systematic increase in RRSE justified the relationship between individual entries on a sampled link, while the RAE defined abnormalities in the data entry [39][43][64]. Other strong statistical variable used for measuring the degree of accuracy was the Kappa statistics, which measures between 0 – 1 for least to high accuracy respectively. Figure 6 presents the distribution of Kappa statistics of the sampled links.

IOP Conf. Series: Materials Science and Engineering 1036 (2021) 012036 doi:10.1088/1757-899X/1036/1/012036

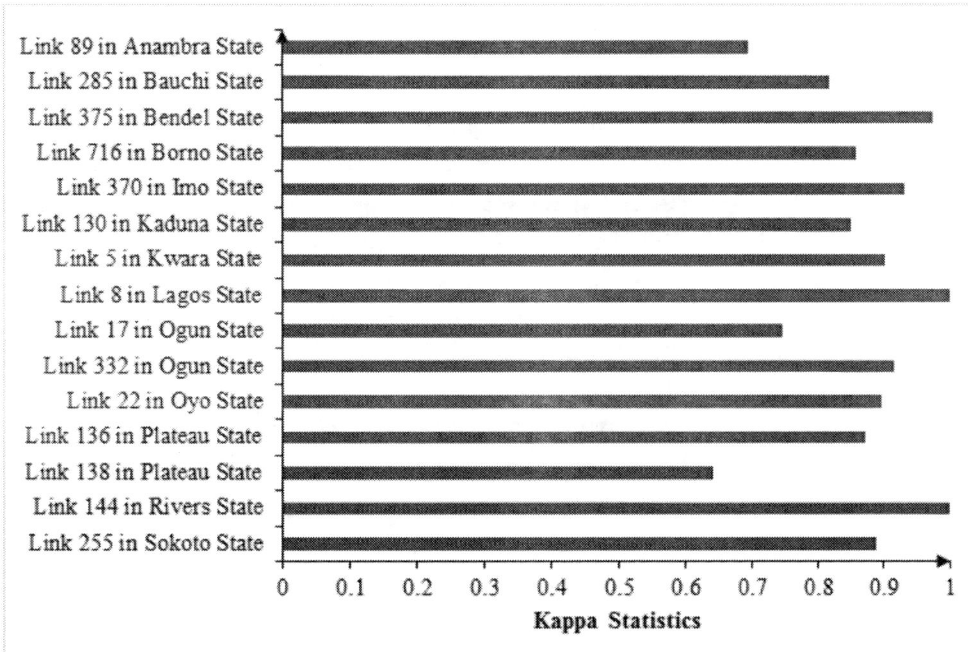

Figure 6. Accuracy of Classifications

Figure 6 revealed that surface condition classifications for links 375, 8 and 144 had relatively high degree of accuracy as compared to links 89, 716, 17 and 138. This trend agreed with the earlier assertion that, missing data and noise elements in the dataset had affected the accuracy of some classification significantly. The analysis of link statuses which grouped classified links as Excellent and Good into 'Worthy' and Fair, Poor an Very Poor into 'Unworthy' statuses indicated that the entire road surface conditions is descripted using Figure 7;

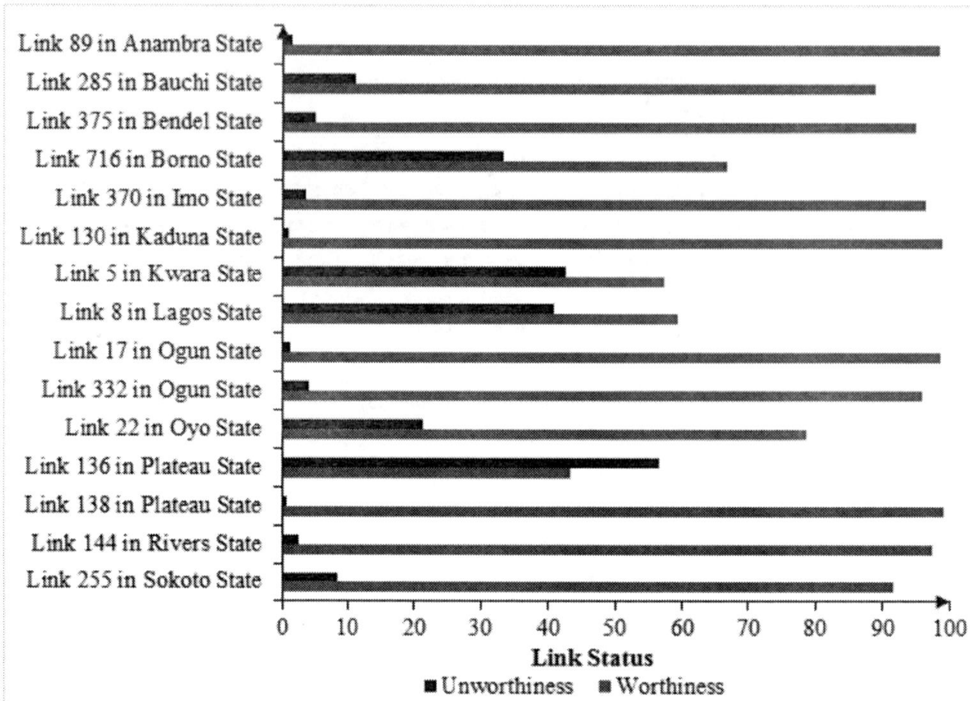

Figure 7. Statuses of Classified Links

Figure 7 showed that links 716, 5, 8, 22 and 136 in Borno, Kwara, Lagos, Oyo and Plateau states respectively had relatively high level of unworthiness for surface condition, which increases safety risk, vehicle wear and level of pollution, hence call for urgent repair actions. On the other hand, links 89, 375, 370, 130, 17, 332, 138, 144 and 255 for Anambra, Bendel, Imo, Kaduna, Ogun, Plateau, Rivers and Sokoto states respectively had worthy surface condition, were safe for driving and do not required major maintenance process as at the time of the field investigation.

4. Conclusion

The study carried out surface condition classification of flexible road pavement using the Naïve Bayes classifier. Surface condition attributes used for the classification included, percentage of fatigue cracks, average rut depth and drainage condition to estimate the Condition Survey Rating Scale (CSRS) of the road pavement segments. Fifteen (15) Federal Highway links across Nigeria were considered for the analysis. The high sensitivity of Naïve Bayes classifier yielded high degree of classification accuracy. Results of the analysis indicated that; links 716, 5, 8, 22 and 136 in Borno, Kwara, Lagos, Oyo and Plateau states respectively had relatively high level of unworthy surface condition for safe travel, which also increased risk and cost of plying the highways, hence call for immediate maintenance and rehabilitations actions. While links 89, 375, 370, 130, 17, 332, 138, 144 and 255 from Anambra, Bendel, Imo, Kaduna, Ogun, Plateau, Rivers and Sokoto states respectively had worthy surface condition for safe driving and do not required immediate maintenance services as at the time of the field survey.

References

[1]. Mallick R B and El-Korchi T 2009 *Pavement engineering principles and practice*, Taylor and Francis Group, USA.
[2]. Taylor M A P and Philip M L 2015 Investigating the impact of maintenance regimes on the design life of road pavements in a changing climate and the implications for transport policy. *Transp. Policy*, **41**, pp 117 – 35.

[3]. Claros G, Carmichael R F and Harvey J 1986 *Development of pavement evaluation unit and rehabilitation procedure for overlay design method.* Lagos: Texas research and development foundation for the Nigeria Federal Ministry of Works and Housing, Nigeria.

[4]. Fwa T F, Tan C Y and Chan W T 1994 Road-maintenance planning using genetic algorithms II: analysis, *J. of Transp. Eng.*, **120 (5),** pp 710 – 22.

[5]. Inkoom S, Sobanjo J, Barbu A and Niu X 2019 Pavement crack rating using machine learning frameworks: partitioning, bootstrap forest, boosted tree, Naïve Bayes, and K-Nearest neighbors, *J. of Transp. Eng., Part B: Pavement,* **145 (3),** pp 1 – 12.

[6]. Prozzi J A 2001 *Modeling Pavement performance by combining field and experimental data,* PhD Thesis at Civil and Environmental Engineering, University of California Berkeley, USA.

[7]. Jeong H, Kim H, Kim K and Kim H 2017 Prediction of flexible pavement deterioration in relation to climate change using fuzzy logic. *J. of Infrastructure System,* American Society of Civil Engineers (ASCE), **23(4),** pp 1 - 11.

[8]. Alharbi F 2018 *Predicting pavement performance utilizing Artificial Neural Network (ANN) Models*, PhD Thesis, Iowa State University capstones, Iowa, Ames, USA.

[9]. Najafi S 2015 *Pavement Friction Management (PFM) – A step towards zero fatalities*, PhD thesis, Faculty of the Virginia Polytechnic Institute and State University, Virginia, USA.

[10]. Kirbas U and Karasahin M 2016 Performance models of hot mix asphalt pavement in urban roads, *Const. and Build. Materials*, **116**, pp 281 – 88.

[11]. Adeke P T, Atoo A A and Joel E 2018a A policy framework for efficient and sustainable road transport system to boost synergy between urban and rural settlements in developing countries: a case of Nigeria, *1st International Civil Eng. Conf. (ICEC 2018),* Department of Civil Engineering, Federal University of Technology, Minna, Nigeria, **1(1),** pp 22 - 30.

[12]. Yin H 2007 *Integrating instrumentation data in probabilistic performance prediction of flexible pavement*, PhD Thesis in the Department of Civil and Environmental Engineering, Graduate School, the Pennsylvania State University, USA.

[13]. Ayed A 2016 *Development of empirical and mechanistic empirical performance models at project and network levels.* PhD Thesis, Department of Civil Engineering, University of Waterloo, Canada.

[14]. Ziliute L, Motiejunas A, Kleiziene R, Bribulis G and Kravcovas I 2016 *Temperature and moisture variation in pavement structures of the test road*, 6th Transport Research Arena, Elsevier, London.

[15]. Murana A A 2016 *Characterisation of subgrade materials from some Nigerian sources for use in the Nigeria empirical-mechanistic pavement analysis and design system*, PhD Thesis, Department of Civil Engineering, Faculty of Engineering, Ahmadu Bello Universiy Zaria, Nigeria.

[16]. Pantuso A, Flintsch G W, Katicha S W and Loprencipe G 2019 Development of network-level pavement deterioration curves using the linear empirical Bayes approach, *Int. J. of Pavement Eng.,* pp 1-14

[17]. Marcelino P, Antunes M, de. L, Fortunato E and Gomes M C 2019 Machine learning approach for pavement performance prediction. *Int. J. of Pavement Eng.*, pp 1 – 14.

[18]. Miradi M 2009 *Knowledge discovery and pavement performance: intelligent data mining,* PhD Thesis submitted to the Section of Road and Railway Engineering, Faculty of Civil Engineering and Geosciences, Delft University of Technology, The Netherlands.

[19]. Abiola O S, Owolabi A O, Odunfa S O and Olusola A 2010 Investigation into causes of premature failure of highway pavements in Nigeria and remedies. *In Proc. of the Nigeria Institution of Civil Engineers (NICE) Conference.* Abuja, Nigeria.

[20]. Chikezie C U, Olowosulu A T, Abejide O S and Kolo B A 2011 Review of application of genetic algorithms in optimization of flexible pavement maintenance and rehabilitation in Nigeria. *World J. of Eng. and Pure and Applied Sc.,* **1(3),** pp 68 – 76.

[21]. Owalabi A and Abiola O S 2011 Development of priority index assessment model for road pavements in Nigeria, *8th Int. Conf. on Managing Pavement Assets,* Abeokuta, Nigeria.
[22]. Murana A A, Olowosulu A T and Otuoze H S 2012 Minimum threshold of Monte Carlo cycles for Nigerian empirical-mechanistic pavement analysis and design system. *Nigerian J. of Tech,* **31(3),** pp 321- 28.
[23]. TAC 2013 *Pavement asset design and management guide,* Ottawa; Transport Association of Canada.
[24]. Adefemi B A and Ibrahim A A 2015 Flexible pavement assessment of selected highways in Ifelodun local government, Ikirun-osun, South – Western Nigeria. *Int. J. of Eng. and Tech.,* **5(8),** pp 475 – 84.
[25]. Mahmood M S 2015 *Network-level maintenance decisions for flexible pavement using a soft computing-based framework.* PhD Thesis, Nottingham Trent University, United Kingdom.
[26]. Huang Y 2017 *Evaluating pavement response and performance with different simulative tests,* PhD Thesis, Virginia Polytechnic Institute and State University, Virginia.
[27]. Saltan M, Terzi S and Kucuksille E U 2011 Backcalculation of pavement layer moduli and Poisson's ratio using data mining. *Expert Systems with Applications*, **38,** pp 2600–08.
[28]. Martin T C 1996 A review of existing pavement performance relationships. *ARRB Transport Research Ltd. Research Report No. 282,* Washington, D.C.
[29]. Shekharan A R 1998 Effect of noisy data on pavement performance prediction by artificial neural networks. *Transportation Research Record 1643,* Washington, D.C., pp 7-13.
[30]. American Society for Testing and Materials (ASTM). 2007 *Standard practice for road and parking lots pavement condition index survey*. D6433-07. Philadelphia.
[31]. Marianingsih S, Utaminingrum F and Bachtiar F A 2019 Road surface types classification using combined of K-Nearest neighbor and Naïve Bayes based on GLCM, *Int. J. of Advances in Soft Comp. and its Applications,* **11(2),** pp 15 – 27.
[32]. Adeke P T, Atoo A A and Orga S G 2018b Assessment of pavement condition index: a case of flexible road pavements on the University of Agriculture Makurdi campus. *Nigerian J. of Tech.* **38(1),** pp 15 – 21.
[33]. Pan Y, Zhang X, Cervone G and Yang L 2018 Detection of asphalt pavement potholes and cracks based on the unmanned aerial vehicle multispectral imagery, *IEEE J. of Selected Topics in Applied Earth Observations and Remote Sensing*, pp 1 – 12.
[34]. Fwa T F and Shanmugam R 1998 Fuzzy logic technique for pavement condition rating and maintenance-needs assessment, *4th Int. Conf. on Managing Pavements, Durban, South Africa*, pp. 465 - 476.
[35]. Hamed R I and Kakarash Z A 2016 Evaluate the asphalt pavement performance of rut depth based on intelligent method. *Int. J. of Eng. and Comp. Sc.,* **5(1),** pp 15474 – 81.
[36]. Munakata T 2008 *Fundamentals of new artificial intelligence, neural evolutionary, fuzzy and more.* 2nd Edition, Computer and Information Science Department Cleveland State University, USA.
[37]. Cubero-Fernandez, A., Rodriguez, F. J., Villatoro, R., Olivares, J. and Palomares, J. M. (2017). Efficient pavement crack detection and classification, EURASIP *J. on Image and Video Processing*, **39,** pp 2 -11.
[38]. Russell S J and Norvig P 2010. *Artificial Intelligence. A modern Approach*. 3rd Edition. Pearson Education, Inc., London.
[39]. Witten I H, Frank E, Hall M A and Pal C J 2016 *The WEKA workbench – data mining practical machine learning tools and techniques,* 4th edition, Morgan Kaufmann, Elsevier, London.
[40]. Fox C 2018 *Data science for transport; self-study guide with computer exercises*. Springer Int. Publishing, Switzerland.
[41]. Arifuzzaman M, Gazder U, Alam M S and Sirin O 2019 Modelling of asphalt's adhesive behaviour using classification and regression tree (cart) analysis. *Computational Intelligence and Neuroscience,* ID 3183050, Hindawi.

[42]. Smadi O G 2000 *Knowledge based expert system pavement management optimisation*, paper based on PhD Dissertation, Iowa State University, Ames, Iowa,United States.
[43]. Witten I H and Frank E 2005 *Data mining – practical machine learning tools and techniques,* 2nd Edition, Morgan Kaufmann, Elsevier, London.
[44]. Gopalakrishnan K, Ceylan H and Attoh-Okine N O 2009 *Intelligent and soft computing in infrastructure systems engineering,* Recent Advances, Springer, Berlin.
[45]. Gopalakrishnan K, Agrawal A, Ceylan H, Kim S and Choudhary A 2013 Knowledge discovery and data mining in pavement inverse analysis, *Transport,* **28(1),** pp 1-10.
[46]. Bal M 2013 Rough sets theory as symbolic data mining method: an application on complete decision table. Information Science Letters, *an Int. J.,* **2(1),** pp 35-47.
[47]. Pawlak, Z. 1982 Rough sets. *Int. J. of Computer and Information Sc.*, **11**, pp 341- 3.
[48]. Pawlak, Z. 2002 Rough Sets and Intelligent Data Analysis. *Information Sc.,* **147**, pp 1 – 12.
[49]. Mahmood M, Rahman M, Nolle L and Mathavan S 2013 A fuzzy logic approach for pavement section classification. *Int. J. of Pavement Research and Tech.,* Chinese Society of Pavement Engineering, **6(5)**, pp 620–26.
[50]. Mahmood M, Rahmood M, Mathavan S and Nolle L 2015 Pavement management: data centric rules and uncertainty management in section classification by a fuzzy inference system, *Bituminous Mixtures and Pavement,* **6**, pp 533 – 41.
[51]. Arhin S A, Williams L N, Ribbiso A and Anderson M F 2015 Predicting pavement condition index using international roughness index in a Danse urban area, *J. of Civil Eng. Research,* **5(1),** pp 10 – 17.
[52]. Alam F and Pachauri S 2017 Comparative study of j48, naïve bayes and one-r classification techniques for credit card fraud detection using WEKA, *Advances in Computational Sc. and Tech.*, **10(6),** pp 1731 – 43.
[53]. Cigsar B and Unal D 2019 Comparison of data mining classification algorithms determining the default risk, *Hindawi Scientific Programming*, **2019**, ID 8706505.
[54]. Tribhuvan A P, Tribhuvan P P and Gade J G 2015 Applying Naïve Bayesian classifier for predicting performance of a student using WEKA. *Advances in Computational Research*, **7(1),** pp 239 – 42.
[55]. Jang W, Lee J K, Lee J and Han S H 2015 Naïve Bayesian classifier for selecting good/bad projects during the early stage of international construction bidding decisions, *Mathematical Problems in Eng.,* **2015**, pp 1 – 12.
[56]. Li, Ningyuan Xie, Wei-chau, Haas, Ralph 1996 reliability– based processing of Markov chains for modeling pavement network deterioration transportation research record: *J. of the Transportation Research Board*, (1524), Transportation Research Board of the National Academies, Washington, D.C. pp 203–213.
[57]. Rose S, Mathew B S and Isaac K P 2017 Development of probabilistic deterioration models and prioritisation of low volume roads for maintenance, *Int. J for Traffic and Transport Eng.,* **7(2),** pp 216 – 31.
[58]. Surendrakuma K, Prashant N and Mayuresh P 2013 Application of Markovian probabilistic process to develop a decision support system for Pavement maintenance management. *Int. J. of Sc. and Tech. Research*, **2(8),** pp 295 – 303.
[59]. Mandiartha P, Duffield C, Thompson R and Wigan M 2012 A stochastic-based performance prediction model for road network pavement maintenance, *Road and Transport Research, ARRB Transport Research Ltd.,* **21(3),** pp 125 – 36.
[60]. Gong H, Sun Y, Shu X and Huang B 2018 Use of random forest regression for predicting IRI of asphalt pavements, *Construction and Building Materials*, **189,** pp 890 – 97.
[61]. Thakur R and Mahajan A R 2015 Pre-processing and classification of data analysis in institutional system using WEKA, *Int. J. of Comp. Applications,* **112(6),** pp 9 – 11.

[62]. Luca M D, Abbondati F, Pirozzi M and Zilioniene D 2016 Preliminary study on runway pavement friction decay using data mining. *6th Transp. Research Area April 18 – 21*, 2016. **14**, pp 3751 – 60.

[63]. Sharma T C and Jain M 2013 WEKA Approach for comparative study of classification algorithm. *Int. J. of Adv. Research in Comp. and Communication Eng.*, **2(4),** pp 1925 – 1931.

[64]. WEKA 2018 *Waikato Environment for Knowledge Analysis: User's Manual*, The University of Waikato, New Zealand.

[65]. Hutter F, Kotthoff L and Vanschoren J 2019 *Automated machine learning – methods, systems, challengers,* The Springer series on challengers in machine learning. Springer, Switzerland.

[66]. Federal Government of Nigeria 2012 *Public private partnership manual for Nigeria*, infrastructure concession regulatory commission, Abuja, Nigeria.

Model uncertainty in concrete breakout resistance of single anchors in shear – assessment of ACI 318 against available alternative

O B Olalusi[1*] and M Akinlolu[1]

[1]Department of Civil Engineering, University of KwaZulu-Natal, Durban, South Africa
*dimejibenlusy@yahoo.co.uk, olalusio@ukzn.ac.za

Abstract. This contribution characterises the model uncertainty related to the shear breakout resistance of anchors located towards the edge of concrete members, considering experiments obtained from previous anchor test. The characterisation includes the computation of the sample moments (mean, standard deviation and skewness) required in reliability analysis. The study analysed the predictive models of current design guidelines (ACI 318 and the modified concrete capacity design method [MCCD] method). The assessment demonstrates that ACI 318 has a higher bias and variability compared to the MCCD method. The influence of the breakout resistance parameters on model uncertainty was evaluated, and significant parameters were identified. The edge distance and the anchor diameter are found to be the most influencing parameters. Also, a suitable theoretical probability distribution is required in reliability analysis for the model uncertainty random variable. Accordingly, it is shown that the log-normal distribution is the most appropriate probability distribution for the derived model uncertainty factors.
Keywords: Model uncertainty, Fasteners, Anchors, Modified CCD, ACI 318

1. Introduction

Fastening/anchorage technology is becoming increasingly important in the construction industry. The increased attention given to the use of fasteners in concrete is motivated by the growing interest in strengthening of existing structures, earthquake retrofitting, preservation of the historic built environment and the use of prefabricated structural elements [1]. Significant efforts are currently invested towards establishing a unified design framework concerning the design of anchors for use in concrete. The ACI 318 [3] and the EN 1992-4 [2] are the most recent and globally accepted design guidelines. Anchors loaded in shear may fail by steel rupture, pry-out failure, pull-out of the anchor, crushing of concrete and concrete breakout. The latter failure mode is investigated in this study.

Concrete related failures pose a critical safety issue since they are quasi-brittle, and as such, may develop swiftly without preceding signs of damage. There is a limited possibility for remediation action and avoidance of complete system failure. Hence, fastening design provisions should be reliable and accurate. This consequently leads to a particular interest in the adequate assessment of the reliability of the current fastening design provisions in the case of concrete breakout failure. To that end, proper characterisation and description of model uncertainty statistics is a necessary prerequisite, since these are required as input in reliability analysis [4]. The presented contribution focuses on the characterisation of model uncertainties in the breakout resistance prediction of shear-loaded anchors, towards the reliability-based design of fastenings to concrete. The uncertainty related to the ACI 318 model [3] is evaluated and compared to that of the modified concrete capacity design (MCCD) method [5].

2. Model uncertainty

Model uncertainty can be understood as a description of the model's deficiency in accurately describing a physical phenomenon (in the present case, the anchor's resistance) due to lack of knowledge in the problem description, or mathematical simplifications [4]. In the realm of structural reliability, it is expressed as a random variable accounting for the abovementioned effects. The model uncertainties related to the ACI 318 and MCCD concrete-edge breakout models are investigated in this section, using the experimental database of 257 anchor tests presented in Section 4. The model uncertainty associated with a single experiment x is obtained as the ratio of the experimental $V_{exp,}$ to predicted shear capacity V_{pred}, expressed in Equation 1 [4]. Apparently, model uncertainty $R_x = 1$ is a condition for an ideal model, $R_x > 1$ implies that the model underpredicts, and $R_x < 1$ that it overpredicts the actual concrete breakout resistance [6].

$$R_x = \frac{V_{exp,x}}{V_{pred,x}(X)} \tag{1}$$

where, $V_{exp,x}$ is the failure load for an anchor test x in shear. X denotes the input variable. $V_{pred,x}$ is the predicted mean breakout resistance for the same anchor test x.

3. Methods for predicting the breakout capacity of anchors in shear

The mean shear capacity of single anchors located towards the edge of non-cracked concrete members according to ACI 318 and MCCD method is obtained based on Equation 2 [7, 8] and Equation 3 respectively [5].

$$V_{pred}(X) = V_{ACI} = \min\left\{\left(\frac{l_f}{d_o}\right)^{0.2} \cdot \sqrt{d_o}\sqrt{f_{cm}}(c_1)^{1.5}, 7.1.\sqrt{f_{cm}}(c_1)^{1.5}\right\}. \tag{2}$$

$$V_{pred}(X) = V_{MCCD} = 3\,.\,d_o^{\alpha}.\,l_f^{\beta}.\sqrt{f_{cm}}.\,c_1^{1.5} \tag{3}$$

where

$$\alpha = 0.1 \cdot \left(\frac{l_f}{c_1}\right)$$

$$\beta = 0.1 \cdot \left(\frac{d_o}{c_1}\right)^{0.2}$$

where d_o is the diameter of the anchor. f_{cm} is the mean concrete strength. l_f is the influence length. c_1 is the edge distance

4. Characteristics of the experimental database of anchors tests

The anchor database considered in this study was assembled from previous studies published in the literature [9-15]. It contains the observed experimental failure load from 257 anchor tests for different types of shear loaded single anchors in uncracked concrete. The database captured concrete strength f_{cm} in the range of 29 – 39 MPa, anchor diameters d_o in the range of 8 – 40 mm and influence length l_f in the range of 40 – 450 mm (Table 1). All anchors in the database failed due to concrete edge breakout. More elaborate details of the database of the anchor test can be found in [6].

Table 1. Range of design parameters for the database.

Parameters	Range				
	Minimum	First Quartile (P25)	Median (P50)	Third Quartile (P75)	Maximum
Concrete cylinder compressive strength f_{cm} (MPa)	29	30.3	32.7	33.6	39.5
Influence length l_f (mm)	50	110	163	313	450
Diameter of anchor d_o (mm)	8	16	24	25	40
Concrete edge distance c_1 (mm)	50	72	100	160	320
Shear breakout capacity V (kN)	8.2	19.1	35	66.9	169

5. Statistical analysis of the model uncertainty

For each experiment, the best estimate/mean resistance is evaluated using Equation 2 and 3. The comparison of the models to experimental results, as shown by Figure 1 indicate the difficulty of the accurate prediction of the breakout capacity for anchors in shear.

The resistance predictions of the different methods vary from the experimentally observed concrete breakout strength. Given the variations, quantification of model uncertainty for the various models is required. The model uncertainty is evaluated from Equation 1 and the main statistical properties such as the mean (μ_R), the standard deviation (dispersion measures) (σ_R), coefficient of variation (Ω_R) and skewness (η_R) are reported in Table 2. As shown in the table, the models generally overestimate the actual breakout resistance. The skewness coefficients of the models are nearer to what is anticipated from a positively skewed distribution such as log-normal distribution.

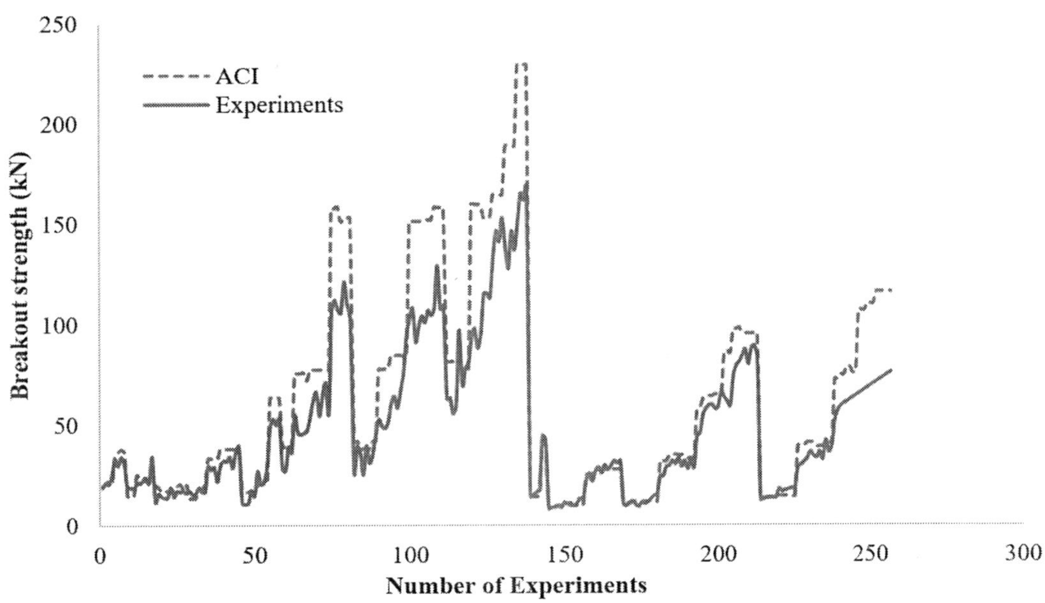

(a)

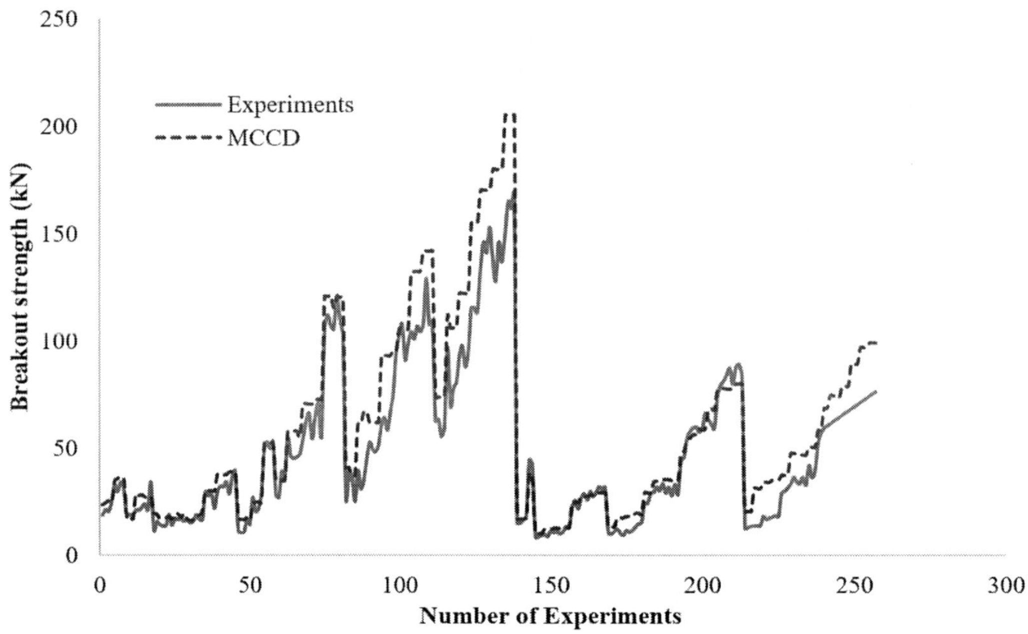

(b)

Figure 1. Comparison of resistance predictions from (a) ACI 318 model and (b) MCCD model to experimental values

Table 2. Statistical properties of model uncertainty

	ACI 318	**MCCD**
Model Uncertainty symbol	R_{ACI}	R_{MCCD}
Mean (μ_R)	0.87	0.93
Standard Deviation (σ_R)	0.18	0.13
Coefficient of Variation	0.21	0.14
Skewness (η_R)	0.65	0.62
Minimum (R_{min})	0.55	0.59
Maximum (R_{max})	1.51	1.47

The trend in the performance of the various models is investigated by correlating the model uncertainty with model parameters (such as f_{cm}, c_1, d_o, l_f) [4]. The correlation coefficients for the relevant parameters are presented in Table 3. The edge distance and the anchor diameter are found to be the most influencing basic variables as indicated by the correlation coefficients.

Table 3. Linear correlation coefficient

	R_{ACI}	R_{MCCD}
d_o $[mm]$	-0.49	-0.19
l_f $[mm]$	0.04	0.18
c_1 $[mm]$	-0.62	-0.25
f_{cm} [MPa]	0.23	0.19

5.1 Probability distribution function

The type of probability distribution has been shown to play a key role in reliability performance assessment [16]. The choice of the appropriate distribution for the model uncertainty random variable was evaluated based on the chi-square goodness-of-fit test. The magnitude of the p-value obtained suggests that the log-normal distribution is a better fit with a p-value higher than 0.05. The Joint Committee on Structural Safety (JCSS) [17] also recommended the log-normal distribution as the most appropriate distribution function for model uncertainty. The histograms and cumulative distribution plots for the model uncertainty variables are presented in Figure 2.

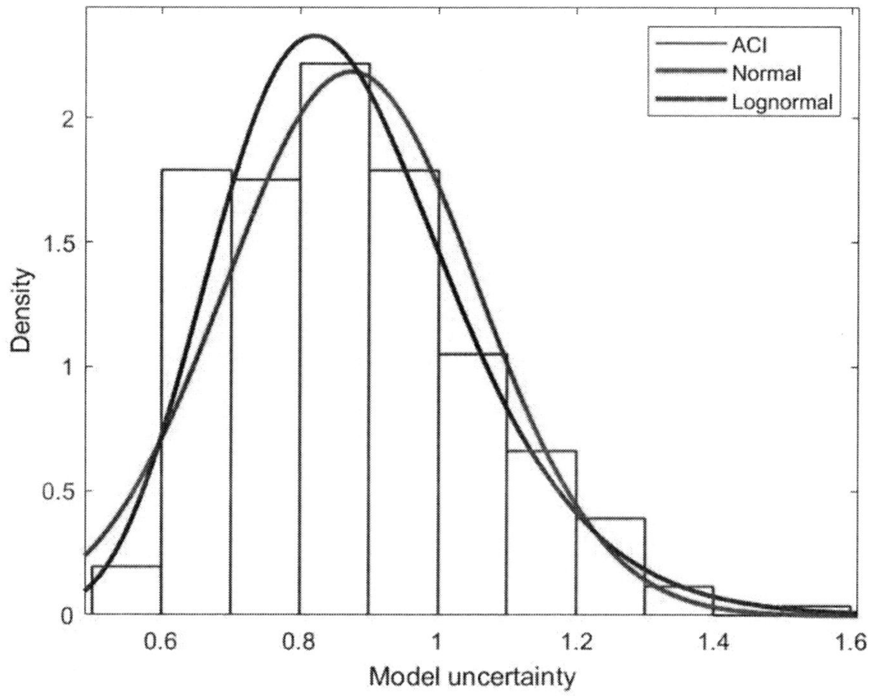

(a) ACI 318

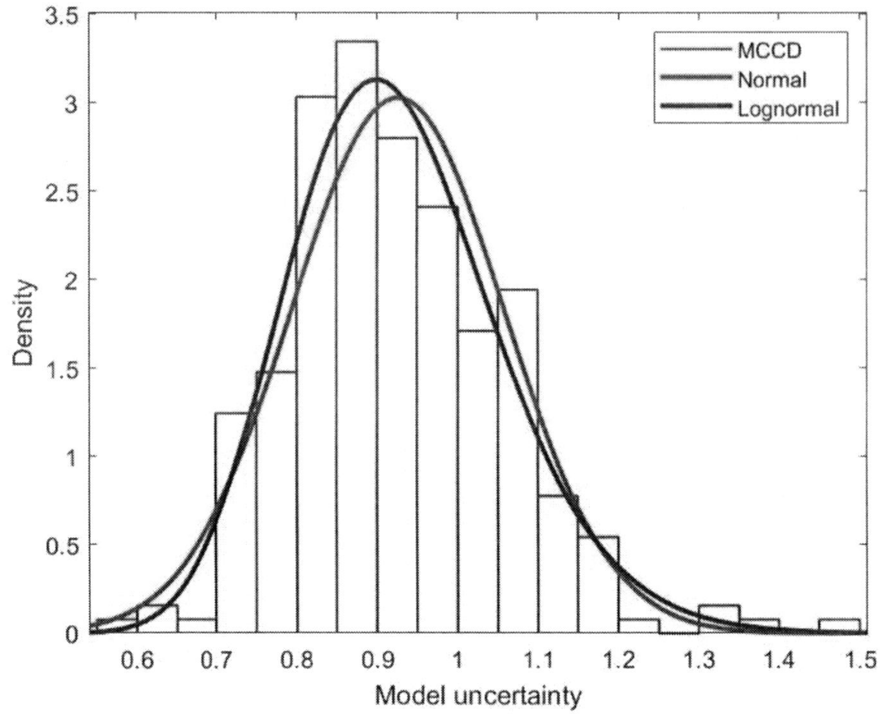

(a) MCCD

Figure 2. Histograms and cumulative distribution plots for model uncertainty

6. Discussion

As presented in Table 2, the model uncertainty associated with the MCCD R_{MCCD} has a bias (mean value) of $\mu_R = 0.93$, indicating that the MCCD model generally overpredicts the shear breakout resistance. MCCD has the closest bias to the value of 1. R_{MCCD} shows small scatter with a standard deviation $\sigma_R = 0.13$. A mild correlation ($r = -0.25$) is observed between R_{MCCD} and concrete edge distance c_1. MCCD did not have any significant correlation with other model parameters.

ACI 318 R_{ACI} was found to have a model uncertainty mean value of $\mu_R = 0.87$ (with 13% unconservative bias) with a standard deviation of $\sigma_R = 0.18$. The model has the largest dispersion. As presented in Table 3, the R_{ACI} has a strong decreasing safety bias with increasing anchor diameter d_o ($r = -0.49$) and concrete edge distance c_1 ($r = -0.62$). The sensitivity analysis identified no significant trend between R_{ACI} and the other basic variables (Table 3).

7. Conclusion

In order to conduct an efficient study of the reliability of fastening to concrete design provisions, characterisation of model uncertainty statistics is a required prerequisite. The model uncertainties related to the ACI 318 and MCCD [5] concrete-edge breakout models were investigated, using the experimental database of 257 anchor tests. The derived model uncertainties statistics were analysed, and suitable probability distribution was recommended. The assessment results in the following conclusions:

- MCCD model displayed the lowest bias (i.e. closest to the mean value of 1) and comparatively smallest dispersion; and mild correlation with shear breakout parameters.
- ACI 318 model displayed the largest dispersion and comparatively highest unconservative bias, with model sensitivity to anchor diameter and concrete edge distance.
- The log-normal distribution was found as the most suitable distribution for the model uncertainties

- The model uncertainty statistics presented in this contribution can be used as basic random variables in reliability investigation and also derive global partial factors for fastening design formulations.

References

[1] Eligehausen R and Gehlen C 2008 eds 7th Int. PhD Symp. in Civil Engineering - Part 3: Fastening University of Stuttgart and fib – Int. Federation for Structural Concrete: Stuttgart Germany

[2] European Committee for Standardization (CEN) EN 1992-4 Eurocode 2: Design of concrete structures - Part 4 Design of fastenings for use in concrete, European committee for standardisation, CEN/TC 250 2019 Brussel

[3] ACI Committee and International Organization for Standardization 2019 Building code requirements for structural concrete (ACI 318-19) and commentary American Concrete Institute 2019

[4] Olalusi O B and Viljoen C 2019 Model uncertainties and bias in shear strength predictions of slender stirrup reinforced concrete beams Structural Concrete. https://doi.org/10.1002/suco.201800273

[5] Hofmann J 2004 Tragverhalten und Bemessung von Befestigungen unter beliebiger Querbelastung in ungerissenem Beton (Load – bearing behavior and design of fastenings under arbitrary shear loading in uncracked concrete – in German) Dissertation (University of Stuttgart)

[6] Olalusi, O B and Spyridis P 2020 Machine learning-based models for the concrete breakout capacity prediction of single anchors in shear. Advances in Engineering Software, **147**, 102832. https://doi.org/10.1016/j.advengsoft.2020.102832

[7] Lee N H, Park K R and Suh Y P 2010 Shear Behavior of Headed Anchors with Large Diameter and Deep Embedments ACI Structural Journal **107** (2)

[8] Dong-Hyun K, Yong-Myung P, Ho-Jung J and Moon-Ki K 2014 Concrete Breakout Capacity of Cast-in-Place Anchor under Shear Loading Advanced Materials Research

[9] Hofmann J, Fuchs W and Eligehausen R 2004 Quertragfähigkeit randnaher Befes-tigungsmittel mit Belastung senkrecht zum Bauteilrand (Behavior of anchorages arranged close to the edge and loaded towards the edge). In: Beton- und Stahlbetonbau 99 No 10 pp 806-7 (in German)

[10] Shaikh A F and Yi W 1985 In-place strength of welded headed studs PCI Journal **30** (2) pp 56-81

[11] Toth M, Bokor B and Sharma A 2019 Anchorage in steel fiber reinforced concrete–concept, experimental evidence and design recommendations for concrete cone and concrete edge breakout failure modes Engineering Structures **181** pp 60-75

[12] Unterweger A 2008 Randnahe Anker unter Querlast: mechanische Modellierung und Bemessung (Anchors close to edge under shear load) Vienna Austria: Universität für Bo-denkultur, Dissertation (In German)

[13] Klingner R E, Muratli H and Shirvani M 1999 A technical basis for revision to anchorage criteria. Division of Engineering Technology, Office of Nuclear Regulatory Research US Nuclear Regulatory Commission

[14] Grosser P 2010 Single Anchors Loaded in Shear close to the Edge in High Strength Concrete Test Report (Germany Institute of Construction Materials University of Stuttgart) Report No. E 10/19 – Bft/17 2010 not published

[15] Senftleben S 2010 Zum Tragverhalten von randnahen Reihenbefestigungen beansprucht durch eine Querlast parallel zum Bauteilrand (Behavior of Multiple Anchor Connections close to the Edge under Shear Loading Parallel to the Edge) Stuttgart Germany: Institute of Construction Materials (University of Stuttgart) Diploma Thesis (in German)

[16] Olalusi, O. B., & Viljoen, C. (2020). Assessment of reliability of EN 1992-1-1 variable strut inclination method of shear design provisions for stirrup failure. Structural Concrete, 21(1), 303-315. https://doi.org/10.1002/suco.201900220

[17] JCSS. JCSS Probabilistic Model Code. (2001). Joint Committee on Structural Safety, Zurich. ISBN 978-3-909386-79-6.

Suitability of using borrow pit soils collected from Southwestern Nigeria as landfill liners

I I Akinwumi, O D Afolayan, C A U Okeke, O O Badiru, K J Jolayemi
Civil Engineering Department, Covenant University, Ota, Nigeria.
*Corresponding author: isaac.akinwumi@covenantuniversity.edu.ng

Abstract. Effective waste disposal system is essential in preventing the infiltration of leachate into the groundwater and in achieving the sustainable development goal of ensuring clean water. One of the ways to provide an adequate disposal system is through the provision and use of engineered landfills for waste containment instead of open dumps. Consequently, this study evaluates the suitability of using soils from six borrow pits collected from six states in Southwestern, Nigeria. To assess the suitability of the soils, laboratory tests were carried out to determine the index properties, mechanical and hydraulic properties of the collected soil samples. The results showed that the soil samples collected from the selected borrow pits in Ogun, Oyo and Ekiti States could be effectively employed for the containment of inert waste. In contrast, the soil samples collected from the selected borrow pits in Osun, Lagos and Ondo States needs to be improved such that their unconfined compressive strengths satisfy the standard requirement (≥200 kN/m^2). This study provides readily accessible soil data for use by engineering professionals involved in selecting suitable soil for use as landfill liners in Southwestern, Nigeria.
Keywords: Engineered landfills; sustainable development; waste containment; waste disposal; waste management

1. Introduction

The rapid growth in population, industrialization and urban development has resulted in an increase in the generation of solid waste [1, 2]. According to Verisk Maplecroft [3], over 2.1 billion tonnes of municipal solid waste (MSW) is generated each year globally. This large volume of generated wastes is a global problem requiring prompt action [4]. In developing countries, the decline in environmental quality and potential loss of valuable resources have been ascribed to the rising rates of waste generation, ineffective waste handling and poor management quality of solid waste [5]. Babayemi and Dauda [6], Babayemi et al. [7] and Minh et al. [8] identified open waste dumping and burning, and landfilling in dug holes as the prevalent waste management types generally employed in developing nations. However, despite the use of these waste management types, Scarlat et al. [9] stated that in 2012, only 54.4% of solid waste generated gets collected, while the rest are indiscriminately discarded [10].

Engineered or sanitary landfills have been proposed in a bid to effectively dispose of this increasingly large volumes of solid waste that are being generated, [11, 12]. Engineered landfills can help to minimize the migration of contaminants from the disposed waste from leaching into the groundwater [11]. In performing this function, landfills are designed with liner and cover to seal off their content from the environment [13]. Landfill liner is an essential part of a landfill, whose material selection is of utmost importance. Ojuri et al. [10] stated that in order to build a cost-effective landfill, there is a need to use locally-available materials as landfill liners. It is in view of this that this study seeks to assess the suitability of locally available soils in Southwestern Nigeria for use as landfill liners.

2. Materials and Methods

2.1 Materials

Soil samples were collected from borrow pits from different locations in Southwestern Nigeria, as depicted in Fig. 1. These soil samples were collected at depths ranging from 1.2-1.5 m, and were packed and kept secured in air-tight polyethylene bags to retain their natural moisture content. Each bag was marked accordingly, indicating soil description, the location of collection and the date of sampling.

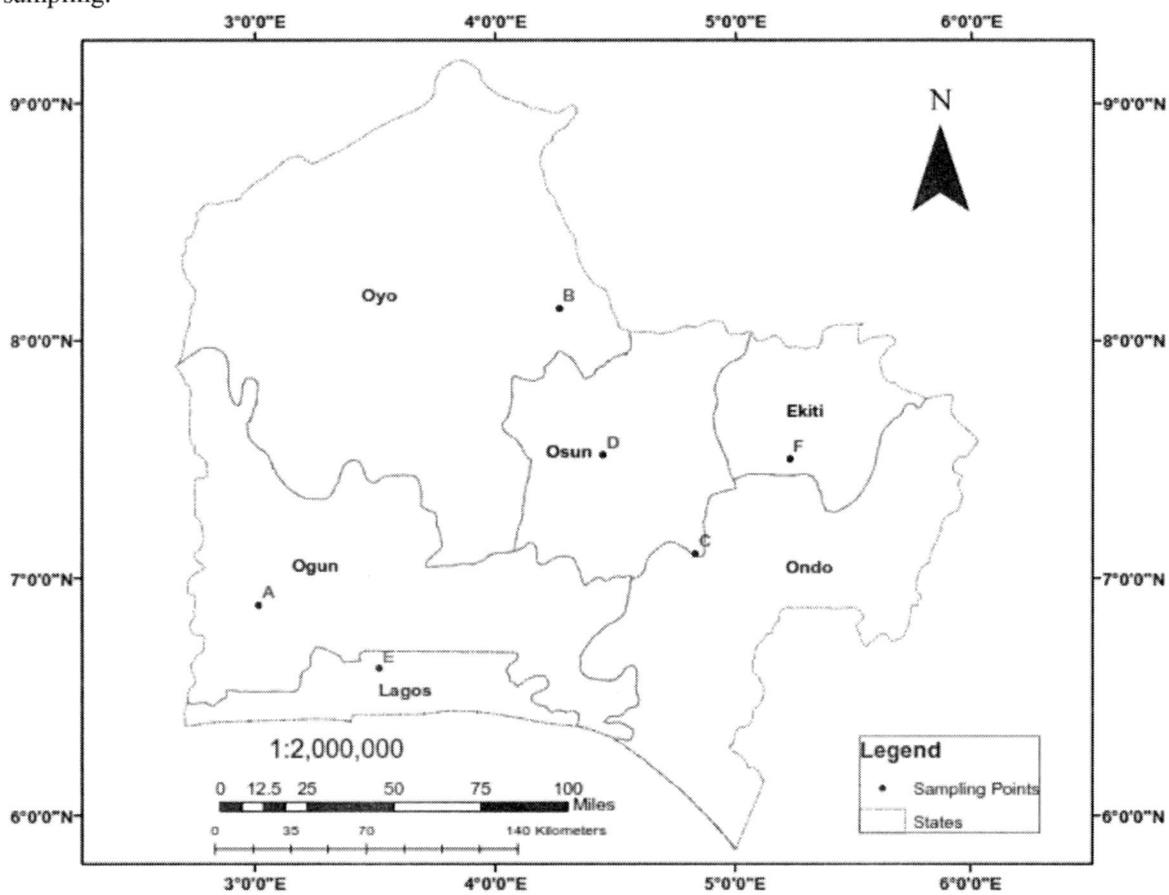

Figure 1. Showing the samples collection location with representative nomenclature

2.2 *Field Survey*

Physical, mechanical and index properties of the soil samples were determined by carrying out colour (visual) inspection, specific gravity, particle size distribution test, compaction test, Atterberg's limit test, unconfined compressive strength and hydraulic conductivity test.

Index properties

Wet sieving was carried out on the samplesin accordance with ASTM D422 [14]. The collected soil samples were weighed on the weighing balance before they were washed through No. 200 (75 micron) sieve. The sample retained on the sieve were oven-dried for 24 hours and their particle size distribution determined by sieving them through the standard set of sieves using a mechanical sieve shaker. Atterberg's limit tests were performed on the collected samples in accordance with ASTM D4318 [15].

Mechanical properties

Moisture-unit weight relationship of the collected samples were conducted and obtained according to ASTM D 698 [16] and Head [17]. This test was carried out to determine the maximum dry density (MDD) and the optimum moisture content (OMC) of the soil samples. Also, the collected samples were compacted and tested to determine their unconfined compressive strength in accordance with ASTM D 2166 [18] and Head [19]. This test involves the use of cylindrical specimens of diameter and length 38 mm and 76 mm, respectively.

Hydraulic properties

The hydraulic conductivity of the samples are essential for determining the ease flow of water or leachates through the samples. Hydraulic conductivity test on the soil samples were performed following Head [19] and ASTM D5084-10 [20].

3. Result and Discussion

The geotechnical properties of the soil samples which were collected from Ogun (A), Oyo (B), Ondo (C), Osun (D), Lagos (E) and Ekiti (F) states, were analyzed and presented in Table 1.

Table 1. Geotechnical properties of the collected soil samples

Properties	A	B	C	D	E	F
Index properties						
Colour	Red	reddish-brown	Grey	reddish-brown	brownish-grey	grey
Specific gravity	2.54	2.57	2.54	2.5	2.51	2.6
Liquid limit (%)	65	38	57	61	46	40
Plastic limit (%)	42.15	25.68	32.86	46.59	26.64	27.16
Plasticity index (%)	22.85	12.32	24.14	14.41	19.36	12.84
% passing 75 micron sieve	62.5	43.3	31.5	41.4	79.5	54.8
USCS	MH	CL	MH	MH	CL	CL
Mechanical properties						
Optimum moisture content (%)	19	17.5	19	22.5	22.5	14.7
Maximum dry density (g/cm^3)	1.7	1.86	1.73	1.68	1.68	1.9
Unconfined compressive strength (kN/m^2)	201.8	214.7	60	39.6	138.6	203.1
Cohesion (kN/m^2)	100.9	107.35	30	19.8	69.3	101.55
Hydraulic properties						
Permeability $x10^{-8}$ (m/s)	7.911	8.345	8.373	7.676	8.065	8.539

3.1 Index Properties

The specific gravity values of the collected samples range from 2.5-2.6, as presented in Table 1. The colour of the samples ranged from reddish-brown to grey to brownish-grey. According to Brady and Weil [21], the reddish-brown colouration of the collected samples are due to the presence of ferric oxide.

Figure 2 shows the plasticity charts for the Unified Soil Classification System (USCS) with the plasticity plots for the six collected samples. It shows that the collected samples range from being classified as sandy elastic silt (MH) for samples A, C and D to being classified as sandy, lean clay (CL) for samples B, E and F as presented in Figure 2. According to unified system of particle size classification and from Figure 3, sample A contains 0.7% gravel, 36.8% sand, 46.29% silt and 16.21% clay; sample B contains 0% gravel, 57.7% sand, 30.32% silt and 11.98% clay; sample C contains 9.4% gravel, 60.1% sand, 20.30% silt and 10.20% clay; sample D contains 6.4% gravel, 52.2% sand, 29.38% silt and 12.02% clay; sample E contains 0% gravel, 20.48% sand, 56.70% silt

IOP Conf. Series: Materials Science and Engineering 1036 (2021) 012038 doi:10.1088/1757-899X/1036/1/012038

and 22.82% clay; and sample F contains 0.5% gravel, 47.7% sand, 38.8% silt and 16% clay. The clay content of >10% in the collected samples contributed to the medium to high plasticity index of the samples, which is an indication of their potential to effectively contain waste when employed as a base liner for landfills [22, 23]. However, soils with high fines content are susceptible to volumetric shrinkage. Rowe *et al.* [24], EPA [23, 25] and Guney *et al.* [26] recommended that for a soil to be effective as a liner material, the particle size must contain at least 15 to20% silt or clay-sized material, gravel-sized particles of less than 10% and plasticity index (PI) should be greater than 10%. The particle size analysis and Atterberg limit requirements for a liner are satisfied by all the collected samples, as shown in Table 1. Hence, classifying them as potential liner materials.

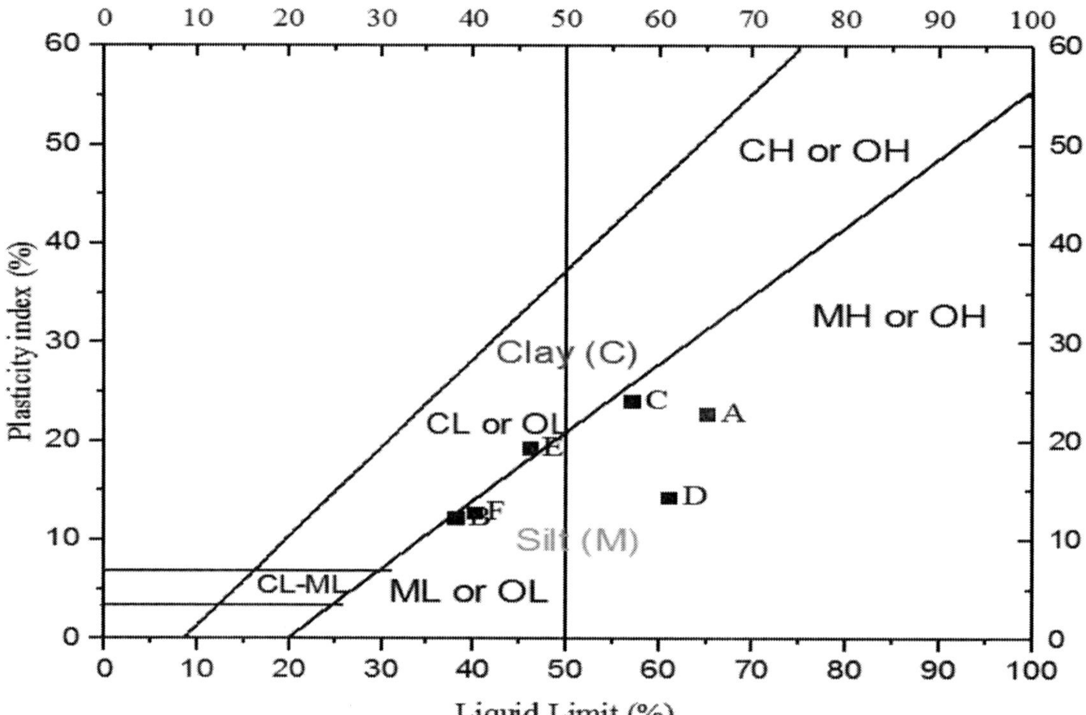

Figure 2. USCS plasticity charts showing the plots for the collected samples

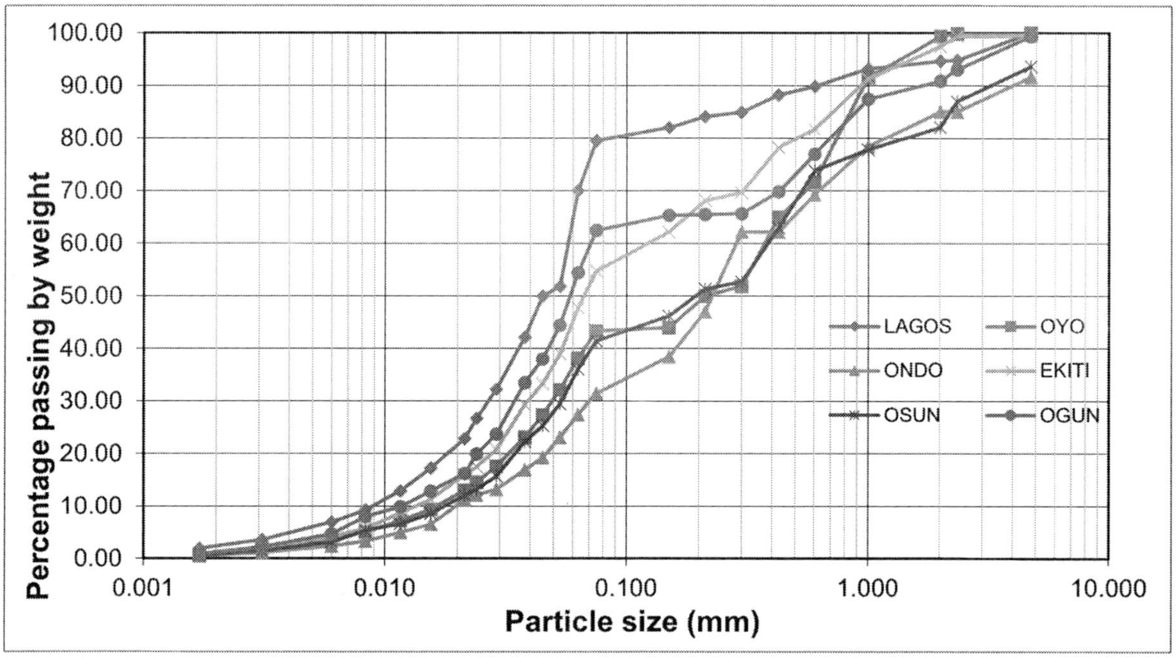

Figure 3. Particle size distribution curves of the collected soil samples

3.2 Mechanical Properties

From Table 1, the maximum dry density and optimum moisture content (OMC) values of the soil samples range from 1.68 to1.90 g/cm^3 and 14.70 to 22.50%, respectively. For practical use, EPA [23] recommended that clay samples are to be compacted to 95% MDD for optimal performance as a base liner in landfills.

Furthermore, Guney *et al.* [26], Amadi *et al.* [27] and Daniel and Wu [28] specified that for a soil to be utilized as a compacted liner material, its minimum unconfined compressive strength must be 200 kN/m^2. From Figure 4, this criterion is only satisfied by samples A, B and F. Samples C, D and E have compressive strength of < 200 kN/m^2. Hence, the samples have to be modified with chemical stabilizers before they can be used as liner materials.

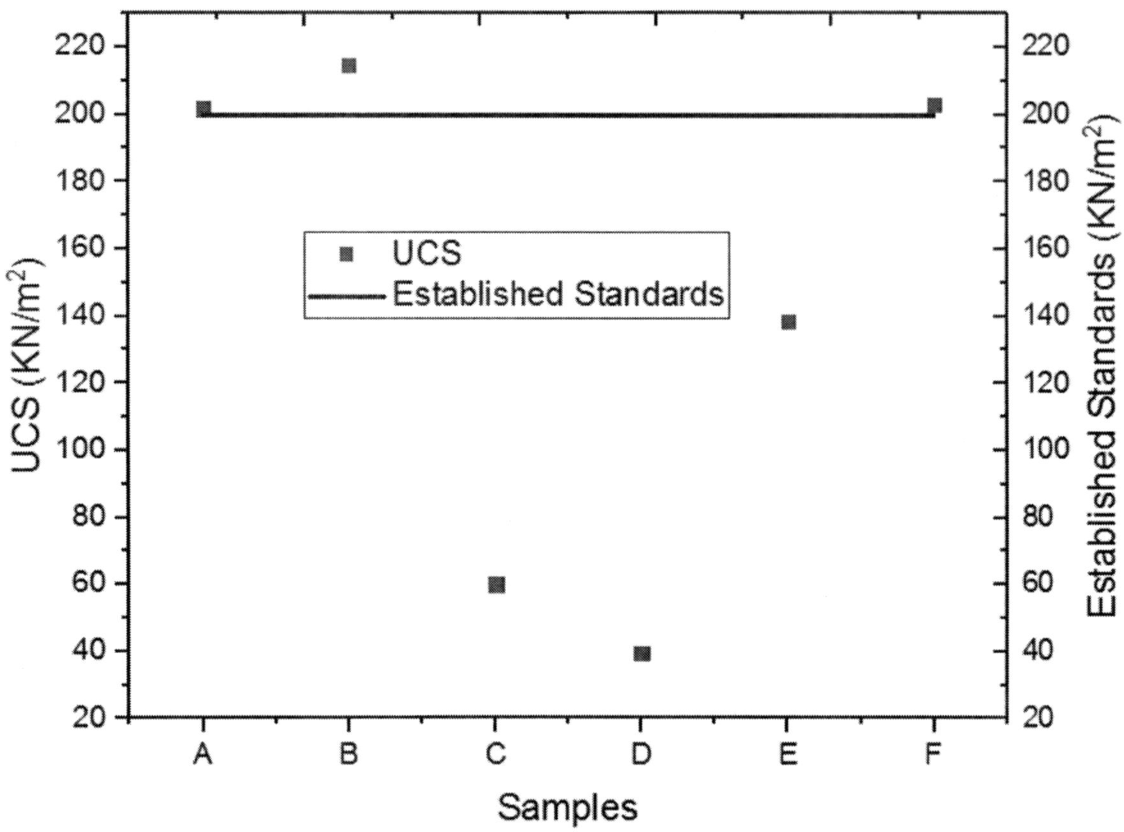

Figure 4. Curve for the unconfined compressive strength of the samples and established standards

3.3 Hydraulic Properties

The hydraulic property of a soil to be used for landfill lining is undoubtedly the most important criterion for assessing the suitability of the soil sample. European Union Council Directive [29], Benson *et al.* [30], Rowe *et al.* [24], Burnley *et al.* [31], Osinubi and Nwaiwu [32] and EPA [23] specified that for a soil sample to be used as a base liner, it must have permeability value of $\leq 1 \times 10^{-9}$ m/s for hazardous and non-hazardous waste, and $\leq 1 \times 10^{-7}$ m/s for landfills to receive inert wastes. From Table 1, the permeability value of the collected samples ranges from 7.676×10^{-8} m/s to 8.539×10^{-8} m/s, hence, they can be used to contain inert wastes in landfills. However, for the samples to be used effectively for the containment of hazardous and non-hazardous wastes, chemical modification of the soil samples may be employed.

4. Conclusion

Selecting suitable soil materials for lining landfills for waste containment is important while planning for the construction of a landfill. Consequently, this study assessed the suitability of using soils collected from six states in Southwestern Nigeria as landfill liner materials. Evaluating the soil samples, relative to standard requirements showed that the soil samples collected at Ogun, Oyo and Ekiti can be effectively used for the containment of inert wastes owing to their index properties, mechanical and hydraulic properties. However, the unconfined compressive strength samples collected from Osun, Lagos and Ondo need to be improved to satisfy the minimum unconfined compressive strength of 200 kN/m^2. The potential implication of this research is that it ascertains the usability of natural soils as landfill liner materials. This is in the quest to protect the environment (mainly groundwater) and ensure the availability of clean water within the study area.

Acknowledgement
The authors acknowledge the contributions of the technical officers of the geotechnical engineering laboratory, Civil Engineering Department, Covenant University and Covenant University Centre for Research Innovation and Discovery for providing a platform for this research work.

References

[1] Rao, AV N. and Chittaranjan, M (2011). Application of Agricultural and Domestic Wastes in Geotechnical Applications: An Overview. *Journal of Environmental Research and Development*, **5** (3).
[2] Akinwumi I I, Booth C A, Ojuri O O, Ogbiye A S and Coker A O 2019 *Containment of Pollution from Urban Waste Disposal Sites*. In Urban Pollution (eds S M Charlesworth and C A Booth), Chapter 17, 223–234, John Wiley & Sons Ltd. doi:10.1002/9781119260493.ch17.
[3] Verisk Maplecroft 2019 Waste Generation and Recycling Indices. https://www.circularonline.co.uk/wp-content/uploads/2019/07/Verisk_Maplecroft_Waste_Generation_Index_Overview_2019.pdf.
[4] Karija, M K, Shihua, Q I, and Lukaw, Y S (2013). The impact of poor municipal solid waste management practices and sanitation status on water quality and public health in cities of the least developed countries: The Case of Juba, South Sudan. *International Journal of Applied Science and Technology*, **3**(4), 87–99.
[5] Babayemi, J O, Ogundiran, B M and Osibanjo, O (2017). Current Levels and Management of Solid Wastes in Nigeria. *Environmental Quality Management*, **23**(4), 29–52. DOI: 10.1002/tqem.21498
[6] Babayemi, J O, and Dauda, K T (2009). Evaluation of solid waste generation, categories and disposal options in developing countries: A case study of Nigeria. *Journal of Applied Sciences and Environmental Management*, **13**(3), 83–88.
[7] Babayemi, J O, Osibanjo, O, and Weber, R. (2014). Assessment of use, reuse, and EoL disposal and X-ray fluorescence analysis screening of waste mobile phones in Nigeria. Environmental Quality Management, **23**(4), 1–12.
[8] Minh, N H, Minh, T B, Kajiwara, N, Kunisue, T, Subramanian, A, Iwata, H and Tanabe, S (2006). Contamination by persistent organic pollutants in dumping sites of Asian developing countries: Implication of emerging pollution sources. *Archives of Environmental Contamination and Toxicology*, **50**(4), 474–481. https://doi.org/10.1007/s00244-005-1087-3.

[9] Scarlat, N, Motola, V, Dallemand, J F, Monforti-Ferrario, F, & Mofor, L (2015). Evaluation of energy potential of Municipal Solid Waste from African urban areas. *Renewable and Sustainable Energy Reviews*, **50**, 1269–1286.
[10] Ojuri, O O, Akinwumi, I I and Oluwatuyi, O E (2017) Nigerian lateritic clay soils as hydraulic barriers to adsorb metals: Geotechnical characterization and chemical compatibility *Environment Protection Engineering*, **43** (4). pp. 209-222.
[11] Akinwumi, I I, Ojuri, O O, Edem, D and Ogbiye, S A (2016) Sawdust Stabilization of Lateritic Clay as a Landfill Liner to Retain Heavy Metals. In: Geo-Chicago 2016: *Sustainable Geo-environmental Systems*, August 14–18, 2016, Chicago, Illinois, United States of America.
[12] Amadi, A A, and Eberemu, A O (2013) Potential Application of Lateritic Soil Stabilized with Cement Kiln Dust (CKD) as Liner in Waste Containment Structures *Geotech Geol Eng* **31:**1221–1230 DOI 10.1007/s10706-013-9645-3,
[13] Amadi A. (2007). Unsafe waste disposal practices in Nigerian cities: geo-environmental perspectives. Nigerian Society of Engineers (NSE) Tech Trans **42**(2):31–44
[14] ASTM D422 (2007). *Standard Test Method for Particle-Size Analysis of Soils*. ASTM International, Philadelphia, USA.
[15] ASTM D 4318 (2010) *Test Methods for Liquid Limit, Plastic Limit, and Plasticity Index of Soils*. ASTM International, Philadelphia, USA.
[16] ASTM D 698 (2012) *Standard Test Methods for Laboratory Compaction of Soil Standard Effort*. ASTM International, Philadelphia, USA.
[17] Head, K H (1994a) Manual for soil laboratory testing, *soil classification and compaction tests*. Halsted Press, New York
[18] ASTM D 2166 (2016). *Standard Test Method for Unconfined Compressive Strength of Cohesive Soil*. ASTM International, West Conshohocken, PA, USA.
[19] Head, K H (1994b) *Manual of soil laboratory testing. Permeability, shear strength and compressibility tests*, 2nd edition. Pentech Press, London.
[20] ASTM D 5084-10 (2010) *Standard Test Method for Measurement of Hydraulic Conductivity of Saturated Porous Material Using a Flexible Wall Permeameter*. ASTM International, Philadelphia, USA.
[21] Brady, N C and Weil, R. R. (2010) *Elements of the Nature and Properties of Soils*, 3rd edition, page 95, Upper Saddle River, New Jersey, Prentice Hall.
[22] NRA (1989). *Earthworks to landfill sites*. North–West Region: National Rivers Authority (NRA).
[23] Environment Protection Agency (EPA) (2014). LFE4 - *Earthworks in landfill engineering: Design, construction and quality assurance of earthworks in landfill engineering*. United Kingdom Environment Agency, Bristol.
[24] Rowe, R K, Quigley, R M, and Booker J R (1995). *Clayey barrier systems for waste disposal facilities*. E and FN Spon, London
[25] Environment Protection Agency (2009). *How to Comply with your Environmental Permit, Additional Guidance for Landfill* (5.02), Environment Agency, Bristol UK.
[26] Guney, Y, Cetin, B, Aydilek, A H, Tanyu, B F and Koparal, S (2014). Utilization of sepiolite materials as a bottom liner material in solid waste landfills. *Waste Management*, **34**(1), 112–124.
[27] Amadi, A A, Eberemu, A O, and Osinubi, K J (2012) Strength consideration in the use of lateritic soil stabilized with fly ash as liners and covers in waste landfills. In: Roman D Hryciw, Adda Athanasopoulos-Zekkos, Nazli Yesiller (Eds) State-of-theart and

practice in geotechnical engineering, *Geotechnical Special Publication (GSP) American Society of Civil Engineers*, **225**: 3835–3844.
[28] Daniel, D E, and Wu, Y K (1993) ASTM (2007) D 422: *Standard Test Method for Particle-Size Analysis of Soils*. ASTM International, Philadelphia, USA.
[29] European Union Council Directive (1999). *Council Directive 1999/31/EC of 26 April 1999 on the landfill of waste. Official Journal of the European Communities*, **16**(7), L182/1– L182/19.
[30] Benson, C H, Zhai, H and Wang, X (1994). Estimating hydraulic conductivity of compacted clay liners. *Journal of Geotechnical Engineering*, **120**(2), 366–387.
[31] Burnley, S, Cooke, D and Gladding, T (2005). T308 Environmental Monitoring, Modelling and Control: *Block 4 Solid Wastes Management*. The Open University, Milton Keynes.
[32] Osinubi, K J and Nwaiwu, C M (2006). Design of compacted lateritic soil liners and covers. *Journal of Geotechnical and Geo-environmental Engineering*, **132**(2), 203–213.

Influence of GIS on sustainable pavement maintenance: a comparative review

B U Ngene, D E Bassey*, A A Busari, G O Bamigboye and A T Nworgu

Department of Civil Engineering, Covenant University, Ota, Ogun State, Nigeria
*Corresponding author: bassey.daniel@gmail.com

Abstract. Road transportation remains dominant in passenger and freight movements across Africa. Yet, a majority of the countries in Africa do not perform proactively in ensuring the sustainability of highway infrastructure. This research reviews recent literature in recognizing the state-of-the-practice on the application of GIS-based techniques in sustainable pavement maintenance, hence promoting a systematic approach for highway maintenance management. Furthermore, it compares contemporary maintenance practices in the five major African regions to highlight the differences and similarities. Most countries have since implemented institutional reforms such as the provision of road funds and agencies, thereby making substantial improvements on road maintenance. However, several challenges, including inadequate financing, lack of technical expertise, poor administration, and lack of continuous research and development still hinder sustainable road maintenance in Africa. Hence, most African countries still depend solely on budget financing while adopting delayed and reactive approaches to highway maintenance activities. The study revealed that sustainable methods of pavement maintenance have neither been incorporated in policy nor practices involving highway pavement management in most African nations. And as such, it encourages respective road agencies to integrate GIS-based techniques into road maintenance management at all functional levels to promote sustainability.

1. Introduction

The serviceability of any nation's roadway network plays a key role in its social and economic advancement since passenger and freight mobility, and land accessibility are major aspects [1, 2]. Gujba *et al.* [3] reported that averagely, road transport in Africa controls about 90% and 80% of passenger and freight movements, respectively. The lack of modal diversity in transportation has resulted in the over-dependence on roads for transportation and consequently, increased carbon emissions, noise pollution, traffic congestion, and ultimately, stress-induced pavement failures. Most African Nations inherited roads initially developed for colonial domination and military ascendancy [4]. An adequately integrated and functional road network system aids in accelerating the fiscal growth of any nation, via tourism, trade, and other feasible developments [5]. Due to limited funding, road networks, especially in the African continent, were only developed and maintained to certain extents. As such, road maintenance practices remain inadequate in many African countries [4].

In response to the United Nations, sustainable development goal 11 for the development of sustainable cities and communities, sustainability in pavement maintenance has gained attention in recent years. To avert failures and guarantee level of service, road infrastructure requires constant Maintenance and Rehabilitation (M and R). In many cases, road management agencies are guilty of allowing roadways to fall into severe conditions before taking the required actions. Nevertheless, proactive preservation actions endorse management systems while the roadways are still in proper condition [6]. As such, highway management agencies search for methods to improve pavement preservation in a quality and cost-efficient manner, such as the advanced Pavement Management Systems (PMS) [7]. PMS institutes organizational and computational programs for pavement assessment, detecting the present conditions, recognizing failure rates, and evaluating several maintenance alternatives taking contemporary and impending costs into account [8].

Several computer programs have been developed employing vital functions of modern PMS alongside spatial data from Geographic Information Systems (GIS) [9]. PAVER™ is the most commonly used PMS software. It conducts series of analyses using pavement condition index (PCI) data with values between zero (failed condition) and 100 (excellent condition) for a precise description of pavement conditions and prediction of contemporary and future M and R requirements [10]. Recently, software has utilized GIS/GPS technology and state-of-the-art graphics processing in simplifying pavement inspection with the use of mobile devices by permitting users to identify device locations on geo-referenced road maps displayed for prompt input of pavement distress information [11]. This paper reviews modern practices on pavement maintenance management systems, highlighting common pavement distresses, M and R categories, and GIS-based methods for sustainable pavement maintenance it further discusses the different M and R practices in the five regions of Africa, taking the most populous country within each region as a case study. The focus is on comparing and contrasting the different contemporary M and R perspectives in the regions.

2. Components of Road Maintenance

2.1 Pavement distresses

Distresses may start during construction and proliferate over time. Poor construction methods most certainly guarantee the poor performance of pavements. Distress may occur in flexible pavements due to reasons like seasonal variations, recurring traffic loadings, poor drainage, aging, or combinations [12]. Some common distresses in asphaltic pavements include rutting, fatigue cracking, roughness, thermal cracking, shoving, bleeding, raveling, and reflection cracking [13].

2.2 M and R categories

M and R involves practices for preserving the performance of road structures. It is achieved by correcting pavement distresses that develop over time while considering durability and environmental sustainability [6]. M and R operations in Nigeria are categorized in Table 1.

Table 1. Road maintenance categories and description.

Category	Description
Routine	Regularly required on roads of all functional classes. Costs are considered fixed in maintenance budgets, activities include lane marking drainage clearing, culverts and bridge maintenance, turf cutting, amongst others.
Recurrent	Conducted at specified intervals within the year. Maintenance operation frequency is reliant on climatic and topographic conditions, and the traffic volume conditions. Activities include repair of cracks, potholes, and other road distresses and releveling or grading for unpaved roads.
Periodic	This comprises major rehabilitation of failed sections along roads. The frequency of occurrence is usually very low as it is performed only when the need arises. Activities include shoulder re-graveling and carriageway resurfacing, amongst others.
Emergency / Special Repairs	Undertaken only in extreme unexpected situations such as momentous landslides, tectonic or seismic movements across roadways, culvert or bridge failures, and pavement explosion due to terrorism amongst other tragedies.

3. Pavement Maintenance Management Systems (PMMS)

PMMS integrates coherent and systematic activities aimed at the effective management and maintenance of road pavements[14]. Benefits of PMMS include: the development of digitized databases for storage of pavement data; encouraging the practice of regular pavement monitoring for collection of pavement distress data; making optimized lists to aid in selecting maintenance and rehabilitation works based on necessity and resources; predicting impending distresses and providing documentation of the required intervention to support requests for project funding.

3.1. Assessment of Current Pavement Conditions

The process begins with taking an inventory of the road network based on visual inspection of the surface as a vehicle with the acquisition system pre-installed on travels along the pavement. The user records the data on spreadsheets based on the objectives of the inspection [15]. A spatial database is developed with collected data representing the road network. ArcGIS® ModelBuilder™ is then used in developing digitized models. Data collected represent locations along the roads and are geo-referenced using the linear referencing system (LRS) during data import. After the database population, analysis is done using codes written in Python® to assess the current state of the pavement expressed in PCI.

3.2. Prediction of Future Pavement conditions

When predicting pavement future PCIs, material quality, climatic conditions, storm and subsurface hydrology, traffic volume, and maintenance policies are governing factors which may become variable. As such, PMS adopts predictive models in deducing future pavement conditions based on contemporary data. Probabilistic and deterministic models such as the Markov chains and AASHTO methods, respectively have been utilized in past studies [16, 17].

3.3. Role of GIS in PMMS

Transportation systems associated data are perfect fits for GIS analysis because of the spatial distribution of features. Hence, GIS has been used in increasing the efficiency of transport planning and M and R systems significantly in terms of cost and time efficiency, and decision making.

3.3.1. Application of PAVER™ and ArcGIS® Programs. The PAVER™ software was developed to aid PMMS activities. Pavement condition ratings are expressed in PCI after parallel assessments of available distress intensities and extents. Figure 1 shows PCI range attainable from PAVER™ 6.5.7

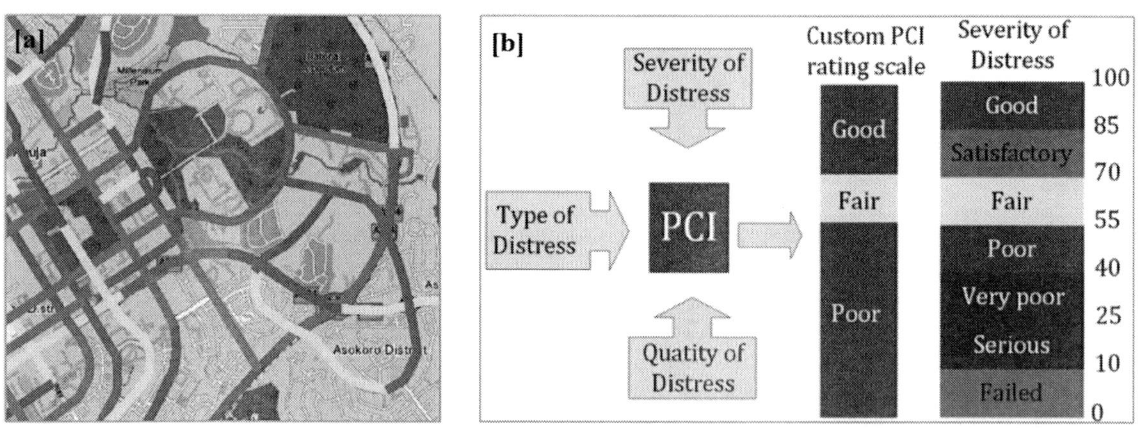

Figure 1. [a] Road network map showing road sections within Abuja, Nigeria with varying PCIs [b] PCI range values (Source: [10])

Characteristically, a scale of seven distinct color-coded classes is used to categorize different pavement conditions. PAVER™ 7.0 receives data collected during initial inspection and assesses using 19 levels of distress, making it a highly thorough PMS software [10]. Output reports from PAVER™ provide summarized charts and tables showing statistics of the selected pavement network such as total area,

IOP Conf. Series: Materials Science and Engineering 1036 (2021) 012039 doi:10.1088/1757-899X/1036/1/012039

number of branches and segments, average PCI of segments, and area-weighted PCI across pavements. The PAVER™ software output data is compatible with most GIS programs. ArcGIS® was used in studies by Bazlamit *et al.* [8] and Wolters *et al.* [9] to further display results with the use of color-coded maps. PAVER™ output file is imported into ArcGIS as a shapefile on a new layer for plotting.

4. Comparison of Contemporary Highway M and R Practices in Africa

4.1 Highway M and R in Western Africa: Nigeria situation

Nigeria, with a population of over 200 million people, highest in Africa, has the largest and second-largest road network in West Africa and Sub-Saharan Africa, respectively. The Nigerian road network is currently estimated at 194,000 km [18]; however, Nigeria's road sector performance has been unsatisfactory, thus negatively affecting socio-economic growth. Continuous funding has been directed towards the development of roads; however, the increasing population and traffic volumes keep the roads in a critical state [4].

The Federal roads maintenance agency (FERMA) is saddled with the duty of inspecting and maintaining all Federal roads within the country. They manage only 17% of the nation's road network while the state and local governments are responsible for the state (16%) and local (67%) roads, respectively[19]. The deterioration rates have outpaced the maintenance rates as a result of the decapacitated levels of finance and technical expertise for M and R efficiency [20].

A few studies have assessed PMS in Nigeria. Owolabi *et al.* [21] developed an assessment model for the performance of the Lagos – Ibadan expressway via pavement condition score (PCS) and international roughness index (IRI) data collected from the evaluation unit of the Federal Ministry of Works. Also, Owolabi and Abiola [22] established a priority index (PI) model for the assessment and forecast of pavement conditions. A linear relationship between equivalent standard axle loading (ESAL), IRI, PCS, and PI, under many forms of pavement distresses, was developed, while the prediction of future pavement conditions was done via artificial neural network and multiple linear regression [22]. Despite endorsements from researches, road pavement maintenance in Nigeria is still done reactively rather than on a proactive basis. As such, the need for a reform of the currently ineffective but adopted M and R schemes in Nigeria is critical. In General, the M and R issue in Nigeria can be addressed through proper funding, research, and development. These can reform the approach of M and R contracting at federal, state, and local government levels to a performance-based approach via public-private participation to ensure cost-efficiency.

4.2 Highway M and R in Eastern Africa: Ethiopia situation

Ethiopia, populated with over 112 million people, second-highest in Africa, and generally categorized as an underdeveloped country has in the past decade increased the reach of its road network by over 100% as a result of its modernized PMS approach to highway M and R. The Ethiopian road network estimated at 6,400 km as of 1951 grew to over 46,000 km in 2010, and about 85,966 km in 2015 [4]. The earlier lack of comprehensive PMS M and R techniques resulted in the accrual of the backlog of road maintenance. Nonetheless, the reform of M and R management initiatives, including ownership, management, financing, and responsibility is progressively leading Ethiopia out of the shadows of poor roadway networks [4].

The Ethiopian Road Transport Authority (RTA) with the aid of the Ethiopian government is responsible for the regulations and controls leading to the road network growth in Ethiopia. RTA was enacted in 1967 to oversee roads and road-using vehicle-related matters, as such, RTA has since taken maintenance as key to achieving its goals [4]. Since road management is continually hindered by funding, the road fund was established in Ethiopia in 1997, with significant consideration towards resolving the maintenance funding issues. This fund is based on a charge-for-service principle. To great extents, the M and R reform has been encouraged. As such, Ethiopia was recently classified as satisfactory performers in terms of M and R. Generally, the M and R culture in Ethiopia has greatly grown in the last few years due to appropriacy in funding via the road fund model and the adoption of PMS in the M and R sector [4].

4.3 Highway M and R in Northern Africa: Egypt situation

Egypt is the third most populous African nation with over 100 million people. Here, road M and R is said to be neglected, as it accounts for only 0.15% of the country's GDP. This is minor compared to Ecuador, Morocco, and Ukraine with 0.23, 0.24%, and 0.45%, respectively. This negligence has been consequential in increased road accident rates which in turn accounts for 1.5% GDP loss [4]. In 2001, the government increased the road funds to improve the transportation sector, However, about 75% of the roads still require maintenance; as approximately 700 of the 1,706 bridges in Egypt were in critical states due to poor M and R in 2009 [4].

In addressing the situation, the General Authority of Roads, Bridges, and Land Transport (GARBLT) was set up to oversee the construction and maintenance of roads in the country. GARBLT recently commenced the bridge maintenance and repair program. However, the program has yielded low success rates as a result of inadequacy in human resources in terms of technical expertise, and equipment for adequate inspection and assessment of bridges. Generally, the state of road M and R in Egypt is still a quandary, and requires rehabilitation rather than maintenance [4]. GARBLT's major issue has been funding, as such an initiative was introduced to generate funds via tolling, advertising, and issuance of fines amongst others.

4.4 Highway M and R in Central Africa: Democratic Republic of Congo (DRC) situation

The DRC, with over 86 million people, is ranked 4th most populous in Africa. DRC's transport infrastructure system is currently one of the most dilapidated across the globe [23]. The DRC road network spans a total of 171,250 km, of which only 2,250 km are paved [24]. Traveling to Kinshasa, the capital, by road is impossible since most of the provincial capitals are not linked. Another major bottleneck in road transportation across the country is the poor state of bridges used in crossing several small rivers. The DRC possesses an extensive inland river network; however, water transport is still hampered by long wait periods at ports resulting from poor infrastructure, high rates of silting, and uneven governance [23]. The Roads Authority of the DRC is responsible for the management of road maintenance and construction operations of all national and regional roads. At the same time, the Direction of Agricultural Roads and Services maintains the secondary and local roads. However, key infrastructure M and R projects are currently being done by international agencies such as the World Bank (3500 km), European Union (2000 km), the UK Department for International Development (1300 km), KfW Development Bank (280 km), and the African Development Bank (150 km) [24].

Presently, the populace suffers from insufficient roadways and are obliged to resort to air transportation for passenger and freight movements. Only a few roadways in the DRC; Kinshasa to Port of Matadi, and Likasi – Kipushi – Lubumbashi in Southern Katanga have remained in good condition [24]. However, since 2005, the United Nations Joint Logistics Centre (UNJLC) has collected and disseminated road condition data in the DRC using the GIS Working Group and its various partners. These data are registered in geo-databases hosted by the UNJLC. They are steadily updated via remote sensing methods from the Belgian University of Louvain [24] in attempts to resuscitate the country's road M and R activities.

4.5. Highway M and R in Southern Africa: South Africa (SA) situation

SA, with over 68 million people, is the sixth most populous country in Africa. SA road network spans about 746,978 km, the longest in Africa, 10th longest globally, and 18th longest in terms of paved roads worldwide (153,719 km) [4]. South-Africa is foremost economically in the continent and as such perform better in road M and R compared to other African Nations. However, about 38% of its roadways are in fair to poor conditions [25]. The South African National Roads Agency SOC Limited (SANRAL) handles road M and R in SA. SANRAL accounts for 92% of the national roadway network, of which 19% are tolled and the remaining, toll-free. The 8% remaining from the total road network are privately managed, hence, tolled. The SANRAL regulated these private bodies based on thirty-year concession contracts [4].

SANRAL alongside Namibia can be categorized as top road M and R performers in Africa even though SA has a road M and R funding backlog between 80 and 149 billion Rand [4]. It has adopted efficient PMS-based routine, periodic and emergency maintenance strategies. Municipal roads are the only fraction not managed adequately due to poor data collection (only 4% of PCI data available), and consequently, poor coordination of municipal M and R responsibilities. However, the performance-based method for contracting M and R works has significantly been instrumental in the success attained in SA [4].

5. Discussion and Recommendations for Road M and R in Africa

In general, M and R activities in most African countries are very similar; however, the methods vary. Nonetheless, the importance of road M and R cannot be overemphasized due to its massive influence on regional amalgamation, economic advancement through efficient public administration and security, and social development. Furthermore, M and R prolongs the serviceability and safety of roadways, thereby minimizing internal and external costs of road usage. However, delay in M and R can be equivalent to its reconstruction as it leads to high direct and indirect costs.

Compared to other African countries, Ethiopia has a working road funding model autonomous of governmental input. However, SA and Namibia also record tremendous success in this regard while other Nations remain reliant on governmental funds. These funds are often constrained by the country's development needs. As such, road M and R always receive the least priority. Moving forward, the development of road funding models suiting specific national and cultural requirements is essential, especially in nations with vast populations such as Nigeria. Figure 2 illustrates the comparison between the countries discussed.

Concentrating on the technical issues related to M and R in Africa, it is clear that the issue of deficiency of data is encountered in all regions. It is therefore of utmost necessity that all African countries adopt GIS-based PMS methods in meeting road M and R requirements effectively. The development of spatial inter-state databases can further help in reducing road construction and M and R costs [4].

	M and R Financing	M and R Agency	Current M and R Practice	Current major problems faced
	Annual Budget Financing	FERMA	Very Poor / No GIS-based PMS	Poor funding Poor Administration
	Road Fund Model	Ethiopian RTA	Good / with GIS-based PMS	Poor funding
	Annual Budget Financing	GARBLT	Very Poor / No GIS-based PMS	Poor funding Poor data services
	Annual Budget Financing	DRC Roads Authority	Terribly Poor / No GIS-based PMS	Poor funding Poor Administration
	Annual Budget and Toll Financing	SANRAL	Satisfactory / with GIS-based PMS	Insuficient data services at municipalities

Figure 2. Comparison of M and R practice in selected African countries

6. Conclusions

This study reviews recent literature in recognizing the state-of-the-practice on the application of GIS-based techniques in sustainable PMMS, hence promoting a systematic approach for highway M and R operations. Additionally, it compares current M and R practices in the five major African regions. The study reveals that compared to individual nation's economic capacity, road M and R efforts are relatively high, however, compared with sustainable standards adopted in developed countries across the globe, these efforts lack significance. In the aspect of road ownership and establishment of management agencies and responsibilities, most African nations have performed well. However, the financing, technicality, and research and development aspects remain deficient as a majority still depend solely on budget financing, while adopting delayed reactive approaches to highway M and R activities. As much as the road agencies may want to perform, the issue of financing has always been a hindrance. In conclusion, there is a great necessity for African countries to:

i. Seek alternative sources for M and R funds,
ii. Integrate GIS-based PMS into M and R works at all functional levels,
iii. Consider the performance-based contracting approach for M and R,
iv. Invest in other modes of transportation such as railways and waterways to reduce pressure on roadways.

Acknowledgements
The authors wish to appreciate the management of Covenant University competently led by the Chancellor, Dr. David O. Oyedepo for the platform made available for this research and the sponsorship.

References

[1] A. M. G. Ribeiro, S. D. Capitão, and R. G. Correia, "Deciding on maintenance of small municipal roads based on GIS simplified procedures," *Case Stud. Transp. Policy*, vol. **7**, no. 2, *pp. 330–337*, Jun. 2019.

[2] A. N. Ede, "Cumulative Damage Effects of Truck Overloads on Nigerian Road Pavement," *Int. J. Civ. Environ. Eng.*, vol. **14**, no. 01, *pp. 21–26, 2014.*

[3] H. Gujba, Y. Mulugetta, and A. Azapagic, "Passenger transport in Nigeria: Environmental and economic analysis with policy recommendations," *Energy Policy*, vol. **55**, *pp. 353–361,* Apr. 2013.

[4] H. M. Mostafa, "Road Maintenance in Africa : Approaches and Perspectives," in *4th International conference on Energy Materials and Environmental Engineering (ICEMEE), E3S Web of Conferences, 01005*, 2018, vol. **38**.

[5] S. A. Ojo, J. O. Olusina, B. U. Ngene, A. A. Busari, J. Adediran, and A. Eletu, "Assessment of road infrastructure using remote sensing and GIS methodology for monitoring the condition of paved and unpaved roads," in *1st International Conference on Sustainable Infrastructural Development, IOP Conf. Series: Materials Science and Engineering, 012099*, 2019, vol. **640.**

[6] Y. Zhang and J. P. Mohsen, "A Project-Based Sustainability Rating Tool for Pavement Maintenance," *Engineering*, vol. **4**, no. 2, *pp. 200–208*, Apr. 2018.

[7] P. C. Anastasopoulos, R. J. G. M. Florax, S. Labi, and M. G. Karlaftis, "Contracting in highway maintenance and rehabilitation: Are spatial effects important?," *Transp. Res. Part A Policy Pract.*, vol. 44, no. **3**, *pp. 136–146*, Mar. 2010.

[8] S. M. Bazlamit, H. S. Ahmad, and T. I. Al-Suleiman, "Pavement Maintenance Applications Using Geographic Information Systems," in *Procedia Engineering*, 2017, vol. **182**, *pp. 83–90.*

[9] A. Wolters, K. Zimmerman, K. Schattler, and A. Rietgraf, "Implementing Pavement Management Systems for Local Agencies: State-ofthe-Art/State-of-the-Practice," Illinois, 2011.

[10] R. R. A. Almuhanna, H. A. Ewadh, and S. J. M. Alasadi, "Using PAVER 6.5.7 and GIS program for pavement maintenance management for selected roads in Kerbala city," *Case Stud. Constr. Mater.*, vol. **8**, *pp. 323–332*, Jun. 2018.

[11] PAVER^{TM}, "PAVER - Pavement Management Software," 2019. [Online]. Available: http://www.paver.colostate.edu/. [Accessed: 11-Jan-2020].

[12] A. Chen, Y. Zhao, P. Li, Y. Li, M. Mohammed, and P. Guo, "Crack propagation prediction of

asphalt pavement after maintenance as a function of initial cracks distribution," *Constr. Build. Mater.*, vol. **231**, p. *117157*, Jan. 2020.

[13] T. F. Fwa, *The Handbook of Highway Engineering*. Boca Raton, FL: CRC Press Taylor & Francis Group, 2006.

[14] J. Farhan and T. F. Fwa, "Pavement maintenance prioritization using analytic hierarchy process," *Transp. Res. Rec.*, no. 2093, *pp. 12–24*, 2009.

[15] L. Picado-Santos *et al.*, "Pavement management system for Lisbon," *Proc. Inst. Civ. Eng. Munic. Eng.*, vol. **157**, no. 3, *pp. 157–165*, 2004.

[16] P. Di Mascio and L. Moretti, "Implementation of a pavement management system for maintenance and rehabilitation of airport surfaces," *Case Stud. Constr. Mater.*, vol. **11**, Dec. 2019.

[17] M. K. Jha and J. Abdullah, "A Markovian approach for optimizing highway life-cycle with genetic algorithms by considering maintenance of roadside appurtenances," *Met. Finish.*, vol. **104**, no. 6, *pp. 404–419*, Jun. 2006.

[18] Federal Road Safety Commision, "Road Accident Statistics in Nigeria," 2013.

[19] Central Bank of Nigeria, "Highway Maintenance in Nigeria: Lessons from other countries," 2013.

[20] Federal Ministry of Works Nigeria, "Compendium report on road infrastructural and related development in Nigeria- An inventors manual," 2013.

[21] A. O. Owolabi, O. M. Sadiq, and O. S. Abiola, "Development of Performance Models for a Typical Flexible Road Pavement in Nigeria," *Int. J. Traffic Transp. Eng.*, vol. **2**, no. 3, *pp. 178–184*, 2012.

[22] A. O. Owolabi and O. S. Abiola, "Development of Priority Index Assessment Model for Road Pavements in Nigeria," in *8th International Conference on Managing Pavement Assets*, 2011, *pp. 1–13*.

[23] World Bank Group., "Transport, Economic Growth, and Deforestation in the Democratic Republic of Congo: A Spatial Analysis.," 2016.

[24] United Nations Joint Logistics Centre (UNJLC), "DRC Snapshot: Roads in the DRC," 2017.

[25] D. Jones, P. Paige-Green, and E. Sadzik, "Development of Guidelines for Unsealed Road Assessment," *Transp. Res. Rec. J. Transp. Res. Board*, vol. **1819**, no. 1, *pp. 287–296*, Jan. 2003.

Effect of iron ore tailings on the plasticity and compaction properties of selected tropical soils

P Yohanna[1], R O Sani[2], K Ishola[3], T S Ijimdiya[4], A O Eberemu[4*] and K J Osinubi[4,5]

[1]Department of Civil Engineering, University of Jos, Plateau State, Nigeria
[2]Department of Civil Engineering, Covenant University, Ota Ogun State, Nigeria
[3]Department of Civil Engineering, Osun State University, Osogbo, Osun State, Nigeria.
[4]Department of Civil Engineering, Ahmadu Bello University, Zaria, Kaduna State, Nigeria.
[5]Department of Civil Engineering and African Centre of Excellence on New Pedagogies in Engineering Education (ACENPEE), Ahmadu Bello University Zaria, Nigeria.

Corresponding Author: reuben.sani@covenantuniversity.edu.ng

Abstract. The workability of any soil, when used as a construction material, is greatly affected by its plasticity and compaction characteristics. This research investigates the effect of iron ore tailings (IOT) on the plasticity and compaction properties of lateritic soil (LTS) and black cotton soil (BCS). Atterberg limit and compaction test using three energies, British standard Light, BSL, British standard Heavy, BSH, and West Africa standard, WAS were carried out. These tests were conducted on mixed ratios of 0, 2, 4, 6, 8, and 10 % of IOT mixed each with LTS and BCS separately. A Statistical conceptual model was developed using Minitab R15 to predict the values of the maximum dry density (MDD) and optimum moisture content (OMC) from the computed laboratory results. Analysis was then carried out on the predicted results using origin pro version 8 to test the efficacy of results obtained. The result of specific gravity test shows increase with IOT concentration. An improvement on the plasticity of the modified soil was noted. The liquid limit for both LTS and BCS decreased with increasing IOT concentration. Values drop from 43.4 to 42.7% for LTS and 56 to 52.6% for BCS at 0 and 10% of IOT content, respectively. Plastic limit increased with increasing IOT concentration for LTS and decreased for BCS. In the case of the Plasticity index, values decreased for LTS and marginally increased for BCS treated with IOT. MDD values increased and thereafter drop for LTS treated with IOT while for BCS values increased from natural up to 10% IOT content. OMC values increased for LTS treated with IOT and decreased for BCS treated with IOT. The effectiveness of the developed model was validated using the values of the correlation coefficient obtained from the statistical analysis, which showed a solid relationship between predicted and measured values. Plasticity index values obtained for IOT stabilized BCS did not meet up to the standard requirement when used as a sub-base and base course material respectively. However, the treatment of LTS and BCS with 8-10% IOT improved the soil properties and can be used for pedestrian walkways.

Keywords: plasticity Index, compaction, conceptual model, predicted results, maximum dry density.

1. Introduction

Soils are important construction materials because they play a strategic role in the construction of roads, embankments and buildings [1]. For any soil to be used as a construction material, it must possess good engineering properties such as high strength and low settlement [1]. The workability of any soil, when used as a construction material, is greatly affected by its plasticity and compaction characteristics [2, 3, 4].

To achieve the United Nations sustainable development goal nine (9) on infrastructure, research is constantly ongoing to improve soils with poor engineering properties. The high cost of materials capable of improving the poor properties of soils to meet the various construction specification requirements, and the harmful effect of these materials to the environment has been a source of concern globally. Therefore, utilizing industrial waste (e.g. iron ore tailings, cement kiln dust, and steel slag) or agricultural waste (e.g. bagasse ash, rice husk ash, and periwinkle shell ash) alone, or in combination with other stabilizing agents have been alternative ways by which researchers use to improve on the poor properties of soils [5 - 10]. Sustainable handling of industrial and agricultural waste that constitutes a waste disposal problem globally requires the discoveries of various options on how to reuse them.

Smelting of iron in iron ore mining industry produces an industrial waste known as iron ore tailings [6]. The global annual production rate of iron ore tailings has been estimated at 14.6 billion tonnes [11]. The presence of high silica contents in most forms of iron ore tailings has drawn much research works towards its utilization as supplementary cementitious materials for construction purposes [12 – 14]. Iron ore tailings from research have been proven to improve on the grading and workability of soils [10, 15]. The plasticity of a soil is its ability to be moulded to any shape [3]. This makes the plasticity of soils very important when used for construction because they would have to take shape to suit their purpose or function during the construction process. This ability of a soil to take shape shows its degree of workability [16].

Stability, increased bearing capacity, resistance to deformation, and improved shear strength are desired properties to be achieved for any soil used for construction purposes [2]. Soil compaction is the process by which these desired properties can be achieved via the densification of the soil aimed at reducing the air void and increasing its density [2]. The higher the density, the greater the bearing capacity and strength of the soil [2]. Desired compaction density needs to be achieved in the field in order to avoid construction failure.

Various research works are available on the improvement of various soil properties using iron ore tailings either as a stone alone stabilizer or in combination with other stabilizing agents [6, 12, 15]. But less emphasis has been placed on how iron ore tailings can enhance the plasticity and the compaction characteristics of soils and the significance of such improvement. Therefore, this research investigates the effect of IOT on the plasticity and compaction properties of selected tropical soils.

2. Materials and Methods

2.1 Materials

2.1.1 Soil

Two soils were utilized for this research work; black cotton soil (Latitude 10°19'N and Longitude 11°30'E) and lateritic soil (Longitude 7°36'E Latitude 11°4'N). Black cotton soil samples were collected from a borrow pit in Gombe state, at depth of 1m, while samples of the lateritic soil were collected from a borrow pit in Zaria, Kaduna State.

2.1.2 Iron ore tailings

For this research, the IOT utilized was sourced from the National mining ore company in Itakpe, Kogi State, Nigeria. The IOT was filtered through sieve number 200 (0.075µm) and kept in air-tight containers in order to avoid pre- hydration of the IOT samples.

2.2 Methods

2.2.1 Preparation of specimens

Batches of the lateritic soil and the black cotton soil utilized for various laboratory tests were prepared using the following mix ratios: 0, 2, 4, 6, 8, and 10% of IOT mixed with soil.

2.2.2 Index properties

Atterberg limits test, (liquid limit, LL, plastic limit, PL and plasticity index, PI) was performed on the natural soils and the various mix ratios of iron ore tailings mixed with the soil. Soil samples that passed through sieve size 425µm was used for this test. Specific gravity test was also conducted on the natural soil, and IOT stabilized soils in accordance to test procedures outlined in British Standards, BS 1377 part 2 [17] and BS 1924 [18], correspondingly.

2.2.3 Compaction test

The compaction characteristics (i.e., MDD and OMC) of the natural and IOT stabilized soils were performed in accordance to test methods outlined in British Standards BS 1377 part 4 [19] and BS 1924 [18] respectively. Compaction was carried out using the BSL, BSH and WAS energy levels. The BSL compaction comprises of a rammer of weight 2.5 kg falling from a 300 mm height, applying onto three (3) layers of soil applying 27 blows in a 1000 cm^3 mould. While the BSH compaction involves a rammer of weight 4.5 kg falling from a height of 457.2 mm, applying 25 blows onto five (5) layers of soil in a 1000 cm^3 mould. WAS compaction involving a rammer of weight 4.5 kg falling from a height of 457.2 mm, applying ten blows onto five (5) layers of soil in a 1000 cm^3 mould.

2.2.4 Procedure for statistical analysis

A statistical conceptual model was established using Minitab R15, which was used to predict the compaction characteristics; OMC and MDD, from computed results obtained in the laboratory. Further analysis was then carried out on the predicted results using origin pro version 8 to validate the results obtained using this developed model.

3. Results and Discussion

3.1 Natural soil properties

Table 1 gives some properties of the LTS and BCS used in their natural form. From the properties obtained, the lateritic soil was classified as an A-7-6 (11) soil, while the black cotton soil was classified as A-7-6 (22) soil based on AASTHO classification [20]. According to AASTHO classification, these soils would have fair to poor performance when used for road sub-grade construction. Based on USCS classification [21], LTS and BCS were classified as CL and CH respectively.

Table 1. Properties of the black cotton soil and laterite soil in their natural form

Natural Soil Property	Lateritic soil (LTS)	Black cotton soil (BCS)
Percentage of soil particles passing sieve No 200	58.8	74.2
Soil natural moisture content (%)	16.24	19.5
The specific gravity of soils	2.59	2.44
Liquid limit of soils (%)	43.4	56
Plastic limit of soils (%)	21.3	25
Plasticity Index of soils (%)	22.1	31
Maximum Dry density of soils (Mg/m^3)		
BSL	1.72	1.68
WAS	1.76	1.64
BSH	1.83	1.56
Optimum moisture content of soils (%)		
BSL	16	19.3
WAS	15	20
BSH	14	23.5
AASTHO classification	A-7-6 (11)	A-7-6 (22)
USCS classification	CL	CH

3.2 Specific gravity.

Figure 1 presents the plots of specific gravities of LTS and BCS against IOT concentration. The values of specific gravity of LTS increased from 2.59 to 3.11 while that of BCS increased from 2.46 – 2.56 when treated with 0 up to 10% IOT content. This increase may be due to the high value of the specific gravity of the IOT (3.29) used for this research when compared to that of LTS (2.59) and BCS (2.46). A continuous increase was observed for both the LTS and the BCS when treated with up to 10% IOT due to the increasing density of the mixtures. Several works of literature reviewed showed a specific gravity range for IOT as always between 2.6 – 3.9 [22 – 25]. Results show that LTS treated with IOT recorded higher specific gravity than that of BCS, this could be linked to the higher specific of LTS (2.59) when compared to BCS (2.46).

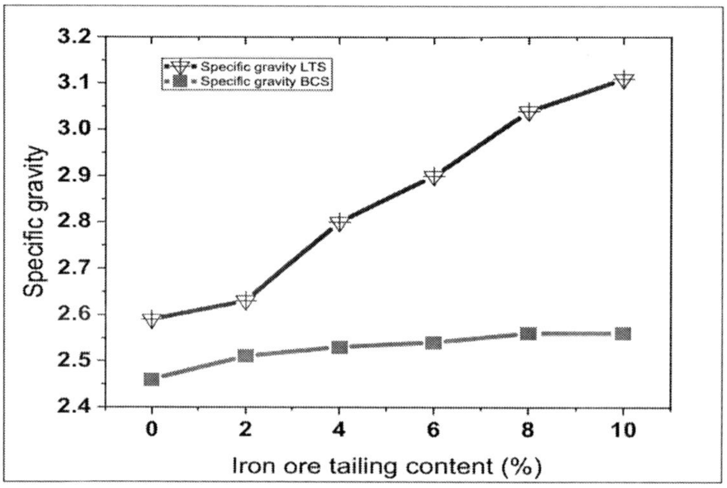

Figure 1. Plot of specific gravity of tropical soils – IOT mixtures

3.3 Atterberg Limits

3.3.1 Liquid limit:
The variation of liquid limit of LTS and BCS are as represented in figure 2 below. A decreasing trend of liquid limits when treated with IOT was observed for both soils. The natural LTS had a liquid limit of 43.4 %. On treatment with stepped concentration of IOT, it had its liquid limit decreased to 42.7% at 10% IOT content. BCS with a liquid limit of 56% for natural soil decreased to 52.6% at 10% IOT content. Flocculation and agglomeration of soil particles resulting from cation exchange reactions might be possible explanations for this decrease in liquid limit observed [6]. A similar trend of decrease in liquid limit was observed in research works carried out by Al-Zoubi [26], Portelinha et al. [27], and Ramesh et al. [28].

3.3.2 Plastic Limit:

The variation of plastic limit of LTS is as represented in figure 2 below. An increasing trend when treated with IOT was observed. The plastic limit of the natural LTS was 21.33%. On treatment with stepped IOT content, it had its plastic limit increased to 27.42% when treated with 10% IOT concentration. The plastic limit values are always subtracted from the liquid limit to give the soils plasticity index, which is often used as a check to the extent of improvement of a soil's properties. The natural BCS plastic limit decreased from 25% to 18.18% on treatment with 10% IOT concentration. The drop in the plastic limit value of BCS when treated with IOT can be explained as the replacement of weakly bonded monovalent cations of Na^+ and K^+ in the clay structures of BCS with divalent cations in the admixtures such as Ca^{2+}. Similar trends of increase in plastic limit were observed in research works carried out by Bayat, Asgari and Mousivand [29] and Khalid et al. [30]. The observed decrease in the plastic limit agrees with results reported by Ramesh et al. [28] and Portelinha et al. [27].

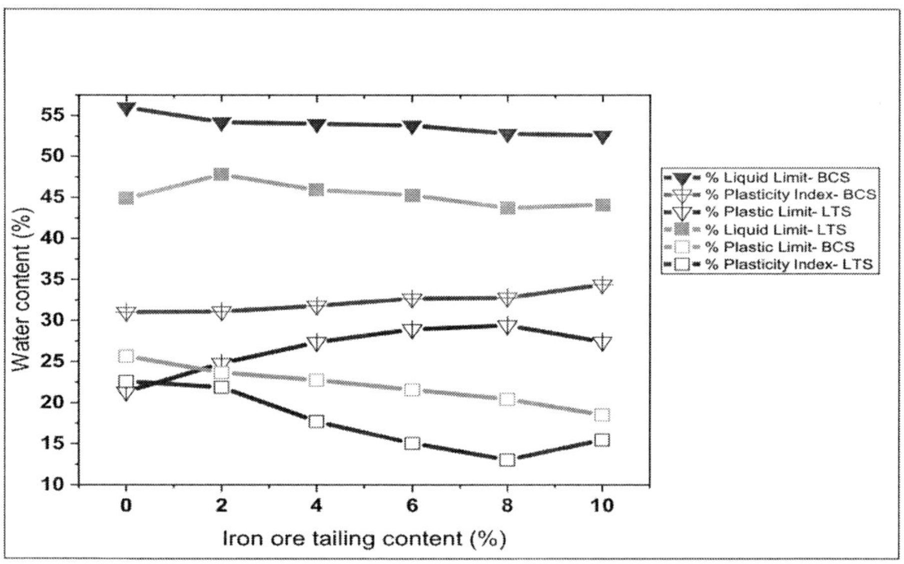

Figure 2. Plot of consistency limits of tropical soils – IOT mixtures

3.3.3 Plasticity Index:

Figure 2 presents the variation in the plasticity index of LTS and BCS. The plasticity index of LTS declined while that of BCS increased. The plasticity index of the natural BCS was obtained as 31% and slightly rose to 34.42% at 10% stepped treatment with IOT. LTS soil had a plasticity index of 22.7% for natural soil, which decreased to 12.9 % at 8 % IOT treatment.

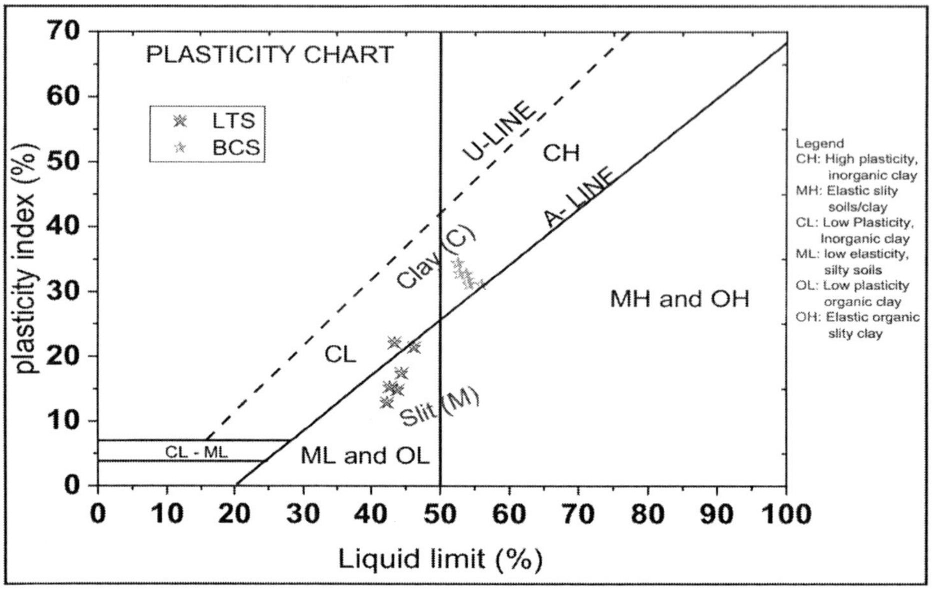

Figure 3. USCS Plasticity Chart for the soils

Figure 3 presents a plot of liquid limit against the plasticity index for both soils on the unified soil classification system USCS. The natural BCS and the IOT modified BCS values plot above the A-line and are characterized as inorganic clay of high plasticity. According to Nigerian standards, [31], all plasticity index values obtained for the natural BCS and the IOT modified BCS did not meet up to the standard values of <16 % when used as sub-base material and ≤ 13 % when used as base course material. However, the modified soils may be suitable for other engineering applications such as dams and embankments.

The plot of the plasticity index of the LTS against its liquid limit on the plasticity chart in figure 3 shows that the natural soil plotted just above the A-line and is characterized as a low plasticity inorganic clay. On modification with IOT, the modified soil samples plotted below the A-Line and are characterized as inorganic silt of medium compressibility and organic silts. This change in soil classification indicates that the soil became more workable when treated with IOT [32, 33]. At 6 – 10 % IOT content, the LTS met the Nigerian standard of <16 % when used as sub-base material while only at 8 % was the standard of ≤ 13 % when used as base course material was met [26].

3.4 Compaction characteristics

Figure 4 and figure 5 presents the variation in the MDD of LTS BCS with an increase in IOT content. The graphs also present the variation of the MDD at different energy levels. The MDD for the LTS at all energy levels was observed to increase to a peak at 6 % IOT content, then decrease on further addition of IOT. In the case of BCS, a continuous rise in MDD values was observed for BCS admixed with IOT. BSH energy level was also observed to have the highest MDDs followed by WAS then BSL energy level for both soils. A careful comparison of the MDD values of the modified soils show that stabilized LTS had the highest peak density values of 1.87 Mg/m^3 to 1.79 Mg/m^3 of BCS when compacted with BSH energy. Similar trends occurred for BSL and WAS energy levels. The trend of increase in MDD can be explained to be due to the IOT occupying void spaces in both the LTS and BCS matrix or the high specific gravity of the IOT [34 – 36]. The need for more moisture during cation exchange reactions or pozzolanic reactions are possible explanations for the decreased MDD observed at 8 and 10 % IOT – LTS treatment.

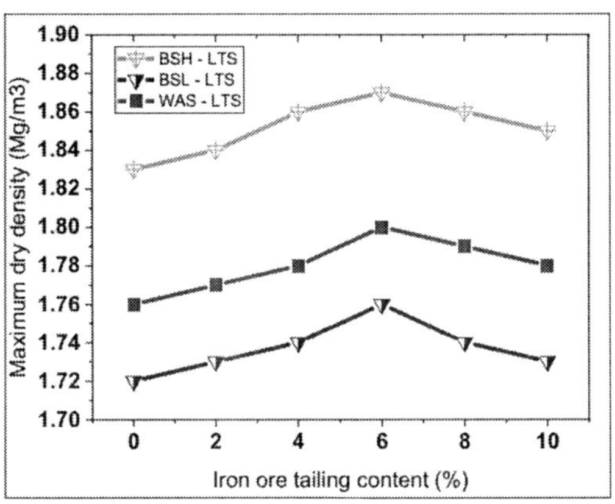

Figure 4. Plot of maximum dry density of lateritic soils – IOT mixtures

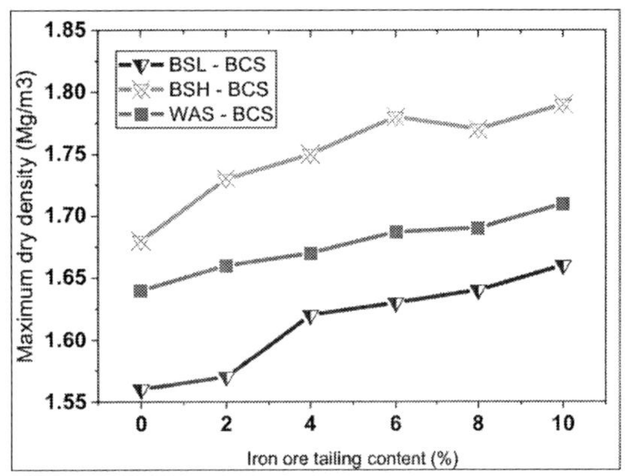

Figure 5. Plot of maximum dry density of black cotton soils – IOT mixtures

Figure 6 and figure 7 present the variation in the OMC of LTS and the BCS with an increase in IOT content. The graphs also present the variation of the OMC at different energy levels. The OMC of LTS at all energy levels are observed to increase to a peak at 6 % - 8 % IOT content then decreased with further addition of IOT. A continuous drop in the OMC values was observed for stabilized BCS. Possible explanations for the increased OMC observed at 0 – 6 % for BSL and at 0 – 8 % for WAS and BSH compacted IOT stabilized LTS, might be due to the need for more moisture during pozzolanic reactions occurring in the stabilized soil [37]. The decreased OMC observed for stabilized BCS can be due to loss in moisture content during the hydration process [38, 39].

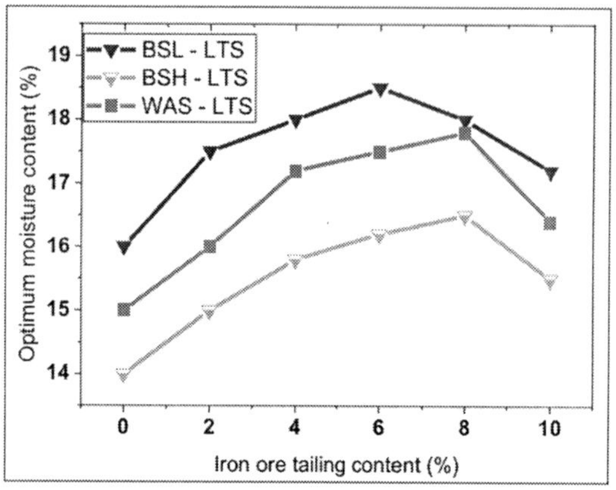

Figure 6. Plot of optimum moisture content of lateritic soils – IOT mixtures

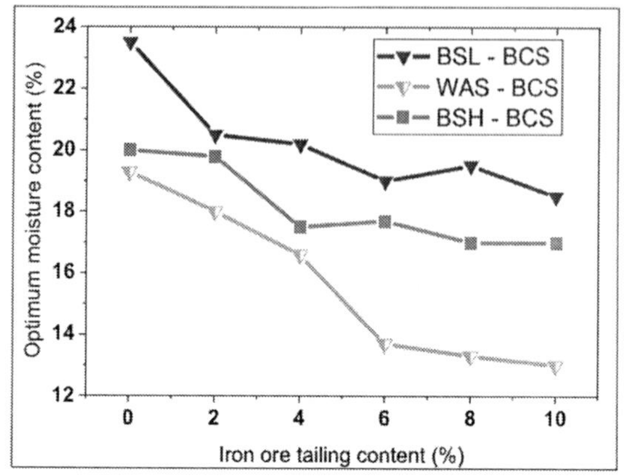

Figure 7. Plot of optimum moisture content of black cotton soils – IOT mixtures

3.5. Conceptual models and comparison between measured and predicted compaction properties (MDD and OMC)

Model equations developed from the laboratory test on both LTS and BCS are shown in equations 1-4. Results show a strong relationship between the compactions characteristics (i.e., MDD and OMC) for both LTS and BCS and the independent parameters considered (i.e., Iron ore tailing content IOT, Specific gravity Gs, Liquid limit, LL, Plasticity index PI and Compactive effort CE). The measured degree of correlation is shown by the coefficient of determination recorded as 96.1% and 97.1% for MDD (LTS) and MDD (BCS), respectively. In the case of OMCs, coefficient of determination of 96.3% and 89.2% were recorded for OMC (LTS) and OMC (BCS), respectively.

From the equations, an increase in variables with positive coefficients considerably increases the compaction properties, while variables with negative coefficients considerably reduce the compaction properties. The predicted compactions properties (MDD and OMC) using the model equations 1-4 are presented in Tables 2 - 5 with their respective percentage errors. MDD (LTS) recorded 0.03-0.76% error (see table 1) while MDD (BCS) recorded 0.03-0.95% error (see table 2). In the case of OMC (LTS) 0.01-1.83% error (see table 3) was noted while OMC (BCS) 0.17-8.28% error (see table 4) was recorded.

$$\mathrm{MDD(LTS)} = 2.42 + 0.0071\mathrm{IOT} - 0.204\mathrm{Gs} + 0.00115\mathrm{LL} - 0.0077\mathrm{PI} + 0.0575\mathrm{CE} \quad (1)$$

$$\mathrm{R}^2 = 96.1\%$$

$$\mathrm{MDD(BCS)} = -3.25 - 0.00498\mathrm{IOT} + 1.32\mathrm{Gs} + 0.0175\mathrm{LL} + 0.0207\mathrm{PI} + 0.0683\mathrm{CE} \quad (2)$$

$R^2 = 97.1\%$

$$\mathrm{OMC(LTS)} = -23.7 - 0.508\mathrm{IOT} + 8.45\mathrm{Gs} + 0.571\mathrm{LL} - 0.362\mathrm{PI} - 1.02\mathrm{CE} \quad (3)$$

$R^2 = 96.3\%$

$$\mathrm{OMC(BCS)} = 282 - 0.111\mathrm{IOT} - 71.4\mathrm{Gs} - 1.35\mathrm{LL} - 0.33\mathrm{PI} - 2.28\mathrm{CE} \quad (4)$$

$R^2 = 89.2\%$

Where MDD(LTS) = Maximum dry density of lateritic soil, MDD(BCS)= Maximum dry density of black cotton soil, OMC(LTS)= optimum moisture content of lateritic soil, OMC(BCS) = optimum moisture content of black cotton soil, IOT = iron ore tailing content, Gs = specific gravity, LL= Liquid limit, PI = plasticity index, CE = Compactive effort.

Table 2. Measured MDD values and predicted MDD values for Lateritic soil from the model

S/No	Compactive effort	Compactive effort index	IOT Content	Measured MDD	Predicted MDD	Absolute error	Percentage Error
1	BSL	-1	0	1.72	1.71	0.01	0.48
2	BSL	-1	2	1.73	1.73	0.00	0.30
3	BSL	-1	4	1.74	1.73	0.01	0.34
4	BSL	-1	6	1.76	1.75	0.01	0.76
5	BSL	-1	8	1.74	1.75	0.01	0.32
6	BSL	-1	10	1.73	1.73	0.00	0.15
7	WAS	0	0	1.76	1.77	0.01	0.52
8	WAS	0	2	1.77	1.78	0.01	0.76
9	WAS	0	4	1.78	1.79	0.01	0.65
10	WAS	0	6	1.8	1.80	0.00	0.23
11	WAS	0	8	1.79	1.80	0.01	0.73
12	WAS	0	10	1.78	1.78	0.00	0.28
13	BSH	1	0	1.83	1.83	0.00	0.18
14	BSH	1	2	1.84	1.84	0.00	0.07
15	BSH	1	4	1.86	1.85	0.01	0.58
16	BSH	1	6	1.87	1.86	0.01	0.44
17	BSH	1	8	1.86	1.86	0.00	0.03
18	BSH	1	10	1.85	1.84	0.01	0.41

Table 3. Measured MDD values and predicted MDD values for black cotton soil from the model

S/No	Compactive effort	Compactive effort index	IOT Content	Measured MDD	Predicted MDD	Absolute error	Percentage Error
1	BSL	-1	0	1.56	1.56	0.00	0.11
2	BSL	-1	2	1.57	1.58	0.01	0.95
3	BSL	-1	4	1.62	1.61	0.01	0.50
4	BSL	-1	6	1.63	1.63	0.00	0.03
5	BSL	-1	8	1.64	1.63	0.01	0.51
6	BSL	-1	10	1.66	1.65	0.01	0.50
7	WAS	0	0	1.64	1.63	0.01	0.81
8	WAS	0	2	1.66	1.65	0.01	0.41
9	WAS	0	4	1.67	1.68	0.01	0.61
10	WAS	0	6	1.69	1.70	0.01	0.70
11	WAS	0	8	1.69	1.70	0.01	0.59
12	WAS	0	10	1.71	1.72	0.01	0.59
13	BSH	1	0	1.68	1.70	0.02	0.89
14	BSH	1	2	1.73	1.72	0.01	0.49
15	BSH	1	4	1.75	1.75	0.00	0.08
16	BSH	1	6	1.78	1.77	0.01	0.72
17	BSH	1	8	1.77	1.77	0.00	0.1
18	BSH	1	10	1.79	1.79	0.00	0.09

Table 4. Measured OMC values and predicted OMC values for Lateritic soil from the model

S/No	Compactive effort	Compactive effort index	IOT Content	Measured OMC	Predicted OMC	Absolute error	Percentage Error
1	BSL	-1	0	16.00	16.02	0.02	0.11
2	BSL	-1	2	17.50	17.18	0.32	1.83
3	BSL	-1	4	18.00	18.02	0.02	0.10
4	BSL	-1	6	18.50	18.42	0.08	0.41
5	BSL	-1	8	18.00	18.44	0.44	2.46
6	BSL	-1	10	17.20	17.39	0.19	1.08
7	WAS	0	0	15.00	15.00	0.00	0.01
8	WAS	0	2	16.00	16.16	0.16	1.02
9	WAS	0	4	17.20	17.00	0.20	1.15
10	WAS	0	6	17.50	17.41	0.09	0.53
11	WAS	0	8	17.80	17.43	0.37	2.10
12	WAS	0	10	16.40	16.37	0.03	0.19
13	BSH	1	0	14.00	13.98	0.02	0.11
14	BSH	1	2	15.00	15.15	0.15	0.97
15	BSH	1	4	15.80	15.99	0.19	1.17
16	BSH	1	6	16.20	16.39	0.19	1.17
17	BSH	1	8	16.50	16.41	0.09	0.55
18	BSH	1	10	15.50	15.35	0.15	0.96

Table 5. Measured OMC values and predicted OMC values for black cotton soil from the model

S/No	Compactive effort	Compactive effort index	IOT Content	Measured OMC	Predicted OMC	Absolute error	Percentage Error
1	BSL	-1	0	23.50	23.22	0.28	1.19
2	BSL	-1	2	20.50	21.76	1.26	6.13
3	BSL	-1	4	20.20	20.17	0.03	0.17
4	BSL	-1	6	19.00	19.21	0.21	1.10
5	BSL	-1	8	19.50	18.95	0.55	2.82
6	BSL	-1	10	18.50	18.38	0.12	0.64
7	WAS	0	0	20.00	20.95	0.95	4.73
8	WAS	0	2	19.80	19.48	0.32	1.61
9	WAS	0	4	17.50	17.89	0.39	2.23
10	WAS	0	6	17.70	16.93	0.77	4.33
11	WAS	0	8	17.00	16.68	0.32	1.91
12	WAS	0	10	17.00	16.11	0.89	5.25
13	BSH	1	0	19.30	18.67	0.63	3.26
14	BSH	1	2	18.00	17.21	0.79	4.41
15	BSH	1	4	16.60	15.62	0.98	5.93
16	BSH	1	6	13.70	14.66	0.96	6.99
17	BSH	1	8	13.30	14.40	1.10	8.28
18	BSH	1	10	13.00	13.83	0.83	6.40

Figures 8 and 9 shows the plot of the predicted MDD values of the stabilized LTS and BCS against the measured MDD values. Using a third-order polynomial relationship to carry out a regression analysis, the correlation coefficient obtained for the predicted MDD against the measured MDD value was $R = 0.9837$ for IOT stabilized LTS and $R = 0.98287$ for IOT stabilized BCS. This shows a very strong relationship between predicted and measured values, thus validating the efficacy of the model.

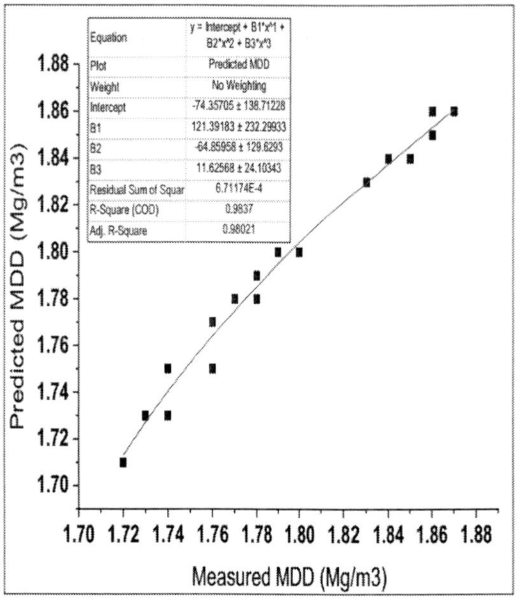

Figure 8. Plot of predicted MDD values against measured MDD values for the lateritic soil

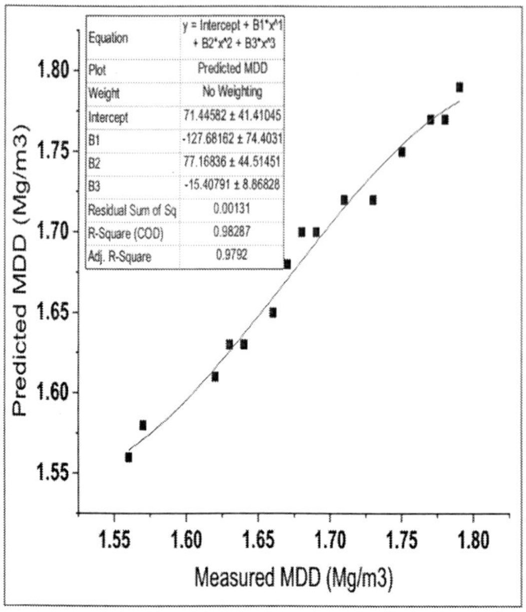

Figure 9. Plot of predicted MDD values against measured MDD values for the black cotton soil

Figures 10 and 11 shows the plot of the predicted OMC values of the stabilized LTS and BCS against the measured OMC values. A third-order polynomial relationship was used to perform the analysis, the correlation coefficient obtained for the predicted OMC against the measured OMC values were R = 0.97518 for IOT stabilized LTS and R = 0.9670 for IOT stabilized BCS. This shows a very strong relationship between predicted and measured values, also validating the efficacy of the model.

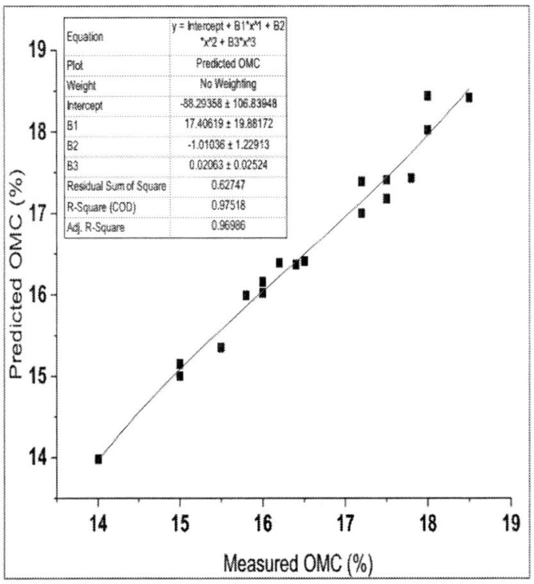

Figure 10. Plot of predicted OMC values against measured OMC values for the lateritic soil

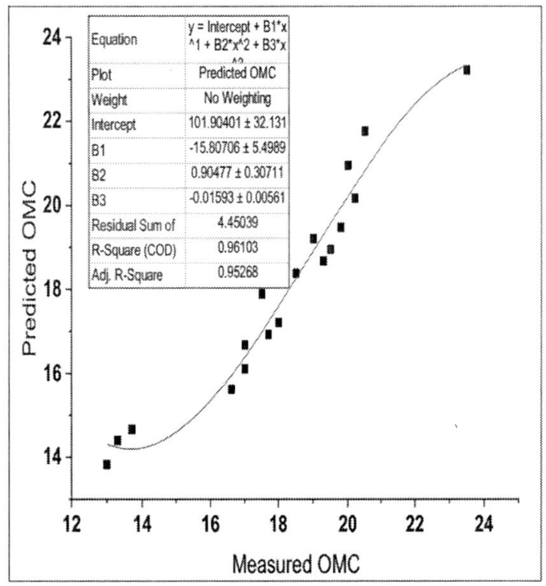

Figure 11. Plot of predicted OMC values against measured OMC values for the black cotton soil

4. Conclusion

The research work investigated the effect of IOT on the plasticity and compaction properties of LTS and BCS. The specific gravity increased with an increase in IOT concentration. The liquid limit for both LTS and BCS declined with an increase in IOT concentration. Values drop from 43.4 to 42.7% for LTS and 56 to 52.6% for BCS was noted at 0 and 10% of IOT content, respectively. Plastic limit increased for LTS and decreased for BCS treated with IOT. In the case of Plasticity index, values decreased for LTS and marginally increased for BCS treated with IOT. MDD values increased and thereafter drop for LTS treated with IOT while for BCS values increased from natural up to 10% IOT content. The OMC of LTS was observed to increase at all energy levels to a peak at 6 % IOT content for BSL compaction and at 8 % IOT content for BSH and WAS compaction, respectively. In the case of BCS, OMC values decreased with increasing IOT content. Possible loss in moisture may have occurred during the hydration process when IOT were used to stabilize BCS. The efficacy of the developed model was validated by the values of the correlation coefficient obtained, which showed a very strong relationship between predicted and measured values. Generally, treatment of LTS and BCS with 8-10% IOT improved the soil properties and can be used for pedestrian walkways.

References

[1] Vincy S S, and Mutharam M, 2009. Delayed Compaction Effects on the Behaviour of Stabilized Soils. India Geotechnical Society, IGC Guntur, India. Pp 419 – 423

[2] Zhang C, Wang H,, You Z, and Xu Yang, P.E. 2016. Compaction characteristics of asphalt mixture with different gradation type through superpave gyratory compaction and X-ray CT scanning. Construction and Building Materials Vol.128, pp 243 – 255.

[3] Moreno-Maroto J M, and Alonso-Azcarate J, 2018. What is clay? A new definition of "clay' based on plasticity and its impact on the most widespread soil classification systems. Applied Clay Science **161**, pp 57 – 63.

[4] Bergaya F, and Lagaly G, 2013. General Introduction: Clays, Clay Minerals, and Clay Science. In: Bergaya, F., Lagaly, G. (Eds.), Handbook of Clay Science, Developments in Clay Science, 2nd Ed. Vol. 5A. Fundamentals, Amsterdam, pp. 1–19

[5] Eberemu A O, and Sada H, 2013. Compressibility characteristics of compacted black cotton soil treated with rice husk ash. Nigerian Journal of Technology (NIJOTECH). Vol. 32(3), pp 507- 521. Faculty of Engineering, University of Nigeria, Nsukka, ISSN 1115-8443.

[6] Osinubi K J, Yohanna P, and Eberemu A O, 2015. Cement Modification of Tropical Black Clay Using Iron Ore Tailings as Admixture. Transportation Geotechnics vol. 5, pp 35– 49. https://doi.org/10.1016/j.trgeo.2015.10.001

[7] Neeraj K S, Swain S K, and Sahoo U C, 2012. Stabilization of a Clayed Soil with Fly ash and Lime: A Micro Level Investigation. Geotech Geol Eng, **30**, pp. 1197-1205. https://doi.org/10.1007/s10706-012-9532.

[8] Nnochiri E S, and Aderinlewo O O, 2016. Geotechnical Properties of Lateritic Soil Stabilized with Periwinkle Shell Ash in Road Construction. International Journal of Advanced Engineering, Management and Science, **2**(5), pp 484-487.

[9] Moses G, Saminu A, and Oriola F O P, 2012. Influence of compactive efforts on compacted foundry Sand treated with Cement Kiln dust. Civil and Environmental Research, **2** (5), pp. 11–24.

[10] Packey D J, 2012. Multiproduct mine output and the case of mining waste utilization. Resources Policy, **37**, pp. 104-108.

[11] Jones H, and Boger D V, 2012. Sustainability and Waste Management in the Resource Industries, Industrial & Engineering Chemistry Research. **51**(30), pp 10057-10065.

[12] Jung M Y, Choi Y W, and Jeong J G, 2011. Recycling of tailings from Korea Molybdenum Corporation as admixture for high-fluidity concrete, Environmental Geochemistry and Health **33** pp. 113-119.

[13] Zhou M K, Zhu Z G, Li B X, and Liu J C, 2017. Volcanic Activity and Thermal Excitation of Rich-silicon Iron Ore Tailing in Concrete, Journal of Wuhan University of Technology-Materials Science Edition **32**(2) pp. 365-372.

[14] Onuaguluchi O, and Eren O, 2012. Durability-related properties of mortar and concrete containing copper tailings as a cement replacement material, Magazine of Concrete Research **64**(11) pp. 1015-1023.

[15] Etim R.K, Eberemu A.O, and Osinubi K J, 2017. Stabilization of black cotton soil with lime and iron ore tailings admixture. Transportation Geotechnics **10**, pp 85 – 95.

[16] Polidori E, 2015. Proposal for a new classification of common inorganic soils for engineering purposes. Geotech. Geol. Eng. **33** (6), pp 1569–1579. http://dx.doi.org/10.1007/s10706-015-9922

[17] British Standard Institution, Methods of Testing Soils for Civil Engineering Purposes (1990), BS 1377, Part 2: classification tests, British Standards Institution: London.

[18] British Standard Institution, Methods of Test for Stabilized Soils (1990). BS 1924, *British Standards Institution*: London.

[19] British Standard Institution, Methods of Testing Soils for Civil Engineering Purposes (1990), BS 1377, Part 4: Compaction related tests, British Standards Institution: London.

[20] AASHTO (1986). Standard Specifications for Transport Materials and Methods of Sampling and Testing. 14th Edition, American Association of State Highway and Transport Officials (AASHTO), Washington, DC.

[21] ASTM (1992) Annual Book of Standards; vol. 04.08. Philadelphia: American Society for Testing and Materials.

[22] Fontes W C, Fontes G G, Costa E C P, Mendes J C, Silva G J B, and Peixoto R A F, 2018. Iron ore tailings in the production of cement tiles: a value analysis on building sustainability. *Ambient. constr*. vol.18 (4). http://dx.doi.org/10.1590/s1678-86212018000400312

[23] Uchechukwu E A, and Ezekiel M J 2014. Evaluation of the iron ore tailings from Itakpe in Nigeria as concrete material. *Advances in Materials* **3**(4): pp 27-32. doi: 10.11648/j.am.20140304.12.

[24] Kuranchie F A, Shukla S K, Habibi D. and Mohyeddin A. 2015. Utilization of iron ore tailings as aggregates in concrete. *Civil & Environmental Engineering, Cogent Engineering* **2**: *1083137* http://dx.doi.org/10.1080/23311916.2015.1083137

[25] Ugamal T I, Ejeh S P, and Amarte Y D, 2014. Effect of Iron Ore Tailing on the Properties of Concrete. *Civil and Environmental Research. Vol.6(10), pp 7 – 13.*

[26] Al-Zoubi M S. 2008. Undrained shear strength and swelling characteristics of cement treated soil. Jordan J Civil Eng **2**(1): pp 53–62.

[27] Portelinha F H M, Lima D C, Fontes M P F. and Carvalho C A B, 2012. Modification of a lateritic soil with lime and cement: an economical alternative for flexible pavement layers. *J Soil Rock*. **35**(1): pp 51–63.

[28] Ramesh H N, Krishnaiah A J, and Supriya M D 2013. Role of moulding water content on the strength properties of red earth treated with mine tailings. *Int J Sci Eng Res* **14**(5): pp 47–50.

[29] Bayat M, Asgari M R, and Mousivand M, 2013. Effects of cement and lime treatment on geotechnical properties of a low plasticity clay. International Conference on Civil Engineering Architecture & Urban Sustainable Development 27&28.

[30] Khalid N, Arshad F M, Mukr, M, Mohamah K, and Kamarudin F, 2015. Influence of Nano-Soil Particles in Soft Soil Stabilization. *Electronic Journal of Geotechnical Engineering*. **20**(2); pp 731 – 738.

[31] Nigerian General Specifications. Roads and bridges. Abuja, Nigeria: Federal Ministry of Works; 1997.

[32] Akinwumi I I, Booth C A, Diwa D, and Mills P, 2016. Cement stabilization of crude-oil-contaminated soil. *Proceedings of the Institution of Civil Engineers—Geotechnical Engineering,* **169**(4): 336–345. DOI: 10.1680/jgeen.15.00108.

[33] Akinwumi I I. and Booth, C.A. 2015. Experimental insights of using waste marble fines to modify the geotechnical properties of a lateritic soil. *Journal of Environmental Engineering and Landscape Management,* **23**(2): 121–128.DOI: 10.3846/16486897.2014.1002843.

[34] Kumar B, and Puri N, 2013.Stabilization of weak pavement subgrades using cement kiln dust. *Int J Civil Eng Technol.* **1**(4): pp 26–37.

[35] Jadhao, P.D, and Nagarnaik, P.B. 2008. Influence of polypropylene fibers on engineering behavior of soil_fly ash mixtures for road construction. Electron J Geotech Eng **13**: pp 1–11.

[36] Phanikumar B R, and Sharma Radhey S, 2004. Effect of flyash on engineering properties of expansive soil. J Geotech Geoenviron Eng **130** (7): pp 764–767.

[37] Joel M, and Edeh J E, 2016. Calcium Carbide Residue-Hydrated Lime Blend as Stabilizing Agent. *International Journal of Science and Technology* **5**(12): 606 – 614.

[38] Hossain K M A, Lachemi M and Easa S 2007 Stabilized soils for construction applications incorporating natural resources of Papua New Guinea. *Resources, Conservation and Recycling,* **51**, 711-731.

[39] Moses G, Saminu A, and Oriola F O P 2012 Influence of compactive efforts on compacted foundry Sand treated with Cement Kiln dust. *Civil and Environmental Research,* **2** (5), 11–24.

IOP Conf. Series: Materials Science and Engineering 1036 (2021) 012041 doi:10.1088/1757-899X/1036/1/012041

Investigating the suitability of *sakasaka* sand as replacement of natural aggregates in concrete for low volume pavement construction

D D Adegoke[1*], R O Sani[1] and K O Daramola[1]
[1] Civil Engineering Department, Covenant University, Ota, Ogun State, Nigeria.

*Corresponding author: dunmininu.adegoke@covenantuniversity.edu.ng

Abstract. The demand for access roads to vital services is on the high side in rural areas. This demand face much challenge due to the cost of construction. However, having the opportunity to use non-conventional materials that are naturally available in meeting this demand will ease people living in such environs. It is on this note that an investigation was carried out on an indigenous, naturally available sand popularly known as Sakasaka by Ilesha indigenes to ascertain its suitability for pavement construction. The Sakasaka collected from Ilesha borrow site, Ilesha West Local Government Area of Osun State, was used to produce concrete cubes that were tested after curing for 7, 14, 21, 28, 56, and 112 days respectively to know the strength of the material and compared to theoretical strength. Prior to the cubes production, some investigations such as chemical analysis, Sieve analysis,Specific gravity and so on were carried out on the sand. The study investigates the physical properties, chemical components of Sakasaka as well as the compressive strength properties of the concrete produced at ratio 1:2:4 using Cements A, B, and C 3x brand. The total numbers of cubes cast were 162 with a concrete cube size of 150x150x150mm. Sieve analysis, specific gravity, slump tests were also carried out for different water-cement ratios for the three cement. The concrete cubes were tested at the ages of 7, 14, 21, 28, 56, and 112 days. The effects of cements (A, B, and C) on the all-in-aggregate was investigated with respect to compressive strength at different curing age. From the results obtained, the compressive strengths of cubes were 17.76 N/mm^2, 19.92 N/mm^2 and 20.5N/mm^2 which was in line with the standard compressive strength of concrete mix 1:2:4 at 28days.
Keywords: Low volume roads, alternative materials, concrete, compressive strength, chemical analysis.

1. Introduction

Low-volume roads (LVR) roads link rural communities with planned road networks, in addition to important general services such as schools, hospitals, farms, and markets. They comprise a significant portion of the African road network, which can be up to 80-90% of the overall network length [1]. These roads are roads with traffic volumes of not more than 400 vehicles per day. They have huge effects on economies, communication, and social interaction [2]. Use of un-conventional materials in constructing these roads depend on the technical, commercial, and environmental state of that neighbourhood [3]. Most rural areas do not have access to conventional due to the cost involved. However, there is a need to connect to important services for ease in living. To create access for locals, the only alternative is the construction of low volume roads at a minimal cost. People living in the suburbs, as well as their governing sectors, possess less finance for sophisticated roads, and this posse a big challenge to such locality.

In emergent nations where concrete is largely employed , the huge and rapid increase in cost of concrete has made construction high-priced. High-pricing as well as negative environmental effects from producing concrete concrete, induced researchs on diverse materials forreplacing constituents of concrete. Many attempts are been made in search for novel materials, which is a combination of two or more materials. Singh, Ransinchung, Debbarma, and Kumar [4] made use of reclaimed asphalt aggregates and wastes from sugarcane mills for asphaltic mix in the construction of low volume roads. It was discovered that 10% of the alternative materials improved the mechanical and durability properties of the concrete. Mahgoub, and Mohamady, and Abdel-Motaleb, [5] worked on recycled building materials for Asphalt Pavement in low-volume agriculture roads and also found out that 25% of the concrete aggregates could be successfully replaced with maintained desired pavement properties. According to Modupe, Olayanju, Atoyebi, Aladegboye, Awolusi, Busari, and Modupe, [6], integrating cow bone ash into asphaltic concrete for hot mix asphalt as filler improved the stability as well as the flow of the asphaltic concrete. However, these studies and many others involve the processing of proposed alternative materials, which might induce additional cost. In other to eliminate additional cost that may be incurred, this study looks at investigating possible utilization of indigenous non-conventional materials like Sakasaka in Ilesha, Osun state, for low volume roads.
Sakasaka is a type of sand that mere looking at it could be classified as all- in aggregate. It is prominent and locally available in areas like Ilesha, Osun State. It has been in use by the locals as materials for the purpose of building construction since sharp sand is not readily available for use. In this study, some engineering properties were checked to determine its suitability for use as a total replacement for sharp sand. To achieve desired target strength, A, B, and cement C were used at different water-contents maintaining a mix design ratio of 1:2:4. The choice of these cements were made with the most suitable and available product in mind.

2. Materials Properties and Preparation

Cement brands A,B and Cwere used as binder material for the various concrete mixes. Concrete mix ratio of 1:2:4 was used with varying water/cement ratios of 0.7, 0.75, and 0.8. Granite aggregate passing through 19mm sieve and retained on 14mm sieve was used as coarse aggregate, and Sakasaka was used as a fine aggregate pass through 4.75 mm and retained on 0.075mm sieve with a maximum size of 12.5mm and potable water was used for mixing. Table 1 shows brands of cement with their manufacturer and grade.

Table 1. Brands of cement with their corresponding manufacturer and grades

Brand	Manufacturer	Type	Grade
Power Max	Lafarge WAPCO Nigeria Plc	CEM II	42.5N
Dangote 3X	Dangote Cement Plc	CEM II	42.5R
Super Set	Lafarge WAPCO Nigeria Plc	CEM II	32.5R

2.1 Aggregate

Sakasaka, found in Ilesha environ, Osun State, was used as fine aggregates, with particle sizes ranging from 75µmm – 4.75mm in diameter, was in place with the requirements of BS 882 [7] and BS EN 12620:[8]; Specific gravity test was performed on the samples in accordance to BS 812; part 2 [9] and EN 12620:2002; A1:2008. The aggregate used in the concrete mix was air-dried to remove all forms of moisture in the sand.

2.1.1. Coarse Aggregate

The crucial requirements for an aggregate to be acceptable for concrete incldes it being strong, durable, and inert. The sole fragment ideally is expected be of equal shape and be uniformly graded from coarse to fine within its particular grading size fraction. Granite with particles passing through the 19mm sieve satisfying these requirements was used for this research work. Specific gravity test was done on the sample in line with ASTM C 128-88, BS 812; part 1 1975 and EN 12620:2002; A1:2008 [7], grain size distribution test was performed on the sample based on BS 1377; part 2 [9], BS EN ISO 17892-1 [10].

2.2 Water
Water is vital as the strength of concrete results from the hydration process which takes place during the cementitious gel formation. Potable water was used in all the works carried out and had no physical impurities. It was also free from acids, organic matter suspended solids, and alkalis, which, when present, may have an adverse effect on the strength of concrete. It conformed to BS 3148 [11] requirements.

2.3 Concrete Mix and Batching
Batching was done by weight, and a concrete mix of 1:2:4 was used for producing concrete cubes. Three water-cement ratio regimes were used. These were; 0.65, 0.7, and 0.75 by weight.In determining the suitability of the aggregates (fine and recycled coarse aggregate) for concrete production, sieve analysis was conducted on the aggregates as prescribed by BS 1377 Parts 1 and 2. The grading for the aggregates and their relative combinations were determined and expressed in terms of grading coefficient: Coefficients of uniformity (C_u) and curvature (C_c). The concrete was used in preparing concrete cubes for compressive test, prior to which slum and compacting factor tests were carried out. Sixty-three cubes each was cast for each cement brand with various percentage of water content. The cubes were tested in sets of nine at 7 days, 14 days, 21 days, 28 days, 56 days, 91days, and 112 days, making a total of one hundred and eighty-nine cubes.

2.4 Experimental Steps
The following experimental steps and preliminary tests were carried out sequentially, and all tests were conducted according to relevant standards and specifications:

- Sample Pre-treatment
- Particle size distribution of coarse and fine aggregates.
- Concrete Workability Tests.
- Casting of Concrete specimens.
- Curing of Specimen.
- Compressive strength Test-Crushing of concrete cubes.

2.4.1 Sample Pre-treatment
The aggregate samples were maintained dry and free from moisture by air drying. The 50kg cement bags were kept on a dry raised platform to keep them dry and avoid caking, and once a bag was opened, it was used up within 24hours after opening. The water used was potable water supplied to the laboratory from the University of Lagos Works department.

2.4.2. Compressive Strength of Concrete Cubes
The compressive strength test was carried out on the cubes for the varying cement brands and water-cement ratios adopted for this study. The cubes were cured for 7, 14, 21, 28, 56, and 112 days respectively.
The compressive strength is computed from the formula below

$$Compressive\ Stress = \frac{applied\,load\ (N)}{cross\ sectional\ area\ of\ specimen\ (mm^2)}\ (\text{N/mm}^2) \quad (1)$$

3. Results and Discussions

3.1. Particle size distribution of sakasaka sand
ASTM C 33[12] and AASHTO M 6 [13 specifies limits based sieve sizes that fine aggregates must meet before they can be used in the production of concrete. Table 2 gives these standard limits.

Table 2. Fine-Aggregate Grading Limits (ASTM C 33/AASHTO M 6).

Sieve sizes (mm)	Percentage passing sieves by mass
9.5	100
4.75	95 - 100
2.36	80 - 100
1.18	50 – 85
600 μm	25 – 60
300 μm	5 – 30 (AASTHO 10 – 30)
150 μm	0 – 10 (AASTHO 10 – 30)

Figure 1 shows the particle size distribution curve of sakasaka sand used for this research. The graph shows the plot of the sieve size analysis and hydrometer analysis.

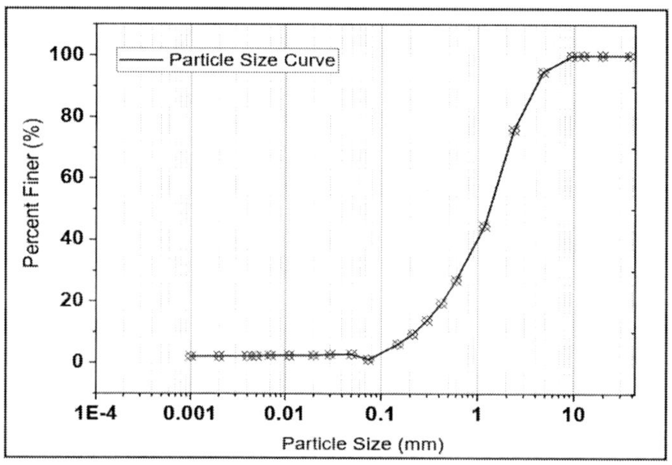

Figure 1. Particle Size Distribution of Sakasaka sand.

From the particle size graph, the fine aggregate just falls within the upper and lower limits specified by the ASTM standard, i.e. 99.9 % passed sieve No. 4 and 6.1 % passed sieve No. 100. Although, 75.9 % passed sieve No. 8, just a little short of the standard specified range of 80 – 100, the remaining six standards where met, making the sand just suitable for use in the production of concrete based on these criteria.

Table 3 shows the percentage gravel, percentage sand, percentage silt, percentage clay, coefficient of uniformity, and coefficient of curvature of the sand.

Table 3. properties of sakasaka sand.

Sand Property	Value
Percentage gravel (%)	37
Percentage sand (%)	54
Percentage silt (%)	7
Percentage clay (%)	2
coefficient of uniformity	9.83
coefficient of curvature	1.25

IOP Conf. Series: Materials Science and Engineering 1036 (2021) 012041 doi:10.1088/1757-899X/1036/1/012041

Based on unified soil classification system, USCS, the sand is classified as a well-graded gravel, GW, or gravel – sand mix having small or none fines since Cu is > 4 and Cc is less than 3. Though the sand is classified as a GW soil it has percentage fines (silt + clay) greater than 5 % maximum allowable value for pure sand classification and percentage gravel greater than 15 % [14]. This excess in percentage fines and gravel may have implications on the workability of concretes produced and can only be improved by the addition of mineral mixtures to mitigate this problem [15].
The specific gravity of sakasaka sand was obtained as 2.76, which is within the standard range set by ASTM of between 2.4 – 3 for sands.

3.2 Aggregate Size
Figure 2 shows the particle size distribution used in the mixing of the concrete.

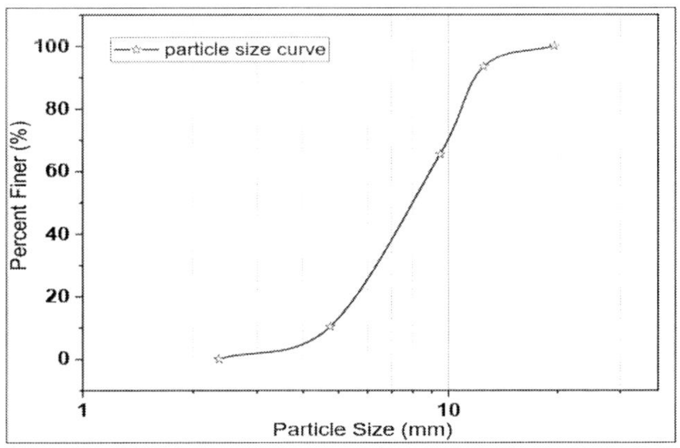

Figure 2. Particle Size Distribution of course aggregate

Aggregates that are retained on sieve No. 4 (4.75 mm) are known as coarse aggregates [16, 17]. 99.97% of the coarse aggregates used for this research were retained on sieve No. 4, meeting the ASTM C 33 [13] and AASTHO M 80 [18] grading requirements for coarse aggregates.

3.3 Concrete Workability Tests.
Concrete workability as defined by ACI Standard 116R-90 [19] is a property of freshly mixed concrete that portrays the ease and homogeneity to which it is mixed, placed, compacted and cured. The workability of a concrete mix is determined by performing the slump test [20] [21]. The slump test results of the freshly prepared concrete using the 3 brands of cement and different water-cement ratios, are presented in Table 4.

Table 4. Slump test results.

Cement brand	Water/Cement Ratio	Slum Value	Slump Type	Workability Class
	0.7	0	True Slump	Low
B	0.75	50	True Slump	Low
	0.8	85	Medium Slump	High
	0.7	70	True Slump	Low
A	0.75	92	True Slump	medium
	0.8	102	True Slump	High
	0.7	10	True Slump	Low
C	0.75	50	True Slump	Low
	0.8	135	True Slump	High

The slump was observed to increase with increasing water-cement ratios for all mixes. Workability is known to be directly proportional to water-cement ratio. As the water content increased, the workability increased. The workability class ranged from low to medium for the mixes and w/c ratio used. All slumps were true slumps as there was no record of shear or collapsed slump. Segregation and bleeding were not experienced as this is usually expected at high workability classes.

3.4 Compacting Factor Test
To further confirm the workability of the fresh concrete at 0.7 water-cement ratio, compaction factor test was done on the fresh concrete. The outcomes obtained are shown in Table 5.

Table 5. Compacting Factor Test @ 0.7wcr.

Weight of cylinder + concrete(kg)	17.11
Weight of cylinder + vibrated concrete(kg)	18.20
Weight of cylinder (kg)	5.96
Compacting Factor of Lafarge C	0.94
Compacting Factor of C	0.89
Compacting Factor of A	0.9

The specified range for compacting factor values are 0.7 to 0.98, value below or above this range is considered improper. At 0.7 water cement ratio that recorded no slump values, the compacting factor for all mix with different cement types were within the specified range making the water cement ratio an appropriate one.

3.5. Test on Hardened Concrete
Figures 3 – 5 show the variation of the compressive strength with time for the three brands of cements used in the casting of concrete at varying water-cement Ratio

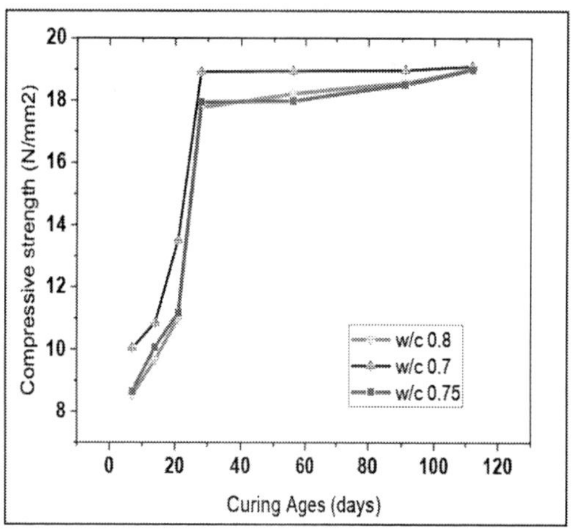

Figure 3. Variation of the compressive strength with Time for cement B samples at varying water – cement Ratio.

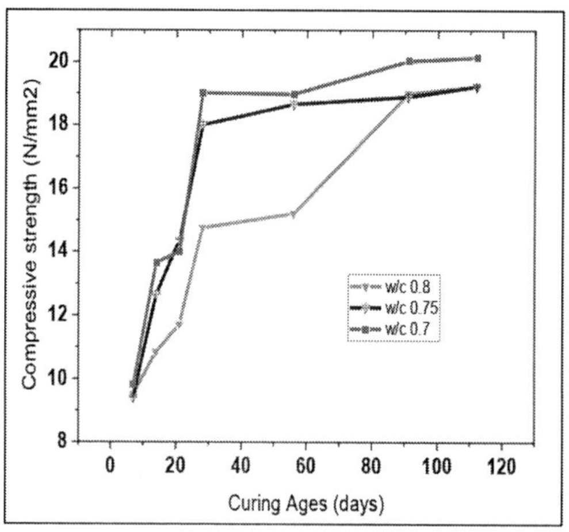

Figure 4. Variation of the compressive strength with Lafarge cement A concrete samples at varying water – cement Ratio.

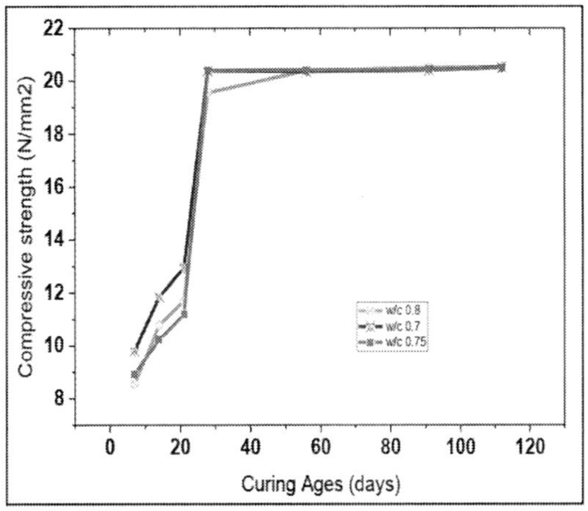

Figure 5. Variation of the compressive strength with Time for Cement C Concrete at varying water-cement Ratio.

From the figures above, it was observed that there was a continual rise in the strength gained for various samples of concrete from 7days to 28 days at various water cement ratio. However, 0.7 w/c appears to produce higher average strength of 20.97N/mm^2 follow by 0.75 w/c which gave average strength of 18.10N/mm^2 and then 0.8w/c gave average strength of 17.76N/mm^2. Cement B for Grade 25 produced higher strength at lower water cement ratio.

Cement A and C seem to follow the same trend with B as shown in the figures, 0.7w/c for all the cement brand produced higher concrete strength compared to strength obtained for water cement ratio 0.75w/c, and 0.8 w/c.

It can be observed from the graph also that cement C samples gained more strength than the rest at 7 days and 28days for various water cement ratio, followed by cement B and then Cement A samples.

Table 6 shows the compressive strength test on cements (N/mm2)The strength produced by Cement A and B is lower than the required strength at 28 days for various water cement ratio while strength produced by Cement C appeared ok at 28 days.

Cement C samples achieved more strength of 102.9% at 28 days for 0.7 w/c and 96.3% at 28 days for 0.8 w/c. Cement C produced the highest strength with lower w/c. The concrete strength decrease with increase in water percentage.

Table 6. Compressive Strength Test on Cements (N/mm^2)

CEMENT TYPE	**CURING AGES(DAYS)**		
	3	7	28
B	12.07	16.5	39.21
C	11.42	13.5	33.85
A	13.09	18.5	41.52

4. Conclusions

From the investigations made during this research, the following conclusions can be deduced:

- Sakasaka is classified as a well Graded gravelly soil (GW), according to the unified classification system.
- Cement C appear to have highest comprehensive strength of 20.5N/mm^2 with Sakasaka only at 28 days and 21N/mm^2 at 112days with minimum water cement ratio of 0.7, based on this research.
- All the cements brands A, B and C used for the purpose of this research, gave strength close to that specified in BS110 and CPU 110 which specified required concrete strength for grade 25 at 7days and 28 days to be 65% and 100% respectively with minimum water cement ratio.
- Both Chemical and metallic composition of Sakasaka justified that if improved upon, it can achieve desired strength.

Recommendations

From this research work done so far, the following recommendations can be made:

- Sakasaka can be used with cement C with minimum water cement ratio of 0.7 for grade 25 concrete where total replacement is required.
- Partial replacement instead of total replacement can be done to improve on the strength.
- Concrete made from Sakasaka can be used for Local Street where the expected wheel load is low.

Acknowledgment

The authors appreciate Covenant University, Centre for Research Innovation and Discovery for providing a platform for this study.

References

[1] Geddes R, and Gongera K. 2016. Economic Growth through Effective Road Asset Management. In *International Conference on Transport and Road Research, 16th–18th March*.

[2] Blazquez D, and Domenech J. 2018. Big Data sources and methods for social and economic analyses. *Technological Forecasting and Social Change*, ***130***, 99-113.

[3] Kurda R, De Brito J, and Silvestre J D. 2019. CONCRETop-A multi-criteria decision method for concrete optimization. *Environmental Impact Assessment Review*, ***74***, 73-85.

[4] Singh S, Ransinchung G D, Debbarma S, and Kumar P. 2018. Utilization of reclaimed asphalt pavement aggregates containing waste from Sugarcane Mill for production of concrete mixes. *Journal of Cleaner Production*, ***174***, 42-52.

[5] Mahgoub H, Mohamady A, Abdel-Motaleb M E, 2019. Recycled Building Materials for Asphalt Pavement in Low-Volume Agriculture Roads *12th International Conference on Low-Volume Roads,* Location: Kalispell Montana, United States, 2019-9-15 to 2019-9-18.

[6] Modupe A E, Olayanju T M A, Atoyebi O D, Aladegboye S J, Awolusi T F, Busari A A, and Modupe O C. 2019. Performance evaluation of hot mix asphaltic concrete incorporating cow bone ash (CBA) as partial replacement for filler. In *IOP Conference Series: Materials Science and Engineering* **640** (1), p. 012082). IOP Publishing.

[7] British Standards Institution 1992. Specification for aggregates from natural sources for concrete BS EN 12620:2002+A1:2008. British Standards Institution, London.

[8] British Standards Institution 2002. Aggregates for concrete BS EN 12620:2002 British Standards Institution, London.

[9] British Standard Institution, Methods of Testing Soils for Civil Engineering Purposes 1990, BS 1377, Part 2: classification tests, British Standards Institution: London.

[10] BS EN ISO 17892-1:2014 Geotechnical investigation and testing. Laboratory testing of soil. Determination of water content.

[11] British Standard Institution 3148: 1980 Methods of test for water for making concrete (including notes on the suitability of the water) British Standards Institution, London.

[12] ASTM (American Society for Testing and Materials). 2003. Standard Specification for Concrete Aggregates. C33. West Conshohocken, PA, USA.
[13] AASTHO (American Association of State Highway and Transportation Officials) M6 2013. Standard specification for fine aggregate.
[14] Thomas F W, 2003. Soil Relationships and Classification. In W. F. Chan and J. Y. Richard (Eds.), the civil engineering handbook (776–794). USA: CRC Press (Taylor and Francis Group).
[15] Neville A M. 2011. Properties of Concrete. (5th ed.), Pearson Education Limited, England.
[16] Duggal S K. 2008, *Building Materials* (3rd edn), New Delhi, New Age International, (P) Limited, Publishers.
[17] Council of Registered Builders of Nigeria (CORBON) and Nigeria Institue of Building (NIOB) 2014, *Improving the CorePractice Area for Builders IV,* 7th MCPDP held at Abuja, Portharcourt, Gasau and Lagos, 20th — 23rd 2014.
[18] AASTHO (American Association of State Highway and Transportation Officials) M80 2013. Standard specification for coarse aggregate.
[19] ACI 116R-90 (American Concrete Institute). Concrete and cement technology. ACI Standards.
[20] Busari A A, Akinmusuru J. O., Dahunsi, B., Ofuyatan, O. M., & Ngene, B. U. 2017. Pavement construction using self-compacting concrete: Mechanical properties. *International Journal of Advanced and Applied Sciences*, *4*(8), 50-55.
[21] Bamigboye G O, Olukanni D O, Adedeji A A, & Jolayemi K J, 2017. Experimental study on the workability of self-compacting granite and unwashed gravel concrete. In *International Journal of Engineering Research in Africa* **31**, pp. 69-76). Trans Tech Publications Ltd.
[22] Ngene B, Bamigboye G, Agomo C, Eneh K, & Olajide O, 2019. Magnetic Field Treated Water Influence On Structural Concrete Elements: Environmental Issues of Concern.

Fresh Characteristics of High-Performance Self-Compacting Concrete using Induction Furnace Slag as Supplementary Cementitious Material

O G Mark[1*], A N Ede[1], C Arum[2], K J Jolayemi[1], I Ojuawo[1], D Ajimalofin[1], J Adediran[1], I Babatunde[1], G B Olimaro[1]

[1]Civil Engineering Department, Covenant University, PMB 1023, Ota, Nigeria
[2] Civil Engineering Department, Federal University of Technology, PMB 704, Akure, Ondo State, Nigeria

*Corresponding author: oluwaseun.mark@covenantuniversity.edu.ng

Abstract. Using wastes products in concrete contributes towards sustainability in civil engineering construction. Producing high-performance-self-compacting concrete (HPSCC), using induction furnace slag (IFS) contributes positively to the sustainability of concrete technology. Many researchers have developed self-compacting concrete (SCC), but there are minimal investigations reported considering the high-performance in the development of SCC. The aim of this paper is to determine the fresh characteristics of HPSCC, using IFS as a supplementary cementitious material. Tests were conducted on filling ability, passing ability, as well as resistance to segregation of the fresh HPSCCs, containing IFS at percentage replacement of 0%, 10%, 20%, 30%, 40% and 50%. Slump flow and T_{50cm} slump flow tests were conducted to determine the filling ability. L-Box test was used to determine the passing ability. V-funnel at $T_{5minutes}$ test was conducted to determine the resistance to segregation. The results showed that the slump flow ranged between 652 and 687 mm, the T_{50cm} slump flow ranged between 2.59 and 3.97 seconds, the L-box value ranged between 0.81 and 0.95, while V-funnel at $T_{5minutes}$ ranged between 1.88 and 3.11. The fresh properties criteria, according to EFNARC were fulfilled, except for few concretes with IFS content greater than 20%. It is therefore recommended that induction furnace slag be incorporated as a supplementary cementitious material (SCM) when producing HPSCC.
Keywords: High-performance-self-compacting concrete, induction furnace slag, supplementary cementitious materials, filling ability, passing ability, resistance to segregation.

1. Introduction

Concrete, mostly limestone Portland cement concrete is a blended substance, which contains coarse and fine aggregates, blended with cement. It is sometimes combined with other hydraulic cements, such as supplementary cementitious materials [1].

IOP Conf. Series: Materials Science and Engineering 1036 (2021) 012042 doi:10.1088/1757-899X/1036/1/012042

Currently, limestone Portland cement, as well as SCMs are the least expensive binding agents that keep the concrete production. Nevertheless, limestone Portland cement manufacture consumes more energy. Each limestone Portland cement (PC) production tonne generates about 1 tonne of carbon dioxide (CO_2). As a result, manufacture of limestone Portland cement cement production causes close to 5% for every carbon dioxide discharge [2].

The principle of high-performance concrete developed recently. At first, it was taken as high-strength concrete. This possesses certain advantages, unfortunately, it is not complete in its sense. It is needful to look into several concrete parameters also, which oftentimes are more important than strength parameter.. High-Performance concrete can be defined as a concrete containing suitable constituents, mixed in accordance to a chose design mix; adequately blended, moved, deposited, compacted and moistened in such a way that concrete produced exhibits an outstanding characteristics in the structure been positioned, in the environmental conditions been subjected to, as well as the loadings it will carry when designed. Therefore, high-performance concrete can be regarded as durable concrete.

Self-compacting concrete can be defined as that unique concrete which has the potential to pass across densely-packed steel bars, fill all parts formwork, as well as compact well without a compactor. It exhibits good passing ability, outstanding filling ability, as well as excellent resistance to segregation [3], [4].

Combining the durability characteristic, as well as the high-strength property of high-performance concrete (HPC) with self-compacting concrete (SCC) and at a lower water-binder proportion, high-performance self-compacting concrete (HPSCC). High-performance self-compacting concrete can be defined as a type of concrete that produces outstanding characteristics, in terms of strength, flowability, transportability, as well as durability in line with the specifications for conditions of service at a specified loading, constituent materials, as well as environmental exposure.

The fundamental HPSCC fresh characteristics are the resistance to segregation, passing ability, as well as filling ability.

The filling ability property of HPSCC can be defined as the potential of HPSCC, in its fresh state to fill all parts of the concrete formwork at its self-weight, without a vibrator [5], [6]. The filling ability property of HPSCC has an important effect on the placing of concrete and its ability to consolidate at its own weight [7]. The passing ability of HPSCC generally refers to the capacity of HPSCC to pass across densely packed reinforcing rods [5], [6]. For densely reinforced structures, the excellent HPSCC passing ability allows for easy placing of the concrete, as well as self-compaction across the heavily reinforced steel rods [7]. The HPSCC's resistance to segregation means the ability of the concrete to continue to be stable without any form of bleeding, settling of the coarse aggregates, as well as disconnection of the concrete paste [6]. Particularly, when HPSCC does not satisfy the adequate resistance to segregation property, there is non-uniformity in the concrete mix.

Supplementary cementitious materials can be defined as those finely grounded materials that add to the characteristics of concrete by their hydraulic and/or pozzolanic behaviour [8]. Mark et al [9] in their review paper found out that SCMs have positive influence on the properties of normally-compacted concrete. Arum and Mark [10] in their study concluded that cupola furnace slag (the old form of induction furnace slag), which is an SCM, has a positive influence on concrete.

In producing cast iron in the modern way, secondary by-products are formed. The most abundant by-product obtained in this modern way is induction furnace slag. When an induction furnace slag is crushed, pulverized, milled and sieved to cement size, it is referred to as ground granulated induction furnace slag (GGIFS). Induction furnace slag is such material that can be used as cement replacement [11]. This material considers being a potential waste material which is dumped near the industrial area. Using a cheaper SCM like induction furnace slag (IFS) will reduce the total cost of producing HPSCC. Likewise, usage of IFS minimizes the cement usage, as a result, lowers the expenses incurred in manufacturing cement as well as reduces the emission of harmful gases to the environment evolving from cement manufacturing industries. Therefore, IFS does not just enhance the durability and properties of concrete. It also brings ample benefits environmentally and economically. Up till now, there is limited research carried out to investigate the capacity of IFS by exploring its influence on durability, fresh and hardened properties of HPSCC.

2. Materials and Methods

2.1. Materials

2.1.1. Induction Furnace Slag (IFS). Locally available induction furnace slag was obtained at Nigeria foundries limited, Sango, Ota, Ogun State, Nigeria. It was crushed, pulverized and milled to fine aggregate size using steel ball rolling mill at Highway/Transportation Laboratory of Covenant University, Ota, Ogun State, Nigeria. The fine aggregate-sized slag was sieved through 45 μm sieve so as to get the powdered IFS that was utilized as the SCM, conforming to cement grading requirement of BS EN 197-1 [12].

2.1.2. Portland Cement. Dangote 3X Portland cement was utilized, and was obtained at a retail outlet in Ota, Ogun State, Nigeria. The cement satisfied ASTM Type 1 specification [13].

2.1.3. Granite. Locally available granite was obtained at a quarry site in the form of a mixture of broken and spherical granite. The nominal size of the granite was 12.5 mm. The spherical granite took half of the whole portion of the granite content.

2.1.4. River Sand. Indigenous river sand was utilized as the fine aggregate. This river sand was obtained as soil deposition.

2.1.5. Potable Water. Drinkable water was utilized for mixing the concrete. The quality of the water was such that it was free from impurities and potable enough for drinking.

2.1.6. Superplasticizer. Polycarboxylate-based HRWR, commercially known as Complast SP 430 superplasticizer was utilized in achieving the expected concrete flow, in accordance to ASTM standard [14].

2.2 Methods

2.2.1. Mixture proportioning of the various types of concrete mixture. A total of six (6) different types of HPSCC mixtures were designed, as shown in Table 1. The concrete mixtures were designated based on the GGIFS content. The mixing ratio of the aggregate contents obtained was based on saturated surface-dry condition. However, the dried aggregates were utilized for the concrete production. Therefore, the aggregates took in substantial quantity of the water for during concrete batching. Also, superplasticizer added to the fluidity of the concrete. Thus, the taking in of water by the aggregates, as well as the addition of fluidity by superplasticizer was considered when determining the water for mixing. Moreover, the aggregate contents were determined based on their absorption and moisture content.

Table 1. Details of concrete mixture proportions.

S/N	Concrete Type (kg//m^3)	Cement (kg/m^3)	GGIFS (kg/m^3)	River Sand (kg/m^3)	Granite (kg/m^3)	Water (kg/m^3)	Superplasticizer (kg/m^3)
1	HPSCC0,100	733.0	0	747.66	696.35	263.88	14.66
2	HPSCC10,90	659.7	73.3	747.66	696.35	263.88	14.66
3	HPSCC20,80	586.4	146.6	747.66	696.35	263.88	14.66
4	HPSCC30,70	513.1	219.9	747.66	696.35	263.88	14.66
5	HPSCC40,60	439.8	293.2	747.66	696.35	263.88	14.66
6	HPSCC50,50	366.5	366.5	747.66	696.35	263.88	14.66

2.2.2. Fresh Concretes Production. Fresh HPSCC concretes were produced according to ASTM standard [15], and were done at the Structures Laboratory, Covenant University, Ota, Ogun State.

2.2.3. Batching procedure. The batch quantities of the fresh HPSCC were calculated before batching. The fresh HPSCC batch quantities taken were at the minimum of 15% higher, in order to cater for the losses at the time of test. The constituent materials were taken on weight basis.

2.2.4. Fresh HPSCC tests. The fresh HPSCC mixtures were tested for filling ability by carrying out slump flow test and T_{500} slump flow test. It was also tested for passing ability by carrying out L-box test and finally tested for segregation resistance by carrying out V-funnel at $T_{5minutes}$ test. These were done at the Structures Laboratory of Covenant University, Ota, Ogun State.

2.2.4..1 Slump flow and T_{500} time Tests. Conduction of slump flow test was done according to EFNARC standard [6], utilizing Abram's slump cone, as seen in Figure 1. Concrete samples were poured inside the apparatus in a single one layer, with no form of compaction. The apparatus was removed uprightly, to make the concrete sample to distort on top of a non-absorbent pan. A stopwatch was started simultaneously while the time required for the flowing concrete to get to 500 mm diameter circle was taken to be the T_{500} time slump flow value. The concrete flow diameter was taken as the slump flow, and was taken at the four (4) points dividing the spread to eight (8) parts.

Figure 1. Slump Flow Test.

2.2.4.2. L-box Test. Conduction of L-box test was done according to EFNARC [6], as seen in Figure 2. The apparatus (L-box) was positioned on a flat ground, while closing the sliding gate with the sliding gate closed. Fresh concrete was poured into the vertical part of the L-box, which was allowed to rest for about a minute. Afterwards, the fresh concrete was made to flow to the horizontal part of the L-box. The vertical height H_1, as well as the horizontal height H_2 of the concrete in the apparatus was taken, immediately the fresh concrete did not flow again. The blocking ratio H_1/H_2 was determined.

Figure 2. L-Box Test.

2.2.4.3. V-funnel at $T_{5minutes}$ test. Conduction of V-funnel at $T_{5minutes}$ test was done according to EFNARC [6], as seen in plate 3. Water was poured into the apparatus (V-funnel), to wet it, and the apparatus was later placed on a flat ground. A container was put under the apparatus, while closing the flap. Filling of the L-box with fresh concrete was done once, without compacting or tapping it with compacting rod. After five minutes, opening of the flap was done, and the concrete was made to flow into the container, by its own weight. Concurrently, the time it took for the fresh concrete to pour into the container was measured as the time of flow at $T_{5minutes}$.

Figure 3. V-funnel at $T_{5minutes}$ Test.

3. Results and Discussion

3.1. Results

The HPSCC mixtures results for slump flow, T_{500} slump flow test, L-box test and V-funnel at

$T_{5minutes}$ test are presented in Figures 4 to 7.

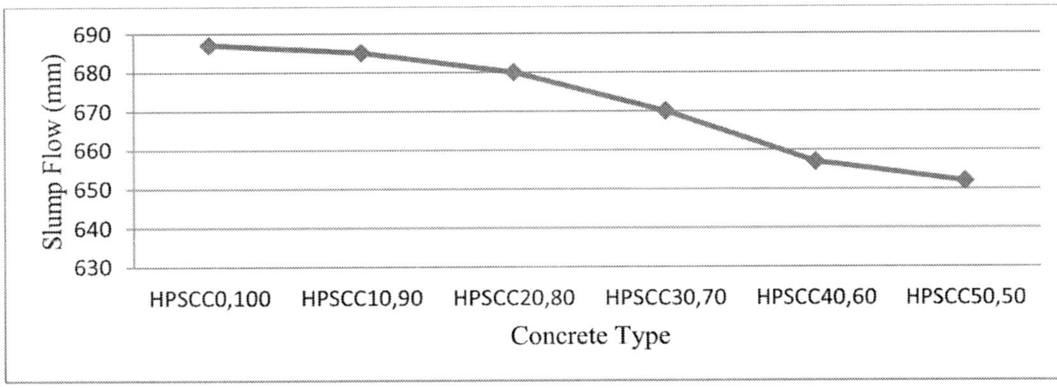

Figure 4. Effect of Induction Furnace Slag on Filling Ability Property of HPSCC.

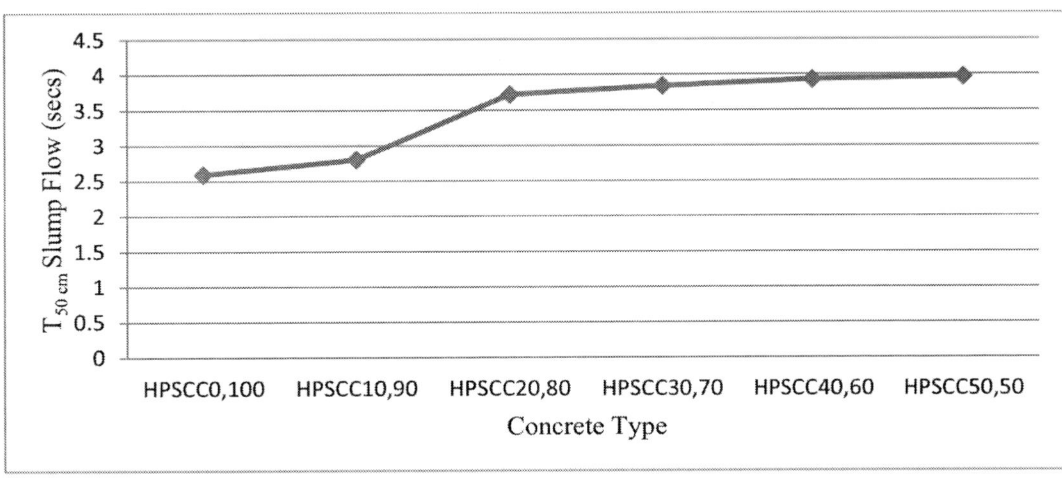

Figure 5. Effect of Induction Furnace Slag on Filling Ability Property of HPSCC.

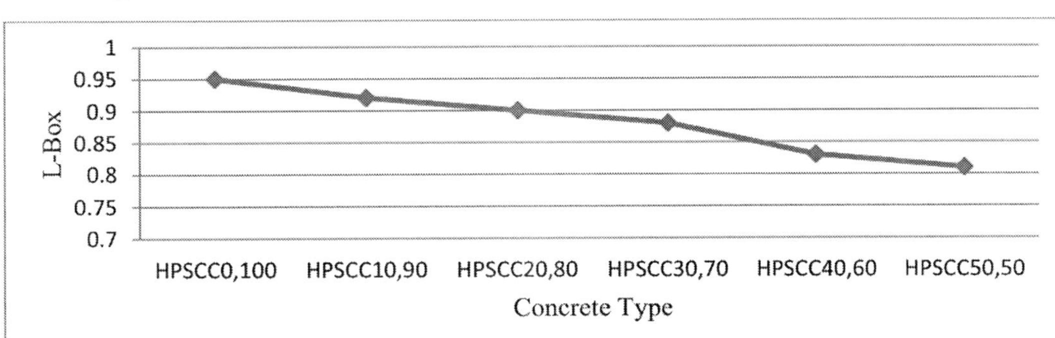

Figure 6. Effect of Induction Furnace Slag on Passing Ability Property of HPSCC.

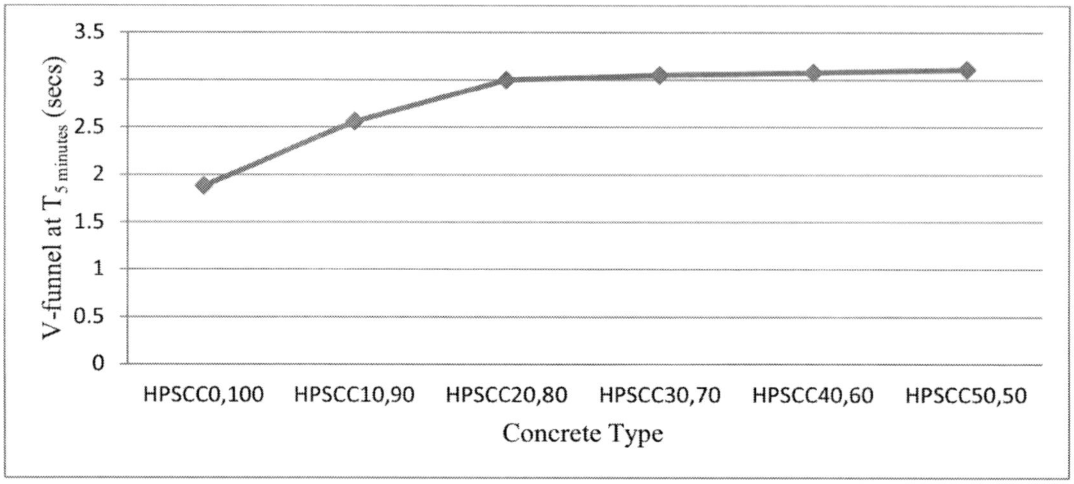

Figure 7. Effect of Induction Furnace Slag on Segregation Resistance Property of HPSCC.

3.2. Discussion

3.2.1 Filling Ability of Concretes. the slump flow value of the fresh HPSCC ranged between 652 mm and 687 mm, indicating an outstanding filling ability of the HPSCC. Usually, HPSCC slump flow values range between 550 and 850 mm [6], [16], [17]. From Figure 4, as IFS quantity increased, the slump flow value reduced considerably. This is caused by the higher contents of concrete paste, reduced quantities of aggregates, as well as higher IFS fineness. Increase in IFS content increased the volume of the fresh concrete paste, as well as reduced the quantities of the aggregates. Slump flow value usually reduces when the volume of concrete paste increases, quantity of aggregates reduce, as well as when the binders fineness increase [18], [19]. The high binders fineness, high volume of fresh concrete paste, as well as reduced quantities of aggregates need additional mixing water for flowability and hence, reduce the deformability of the concrete, as the collision between aggregates is greatly increased with lower inter-particle distance. Therefore, HPSCC filling ability can be assessed by slump flow test.

The T_{50cm} slump flow of different HPSCC mixtures ranged between 2.59 and 3.97 seconds, as seen in Figure 5. The acceptable range is between 2 and 5 seconds [6]. The T_{500} slump flow of concretes with higher IFS content was above the control mix. This was because there was increase in kinetic energy. The flow times increased with lower W/B ratio and greater IFS content, thus indicating an increase in the plastic viscosity of the concretes. The volume fraction and surface area of binder were increased with higher IFS content. The increased volume fraction and surface area of the binder increase the viscosity of concretes, and hence increase the flow time of concretes [20]. However, the aggregate content of concretes was reduced and the paste volume was increased simultaneously with the greater IFS content. The reduced aggregate content and the increased paste volume increased the plastic viscosity of the concretes [21], [6].

3.2.2. Passing Ability of Concretes. The ratio h_2/h_1 varied in the range of 0.81 to 0.95 as can be seen from figure 3. Usually, the ratio of h_2/h_1 for HPSCC varies from 0.8 to 1.0 [6]. The minimum ratio h_2/h_1 was observed for the HPSCC with 50 % IFS. But the highest ratio h_2/h_1 was attained for the HPSCC with 0%

IFS. The trend was similar as observed in case of the filling ability properties of the concretes. The ratio h_2/h_1 exhibited a good passing ability of the concretes.

3.2.3. Segregation Resistance of Concretes. The V-funnel at $T_{5minutes}$ varied in the range of 1.88 to 3.11 seconds as can be seen from figure 4. Usually, the V-funnel at $T_{5minutes}$ for HPSCC varies from 0 to 3 seconds [6]. The minimum value was observed for the HPSCC with 0% IFS. But the highest value was attained for the HPSCC with 50% IFS. The trend was similar as observed in case of slump flow. The V-funnel at $T_{5minutes}$ exhibited a good segregation resistance of the concretes.

4. Conclusions and Recommendations

a. The test results obtained for the fresh properties showed that the slump flow reduced with greater IFS content. In addition, the T_{50cm} slump flow increased considerably with greater IFS content. Also, the L-Box decreased with higher IFS content. Likewise, the V-funnel at $T_{5minutes}$ increased considerably with greater IFS content.
b. The filling ability, passing ability and segregation resistance criteria were fulfilled for all HPSCCs except few concretes with IFS content greater than 20% that exhibited high viscosity mostly due to the excessive surface area of IFS.
c. It is recommended that induction furnace slag should be used as a supplementary cementitious material, since the fresh properties of the resulting IFS-based HPSCC were within the standard limits specified by EFNARC.

Acknowledgement
The authors would like to appreciate Covenant University Centre for Research, Innovation and Discovery for the financial support.

References

[1] Ede A N, Olofinnade O M, Bamigboye G O, Shittu K K and Ugwu E I 2017 prediction of fresh and hardened properties of normal concrete via choice of aggregate sizes, concrete mix-ratios and cement. *International Journal of Civil Engineering and Technology* **8** 288-301.
[2] Tatem A J 2003 Increasing the spatial resolution of agricultural land cover maps using a Hopfield neural network. *International Journal of Geographical Information Science* **17** 647-72.
[3] Khayat K H 2009 Workability, testing, and performance of self-consolidating concrete", *ACI Materials Journal* **96** 346-53.
[4] Ede A N, Oshogbunu O, Olofinnade O M, Mark O G and Awoyera P O 2019 Effects of bamboo fibres and limestone powder on fresh properties of self-compacting concrete, *10th International Structural Engineering and Construction Conf.,* **42**-1-6.
[5] Bartos P J M 2000 Measurement of key properties of fresh self-compacting concrete", *Paper presented in the CEN/STAR PNR Workshop on Measurement, Testing and Standardization: Future Needs in the Field of Construction Materials*, Paris, France, 5-6 6pp.
[6] EFNARC 2002 *Specifications and Guidelines for Self-Consolidating Concrete*, European (Surrey, UK: Federation of Suppliers of Specialist Construction Chemicals), p 32.
[7] ACI 237R 2007 *Self-Consolidating Concrete, ACI Manual of Concrete Practice, Part 1*, (Farmington Hills, Michigan, USA: American Concrete Institute), p 30.
[8] CAN/CSA A3001 2003 *Cementitious Materials for Use in Concrete, Cementitious Materials Compendium*, (Ontario, Canada: Canadian Standards Association)
[9] Mark O G, Ede A N, Olofinnade O, Bamigboye G, Okeke C, Oyebisi S O and Arum C 2019

Influence of some selected supplementary cementitious materials on workability and compressive strength of concrete – a review. *IOP Conf. Series, ist international conference on sustainable infrastructural development.* **640** 1 – 18.

[10] Arum C and Mark O G 2014 Partial replacement of Portland cement by granulated cupola slag sustainable option for concrete of low permeability *Civil and Environmental Research* J **6** 17 – 26.

[11] Alizadeh R, Chini M, Ghods P, Hoseini M, Montazer P and Shekarchi M 2013 Utilization of electric arc furnace slag as aggregates in concrete, 6th ACI international conference on recent advances in concrete technology, Bucharest, Romania.

[12] BS EN 197-1 2011 *Cement Composition Specifications and Conformity Criteria for Common Cements.* (USA: British standard institution), p 56.

[13] ASTM C 150 2004 *Standard Specification for Portland Cement, Annual Book of ASTM Standards*, Vol.04.01, (Philadelphia, USA: American Society for Testing and Materials).

[14] ASTM C 494/ C494M 2004 *Standard Specification for Chemical Admixtures for Concrete, Annual Book of ASTM Standards*, Vol.04.02, (Philadelphia, USA: American Society for Testing and Materials).

[15] ASTM C 192/C 192M 2004 *Standard Practice for Making and Curing Concrete Test Specimens in the Laboratory, Annual Book of ASTM Standards*, Vol.04.02, (Philadelphia, USA: American Society for Testing and Materials).

[16] SCCEPG 2005 *The European Guidelines for Self-Compacting Concrete: Specification, Production and Use*, Self-compacting Concrete European Project Group, (West Midlands, UK: The European Federation of Concrete Admixtures Associations), p 63.

[17] Khayat K H 2000 Optimization and performance of air-entrained, self-consolidating concrete, *ACI Materials Journal.* **97** 526-35.

[18] Bonen D and Shah S P 2005 Fresh and hardened properties of self-consolidating concrete", *Progress in Structural Engineering and Materials* **7** 14-26.

[19] Yen T, Tang C W, Chang C S and Chen K H 1999 Flow behaviour of high strength high-performance concrete", *Cement and Concrete Composites.* **21** 413-24.

[20] Kim H, Park Y D, Noh J, Song Y, Han C and Kang S 2007 Rheological properties of self-compacting high-performance concrete", *Proceedings of the Third CANMET/ACI International Conference* **172** 653-68.

[21] Struble L and Hawkins P 2004 *Hydraulic Cements – Physical Properties", Tests and Properties of Concrete.* American Society for Testing and Materials, (Philadelphia; ASTM STP 169C) pp.449-61.

Review of the mechanical and durability properties of natural fibre laminate-strengthened reinforced concrete beams

A N Ede[1,*], B Ben-Ejeagwu[1], A I Akpabot[1], S O Oyebisi[1], O M Olofinnade[1], O G Mark[1]

[1]Department of Civil Engineering, Covenant University, Ota, Ogun State, Nigeria

*Corresponding author: anthony.ede@covenantuniversity.edu.ng

Abstract. The brittle nature of concrete has always necessitated the need for various forms of reinforcement that will absorb the tensile stress that is responsible for cracking. Conventionally, steel rebar embedded in concrete, giving rise to reinforced concrete structures has turned out to be a composite material that has been most used for infrastructural development around the world. The ductility impacted on concrete by steel rebar has been very efficient, though reinforced concrete structures are prone to durability challenges when exposed to adverse environmental conditions. Most times, repairing deficient concrete structures is more economical and sustainable than total replacement. Various approaches have been successfully implemented around the world in strengthening weakening reinforced concrete structures, nevertheless, bonding of fiber reinforced polymer (FRP) composites to the soffit of weak beams is turning out to be a viable option. Of particular interest is the adoption of naturally occurring fibers for laminates used for beam strengthening. In this era of Sustainable Development Goals (SDGs), in which the world is striving to reduce the effects industrial productions on the environment, advances in adopting renewable materials is highly welcomed. This research reviews the past works on the use of natural fibre-based composites in strengthening deficient reinforced concrete beams. The outcome of this research will go a long way in contributing towards the achievement of the SDGs.
Keywords: Natural fibre, FRP, Strengthening, Reinforced concrete beam, SDGs

1. Introduction

Fibres have been utilized by mankind for various purposes since the dawn of time. They have been used to make clothing, to make tools and household implements, and even medicines. Natural fibres however, are greatly preferred over other types of fibre due to their abundance in nature, low cost and sustainability. Natural fibres have already made quite a significant impact on the infrastructural and commercial products market with examples such as jute, flax, kenaf, and hemp [1]. While these fibres are renewable and ecologically meritorious, they have some disadvantages like lower durability and smaller strength. Fortunately, recent research has led to the development of several treatment procedures to enhance the durability of natural fibres [2]. The use of natural fibres in the form of a laminate to strengthen reinforced concrete members is a relatively recent and more sustainable method of retrofitting and upgradation which has been made necessary by numerous factors that ultimately lead to the gradual degradation of reinforced concrete structures. It is only one of numerous techniques utilized some of which include; Polymer injection, Steel plate bonding, Concrete jacketing and use of advanced composite materials like Fibre Reinforced Polymers and Ferro cement [3]. The properties of

IOP Conf. Series: Materials Science and Engineering 1036 (2021) 012043 doi:10.1088/1757-899X/1036/1/012043

natural fibre reinforced polymer laminates rely heavily on the properties of the particular fibre acting as the reinforcement therefore it is vital to understand the behaviors, strengths and weaknesses of these fibres in order to produce a laminate that will effectively serve its purpose of strengthening a concrete beam.

2. Strengthening and repairing concrete

Concrete is one of the most common building materials utilized by man due to its various favorable characteristics and properties that make it an ideal option for construction purposes. It is a composite material that possesses many qualities like strength, versatility, durability and economy, along with the extremely useful ability to be molded into almost any shape or form [4]. Despite its advantages, concrete is notoriously weak in tension, brittle and prone to cracking. Thus, these concrete structures gradually deteriorate over time for various reasons such as; inadequate design and construction, increase in service loads, poor quality of building materials, environmental or external factors, corrosion of steel reinforcement used in the concrete structure, natural disasters like earthquakes, tsunamis, tornados etc. [5]. This necessitates the implementation of retrofitting techniques such as NFRP laminate strengthening.

2.1. Concrete strengthening methods

Concrete members can be strengthened in numerous ways with different materials and varying degrees of success. One of the oldest and most common ways to strengthen or repair concrete structures is by employing sprayed concrete or shotcrete. Sprayed concrete is characterized into wet and dry spraying methods and it is usually implemented when large areas of concrete or mortar have been damaged corrosion or fire, access to damaged area is difficult, and when utilization of formwork is unrealistic [6]. Steel plate bonding is another very popular concrete strengthening procedure. It usually involves the adhesion of steel plates to the tension face of a concrete beam. It effectively increases both the shear and flexural strength capacities of the beam and is a well acknowledged technique because it is readily available, high ductility, low cost, and high fatigue strength. However, it suffers the same disadvantages associated with all forms of plate bonding techniques which is premature debonding [7].
Polymer grouting or polymer injection is yet another technique which is usually employed in underground conditions with the presence of water due to the good reaction time and plugging effect of the polymers [8]. Polymers such as epoxy, vinyl ester, polyester, polyurethane and acrylic resins are used. The commonest polymer used for injection is epoxy. Polyurethane and acrylic resin-based polymers are usually utilized to treat water retaining structures, underground structures and also limit water seepage [9].
The procedure of jacketing has been proven to improve the bending strength, ductility, stiffness, shear capacity and axial load carrying capacity of concrete members. The technique entails wrapping concrete structural elements with jackets which can be fabricated using concrete, steel and even fibre reinforced polymer composites [10].
The use of ferro cement which is an advanced composite is another technique for strengthening concrete. Ferro cement is a kind of thin wall reinforced concrete usually made using hydraulic cement mortar, with minimally spaced layers of continuous and relatively small size wire mesh as reinforcement [11]. Ferro cement is an excellent choice for the strengthening of concrete structures due to its high strength-to-weight ratio, raw materials availability, malleability and its superior cracking behaviour to regular cement [12].

Another commonly used advanced composite is the Fibre reinforced polymer (FRP) composite. This is a composite compound between a polymer and a fibrous material. The most common synthetic fibres utilized in FRP composites are glass, carbon, acrylic and aramid. Even paper and wood have been used on rare occasions with varying degrees of success. These synthetic fibres are products of well-informed research and development of materials in the textile and petrochemical industries to meet precise needs and standards required to create the most effective composite materials. These fibres provide high tensile strength to support the tensile and bending moment stiffness of the composite structures they are used to construct. However, they have some significant disadvantages which deters engineers from using them in structural construction. The advanced composites created with synthetic materials are relatively expensive to be mass produced for domestic materials and therefore are mostly used for small, unique job descriptions. Also, these composites are non-biodegradable and thus unsustainable. This presents a problem in this recent green age. Some of the composites fabricated with these synthetic fibres have been proven to be detrimental to the health of humans. They have caused adverse reactions on the skin and also in the human respiratory system [13]. These challenges have led to researchers looking to other types of fibres to strengthen concrete. These fibres would have to undoubtedly improve the structural integrity of the concrete, control the effects of cracking, be sustainable and environmentally friendly. The obvious solution was to turn to natural fibres.

2.2. The Natural fibre approaches

Natural fibres can be grouped into Organic and Inorganic fibres. Organic fibres are then further categorized into plant fibres and animal fibres while inorganic fibres consist of mineral fibres. Plant fibres can be procured from leaves, bark, stem, husk, seeds and grass. However, Animal fibres are obtained from the skin and coats of different animals [14].

It has been established that these natural fibres have certain characteristics and properties that give them the ability to counteract different problems experienced by concrete during construction. These unique properties determine the areas of applicability for each fibre and therefore makes them one of the first factors to consider when choosing a particular fibre for a job. These fibres can be used to strengthen concrete in different ways. They can be directly integrated into the concrete mix itself as short discontinuous fibres and they can also be used to make polymer laminates which will then be externally bonded to a concrete member. The advantages of utilizing natural fibres to strengthen concrete are numerous and significant enough to make them a well sought-after commodity in the construction industry. The fibres are readily available and abundant in nature, their extraction from nature requires low energy consumption, they are cost-effective, they are durable have good workability and decent strength. However, they have poor resistance to fire, they are highly susceptible to water and humidity damage, their strength is easily compromised, there is no direct or specific local design code for them and they have poor resistance to most forms of chemical attack. Table 1 shows the properties of some natural fibres.

Table 1. Properties of some natural fibres [15]

Fibre	Fibre type	Density (g/cm^3)	Elongation at break (%)	Tensile strength (MPa)	Elastic Modulus (GPa)
Jute	Bast	1.3	1.5 – 1.8	393 - 773	26.5
Flax	Bast	1.5	2.7 – 3.2	500 – 1500	27.6
Hemp	Bast	1.47	2 – 4	690	70
Kenaf	Bast	1.45	2.7 – 6.9	295 – 930	53
Ramie	Bast	1.44	4 – 4	400 – 938	61.4 – 128
Sisal	Leaf	1.5	2.0 – 2.5	511 – 635	9.4 – 22
Pineapple	Leaf	1.5	1 – 3	170 – 1627	34.5 – 82.5

Banana	Leaf	1.35	3 – 10	529 – 914	8 – 32
Cotton	Seed	1.5 – 1.6	7.0 – 8.0	400	5.5 – 12.6
Coir	Fruit	1.2	30.0	593	4.0 – 6.0
Wheat	Stalk	0.6 – 0.8	2.7	10 – 200	1 – 12
Rice	Stalk	1.65	2.2	449	1.21 – 1.25
Bamboo	Cane	0.6 – 1.1	4.0 – 7.0	140 – 230	11 – 17
Spider silk	Silk		17 – 18	875 – 972	11 – 13

2.3 Some types of natural fibre

2.3.1 Coconut fibre. Identified as the most durable and most resistant of all commercial natural fibres, coconut fibre or coir as it is popularly called is the short fibre extracted from the outer shell of coconuts otherwise called husks [16]. The fibre is obtained from the tissue surrounding the seed of the palm (Cocos nucifera) Coconut fibres procured from the tissues of the outer shell, belongs to the palm fibres family. They are agricultural waste products gotten from process of coconut oil. They are available in vast amounts in the tropical areas like in Africa, Asia and southern America. They are found in commercial amounts in the southern parts of Ghana [17]. The lengths of natural coir ranges from 60 to 230mm. Its diameter is about 0.17 to 0.24mm. The color of the fibre varies from pale yellow to dark brown. The chemical components of coconut fibre are similar to the composition of other vegetable fibres. Its main components are cellulose, hemi-cellulose, pectin and lignin. paragraph text follows on from the subsubsection heading but should not be in italic [3].

2.3.2 Sugarcane fibre

Sugar cane fibre also known as "bagasse" is the fibrous leftover material that is gotten after the juices of the sugar cane plant has been extracted. Sugarcane (Saccharum officinarum), is a species of tall perennial true grass and belongs to the grass family known as Poaceae. The sugar cane plant is mainly grown for its sweet tasting juice and is primarily used as an additive in different foods, drinks and other products. The cultivation of sugarcane fibres is an ancient practice as the Chinese and the Indians grew them as far back as 2,500 years ago [18]. There are two main types of sugarcane fibre. The Factory Bagasse which is obtained by industrial processes through repeated steps of extraction and the Farm Bagasse or cane stalks which are procured from small farms or factories. Its quality and quantity are significantly lower than that of the factory bagasse [21]. Bagasse is a lignocellulosic material and so is mostly constituted by cellulose, hemicellulose and lignin. These three constituents contribute more than 90% by dry weight of the fibre [20].

2.3.3 Kenaf fibre.

Kenaf fibre is obtained from the kenaf plant (Hibiscus Cannabinus L) which is a warm season annual crop significantly similar to jute and cotton. Recently, there has been a lot of commotion surrounding the use of kenaf fibres especially for composite reinforcement. This is due to the fact that kenaf possesses good mechanical properties and they grow quickly. They can rise to a height of 4-5 m in about 4-5 months of the growing season with the kenaf's stalk diameter reaching about 25-35 mm [21]. It is native to Africa where it has been cultivated for use in making ropes and animal feed [22]. Kenaf has a single unbranched stem which consists of two main types of fibre; the outer bast fibre which makes about 35% of the dry stem's weight and is considered the stronger and the more important fibre and the inner core fibre which amounts to about 65% of the stem's dry weight [23].

2.3.4 Banana fibre.

Banana fibre is a ligno-cellulosic fibre that is gotten from the pseudo-stem of the popular banana plant (Musa Sepientum) Its scientific name is Musa acuminate. It is a naturally occurring bast fibre that is commonly used in the production of hand-crafted works like mats, bags, hats, ropes and twine [24]. It should be noted that the main constituent of the fibre is cellulose which is then followed by hemi-cellulose. The cellulose is also the reinforcement for most of the other constituents present in the fibre. A high cellulose content and low microfibrillar angle are well sought-after properties for bast fibres and therefore make the banana fibre an ideal candidate [25]. Figure 1 shows various fibers recently adopted for improving concrete strengths. They are Momordica angustisepala fibers [26], Kenaf fibers [15] and Bamboo fibers [27].

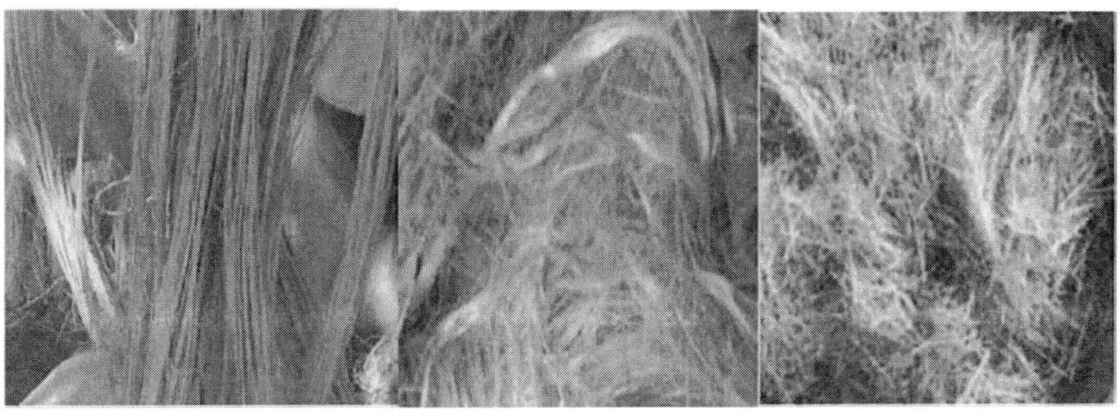

Figure 1. Different types of natural fibres [26, 15, 27]

3. Synthetic and natural FRP flexural strengthening systems

FRPs can be utilised internally as reinforcement for concrete building elements like beams, slabs, girders foundations and walls and also externally to upgrade or repair an existing structure made of concrete, steel or masonry [29]. There have been numerous research works undertaken on the strengthening and repair of reinforced concrete structural elements, bridge components and existing reinforced concrete structures using both synthetic and natural FRPs and some of their various strengthening systems will be discussed in this section. Stiffness enhancement of FRP-strengthened reinforced concrete beams via dynamic-based non-destructive tests were shown in the works [29] for static loads and in [30] for cyclic loads. In these two researches, it was proven that FRP composites are very essential in the restoration of damaged structures to functional state.

Nayak *et al.* [31] investigated the effect of externally strengthening reinforced concrete beams with glass fibre reinforced polymer (GFRP) fabrics on the flexural strength of the beams. They prepared nine beams wrapped externally in different number of layers and different lay-up patterns then one unwrapped control beam. All the beams were subjected to two-point loading. The first four beams were wrapped with the GFRP fabric fully at the bottom and on the sides at a height of 25mm from the bottom. The next four were wrapped at the bottom and 75mm from the bottom in the sides. The remaining one was wrapped completely at the bottom and on the two sides. It was observed that all ten beams showed similar behaviours initially. It was then concluded that the flexural strength of the concrete beams increased with an increase in the number of layers of GFRP fabric bonded to it in different laying patterns. The increase in layers was all observed to accompany a gradual change in the failure mode from ductile to brittle behaviour. It was also discovered that of the three cases of GFRP wrapping, the model in which the bottom portion and half of the two sided were wrapped exhibited the best performance in strength and ductility when compared to the beam without GFRP sheets. Using multiple layers of synthetic FRP reinforcement has evidently been proven to enhance the mechanical properties of concrete elements as also discovered by Karzad *et al.* (2019) [32]. They found that the strengthening of reinforced concrete beams with shear deficiency using one layer of CFRP resulted in average of 65% improvement of shear capacity which can be further enhanced by adding another layer of CFRP to gain an additional 30% increase in shear capacity.

Despite the effectiveness of synthetic FRP systems, they are less preferred than natural FRPs due to their cost of manufacturing and their non-biodegradability. Natural FRPs have been observed to be cost effective, biodegradable and can achieve better strengthening performance in terms of cost and effect [33]. Chen et al. [33] conducted a study to explore the feasibility of replacing synthetic FRPs with natural FRPs in structural strengthening works and highlighted the advantages they had over synthetic FRP. Six beams prepared differently were subjected four-point bending loads. A typical beam was 150mm wide, 250mm in depth and 2400mm in length. Four-point bending was applied thereby equally dividing the shear and pure bending spans into 700mm. One of the beams was unstrengthened (control), four beams were strengthened using jute or flax FRP laminates and one was strengthened with carbon fibre reinforced polymer (CFRP) laminate. The NFRP laminates were formed using the wet lay-up method and left to cure for seven days at the least. They were found to provide better strengthening than the CFRP did when the NFRP laminates were used in large amounts. A cost analysis was further conducted and it exposed the fact that the NFRP laminates were 20% and 40% cheaper in flax and jute respectively than the CFRP laminates for a unit increase of load-carrying capacity. Di Luccio et al. [34] performed an experimental study to explore the cyclic response of reinforced concrete walls strengthened externally using carbon and Flax FRP strips in a bid to compare their overall performances. One RC wall was to serve as a reference so it was not strengthened, two walls were strengthened with strong bidirectional strips of CFRP utilising the wet lay-up system. The last three walls were retrofitted with unidirectional flax fibre reinforced polymer (FFRP) strips. One of these three walls however, was additionally strengthened with a bi-directional glass fabric. The test results showed an increase in resistance of about 100% in the FFRP strengthened walls. Also, the RC walls upgraded with the CFRP strips failed at displacement levels lower than those of the control or reference beam, while there was a significant 30% increase of displacement identified in the FFRP strengthened walls. The CFRP strips failed due to concrete peeling or pulling of the external anchorage.

The most common form of FRP reinforcement for concrete and masonry structures is the external bonding reinforcement (EBR) method therefore it is used in most research works and studies. However, there is another relatively recent strengthening method called the near surface mounting method (NSM) which involves cutting out a hole or groove in a concrete element, applying the proper bonding adhesive and placing FRP rods, bars or strips into the hole. The NSM technique has some advantages over the EBR method such as better strain distribution leading to higher utilisation of potential strain and its higher resistance to environmental factors by virtue of its embedment in the concrete cover [35]. In a critical review performed by Parvin and Shah [35], it was deduced that not only does using the NSM technique eliminate the notorious issue of debonding experienced in the EBR method, it also possesses higher fire resistance properties than the EBR method thereby reducing insulation length and thickness required to hinder the adhesive from reaching its glass transition temperature. They additionally observed that prestressing of NSM-FRP greatly improves the structural performance of reinforced concrete structures.

4. Conclusion

Natural fibres present a more sustainable and ecofriendly substitute to synthetic fibres as reinforcement in polymer composites. While the natural fibres have to be used in more quantities to provide as much strength as synthetic fibres would, they are a far cheaper and reasonable option. Concrete members strengthened with NFRP laminates were found to offer up significantly more resistance when subjected to cyclic loading and they also failed at higher displacement levels than CFRP laminates. The adoption of the near surface mounting method (NSM) has eliminated the greatest opposition to the external bonding method (EBR) thereby offering a promising future for the concept of NFRP laminate strengthening of concrete.

References

[1] Tumolva, T. P., Vitoa, J. K. R., Ragasaa, J. C. R., & Cruza, R. M. C. Dela. (2017). Mechanical properties of natural fiber reinforced polymer (NFRP) composites using anahaw (Saribus rotundifolius) fibers treated with sodium alginate. 8(October), 66–73.
[2] Sonar, I. P. (2013). Natural fibre reinforced cement concrete : Avenue through some investigations, **2**(2), 6–10.
[3] L.S, S. kumar, Reddy, H. N. J., & Nizar, R. (2013). Retrofitting of RC beams using natural FRP wrapping (NSFRP). International Journal of Emerging Trends in Engineering and Development, **5**(3), 168–178.
[4] Shelton, D. P., & Harper, J. M. (1982). G82-623 An Overview of Concrete as a Building Material An Overview of Concrete as a Building Material.
[5] Sen, T., & Jagannatha Reddy, H. N. (2013). Strengthening of RC beams in flexure using natural jute fibre textile reinforced composite system and its comparative study with CFRP and GFRP strengthening systems. International Journal of Sustainable Built Environment, 2(1), 41–55.
[6] Austin, S. A., Robins, P., & Goodier, C. I. (1998). Construction and repair with wet-process sprayed concrete and mortar.
[7] Jumaat, Z., & Alam, A. (2006). Problems associated with plate bonding methods of strengthening reinforced concrete beams. Journal of Applied Sciences Research, **2**(10), 703–708.
[8] Lin, Z., Guo, C., Cao, D., Ni, P., & Wang, F. (2020). An experimental study on the cutting failure of polymer grouting. Construction and Building Materials, **258**, 119582. https://doi.org/https://doi.org/10.1016/j.conbuildmat.2020.119582
[9] The Constructor. (2014). Injection Grouting for Concrete Crack Repair - its Types and Materials. Retrieved June 25, 2020, from https://theconstructor.org/concrete/injection-grouting-concrete-repair-types-materials/9111/
[10] Vandoros, K. G., & Dritsos, S. E. (2008). Concrete jacket construction detail effectiveness when strengthening RC columns. Construction and Building Materials, **22**(3), 264–276. https://doi.org/https://doi.org/10.1016/j.conbuildmat.2006.08.019
[11] Sirimontree, S., Witchayangkoon, B., & Lertpocasombut, K. (2015). Strengthening of Reinforced Concrete Column via Ferrocement Jacketing. Construction Innovation, **4**(1), 39–47.
[12] Jaafer, A. (2015). Experimental investigation on the ferrocement slabs with a sifcon matrix. Wasit Journal of Engineering Science, **3**(1). https://doi.org/10.31185/ejuow.Vol3.Iss1.34
[13] Lau, K. tak, Hung, P. yan, Zhu, M. H., & Hui, D. (2018). Properties of natural fibre composites for structural engineering applications. Composites Part B: Engineering, **136**(October 2017), 222–233. https://doi.org/10.1016/j.compositesb.2017.10.038
[14] TextileSchool. (2018). History of Fibres Natural and Manmade Fibres. Retrieved January 5, 2020, from https://www.textileschool.com/345/history-of-fibres-natural-and-manmade-fibres/
[15] Nwankwo, C. O., Ede, A. N., Olofinnade, O. M., & Osofero, A. I. (2019). NFRP strengthening of reinforced concrete beams. IOP Conference Series: Material Science & Engineering, **640**(1).
[16] Food and agriculture organization. (2019). Future Fibres: Coir. Retrieved from

http://www.fao.org/economic/futurefibres/fibres/coir/es/
[17] Yalley, P. P., & Kwan, A. S. . (2012). Coconut Fibre as Enhancement of Concrete. Journal of Engineering and Technology, **2**, 54–73.
[18] Grant, A. (2018). What Is Sugarcane Used For – Learn What To Do With Sugarcane Plants. Retrieved December 4, 2019, from https://www.gardeningknowhow.com/edible/herbs/sugarcane/common-sugarcane-uses.htm
[19] Heuzé, V., Tran, G., & Archimède, H. (2012). Sugarcane bagasse | Feedipedia. Retrieved from http://www.feedipedia.org/node/559
[20] Carvalho, W. (2009). Chemical composition of the sugarcane bagasse. Society for Industrial Microbiology, p. 81232. Retrieved from https://sim.confex.com/sim/31st/techprogram/P8845.HTM
[21] Ishak, M. R., Leman, Z., Sapuan, S. M., Edeerozey, A. M. M., & Othman, I. S. (2010). Mechanical properties of kenaf bast and core fibre reinforced unsaturated polyester composites. IOP Conference Series: Materials Science and Engineering, **11**, 012006. https://doi.org/10.1088/1757-899x/11/1/012006
[22] Abdul Rahman Khan. (2011). Kenaf, a fibre for the future: the Harusmas experience. JEC Composite Magazine, 21–23. Retrieved from http://www.jeccomposites.com/knowledge/international-composites-news/kenaf-fibre-future-harusmas-experience
[23] Akubueze, E. U., Ezeanyanaso, C. S., Orekoya, E. O., Akinboade, D. A., Oni, F., Muniru, S. O., & Igwe, C. C. (2014). Kenaf Fibre (Hibiscus cannabinus L .): A Viable Alternative to Jute Fibre (Corchorus genus) for Agro-Sack Production in Nigeria. World Journal of Agricultural Sciences, **10**(6), 308–313. https://doi.org/10.5829/idosi.wjas.2014.10.6.1824
[24] Alam, F. (2014). Properties of Banana Fiber | Manufacturing Process of Banana Fiber | Application of Banana Fiber - Textile Learner. Retrieved from http://textilelearner.blogspot.com/2014/01/properties-of-banana-fiber.html#ixzz4U9lMrj9u
[25] Bhatnagar, R., Gupta, G., & Yadav, S. (2015). A Review on Composition and Properties of Banana Fibers. Int. J. Sci. Eng. Res., **6**(5), 143–148. Retrieved from www.ijser.org
[26] Ede, A. N., Olofinnade, O.M., Ugwu, E.I., Salau, A.O. (2018). Potentials of Momordica angustisepala fiber in enhancing strengths of normal portland cement concrete Cogent Engineering, **5**(1),1431353. pp. 1-17
[27] Ede, A.N. Olofinnade, O.M., Joshua, O., Nduka, D.O., Oshogbunu, O.A., (2020). Influence of bamboo fiber and limestone powder on the properties of self-compacting concrete, Cogent Engineering, **7** (1), 1721410
[28] Jawdhari, A., & Harik, I. (2018). Finite element analysis of RC beams strengthened in flexure with CFRP rod panels. Construction and Building Materials, **163**, 751–766. https://doi.org/10.1016/j.conbuildmat.2017.12.139
[29] Ede A N and Pascale G 2016 Structural Damage Assessment of FRP Strengthened Reinforced Concrete Beams under Cyclic Loads Mater. Sci. Forum, **866** 139–42
[30] Ede A N Bonfiglioli B Pascale G and Viola E (2004) Dynamic assessment of damage evolution in FRP strengthened RC beams, International Conference on Restoration , Recycling and Rejuvenation Technology for Engineering and Architectural Application (Italy) pp 155–163.
[31] Nayak, A. N., Kumari, A., & Swain, R. B. (2018). Strengthening of RC Beams Using Externally Bonded Fibre Reinforced Polymer Composites. Structures, **14**, 137–152. https://doi.org/10.1016/j.istruc.2018.03.004
[32] Karzad, A. S., Leblouba, M., Toubat, S. Al, & Maalej, M. (2019). Repair and strengthening of shear-deficient reinforced concrete beams using Carbon Fiber Reinforced Polymer. Composite Structures, **223**(January), 110963. https://doi.org/10.1016/j.compstruct.2019.110963
[33] Chen, C., Yang, Y., Yu, J., Yu, J., Tan, H., Sui, L., & Zhou, Y. (2020). Eco-friendly and mechanically reliable alternative to synthetic FRP in externally bonded strengthening of RC

beams: Natural FRP. Composite Structures, 241, 112081.
[34] Di Luccio, G., Michel, L., Ferrier, E., & Martinelli, E. (2017). Seismic retrofitting of RC walls externally strengthened by flax–FRP strips. Composites Part B: Engineering, **127**, 133–149. https://doi.org/10.1016/J.COMPOSITESB.2017.06.017
[35] Parvin, A., & Shah, T. S. (2016). Fiber reinforced polymer strengthening of structures by near-surface mounting method. Polymers, 8(8). https://doi.org/10.3390/polym8080298

Assessing the low-cost buildings in Nigeria using Hydraform blocks as walling materials: the journey so far

S Oyebisi[1]*, A Ede[1], O Afolayan[1] S Oladeji[2], B Oniemayin[1], J Lazarus[1]

[1]Civil Engineering Department, Covenant University, Ota, Nigeria
[2]Civil Engineering Department, Lead City University, Ibadan, Nigeria
*Corresponding author: solomon.oyebisi@covenantuniversity.edu.ng

Abstract. Man's growing desire for affordable and comfortable buildings have necessitated the diversification of materials and methods in the built sector. Thus, this paper assessed the hydraform blocks as walling materials to examine its current level of usage in the Nigerian built sector. Information was generated via literature reviews, newspapers, and professional perceptions. The findings revealed that despite the concerted feats of private developers in the use of hydraform blocks as walling materials for the low-cost building construction in Nigeria, efforts by the Federal Government has yielded limited success in this regard. Also, State Governments have not fared better. Therefore, this paper recommended and proffered solutions to bridge this gap.
Keywords: Housing; sustainability; environment; low-cost buildings; local materials; building construction; hydraform blocks.

1. Introduction

The physical survival of mankind depends on housing, besides food and clothing [1, 2]. Adequate and comfortable housing influences the physical, social, and moral health, and contributes to the works' stability, efficiency, and development [3]. In 2019, the Federal Ministry of Works and Housing (FMW&H) proposed to build 2,383 housing units in Nigeria under the National Housing Programme in 36 states of the Federation and the FCT. Besides, in 2020, the FMW&H plans to build just about 2,000 homes [4]. However, to keep up with growing population and urban migration, the World Bank recommends a yearly construction of 700,000 housing units to breach a huge housing deficit in Nigeria. The World Bank estimates Nigeria's total housing deficit to be 17-23 millions [4]; though the present demand is estimated at 37 million houses [5]; and this figure increases yearly by 20%. Efforts by the previous governments to tackle housing situation in Nigeria have yielded limited success after 59 years of independence [5]. Therefore, it was added that the private sector is critical in making housing available and affordable in Nigeria, as the government cannot do it alone.

Globally, housing problems are summarized by the houses needed by the poor and low-income earners [6]. The housing deficits, particularly targeted at the poor and low-income earners have increased due to weak housing policy, rapid urbanization, and high cost of conventional building materials [6]. Report from a previous research revealed that no civil servant in the Nigerian Federal Civil Service, under grade level 13, can afford a house costing ₦4.5 million at 6% salary on a 25 years of mortgage loans [7]. This signifies that affordable housing will become ever more elusive for low-income earners. Low-cost housing is an arrangement by which the prospective house owners fulfill the periodic mortgage needs without compromising the health and other basic needs of the family members [8].

On the account of the foregoing, this paper aims to examine the usage level of hydraform blocks as walling materials in the Nigerian built sector. Data was obtained through the reviews of literature, newspapers, and professional views. The benefits of hydraform interlocks as low-cost building blocks were appraised, while the Government efforts vis-à-vis professional perceptions of the low-cost housings were also examined. Finally, the paper concludes and recommendations are made.

2. Literature Review

2.1 The Concept of Hydraform Blocks in Nigeria

In the 20th century, traditional adobe or unfired laterite bricks were prevalent in some African countries [9]. However, in recent time, the search for more efficient, versatile, affordable, and eco-friendly blocks led to the development of hydraform blocks, otherwise called laterite interlocking blocks (LIB) [10]. Hydraform block, as shown in Figure 1, is produced from laterite, cement, and water, and compressed hydraulically with the aid of hydraform hydraulic machine (Figure 2), resulting in a high quality interlocking solid block [11], which is largely dry-stacked. Consequently, the *Nigerian Building and Road Research Institute* [NBBRI] designed an interlocking block machine, as shown in Figure 3, to produce LIBs, mainly comprised cement and laterite of ratio 1:20, respectively [12]. The designed concept was based on the LIBs with a geometric size 225 x 225 x 112 mm^3 [13]. The geometry is assembled without the application of mortar, hence making the laying concept to be differed from conventional blocks [9]. Hydraform blocks satisfied the specification of Industrial Building System [14], and have been used in many parts of the World for load bearing masonry structures [15].

Figure 1. Stacked hydraform blocks
Source: Nairaland Forum, 2020 [16].

Figure 2. Hydraform hydraulic machine
Source: Nairaland Forum, 2020 [16].

Figure 3. Hydraform hydraulic machine designed by NBRRI
Source: NBRRI Newsletter, 2006 [12].

2.2 Benefits of Hydraform Blocks in Building Construction

The vital element in building construction is building materials. About 60-80% of building materials account for the entire cost of building construction [17]. This signifies that access to efficient, affordable, and eco-friendly buildings in Nigeria would continue to be difficult except the high cost of conventional building materials is checked and alternative materials are harnessed. However, hydraform blocks have been predominantly used in South Africa, Kenya, Uganda, Argentina, India, Mexico, and United States of America, to construct houses, schools, and general single and multi-storey buildings [18].
Hydraform blocks offer substantial cost saving by harnessing 90-95% of laterite, a locally available and subsoil material, and 5-10% cement for their production [18]. Besides, the cost of burning is eliminated. Hydraform blocks are produced on site, thus reducing the transport cost. Except in lintels and high gables, LIBs are dry-stacked, as shown in Figure 4, hence eliminating mortar in superstructure walls [18]. Confirming the foregoing assertion, previous studies revealed that about a 30% of cost and time are saved [19-21]. Also, following the interlocking concept of hydraform blocks, about a 75% of the building requires no mortar [22].

Figure 4. Dry-stacking of LIBs without mortar
Source: Nairaland Forum, 2020 [16].

Hydraform blocks offer faster rate of construction than conventional blocks [18]. In a day, about 21 m^2 of walls can be laid with 800 hydraform blocks by a bricklayer, hence making LIBs to be laid three times faster than conventional blocks; this also reduces the labour time per hour [18].

The use of LIBs as walling materials exhibit excellent thermal capacity (the ability to absorb and hold heat). According to *John Morris*, hydraform blocks offer three times of thermal insulation efficiency as concrete and about two times as fired clay bricks [18]. This signified that LIBs are energy-efficient building materials and can be used as walling materials in hot or arid climatic weather conditions [23, 24], and consequently, contribute to the sustainable development agenda.

Hydraform blocks yield higher impact strength compared with concrete blocks [18]. After 14 days curing, hydraform blocks attain 7 MPa of compressive strength, consisting a mix ratio 1:20 of cement and laterite, respectively. Moreover, LIBs are eco-friendly materials because they are produced under high compression from laterite, without using fuel-wood to burn or fire them [18]. It was reported that about 657 MJ/ton of energy is required to produce LIBs compared with 4187 MJ/ton for fired bricks [25]. This signified a reduction of about 85% energy input when LIBs are produced compared with fired bricks. Also, about a 41 kg CO_2/ton is emitted during the production of LIBs compared with 202 kg CO_2/ton associated with fired bricks [25]; this indicated that there was about 80% reduction in CO_2 emission during LIBs' production compared with fired bricks.

Hydraform blocks exhibit attractive surface, in terms of aesthetics, as shown in Figures 5 and 6 such that rendering, plastering, and painting may not be necessary; these also reduce the cost of finishes. Also, production and erection are simple to lay under the supervision of hydraform-trained personnel.

Figure 5. Construction of Heart Mega School using LIBs, Famese, Akure, Nigeria
Source: Aghimien, Aghimien, 2015 [26].

Figure 6. A residential building built with LIBs in Lekki, Lagos, Nigeria
Source: Adegboye, 2012 [2].

3. Research Methods

On the basis of methodology used in carrying out this study, qualitative investigation was conducted to obtain knowledge as well as to collect data via reviews of literature, website/internet accesses, newspapers, and professional perceptions of the built industry.

4. Results and Discussion

4.1 Application of hydraform blocks in the Nigerian built sector: the journey so far

In 1991, compressed stabilized interlocking earth blocks (CSIEB) were reportedly first used to construct 60 housing units estate at the University of Lagos, Lagos, Nigeria [27]. Afterward, the use of CSIEB has continued to enjoy patronage as low-cost building materials; this ultimately resulted in the introduction of CSIEB as hydraform blocks in the Nigerian built sector [28]. Although, hydraform innovation is South African technology, which has been in existence 30 years ago [22].

In 2005, the Federal Government of Nigeria, through the Association of Housing Corporation of Nigeria (AHCN) in collaboration with the NBRRI, flagged-off the construction of 500 low-cost housing units in each state of the federation and the Federal Capital Territory (FCT) using LIBs [29]. NBRRI provided the fabricated interlocking block-making machines, while the Federal Mortgage Bank OF Nigeria (FMBN) provided the funds needed for the construction. Consequently, Ekiti State, through the Ekiti State Housing Corporation, executed the project, completed, and named it as Olusegun Obasanjo Housing Estate, Ado-Ekiti, Nigeria, as shown in Figure 7. The housing units were allocated to civil servants, mostly, the low-income category.

Figure 7. Pictorial view of Olusegun Obasanjo Housing Estate, Ado-Ekiti, Nigeria
Source: Oyebamiji [30].

Kwara State Government, through the Private Developer (International Business Bureau Limited), executed the housing units, comprising two-bedroom standard, two-bedroom luxury, and three-bedroom standard bungalows located at Akerebiata, Sobi Road, Ilorin, Nigeria. In the same vein, FMBN provided the funds, while the Kwara State Government provided the land. The phase 1 (250 units) of the buildings was completed and commissioned in 2009, while the other phase remains undeveloped, not only in Kwara State but also, other states of the Federation [31]. The low-cost housing units, two-bedroom standard, two-bedroom luxury, and three-bedroom standard bungalows were allocated to the low-income earners and sold for two millions (₦2,000,000.00), two millions and five hundred thousand (₦2,500,000.00), and three millions (₦3,000,000.00) naira only, respectively [21, 31].

In 2010, Osun State Government, through the Osun State Property Development Corporation (OSPDC) purchased hydraform block-making machines, purposely, to construct low-cost housing for low-income earners, and to build elementary schools for teaming students in the state [32]. Moremi housing estate at Ile-Ife and Atakunmosa West Local Government elementary school, Osun State, Nigeria, were built. However, despite all efforts by the OSPDC, the existing reality revealed that the mandate is far from being attained [32].

In 2017, chairman, Hydraform Nigeria, *Platter* [33] asserted that over 5000 low-cost housing units have been built in Nigeria by the Hydraform Nigeria using hydraform blocks. It was further reported that over 4000 low-cost housing units have been built in Abuja (FCT) alone, and the Hydraform Nigeria had also worked in Jigawa, Anambra, Enugu, Yobe, Lagos, and some other states in Nigeria in building affordable houses for low-income earners. However, it was discovered that Government at all levels failed to take the lead in tackling the problems of affordable buildings in Nigeria using hydraform blocks [33]. Moreover, in 2016, Hydraform Nigeria sold a three-bedroom bungalow built with hydraform blocks for four millions and five hundred thousand naira only (₦4,500,000.00) and was the cheapest in the built sector at that time [34].

IOP Conf. Series: Materials Science and Engineering 1036 (2021) 012044 doi:10.1088/1757-899X/1036/1/012044

4.2 Government efforts vis-à-vis professional perceptions

The investigations on the promulgated National Housing Scheme of 1991 revealed that the programme was faulty and weak [34]. Most estates constructed under the low-cost housing programme of the previous regimes, have been in moribund owing to the fact that those housing units were unacceptable to the prospective users [34]. The reports showed that the end-users were neither asked nor consulted for any contribution during the planning stages. Some estates were built in a vulnerable area, while some were built under the low-cost housing programme but were bought by the rich and rented to the low-income earners at high cost, whom the scheme was meant for in the first place [34]. Thus, housing affordability becomes a challenge. Consequently, the Federal Ministry of Works and Housing, in 2015, adopted a novel approach to address the deplorable situation of low-cost housings in Nigeria [34]. A team of Architects, Surveyors, and Town Planners were selected across Nigeria and inaugurated with a responsibility to consult Nigerians in the six geopolitical zones on their acceptable kinds of houses. The consultations were conducted between 2016 and 2017. The results revealed that most Northern States and Southern States of Nigeria preferred bungalows with courtyards and flat blocks, respectively [34]. These results have been incorporated into the current National Housing Scheme, presently undertaking in 34 of the 36 states of the federation and the FCT. Moreover, FMBN, a facilitator and guarantor, was empowered by the Federal Government of Nigeria to grant mortgage loans to loan seekers at very affordable interest rates. Many in-built innovations were also added to the programme to make it more inclusive and sustainable [34].

Reacting to the foregoing efforts by the Federal Government of Nigeria, many professionals in the Nigerian built sector affirmed that the programmes were not meant for the poor, house seekers, and low-income earners in the society. In 2019, an expert in the low-cost housing development, *Plattner,* who also doubled as Chairman, Hydraform Nigeria, stated that most housing projects initiated by the Federal and State Governments were not designed to cater for low-income earners due to the uses of conventional blocks as walling materials without recourse to the locally made available materials, LIBs [33]. Therefore, Government at all levels must adopt new technologies and local materials as alternative means of providing affordable houses. Moreover, *Fatoki,* an estate surveyor and valuer and former chairman, Lagos State Branch of the Nigerian Institution of Estate Surveyors and Valuers (NIESV) [2], and *Eleh,* a former president, NIEV [17], stated that Nigerian housing delivery was a failure and would remain a thorny challenge if the Land Use Acts and the cost of building materials were not reviewed. Furthermore, *Zubairu,* project manager, Niger State Urban Support Programme [5] asserted that both Federal Government and the State Government are yet to formulate an effective and sustainable programme for affordable houses of low carbon footprint to Nigerians, particularly, the poor and low-income earners. In 2019, it was also reported that the National Housing Policy Formulation was not integrative in that State and Local Governments were not involved in the housing provisions; this makes the affordable houses unrealizable, despite high promises by the Federal Government [5]. In the same vein, *Lufadeju,* Chairman, Board of Trustees, Real Estate Developers Association of Nigeria (REDAN), reported that policy failure and lack of priority have undermined the provision of affordable houses [5]. Besides, lack of political will and lukewarm and penchant attitudes of Nigerians to the patronage of cheaper building materials, LIBs, are not helpful. The president, Association of Housing Corporation of Nigeria (AHCN), *Adamu* [5] posited that the challenges of mass and low-cost housing provisions was due to usurpation of statutory responsibilities by the Governments agencies governing the housing provisions. On the other hand, *Awobodu,* President, Nigerian Institute of Building (NIOB), substantiated that low-cost housing delivery in Nigeria still remains a mirage because Government fails to strike balance between the building material cost and the purchasing power of Nigerians [5]. It was further noted that real estate developers have resorted to obtaining loans from the commercial banks at about 24-28% interest rates since 2012 till date when Estate Development Loan (EDL) was suspended by the FMBN; this impacts negatively on the housing cost in Nigeria.

5. Conclusions and Recommendations

This paper examined the low-cost buildings in Nigeria using LIBs as walling materials. It was shown that harnessing the naturally occurring subsoil material, laterite, to build the sustainable housing units would alleviate suffering, remove cost burden on the beneficiary units, and make houses affordable for the less privileged and low-income earners. For effective and efficient low-cost housing provisions in Nigeria, the following recommendations are made:

i. Government should take the leading roles in creating an enabling environment for the provision of low-cost housing units using LIBs as alternate to concrete blocks.
ii. NBRRI should be more proactive in its partnership with the Hydraform Nigeria to train more Nigerians on the production and laying mechanisms of hydraform blocks; this would not only reduce the problems of housing affordability in Nigeria but also create jobs for the teaming jobless Nigerian youths.
iii. Stakeholders in the Nigerian built sector must collaborate with strong advocacy and synergy to market and recommend hydraform blocks as energy-efficient, sustainable, and affordable building blocks by influencing policy formulation and implementation at all levels of government.
iv. National Housing Programme should be integrated such that State and Local Governments would also be actively involved in low-cost housing provisions using LIBs. This would allow the Governments to focus on implementation, monitoring and evaluation, and impact assessment while eliminating direct construction by all tiers of government.
v. All tiers of government must create a conducive atmosphere and a favorable condition for the manufactures of hydraform blocks to reduce the cost of building material while competing with the cheaper and substandard imported brands and providing employment opportunities.

Acknowledgement

The authors acknowledge the Covenant University Centre for Research, Innovation and Discovery (CUCRID) for the financial support and conducive environment to conduct the research.

References

[1] Kremer, William, Claudia H. Abraham Maslow and the pyramid that beguiled business. BBC (2013, Sep. 1). Available at https://www.bbc.com/news/magazine-23902918/ [accessed 20 June 2020].
[2] Adegboye K. 52 years of homelessness: experts proffer solutions to dearth of affordable housing-Nigerian Vanguard news, 2012. Available at https://www.vanguardngr.com/2012/10/52-years-of-homelessness-experts-proffer-solutions-to-dearth-of-affordable-housing/ [accessed 25 April 2020].
[3] Baumeister, Roy F, Mark R L. The need to belong: desire for interpersonal attachments as a fundamental human motivation. Psychological Bulletin, 1995, **117**(3), 97-529.
[4] Premium times. Nigeria needs 700,000 homes yearly-but Government plans to build only 2,383. The premium times, 2019. Available at https://www.premiumtimesng.com/news/headlines/359274-nigeria-needs-700000-homes-yearly-but-govt-plans-to-build-only-2383.html/ [accessed 25 April 2020].
[5] The Guardian. The property: Not yet affordable housing despite high promises. The Guardian, 2019. Available at https://guardian.ng/property/not-yet-affordable-housing-despite-high-promises/ [accessed 26 April 2020].

[6] Iheme J O, Ekung S B, Effiong J B. The effect of government policy on housing delivery in Nigeria: a case study of Port Harcourt low Income housing programme. International Letters of Social and Humanistic Sciences, 2015, **61**, 87-98.
[7] Onyike J A. Addressing the urban Housing problems of Nigeria in the 21st century. Department of Estate Management, Imo state University Owerri Nigeria, 2009.
[8] Adediji Y M D. Affordable and functional housing in a developing economy: A case study of Nigeria. Journal of Land Use and Development Studies, 2006, **2**(1), 113-120.
[9] Anand K B, Ramamurthy K. Development and performance evaluation of interlocking block masonry. Journal of Architecture Engineering, 2000, **6**(2), 45-51.
[10] Hydraform. Hydraform building system. A solution to eco-friendly and energy Efficient Building. 2012. Available at www.hydraformindia.com [accessed 2 May 2020].
[11] Hydraform. Hydraform training manual. 2013. Available at www.hydraformindia.com [accessed 2 May 2020].
[12] Nigerian Building and Road Research Institute. NBRRI interlocking block making machine. NBRRI Newsletter, 2006, 1(1): 15-17.
[13] Adedeji Y M D. Materials preference options for sustainable low-income housing in selected cities in Nigeria. Unpublished PhD Thesis, submitted to The Federal University of Technology, Akure. 2007.
[14] Ahmad Z, Othman S Z, Yunus B, Mohaed B. Behaviour of masonry wall constructed using interlocking soil cement bricks. World Academy of Science and Technology, 2011, **60**, 1263-1269.
[15] Ventakarame B V, Gupta A. Characteristics of soil cement blocks using highly sandy soils. Materials and structures, 2005, **38**(6), 651-658.
[16] Nairaland Forum. Hydraform interlocking block and brick making machine for hire/rent - business to business. Nairaland, 2020. Available at https://www.nairaland.com/3232610/hydraform-interlocking-block-brick-making [accessed 25 June 2020].
[17] Njoku J. Dearth of affordable building materials: Experts proffer solutions. The Vanguard, 2012. Available at https://www.vanguardngr.com/2012/07/dearth-of-affordable-building-materials-experts-proffer-solutions/ [accessed 23 June 2020].
[18] Hydraform Engineering Nigeria Limited. African business. African business directory, 2020. Available at https://directory.africa-business.com/business/13d463157e/Hydraform-Engineering-Nigeria-Ltd [accessed 25 June 2020].
[19] Oyebisi S O, Olutoge F A, Ofuyatan M O, Abioye A A. Effect of corncob ash blended cement on the properties of lateritic interlocking. Progress in Industrial Ecology-An International Journal, 2017, **11**(4), 373-387.
[20] Raheem A A, Bello O A, Makinde O A. A comparative study of cement and lime stabilized laterite interlocking blocks. Pacific Journal OF Science and Technology, 2010, **11**(2), 27-34.
[21] Oyebisi S O, Ede A N, Ofuyatan M O, Oluwafemi J O, Akinwumi I I. Comparative study of corncob ash-based lateritic interlocking and sandcrete hollow blocks. International Journal of Geomate, 2018, **15**(51), 209-216.
[22] Hydraform. Hydraform training manual. 2004. Available at www.hydraform.com [accessed 23 October 2019].
[23] Oyedepo S O. Energy efficiency and conservation measures: Tools for sustainable energy development in Nigeria. International Journal of Energy Engineering. 2012, 2(3): 86-98.
[24] Ede A N, Douglas K K, Oyebisi S O, Oluwafemi J O. Study of energy efficient building design techniques: Covenant University Health Centre. Journal of Physics: Conference Series 1378 (2019) 032037. Doi:10.1088/1742-6596/1378/3/032037.

[25] Oti J E, Kinuthia J M, Bai J. Engineering properties of unfired clay masonry bricks. Engineering and Geology, 2009, **107**, 130-139.

[26] Aghimien D O, Aghimien E I. Assessment of the use of compressed stabilized interlocking earth block for building construction in Nigeria. Civil and Environmental Research, 2015, **7**(12): 1-7.

[27] Olusanya O. Sustainable industrialization: a cost-efficient approach to cooperative housing". In O. Jadesola (Ed), Women and Housing in Nigeria, Issues, Problems and Prospects, 2001: 52-58.

[28] Adedeji Y M D, Fasakin J O. Acceptability indices of interlocking masonry for housing construction in Nigeria. Pakistan Journal of Social Sciences, 2008, **5** (7): 696-700.

[29] Olotuah A O, Taiwo A A. Housing the urban poor in Nigeria through low-cost housing schemes. International Journal of Physical and Human Geography, 2013, **1**(3), 1-8.

[30] Afolami A J, Oyebamiji I O. Thermal perception of residents in housing developments built with laterite interlocking blocks in Ado-Ekiti, Nigeria. FUTY Journal of the Environment, 2017, **11**(1), 120-135.

[31] Njoku Jude, Adegboye Kingsley. Affordable housing: With bricks, you can reduce cost by 40%. The Vanguard, 2012. Available at https://www.vanguardngr.com/2012/04/affordable-housing-with-bricks-you-can-reduce-cost-by-40/ [accessed 24 June 2020].

[32] Adejumo M A. Hydraform brick: a cost saving benefit with good aesthetics for housing development in Osun State of Nigeria (case study of Osun State Property Development Corporation), 2016.

[33] Daily Trust. Housing Projects: Most of them not designed for masses – Hydraform MD. The PropertyProinsider, 2017. Available at https://www.propertypro.ng/blog/housing-projects-most-of-them-not-designed-for-masses-hydraform-md/ [accessed 24 June 2020].

[34] Durugbo M. The Myth of a National Housing Deficit. The PM News, 2020. Available at https://www.pmnewsnigeria.com/2020/01/24/the-myth-of-a-national-housing-deficit/ [accessed 9 April 2020].

Management of construction & demolition waste: barriers and strategies to achieving good waste practice for developing countries

O M Olofinnade[1,*] I Manda[1] and A N Ede[1]

[1]Department of Civil Engineering, Covenant University, Ota, Ogun State, Nigeria

*Corresponding author: rotimi.olofinnade@covenantuniversity.edu.ng

Abstract. Environmental contamination due to construction and demolition (C&D) waste management is a global issue. It is an even bigger issue specifically for developing countries that encounter different obstacles due to the construction industry's practices caused by a lack of knowledge and the economic status of these countries. The aim of this study is to review various research conducted in developed and developing countries so as to identify the significant problems and to seek appropriate ways to implement the management of C&D waste in developing countries. The findings of this research were grouped into seven (7) areas, which include the economy, knowledge across the industry, storage and cross-contamination of wastes, technology, design for deconstruction, legislation, and environmental health. Each barrier is attached with a strategy appropriate for developing nations, which would improve the C & D waste management industry if implemented. This study's results can be used as a reference for further research into these barriers to develop comprehensive strategies that will significantly improve the C&D waste management industry.
Keyword: Construction and demolition wastes, recyclable waste, construction waste management, sustainability

1. Introduction

Demolition waste is the rubble obtained from destroying building structures. Despite being destructive, demolition is vital for developing megacities, with space scarcity a significant obstacle. Many buildings are demolished, but they mainly occur to buildings that have approached or exceeded their lifespan and buildings undergoing renovations. Waste on a building construction can unnecessarily be produced due to the excessive ordering or quantification of materials and the mishandling of the materials by unskilled laborers. Natural disasters can also generate construction and demolition (C&D) waste. These few activities result in several significant problems, e.g., how to transport the waste, store it temporarily before processing, and finally, where to dispose, mostly landfill sites. C&D waste has chemical properties that make it difficult to be broken down or changed by waste processing techniques like other forms of waste [1]. The construction industry contributes highly to the depletion of natural resources and degradation of the environment. An estimated 3 billion tons was observed by Saghafi & Teshnizi [2] as the number of raw materials used for the manufacturing of construction materials worldwide. Despite being essential in the construction process, the process of extraction, processing, handling, and uses amount to 7% of the total global energy consumption. Furthermore, the transportation of primary materials alone is responsible for the usage of 40% of the total energy consumed by the industry. The construction industry has begun to use different methods for its sustainability by reducing the generation of waste, making sure energy is used efficiently, and reusing construction and demolition waste as recycled aggregates [3].

Furthermore, there has been a rapid increase in urban populations throughout the world, which has led to an increase in construction activities, and this has developed a problem due to the increase in construction and demolition waste as a result of that. In Hong Kong, the amount of C&D waste represented six times that of municipal solid waste. The construction industry is in pressing need of an alternative and sustainable source of raw construction materials. It has become an increasingly important issue to study the effective management of these wastes and their potential to be recycled globally due to their effect on our environment [4]. Many different researchers have proposed ways in which the use of natural resources can reduce with the use of supplementary cementitious materials. These include silica fumes, fly ash, quick lime, or waste products derived from ceramics, palm oil waste, etc. The addition of these materials significantly changes the properties of the resultant concrete, which can be used as an outlet for other waste products in different industries, which also reduces the handling cost of waste [1]. C&D waste is less likely to be used as a different product, which increases the need for landfills, but this is not a sustainable management option due to its immediate effects on the environment. This has made it a necessity to recycle and reuse these wastes as construction materials. Reduce, reuse, and recycle are no longer just a term but are beginning to be realized quickly in some countries. This was observed as the EU in 2010 who set a target of increasing the recycling rate of C&D waste from 10-90% to a minimum of 70% average recycling rate by 2020. This was achieved by a few countries like Belgium, the Netherlands, Switzerland, and Austria by 2013 [4].

Sustainability measures are unavoidable in the construction industry, which means that it is not just meant to reduce the carbon footprint, but it is also meant to contribute to the conservation of natural resources, which are crucial for the constant development of our society and the societies to come. Many developments are underway to reduce the depletion of natural resources, but completely stopping the use of natural resources is practically impossible. It has been discovered that as C&D waste is disposed of and stored, it tends to mix with materials that are not reusable for construction work. These materials alter the properties of the C&D waste and reduce their reusability, making it hard for them to be treated, and some are rendered unusable. The management of C&D waste is crucial in increasing the amount of C&D waste that can be recycled. Developing countries, also known as third world countries, are a large number of dissimilar societies, cultures, and civilizations not part of the Western world with economies that are less industrialized or modernized. A common fact in these countries is that most of their people have a low standard of living and regularly fight wars, they are unstable politically and experience population explosions, which have led to the destabilization of their economies and led to an inadequate financial standing [5]. Although waste reuse and recycling are practiced in developing countries, they only apply to a specific type of waste, mainly steel, paper, wood, and plastic. Inert wastes like concrete and broken bricks are not recycled. As the real estate industry and infrastructure development continue to expand, C&D waste generation is likely to increase further. To improve the sustainable construction in developing countries, we can do this by the effective utilization of resources in construction, material recovery, an improved system for C&D waste management, and energy savings [6].

The C&D waste is a significant problem in both developed and developing countries; what differs is the nature of generation and handling of the waste. Developing countries are beginning to engage in waste management practices, but still, there is a culture of insufficient collection and improper disposal that still reigns supreme in the industry. This waste constitutes a large amount of waste dumped in river banks, roadsides, and many other open spaces. It has resulted in health and environmental impacts coupled with negative economic impacts on nearby properties. There is a need for urgent improvement in construction waste disposal practices to meet the demand for improved construction waste management [7]. In order for C&D waste management's goals to be achieved, significant barriers hindering its implementation need to be recognized and considered. There is a strong need for a long-term plan combined with adequate development and research. Very few initiatives involving legislation and regulation are followed, and this has failed the necessary economic, technical, and administrative instruments.

2. Methodology

This review aims to identify barriers significantly affecting the growth of the C&D waste management industry by looking into various research conducted in both developed and developing nations and seeks appropriate strategies that can be implemented in developing countries.

3. Discussion

3.1 Economy

Developing countries are known to be low-income countries with a lack of adequate finances to expand and sustain the C & D waste management industry. Therefore, economic instruments can be used to improve waste management, but the framework for regulation and their application are a hindrance to the implementation of these measures. There needs to be sure things set in place before it is implemented like an appropriate public awareness about the environment, excellent management expertise, innovations, and proper accounting practices because the technical, managerial, and social point of view are difficult to manage. There needs to be an application of new charges and it should be a system programed to support itself. The methods of reducing, reusing, and recycling are widely used in municipal solid waste practices but less practiced when it comes to C&D waste. One reason could be that the recycling of building materials is not cost-effective. There is a severe lack of demand for particular wastes of construction like concrete and broken bricks [6], [17-18].
According to Akhtar & Sarmah [1], the recycling of waste and its management can minimize the volume of C&D waste in the ever-growing construction industry. This process increases the prosperity of a country by creating many new jobs, which can be referred to as a circular waste economy. Ferronato & Torretta [8] noted that we need to include the informal sector by finding ways of recycling and reducing waste inflow into final disposal by developing methods that require low technological requirements and economic investments in developing countries. Although to do this, we will need to understand the most appropriate strategy for their involvement by further investigating various methods of involving the informal sector. Mulenga [9] suggested that Zambia with its neighboring countries need to create sustainable practices that will include Climate Change and Green construction principles. Countries and regional organizations such as SADC, SATTC, and ASANRA are encouraged to be frontrunners in providing funds for these attempts. High risk attempts are required to establish sustainable C&D waste management, and there needs to ne coordination between education, training, research and development institutions, and the construction industry at large.

3.2 Industrial knowledge and training

The next challenge would be to change the perception of C&D waste being considered as solid waste and to re-frame it as a raw material in the construction business. The uncontrolled disposal of C&D waste into landfills continues to be a significant problem in developing countries, and it is the main reason for the increased waste of land and natural resources [10]. Excessive use of natural resources and the large amount of C&D waste observed in our environment is due to a lack of awareness of source efficient construction practices. A study by Muleya & Kamalondo [11] on the waste management practices in Zambia, targeted groups which were separated into four: contractors, consultants, government units, and regulators. This

IOP Conf. Series: Materials Science and Engineering 1036 (2021) 012045 doi:10.1088/1757-899X/1036/1/012045

research illustrated that what was practiced in waste management did not align fully with the Zambia construction industry (ZCI) waste management hierarchy. There was an agreement between all groups that there was a need for a Site waste management plan (SWMP), which would be a successful and safe way of managing construction waste in the ZCI. Significant challenges identified include; Traditional habits of construction and behavior, inadequate exposure or training, and a lack of funds.

There needs to be an introduction of awareness and information campaigns, implementation of systems for treatment and regulation and control agencies, the improvement of sites of disposal and their management, the enhancement of financial sustainability of the systems and the introduction of future management plans in order to improve the integrated solid waste management of a country, region, municipality or rural area [8]. Zulu [12] evaluated exercises, techniques, and awareness of solid waste management (SWM) in chosen townships in the capital of Zambia, Lusaka. It was found that 91% of the respondents were not aware of solid waste management and did not know the appropriate disposal methods. It was concluded that people were not knowledgeable enough on solid waste management and practiced inferior ways of getting rid of waste among all respondents regardless of their backgrounds. This study signifies how big a task it will be to raise awareness to achieve sustainable waste management and disposal in the construction industry. The municipal solid waste management sector has gone great lengths in raising awareness on the disposal of waste and how to manage it, but even in recent times, serious problems are faced, and a majority of the people still do not engage in these practices. There is a need to change the attitude and perception of construction project stakeholders to show that the waste recycled products fit their needs to influence their behavior and decision-making. The Institute of Public Works Engineering Australia Limited has come up with courses in which all stakeholders involved in public works understand the specific requirements of newly introduced products like recycled aggregates and substitute materials and grasp how they can be incorporated in projects. Even as the range of new recycled materials and products increases on the market, the requirement for courses recognizing them will be more critical and the ways in which they can be utilized efficiently [13].

3.3 Storage and Cross-contamination of wastes

Construction material's recovery rates are severely impeded due to cross-contamination with other materials, which is prevalent in the construction and demolition waste industry. It must be ensured that materials are obtained closer to the source, which reduces the likelihood of mixing it with other wastes and obtain high recovery rates. As noticed from the case studies, the highest recovery rates were obtained from waste materials that were captured and segregated early [13]. Segregation of plasterboard increases the total cost of a construction project, but this cost is recouped by gate fees being reduced for waste without plasterboard sort out and the income generated from the recycling of low sulphate C&D waste [14]. There is a challenge in the categorization of C&D waste faced by many developing economies because no specific laws are governing the production, management, and discarding of C&D waste. In most of these countries, it has been considered municipal solid waste (MSW) and planned for in that way. This poses to be a problem for recycling because there are plenty of chemical and physical differences between the two, which means it would need additional treatment if it were to be used for construction [10].

Policies must be set in place to mandate contractors to implement the calculation of waste indices, which will ascertain the quantity of waste generated on-site. This will help plan for the right measure to control it. Much should be done to encourage contractors to make use of prefabricated materials. There should be a correlation between specific materials purchased for use and their design specifications which can guide site personnel on ways in which they can avoid the wastage as they carry out construction works [15]. Material use efficiency can be achieved by the creation of space and the availability of consolidation centers within reach, which are a vital aspect of the adequate sorting and stocking of materials on site. The use of consolidation centers or availability of impromptu deliveries allows for preferable organization of the area; they reduce the waste and increase the efficiency of works on site, which makes savings achievable. Waste sorting and processing is a technology well spread in developed countries and is standard for waste treatment facilities. This practice needs a way in which waste can be segregated and a constant demand for these products which keeps it away from gathering subsidiary products. Recycled products are relatively cheaper than natural materials. With the implementation of this method, it is possible to raise the confidence of all contributors in the industry as it is also possible to achieve increased amounts of recycled products (higher than 90% for some frontrunners) and it yields notable benefits from the life-cycle perspective on the environment [14].

3.4 Technology

According to a literature review by Mulenga [9], sustainable C&D waste management is possible only if every phase of construction has been considered. It is crucial without attention to the degree of mechanization. C&D waste production and disposal need to be studied in depth in order to benchmark good practice. A great number of studies concerning waste have been overseen in developed countries and tend to focus on municipal waste compared to C&D waste. This has left a gap for the developing world to address and develop models to manage C&D waste.

A significant obstacle to recycling landfilled materials is finding the right equipment to sufficiently clean and segregate these waste products so that they can be reused. This has created an opportunity for a market for recycling plants that could be used for this cause. This study showed that this is specifically a problem for waste produced by timber where the contamination happens due to the treated timber and funding of extensive investigation is required to conquer these barriers [13]. Another challenge is in contingency C&D waste management. This can happen due to a destructive disaster that would need massive scale management in an emergency situation. In an ideal situation, rescue and cleanup of rubble must be done simultaneously because they complement each other. It helps prevent the spread of disease and also assist in returning back to normal life as quickly as possible.

Reuse of materials is viable [20], but there isn't a large market for recycled products, and this has meant that there is a large stockpiling up, which hinders the applicability of these materials. The reuse of materials is a measure applied scarcely, and this is because of its high demand for large sums of funding. However, despite how much is being done, its use as a standard product in construction is far from being achieved. A good case in point is the price of recovered bricks, which can sometimes cost more than standard bricks by 100%, although it is possible to save about 50% of the total cost used for investment by the reclamation of steel frames. It is also possible completely stop the dumping of waste in landfills if reuse is integrated with other best practices [14]. There are many benefits to adopting an SWMP based on findings, which include: reduced amount of waste in landfills, reduced waste on-site, and prevention of pollution. It can provide a prescribed approach to on-site waste management and helps provide an efficient way to allocate resources. Regulators showed a willingness to support SWMP in giving fines for non-compliance and halting of projects that do not meet the guidelines with deregistration as a last resort [11].

3.5 Design for deconstruction

A way in which construction materials can be reused better is by the adoption of deconstruction rather than demolition, which would help in the selection of materials needed for reuse. De-construction demands a lot of manual labor, time, and very specialized light machinery, making it a costly option [10]. Designing for deconstruction is a process by which resource recovery is of the highest priority and is slowly gaining prominence. It is designed so that as it approaches the building's design life, it can be taken apart and offers more excellent resource recovery opportunities. It may influence the design of materials to suit better this new form of design [13]. Building deconstruction can be used when waste management is a costly option, and some materials or components are scarce, which would drive the demand for it. Despite being very expensive, it is efficient with recovery rates of up to 95-99% [14]. De-construction of buildings has been studied by many, and the reuse of materials has yielded positive results. The reuse of construction materials has many benefits. However, the price and time factor related to it, and the structured operation are noteworthy barriers for this particular technology to be adopted [1].

3.6 Legislation

There is a deficiency in legislation pertaining to C&D waste management for it to be sustainable, a common issue in developing or transitory economies like India, Malaysia, and the Czech Republic. This means that due to the lack of legislation to govern and monitor C&D operations by these governments, there is a high likelihood of experiencing the development of compromised structures due to reduced quality. This becomes a hazard to the environment and is not safe for its occupants, which may lead to the demolition of inferior quality structures, which may lead to non-recyclable materials and materials harmful to the environment. The lack of enforcement in these countries also results in the illegal disposal of this waste into landfills [10]. Site waste management and prevention can be achieved by adequate enforcement of legislation, improvement of facilities, and increasing the accessibility to waste management services because the reduced rate of recovery of waste materials in some countries is not because of below par management. Generally, waste management of construction worst don't exceed 3% of the entire cost. This means that preventing waste will only result in a small amount of savings. Ideally, about 99% of all waste is diverted from landfills [14]. The policy is needed in order to mandate contractors to include a waste management plan in the documents expected to be submitted during tendering processes. Contractors need to form a department that evaluates the plan in order for it to comply with the provisions by building the capacity of the site workers and workforce development. The site workers need to be educated on the environmental and health risks that accompany waste generation from materials that have been used during the construction process [15].

3.7 Environmental Health

The production and transportation of cement and aggregates yield a high carbon footprint with cement having the highest carbon footprint. This is shown in a study that observed that cement production emits 7% of the world's CO_2 emission [3]. There are particular elements of demolition waste like plasterboard that are hazardous to the environment when it is landfilled because it is run down when deposited in landfills and emits sulfide gas, which is toxic. The waste obtained from individual construction and demolition ends up being deposited in municipal bin/waste storage deposits, which makes the municipal waste heavy. It makes the municipal waste quality degrade and also makes it difficult for further treatments like composting. Furthermore, about 10-20% of this waste finds its way into surface drains which chokes them [16]. Ferronato & Torretta [7] reviewed environmental contamination and social issues in developing countries due to solid waste (SW) mismanagement. The results of this review showed that there is a need for solid waste management (SWM) system in order to get through the reduction of the environmental footprint and to improve the targets of sustainable development goals (SDGs). This view was also echoed in the study of [19, 20] on achieving some of the SDGs goals. In developing countries, too often, SWM is considered as a single stream disposed of in open dumpsites. However, it is required that there should be an application of impromptu collection and treatment solutions for each waste flow in the implementation of future management plans in municipal areas: Municipal Solid Waste (MSW) (organic and inorganic), healthcare waste (HW), C&D waste, waste electric and electronic equipment (WEEE) and used batteries, industrial and toxic waste and used tires. Models for quantities of waste to be quantified need to be determined, which can be rooted on the entire price of a construction project or on the amount of space on the floor depending on the building classification [9].

4. Conclusion and Recommendation

The following conclusions can be drawn from these reviews:

i. Developing countries are neglecting the implementation of sustainable C & D waste management practices, and it is required that there should be cooperation amongst the educational, training, research, and development institutions.
ii. The implementation of C&D waste management practices offers a new market for the private sector, which could generate a large source of income as well as increase the jobs for a country's citizens.
iii. The best way to implement sustainable C&D waste in developing countries is by the enforcement of legislation that all parties in the construction industry need to follow. C&D waste management guidelines need to be set in place from tendering to ensure that it is included in all projects' total costs.
iv. De-construction is currently not a viable option due to the current state of C&D waste management in developing countries. This is because De-construction is very expensive to implement and will require the financial input and cooperation of many institutions.
v. Funding is required to push C&D waste management as it will require huge investments before the industry can generate profits. This can be done with the help of governments and regional organizations (ECOWAS, SADC, etc.)
vi. Developing countries can achieve C&D waste management sustainability by developing methods that require low technological requirements and economic investments and not by replicating developed nations' strategies due to their inferior financial status.

Most research on the management of construction and demolition waste has been conducted in developed countries and therefore the solutions obtained are only suitable for these countries. Therefore, there is a need for research in the technologies and strategies that are suitable for implementation in developing countries.

Acknowledgment

The authors would like to acknowledge the support provided by the Centre for Research, Innovation and Development (CUCRID), Covenant University, Ota towards the publication of this research work.

Reference

[1] Akhtar, A., & Sarmah, A. K. 2018. Construction and demolition waste generation and properties of recycled aggregate concrete: A global perspective. *Journal of Cleaner Production*, 262-281.
[2] Saghafi, M., & Teshnizi, Z. 2011. Recycling value of building materials in building assessment systems. *Energy Build*, 3181-3188.
[3] Oh, D., Noguchi, T., Kitagaki, R., & Park,, W. 2014. CO_2 emission reduction by reuse of building material waste in the Japanese cement industry. *Renew. Sustain.*, ***38***, 796-810. Retrieved from https://doi.org/10.1016/j.rser.2014.07.036.
[4] Hossain, U., Poon, C. S., Lo, I. M., & Cheng, J. C. 2016. Comparative environmental evaluation of aggregate production from recycled waste materials and virgin sources by LCA. *Resources, Conservation and Recycling*, 67-77.
[5] Owusu, M. (2001). Third World. *International Encyclopedia of the Social & Behavioral Sciences*.

[6] Manowong, E., & Brockmann, C. 2012. Construction waste management in Newly Industrialized countries. *Semantic Scholar*, 1-13.
[7] Mahayuddin, S. A., Badaruzzaman, W. H., Pereira, J. J., & Mokhtar, M. 2008. Construction waste management in a developing country: Case study of Ipoh, Malaysia. *WIT Transactions on Ecology and the Environment*, 481-489.
[8] Ferronato, N., & Torretta, V. 2019. Waste Mismanagement in Developing Countries: A Review of Global Issues. *International Journal of Environmental Research and Public Health*, 1-28.
[9] Mulenga, M. N. 2018. Towards Sustainable Construction Waste Minimization and Management in Zambia and Beyond. *Rwanda Journal of Engineering, Science, Technology and Environment, 1*(1), 1-10.
[10] Periathamby, A. 2009. Challenges in sustainable management of construction and demolition waste. *Waste Manag Res 2008; **26**; 491* 491-492.
[11] Muleya, F., & Kamalondo, H. 2017. An Investigation of Waste Management Practices in the Zambian Construction Industry. *Journal of Building Construction and Planning Research,* ***5***, 1-13.
[12] Zulu, A. 2019. Assessment of Solid Waste Management in Residential Townships in Lusaka Zambia. *ACTA scientific agriculture,* ***3***(11), 62-66.
[13] Edge Environment, L. P. 2011. *Demolitio waste guide - recycling and reuse across the supply chain.* Australia: Department of Sustainability, Environment, Population and communities, Australian Government.
[14] Gálvez-Martosa, JL., Stylesb, D., Schoenbergerd, H., & Zeschmar-Lahl, B. 2018. Construction and demolition waste best management practice in Europe. *Resources, Conservation & Recycling*, 166-178.
[15] Wahab, A. B., & Lawal, F. A. 2011 An evaluation of waste control measures in construction industry in Nigeria. *African Journal of Environmental Science and Technology, 5(3)*, 246-254.
[16] Ponnada, M. R., & Kameswari, P. 2015. Construction and Demolition Waste Management – A Review. *International Journal of Advanced Science and Technology*, 19-46.
[17] Coelho, A., & Brito, J. D. 2013. Economic viability analysis of a construction and demolition waste recycling plant in Portugal e part I: location, materials, technology and economic analysis. *Journal of Cleaner Production*, 338-352.
[18] Ede AN, Olofinnade OM, Bamigboye GO, Shittu KK & Ugwu EI. 2017. *Prediction of fresh and hardened properties of normal concrete via choice of aggregate sizes, concrete mix-ratios and cement.* International Journal of Civil Engineering and Technology **8** (10), 288–301
[19] Davies IEE, Nwankwo CO, Olofinnade OM, & Michaels TA 2019 Insight review on impact of infrastructural development in driving the SDGs in developing nations: a case study of Nigeria *IOP Conf. Series: Materials Science and Engineering*, **640**, 012112 doi:10.1088/1757-899X/640/1/012112
[20] Olofinnade OM, Busari AA, Akinwumi II, Awoyera PO & Ekanem J 2019 Perception of professionals on causes of structural cracks in concrete buildings *IOP Conf. Series: Materials Science and Engineering* **640**, 012036 doi:10.1088/1757-899X/640/1/012036

Suitability of using post-consumer polyethylene terephthalate wastes in cement-based hollow sandcrete blocks

I E E Davies[1] and O M Olofinnade[1]
[1]Department of Civil Engineering, Covenant University, Ota, Ogun Sate, Nigeria

*Corresponding author: rotimi.olofinnade@covenantuniversity.edu.ng

Abstract. In recent times, efforts are channeled towards limiting the impact of waste plastic pollution on the environment. This study examines the beneficial utilization of polyethylene terephthalate (PET) wastes, which is a non-biodegradable waste materials as a partial substitute for sand in the production of non-load bearing hollow sandcrete blocks. The physical and mechanical properties such as density and compressive strength of the hollow sandcrete blocks using PET waste as a replacement for fine aggregate at 5%, 10%, 30% and 50% were determined. Tests results show that the density of the produced sandcrete blocks decreases as the waste PET content increases. Results showed that the density varied from 1391.33 kg/m^3 for 50%PET content to 1954.54 kg/m^3 for 5%PET content, while the density of the reference sandcrete was 2002.15 kg/m^3 for 7 days. The values of densities ranged between 1453.88 kg/m^3 for PET 50% and 1851.09 kg/m^3 for PET 5% and the density of reference sandcrete was 1991.77 kg/m^3 at 28 days. The sandcrete blocks also demonstrate a similar reduction trend in compressive strength as the PET content increases. The compressive strength ranged between 1.140 N/mm^2 for 50%PET and 2.564 N/mm^2 for 5%PET content and 1.510 N/mm^2 for 50%PET to 2.991 N/mm^2 for 5%PET respectively, as the curing age increases from 7 to 28 days. The hollow sandcrete block containing an optimum 5% replacement percentage of sand with PET waste showed comparable strength to the conventional sandcrete blocks. This will help to mitigate the menace of marine plastic pollution and promote responsible consumption.

Keywords: Sandcrete blocks, Natural aggregate, Compressive strength, Density, Coefficient of uniformity, Specific gravity

1. Introduction

Annually, global waste generation both organic and inorganic materials have been on the increase due to rise in world population and urbanization. Polyethylene terephthalate (PET) is rated as the most produced plastic types and are used globally in packaging of different products such as water, beverage, foods, soft drinks and other consumer's goods [1, 2]. The consumption of PET used in bottling of drinks was reported to be about 10 million tons and estimated to rise at about 15% annually [1]. The PET waste is the most recycled plastic waste, a study by [3] reported that PET is the most abundant type of plastics found in urban municipal waste streams. Polyethylene terephthalate (PET), is one of the major sources of environmental pollution, thus the need for an urgent attention in the management of this non-biodegradable waste materials. One of the ways advocated of addressing this menace is to find an alternative re-use or recycling approach of the PET wastes [4, 5]. Huge quantities of plastic wastes are generation and indiscriminate dumping of this hydrocarbon byproduct, has spark a lot of research towards the possibility of recycling of these wastes in concrete, in the production of sandcrete block as partial replacement to natural aggregate particularly where strength may not be a necessary criteria under consideration and also as a constituent pavement construction [6]. Many researchers have researched on the effect of replacement of PET on the density of sandcrete block such as [7-10]. A study by Safi et al. [9] opined that production of lighter density or light mortar could be achieved by partially replacing sand with PET waste aggregate [11-15]. The focus of this study is to show the suitability of utilizing the enormous quantity of PET in the environment as an alternative aggregate material in the production of hollow sandcrete blocks. This work will examines the use of PET waste, as partial replacement to fine aggregates through the determination of density and compressive strength properties of the produced hollow sandcrete blocks.

2. Materials and Method

2.1 Materials

The ordinary Portland cement used in this research was produced by Dangote cement. This cement type is the most widely used binder material in the production of hollow sandcrete blocks in Nigeria. Hence, it was used as adhesive in the mix design of the hollow blocks. The cement grade was 42.5 and complied with the specification prescribed in BS EN 197-1:2000 [16] and NIS 444-1:2003 [17]. The river sand was used as the natural fine aggregate in the hollow sandcrete block mixes. The sand was purchased in Ota, Ogun state and brought to Civil Engineering Laboratory. The bulk density, specific gravity and water absorption capacity were carried out on the sand according to BS EN 1097-3:1998 [18] and is presented in Table 3, while the grading size distribution of the sand is shown in figure 1.
In addition, the PET plastic bottles were collected, gathered and shredded into granulated form in Covenant University. The bottle used were pre-washed with water, dried, melted at about 280 ^{0}C and 320 ^{0}C and allowed to solidify before being grounded into fine granulated form (below 4.75mm sieve size) by a crushing machine. The acquired PET plastic were smooth finely granulated PET Aggregate. Physical properties such as density, specific gravity and water absorption capacity were determined according to BS EN 1097-3:1998 [18]. These results are tabulated in Table 3. The grade size distribution of the PET Aggregate is describe in figure 1. Potable water was employed for the mixing of the matrix composite in accordance to BS 3148:1980 [19].

2.2 Fabrication procedure of the hollow sandcrete blocks

Manual mould was used to produce the sandcrete block. The internal dimension of the block mould was 450 × 150 × 225mm (Length × Width × Height). The mould had a void quantifying about half the volume of the block. In this study, cement-to-aggregate mix ratio was 1:6 based on the specification given by the Nigerian National Building Code [20]. The granulated PET was used to replace sand as aggregate at 5%, 10%, 30% and 50% by weight in the aggregate mixture. The experimental design of the aggregate is shown in Table 1. Hand mixing was employed in the mixing of the cement and aggregate. After mixing the constituent materials, potable water was supplied to the mixture to secure a fluid slurry. The water/cement ratio used was 0.55. After mixing, the fluid slurry was poured into the hand block mould. After which, the blocks were removed from the mould and were placed carefully in separate rows with adequate spacing between each successive blocks. Thereafter, the block samples were openly cured for 7 days and 28 days, by sprinkling water once daily and covering with polyethylene nylon after 24 hours of moulding. Figure 2 depicts the produced hollow sandcrete block samples with PET before covering with polyethylene nylon after water sprinkling.

2.3. Tests

i. Chemical composition of PET-aggregate - oxides composition of PET-aggregate was determined by using X-ray fluorescence (XRF).
ii. Physical properties of natural aggregate and fine granulated PET aggregate test - physical properties such as density, particle size distribution, specific gravity and water absorption capacity were determined according to BS EN 1097-3:1998 [18].
iii. Density of the hollow sandcrete block samples were measured in accordance to BS EN 772-11:2011 [21].

iv. Compressive strength - the compressive strength test was carried out on the hollow sandcrete blocks according to BS 6073-2:2008 [22]. Figure 2 shows a block sample under compressive load.

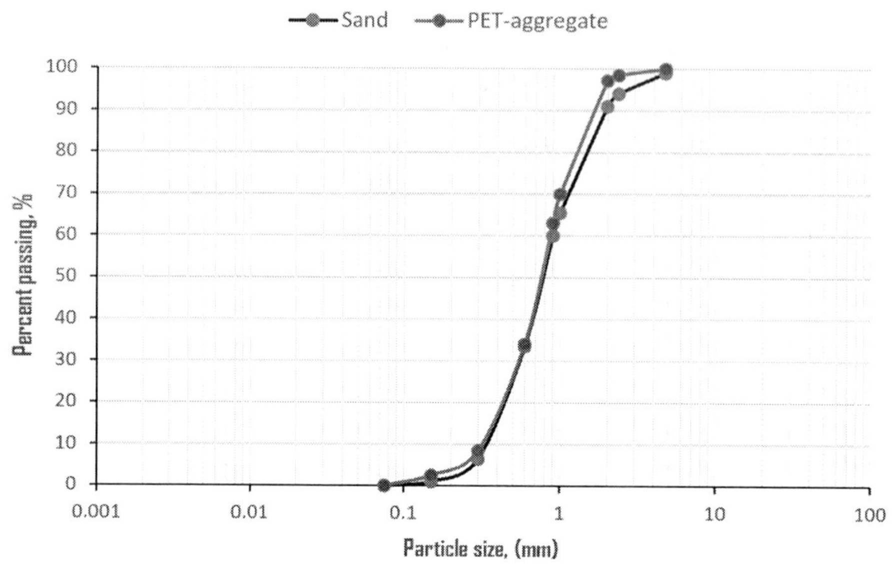

Figure 1. Particle size distribution of sand and PET aggregate

Figure 2. Hollow sandcrete block after water sprinkling; and under compressive load

Table 1. Experimental Design

Samples	% cement content	% sand content	% PET-aggregate	Water/ cement ratio	Remarks
AB	100	100	0	0.55	Reference sandcrete
A1	100	5	95	0.55	5% sandcrete block
A2	100	10	90	0.55	10% sandcrete block
A3	100	30	70	0.55	30% sandcrete block
A4	100	50	50	0.55	50% sandcrete block

3. Results and Discussion

3.1 Particle size distribution

The results of the particle size distribution of PET wastes bottle are shown in figure 1. The coefficient of uniformity of sand and PET wastes bottle were 2.66, and 2.61, respectively. The coefficient of curvature of sand and PET were 1.02 and 1.16 respectively. This is an indication that the sand is well graded. The particle sizes of PET waste aggregate fell within the nominal sizes of 0.150 to 4.75 mm, while grain distribution curve shows that 1.0% of the river sand particle sizes were bigger than 4.75 mm. The coefficient of curvature of the PET aggregate is close to the reported coefficient of curvature reported for palm kernel shell and periwinkle shell [23].

3.2 Physical properties and chemical composition of aggregate

Some of the physical properties of natural aggregate and PET waste plastic are shown in Table 2. The specific gravity of river sand and PET waste plastic were 3.00 and 1.42. Bulk densities of uncompressed river sand and PET waste plastic 1292 kg/m^3 and 846.17 kg/m^3. The water absorption capacity of river sand and PET waste plastic were 0.95%, and 0.02% respectively. Thus implies that the PET aggregates is of lighter weight and low water absorption capacity compare to the conventional sand materials.

Table 2. Physical properties of aggregates

Properties	Sand	PET
Specific gravity	3.00	1.42
Water absorption, %	0.95	0.02
Bulk density, kg/m^3	1292	846.17

Meanwhile, the results of the chemical composition of the granulated PET aggregate materials showed that the main oxide compositions of the material include; SiO_2 (56.4%), Al_2O_3 (23.71%) and Fe_2O_3 (7.38%), MgO (2.95), Na_2O (2.7%) and CaO (2.12). These are the major chemical oxides in the granulated PET material. The recorded loss on ignition (LOI) for the PET plastic is 1.2. Aggregate with high quantity of reactive silica has the potential to react with alkali in cement. The possibility of Alkali-silica (ASR) compound in the sandcrete blocks was not investigated due to the moderate percentage of Na_2O. A likely occurrence of ASR can be detrimental to the long term performance of the sandcrete blocks [24, 25]. Meanwhile, the plastics contain some severe and toxic organic and inorganic chemical compounds that are dangerous to human health such as nickel, lead, cadmium, chromium, mercury, bromine, tin, antimony, bisphenol A, and chloro-ethane monomer.

3.3 Density of sandcrete blocks

The results of densities for the different weight percentages of waste PET plastics for the production of sandcrete blocks are shown in figure 3. The density varied from 1391.33 kg/m^3 for PET 50% to 1954.54 kg/m^3 for 5% PET, while the density of reference sandcrete was 2002.15 kg/m^3 at 7 days curing age. The values of densities ranged between 1453.88 kg/m^3 for 50% PET and 1851.09 kg/m^3 for 5% PET, while the density of reference sandcrete was 1991.77 kg/m^3 at 28 days curing age. It was observed that the recorded density for all the hollow sandcrete block samples achieved the minimum requirements for density and compacted bulk density of lightweight coarse aggregate, which are 880 kg/m^3 [26] and 1000 kg/m^3 [27]. According to BS 2028 [28], hollow sandcrete blocks with density of 1500 kg/m^3 are grouped as Type A: dense aggregate concrete blocks. However, the density decreased significantly with increased incorporation of PET waste aggregates as partial replacement for fine aggregate. The observed reduction in density of sandcrete blocks could be adduced to the replacement of heavier material (sand) with the lightweight material (crushed PET waste aggregate). A similar reduction in density was reported for sandcrete blocks produced from crushed plastic at replacement percentages of 0%, 5%, 10%, 15%, 20% and 100% at the end of 28 days [29].

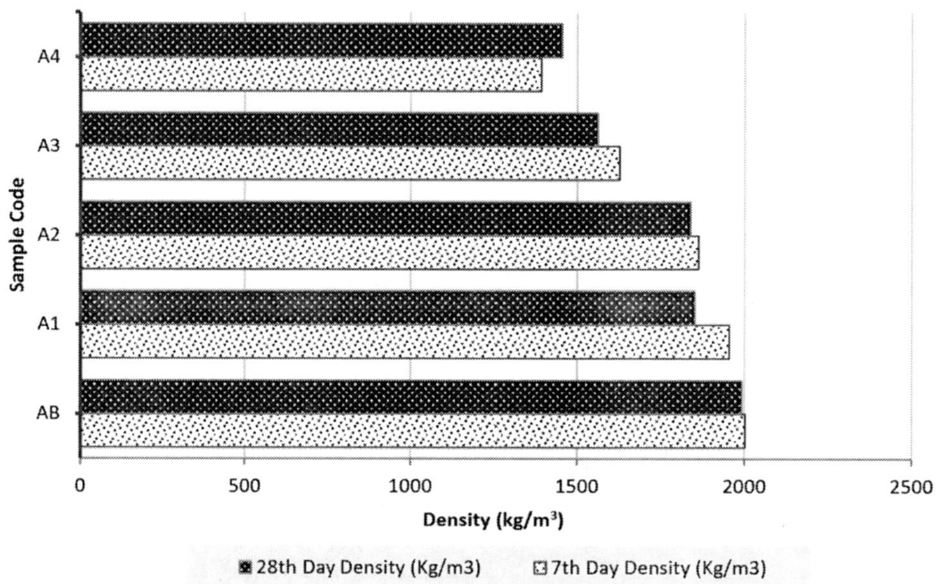

Figure 3. Density of hollow sandcrete blocks

3.4 Compressive strength of sandcrete block

The compressive strength of sandcrete blocks at different PET waste percentages for 7 days is presented in Figure 4. The compressive strength ranged between 1.140 N/mm^2 for PET 50% and 2.564 N/mm^2 for 5% PET and 1.510 N/mm^2 for 50% PET to 2.991 N/mm^2 for 5% PET respectively, as the curing age increases from 7 to 28 days. The compressive strength of reference sandcrete blocks increased from 2.707 N/mm^2 to 3.419 N/mm^2 as the curing age increases from 7 to 28 days. The results obtained from the tests were compared relatively with the recommended strength values of Nigerian Building and Road Research Institute [30], Nigerian Building Code [20], and Nigerian Industrial Standards [31] for manually produced non-load bearing hollow sandcrete blocks. The results indicated that the reference hollow sandcrete block samples and hollow sandcrete block samples produced from 5% and 10% PET waste replacement for sand satisfied the minimum requirement of 2.5 N/mm^2 recommended for compressive strength of hollow sandcrete blocks at 28 days curing age by NIS 87: 2000 [32] for non-load bearing walls application. These results are higher than those reported in the work of [33] with values ranging from 1.01 N/mm^2 to 1.68 N/mm^2 and 0.53 N/mm^2 to 1.59 N/mm^2 for sandcrete blocks containing sawdust as replacement for cement at 28-days curing age. Study by [34] reported a decreasing trend in the compressive strength of cement-based composites as the percentage levels of waste plastic increases. Also, a related study by [35] showed that the compressive strength of the manually compacted sandcrete block samples ranges between 2.61 N/mm^2 to 2.89 N/mm^2 compare to mechanically compacted sandcrete block samples with varied between 2.26 N/mm^2 to 3.03 N/mm^2 as the curing age increases.

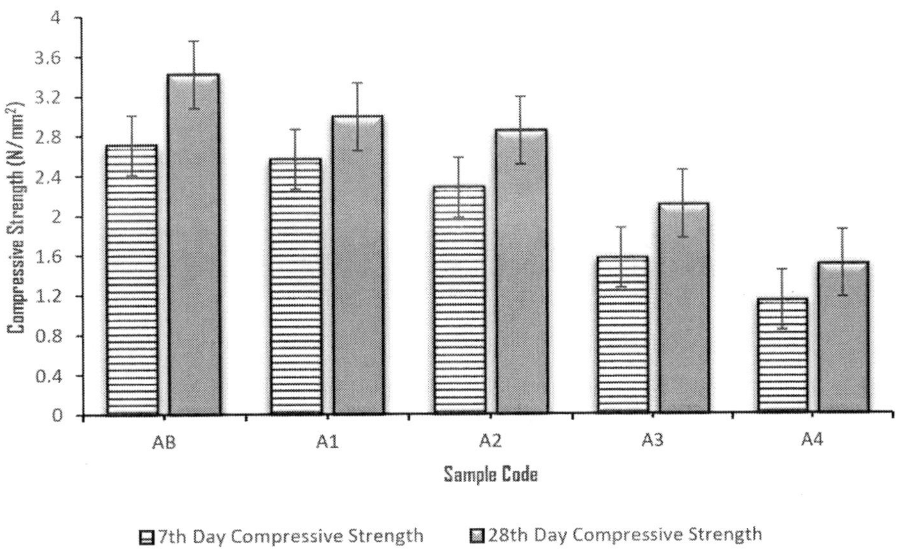

Figure 4. Compressive strength of hollow sandcrete blocks

4. Conclusion

This study showed that the control sandcrete block sample had highest density compare to all other sandcrete blocks containing the crushed PET waste as replacement for sand. In addition, the compressive strength of sandcrete blocks reduced with increasing percentage replacement of the natural aggregate content with PET aggregate waste in the mixes for all the samples. However, reference sandcrete block sample had the highest compressive strength. However, hollow sandcrete blocks containing 5% granulated PET aggregate showed comparable strength to the reference sandcrete blocks, thus implies the possibility of using the granulated PET at an optimum of 5% to partially substitute sand in mix for the production of non-load bearing hollow sandcrete blocks for sustainable construction.

References

[1] Frigione M. 2010. Recycling of PET bottles as Fine aggregate in concrete. Waste Manag. **30** 1101-1106.

[2] Appiah K J, Berko-boateng V N, and Tagbor A T 2017 Use of waste plastic materials for road construction in Ghana *Case Studies in Construction Materials* **6** 1–7 https://doi.org/10.1016/j.cscm.2016.11.001

[3] de Mello D, Pezzin S H and Amico S C 2009 The effect of post-consumer PET particles on the performance of flexible polyurethane foams. *Polymer Testing 28* 702–708

[4] Sukontasukkul P, and Chaikaew C 2006 Properties of Concrete Pedestrian Block Mixed with Crumb Rubber *Construction and Building Materials* **20** 450–457.

[5] Choi Y W, Moon D J, Chung J S, Cho S K, 2005 Effects of waste PET bottles aggregate on the properties of concrete *Cem. Concr. Res.* **35** (4) 776–781

[6] Avila A F, and Duarte M V 2003 A mechanical analysis on recycled PET/HDPE Composites *Polym. Degrad. Stabil.* **80** (2) 373–82.
[7] Baboo R, Tabin-Rushad S, Bhavesh KR, Duggal SK (2012). Study of waste plastic mix concrete with plasticizer, International Scholarly Research Network, ISRN Civ Eng; 2012. 1–5.
[8] Ferreira L, Brito J D, Saikia N 2012 Influence of curing conditions on mechanical performance of concrete containing recycled plastic aggregate *Constr Build Mater* **36** 196–204
[9] Safi B, Saidi M, Aboutaleb D, and Maallem M 2013 The use of plastic waste as fine aggregate in the self-compacting mortars: Effect on physical and mechanical properties *Construction and Building Materials* **43** 436–442
[10] Siddique R, Khatib J, Kaur I 2018 Use of recycled plastic in concrete: a review *Waste Manage.* **28** 1835–52
[11] Malagaveli V 2011 Strength characteristics of concrete using solid waste an experimental investigation *Int. J. Earth Sci. Eng.* **4** 937-940
[12] Ramadevi K, Manju R 2012 Experimental investigation on the properties of concrete with plastic PET (bottle) fibers as fine aggregate *Int. J. Emerg. Technol. Adv. Eng.* **2** 42-46
[13] Pelisser F, Montedo, R K O, Gleize J P P and Roman R H. 2012 Mechanical properties of recycled PET fibers in concrete. *Mater. Res.* **15** (4), 679-686.
[14] Rahmani E, Dehestani M, Beygi M H A, Allahyari H, Nikbin I M 2013. On the mechanical properties of concrete containing waste PET particles. Constr. Build. Mater. 47, 1302-1308
[15] Prahallada M C and Parkash K B 2013 Effect of different aspect ratio of waste plastic fibers on the properties of fiber reinforced concrete an experimental investigation *Int. J. Adv. Res. IT Eng.* **2** 1e13
[16] BS EN 197-1 2000. Composition, Specification and Conformity Critaria for Common Cements, British Standard Institute, 389 Chiswick high Road, London, W4 4AL
[17] NIS 444-1 2003. Composition, specification and conformity criteria for common cements. Standards Organisation of Nigeria.
[18] BS EN 1097-3 1998 Tests for mechanical and physical properties of aggregates. Determination of loose bulk density and voids
[19] BS 3148 1980. Methods of test for water for making concrete (including notes on the suitability of the water)
[20] Nigerian Building Code 2006 Federal Republic of Nigeria: National Building Code First Edition. Nigeria: National Building Code.
[21] BS EN 772-11 2011. Methods of test for masonry units. Determination of water absorption of aggregate concrete, autoclaved aerated concrete, manufactured stone and natural stone masonry units due to capillary action and the initial rate of water absorption of clay masonry units
[22] BS 6073-2 2008. Precast concrete masonry units. Guide for specifying precast concrete masonry units
[23] Eziefula U.G., Opara H.E and Anya C.U. 2017. Mechanical properties of palm kernel shell concrete in comparison with periwinkle shell concrete *Malaysian Journal of Civil Engineering* **29**(1):1-14
[24] Dhir R K., and Jackson N 1996 Concrete. In Jackson N. and R.K. Dhir (eds.) *Civil Engineering Materials (5th ed.).* Basingstoke: Palgrave 161-298
[25] Islam M S and Ghafoori N 2013. Evaluation of Alkali-Silica Reactivity Using Aggregate Geology, Expansion Limits of Mortar Bars and Concrete Prisms, and Kinetic Model. Journal of Materials Science Research **2**(2) 103-117
[26] ASTM C330 – 04 2004. Standard Specification for Lightweight Aggregates for Structural Concrete
[27] Prusty J K and Patro S K 2015. Properties of fresh and hardened concrete using agro-waste as partial replacement of coarse aggregate – a review. *Construction and Building Materials,* **82**, *101–113.* doi:10.1016/j.conbuildmat.2015.02.063
[28] British Standard (BS 2028), Specification for Precast Concrete Blocks. British Standard Institution, Gayland and Sons Ltd. London. (1968).

[29] Akinyele, J. O., & Toriola, I. O. 2018. The effect of crushed plastics waste on the structural properties of sandcrete blocks. *African Journal of Science, Technology, Innovation and Development, 1–5.* doi:10.1080/20421338.2018.1496614
[30] Nigerian Building and Road Research Institute (NBRRI). 2006. NBRRI Interlocking Blockmaking Machine, *NBRRI Newsletter*, Vol. **1** (1) pages 15 – 17
[31] Nigerian Industrial Standard, NIS 87:2007 2007 Standard for sandcrete blocks Lagos, Nigeria: Standard organization of Nigeria. Lagos, Nigeria: Standard organization of Nigeria
[32] Nigerian Industrial Standard (NIS) 87 2001. Standard for Sandcrete Blocks
[33] Raheem A A. and Sulaiman O K 2017 Sawdust ash as partial replacement for cement in the production of sandcrete hollow blocks. *International Journal of Engineering Research and Applications* **3** (4) 713–721
[34] Naik, T.R., Singh, S.S., Huber, C.O., Brodersen, B.S., 1996. Use of post-consumer waste plastic in cement-based *composites. Cem. Concr. Res.* **26** 1489-1492
[35] Odeyemi, S O, Otunola O O, Adeyemi A O, Oyeniyan W O, Olawuyi M Y 2015. Compressive strength of manual and machine compacted sandcrete hollow blocks produced from brands of Nigerian cement *American Journal of Civil Eng*. **3** 2015 2-3

IOP Conf. Series: Materials Science and Engineering 1036 (2021) 012047 doi:10.1088/1757-899X/1036/1/012047

Influx potential and sequestration of CO_2 in Cement based material towards establishing a proactive measure for combatting structural deformations

S E Sanni[1], J O Odigure[2], O Agboola[1], M E Emetere[3], E E Okoro[4] and O Aderinsola[1]

[1] Department of Chemical Engineering, Covenant University, P.M.B 1023, Ota, Ogun State, Nigeria.
[2] Department of Chemical Engineering, Federal University of Technology, P.M.B 65, Niger State, Nigeria.
[3] Department of Physics, Covenant University, P.M.B 1023, Ota, Ogun State, Nigeria.
[4] Department of Petroleum Engineering, Covenant University, P.M.B 1023, Ota, Ogun State, Nigeria.
samuel.sanni@covenantuniversity.edu.ng

Abstract. The effect of carbon dioxide sequestration by Sandcrete structures was investigated as a means of reducing CO_2 emissions. Experiments were conducted to determine the concentration of trioxocarbonate IV (CO_3^{2-}) in sandcrete samples as well as its effect on soil pH. The results showed that the presence of CO_3^{2-} in sandcrete samples evidenced the process of carbonation in sandcrete structures and other cement-based material which provides an alternative means of CO_2 Sequestration. Also, the concentration of CO_3^{2-} varied along heights and horizontal distance within sections of the sandcrete structure. The acidic nature of the soil close to some sections of the sandcrete structure experienced varying pHs of 6.14 - 6.35 which gave room for enhanced leaching which seriously undermines the strength of the sandcrete structure. The percentage concentration differential in the horizontal and vertical directions, were found to vary widely from 5-46%. This constitutes a potential danger i.e. cracking and possible collapse of the structure. The results of calculated diffusivity values and concentration gradients of the carbonate ion conformed with results obtained using a predictive model which helped in monitoring the migration patterns of CO_3^{2}- in the sandcrete structure.

Keywords: Carbon sequestration, sandcrete structures, influx potential, structural deformation, CO_3^{2-}-migration pattern

1. Introduction

In developing nations such as Nigeria, ninety percent of buildings including industrial structures comprise of sandcrete blocks [1]. Currently, efforts to tackle climate change are on the increase globally [2]. Anthropogenic operations (gas flaring, bush burning and agricultural activities) within such proximities release CO_2 into the atmosphere which affect these structures as well as cause global warming [3-5]. Cement-based structures consists of materials such as sand, stones and gravels which use cement as a binder to create a structure suitable for application in construction works. However, it has been proven that cement-based structures have the potential for atmospheric carbon dioxide sequestration [6]. Upon hydration, cement produces hydroxides that serve as binding agents which help to maintain the strengths of cement based structures [7]. An increase in atmospheric CO_2 leads to ocean acidification, which may impose drastic consequences in marine ecosystems and according to literature, carbon mitigation is one of the newest strategies of the world's energy transition schemes [8].

Carbon dioxide sequestration is a geo-engineering technique for long term storage of CO_2 and other forms of carbon [6]. CO_2 sequestration includes the storage and capture of carbon/its compounds. This refers to large-scale, permanent artificial capture and sequestration of industrially produced CO_2 using subsurface saline aquifers, reservoirs, ocean water and aging oil fields [9]. Pre- and post-combustion systems for capturing CO_2 were previously discussed [6,10]. Carbonation reactions are exothermic reactions involving simple reactions between binary oxides such as MgO and CaO [11]; carbonates have a lower energy state than CO_2. According to Huntzinger et al. [12], mineral carbonization is one of the only forms of carbon sequestration that is permanent/stable. The disadvantage of other forms such as ocean, geologic and terrestrial sequestration, is largely due to the potential for leakage over a long period of time. This makes the sequestration process ineffective if the trapped CO_2 rapidly escapes back into the atmosphere.

Several models were discussed by Ramezaniapour et al. [13] and Odigure [7] on carbonation of cement-based structures. However, most of these works focused on determining the depth of carbonation, physical as well as chemical changes of the cement-based structures. In an earlier investigation, Odigure [7] presented a diffusion model for determining the diffusion of gaseous pollutants in sandcrete blocks. The shortcoming of the model is that other reactions were not taken into consideration (i.e. the formation of calcium bicarbonate from $CaCO_3$ and its subsequent leaching from the wall to the ground.

2. Theory

Carbonation is a chemical reaction in which solid products of cement hydrates, primarily calcium hydroxide or calcium silicate hydrates (CSH), calcium aluminate hydrates and calcium sulfoaluminate hydrates (mainly ettringite) in cement-based materials, react with carbonic acid to form carbonates [6]. A technique on how to sequester carbon dioxide was developed by Klaus Lackner [14]. This involves the combination of Carbon dioxide with Fe and Mg-rich serpentine rocks to produce CO_3^{2-} by the reaction in (1).The chemical reactions involved in this process are as given in (2) and (3):

$$Mg^{2+} + CO3^{2-} \rightarrow MgCO_3 \quad (1)$$

$$Ca\ (OH)_2 + H_2O + CO_2 = Ca\ (HCO_3)_2 \quad (2)$$

$$Ca\ (HCO_3)_2 + CO_2 = CaCO_3 + H_2O \quad (3)$$

The formed cement hydrates are brought about by hydration in sandcrete structures when they come in contact with acidic gases such as CO^2, SO^3 in the atmosphere. Hydrated cement contains calcium hydroxide, calcium silicate hydrate, calcium aluminate hydrate, hydrated ferrites and calcium monosulphoaluminate. The rate is dependent on the concentration of these gases in the atmosphere and their relative atmospheric humidity. Long exposure of sandcrete structures to aggressive acids, salts and alkalis, enhance their physicochemical as well as their mechanical deterioration. Therefore, this research investigates the possibility of carbon dioxide sequestration in sandcrete structures as well as the effects of carbonation on sandcrete structures. Products arising from carbonation reactions induce important physicochemical changes on these sandcrete structures [9]. One major effect of carbonization in sandcrete structures, is the loss of aesthetic value. Carbonized spots are not localized but spread across the structures of interest. An increase in atmospheric CO_2 contributes significantly to climate changes and global warming as well as the formation of acid rain which enhances corrosion of reinforced steel in concretes and roofing sheets [15,16]. The potential of this process is yet to be exploited, perhaps due to lack of understanding of the nature and possibility of CO_2 sequestration in cement-based materials [7], hence, the motivation for this research which entails the investigation of the mechanism of carbonation of sandcrete structures in relation to the deterioration of sandcrete structures.

IOP Conf. Series: Materials Science and Engineering 1036 (2021) 012047 doi:10.1088/1757-899X/1036/1/012047

Cement hydrates formed as products of hydration in sandcrete structures can react with acidic gases such as CO_2, SO_3 present in the atmosphere. This reaction occurs internally and may have significant adverse effects. The rate of reaction is dependent on the concentration of these atmospheric gases and their relative humidity. Long exposure of sandcrete structures to aggressive media such as acids, salts and alkalis enhance physicochemical and mechanical deterioration of the structures [17,18]. Consequently, this research investigates carbon dioxide sequestration potential and its effects on the chemical and mechanical properties of sandcrete structures at Covenant University by using a model approach. Also, an estimation of the percentage concentration of CO_3^{2-} in sandcrete samples of cracked walls of equal height and depth at a mapped location (Covenant University fence) was carried in order to predict the levels of structural deterioration at such sites. The concentrations of CO_3^{2-} and calcite present in each of the samples as well as the pH of the soil at different sections of the sandcrete structure were also determined.

$$Ca(OH)_2 + CO_2 \rightarrow CaCO_3 + H_2O \quad (4)$$

$$NaOH(aq) + HCl(aq) \rightarrow NaCl(aq) + H_2O \quad (5)$$

$$CaCO_3(aq) + 2HCl(aq) \rightarrow CaCl_2(aq) + H_2O + CO_2(aq) \quad (6)$$

Acid concentration in pulverized sandcrete was calculated using:

$$C_AV_A/C_BV_A = n_a/n_b \quad (7)$$

Concentration of HCl (g/l) = Molar mass of HCl * molar concentration of HCl (8)

Mass of HCl that reacted with pulverized sandcrete is: volume HCl (l) 3.65g/l * concentration

(9)

Mass of calcite = mass of HCl used * molar mass of $CaCO_3$ in g/mol / (molar mass of HCl) (10)
The mass of CO_2 adsorbed (g/l) calculated mathematically as:
mCO_2 = mHCl* mCO_2/ molar mass of HCl
Where: mCO_2 = molar mass of CO_2 (g/mol), molar mass of HCl = 35.5
Equation 6 was used to determine the parameters for the carbonation process.

$$k^2 - Pk - q = 0 \quad (11)$$

where P (concentration gradient) = u / (De) L and u = z/t; z = 0.94 m t = time (h) and q = De / (De)L ln r.

Solving equation (11) as a quadratic expression gives the possible values of k.

3. Materials and Methods

3.1. Raw Materials, Equipment and Reagents

Sandcrete samples at sections of a fence at Covenant University, 0.5 M NaOH, 0.1 M (HCl) (Sigma Aldrich), phenolphthalein indicator, distilled water, conical flask, burette, pipette, electronic weighing balance (Stwart systems), spatula, retort stand, 1000 and 50 ml measuring cylinders, laboratory test sieve (300 μm), 250 ml beakers, hammer, measuring tape, ceramic mortar and pestle and pH Meter (Hanna 211 meter).

3.2. Methods

Sandcrete samples were obtained from six sections of a fence at Covenant University, Ota, Ogun state, Nigeria. The samples were obtained from long-exposed sandcrete blocks of the fence by striking a hammer into the blocks that make up the fence at the top (2.82 m section), middle (1.88 m section) and ground level (0.94 m section); the particles falling off were collected in containers. These samples were from six sections of the sandcrete fence at equal height interval and depth (i.e. 20 mm). By means of a measuring tape, horizontal, middle and ground level points of each section of the fence were measured; the sandcrete fence has aged 14 years. Control samples i.e. five months old samples were obtained from a construction site at Covenant University. Soil samples were also obtained at different wall sections of the sandcrete fence to determine the pH of the soil.

3.2.1. Sample preparation

Each of the sandcrete samples discussed in section 3.2 was ground to fines and sieved with a 300 µm mesh to obtain very fine particles. In order to avoid aeration or absorption of moisture, the samples obtained were stored in transparent airtight moisture-proof containers prior testing.

3.2.2. Estimation of calcite and CO_3^{2-} concentration by quantitative measurement

To estimate the concentration of carbon dioxide and calcite in the samples, back titration was employed. 0.1 M hydrochloric acid was prepared by dissolving 8.33 ml of concentrated HCl to 1 L of distilled water in an Amber bottle. 0.5 M NaOH was prepared by mixing 20 g of Na-pellets in 1 L of distilled water in a reagent bottle. 3 g of the pulverized carbonated sandcrete sample was taken and transferred into a conical flask. 20 mL of 0.1 M HCl was added. The mixture was shaken thoroughly and observed for CO_2 release. 3 drops of phenolphthalein indicator was then added. At the end of CO_2 release, 0.5 M NaOH was measured and titrated against a standard solution of carbonated sandcrete so as to neutralize the HCl therein. Titration of other samples was done thrice in order to abate inaccuracies. The appearance of a pale permanent pink color signified the end point and the titre values were recorded.

3.2.3. Soil-pH at the wall sections of Sandcrete fence

20 g of each soil sample was weighed and put in a beaker. 30 mL of distilled water was added and the mixture was stirred properly. The measuring probe of the pH meter which was connected to an electronic meter was then inserted in the mixture. Data were read off from the panel of the pH meter.

4. Results

At the 0 and 8.3 m points along the horizontal axis of the fence, the concentration of CO_3^{2}- is highest at the middle section of the fence with values being 1.38 and 1.38 g/L respectively (Figure 1). At the 3.8 m point along the horizontal, the highest CO_3^{2}- concentration (1.34 g/L) was observed at the top section of the fence while the ground section at the lowest concentrations of carbonates at the aforementioned sections with values being 1.03, 1.05 and 0.83 g/L at all the 0, 3.8 and 8.3 m sections respectively. At the 10.48, 14.25 and 17.21 m points along the axis of the fence, the CO_2 concentrations were highest at the ground with corresponding values of 0.63, 1.38 and 1.5 g/L respectively with the lowest carbonate concentrations at such points being 0.26 (ground level), 0.22 (at the top level) and 1.43 at the top section of the fence above its horizontal axis. The high, medium and low carbon contents recorded at several sections of the fence are as a result of low influx of and dissolution of carbon in precipitation or moisture in the atmosphere which may have been transported as acid water or trioxocarbonate IV acid via diffusion into the walls of the fence. Also, the ease with which carbonation occurred in the fence is a measure of the type of cement used in the construction of the fence. Also, the manufacturer may have produced blended Portland cement using carbonaceous materials like limestone as additive. This may also result from the difference in the rate of leaching which is connected to porosity of the blocks. Furthermore, the production of $CaCO_3$ by carbonation can result in the gradual clogging of the pores of the sandcrete [6].

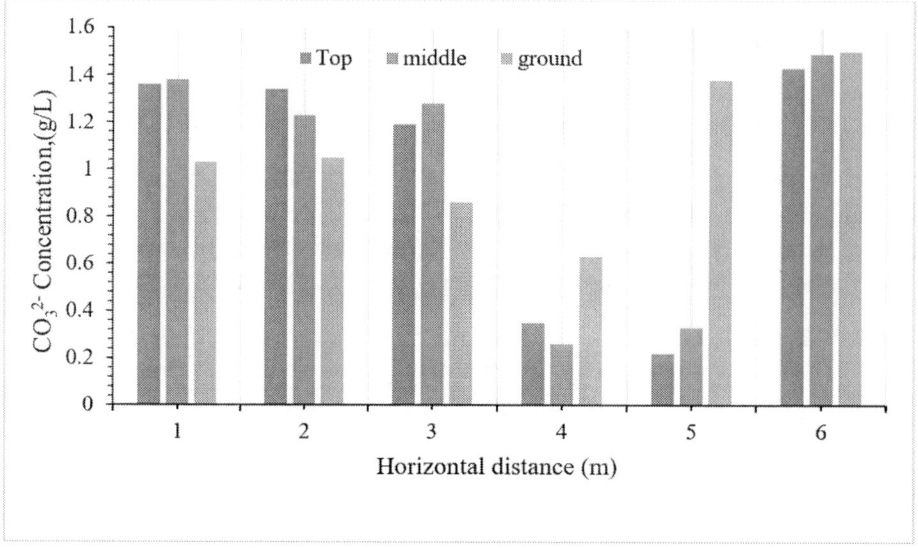

Figure 1. Concentration variation of carbonate at different sections of the fence. Key codes: 1 = 0 m, 2 = 3.8 m, 3 = 8.3 m, 4 = 10.48 m, 5 = 14.25 m, 6 = 17.21 m along the horizontal axis of the fence

Leaching occurred faster at more porous sections than less porous sections of the fence, hence, the extent of block-porosity influenced the rate of carbonation. Hence, this informs that porosity is not the same at all points in the sandcrete wall.

4.1. Concentration and pH of the soil on which the fence rests

Figure 2 gives an illustration of the soil pH at different sections of the pipe. The highest pH (7.63) was recorded at the 3.8 m section of the soil holding the fence in place. The soil pH was found to be lowest i.e. 6.24 at the 0 and 8.3 m sections, 6.75 at the 17.21 m section, 6.35 at the 10.48 m section and 7.5 at the 14.24 m section. This is an indication that the soil pH was either weakly acidic or weakly alkaline. Since carbonic acid is a weak acid, it justifies the recorded pHs in the acidic range, however, owing to the presence of minerals such as K, Ca, Ph, etc, the salts of these elements may have resulted in the basic pHs recorded above the pH of 7 (neutrality).

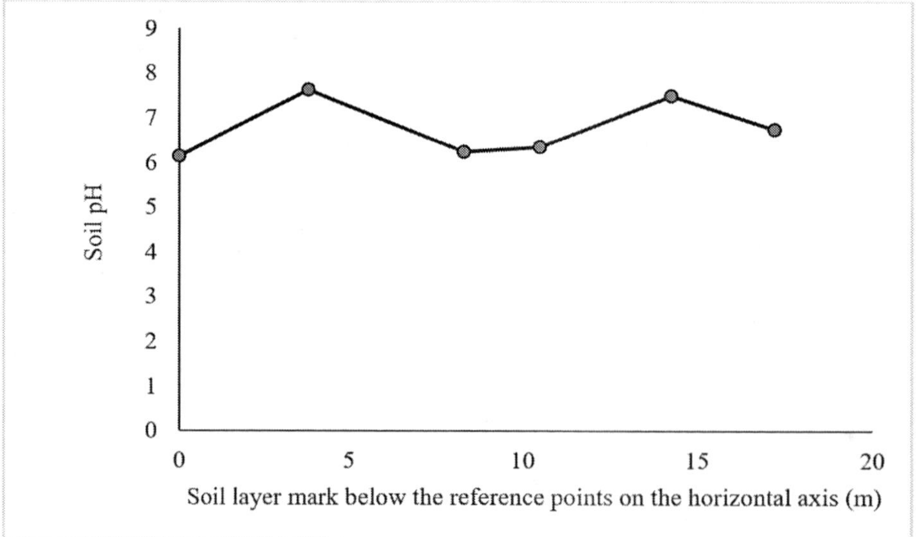

Figure 2. Variation of soil pH at different sections of the fence.

A close look at the pH trend observed at the 0, 7.61, 8.3 and the 10.48 m sections is a clear confirmation that the soil's acidity is a function of the carbonate concentration at those points of the soil holding the fence in place. Owing to the presence of minerals in the soil and conversion of the dissolved carbonate to salt, pH variation possibly set-in which gave rise to soil pH values > 7. The migration pattern of the CO_3^{2-} horizontally showed that there are sections that tend to allow more transfer of the carbonate to the soil. This might be due to difference in the soil chemistry. If the pH of the surrounding soil is acidic, leaching is enhanced and if it is alkaline, rate of leaching is reduced thereby retaining more of the carbonates in the sandcrete wall.

4.2. Estimated CO_3^{2-} concentration gradient and diffusivity in the sandcrete wall

CO_3^{2-} concentration gradient

The estimated carbonate concentration gradients at selected sections/heights above the ground level of the fence are shown in Table 1. The highest carbonate-concentration gradient was obtained at the lowest height above ground level. This shows that, the carbonate concentration gradient or driving force is highest towards the bottom of the fence giving a value of 1.5 g/L.m at the 1.88 m height above ground level/soil holding the fence in place. The lowest carbonate concentration was seen at the highest of all 3 heights considered at the different (6) horizontal sections considered hence, this justifies the fact that carbonate concentration gradient dropped at from the 0-14.24 m points whereas, beyond the 17.21 m point, the value increased significantly. A close look at the extreme values of table 1 shows that the terminals/extremes of the fence stored more carbonate that any other part of the fence because the values are higher at both ends (i.e. the 0 and 17.21 m points) than at any other point on the fence. This is due to those areas being closed ends where carbonate radical diffuses and thus gets stored because there are no extensions/points beyond those regions.

Based on Fick's law of diffusion, concentration gradient is inversely proportional to diffusivity hence, the extreme points of highest concentration gradients have the least diffusivities (Table 2). The extent of carbonation was investigated by calculating diffusivity (De) L values according to the mathematical model. (De)L represents the dispersion of the dissolved $CaCO_3$ as Ca $(HCO_3)_2$ in the block-pore volume. It also represents the effective diffusivity of CO_3^{2-} ions into the sandcrete fence. It was approximately determined from the concentration gradient "P". The estimated negative values are indicative of a decrease in concentration of carbonates in the reference sections of the sandcrete fence. The higher the diffusivity, the lesser will be the expected mechanical strength of the sandcrete fence [7].

Table 1. Concentrations of CO_3^{2-} at the various sections of the fence.

Vertical Height, m	Estimated concentration gradient g/L.m					
	0 m	3.8 m	8.3 m	10.48 m	14.24 m	17.21 m
2.82	1.36	1.34	1.19	0.35	0.22	1.43
1.88	1.38	1.23	1.28	0.26	0.33	1.49
0.94	1.03	1.05	0.86	0.63	1.38	1.5

Table 2. Estimated diffusivities at the different sections of the pipe.

Vertical Height, m	Estimated Diffusivities * 10^{-6} m^2/s					
	0 m	3.8 m	8.3 m	10.48 m	14.24 m	17.21m
	-				-	-
1.88-2.88	0.36	5.46	-6.56	6.69	5.46	1.09
0.94-1.88	1.73	3.31	1.42	-1.63	-57.46	-0.33

A high rate of carbonate diffusivity is aided by the temperature of the environment, hence, an environment of high temperature increases the rate of diffusion. Also, Sections with high diffusivity gave lower CO_3^{2-} concentration gradients or lower diffusion rates of the carbonate ion. Furthermore, the diffusivity of CO_2 is 10,000 times smaller in water than in air. This means that diffusivity is faster in air than in water. Therefore the rate of carbonation occurs faster in air than water. According to Odigure [14], this parameter gives the extent of reaction between the cement mineral hydrates and CO_2 and their leaching from the structure. De represents the actual values of CO_2 molecules that diffused into and reacted with the cement minerals in the sandcrete matrix. The parameter r represents depth. This varies but for this research work, q was assumed to be zero since samples were collected at equal height and at constant depth of 20 mm.

The depletion of CO_3^{2-} within the sandcrete will lead to a fall in the binding properties of any formed $Ca(OH)_2$ and may cause a defect or crack in the structure. The observed horizontal changes in carbonate concentration differs with sections of the fence which will exhibit a variation in pH and %composition of its carbon constituents. Furthermore, these changes can lead to internal stress within and along the fence wall which result in crack-formation in the fence structure. Based on the investigation carried out by Odigure [14], the results obtained in this study in terms of calculated diffusivities and concentration gradients of CO_3^{2-} are in close agreement.

Figure 3 is an illustration of the irregular/random variation of soil pH and heights across the fence. At the 1.88- 2.88 m height above the ground, within the soil pH of 6.24-7.5, the diffusivity of CO_3^{2-} is lowest with corresponding values in the range of -6.56-5.46 m^2/s and hence, based on Fick's second law of diffusion, the concentration of the carbonate is highest whereas, it is lowest at pHs of 6.35 and 7.63, the resultant diffusivities are 6.69 and 5.46 m^2/s. However, within 0.94-1.88 m height along the fence, the pHs are 6.35 and 7.5 with corresponding diffusivities of -1.65 and 57.46 m^2/s respectively, which are the least recorded values hence, the concentrations of CO_3^{2-} are expected to be highest here; also, within the measured pHs of 6.14 and 7.63, the estimated diffusivities are 1.73 and 3.31 respectively, which are expected to give the lowest CO_3^{2-}concentrations. Based on the results, the soil pH range is weakly acidic and alkaline with varying values which influence the estimated diffusivities.

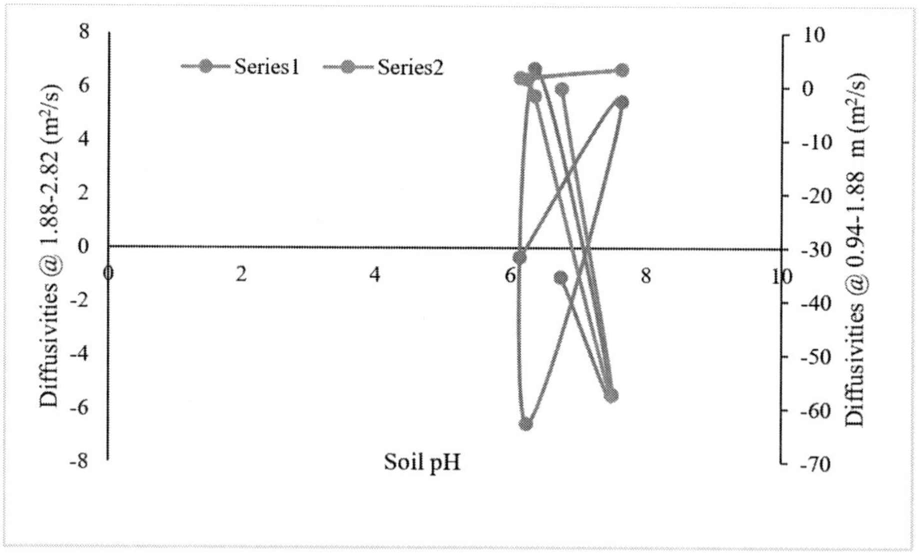

Figure 3. Soil pH and diffusivities at different heights.

5. Conclusion

The presence of CO_3^{2-} in sandcrete samples taken from the fence actually confirm the effect of carbonation in sandcrete structure which gives a clue to the reason for structural failures in fences caused by carbon sequestration. CO_3^{2-} concentration varied with heights and horizontal distances within sections of the sandcrete fence. pH ranges were judged to be in the weakly acidic and weakly alkaline ranges thus encouraging leaching of the soil/weakening of the fence which resultantly undermines the strength of the fence. The estimated percent carbonate concentration differential in the horizontal and vertical directions were found to vary widely from 5 to 46% which constitute a potential danger to the fence in terms of being subsequently probable to cracking and collapse. The predictive model gave comparable diffusivity values with those obtained in literature. Furthermore, at the 1.88- 2.88 m height above the ground, the concentration of the carbonate was highest whereas, it was lowest at pHs of 6.35 and 7.63, with resultant diffusivities of 6.69 and 5.46 m^2/s respectively while at the 0.94-1.88 m height along the fence, the recorded pHs are 6.35 and 7.5 with corresponding diffusivities of -1.65 and 57.46 m^2/s respectively; these represent the least recorded diffusivities and are expected to be points having the highest concentrations of CO_3^{2-}

Recommendations

Based on the results of this study, it is necessary to carry out routine CO_2 measurements around Covenant University fences in order to ascertain their deformation potentials; this is aimed at developing adequate carbon capture strategies (adsorption systems/CO_2-traps) for CO_2 mitigation. Also, in the year 2019, a collapse of one of the Covenant University fences occurred and this is a clear evidence that the fence had been weakened at some of its sections owing to the influx of acid rain and CO_2. Furthermore, an investigation of the mechanism and kinetics of the carbonation process of cement needs be carried out in order to clearly understand the sequestration process of cement or sandcrete structures. Analysis involving the use of Scanning Electron Microscopy, Thermal Gravimetric Analysis and X-ray Diffraction can be carried out to investigate the carbonation profile, the mineralogy and morphology of the understudied sandcrete structure. The effect of temperature on cement-carbonation should be investigated. Estimation of Analysis on deposition and penetration rates of other gaseous pollutants such as SO_2 and NO_2 may be considered in order to ascertain their influence on the carbonation processes of sandcrete structures. The carbonation of other cement-based materials such as concrete can be investigated with the results compared with those of these study. The use of Atomic Absorption Spectrophotometer (AAS) is necessary so as to ascertain the types of metal ions especially Ca^{2+} ions in sandcrete samples as this helps to predict the extent of carbonation.

Acknowledgements

The authors would like to appreciate The Covenant University Centre for Research and Innovation Development for her support in ensuring the publication of this manuscript.

References

[1] Oyekan G L and Kamiyo O M 2011 *J. Eng. Technol. Res.*, **3,** 3, 88-98.
[2] Li H, Jiang H-D, Yang B and Liao H 2019 *Sci. Tot. Environ.* **687**, 687–701.
[3] Zarate-Barrera T G and Maldonado J H, 2015 *PLoS ONE* **10,** 5 e0126627. Doi: 10.1371/journal. pone.0126627.
[4] Sood A and Vyas S 2017 IOP Conf. Ser.: Earth Environ. Sci. **83** (Rome, Italy) 012024.
[5] Mal S, Singh R B, Huggel C and Grover A 2018 *Introducing linkages between climate change, extreme events, and disaster risk reduction*. (Eds.) S Mal, R B Singh and Cham C Huggel, Springer International Publishing, pp. 1–14.
[6] Kumar R and Naik R K 2010 *Global warming and cement-based materials. Milwaukee,* Wisconsin USA: Tarun R Naik, UWM Center for By-Products Utilization pp 1–70.
[7] Odigure J O 2002 *Cement and Concrete Research, vol.* **32**, 1451 – 1455.
[8] Oelkers E H and Cole D R 2008 Carbon dioxide sequestration: *a solution to a global problem,* Elements Magazine pp. 305-310. DOI: 10.2113/gselements.4.5.305.
[9] Fernández B M, Simons S J, Hills C D, Carey P J 2004 *J. Haz. Mat.,* **112**, 3, 193–205.
[10] James C F 2009 *The Reduction of CO_2 Emissions via CO_2 capture and solid oxide fuel cells*, PhD thesis, University of Akron USA.

[11] Yegulalp T M, Lackner, K S and Ziock H J 2000 Proceedings of 6th Int. Conf. on Environmental Issues and Management of Waste in Energy and Mineral Production. Calgary Alberta, Canada, Balkema, Rotterdam, pp. 375 –382.
[12] Huntzinger D N, Gierke J S, Kawatra K S, Eisele T C and Sutter L L 2009 *Environ. Sci. Technol.*, **43**, 6, 1986-1992.
[13] Ramezanianpour A A, Mahdikhani M and Ahmadibeni G 2009 *Int. J. Civil Eng.* 7, 2, 83–91.
[14] Odigure J O 2008 Migration model of carbonized cement mineral hydrates in sandcrete. Proc. Int. Conf. on Concrete Durability: Achievement and Enhancement, 8 – 9 July, University of Dundee, Scotland, UK. pp. 249 – 257.
[15] Osigwe O and Nwanakoala H 2013 *Int. J. Sust. Energy Environ.*, pp. 158 - 166.
[16] Song K, Zang S, Zhao Y, Li L, Du J, Zhang N et al. 2013 *Hydrology and Earth System Sciences*, **17**, 10, 4269 – 4281. https://doi.org/10.5194/hess-17-4269-2013.
[17] Wang X, Akhmedov N, Duan Y, Luebkea D and Li, B 2013 *J. Mater. Chem. A*, **1**, 2978–2982.
[18] Majumder S C, Islam K and Hossain M H 2019 *Geol., Ecology, Landscapes* vol. **3**, 1, 29–36. https://doi.org/10.1080/24749508.2018.1481656.

IOP Conf. Series: Materials Science and Engineering 1036 (2021) 012048 doi:10.1088/1757-899X/1036/1/012048

Protection performance of orange oil distillates on low carbon steel in dilute acid electrolytes

R T Loto[1*], E H Mbah[1] and J I Ugada[1]
[1]Department of Mechanical Engineering, Covenant University, Ogun state, Nigeria
Email: tolu.loto@gmail.com

Abstract. Orange essential oil distillate (OOE) was studied for its inhibition performance on low carbon steel in 0.5 M H_2SO_4 and HCl aqueous electrolyte by electrochemical method and open circuit potential measurement. OOE performance was significantly proportional to it concentration with highest protection efficiency of 76.93% and 94.90% at optimal OOE concentration (10% OOE) in H_2SO_4 and HCl electrolyte. The performance of OOE at low concentration in H_2SO_4 was quite poor with inhibition efficiency values below 40%. OOE exhibited greater inhibition action in HCl solution with effective performance beginning at 4% OOE concentration compared to 6% concentration in H_2SO_4 electrolyte. Corrosion potential deviated basically in the anodic direction signifying the dominant inhibition effect to be through surface coverage of the steel surface. OOE exhibited mixed type corrosion inhibition. Open circuit potential graphs show OOE increases the thermodynamic fluctuation of the carbon steel in H_2SO_4 solution relative to the plots at 0% OOE, despite effective protection. In HCl, the open circuit potential plots were more electropositive relative to the plots at 0% OOE signifying decreased tendency to corrosion. Optical images of the carbon steel after evaluation with and without OOE significantly contrast each other.

1. Introduction

The corrosion dilemma is prevalent worldwide and accounts for a significant portion of the gross domestic product of countries. Most of the cost resulting from corrosion damage goes to maintenance and repair, replacement of damage parts and, corrosion control and monitoring techniques [1-3]. There are cases where damage due to corrosion of metallic parts results in leakage of toxic substances to the environment, injury to industry personnel due to industrial accidents, financial cost from plant shutdowns and unforeseen circumstances [4, 5]. One of the important focal points of material degradation is the universal utilization of carbon steels due to their inadequate corrosion resilience and inability to passivate against aqueous reactive species [6, 7]. Carbon steels are lowly priced, reusability, easily obtainable and desirable mechanical properties which can be easily altered to meet specific industrial requirements [8]. They are applied in petrochemical industry, automobile industry, mining industry, marine industry, desalination plants and energy generating industry. There are a number of effective methods of corrosion prevention and control. However, utilization of corrosion inhibitors remains the most competitive and proven method of corrosion control to extend the lifespan of carbon steels [9]. Chemical compounds of organic origin are widely used to enhance the corrosion resilience of carbon steel. They perform through evolution of a shielding film on the steel's exterior. However, conventional inhibiting fluid derivatives are unsustainable due to the damage caused to the environment and personnel involved in their application [10-18]. Utilization of plant distillates for corrosion inhibition has seen increasing use of recent due to their non-toxic nature [19-24]. This manuscript assesses the protection performance of citrus sinensis oil distillate in aqueous H_2SO_4 and HCl aqueous electrolyte.

2. Experimental analysis

Orange essential oil distillate (OOE) was concocted in cubic concentrates of 2%, 4%, 6%, 8% and 10% per 200ml of 0.5 M H_2SO_4 and HCl solution. Low carbon steel (LCS) rods were sliced and partitioned into 6 test exhibits for each experimental study. Their general dimension is 0.7 cm radius and 0.7 cm length. The circular exterior ends of the rods were smoothened with coarse papers of different grits. Cu wires were attached to one end of the steel surface and mounted in prehardened resin before hardening after 5 minutes of encasement. Potentiodynamic polarization was executed with a potentiostatic instrument. The mounted LCS working terminal, Pt wire counter terminal and Ag/AgCl reference terminal were immersed within a container with the acid-inhibitor solution at specific inhibitor concentration per test. The electrodes are attached to the potentiostat-computer system to observe the output polarization plots from -1.2V to 0.5 V at sweep rate of 0.0015 v/s. Corrosion rate and other electrochemical parameters were gotten from the polarization test which were discussed in the subsequent sections. Open circuit potential measurement of the alloy in the aqueous acid/inhibitor electrolyte was executed with Digi-Ivy potentiostat for 4000 s at step potential of 0.1 V. Optical illustrations of the protected and corroded steel surface were obtained and assessed in the subsequent sections of the manuscript.

3. Results and discussion

3.1. Potentiodynamic polarization

Potentiodynamic polarization graphs of LCS in OOE/H_2SO_4 and OOE/HCl solution at exact OOE concentration are shown in Fig. 1(a) and (b). Outputs from the test is shown in Table 1. Results in Table 1 shows OOE strongly influenced the corrosion of LCS in both acids. However, general observation of the results shows OOE exhibited higher efficiency in HCl solution compared H_2SO_4 solution. The higher inhibition efficiency in HCl solution is because of the higher dissociation and reaction potential of H_2SO_4 in H_2O as against HCl where H_2SO_4 releases two protons whereas HCl releases on proton. Secondly the size of Cl^- ions is much smaller than the corresponding SO_4^{2-} ions as such the reaction effect of Cl^- ions tends to be limited as against the general surface deterioration caused by the action of SO_4^{2-} ions. Effective inhibition performance was attained at 4% OOE concentration in HCl compared to 6% OOE concentration in H_2SO_4. Nevertheless, the progressive rise in OOE inhibition efficiency analogous to OOE concentration depicts the protection effect of the oil distillate is concentration dependent wherewith surface coverage of the steel surface by the inhibiting ions suppresses the oxidation of the steel. Optimal inhibition efficiency of 76.93% was attained at 5% OOE concentration in H_2SO_4 and 94.90% at 5% concentration in HCl. Variation of corrosion potential values analogous to the potential at 0% OOE shows OOE exhibits mixed type inhibition reaction mechanism. This is proven from the variation of the inclines of the anodic-cathodic polarization plots where significant decrease in the slopes occurred with increase in OOE concentration. However, the cathodic plots in Fig. 1(b) shows at 6% - 10% OOE concentration shows dominant cathodic inhibition effect whereby the O_2 reduction and H_2 evolution reactions are hindered.

3.2. Open circuit potential measurement

Open circuit potential (OCP) plots for LCS corrosion in H_2SO_4 and HCl solution at 0%, 2% and 10% OOE concentration are shown in Fig. 2(a) and (b). The plots in Fig. 2(a) shows OCP at 0% OOE is more electropositive than the plots at 2% and 10% OOE concentration. The plot at 0% OOE started at -0.478 V and sharply deviated electropositively to -0.450 V at 400 s before gradually progressive in the same direction. The plots at 2% and 10% OOE transited from onset to electronegative potentials culminating at -0.475 V and -0.499 V at 4000 s. The negative shift at 2% and 10% OOE shows that OO increases the thermodynamic tendency LCS to corrode despite effective inhibition performance at 10% OOE concentration. The phenomenon is probably due to lateral interaction effect among inhibitor molecules which strongly influences their inhibition performance. This is evident in the significant potential transient at 10% OOE concentration. The OCP graphs in Fig. 2(b) shows the plot at 2% OOE is more electropositive than the plots at 0% and 10% OOE concentration despite gradual progression to electronegative values coupled with visible potential transients. The potential fluctuations signify contentious adsorption among OOE ionic species and Cl^- ions in the electrolyte.

Table 1. Outputs from potentiodynamic polarization of LCS in 0.5M H_2SO_4 and HCl solution at exact OOE concentration (n=1)

Sample	OOE Conc. (%)	LCS C_R (mm/y)	OOE I_E (%)	C_I (A)	C_J (A/cm^2)	C_P (V)	R_p (Ω)	B_c (V/dec)	B_a (V/dec)
H_2SO_4 Solution									
A	0	9.47	0	9.38E-04	8.30E-04	-0.429	17.40	-7.203	1.451
B	2	7.58	19.95	7.51E-04	6.64E-04	-0.427	29.49	-7.503	0.138
C	4	6.12	35.35	6.06E-04	5.36E-04	-0.402	43.26	-10.480	2.969
D	6	3.52	62.82	3.49E-04	3.08E-04	-0.390	75.64	-7.943	1.054
E	8	2.84	69.97	2.82E-04	2.49E-04	-0.420	91.24	-2.475	4.046
F	10	2.18	76.93	2.16E-04	1.91E-04	-0.428	118.80	-6.364	9.240
HCl Solution									
A	0	5.52	0	5.47E-04	4.84E-04	-0.453	47.18	-6.646	13.15
B	2	2.76	49.92	2.74E-04	2.42E-04	-0.445	61.19	-5.807	-0.064
C	4	2.19	60.41	2.16E-04	1.92E-04	-0.458	83.81	-5.140	4.930
D	6	0.34	93.83	3.37E-05	2.99E-05	-0.461	761.40	-7.394	10.500
E	8	0.34	93.86	3.35E-05	2.97E-05	-0.458	766.00	-5.107	11.180
F	10	0.28	94.90	2.79E-05	2.47E-05	-0.458	820.80	-5.130	10.730

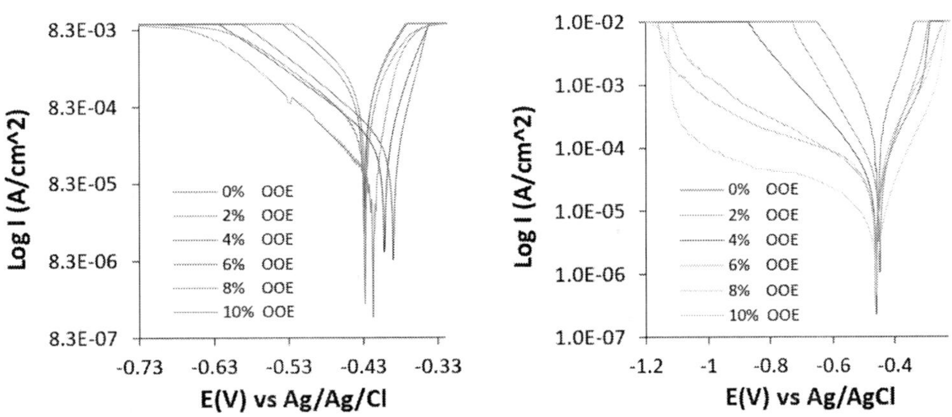

(a) (b)

Figure 1. Potentiodynamic polarization plots LCS corrosion at 0% - 6% OOE concentration within (a) H_2SO_4 solution and (b) in HCl solution

The plot at 0% OOE shifted to electropositive values, but its final potential is more electronegative than the final value at 2% OOE concentration due to the inhibition effect of protonated OOE molecules aided by the electronegative steel surface due to initial adsorption by Cl^- anions. The plot at 10% OOE concentration in Fig. 2(b) exhibited similar plot configuration to the plot at 10% concentration in Fig. 2(a) due to reasons earlier discussed. Nevertheless, despite effective inhibition performance by OOE on LCS, OOE increases the thermodynamic instability of LCS surface and its tendency to corrode due to instability of of the protected steel surface.

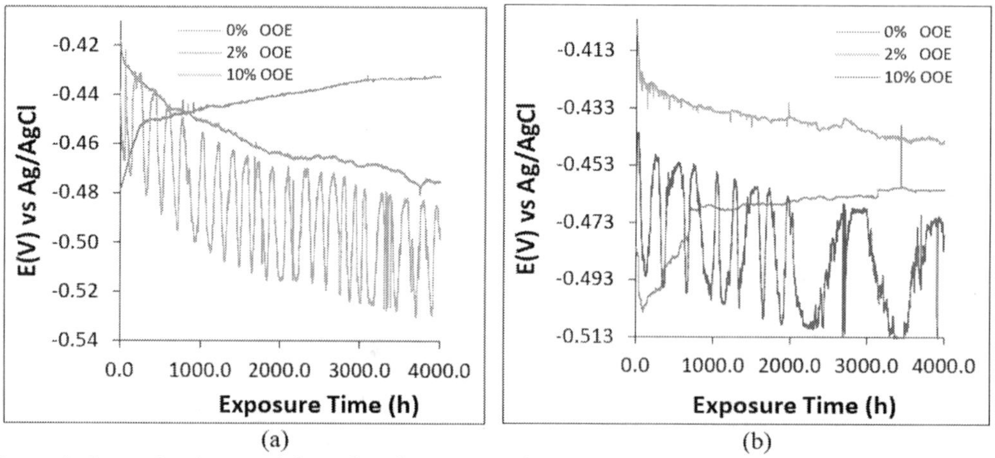

Figure 2. Open circuit potential graphs of LCS corrosion in (a) H_2SO_4 solution and (b) HCl solution at 0%, 2% and 10% OOE concentration

3.3. Optical microscopy study

Optical illustrations of LCS prior to corrosion, after in H_2SO_4 and HCl solution without OOE and after corrosion from the electrolytes in the presence of OOE are laid out from Fig. 3 to 5(b). Severe deterioration is visible on LCS surface in Fig. 4(a) and (b) due to the absence of OOE to counteract the degenerative reactions of SO_4^{2-} and Cl^- anions in the acid media. As a result, LCS surface was oxidized undergoing anodic dissolution. The degree of deterioration is greater on Fig. 4(a) which contrast to Fig. 4(b) because of the higher dissociation constant of H_2SO_4 in H_2O compared to HCl. The images in Fig. 5(a) and (b) significantly differs to the images earlier discussed due to suppression of the reduction-oxidation reaction mechanism. The mild surface deterioration on the images [Fig. 5(a) and (b)] is due to preadsorbed corrosive species on the steel surface which increased its ionization potential and electronegativity before subsequent attraction of cationic OOE molecules unto the steel which blocked further diffusion of corrosive species.

Figure 3. Optical illustration of LCS before corrosion

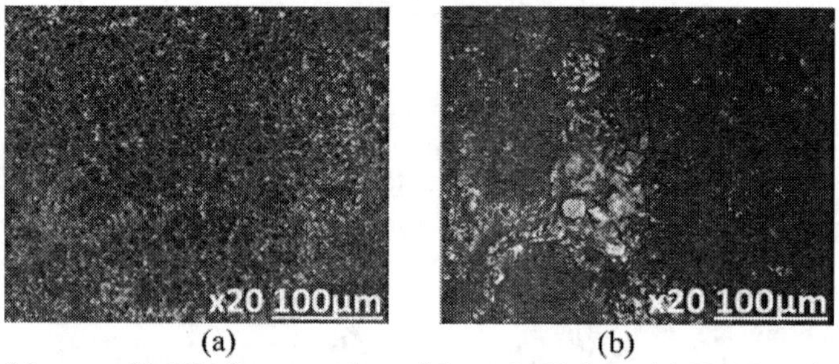

Figure 4. Optical image of LCS after corrosion without OOE from (a) H_2SO_4 solution and (b) HCl solution

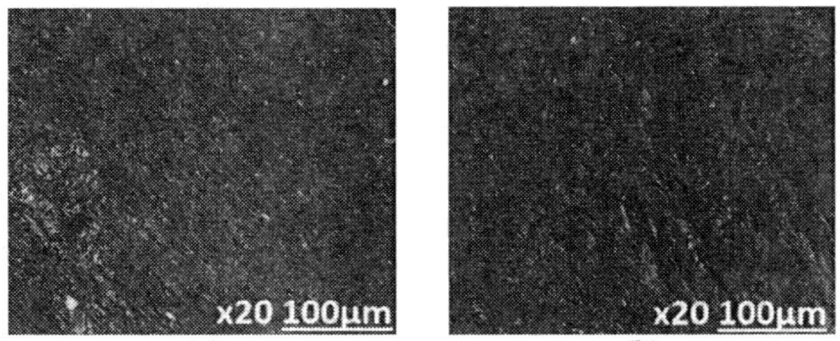

(a) (b)

Figure 5. Optical image of LCS after corrosion from (a) H_2SO_4/OOE solution and (b) HCl/OOE solution

4. Conclusion

Orange essential oil distillate sufficiently stifled the degeneration of the carbon steel in H_2SO_4 and HCl aqueous electrolyte. The oil distillate exhibited higher efficiency at all concentrations in HCl solution compared to H_2SO_4 solution were effective performance occurred at higher concentrations of the distillate. The distillate exhibits mixed type protection characteristics with dominant cathodic inhibition effect at higher distillate concentrations in HCl solution. The morphology of the alloy after corrosion in the electrolyte in the absence of the distillate showed severe deterioration compared the improved morphology with the distillate. Corrosion potential measurement showed the distillate induced the thermodynamic instability of the steel in both acids despite effective corrosion inhibition performance.

Acknowledgement

The author declares the support of Covenant University Ota, Ogun State, Nigeria for this investigation.

References

[1] Corrosion Costs and Preventive Strategies in the United States, NACE Publication NO. FHWA-RD-01-156. https://www.nace.org/uploadedFiles/Publications/ccsupp.pdf.
[2] Akinyemi OO, Nwaokocha CN and Adesanya AO 2012 Evaluation of corrosion cost of crude oil processing industry *J. Eng. Sci. Technol.* **7(4)** 517 – 528.
[3] Jekayinfa SO, Okekunle PO, Amole IG, J.A. Oyelade JA 2005) Evaluation of corrosion cost in some selected food and agro-processing industries in Nigeria *Anti-Corros. Method. M.* **52(4)** 214-218.
[4] The effects and economic impact of corrosion, Corrosion: Understanding the Basics, ASM International, 2000. www.asminternational.org/documents/10192/1849770/06691g_chapter_1.pdf.
[5] Perumal KE 2014 Corrosion risk analysis, risk based inspection and a case study concerning a condensate pipeline Procedia Eng. **86** 597–605.
[6] Loto RT and Loto CA 2018 Corrosion behaviour of S43035 ferritic stainless steel in hot sulphate/chloride solution *J. Mater. Res. Techn.*7(3) 231-239.
[7] Loto RT and Babalola P 2017 Corrosion polarization behavior and microstructural analysis of AA1070 aluminium silicon carbide matrix composites in acid chloride concentrations *Cogent Eng.* **4(1)** 1422229
[8] Dwivedi D, Lepková K and Becker T 2017 Carbon steel corrosion: a review of key surface properties and characterization methods *RSC Advances* 8 4580-4610.
[9] Dariva CG and Alexandre F. Galio AF 2013 Corrosion inhibitors – principles, mechanisms and applications, Developments in Corrosion Protection, Intechopen, pp. 365-379. http://dx.doi.org/10.5772/57255.
[10] Kuznetsov YI 1996 Organic inhibitors of corrosion of metals, Springer US. Doi.org/10.1007/978-1-4899-1956-4.
[11] Sanyal B 1981 Organic compounds as corrosion inhibitors in different environments — A review *Prog. Org. Coat.* **9(2)** 165-236.
[12] Fouda AS, Diab MA, and Fathy S 2017 Role of some organic compounds as corrosion inhibitors for 316L stainless steel in 1 M HCl *Int. J. Electrochem. Sci.* **12** 347–362. doi.org/ 10.20964/2017.01.60.
[13] Rivera-Grau LM, Casales M, Regla I, Ortega-Toledo DM, Ascencio-Gutierrez JA, Porcayo-Calderon J and L. Martinez-Gomez L 2013 Effect of organic corrosion inhibitors on the corrosion performance of 1018 carbon steel in 3% NaCl solution *Int. J. Electrochem. Sci.* **8** 2491 -2503.
[14] Winkler DA 2017 Predicting the performance of organic corrosion inhibitors *Metals MDPI* **7(12)** 553; https://doi.org/10.3390/met7120553.
[15] Singh WP and Bockris JO 1996 Toxicity Issues of Organic Corrosion Inhibitors: Applications of QSAR Model NACE International, Houston, Texas.
[16] Winkler DA, Breedon M, Hughes AE, Burden FR, Barnard AS, Harvey TG and I. Cole I 2014 Towards chromate-free corrosion inhibitors: structure–property models for organic alternatives *Green Chem.* **16** 3349-3357. 10.1039/C3GC42540A.

[17] Cicek V 2017 Chromates: Best corrosion inhibitors to date, Corrosion Engineering and Cathodic Protection Handbook, Scrivener Publishing LLC, Inc, pp.27-30. doi.org/10.1002/9781119284338.ch11.
[18] Loto RT and Loto CA 2012 Effect of P-phenylediamine on the corrosion of austenitic stainless steel type 304 in hydrochloric acid *Int J Elect Sci* 7(10) 9423-9440
[19] Verma DK and Khan F 2015 Corrosion inhibition of mild steel by extract of bryophyllum pinnatum leaves in acidic solution *Chemistry and Material Research* 7 2225-0956, 2015.
[20] Subhashini S, Rajalakshmi R, Prithiba A, Mathina A 2010 Corrosion mitigating effect of Cyamopsis Tetragonaloba seed extract on mild steel in acid medium *E-J.Chemistry* **7(4)** 1133-1137.
[21] Obot IB, Obi-Egbedi NO, Umoren SA and Ebenso EE 2010 Synergistic and antagonistic effects of anions and ipomoea invulcrata as green corrosion inhibitor for aluminium dissolution in acidic medium *Int. J. Elect. Sci.* 5(7) 994–1007.
[22] Loto RT, Loto CA, Ayozie B and Sanni T 2018 Anti-corrosion properties of rosemary oil and vanillin on low carbon steel in dilute acid solutions, The Minerals, Metals & Materials Society, TMS 147th Annual Meeting & Exhibition Supplemental Proceedings, The Minerals, Metals & Materials Series, pp. 883-890. https://doi.org/10.1007/978-3-319-72526-0_84.
[23] Loto RT 2017 Corrosion inhibition effect of non-toxic α-amino acid compound on high carbon steel in low molar concentration of hydrochloric acid *J. Mater. Res. Techn.* doi.org/10.1016/j.jmrt.2017.09.005.
[24] Loto CA and Loto RT 2013 Effect of dextrin and thiourea additives on the zinc electroplated mild steel in acid chloride solution *Int. J. Elect. Sci*. 8(12) 12434-12450.

Strength Evaluation of Aluminium Fibre Reinforced Particle Board made from Sawdust and Waste Glass

O D Atoyebi[1], S O Ajamu[1], S O Odeyemi[2], D J Ojo[1] and J A L Ramonu[3]

[1]Department of Civil Engineering, Landmark University, Omu Aran, Kwara State, Nigeria.
[2]Department of Civil Engineering, Kwara State University, Malete, Kwara State, Nigeria.
[3]Department of Physical Planning and Development, Landmark University, Omu Aran, Kwara, State, Nigeria

Corresponding Author: atoyebi.olumoyewa@lmu.edu.ng

Abstract. This research investigated the effects of aluminium shavings on particleboards produced from sawdust and waste glass using Urea formaldehyde as binder. Twelve (12) boards were produced and each had different variations of sawdust, waste glass and aluminium shavings with a constant volume of adhesive used. The boards were tested for physical properties such as density, Thickness swelling (TS) and Water absorption (WA) and mechanical properties such as modulus of rupture (MOR) and modulus of elasticity (MOE). The effects of the aluminium shavings on the boards were observed in the MOR and MOE values such that these values for board A1 were 3.30 N/mm^2 and 10.45 N/mm^2 for 20% aluminium shavings content but reduced to 3.16 N/mm^2 and 8.97 N/mm^2 respectively for board A2 with 10% aluminium shavings content. This trait was observed in other boards with the same content except in boards B1 and B2 where the MOR and MOE values increased with the values for B1 being 2.06 N/mm^2 and 6.98 N/mm^2 and 2.91 N/mm^2 and 10.17 N/mm^2 respectively. The values for density of the boards increased as the sawdust decreased and also as the glass content increased. The boards with the highest values for thickness swelling and water absorption were the first three as they possess the highest quantity of sawdust. From the results, none of the boards produced met up to the requirements for use for structural purposes, board B3 possesses the highest MOR value while board D3 possesses the highest MOE value. Therefore, the boards can serve non-structural purposes such as partitions and indoor uses.

1. Introduction

Due to the rising need for cheaper construction materials, search for solution to housing problems experienced across the world by mostly developing countries and also to provide a way of effectively recycling waste materials, researches have been taken up to explore the possibilities of utilising unconventional materials that bear close similarities with the materials normally used for structural development[1,2]. One of these construction materials is Particle board. Particle board is a construction material that has vast applications and is also eco-friendly as it usually makes use of waste materials thereby providing a way of recycling such materials [3–9]. The adhesive used in bonding the particles together could be any binding agent but there are some synthetic resins which are used frequently, these include urea formaldehyde, phenol formaldehyde etc. However, urea formaldehyde is the most commonly used due to its cheap price and easy production [10,11]. The particle board is an engineered material made from wood-based products and during the development of this board, several factors affect the properties of the board. Such factors could include the particle size and the adhesive content. These factors could directly affect properties such as board density, modulus of rupture, modulus of elasticity etc. Different kinds of materials would have different interactions with the adhesive used, so the suitability of each material to the adhesive used would be different for several materials which would ultimately have either positive or adverse effects on the particle board produced [12].

Industrial waste management refers to the ways through which waste materials could be reused and recycled to serve a functional purpose. This shows that the mentality of regarding waste as being useless is being dismissed and more value is being attached to waste materials as more environmental awareness is created to control environmental pollution. There are various forms in which waste could present itself, some of which include; municipal solid waste such as household refuse, wastewater such as sewage, radioactive waste [13]. Sawdust is an industrial waste that is available in large quantities in sawmilling factories, paper industries and also wood processing industries and it is usually dumped in heaps to be burnt thereby causing a major concern for environmental well-being. There are several applications of sawdust as it is a raw material which could be used in the manufacture of construction materials [14,15]. Such applications include: as a material for the construction of notice boards, furniture, as a fuel source, it could also serve as an insulating material in electronic devices like refrigerators. More applications of sawdust include to produce briquettes, mixing sawdust with wood ash and calcium carbonate forms fertilizers and also sawdust could be used as a composing material for particleboard production and it can be used in the production of oil [4,16,17]. Most of the research done on the recycling of waste glass has been on the use of waste glass either as aggregate in concrete construction or as a replacement for cement in concrete construction [18–20]. A replacement of cement with waste glass powder would also mean a reduction in the production cost of concrete thereby providing a way of spending less and also promoting environmental growth by recycling waste materials. However, the use of waste glass as an aggregate in concrete production would cause a few drawbacks such as reduced bond between the aggregate and cement paste which would be stronger with a natural aggregate and the major hindrance to the progress of utilising waste glass in concrete is the Alkali-Silica Reaction. The outcome of the Alkali-Silica reaction is the alkali-silica gel which has the attribute of absorbing water with the inclination of swelling and exhibiting an increase in volume and due to this volatile behaviour, damage occurs within the structure of the concrete amid the generation of internal stresses. ASR is a problem that might not manifest for years and so, therefore, is a potential risk [21].

2. Materials and Methods

The materials used for the production of the particle boards are sawdust, waste glass, aluminium shavings and Urea Formaldehyde. These materials were sourced locally in the Omu-aran community and were varied in proportions for each board following the experimental design designated for them as depicted in the table 1. The sawdust used for the project was acquired from the Omu-aran community at a moisture content of 9.46%, which was determined with the aid of an AND MX50 moisture analyser and room-dried for 7 days. The waste glass used were washed and dried to remove any impurities that may be present in it. The waste glass used for the research work was gathered and collected for the purpose of the experiment and the particle size used for the experimental procedure was sizes passing through sieve size 2mm and retained on 300 microns. The glass particles were later washed and dried to remove any impurities that may be present in it.
Fabricated steel mould with internal dimensions of $350\,x\,350\,x\,30\,mm$ were used, the mould cover was made in such a way that it extended 15 mm downwards, this provided the assurance that the board was actually compressed to the required thickness (15mm) when placed on the hydraulic press.

Table 1. Mix design for each Board Sample.

Board Sample	Sawdust	Waste Glass	Aluminium Shavings
A	80%	0%	20%
B	90%	0%	10%
C	100%	0%	0%
D	64%	16%	20%
E	72%	18%	10%
F	80%	20%	0%
G	48%	32%	20%
H	54%	36%	10%
I	60%	40%	0%
J	40%	40%	20%
K	45%	45%	10%
L	50%	50%	0%

The quantity of Urea formaldehyde was constant in all the boards. The adhesive used was 30% of the board volume.

2.1. *Production of Particleboard*

The quantity of material required for each board was measured out in volume into a head pan and a dry mix was done before the required quantity in volume of Urea formaldehyde (top bond) was poured into the head pan and then mixed till a completely uniform and homogenous paste was attained. The paste was then poured into cut-out polythene placed in the mould and then the paste was spread uniformly in the mould down to the edges. Another cut out polythene nylon was placed on the surface of the composite and the cover was placed on it. The mould was then transported to the hydraulic press which is used to compress the composite to the required thickness and to also attain compaction. This was done for 10 minutes after which the mould was then taken to the oven to be heated for 1 hour at 80°C. The mould was then brought out to cool for 10 minutes after which the mould cover was removed and left open for 24 hours. The board was then removed from the mould and placed in the oven to be heated at 130°C for 3 hours. The board was then allowed to slowly cool in the oven for 30 minutes before it was brought out of the oven and placed on a flat surface to finally cool for an hour after which it was then stacked.

Each board specimen was cut into the required dimensions for the tests; 50×50×12 mm for density test, water absorption and thickness swelling while 290×50×12 mm was used for the Modulus of Rupture and Modulus of Elasticity tests using the Testometric Tensile testing Machine (M500-50AT).

3. Results and Discussion

3.1. *Physical Properties*

3.1.1. Density Test: This is the ratio of the mass of a material to its volume. Density is a physical property for particle boards this property is dependent on the particle size of the material in use. For commercial purposes, the standard density for particle board is 37-50 lb/cubic ft. (593-801 kg/m^3). It is taken as;

$$\text{Density (kg/m}^3\text{)} = \frac{W_a}{V_a} \quad (1)$$

Where; W_a- air dried weight, V_a– air dried volume

The density for three samples of each board produced is determined and recorded as shown in Table 2. For this research work, the mean densities obtained ranged from 568.89 kg/m^3 to 2004.45 kg/m^3 and these are as shown in table below.

Table 2. Mean density for Particleboards produced.

Board Sample	Mean Density (Kg/m^3)
A	808.89
B	568.89
C	577.78
D	1004.44
E	1044.44
F	1048.89
G	1662.22
H	1524.44
I	1671.11
J	1862.22
K	1817.78
L	2004.45

It can be observed from the table that board L had the highest density, this might be as a result of the trend in the table which is an increase in the density as there is a decrease in the sawdust content and increase in the glass content. Boards A, B and C have the highest composition of sawdust and contain no waste glass but possess the lowest values for density while boards J, K and L have the lowest composition of sawdust and contain the highest content of waste glass but possess the highest values for density. This gives a better density compared to boards made from sawdust and plastic waste as reported by [6]. According to ANSI 208 [22] standard, board B and C can be classified as low density particle boards with densities less than $640kg/m^3$, while all other boards are classified as high densities particle board with densities above $800kg/m^3$ [22].

3.1.2. Thickness Swelling: This is done through a test which determines the alteration of the thickness of the particle board after immersion in water for a specified period of time. This property depends on the moisture and absorption nature of the board specimen. It should not be more than 2-3% and its value decreases after the specimen dries.

$$\text{Thickness swelling (\%)} = \left(\frac{T_f - T_i}{T_i}\right) \times 100 \qquad (2)$$

Where; T_i- initial thickness, T_f - final thickness

For one of the physical properties for particle boards, the thickness swelling for each board is determined and each board has three samples. The mean thickness swelling percentage value is then obtained and recorded. The test involves submerging the test samples in water at room temperature and recording the variations in the thickness after 2 hours and 24 hours. The mean values for the percentage thickness swelling are as shown in the table 3.

Table 3. Mean values for thickness swelling

Board Sample	% TS after 2 hours	%TS after 24hours
A	3.38	21.41
B	4.44	33.41
C	5.5	32.52
D	2.95	3.46
E	1.56	1.89
F	1.39	2.41
G	1.25	7.95
H	0.77	8.9
I	2.46	9.82
J	0.44	1.09
K	0.58	0.95
L	0.61	0.94

It can be observed that boards A, B and C have the highest values for thickness swelling, this is a result of the high sawdust content in them. The trend, therefore, observed in the table is that decrease in the sawdust content leads to decrease in the thickness swelling. It is also observed that due to the high sawdust content, the difference between the thickness swelling after 2 hours and that of 24 hours in the first three boards is quite large. Board J has the lowest thickness swelling value after 2 hours but not the lowest after 24 hours, board L has the lowest thickness swelling value after 24 hours but not the lowest after 2 hours. The American National Standards Institute [22] stipulated that the maximum thickness swelling is 8%. Thus, boards A, B, C and I does not fulfill the thickness swelling requirement.

3.1.3. Water Absorption: Water absorption is the test done to determine the ability of the particle board to absorb water at room temperature until constant weight is attained. This procedure puts the other properties to test as it causes a reduction in the strength of the particle board.

$$\text{Water absorption (\%)} = \left(\frac{W_f - W_i}{W_i}\right) \times 100 \qquad (3)$$

Where; W_f – final weight and W_i – initial weight

Water absorption is a test performed to determine the ability of the particle board to absorb water until a constant weight is attained. Board samples of dimensions 50×50×15 mm were submerged in water at room temperature and the weight was measured and recorded after 2 hours and 24 hours. The values for the water absorption percentage are as shown in table 4.

IOP Conf. Series: Materials Science and Engineering 1036 (2021) 012049 doi:10.1088/1757-899X/1036/1/012049

Table 4. Mean values for Water absorption percentage.

Board Sample	% WA after 2 hours	%WA after 24 hours
A	63.88	77.47
B	74.95	93.79
C	98.52	121.53
D	40.13	49.2
E	41.64	50.91
F	40.33	49.08
G	31.82	35.68
H	30.05	37.05
I	26.38	33.7
J	31.85	37.75
K	35.45	39.25
L	31.97	34.58

It can be observed that boards A, B and C have the highest values for water absorption due to their high sawdust content as sawdust absorbs water easily. It can also be observed that board I has the lowest value for water absorption while board C has the highest value for water absorption owing to the fact that it comprises of only sawdust. It can be seen the addition of aluminium shavings to the particle board aids in the resistance of the board to water absorption as aluminium shavings does not absorb water.

3.2. Mechanical Properties

3.2.1. Static Bending: The mechanical properties of the particle boards produced were tested and the various values for the Modulus of elasticity and the Modulus of rupture were determined. Modulus of elasticity is the ability of a material that describes the resistance offered by the material to deformation upon the application of stress. It is the property of a material that describes the resistance offered by the material to deformation upon the application of stress. This is a mechanical property for particle board.

$$MOE = \frac{P_{bp}L^3}{4bh^3Y_p} \quad (4)$$

Where; P_{bp} – load at the proportionality limit, L – Span length in mm, b – Width of the specimen in mm, h – Thickness of the specimen in mm, Y_p – deflection corresponding to P_{bp} (mm)

3.2.2. Modulus of Rupture: Modulus of rupture is the property of a material that measures the stress applied in the material and its structural components. Just like MOE, this is a mechanical property of a particle board which measures the stress applied in a material and its structural components. This material property is dependent on the density of the specimen material.

$$MOR = \frac{3P_bL}{2bh^2} \quad (5)$$

Where; P_b – Maximum load, L – Span length in mm, b – Width of the specimen in mm, h – Thickness of the specimen in mm

The mean values of three samples each for the Modulus of rupture and Modulus of elasticity are as shown below.

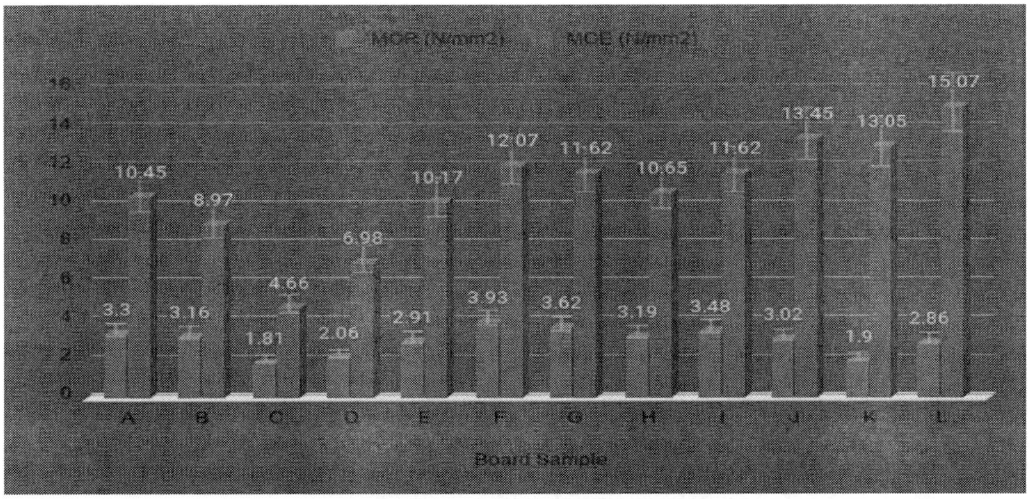

Figure 1. Mean values for MOR and MOE.

It can be observed that board L has the highest value for Modulus of elasticity while board F has the highest value for Modulus of rupture. The board with the lowest MOR and MOE is C. The BS EN 13353 [23] states that panels intended for structural use should have densities of minimum 420 Kg/m^3 and also have minimum values of MOR and MOE to be 5 N/mm^2 and 600 N/mm^2 respectively. This means that none of the boards produced can serve the purpose of structural use. It was observed that, apart from the case of boards D, E & F, there was a higher MOE and MOR value when the composition of aluminium shavings increased in the boards. Board A has higher MOE and MOR compared to B, board G has higher MOE and MOR values when compared to H and J has higher values than K.

4. Conclusion

The results of this research work show that the particle boards produced cannot be used for structural or load bearing purposes. It was also observed that the only noticeable effect aluminium shavings had on the panels was in the Modulus of elasticity and Modulus of rupture values and the impact it had was minimal. The properties the panels exhibited make it possible for them to be used as partition walls. The addition of glass to the boards brought about an increase in the densities of the boards. Therefore the boards produced can serve non-structural purposes such as partitions etc.

Acknowledgements

The authors wish to acknowledge the financial support offered by Landmark University Centre for Research, Innovation and Development (LUCRID) in the actualization of this research work for publication

References

[1] Akinyemi A B, Afolayan J O and Oluwatobi E O 2016 Some properties of composite corn cob and sawdust particle boards. *Constr Build Mater,* **127** 436–41.

[2] Da Silva F M, Gachet Barbosa L A, Lintz R C C and Jacintho A E P G A 2015 Investigation on the properties of concrete tactile paving blocks made with recycled tire rubber. *Constr Build Mater* **91** 71–9.

[3] Abdulkareem S A, Raji S A and Adeniyi A G 2017 Development of Particleboard from Waste Styrofoam and Sawdust. *Niger J Technol Dev* **14** 18–22.

[4] Atoyebi O D, Odeyemi S O, Azeez L O and Modupe A E 2019 Physical and Mechanical Properties Evaluation of Corncob and Sawdust Cement Bonded Ceiling Boards. *Int J Eng Res Africa* **42** 65–75.

[5] Atoyebi O D, Osueke C O, Badiru S, Gana A J, Ikpotokin I, Modupe A E and Tegene G A 2019 Evaluation of Particle Board from Sugarcane Bagasse and Corn Cob. *Int J Mech Eng*

Technol **10** 1193–200.
[6] Atoyebi O D, Adediran A A and Adisa C O 2018 Physical and Mechanical Properties Evaluation of Particle Board Produced From Saw Dust and Plastic Waste. *Int J Eng Res Africa* **40** 1–8.
[7] Odeyemi S O, Abdulwahab R, Adeniyi A G and Atoyebi O D 2020 Physical and Mechanical Properties of Cement-Bonded Particle Board Produced from African Balsam Tree (Populous Balsamifera) and Periwinkle Shell Residues. *Results Eng* **6** 100126.
[8] Mantanis G I, Athanassiadou E T and Barbu M C 2017 Adhesive systems used in the European particleboard , MDF and OSB industries. *Wood Mater Sci Eng* **13** 2 101–16.
[9] Klímek P, Wimmer R, Kumar P and Kúdela J 2017 Utilizing brewer's-spent-grain in wood-based particleboard manufacturing. *J Clean Prod* **141** 812–7.
[10] Nasser R A 2012 Physical and Mechanical Properties of Three-Layer Particleboard Manufactured from the Tree Pruning of Seven Wood Species. *World Applied Sciences* **19** 741–53.
[11] Moslemi A, Zolfagharlou koohi M, Behzad T and Pizzi A 2020 Addition of cellulose nanofibers extracted from rice straw to urea formaldehyde resin; effect on the adhesive characteristics and medium density fiberboard properties. *Int J Adhes Adhes* **99** 102582.
[12] Astari L, Prasetiyo K W and Suryanegara L 2018 Properties of Particleboard Made from Wood Waste with Various Size. *IOP Conf Ser Earth Environ Sci* **166**.
[13] Awuchi C G and Igwe V S 2017 Industrial waste management: Brief survey and advice to cottage , Small and medium scale industries in Uganda. *Int J Adv Acad Res* **8** 26–43.
[14] Osuolale O M, Atoyebi O D and Tunde O V 2019 Experimental Study of the Strength Performance of Sawdust Ash Pervious Concrete. *J Eng Appl Sci* **14** 8321–8.
[15] Mirski R, Dukarska D, Derkowski A, Czarnecki R and Dziurka D 2020 By-products of sawmill industry as raw materials for manufacture of chip-sawdust boards. *J Build Eng* **32** 101460.
[16] Rominiyi O L, Adaramola B A, Ikumapayi O M, Oginni O T and Akinola S A 2017 Potential Utilization of Sawdust in Energy, Manufacturing and Agricultural Industry; Waste to Wealth. *World J Eng Technol* **05** 526–39.
[17] Atoyebi O D, Awolusi T F and Davies I E E 2018 Artificial neural network evaluation of cement-bonded particle board produced from red iron wood (Lophira alata) sawdust and palm kernel shell residues. *Case Stud Constr Mater* **9**.
[18] Atoyebi O D and Sadiq O M 2018 Experimental data on flexural strength of reinforced concrete elements with waste glass particles as partial replacement for fine aggregate. *Data Br* **18** 846–59.
[19] Sadiq O M and Atoyebi O D 2015 Flexural Strength Determination of Reinforced Concrete Elements with Waste Glass as Partial Replacement for Fine Aggregate. *NSE Tech Trans J Niger Soc Eng* **49** 74–81.
[20] Atoyebi O D, Gana A J and Longe J E 2020 Strength assessment of concrete with waste glass and bankoro (Morinda Citrifolia) as partial replacement for fine and coarse aggregate. *Results Eng* **6** 100124.
[21] Tamanna N, Sutan N M, Lee D T C and Yakub I Bin 2013 Utilization of waste glass in concrete. *6th Int. Eng. Conf. Energy Environ. (ENCON 2013)*, 323–9.
[22] American National Standard Institute. ANSI A208.1-1999 PB.
[23] British, Standard, Institution. BS EN 13353.2003 British Standard Institution Solid wood panel (SWP)-Requirements. 2003.

Evaluation of Strength and Moisture Absorption Characteristics of Lime Treated Clay Interlocking Bricks

O G Fadugba[1,*], A A Busari[2], A O Adeboje[3], A E Modupe[4], J N Falana[5] and B D Oluyemi-Ayibiowu[1]

[1]Department of Civil Engineering, The Federal University of Technology, Akure, Nigeria.
[2]Department of Civil Engineering, Covenant University, Ota, Ogun State, Nigeria
[3]Department of Civil and Environmental Engineering, University of Lagos, Lagos, Lagos State, Nigeria
[4]Department of Civil Engineering, Landmark University, Omu Aran, Kwara State, Nigeria
[5]Department of Civil Engineering, The Federal Polytechnic Ile-Oluji, Ondo State,Nigeria

*Corresponding email: ogfadugba@futa.edu.ng

Abstract. The durability and stability of masonry structures throughout usage can be affected by producing process, material properties and environmental conditions such as moisture hence finding out the factors affecting durability of clay bricks is vital so as to produce bricks of high strength which will absorb less water. The effect of stabilizing clay for the production of interlock bricks using lime at various percentages was investigated. Soil type used in the production of interlocking bricks was A-6 soil using the AASHTO system of soil classification. Maximum reduction in Liquid Limit and Plastic Limit of the clay material was observed at 8% addition of lime. The Specific Gravity rose from initial of 2.71 at 0% to 2.85 at 10% addition of lime. The maximum reduction of Average Water Absorption (AWA) of clay modified with lime was observed at 8% and the value of AWA recorded was 13.3%, which brought the clay brick to the first class designation from the second class designation of 0% of lime addition. The compressive strength reduced at 8% for all the sampling period of 28 days, the compressive strength test is lower than specification as such the bricks can be used in non-structural member in construction

Keywords: Interlocking Bricks; moisture absorption; Lime; compressive strength

1. Introduction

Around the globe, the usage of cement for infrastructural progression is a key to the advancement of various economies [1]. Regardless, production of cement is the second biggest wellspring of CO_2 generated around the world. It contributes around 5-8% of overall CO_2 emissions, generally from the usage of non-sustainable energy source for calcination [2]. One ton of cement production is obligated for one ton of CO_2 spread: half of the CO_2 is from the chemical process of clinker production and the rest on account of the calcinations of limestone [3]. A critical source is lime-stone for the production of cement which creates CO_2 has essential depleting factor in the ozone layer.

To help decrease the shortfall in construction of low cost housing for the masses, in Ghana, the Building and Road Research Institute (BRRI) of the Council for Scientific and Industrial Research (CSIR) has made tremendous progress in research towards the development and advancement of locally–developed building materials. Among the materials are burnt clay bricks ([4]; [5]). To this end, broad research has been carried out to establish bricks and tile processing plants and the preparation of craftsmans on brick production ([4]; [6]). Burnt clay bricks have been acclaimed to have a few specialized, financial and ecological favorable circumstances. They are sturdy and have great aesthetic intrigue.

Lime is the by-product from burnt limestone. It is the most established created material utilized for development works, being utilized by human for over 2000 years. Additionally, Lime has been utilized by Romans for road construction. Soils stabilization with lime is broadly utilized in different construction works, for example, foundation base, highways, railroads, air terminals, embankments, slope protection, trench lining etc. Lime is utilized tremendously for stabilizing soil as it is economic, easy construction just as innovation is straightforward that gives attraction for the specialists in building and road constructions.

The sturdiness or stability of brick work structures during use can be influenced by manufacturing process, material properties, and environmental conditions, for example, dampness and salt attack just as upkeep culture ([7]; [8]). One of such extreme conditions which can cause intense disintegration of properties in brick work structures is marine environment. Therefore studying these variables influencing durability of burnt clay bricks produced and utilized locally is imperative. Durability properties incorporate physical behaviour like porosity, pore size distribution, bulk density and mechanical strength. These are subject to the kinds of raw material, the forming process and firing condition utilized in assembling. These when assessed will inform proper choice regarding building materials in construction, guarantee quality and performance during serviceability life when subjected to aggressive and unfavourable environmental condition

2. MATERIALS AND METHODOLOGY

The apparatus used were compression machine, Oven, electric weighing balance and tray. The soil type in this experiment is laterite and was gotten from a new site under construction behind SEMS building in FUTA. The bigger lumps were crushed into smaller particle sizes and the soil sample was sieved using 6mm sieve to obtain finer particles. The lime was gotten locally from a laboratory specifically Vectra Scientific Laboratory, Ilesha Road, Akure. Treatment of soil sample with different quantity of lime varied from 2% to 10% of the total dry weight of the soil sample-lime mixture. All laboratory experiments were performed in the Kure Geotechnical laboratory, Federal University of Technology, Akure.

2.1. METHODOLOGY

Physical property tests were carried out on the lateritic soil in conformity with relevant standards. Laboratory index properties tests were used to categorise the sample. The tests are the moisture content [9], Atterberg limit [10], grain size distribution [10], specific gravity [9] and linear shrinkage [11].

2.1.1. *Water Absorption Test (WAT)*

The water absorption test (WAT) was performed according to specifications by [12] and [13]. Samples were selected randomly and prepared by drying at a temperature of 105^0C to 115^0C for 48 hours in a ventilated oven till it achieves constant mass and then cool to room temperature, the weight of the bricks were measured and recorded as (M_1).
The samples were then immersed in clean water at room temperature for 24 hours. After 24 hours, samples were removed with a damp cloth and weighed after 3 minutes of removing from the water and recorded as (M_2). Equation 1 was used to calculate the percentage of water absorption (PWA) of the brick by its mass.

$$\mathrm{PWA} = \frac{(\mathrm{M_2} - \mathrm{M_1})}{\mathrm{M_1}} \times 100 \quad (1)$$

Where: PWA is Percentage water absorption, M_1 is the mass of brick after drying, M_2 is the mass of brick after immersion in water.

2.1.2. *Compressive Strength Test*
The strength test was directed according to the specifications of BS EN 772-1 [14]. The sample two sides were topped with steel plates to guarantee that the two faces have a level surface for pressure from the compression machine. All samples were oven dried for 24 hours to accomplish a consistent mass in the samples before curing was done for 28 days. Tests were done on selected samples periodically for 7, 14, 21 and 28 days. Compressive strength was determined utilizing the total load applied to the brick divided by the net gross area

3. RESULTS AND DICUSSION
3.1. Chemical Composition of Lime
Table 1 presented the chemical composition of the lime used in this research, it was observed the lime contained about 74% CaO as highest composition of the material.

Table 1. Geotechnical Index Parameters

Lime Composition	Amount
CaO, (%)	73.77
MgO, (%)	3.300
SiO_2, (%)	0.291
P_2O_5, (%)	0.202
Al_2O_3, (%)	0.170
Cl, (%)	0.131
K_2O, (%)	0.040
SrO, (%)	0.029
MnO, (%)	0.031
SO_3, (%)	0.022
TiO_2, (%)	0.012
ZnO, ppm	91.20
CuO, ppm	47.00
ZrO_2, ppm	3.00

3.2. Geotechnical index characteristics of Unmodified and Modified Sample

From the Atterberg test carried out on the unmodified sample shows that the soil sample has an average plastic limit of 27.36%, the Liquid Limit for the sample is 40%, and the plasticity index recorded was 12.64%. The specific gravity and natural moisture content recorded was 2.71 and 9.85% respectively. In accordance with AASHTO M145, the soil can be classified as A-6 soil because more than 36% of the soil particles passes sieve no 200 (Figure 2). The liquid limit of the soil is <40% while the plasticity index is >11%. This signifies that the soil is a clayey soil and the general subgrade rating is fair to poor. According to ASTM D854, the inorganic clay generally range from 2.70 to 2.80. The above specific gravity test result of 2.71 therefore signifies that the soil sample is a clay soil.

Figure 1 displayed the atterberg limits and specific gravity results of soil sample with the addition of lime at varying percentages. The declined in liquid limit within 2% to 8% addition of lime is as a result of exchanges among the free calcium of the lime and the absorbed cations of the clay mineral. This will cause a decrease in size of the diffused water layer encircling the clay particles. The decrease in size of the diffused water layer makes closer contact among the clay particles resulting in flocculation/agglomeration of these particles.
The specific gravity (SG) for the control (untreated) and treated samples is displayed in Figure 1, the S.G of the sample rose from 2.71 at 0% of lime (untreated) to 2.85 at 10% of addition of lime content which is

in agreement with [15] and this is as a result of the molecular reshuffling of soil matrix due to the higher density of lime over the clay sample.

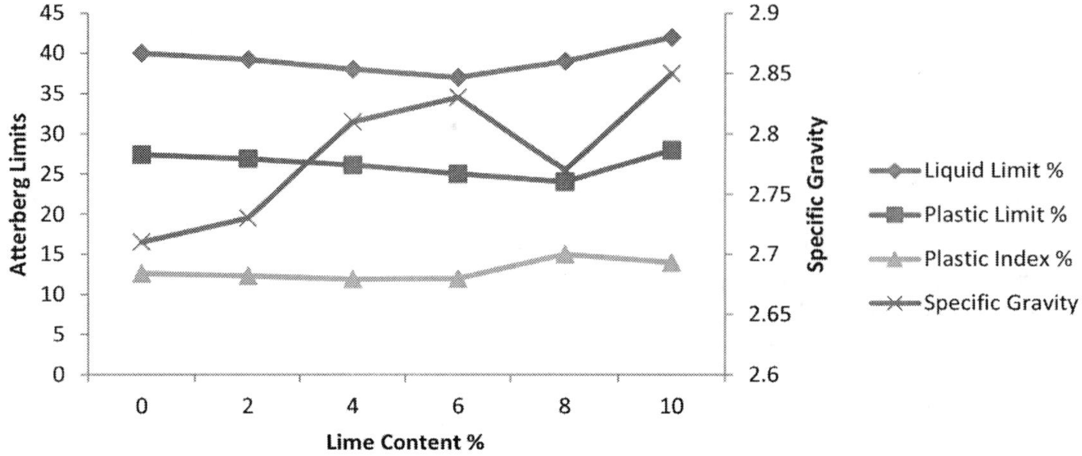

Figure 1. Plot of Atterberg limits, Specific Gravity against lime content

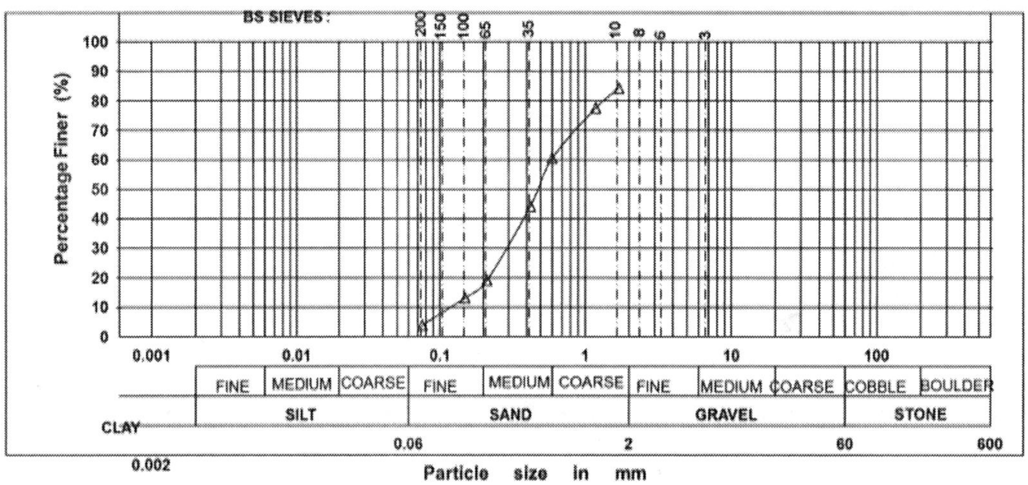

Figure 2. Particle size distribution of laterite used in brick production

3.3. Absorption Characteristics

Figure 3 presented the average water absorption characteristics of the brick with different proportion of lime content. It was observed that at 0% of lime the brick average water absorption recorded was 16.5% which falls within the second class brick designation (IS 1992)[12]. The addition of lime to the brick reduced the average water absorption steadily to fall within the first class brick category from 6% to 10%. The maximum reduction in the water absorption was observed at 8% addition of lime. From the results, increase in lime percentage reduced the absorption of water by the bricks which displayed the hydrophobic characteristics of lime.

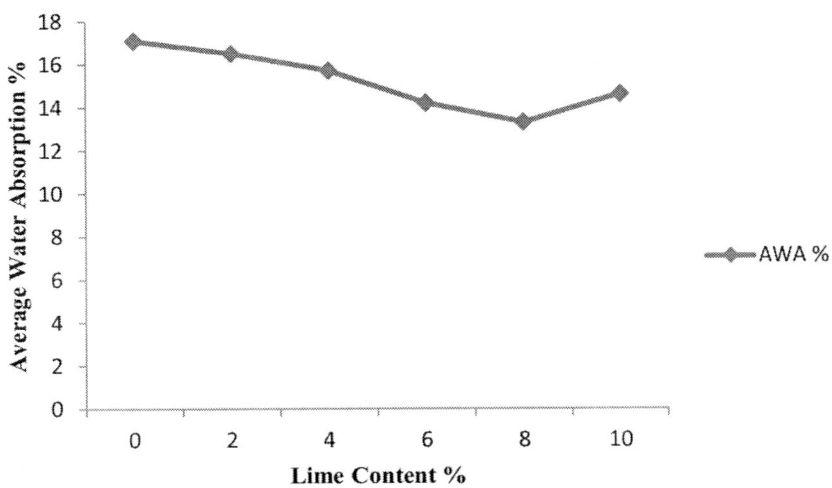

Figure 3. Average Water Absorption of Bricks

3.4. Compressive Strength Characteristics

Figure 4 displayed the compressive strength of the bricks with different proportion of lime and curing days. At 7th day curing of brick, the compressive strength increased from 1.43 N/mm^2 at 0% lime content to 1.99 N/mm^2 at 6% lime content, a slight decreased was observed at 8% addition of lime, the strength reduced to 1.81 N/mm^2. The same trend was observed in 14, 21 and 28 day compressive strength. The results from the compressive strength are contrary to the results from the experiments of [16], that the minimum compressive strength of load-bearing concrete masonry is 5 N/mm2. The results obtained from the compressive test is lower than the specification as such the bricks can be used in non-structural member in construction

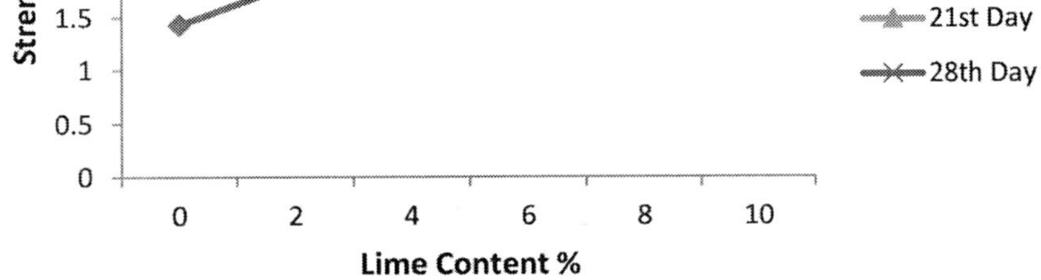

Figure 4. Average Compressive Strength Characteristics of Bricks

4. CONCLUSION

During the investigation, clay material was mixed with varying percentages of lime in order to recognize the behaviour of the material. It was established that:

1. reduction in the liquid limit, plastic limit was observed up to 8 % addition of lime.
2. the specific gravity level rose from the initial 2.71 at 0 % of lime content to 2.85 at 10% content of lime
3. addition of lime to the brick reduced the average water absorption to fall within the first class brick category from 6% to 10%. The maximum reduction in the water absorption was observed at 8% addition of lime.
4. the compressive strength test results is lower than specification as such the bricks can be used in non-structural member in construction

REFERENCES

[1] Mehta P. K. Monteiro P. J. M. (2006) Concrete—Microstructure, Properties, and Materials. A New York, NY: McGraw Hill
[2] Scriverier K. L. and Kirkpatrick R. J (2008): Innovation in use and Research on Cementations material; Cement and Concrete Research 38(2), 128-136. Doi:10.1016/j.cemconres.2007-09-025
[3] Taylor M. Tam C. Gielen D. (2006) Energy efficiency and CO2emissions from the global cement industry. In: International Energy Agency -World Business Council for Sustainable Development Workshop Energy Efficiency and CO_2 emission Reduction Potentials and Policies in the Cement Industry, Paris, France
[4] Ayetey J. K. (1977) Clays and their evaluation for bricks manufacture; Ghana clay report, BRRI current note 11, p.42
[5] Kesse G. O. (1985) The mineral and rock resources of Ghana. AA Rotterdam, p.437-439
[6] Obeng K. (1993) Study of properties of some Ghanaian clays used for the production of bricks, Journal of Building and Road Research, 1&2. B.R.R.I, Kumasi, p.12-14
[7] Surej R. K. (1998) Development of New Durability Index for Clay Brick, Journal of Architechtural Engineering, pg 87-93T. Stambolov, "The Corrosive Action of Salts," Lithoclastia 1976, p
[8] Zsembery S. (2001) Manual 2: The Properties of Clay Masonry Units, Clay Brick and Paver Institute (CBPI), Australia
[9] British Standard Institution (1990); Soils for Civil Engineering Purposes – Classification Tests, BS 1377: Part 2, London, British Standard Institution (BSI).
[10] British Standard Institution (1997); Test for Geometrical Properties of Aggregates – Determination of Particle Size Distribution (Sieving Method), BS EN 933: Part 1, London, BSI.
[11] Madu, R. M. (1984); The Influence of Soil Chemical Composition on Cement Treated Laterite. Proceedingd 8th Regional Conference for Africa on Soil Mechanics and Foundation Engineering, Harare, Zimbabwe, 4-7 June 1984, Pp. 391-402.
[12] IS No.3495 (1992); Methods of Tests of Burnt Clay Building Bricks: Part 2, Determination of Water Absorption (Third Revision)
[13] Annual Book of ASTM Standards, ASTM International, West Conshohocken, PA, 2007: Volume 4.05, C67 "Standard Test Methods for Sampling and Testing Brick and Structural Clay Tile".
[14] EN, BS. 772-1 (2011): Methods of test for masonry units. Part 1: Determination of compressive strength (British Standards Institute).
[15] Ali, J. A., Muzamir, M and Mohammed, Y. F. (2016): Engineering Properties of Clayey Soil Stabilized with Lime, ARPN Journal of Engineering and Applied Sciences 11: 4, 2434-2441
[16] Hendry, A. W., Sinha, B. P. , and. Davies, S. R (2004): Design of masonry structures (E & FN Spon, 2004).

Electrochemical control of high carbon steel corrosion using rosemary oil in citric acid medium

O A*Odunlami[1], O G Abatan[1], A A Busari[2], G T Alao[3], R T Loto[4], O T Olomukoro[1]
[1]Department of Chemical Engineering, Covenant University, Ota, Ogun state, Nigeria
[2]Department of Civil Engineering, Covenant University, Ota, Ogun state, Nigeria
[3]Department of Mathematics, Covenant University, Ota, Ogun state, Nigeria
[4]Department of Mathematics, Covenant University, Ota, Ogun state, Nigeria
*Corresponding olayemi.odunlami@covenantuniversity.edu.ng

Abstract. The Inhibition efficiency of Rosemary oil (*Rosmarinus officinalis*) for corrosion control of high carbon steel was done in citric acid medium by weight loss method. The *Rosmarinus officinalis* was administered in the proportion of 0%, 0.50%, 1.00%, 1.50%, 2.00% and 2.50% concentration. This study unveiled that the inhibition efficiency performed excellently with maximum percentage of 88.3% at room temperature in 0.5M citric acid solution. This has shown the effectiveness of *Rosmarinus officinalis* inhibitor in corrosion analysis of high carbon steel.

Keywords: High Carbon Steel, corrosion inhibitors, weight loss method, rosemary oil.

1. Introduction

Corrosion, the gradual destruction of materials due to chemical or electrochemical reaction with the environment. The easily identified occurrence of corrosion is the oxidation of Iron known as Rust. It has brought about great challenges to the applications of metallic material mostly in industries, such as petrochemical, food, automobile etc. [1-3] causing the destruction of industrial infrastructures. Therefore, corrosion control is required. The damage of metals by acids can be greatly controlled by the corrosion inhibitors. Researches have been carried out on both organic (nitrogen, Sulphur, oxygen) and inorganic (chromates, phosphates, molybdates) compounds as corrosion inhibitors [4-8]. Extracts of plants are found to be biodegradable, non-toxic, environment friendly, easily available and relatively cheap unlike the pure synthetic chemicals. Many of the naturally occurring substances like black pepper [9], onion, garlic [10] among others prevented corrosion of steels in different media from recent researches, also rosemary oil extract yielded 90% inhibitor efficiency on mild steel in water-base and 85% on 439LL ferritic stainless steel in 1M H_2SO_4 solution in previous researches[11,12]. As a result of these, rosemary oil was used to confirm the behavioral effect on high carbon steel in citric acid solution for this research.

This steel is among the commonest types of steel because of its high strength and low cost. Its material characteristics is widely preferred in several applications [13]. High carbon steels have 0.55 to 1.4% carbon. They are strong and hard, and the least ductile of the carbon steels [14]. They possess higher carbon content than mild carbon steel. They are used mostly in metal manufacturing such as drills, axes, saws, hammers, knives and wood [15]. This study investigates the effectiveness of *Rosmarinus officinalis* on high carbon steel in 0.5M citric acid at every 24 hours for 21days at room temperature by weight loss method.

2. Materials and method

The steel has the following composition: 0.9% Carbon, 0.5% Manganese, 0.04% Phosphorus, 0.05% Sulfur and 98.51% Iron. The samples were machined into cylindrical shapes of these dimensions: 1cm (L), 1.2cm (diameter) and 6.03cm^2 (A_S). Abrasive papers were used for the preparation of the metal surface. *Rosmarinus officinalis*, a leaf extract of a woody shrub-type plant which is a transparent yellowish liquid was used for corrosion inhibition in citric acid medium. It was purchased from Jannys beauty store in Ikoyi, Lagos.

2.1. Experimental procedure

The weight loss analysis of the cylindrical shaped carbon steel coupons was done using weighing balance. The samples were immersed in 400ml of 0.5M citric acid solution. The experiment was exposed to the atmosphere and the coupons fully immersed at ambient temperature for 21 days. Each sample was analyzed every 24 hours after being washed with distilled water and dried.

3. Result

3.1. Weight loss plot

The plots of weight loss against exposure time for high carbon steel in citric acid solution is shown below:

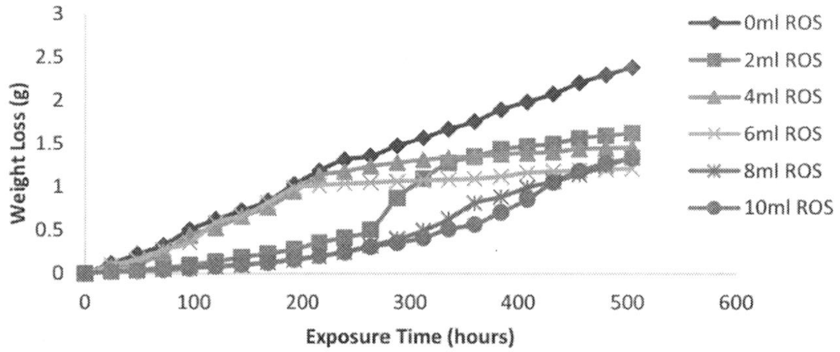

Figure 1. WL against exposure time of high carbon steel in 0.5M citric acid solution

3.2. Corrosion rate plot

Rate of corrosion of carbon steel in citric acid solution is shown in the figure below:

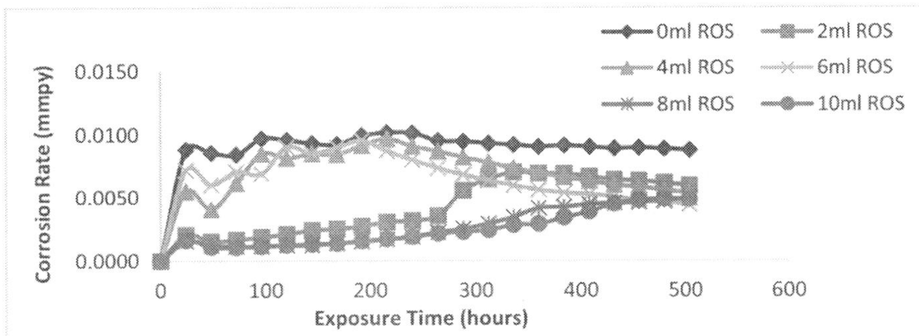

Figure 2. Corrosion rate against exposure time of carbon steel in 0.5M citric acid solution

3.3. Corrosion rate plot

The inhibitor efficiency of *Rosmarinus officinalis* from weight loss experiment in the corrosion of high carbon steel in citric acid is shown below:

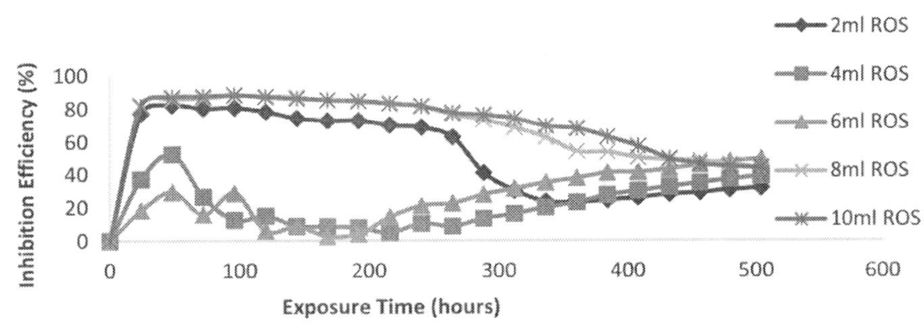

Figure 3. Inhibition efficiency of *Rosmarinus officinalis* against exposure time in 0.5M citric acid.

4. Discussion of results

The result showed that the maximum inhibition efficiency was obtained as 88.3% at 10ml inhibitor concentration. This could be traceable to the high performance of film protection formation of *Rosmarinus officinalis* on the surface of the steel sample to prevent further attack on the steel. Furthermore, the inhibitor also displayed an excellent level of adsorption process by protonating the steel samples from continuous corrosion effect. Adsorption process actually occurs by the interactive force between the molecular properties of rosemary oil and the surface of the high carbon steel samples.

5. Conclusion

Rosmarinus officinalis performed successfully well on high carbon steel with 88.3% maximum inhibitor efficiency in 0.5M citric acid solution. Therefore, rosemary oil displayed an excellent inhibition of high carbon steel which makes it suitable for industrial applications. This further explains the adsorption reaction between the surface of the high carbon steel and the *Rosmarinus officinalis* inhibitor. It also a high level of eco-friendly inhibitor for a sustainable world.

Acknowledgement

The authors acknowledge Covenant University for the financial support offered in the publication of this research.

References

[1] Fajobi M, Loto R and Oluwole O 2019 Corrosion in Crude Distillation Overhead System: A Review. *Journal of Bio-and Tribo-Corrosion* **5**(3) 67

[2] Fajobi M, Fayomi O, Akande I and Odunlami O 2019 Inhibitive Performance of Ibuprofen Drug on Mild Steel in 0.5 M of H_2SO_4 Acid. *Journal of Bio-and Tribo-Corrosion*, **5**(3) 79
[3] Odunlami O, Elehinafe F, Oladimeji T, Fajobi M, Okedere O and Fakinle B 2018 Implications of lack of maintenance of motorcycles on ambient air quality. In *IOP Conference Series: Materials Science and Engineering* **1** (413) 012-055 IOP Publishing
[4] Refaey S *Appl surf Sci.*, 2005, **240** (1-4), 396-402
[5] Quiraishi M and Sharma H *J Appl Electrochem.*, 2005 **35**(1), 33-39
[6] Ashassi-Sorkhabi A, Shaabani B and Seifzadeh D *Appl Surf Sci.* 2005 **239**(2) 154-164
[7] Matheswaran P and Ramasamy A *E-J Chem.* 2010 7(3), 1090-1094
[8] Ali A, Saeed M and Rahman S *Corros Sci.* 2003 **45**(2), 253-266.
[9] Pandian B and Mathur G *Mater Lett.* 2008 **62**(17-18) 2977-2979
[10] Parikh K and Joshi K *Trans SAEST* 2004 **39**(3-4), 29-35
[11] Loto R and Oghenerukewe E (2016) Inhibition studies of rosmarinus officinalis on the pitting corrosion resistance 439LL ferritic stainless steel in dilute sulphuric acid. *Oriental Journal of Chemistry,* 32(5), 2813-2832.
[12] Loto R and Busari A (2020) Inhibition effect of rosemary oil on mild steel corrosion in a water-based petrochemical drilling fluid. *MS&E,* 770(1), 012046.
[13] Loto R 2018 Electrochemical analysis of the corrosion inhibition properties of L-leucine and trypsin complex admixture on high carbon steel in 1M H_2SO_4 solution *Quimic Aplicada y Analitica*, **47** (2), 12-20.
[14] Davis J 2001 Surface Engineering for Corrosion and Wear Resistance. *The Materials Information Society ASM International, Materials park Ohio*
[15] Osarolube E, Owate I, and Oforka N 2008 Corrosion behaviour of mild and high carbon steels in various acidic media. *Scientific Research and Essay* **3** (6) 224-228.

Assessment of the Engineering Properties of Modified Asphalt Using Aluminium Dross as a Filler

A A Busari[1,2]*, A Adeboje[3], A E Modupe[4], G O Fadugba[5], R T Loto[6], G O Abatan[7]

[1]Department of Civil Engineering, Covenant University, Ota, Ogun State, Nigeria
[2]Department of Civil Engineering, Tshwane University of Technology, Pretoria, South Africa
[3]Department of Civil Engineering, University of Lagos, Lagos State, Nigeria
[4]Department of Civil Engineering, landmark University, Omu-Aran, Kwara State, Nigeria
[5]Department of Civil Engineering, Federal University of Technology, Akure, Ondo State, Nigeria
[6]Department of Mechanical Engineering, Covenant University, Ota, Ogun State, Nigeria
[7]Department of Chemical Engineering, Covenant University, Ota, Ogun State, Nigeria

Corresponding Author; ayodebamiakinbode@gmail.com, +2348035419504

Abstract. Aluminum dross waste creates environmental burden for manufacturers and hence the need for its re-use. This experimental research assessed the engineering properties of modified asphalt using aluminium dross as a filer material. The aluminium dross used was obtained from a steel manufacturing industry. Elemental composition of the oxides was done using XRF equipment. Sieve analysis of the aluminium dross and the coarse aggregate was also assessed. Additionally, ductility, penetration, ball and ring and Marshall stability test were used in achieving the aim of this research. The outcome of the research revealed that 10-20% addition of the aluminium dross improved the stability of the modified asphalt. Physical observation revealed that the utilization of the waste stiffened the modified asphalt. The result showed that the addition of aluminium dross reduced the VMA. This will also increase the asphalt mix flushing. Additionally, at higher percentage addition of the waste the stiffness of the asphalt mix increases especially at a higher temperature. The use of this material will help in the reduction of solid waste and reduce the cost of management while reducing the cost of pavement construction in track with the trash to treasure drive.

Keywords: Asphalt; Aluminium Dross; Pavement; Bitumen; Flexible Pavement; Transportation

1. Introduction

1.1 Motivation

The need for the effective disposal of aluminium dross waste is a great burden for most steel and aluminium producing companies globally. This is due to the myriad effect of the improper disposal of this waste to human and animal health. Additionally, it has serious environmental and health implications too such as Alzheimer's disease, silicosis and bronchitis as avowed by [1-15]. Though aluminium dross is not totally a waste, a lot of energy and money is involved in extraction aluminium from the solid waste. Based on the energy requirement and the cost involved in the extraction, this waste is usually landfilled. [16] Affirmed that about one million tonnes of black dross and about four million tonnes of white dross are being produced globally from steel and aluminium producing company.

According to the author, 95% of these materials are openly dumped with serious health and environmental effect. Hence this research attempts to use solid waste as a filler in the production of a sustainable, cost-effective asphalt.

1.2 Background

The urge to provide sustainable material in road construction has birthed the need for the development of alternative materials in asphalt production, which include the use of wastes both processed and recycled. These wastes ranges from agricultural to industrial waste [17-22]. The use of aluminium dross in pavement construction was adopted in a bid to reduce solid waste from the steel production industry. Aluminium dross a light weight metal, ductile material having is the by-product of aluminium production [23]. Its abundance is due to the wide usage industries and homes [24]. Aluminum dross as defined by [25] is a by-product of the aluminium smelting process. It is an inevitable waste product from the primary and secondary aluminium processing company [26,27]. Aside from aluminium metal, dross may likewise contain other chemical compounds e.g. MgF_2, SiO_2 and MgO and so on.

This solid waste is divided into white and black dross based on the coloration and the content of the aluminum in it. However, the white dross is richer in aluminum than the black dross. Additionally, the white dross is known as the non-salt containing dross while the black dross is the salt containing dross [28-32]. Additionally, [29] affirmed that aluminium dross are classified according to their content.

Recent review revealed that the recovery of the aluminum from the dross is been processed in the rotary kilns, while the salt cake is usually land filled. Much energy is devoured to recuperate the aluminium from the dross; this energy can be spared if the dross could be redirected and used as a construction material. (Reddy et al., 2014). Other application of aluminium dross as a construction material as avowed by [16, 33]. A lot of energy is consumed in the recovery process which can be conserved if the waste is re-used for other purposes and hence its utilization in pavement construction which is an important component of effective transportation [34-42]. This research assessed the engineering properties of aluminium dross as a filler in asphaltic concrete pavement.

2. Materials and Methods

2.1 Materials

The aluminium dross waste used for this research was obtained from a steel rolling industry in Ota, Ogun State, Nigeria is the capital of the Ado-Odo/Ota local government area. Municipal waste generated in the state is quite high due to the large number of industries. One of such waste is aluminium dross. Based on this research, the effect of aluminium dross in asphaltic concrete is determined using 0%, 10%, 20% and 30% of aluminium dross as a filler in the modification of asphaltic concrete sample. The coarse and fine aggregate used was purchased from a quarry in Ota, Ogun State, Nigeria. Bitumen of grade 60/70 was purchased in Lagos, Nigeria and used all through the research.

2.2 Sample preparation

The aluminium dross was obtained and packed in air-tight polythene bags. It was air-dried for forty-eight (48) hours. The lump aluminium dross was crushed and sieved into 106μm and 184μm. Other wastes in the dross sample were removed before pulverization. The aluminium dross was used at 0,10,20,30% addition of the bitumen mixture. Fine aggregates passing through a 4.25 mm sieve, coarse aggregates retained on the sieve of mesh size 4.75 mm was used for the research.

2.3 Test and Methods

*E*ngineering tests were carried out on the bitumen to determine the rheology. The softening point and viscosity test were done in accordance with ASTM standards [43]. The bitumen penetration test was done in accordance with AASHTO specification. To achieve the aim of the research, the bitumen was modified with aluminium dross as a filler at at 0, 10, 20 and 30%. The rheology of the modified bitumen was also subjected to the rheological test. Finally, marshal test was conducted to ascertain the stability of the modified asphalt using AASHTO specification.

IOP Conf. Series: Materials Science and Engineering 1036 (2021) 012052 doi:10.1088/1757-899X/1036/1/012052

3. Results and Discussion

3.1 Preliminary test on the Aluminium Dross

The chemical composition of the oxides in the aluminium dross waste was assessed using X-ray fluorescence which is a non-destructive analytical technique. The result of the experiment is as shown in Table 1, with Al_2O_3 having the highest percentage of oxides and SrO as the least oxide in the aluminium dross sample.

Table 1: Oxide Concentration Table

ELEMENT	CONCENTRATION (Wt. %)
Magnesium Oxide	0.93
Aluminium Oxide	77.17
Silicon Oxide	13.48
Sulphur Oxide	0.70
Chlorine	0.27
Potassium Oxide	0.16
Calcium Oxide	1.12
Chromium Oxide	0.07
Manganese Oxide	0.20
Iron Oxide	4.18
Zinc Oxide	0.23

The sieve analysis of the aggregates used in the modification of the asphalt is as shown in Figure 1. The result conformed with AASHTO specification for the design of asphaltic concrete for road construction.

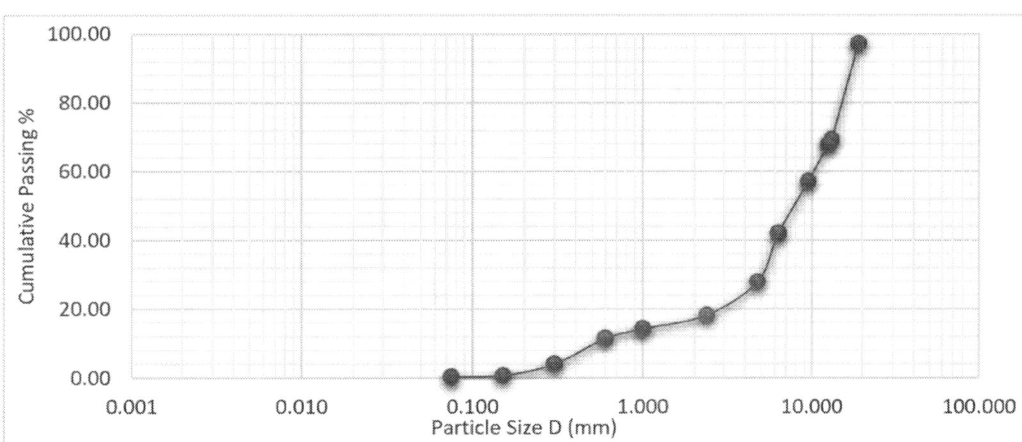

Figure 1: Sieve Analysis of the Coarse Aggregate

3.2 Bitumen Tests

The tests on the purchased bitumen are as shown in Table 2. The result showed the softening point test, penetration test, softening point test and ductility test.

Table 2: Preliminary test on the bitumen

Tests	Sample 1	Sample 2	Sample 3	Average
Penetration Test	5.5	5.7	5.55	5.6
Ductility Test	49	47	52	49.3
Softening Point Test	43	44	43.5	43.5
Ball and Ring Test	5	5.3	5.7	5.3

3.3 Asphalt Test

Stability verifies the performance of the asphalt mixture under loading. The stability at 20% is the highest with 14.97kN see Table 4. As the percentage of the filler in the asphalt increases, the stability. With about a 6.7% decrease in the Marshall stability. However, a lower percentage addition of this waste gave a better result than 30% addition.

Table 3: Marshall and Volumetric Properties of Bitumen Content Mix

Tests	**10%**	**20%**	**30%**
Marshall Stability (kN)	14.07	14.97	13.96
Flow (mm)	3.15	3.24	4.18
Density (g/cm$^{3)}$	2.35	2.35	2.33

The various volumetric results are tabulated in Table 4. From the research, the higher the aluminium dross in the sample, the lower the volume of void. This indicates that at a higher percentage of the aluminium dross, the pavement became less crack sensitive due to the oxidization of the asphalt, which will fill the voids.

Additionally, the Voids in Mineral Aggregates (VMA) was also assessed, as shown in table 5. This was obtained by the addition of the air voids and the asphalt content. This parameter is a function of the aggregate size used in the asphalt mixture. This buttress the assertion of Haydar, (2011) that the addition of mineral filler will reduce the VMA. The result showed that the addition of aluminium dross would reduce the VMA. The implication of this that the addition of this waste may impede the absorption of bitumen by the aggregates, thereby reducing the durability of the asphalt and the pavement. This will also increase the asphalt mix flushing. It also implies that at higher percentage addition of the waste in the asphalt mix. This will invariably, increase the stiffness of the asphalt mixture, especially at a higher temperature. Hence the use of this material may not be encouraged in the temperate region because of the high temperature in this zone so to reduce the incidence of thermal cracks (Haydar, 2011).

Table 4: Asphalt Volumetric Analysis

Tests	10%	20%	30%
Vv (%)	4.52	4.05	3.87
VFB (%)	67.02	69.4	70.2
VMA (%)	13.7	13.2	12.97

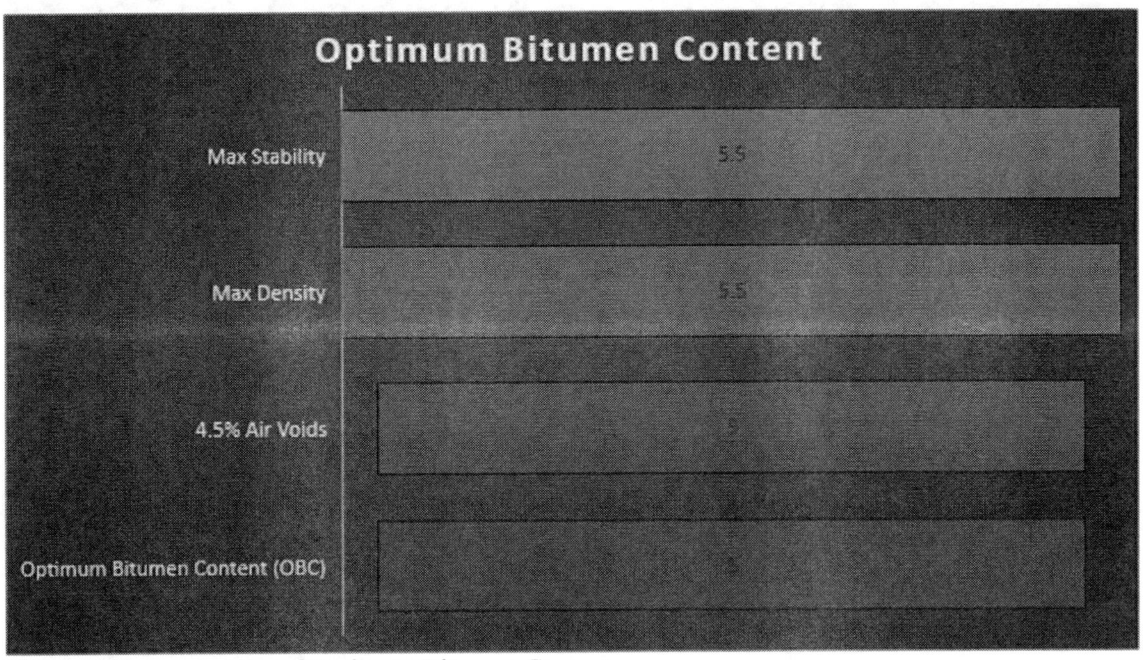

Figure 2: Determination of Optimum Bitumen Content

3.4 Marshall Stability

Marshal stability was conducted to assess the effect of the modified asphalt sample under loading. The result of the Marshall stability is as shown in figure 3. From the figure, the Marshall stability result obtained was higher than the control at 10 and 20% addition of the solid waste. The mix with 30% aluminium dross gave the lowest stability result. The increase in the stability may be attributed to the improved adhesion in the mixture (Sabina et al., 2009; Chen et al., 2009) because of the addition of the waste. Additionally, the dry method of mixing also aided the improved Marshall result, as the natural crystalline nature of the mixture was retained. The had a minimal change in the shape property.

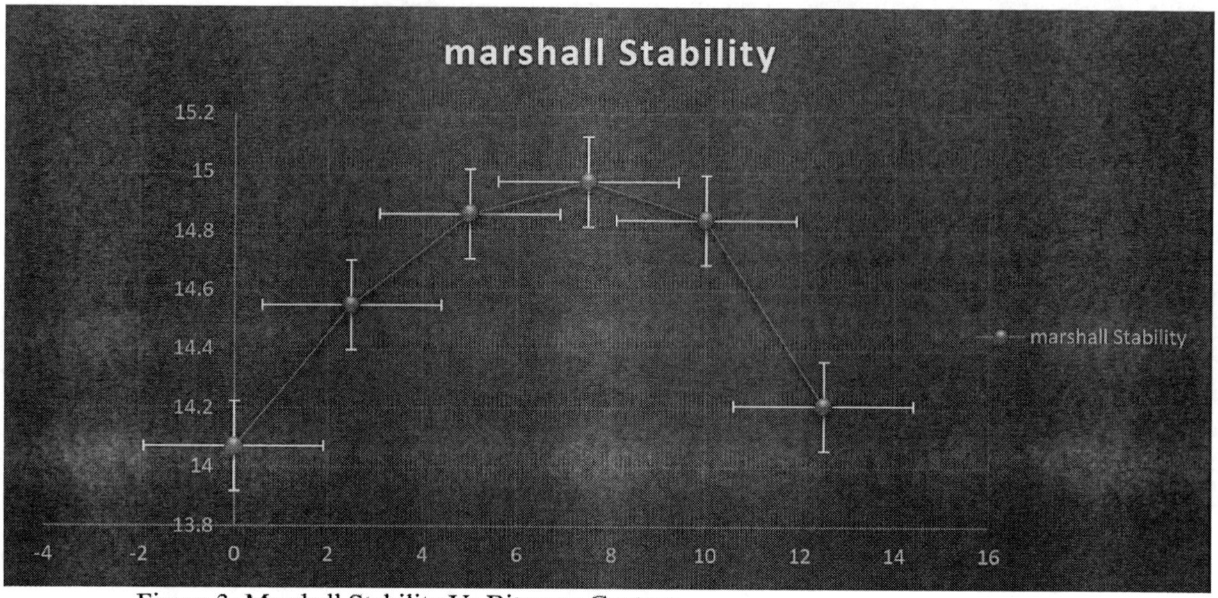

Figure 3: Marshall Stability Vs Bitumen Content

4. Conclusions

This experimental research assessed the utilization of aluminium dross as a filler in the modification of asphalt for the design and construction of a sustainable pavement. To this end, a preliminary test was carried out on the purchased bitumen sample before modification. Stability and volumetric analysis of the asphalt was also conducted. The findings of the research are as listed:

i. Physical observation revealed that the utilization of the waste stiffened the modified asphalt.
ii. The result of the Marshall stability is as shown in figure 4. From the figure, the Marshall stability result obtained was higher than the control at 10 and 20% addition of the solid waste.
iii. The increase in the stability may be attributed to the improved adhesion in the mixture.
iv. The result showed that the addition of aluminium dross would reduce the VMA.
v. The addition of this waste may impede the absorption of bitumen by the aggregates, thereby reducing the durability of the asphalt and the pavement.
vi. This will also increase the asphalt mix flushing. It also implies that at higher percentage addition of the waste in the asphalt mix.
vii. The use of the waste will increase the stiffness of the asphalt mixture especially at a higher temperature

Ultimately the use of aluminium dross is useful in wearing course construction in a bid to reduce solid waste and reduce the cost of road construction.

Recommendation

Future research should investigate the effect of this using waste at a higher percentage on the stability of asphalt.

Acknowledgement

The authors appreciate the management of Covenant University for the sponsorship of this article. The help and assistance offered by Mr. Adeniran Joel in the laboratory is appreciated.

References

[1] Lukiw W, Bazan N 2000 Neuroinflammatory signaling upregulation in Alzheimer's disease. Neurochem Res 25:1173–1184

[2] Campbell A, Yang E, Tsai-Turton M, Bondy S 2002 Proinflammatory effects of aluminum in human glioblastoma cells. Brain Res 933:60–65

[3] McLachlan D, Lukiw W, Mizzen C, Kruck T 1989 Chromatin structure in Alzheimer's disease: effect on 5′ leader sequence for NF-L gene and role of aluminum. Prog Clin Biol Res 317:1061–1075

[4] McLachlan D, Kruck T, Lukiw W, Krishnan S 1991 Would decreased aluminum ingestion reduce the incidence of Alzheimer's disease? CMAJ 145:793–804

[5] Lukiw W, Percy M, Kruck T 2005 Nanomolar aluminum induces pro-inflammatory and pro-apoptotic gene expression in human brain cells in primary culture. J Inorg Biochem 99:1895–1898

[6] Walton J 2013 Aluminum involvement in the progression of Alzheimer's disease. J Alzheimers Dis 35:7–43. https://doi.org/10. 3233/JAD-121909

[7] Walton J 2014 Chronic aluminum intake causes Alzheimer's disease: applying Sir Austin Bradford Hill's causality criteria. Journal of Alzheimers Dis 40:765–838. https://doi.org/10.3233/JAD-132204

[8] Pogue A, Lukiw W 2014 The mobilization of aluminum into the biosphere. Front Neurol 5:262. https://doi.org/10.3389/ fneur201400262

[9] Pogue A, Dua P, Hill J, Lukiw W 2015 Progressive inflammatory pathology in the retina of aluminum-fed 5xFAD transgenic mice. J Inorg Biochem 152:206–209. https://doi.org/10.1016/ jjinorgbio201507009

[10] Pogue A, Jaber V, Zhao Y, Lukiw W 2017 Systemic inflammation in C57BL/6J mice receiving dietary aluminum sulfate; upregulation of the pro-inflammatory cytokines IL-6 and TNFα Creactive protein (CRP) and miRNA-146a in blood serum. J Alzheimers Dis Parkinsonism 7(6). https://doi.org/10.4172/2161- 04601000403

[11] Garza-Lombó C, Posadas Y, Quintanar L, Gonsebatt M, Franco R 2018 Neurotoxicity linked to dysfunctional metal ion homeostasis and xenobiotic metal exposure: redox signaling and oxidative stress. Antioxid Redox Signal 28:1669–1703. https://doi.org/10. 1089/ars20177272

[12] Krishnan S, Gillespie K, McLachlan D 1972 Determination of aluminum in biological material by atomic absorption spectrophotometry. Anal Chem 44:1469–1470

[13] Kruck T, Cui J, Percy M, Lukiw W 2004 Molecular shuttle chelation: the use of ascorbate desferrioxamine and Feralex-G in combination to remove nuclear bound aluminum. Cell Mol Neurobiol 24:443–459

[14] Kandimalla R, Vallamkondu J, Corgiat E, Gill K 2016 Understanding aspects of aluminum exposure in Alzheimer's disease development. Brain Pathol 26:139–154. https://doi.org/10. 1111/bpa12333

[15] Kawahara M 2016 Link between aluminum neurotoxicity and neurodegenerative disorders. Nihon Rinsho 74:1176–1185

[16] Pickens, J, Morris, E 2001 Process for preparing calcium aluminate from aluminium dross, In Office, U. S. P. T.O. (Ed.) US Patent No: 6238633.

[17] Elzea K, Nikhil T, James M, Stanley T 2006 Industrial Minerals, and rocks: Commodities, Markets and uses (7th ed.), Society for Mining, Metallurgy and Exploration p.1406.

[18] Hollins O. 2007 Aluminum industry could dramatically reduce the land filling of furnace waste. URL< http://www.ohlsti.co.uk/ohl/newsletter/ohl_wmr312.pdf

[19] Dunster A, Moulinier F, Abbott A, Conroy K, Adams D, Widyatmoko B 2005 Added value of using new industrial waste streams as secondary aggregates in both concrete and asphalt". Aggregates Research Programme STBF 13/15C.

[20] Mukhopadhyay Y, Ramana S, Upendra, J 2004 The Minerals, Metals & Materials Society Jawaharlal Nehru Aluminum Research development and Design Centre Light Metals.

[21] Shinzato M, Hypolito R 2005 Solid waste from aluminum recycling process: Characterization and reuse of its economically valuable constituents. Waste Management, 25 pp. 37-46

[22] Chen H, Xu Q, Chen S, Zhang Z 2009 Evaluation and design of fiber-reinforced asphalt mixtures. Mater Design 30:2595–603.

[23] Oluropo S Adeosun, O Sekunowo I, Omotayo O, Wasiu A, Adebowale M 2014 Physical and Mechanical Properties of Aluminum Dross, Advances in Materials. Vol. 3, No. 2, 2014, pp. 6-10

[24] Elzea K, Nikhil T, James M,Stanley T, Industrial Minerals and rocks: Commodities, Markets and uses (7th ed.), Society for Mining, Metallurgy and Exploration (SME), 2006, p.1406.

[25] Tsakiridis P, 2012 Aluminium salt slag characterization and utilization A review. Journal of Hazardous Materials, Vol. 217-218. pp. 1-10, https://doi.org/10.1016/j.jhazmat.2012.03.052

[26] Reddy M, Neeraja, D 2016 Mechanical and durability aspects of concrete incorporating secondary aluminium slag. Resource-Efficient Technologies, Vol. 2. No. 4. pp. 225-232, https://doi.org/10.1016/j.reffit.2016.10.012

[27] Mailar G, Raghavendra N, Sreedhara B, Manu D, Hiremath P, Jayakesh K 2016 Investigation of concrete produced using recycled aluminium dross for hot weather concreting conditions. Resource-Efficient Technologies, Vol. 2. No. 2. pp. 68-80.

[28] Kulik G, Daley J 1990 The Minerals, Metals & Materials Society-AIME 427 https://pdfs.semanticscholar.org/c0fe/c7f460d79a5406c575a47032a3d4f024f455.pdf

[29] Peterson R and Newton L 2002 TMS Annual Meeting (Warrendale, Pennsylvania) 1029 L S Carvalho, 1991 III Seminário de Tecnologia da Indústria do alumınio 169

[30] Gwinner D 1996 Environmental Issues in the Aluminum Reclamation Industry ABAL (Associação Brasileira do Alumınio) 40

[31] Reddy M, Neeraja D 2018 Journal of Material Science 43 124 https://doi.org/10.1007/s12046-018-0866-2

[32] Hwang J, Huang X, Xu Z 2006 Journal of Minerals & Materials Characterization & Engineering 5 (1) 47 https://doi.org/10.4236/jmmce.2006.51003

[33] Brough M 2002 Aluminum Lightens the Environmental Load. Vision-The newsletter of the Foresight and Link Initiative. Winter,URLhttp://www.berr.gov.uk/files/file30193.pdf (accessed November 11, 2019).

[34] Busari A 2020 Active Transportation in Sub-saharan Africa: Walking as a Modal Choice. The Open Journal of Transportation Engineering.

[35] Busari A, Adeyanju E, Loto T, Ademola D 2019 Recycled Aggregate in Pavement Construction: Review of Literatures. Journal of Physics. Volume 1378, Issue 2(18).

[36] Busari A, Oluwafemi D, Ojo A, Oyedepo J, Ogbiye A, Adegoke D 2019 Mobility Dynamics of the Elderly: Review of Literatures (2019). Materials Science and Engineering Volume 640, Issue 1, 13 November 2019,

[37] Busari A, Dahunsi B, Akinmusuru J Loto T, Ajayi S 2019 Response Surface Analysis of the Compressive Strength of Self-Compacting Concrete Incorporating Metakaolin. Advances in Science and Technology Research Journal. Volume 13, Issue 2, June 2019, pages 7–13. (ISI). (Scopus).

[38] Busari A, Ojo O 2018 Survey Data on The Effect of Demography and Socio-Economic Parameters on Non-Motorized trip. Data in Brief. (Scopus, Elsevier).

[39] Busari A, Akinwumi I, Awoyera P, Olofinnade O, Tenebe T, Nwanchukwu. J. 2018 Stabilisation Effect of Aluminum Dross on Tropical Lateritic Soil. Journal of Engineering Research in Africa. Vol. 39, pp 86-96 (Indexed in Scopus).

[40] Busari A, Oyedepo J, Modupe A, Bamigboye G, Olowu O, Adediran J, Ibikunle F 2017 The trip pattern of low-density residential area in semi-urban industrial cluster: predictive modelling. International Journal of Human Capital Urban Management, 2(3): 211-218, Summer 2017 DOI: 10.22034/ijhcum.2017.02.03.00

[41] Busari A, Joseph F, Ajayi S, Alayande T, Nwachukwu J, Agbama D 2019 Index Properties of Aluminum Dross Modified Pavement Geo-material. (2019). Journal of Physics. Volume 1378, Issue 2(18).

[42] Busari A, Dahunsi B, Akinmusuru J 2019 Sustainable concrete for rigid pavement construction using de-hydroxylated Kaolinitic clay: Mechanical and microstructural properties. Construction and Building Materials. Volume 211, 30 June 2019, Pages 408-415 (ISI, Scopus).
[43] ASTM D4402/D4402M–15. Standard Test Method for Viscosity Determination of Asphalt at Elevated Temperatures Using a Rotational Viscometer. ASTM International, West Conshohocken, PA, USA.

Predicting the Compressive Strength of Concrete By Ultrasonic Pulse Velocity

[1]O Ofuyatan, [2]A Olowofoyeku, [3]J Oluwafemi, [4]J Ighalo
[1,3] Department of Civil Engineering, Covenant University, Ota, Ogun State, Nigeria
[2]Department of Civil Engineering, Yaba College of Technology, Yaba, Lagos.
[4]Department of Chemical Engineering, Faculty of Engineering and Technology, University of Ilorin

Corresponding Author: olatokunbo.ofuyatan@covenantuniversity.edu.ng
Olatokunbo M. Ofuyatan, http://orcid.org/0000-0001-9052-2758

Abstract. The key resolution of non-destructive methods applied to concrete structures is to offer a suitable estimate of the quality of the material. One of the modern methods used in the test of concrete strength is the non-destructive testing applying the ultrasonic pulse velocity (UPV). In this study, the ultrasonic pulse velocity technique as non-destructive testing of concrete was used to distinguish concrete mixture design and show the predictive relationship between compressive strength and UPV. The effect of various grades of concrete (M15, M20, M25, M30, and M35) with water to cement ratio (w/c) of 0.45 and 0.5 were examined. It was observed that grade M15 cubes had the lowest transit time (37.6 – 41.2 μs) and grade M35 cubes had the highest transit time (48.7 – 49.9 μs) showing that the concrete with short transit time generates the highest pulse velocity than that with low transit time. Grade M15 cubes had the highest pulse velocity (3.6 – 4.0 km/s) and grade M35 cubes had the lowest pulse velocity (3.0 – 3.1 km/s). M15 cubes had the lowest actual compressive strength (18.1 – 19.3 N/mm2) and grade M35 cubes have the highest actual compressive strength (34.6 – 36.0 N/mm2). There was a good correlation between compressive strength and UPV. The study furthermore revealed that UPV can be used to accurately predict the compressive strength of concrete. For all concrete grades, the percentage variation did not exceed ± 2.5%.

1. Introduction

The use, strength and safety of concrete are determined by the characteristic strength [1-4]. Each structural element has its quality strength that determines its life and usage. Compressive strength is one of the valuable factors in concrete construction and it is dependent on other factors. One of the techniques used in assessing the strength of existing structures and durability is non-destructive testing (NDT) [5]. The ultrasonic pulse velocity method is a dynamic method that measures the time of travel of pulse generated by cracks and shock [6].
The development of UPV (ultrasonic pulse velocity) began in England and Canada [7], to the stage of laboratory research where it passed to the application on-site, to become an essential tool for non-destructive testing. Acoustic methods with visual inspection are the oldest form of non-destructive testing [8]. Cracks, voids and delamination are detected through sound. In 1920, Russian scientist Sergi Y. Sokolov Institute of Electrical Leningrad in the USSR was the first who proposed to use the ultrasonic wave velocity. However, until 1942 that the real progress was made NDT Resource Centre (the University of Michigan by Sproule in England). It was in the early 1970s with the improvement of

technology, the onset of the fracture mechanics and development of new law that a real significant change occurs in the field of non-destructive testing [7].
The excitation that is employed in the receiver is shocked by the waves which are generated by piezoelectric crystals as applicable in the case of ultrasonic testing of concrete [9]. The broadcasters encountered in the path of these waves causes diffusion and diversion in that exist as dissipation f heat of heat. This is due to the viscosity of the material. The waves also get modified in speed and this is likely because of the attenuation of the waves through the several processes undergone [10]. The diffusion that take place in the process of this wave propagation is always caused through sand grains, cracks, reinforcement bars or micro-cracks. The principle binding in the emission of these waves frequencies makes the transference of beam energy to coherent waves possible and these include backscattered and incoherent waves. The principle adopted in this process aids the determination of the non-uniform characteristics of the concrete through the measurement of the velocities of the waves. The adoption of this principle also helps in structural crack detection [11]. It can also be used to determine the coefficient of elasticity or Poisson coefficient.

Since there can be no direct measurement of the strength properties and quality of concrete without destructive stress (compressive, flexural and tensile), non-destructive methods have been developed for the assessment [12, 13]. Ultrasonic pulse shows potential for concrete in-situ testing by measuring the speed of ultrasound through materials. It measures the high frequency of ultrasonic pulse through concrete member between the transducers. High pulse reading is the quality of good concrete. Fairly good correlation is obtained between cubes compressive strength and pulse velocity. It enables structural concrete to be predicted within 20% if the aggregate mix is constant [14]. The test can be carried out on both laboratory size specimen and completed concrete structures.

Yılmaz and Ercikdi [15] used UPV to determine the strength in cemented paste in backfill. Wang and Wang [16] reported the use of UPV in self-consolidating waste glass concrete had an increase in strength due to the age and a decrease as the water/binder increases. Al-Nu'man et. al [17] a compressive strength formula for a single grade of concrete using ultrasonic pulse velocity. Shariati et. al [18] utilised UPV and Schmidt rebound hammer as non-destructive tests for evaluating the compressive strength of reinforced concrete. Quite recently, Fatahi and Jafari [19] studied the prediction of compressive strength for lightweight aggregate concrete. Also recently, the effects of the water-to-cement (W/C) ratio, curing conditions and aggregate were experimentally examined by Lee and Lee [20] based on early-stage direct transmission UPV.

This study is a continuation of these lines of investigation. In this study, the ultrasonic pulse velocity technique as non-destructive testing of concrete was used to distinguish concrete mixture design and show the predictive relationship between compressive strength and UPV. Besides the evaluation of cement properties and concrete flow properties, the effect of various grades of concrete with water to cement ratio was examined.

2. Materials and Method

Please follow these instructions as carefully as possible so all articles within a conference have the same style to the title page. This paragraph follows a section title so it should not be indented.

2.1. Materials

A natural aggregate of both fine and coarse aggregate was used in this study. The fine aggregate used was sharp-sand cleaned, air-dried, sieved with 5 mm sieve and free from waste and larger aggregate. The coarse aggregate used was crushed granite. Sieve analysis was carried out on both aggregates to know their size distribution. The cement used was ordinary Portland cement (Dangote 3x cement) with an initial setting time of 110 minutes, final setting time of 280 minutes, soundness of 0.2, fines of 5% and standard consistency of 37.5%. The water was clean and free from chemical, oil and other impurities.

2.2. Concrete Mixes

The constituent shown in Table 1 were used to make the concrete mix at varying concrete grade as seen in Table 2. The fresh test was carried out on the varying grades and the mix was used to cast 15 concrete cubes specimen which was used in this study. The varying concrete mix proportion was used according to BS EN 206-1. Each strength of cube range from 15 – 35 MPa and UPV ranges from 3 – 4.5 km/s the mixes were cast in wood cube mould of 150 mm by 150 mm by 150 mm for approximately 24 hours then cured for 28 days before the test was carried out.

Table 1. Concrete constituent quantity and water-cement ratio

Grade/Constituents	15 MPa	20 MPa	25 MPa	30 MPa	35 MPa
Cement (kg)	2.552	3.003	3.6435	4.641	6.379
Fine Aggregates (kg)	7.656	7.5075	7.287	6.962	6.379
Coarse Aggregates (kg)	15.312	15.015	14.574	13.923	12.76
Water (kg)	1.276	1.5024	1.8218	2.089	2.871
W/C Ratio	0.5	0.5	0.5	0.45	0.45

Table 2. Grade designation and velocity criteria

Mix	Concrete Grade	Ratio	Pulse Velocity (km/sec)	Concrete Quality Grade
1	15	1: 3: 6	Above 4.5	*Excellent*
2	20	1: 2.5: 5	3.5 – 4.5	*Good*
3	25	1: 2: 4	3.0 – 3.5	*Medium*
4	30	1: 1.5: 3	Below 3.0	*Doubtful*
5	35	1: 1: 2	-	–

2.3. Fresh and Hardened Concrete Tests
The slump tests (Abrams cone) were conducted on fresh samples of each grade mix respectively according to the EN 12350-2 (2000) [21]. The hardened concrete test was conducted: compressive strength-EN 12390-3 (2003) [22], ultrasonic pulse velocity-BS 1881 (1986): part 203 [23].

2.4. Ultrasonic Pulse Velocity Method
The ultrasonic pulse velocity technique as non-destructive testing of concrete was used to distinguish concrete mixture design and show the predictive relationship between compressive strength and UPV. This was done in accordance with BS 1881 (1986): part 203.

3. Results and Discussion

3.1. Cement Properties
For the fineness of cement, three samples of the same weight (100 g) and was sieved exhaustively with a sieve of size 90 micron. It was observed the % weight of residue of each sample do not exceed 10% which makes the cement fineness for each sample moderate. The results obtained are summarized in Table 3.

Table 3. Summary of results for the fineness of cement

Samples	A	B	C
Weight of cement (g)	100	100	100
Size of sieve (micron)	90	90	90
Time (minutes)	15	15	15
Weight of retained sample (g)	70	58	83
% weight of residue	0.7	0.58	0.83

For the soundness of cement, it was observed that the distance of the indicator before submerging to boiling is 45 mm and distance after boiling and cooling is 46.5 mm. The difference in distance between both periods turns out to be moderate (1.5 mm) which does not exceed 10 mm. It was also

observed that the standard consistency of the ordinary Portland cement (OPC) used was in range with normal consistency of OPC which is within the range of 26 – 33% according to IS:4031 (Part 4):1988. And also, the standard consistency result made water content for other tests like initial and final setting time achievable.
Specific gravity defines that the substance is how much heavier than water or reference substance of the same volume [24]. The observation was made that the specific gravity of the cement used is 2.72 times heavier than water, and it is also less than 3.15 g/cc which is the range for specific gravity of Portland cement. It was observed that the weight of cement for grade M35 was the highest and this was due to the mix design specified for the grade. M15 had the highest value in weight for fine aggregate and this was due to the value of the fine aggregate in the mix ratio which supersedes other fine aggregate values. The final weight of the concrete cubes for all grades was all-around 8.1 kg.

3.2. Cement Properties

According to the test carried out and result generated from slump test, it was observed that the water-cement ratio used for mixing each grade has an effect on the level of shear that occur in slump test for each concrete grades. The results obtained are summarized in Table 4. The slump of the concrete increased from 15 mm for grade M15 to 40 mm for grade M35. The relationship between the penetration and setting time is shown in Fig. 1. It can be observed that cement penetration increased with increasing setting time.

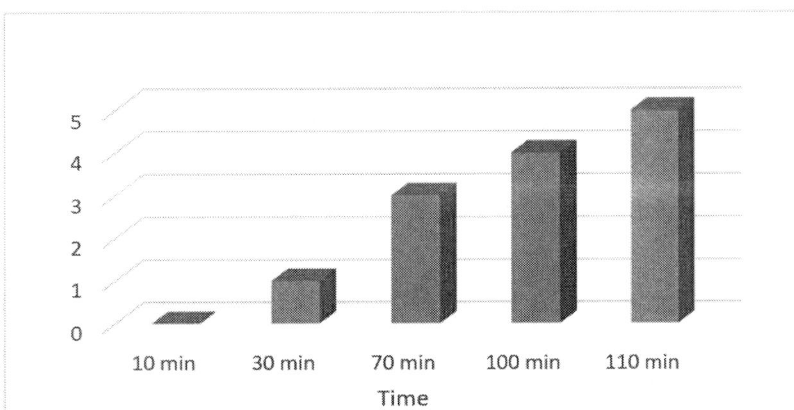

Figure 1. Cement penetration against setting time.

Table 4. Slump experiment result

Grade (MPa)	Slump Height (mm)	Collapse Height (mm)	Slump (mm)
M15	300	285	15
M20	300	280	20
M25	300	275	25
M30	300	265	35
M35	300	260	40

The observation was made for the transit time recorded from the ultrasonic pulse velocity test that was done using the direct transmission method on a concrete cube. This was implemented three times to achieve accurate transit time on a particular concrete cube. The results obtained are shown in Fig. 2. It was also observed that grade M15 cubes have the lowest transit time (37.6 – 41.2 μs) and grade M35 cubes have the highest transit time (48.7 – 49.9 μs). It is now understood that materials with short transit time generate the highest pulse velocity than that with low transit time. Reason for high transit is defect in concrete or failure to add enough coupling agent that can eliminate diffraction in wave [25].

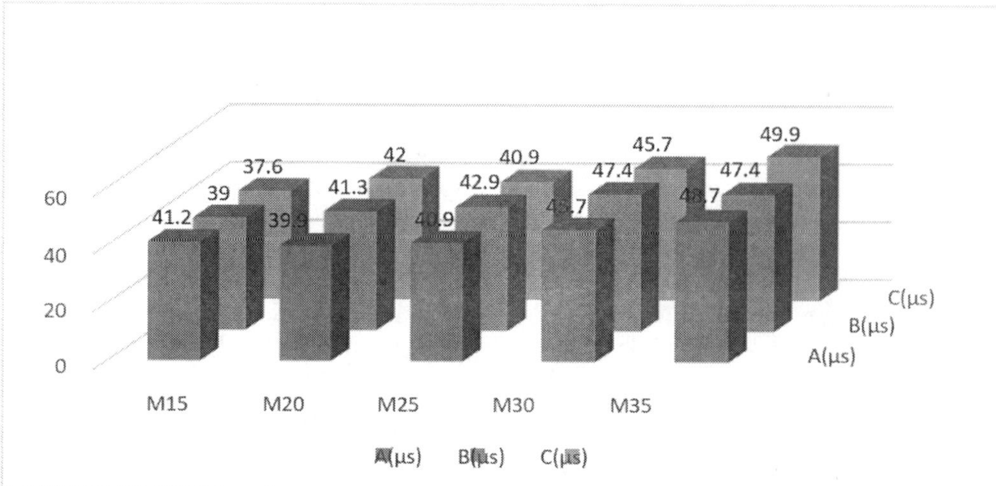

Figure 2. The transition time of the different concrete grade

Practically, it is known that pulse velocity is a dependent parameter which depends mainly on the ratio between path length and transit time [26]. The results obtained for the pulse velocity and actual compressive strength are summarized in Table 5. It was observed that the value with highest transit time has the lowest pulse velocity and that with the lowest transit time has the highest velocity and that fall on grade M35 and M15 cube C respectively. Grade M15 cubes have the highest pulse velocity (3.6 – 4.0 km/s) and grade M35 cubes have the lowest pulse velocity (3.0 – 3.1 km/s). According to the test carried out using UPV, it was observed that the pulse velocity increases as the grade changes and the velocity generated from the transit time and path length were a good pulse velocity value. Al-Nu'man et. al [17] recommended a UPV >3.5 km/s as the threshold for the concrete to be considered good. Also from the results in Table 5, M15 cubes have the lowest actual compressive strength (18.1 – 19.3 N/mm2) and grade M35 cubes has the highest actual compressive strength (34.6 – 36.0 N/mm2).

Table 5. Result of pulse velocity and compressive test of concrete grades cubes

	Pulse Velocity			Actual Compressive Strength		
Grade	A (km/s)	B (km/s)	C (km/s)	A (N/mm^2)	B (N/mm^2)	C (N/mm^2)
M15	3.6	3.9	4.0	18.1	18.6	19.3
M20	3.8	3.63	3.57	21.3	20.1	20.9
M25	3.67	3.5	3.7	25.9	26.2	25.8
M30	3.3	3.2	3.3	28.7	30	28.8
M35	3.1	3.2	3	34.6	35.1	36

3.3. Cement Properties

The prediction of compressive strength is shown in Tables 6 – 10 and the corresponding correlation plots shown in Fig. A – E (in the Appendix) respectively. For the predicted compressive strength of concrete cubes for different grades, it was observed that the strength of each grades cube conformed to the expected strength of a particular grade, except cube A of grade M35 which falls below the estimated strength. The Compressive strength derived from crushing inclines accurately with the grade Strength and also grades having a water-cement ratio of 0.5 has a high pulse velocity than those grades having a water-cement ratio of 0.45, this means that the increase in moisture content in concrete critically affects the ultrasonic pulse velocity of a particular structure, the coefficient of determining R2 for M15, M25 & M35 seems uniform while that of M20 and M30 falls. For both the actual and predicted strength it was observed that as the pulse velocity increases the strength of the concrete cube decreases and vice versa. This relationship has also been observed by other researchers [19, 20]. It was also observed that the actual and predicted strength of concrete cubes of M25 are the same and has no percentage variation and also cube B & C of M30 also share from the features. For all concrete grades, the percentage variation did not exceed ± 2.5%. this interval is relatively better than that obtained in an early study [13]. Based on a comparison between the actual and predicted compressive strength it can be concluded that UPV can be used to accurately predict the compressive strength of concrete.

Table 6. Result of pulse velocity and compressive test of concrete grades cubes

Samples	Transit Time (μs)	Pulse Velocity (Km/s)	Actual Compressive Strength (N/mm^2)	Predicted Compressive Strength (N/mm^2)	Percentage Variation	Linear Regression Equation	R^2
A	41.2	3.6	18.1	18.04	-0.33		
B	39	3.9	18.6	18.8	1.08	C= 2.6923V + 8.3462	$R^2 = 0.864$
C	37.6	4.0	19.3	19.12	-0.93		

Table 7. Concrete grade M20 experiment and prediction results

Samples	Transit Time (μs)	Pulse Velocity (Km/s)	Actual Compressive Strength (N/mm^2)	Predicted Compressive Strength (N/mm^2)	Percentage Variation	Linear Regression Equation	R^2
A	39.9	3.8	21.3	21.2	-0.45		
B	41.3	3.63	20.1	20.6	2.49	C = 2.904V + 10.119	$R^2 = 0.32$
C	42	3.57	20.9	20.5	-1.93		

Table 8. Concrete grade M25 experiment and prediction result

Samples	Transit Time (μs)	Pulse Velocity (Km/s)	Actual Compressive Strength (N/mm^2)	Predicted Compressive Strength (N/mm^2)	Percentage Variation	Linear Regression Equation	R^2
A	40.9	3.67	25.9	25.9	0		
B	42.9	3.5	26.2	26.2	0	C = -1.9198V + 32.923	$R^2 = 0.9894$
C	40.5	3.7	25.8	25.8	0		

Table 9. Concrete grade M30 experiment and prediction result

Samples	Transit Time (μs)	Pulse Velocity (Km/s)	Actual Compressive Strength (N/mm^2)	Predicted Compressive Strength (N/mm^2)	Percentage Variation	Linear Regression Equation	R^2
A	45.7	3.3	28.7	28.8	0.35		
B	47.4	3.2	30	30	0	C = -12.5V + 70	$R^2 = 0.9952$
C	45.7	3.3	28.8	28.8	0		

Table 10. Concrete grade M35 experiment and prediction result

Samples	Transit Time (μs)	Pulse Velocity (Km/s)	Actual Compressive Strength (N/mm^2)	Predicted Compressive Strength (N/mm^2)	Percentage Variation	Linear Regression Equation	R^2
A	48.7	3.1	34.6	35.2	1.73		
B	47.4	3.2	35.1	34.8	-0.85		
C	49.9	3.0	36.0	35.7	-0.83	C = -4.5V + 49.183	R^2 = 0.4023

4. Conclusion

In this study, the ultrasonic pulse velocity technique as non-destructive testing of concrete was used to distinguish concrete mixture design and show the predictive relationship between compressive strength and UPV. Several conclusions can be drawn from the results of the study. Grade M15 cubes have the lowest transit time (37.6 – 41.2 μs) and grade M35 cubes have the highest transit time (48.7 – 49.9 μs) showing that the concrete with short transit time generates the highest pulse velocity than that with low transit time. Grade M15 cubes have the highest pulse velocity (3.6 – 4.0 km/s) and grade M35 cubes have the lowest pulse velocity (3.0 – 3.1 km/s). M15 cubes have the lowest actual compressive strength (18.1 – 19.3 N/mm2) and grade M35 cubes have the highest actual compressive strength (34.6 – 36.0 N/mm2). For both the actual and predicted strength it was observed that as the pulse velocity increases the strength of the concrete cube decreases and vice versa. There was a good correlation between compressive strength and UPV. The study revealed that UPV can be used to accurately predict the compressive strength of concrete. For all concrete grades, the percentage variation did not exceed ± 2.5%. From the observations of this study, it is recommended that a non-destructive test should be implemented for the concrete test due to its effectiveness, accuracy and rapidity. Furthermore, this study has also opened up some other areas for future work. Result generated from UPV test if not satisfactory could be modelled with an artificial neural network which can aid in generating more accurate predictions for the properties. For better assurance on concrete strength result, the combined test can be done to make the result more satisfactory.

5. Acknowledgement

The authors appreciate and thank Covenant University for their support.

References

[1] Steenbergen R and Vervuurt A 2012 Determining the in situ concrete strength of existing structures for assessing their structural safety *Structural Concrete (Electronic Materials* vol 13*)* pp 27-31

[2] Oluwafemi J, Ede A, Ofuyatan O, Oyebisi S and Bankole D 2019 Recycling of Concrete Demolition Waste: pathway to sustainable development *IOP Conf. Series: Materials Science and Engineering (Electronic Materials* vol 640*)* pp 1-6

[3] Ofuyatan O, Adeniyi A, Ijie D, Ighalo J and Oluwafemi J 2020 Development of high-performance self compacting concrete using eggshell powder and blast furnace slag as partial cement replacement *Construction and Building Materials (Electronic Materials* vol 256*)* pp 1-10

[4] Ofuyatan O, Olowofoyeku A, Adaramaja G, Oluwafemi J and Edeki S 2020 Development of high-performance self compacting concrete using eggshell powder and blast furnace slag as partial cement replacement *Construction and Building Materials (Electronic Materials* vol 13*)* pp 1-7

[5] Ali-Benyahia K, Sbartaï Z M, Breysse D, Ghrici M and Kenai S 2019 Improvement of nondestructive assessment of on-site concrete strength: Influence of the

selection process of cores location on the assessment quality for single and combined NDT techniques *Construction and Building Materials (Electronic Materials* vol 195*)* pp 613-622

[6] Boccaccini D, Romagnoli M, Kamseu E P, Veronesi C, Leonelli, and Pellacani G C 2007 Determination of thermal shock resistance in refractory materials by ultrasonic pulse velocity measurement *Journal of the European Ceramic Society (Electronic Materials* vol 27*)* pp 1859-1863

[7] Hannachi S and Guetteche M 2014 Review of the ultrasonic pulse velocity evaluating concrete compressive strength on site in *Proceedings of Scientific Cooperation International Workshops on Engineering Branches, Istanbul, Turkey*

[8] Cawley P 2001 Non-destructive testing—current capabilities and future direction *Proceedings of the Institution of Mechanical Engineers, Part L: Journal of Materials: Design and Applications (Electronic Materials* vol 215*)* pp 213-223

[9] Lim Y Y, Smith S T and Soh C K 2018 Wave propagation based monitoring of concrete
curing using piezoelectric materials *Review and path forward. NDT & E International (Electronic Materials* vol 99*)* pp 50-63

[10] Krautkrämer J and Krautkrämer H 1990 Ultrasonic testing by determination of material
properties, in Ultrasonic Testing of Materials *Springer* pp 528-550

[11] McGovern M E 2016 Non-destructive quantitative assessment of oxidation in aged and
rejuvenated asphalt concrete pavements using a non-collinear wave mixing approach. *University of Illinois at Urbana-Champaign*

[12] Tsioulou O, Lampropoulos A and Paschalis S 2017 Combined non-destructive testing (NDT) method for the evaluation of the mechanical characteristics of ultra high performance fibre reinforced concrete (UHPFRC *Construction and Building Materials (Electronic Materials* vol 131*)* pp 66-77.

[13] Qasrawi H Y 2000 Concrete strength by combined nondestructive methods simply and reliably predicted *Cement and concrete research (Electronic Materials* vol 30*)* pp 739-746

[14] Hernández M, Anaya J, Ullate L and Ibañez A 2006 Formulation of a new micromechanic model of three phases for ultrasonic characterization of cement-based material. *Cement and Concrete Research (Electronic Materials* vol 36*)* pp 609-616

[15] Yılmaz T and Ercikdi B 2016 Predicting the uniaxial compressive strength of cemented paste backfill from ultrasonic pulse velocity test *Nondestructive Testing and Evaluation (Electronic Materials* vol 31*)* pp 247-266

[16] Wang C C and Wang H Y 2017 Assessment of the compressive strength of recycled waste LCD glass concrete using the ultrasonic pulse velocity *Construction and Building Materials (Electronic Materials* vol 137*)* pp 345-353

[17] Al-Nu'man B S, Aziz B R, Abdulla S A and Khaleel S E 2015 Compressive strength formula for concrete using ultrasonic pulse velocity *International Journal of Engineering Trends and Technology* (IJETT) *(Electronic Materials* vol 26*)* pp 8-13

[18] Shariati M, Ramli-Sulong N H, Arabnejad M M, Shafigh P, and Sinaei H 2011 Assessing the strength of reinforced concrete structures through Ultrasonic

Pulse Velocity and Schmidt Rebound Hammer tests *Scientific Research and Essays (Electronic Materials* vol 6*)* pp 213-220

[19] Fatahi O and Jafari S 2020 Prediction of Lightweight Aggregate Concrete Compressive Strength *Journal of Rehabilitation in Civil Engineering (Electronic Materials* vol 8*)* pp 45-57

[20] Lee T and Lee J 2020 Setting time and compressive strength prediction model of concrete by nondestructive ultrasonic pulse velocity testing at early age ` *Construction and Building Materials (Electronic Materials* vol 252*)* pp 119027

[21] EN B *12350-2* 2000 Testing fresh concrete: slump test *London: British Standards Institution*

[22] EN T *12390–3 2003* Beton-Sertleşmiş Beton Deneyleri-Bölüm 3: Deney Numunelerinde Basınç Dayanımının Tayini *TSE, Ankara*

[23] British Standard B *Part 203 1986.* Recommendations for measurement of velocity of ultrasonic pulses in concrete. BSI, London, 1881

[24] Blake G 1965 Bulk density *Methods of Soil Analysis: Part 1 Physical and Mineralogical Properties, Including Statistics of Measurement and Sampling (Electronic Materials* vol 9*)* pp 374-390

[25] Jones R and Fąçąoaru I 1969 Recommendations for testing concrete by the ultrasonic pulse method *Matériaux et constructions (Electronic Materials* vol 2*)* pp 275-284

[26] Reusz G S, Bárczi A, Dégi A, Cseprekál O, Kis É, Szabó Á, Csóka M, Rudas G, Végh A and Temmar M 2020 Distance measurement for pulse wave velocity estimation in pediatric age: Comparison with intra-arterial path length *Atherosclerosis*

Effect of sorghum husk ash and calcium chloride on compressive strength of grade 20 concrete

M A Tijani[1*], S O A Olawale[1], L Abdullahi[1], M A Aliyu[1] and W O Ajagbe[2]

[1]Department of Civil Engineering, Osun State University Osogbo, Nigeria
[2]Department of Civil Engineering University of Ibadan, Ibadan, Nigeria

*Corresponding author: murtadha.tijani@uniosun.edu.ng

Abstract. This study examined the influence of sorghum husk ash (SHA) and calcium chloride ($CaCl_2$) as fractional substitution of cement in grade 20 concrete. Sorghum husk Ash (SHA) was gotten by burning of sorghum husk at a regulated temperature. The substitution of cement by SHA were 0, 5, 10, 15, 20 and 25%. A fixed amount of 1% $CaCl_2$ was mixed with Cement/SHA in all the test samples apart from control mixture. Sample sizes of 150 x 150 x 150 mm concrete cubes were cast and healed in water respectively for 7 as well as 28 days. The tests carried out on prepared fresh and hardened concrete were slump and compressive strength. The result of slump showed that the concrete turn out to be stiffer with increase in proportion of SHA. The 28 day compressive strengths result shows that 5%SHA/1%$CaCl_2$ have the maximum strength of 26.45 N/mm^2 followed by 10% (25.01 N/mm^2), 0%SHA/1%$CaCl_2$ (23.11 N/mm^2), 15%SHA/1%$CaCl_2$ (22.45 N/mm^2), 20%SHA/1%$CaCl_2$ (20.90 N/mm^2) and 25%SHA/1%$CaCl_2$ (18.13 N/mm^2). This implies that the best addition of SHA/$CaCl_2$ as fractional substitution for cement in concrete is in the range 0 - 20% since their compressive strength results were above 20 N/mm^2 specified for grade 20 concrete. Incorporation of 5%SHA/1%$CaCl_2$ and 10%SHA/1%$CaCl_2$ substitution would yield a concrete of greater compressive strength than standard grade 20 concrete.

1. Introduction

Increasing buildings cost in developing countries like Nigeria have been ascribed to several factors [1], among which are incapability of manufacture companies to meet the enormous building materials demand, high cost of transportation and national currency depreciation. Others include overwhelming building materials prices and over-dependency on building materials importation that is frequently proportional to inflation. The most dominant construction material is concrete. It be made up of cement, coarse aggregate, fine aggregate, water and at times admixture in a certain prescribed proportion [2], [3]. Research inclinations in materials engineering have recently concentrated on giving choices to subdue the significant expense of conventional concrete materials. The most recently trending research exercises on proffering the alternatives are centered on the utilization of pozzolanas from agricultural wastes as admixtures or additives in concrete [4].

Some of the agricultural waste products that has been confirmed as pozzolanic materials include palm oil fuel ash, rice husk ash, palm kernel shell ash, sawdust ash, sorghum husk ash as well as guinea corn husk ash. These materials are readily available in Nigeria to partially replace cement without economic implications [5]. They are supplementary cementitious materials (SCMs) which improves the properties of hardened concrete when used with Portland cement through hydraulic or pozzolanic operation or both [5, 6, 7, 8, 9, 10, 11, 12, 13, 14].

USDA [15] stated that Nigeria is the world's second-largest manufacturer of sorghum, with almost 6.6 million tons of sorghum manufactured annually in the country. Approximately 11 per cent SHA content is obtained from proper sorghum husk burning [6]. Studies have shown that a mixture of SHA's chemical constituents suits it as a strong pozzollana [14, 16]. Most authors who worked on suitability of supplementary cementitious materials recommended high range water reducing admixture (e.g superplasticizer) to reduce water/cement ratio and improve concrete strength. However, this study adopted an accelerating admixture (calcium chloride) as addition with SHA to improve the concrete strength.

2. Materials and methods

2.1. Materials

Portland lime cement of grade 42.5R with specific gravity 3.14 was adopted for this study. Natural sand that passes 4.75 mm sieve having 2.57 specific gravity and 0.57 water absorption capacity was utilized as fine aggregate. The crumpled stone (granite) utilized as concrete coarse aggregate has 19 mm maximum size, 2.72 specific gravity and 0.3% water absorption respectively. Individually, the fine and coarse aggregates utilized conformed to [17] while the calcium chloride conformed to [18]. The sorghum husk utilized was collected from Amoloko, Minna, Niger State, Nigeria. It was calcinated into ashes (SHA) in regulated furnace for 3 hours at 700°C. The SHA was then powdered after allowing to cool utilizing ceramic mortar and pestle. The ash was sieved using 75 µm sieve to obtain ash that is fine enough to react perfectly with ordinary cement. Ash specific gravity was tested to be 2.2. The chemical constituents of SHA as presented by Tijani *et al.* [16] is given in Table 1. The combined fraction of silica, alumina and ferric oxide is 77.3 percent, which fulfills the pozzolana condition of ASTM [19].

Table 1. Chemical constituents of SHA

Chemical components	%
SiO_2	55.3
Al_2O_3	10.1
Fe_2O_3	11.9
SO_3	0.5
MgO	1.2
K_2O	4.5
Na_2O	0.7
CaO	10.4
$SiO_2 + Al_2O_3 + Fe_2O_3$	77.3

2.2. Methods

The weight method of batching was adopted using design mix for normal strength concrete of proportion 1:2:4 and 0.6 water/cement ratio. The weight of concrete was calculated based on the volume of mould that was used and the number of cubes for each test sample. The SHA was varied with cement in the order 0, 5, 10, 15, 20 and 25%. 1% $CaCl_2$ by weight of cement was blended with

IOP Conf. Series: Materials Science and Engineering 1036 (2021) 012054 doi:10.1088/1757-899X/1036/1/012054

cement/SHA mixes. A total of 36 concrete cubes of 150 x 150 x 150 mm were produced and preserved in water for 7 as well as 28 days. The slump test which conformed to [20] was carried out on fresh concrete while the hardened concrete were tested for density and compressive strength in accordance with [21] and [22] respectively. Figure 1 displays the compressive strength test setup.

Figure 1. Test setup for Compressive strength

3. Results and Discussion

3.1. Slump of SHA-$CaCl_2$ Concrete

Figure 2 displayed the outcomes of slump test. The figure shows that slump of concrete becomes lower as the SHA percentage risen. The reduction in slump when likened to the slump of control mix was substantial. The slump falls between 38.8 and 2.5 mm. The concrete was observed to become less workable as the proportion of SHA becomes higher. The implication of this is that a workable mixture will requires additional water. The amplified amount of silica in the mix might be the reason for additional water demand as SHA proportion becomes high. The silica-lime reaction necessitates more addition of water to the water needed in the course of hydration of cement [6, 16, 23].

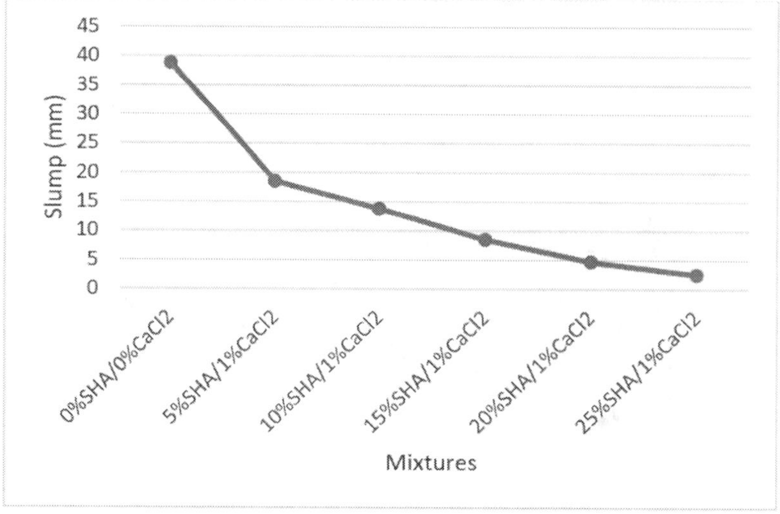

Figure 2. Slump of SHA-$CaCl_2$ Concrete

3.2. Density of SHA-$CaCl_2$ Concrete

Figure 3 presents the results of 7 and 28 day density of SHA-$CaCl_2$ concrete mixtures at different substitution levels. It could be detected that hardened densities (both 7 and 28 day) slightly decreased as the amount of SHA addition increased from 0 to 25%. The respective values obtained at 7 as well as 28 days were 2577, 2502, 2485, 2477, 2450 and 2397 kg/m^3 and 2554, 2492, 2479, 2450, 2419 and 2360 correspondingly for 0%SHA/0%$CaCl_2$, 5%SHA/1%$CaCl_2$, 10%SHA/1%$CaCl_2$, 15%SHA/1%$CaCl_2$, 20%SHA/1%$CaCl_2$, and 25%SHA/1%$CaCl_2$. This implied a decrease of 2.91, 3.57, 3.88, 4.93 and 6.98% at 7-day with reference to control and 2.43, 2.94, 4.07, 5.29 and 7.60% at 28-day with reference to control. The decrease in densities with increase in the amount of SHA might be credited to the lesser specific gravity of SHA (2.20) than that of cement (3.14). Furthermore, a slight decrease in densities could be observed at 28 day when compared with 7 day results. This decreased could be due to the fact the some water are still present in the concrete which have not been fully used for hydration at 7 day of curing.

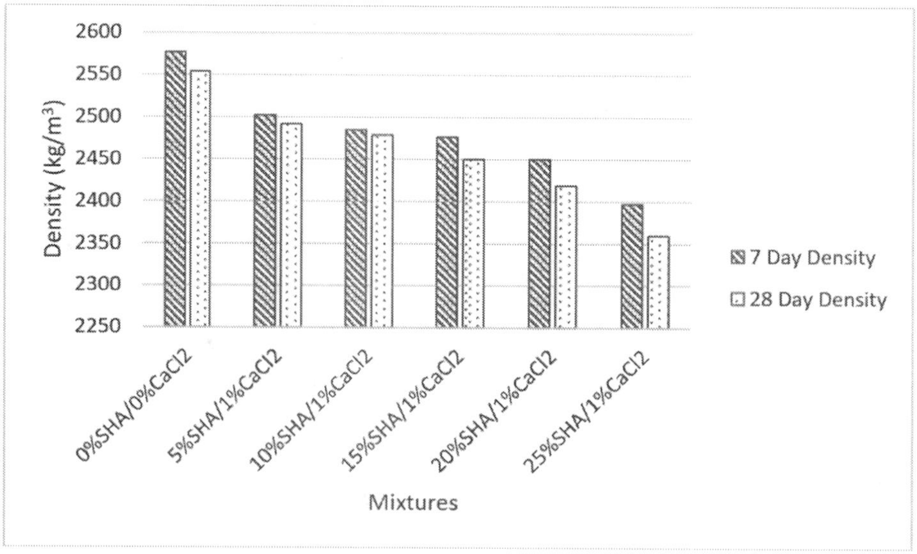

Figure 3. Densities of SHA-$CaCl_2$ Concrete

3.3. Compressive Strength of SHA-$CaCl_2$ Concrete

The 7 as well as 28-day compressive strength outcomes of SHA-$CaCl_2$ concrete are represented in Figure 4. The result indicated that the compressive strengths rises as the SHA/$CaCl_2$ percentage upsurges from 0% to 5% and reduces as the percentage increases from 5% to 25%. Conversely, as the amount of curing days increased, the compressive strengths increased for each percentage of SHA/$CaCl_2$ replacement. The figure shows that for the control cube (plain concrete i.e., 0% replacement), the compressive strength amplified from 15.20 N/mm^2 at 7-day to 23.11 N/mm^2 at 28-day. The 28-day compressive strength was more than the stated value of 20 N/mm^2 for concrete grade 20 [24, 25]. The compressive strength of the 5% replacement by SHA/1% $CaCl_2$ revealed rise in compressive strength from 18.46 N/mm^2 at 7-day to 26.45 N/mm^2 at 28-day. The 28-day compressive strength was more than the stated value of 25 N/mm^2 for grade 25 concrete [24, 25].

The compressive strength of the 10% substitution by SHA/1% $CaCl_2$ revealed rise in compressive strength from 16.25 N/mm^2 at 7-day to 25.01 N/mm^2 at 28-day. The 28-day compressive strength was more than the stated value of 25 N/mm^2 for concrete of grade 25 [24, 25]. The compressive strength of the 15% substitution by SHA/1% $CaCl_2$ indicated rise in compressive strength from 15.56 N/mm^2 at 7-day to 22.45 N/mm^2 at 28-day. The 28-day compressive strength was more than the stated value of 20 N/mm^2 for concrete grade 20 [24, 25].

The strength of the 20% substitution by SHA/1%$CaCl_2$ revealed compressive strength rise from 13.31 N/mm^2 at 7-day to 20.90 N/mm^2 at 28-day. The 28-day compressive strength was more than the stated value of 20 N/mm^2 for concrete grade 20 [24, 25]. The strength of the 25% substitution by SHA/1%$CaCl_2$ revealed compressive strength rise from 11.95 N/mm^2 at 7-day to 18.13 N/mm^2 at 28-day. The 28-day compressive strength was less than the stated value of 20 N/mm^2 for concrete grade 20 [24, 25].

At 7-day, the results of compressive strength of 0%SHA/0%$CaCl_2$, 5%SHA/1%$CaCl_2$, 10%SHA/1%$CaCl_2$ and 15%SHA/1%$CaCl_2$ met the lowest indicated compressive strength of 13.5N/mm^2 for grade 20 concrete stated by [25]. Nonetheless, the compressive strengths result at 28-day shows that 5%SHA/1%$CaCl_2$ have the maximum strength of 26.45 N/mm^2 followed by 10% (25.01 N/mm^2), 0%SHA/1%$CaCl_2$ (23.11 N/mm^2), 15%SHA/1%$CaCl_2$ (22.45 N/mm^2), 20%SHA/1% $CaCl_2$ (20.90 N/mm^2) and 25%SHA/1%$CaCl_2$ (18.13 N/mm^2). This implies that the optimal addition of SHA/$CaCl_2$ as fractional substitution for cement in concrete is in the range 0 - 20% since their compressive strength values were above 20 N/mm^2 specified for concrete grade 20.

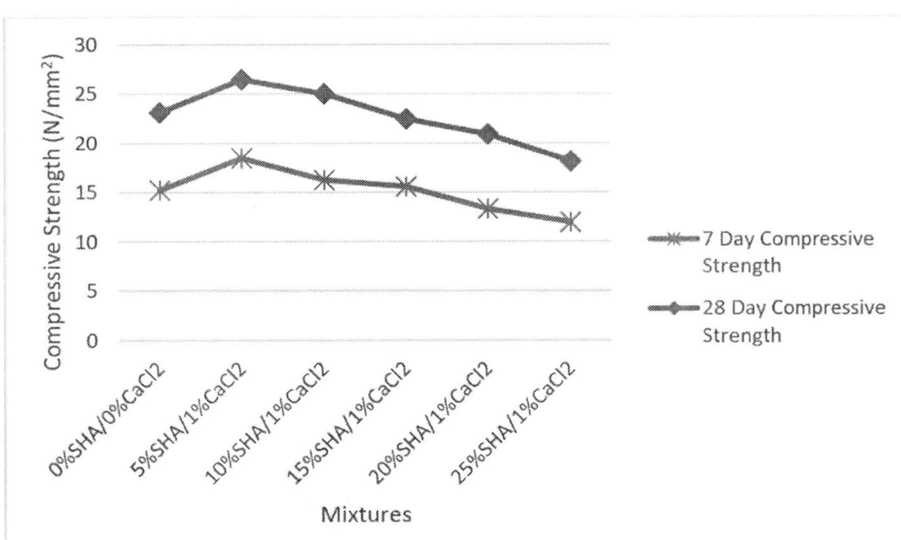

Figure 4. Compressive Strength of SHA-$CaCl_2$ Concrete

4. Conclusion

The influence of SHA and $CaCl_2$ as fractional substitution of cement in concrete grade 20 was investigated in this study. Sorghum husk Ash was gotten by calcination of sorghum husk at a regulated temperature. The substitution of cement by SHA were 0, 5, 10, 15, 20 and 25%. A fixed amount of 1% $CaCl_2$ was mixed with Cement/SHA in all the test samples apart from control mixture. Sample sizes of 150 x 150 x 150 mm concrete cubes were cast and preserved in water respectively for 7 as well as 28 days. The tests conducted on prepared fresh and hardened concrete were slump and compressive strength. The slump values indicated that concrete becomes less workable as the SHA/$CaCl_2$ proportion rises. The compressive strengths of concrete declined as the proportion of SHA/$CaCl_2$ substitution upgraded at all curing ages. The compressive strengths result at 28 days shows that 5%SHA/1%$CaCl_2$ possess the maximum strength of 26.45 N/mm^2 followed by 10% (25.01 N/mm^2), 0%SHA/1%$CaCl_2$ (23.11 N/mm^2), 15%SHA/1%$CaCl_2$ (22.45 N/mm^2), 20%SHA/1%$CaCl_2$ (20.90 N/mm^2) and 25%SHA/1%$CaCl_2$ (18.13 N/mm^2). This implies that the most favorable addition of SHA/$CaCl_2$ as partial substitution for cement in concrete is in the range 0 - 20% since their compressive strength values were above 20 N/mm^2 specified for concrete grade 20. Combination of 5%SHA/1%$CaCl_2$ and 10%SHA/1%$CaCl_2$ substitution would yield concrete of greater compressive strength than standard grade 20 concrete.

References

[1] Fasakin J O and Ogunsemi D R 2003 Modelling the selling prices of residential buildings in Nigeria. *Journal of Financial Management of Property and Construction* 8 (2) 111 – 114.
[2] Ajagbe W O, Tijani M A and Agbede O A 2018 Compressive strength of concrete made from aggregates of different sources. *USEP: Journal of Research Information in Civil Engineering* 15 (1) 11963 – 1974.
[3] Neville A M 2011 Properties of concrete, 5th Edition. Pearson Education Limited, Edinburgh Gate, Harlow, Essex CM20 2JE, England.
[4] Ogork E N and Ibrahim T S 2017 Properties of cement paste and concrete containing calcium carbide waste as additive. *Nigerian Journal of Technology (NIJOTECH)* 36 (1) 26 – 31.
[5] Tijani M A, Ogunlade C A, Ajagbe W O, Olawale S O A, Akinleye M T and Afolayan O D 2018 Development of green concrete using agricultural and construction wastes in Nigeria: A review. *Adeleke University Journal of Engineering and Technology* 1 (1) 40 – 50.
[6] Tijani M A, Ajagbe W O and Agbede O A 2018a Modelling the Effect of burning temperature and time on chemical composition of sorghum husk ash for optimum pozzolanic activity. *Journal of Engineering and Engineering Technology (FUTAJEET)* 12 (2) 273 – 281.
[7] Ephraim M E, Akeke G A and Ukpata J O 2012 Compressive strength of concrete with rice husk ash as partial replacement of ordinary Portland cement. *Scholarly Journal of Engineering Research* 1(2) 32 - 36.
[8] Olutoge F A, Quadri H A and Olafusi O S 2012 Investigation of the strength properties of palm kernel shell ash concrete. *Engineering, Technology and Applied Science Research* 2 (6) 315 - 319.
[9] Akeke G A, Ephraim M E, Akobo I Z S and Ukpata J O 2013 Structural properties of rice husk ash concrete. *International Journal of Engineering and Applied Sciences* 3 (3) 57 - 62.
[10] Williams F N, Anum I, Isa R B and Aliyu M 2014 Properties of sorghum husk ash blended cement laterized concrete. *International Journal of Research in Management, Science and Technology* 2 (2) 73 - 79.
[11] Ndububa E E and Nurudeen Y 2015 Effect of guinea corn husk ash as partial replacement for cement in concrete. *IOSR Journal of Mechanical and Civil Engineering (IOSR-JMCE)* 12 (2) 40 - 45.
[12] Oyejobi D O, Abdulkadir T S and Ahmed A T 2016 A study of partial replacement of cement with palm oil fuel ash in concrete production. *International Journal of Agricultural Technology* 12 (4) 619 - 631.

[13] Joshua O, Olusola K O, Ogunde A O, Ede A N, Olofinnade R M and Nduka D O 2017 Investigating for pozzolanic activity in palm kernel nut waste ash (PKNWA) with cement towards a sustainable construction. *International Journal of Applied Engineering Research* 12 (23) 13959 - 13565.

[14] Ogork E N and Danja Y N 2018 Evaluation of properties of cement and concrete containing sorghum chaff ash. *Bayero Journal of Engineering and Technology* 13 (1) 45 – 53.

[15] United States Department of Agriculture (USDA) 2017 Nigeria now second world's leading producer of Sorghum. Vanguard, *28 September, 2017.* Available at https://www.vanguardngr.com/2017/09/nigeria-now-second-worlds-leading-producer-sorghum/ Assessed on 24/11/2017.

[16] Tijani M A, Ajagbe W O, Ganiyu A A and Agbede O A 2019 Sustainable pervious concrete incorporating sorghum husk ash as cement replacement. IOP Conference Series: Materials Science and Engineering 640: 012051.

[17] British Standard BS 8110 Part 2, Structural Use of Concrete. Code of Practice for Design and Construction, British Standards, BSI Group Headquarters 389 Chiswick High Road, London, W4 4AL, UK, Standards Policy and Strategy Committee, (1985), pp.7/1.

[18] ASTM Standard ASTM C494 1999 Standard Specification for Chemical Admixtures for Concrete, ASTM International, West Conshohocken, PA.

[19] ASTM Standard ASTM C 618 2005 Specification for coal fly ash and raw or calcined natural pozzolana for use as a mineral admixture in Portland cement concrete. ASTM International, West Conshohocken, PA.

[20] ASTM Standard ASTM C192/C192M 2006 Standard Practice for Making and Curing Concrete Test Specimens in the Laboratory, ASTM International, West Conshohocken, PA.

[21] British Standard BS EN 12390-7 2009 Testing Hardened Concrete - Density of Hardened Concrete, British Standards, BSI Group Headquarters 389 Chiswick High Road, London, W4 4AL, UK, Standards Policy and Strategy Committee.

[22] British Standard BS EN 12390-3 2009 Testing Hardened Concrete - Compressive Strength of Test Specimens, British Standards, BSI Group Headquarters 389 Chiswick High Road, London, W4 4AL, UK, Standards Policy and Strategy Committee.

[23] Tijani M A, Ajagbe W O, Ganiyu A A, Aremu A S and Ojewole Y N 2020 Strength and Absorption of Sorghum Husk Ash Sandcrete Blocks. *Premier Journal of Engineering and Applied Sciences* 1 (1) 1 – 7.

[24] CP, 2013 Code of Practice for Structural Use of Concrete, February 2013. The Government of the Hong Kong Special Administrative Region, pp.14.

[25] British Standard *BS882,* 1992. Specification for aggregates from natural sources for concrete. British Standard Institution, London, UK.

Performance evaluation of the mechanical and durability properties of concrete reinforced with hybridized fibers

T F Awolusi[1,4], J N Falana[2], P O Ajayi[3], O O Akinkurolere[3], O G Aluko[3]

[1]Civil Engineering Department, Landmark University, Omu Aran, Nigeria
[2]Civil Engineering Department, Federal Polytechnic Ile Oluji, Ondo State, Nigeria
[3]Civil Engineering Department, Ekiti State University, Ado Ekiti, Nigeria

Abstract. The hybridization of two different elements is done for synergy and better performance. This study investigates the effects of combining bamboo, steel, and carpet fibers, as reinforcement in concrete. The steel and carpet fibers were obtained from discarded tires, and carpets, respectively. The addition of fibers to the mixes was kept constant at 2% of the concrete volume. The first category concrete mix contained a hybrid of steel and carpet fibers at a fraction of 1.75% and 0.25%, respectively, the second category combined steel and bamboo fibers at a fraction of 1% - 1%. The third category contained only steel fibers at a 2% volumetric fraction. The mechanical property of the concrete was evaluated through compressive and flexural strengths, while the durability was examined through water absorption and the acid attack. The compressive strength results obtained for the three categories of fiber reinforced concrete ranged between 7.2 and 12.8 N/mm^2, with the third category performing better. Also, the flexural strength results ranged between 5.37 and 7.47 N/mm^2, with the first category having the best results. The second category had the least water absorption capacity, and the concrete reinforced with only steel fibers showed the best acid attack resistance. The results also confirmed that short discontinuous fibers are suitable strength and durability enhancers in producing durable eco-friendly concrete.

Keywords: steel fiber reinforced concrete; steel fibers; bamboo fibers, carpet fibers

1. Introduction

The introduction of short discontinuous fibers in concrete is referred to as fiber reinforced concrete. Typically, fibers have been found suitable for enhancing the performance of concrete [1]. In this regard, quite several fibers have been incorporated into the concrete mix, and these fibers range from natural to industrially manufactured fibers [2]. The process of combining two or more fibers in concrete is referred to as hybridization. In hybridization, the performance of the composite material named concrete is improved with benefits from each of the individual fibers added to the concrete mix. The advantage of incorporating these fibers into concrete includes enhanced tensile strength, ductility, toughness, and durability [3]. Natural occurring fibers such as bamboo fibers are usually obtained at little or no cost, while the industrially manufactured fibers increase the cost of producing concrete. The addition of 1% of steel fibers content approximately doubles the material costs of concrete [4]. Most developing countries, such as Nigeria, lack the presence of viable steel manufacturing industries to meet the demand for steel fibers if the material were to be incorporated into the concrete mix. Steel fibers waste tires are now being considered as an alternative source of steel fiber for use in concrete. Due to the high volume of waste tire generated annually, they are readily available, and a few pieces of research have also been carried out in this area [5], [6], [7], [8] and [9].

It is estimated that 259 million tires are discarded annually across Nigeria [9], and it is expected that this number will increase, considering the ever-growing population and rising interest in personal mobility. Furthermore, most of the populace depends on used tires, also known as 'Tokunbo Tires,' as a replacement for worn-out tires. This, coupled with the deplorable states of most Nigeria roads, causes these tires to wear quickly and, in turn, increases the rate at which waste tires are generated. Tires are made of elastomers (synthetic rubber and natural rubber), fillers (carbon black and silica), reinforcing metals (wires and bead wires), plasticizers (resins and oils), textile reinforcement (rayon, aramid, nylon, and polyester) and chemical (sulfur) [10]. It is most effective to make use of the metal reinforcement derived from tires as fibers to reinforce concrete, while other materials from tires can be used for other purposes such as making surfaces and into chips for use as solid fuel, considering the cost of purchasing steel fibers, the amount of landfill space occupied by waste tires and the eventual environmental pollution or hazard that results from its disposal [11].

A carpet is a form of polypropylene fiber that has a tensile strength of 40-60N/mm^2 [12]. Carpets are made of polyvinyl chloride (or vinyl) resin, plasticizer (high molecular solvents), pigment and trace stabilizers, and a carrier sheet or backing. Due to the low standard of living of Nigerians, carpets seem the cheapest flooring material, and depending on the maintenance and usage, carpets last for an average of 3-5 years before ending up as waste. This waste carpet comprises different polymeric materials, which makes it suitable for use as fiber in concrete. Waste carpets are relatively readily available, and fibers obtained from shredding these carpets are flexible with high resistance to corrosion, judging by the presence of polyvinyl chloride (or vinyl) resin. This provides another avenue to reuse this carpet and also to reduce the problems associated with its disposal. Since the material is non-biodegradable, it seldom decays, except by burning. This releases pollutants that are hazardous to the environment, especially in the case of open burning as often practice in the country.

Bamboo fibers, as mention earlier, are natural fibers that are environmentally friendly and have an excellent mechanical property that makes them a material of choice in reinforcing concrete. The bamboo fibers can be used in place of the non-degradable synthetic fibers, which are capable of emitting poisonous gases into the atmosphere. The bamboo fibers have tensile strength ranging between 150 and 810MPa, which makes it suitable for use as conventional fiber [13]. The bamboo plant is readily available in most parts of Nigeria due to its fast-growing nature, and it has a growth rate of 3 cm per hour. Besides, bamboo has been found useful for a variety of applications in the construction industry ranging from making of scaffolding, bridges, walls, partitions, roofing, and houses [14].

A study of the mechanical behavior of concrete reinforced with different types of fibers such as polypropylene, glass, and steel at different volumetric fractions ranging between 0.5 and 2% was conducted. The results showed that the compressive strength generally increased with an increase in fiber volumetric fraction [14]. Another investigation into the mechanical properties of hybrid steel polypropylene (ST-PP) Fiber Reinforced Concrete (FRC) at a low fiber volume fraction of 0.5% shows that the hybrid fibers can enhance composite performance when compared to FRC [15]. The mechanical properties of different mixes incorporating various volumetric fractions of ST and PP fibers in hybrid form were also investigated. The results indicate that a hybrid fiber system can lead to similar significant improvements that can also be achieved with the mono fiber system at higher fiber content [16]. Moreover, a study of the mechanical properties and water permeability of fiber reinforced concrete containing hybrid steel- polyvinyl alcohol (ST-PVA) incorporating both macro and micro-fibers showed that the hybridization of macro and micro-fibers in mortar significantly improved the resistance to water permeation. And it also enhanced the mechanical properties of the concrete [17].

The use of short discontinuous fibers as a means for reinforcing concrete for improved performance has been known for several years. These short discontinuous fibers are either used solely or in combination with other industrially manufactured fibers. From the environmental viewpoint, the study examines the potentials of recycling waste materials as a source of fiber for reinforcing concrete. The steel fibers extracted from waste tires, as well as polypropylene fibers extracted from waste carpet, provides an opportunity to reduce the problems associated with waste generation and handling. Besides, the study examines the performance of hybrid short discontinuous fibers obtained from waste and natural means on the mechanical and durability properties of concrete. To this end, the combinations of metallic and natural fibers, as well as metallic and polypropylene fibers, were considered.

2. Materials, Methods, and Testing

2.1. Material

- *Cement*

The cement used was Ordinary Portland Cement of grade 32.5, with a consistency of 27%. It had an initial setting time of 30mins and a final setting time of 450mins. It was purchased from Ado Ekiti local market. The cement properties conformed to British Standard for Portland Cement [18].

- *Steel fibers*

The steel fibers were obtained from waste tires by manually shredding, after which they were cut to a length of 30 mm using the cutting disc. The addition of fibers to the concrete mixes was kept constant at a volumetric fraction of 2% of the concrete volume. Three different categories of the concrete mix were considered in the study. The first category contained a combination of steel and carpet fibers at a volumetric fraction of 1.75% and 0.25%, respectively, while the second category of the concrete mix contained a combination of steel and bamboo fibers at a volumetric fraction of 1% - 1% respectively. Lastly, the third category contained only steel fibers at 2% volumetric fraction

- *Carpet fibers*

Waste carpets were taken from dumpsites located at the Satellite Area of Ekiti State University (EKSU) campus. The shredding of the waste carpet into fibers of length 18mm was done using blade and scissors.

- *Bamboo fibers*

Bamboo trees are located at the Satellite Area of EKSU. Selected trees were cut, after which they were allowed to dry under ambient temperature for weeks. After drying, they were cut into pieces at the sawmill before being shredded into small sizes and length of 30 mm.

- *Superplasticizer*

Superplasticizer was added at 1.5% by weight of cement to improved workability

- *Silica sand*

Silica sand was obtained from Laguna Beach in Lagos state and transported in sacks bags to Ado Ekiti. It was sun-dried and made fine. The silica sand was used as filler in the matrix. A constant proportion of 10% by weight of cement was used.

- *Aggregates*

The fine and coarse aggregates used are stone dust and granite, respectively. The stone dust passed through sieve 4.75mm while passed through sieve size 20mm.

- *Water*

Potable water was used

2.2. *Methods*

The fine, coarse aggregates, filler, cement, superplasticizer, and fibers were weighed using a weighing balance. The casting was done in segments to accommodate the different types of fiber combinations, as presented in Table 1. In the sequence of mixing, cement was mix with 10% by weight of silica sand before the addition of the fine aggregate and coarse aggregate. A constant water-cement ratio of 0.42 was used, and the mixing time was between 3-5mins. The freshly prepared concrete was placed in appropriate molds and compacted in 3 layers.

After 24 hours of casting, the concrete cubes and beams were removed from their respective molds. The weight of each specimen was taken before the concrete cubes and beams were submerged in water for curing. The workability of the concrete containing steel-polymer hybridized fiber matrix had better workability when compared to the steel-bamboo hybridized fiber matrix despite the presence of superplasticizer. This was attributed to the water-absorbing nature of the bamboo fiber. Besides, concrete specimens containing only steel fiber required more setting time for adequate bonding than their hybridized counterpart.

Table 1. Details of concrete mix.

Samples	Cement (kg)	Silica sand(kg)	Water (kg)	SP (kg)	Stone dust (kg)	Granite (kg)	Steel fiber (kg)	Carpet fiber (kg)	Bamboo fiber (kg)
C	3.66	0.37	1.54	0.06	4.77	14.28	0.32	0.11	---
B	3.66	0.37	1.54	0.06	4.77	14.28	0.22	---	0.07
S	3.66	0.37	1.54	0.06	4.77	14.28	0.43	---	---

C = Concrete cubes hybridized with steel and carpet fiber at 1.75%-0.25%, respectively.
B = Concrete cubes hybridized with steel and bamboo fiber at1%-1%, respectively.
S = Concrete cubes reinforced with 2% steel fiber.

2.3. *Testing*

The performance of the hybridized fiber specimens in comparison with the control specimen was done using compressive strength test, flexural strength test, sorptivity test, and susceptibility to acid attack. The compressive strength test was carried at 28 days using the 100 mm cubes, and the average maximum crushing load divided by the area was used to obtain the compressive strength values. The third point bending test with beam size 400 mm by 100 mm by 100 mm was used to determine the flexural behavior of the test specimen. The sorptivity test was done to ensure water penetration resistance. Cubes of 100 mm sizes were coated at the four edges to ensure unidirectional absorption, and after that, the test specimen was dried before being exposed to 10± mm of water. The test specimens were removed from the water and dried with a towel before weighting at time intervals of 15 min, 30 min; 1h, 1.5h, 3 h, 5 h, 8 h, 24 h, 72 h, 96 h, 120 h, 144 h, and 168 h. The test specimens were submerged in water immediately after their weights were taken. The susceptibility to acid attack was done using 100 mm cubes, and the acidic water contained 3% concentrated on hydrochloric acid. Test specimens were in the acidic concentration for 90 and 120 days before testing.

3. Results and Discussions

3.1. *Compressive Strength*

The compressive strength at twenty-eight (28) days for steel-bamboo fibers (B_1- B_3) and steel-carpet fibers (C_1 - C_3) was compared with concrete reinforced with only steel fibers (S_1- S_3) as presented in Table 2. Obviously, the strength values of specimens reinforced with steel-bamboo fibers performed better than the control specimens. Besides, the control specimen showed better behavior than specimens reinforced with steel-carpet fibers. From these results, it can be inferred that the synergy between steel and bamboo was better than that between steel and carpet, which demonstrates the positive effect of hybridization on the performance of concrete. The percentage combination of steel-bamboo fibers was equal; that is 1% each. From this result, it could be inferred that increasing the percentage of the steel fiber beyond 1% may not be necessary because a more significant proportion of the steel fiber, as observed with the control specimen (2%), does not translate to an increase in compressive strength. This assertion is supported by previous research [19]. It has been previously observed that there is a reduction in compressive strength of steel fiber reinforced concrete when the percentage of fiber is increased above 1% [20]. It has also been reported that the maximum increase in compressive strength of concrete reinforced with bamboo fibers is at 1% [21]. Therefore, by combining the optimum steel fibers and bamboo fibers can enhance the compressive strength property of the concrete.

Table 2. Compressive strength of cubes at 28 days.

Sample Number	Mass before immersion (kg)	Mass after immersion (kg)	Volume of cube (m^3)	Density of cube (Kg/m^3)	Crushing load (KN)	Area of cube (mm^2)	Compressive strength (N/mm^2)	Average Compressive strength (N/mm^2)
B1	2.44	2.50	0.001	2500	140.4	10000	14.04	12.8
B2	2.56	2.62		2620	130.4		13.04	
B3	2.48	2.54		2540	113.3		11.33	
C1	2.36	2.46	0.001	2460	60.2	10000	6.02	7.21
C2	2.48	2.56		2560	87.0		8.70	
C3	2.60	2.68		2680	69.1		6.91	
S1	2.66	2.72	0.001	2720	85.1	10000	8.51	8.00
S2	2.56	2.62		2620	78.7		7.87	
S3	2.52	2.58		2580	76.2		7.62	

3.2. *Flexural Strength*

The flexural behavior of the three categories of fiber reinforced concrete considered in the study are presented in Table 3. From this table, it was observed that concrete reinforced with hybrid steel–carpet fibers had the best flexural strength, which is an improvement on the steel fiber reinforced concrete. However, concrete reinforced with hybrid steel–bamboo fibers had the least strength when compared with the others. The enhanced performance may be attributed to the malleable nature of both the steel and carpet fibers since the bamboo fiber is stiffer. Carpet fiber is usually flexible, and this has appeared to affect the bending strength of the hybrid fiber reinforced concrete. Also, the carpet fibers being of a shorter length acted as the micro-fiber to take care of mini-cracks, thereby improving the ductility of the concrete. A previous study on the flexural strength of concrete reinforced with industrial manufactured steel fiber at 1.5% volumetric fraction of concrete gave flexural Strength values less than 8 N/mm^2 at 28 days [22]. The result is similar to that achieved in this study with the hybrid steel- carpet fibers. The advantage presented here is that both hybrid fibers obtained from waste products can be used to positively enhance the flexural behavior of fiber reinforced concrete, which promotes sustainability in the construction industry. Since both waste products are non-biodegradable, their recycling provides an avenue for reducing the environmental challenges associated with its generation and handling.

Table 3. Flexural strength of beams at 28 days.

SAMPLE NO	Mass before immersion (kg)	Mass after immersion (kg)	Depth and width of specimen (mm)	Length (mm)	Maximum load (KN)	Position of fracture "a" (mm)	Flexural strength (N/mm^2)	Average Flexural strength (N/mm^2)
B1	10.78	10.84	100x100	400	11.0	13.9	4.95	5.37
B2	10.94	11.00			14.0	14.0	6.30	
B3	10.34	10.40			10.8	12.8	4.86	
C1	11.06	11.14	100x100	400	16.8	12.7	7.56	7.47
C2	11.07	11.16			16.9	14.5	7.61	
C3	11.58	11.66			16.1	14.4	7.25	
S1	11.08	11.14	100x100	400	15.0	14.1	6.75	6.03
S2	11.84	11.90			12.0	13.0	5.40	
S3	11.32	11.38			13.2	10.8	5.94	

3.3. Water absorption property

The water absorption property of the three categories of fiber reinforced concrete considered in the study is graphically presented in Figure 1. From this graph, it was generally observed that all fiber reinforced concrete attained optimum water absorption at 5 hours, after which a decline in the rate of absorption was observed. More specifically, it was noted that the hybrid steel - bamboo fiber reinforced concrete had better resistivity to water penetration than the hybrid steel – carpet fiber reinforced concrete as well as the steel fiber reinforced concrete. It has been observed that the nature of the concrete pore structure determines the durability of the concrete [23]. Concrete with a more open pore structure is likely to be susceptible to degradation resulting from infiltrating substances. These substances, which may be gases and/or liquids, can penetrate the concrete from the surrounding environment, resulting in physical and chemical reactions within its internal structure, probably leading to irreparable damage [24]. The Water absorption test has also demonstrated the positive effect of hybridization in fiber reinforced concrete since there is a linear relationship between water absorption and porosity of concrete [25]. The ability of hybrid steel - bamboo fiber to outperform steel – carpet fiber may be attributed to the low volumetric fraction of the carpet fiber considered in the study. The results of the study have also indicated that concrete reinforced with only steel fibers or high volumetric fractions of steel fiber is likely to absorb more water.

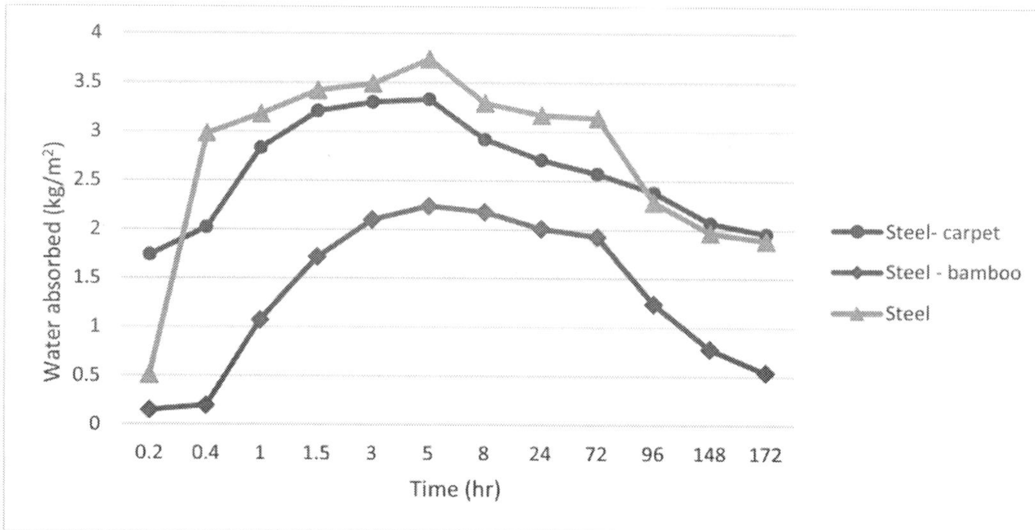

Figure 1. Water absorption of fiber reinforced concrete.

3.4. *Susceptibility to acid attack*

The compressive strength determined after 90 days of submerging the cubes into water containing hydrochloric acid is presented in Table 4. A general trend of an increase in compressive strength for all categories of fiber reinforced concrete was observed when compared to the strength values obtained at 28 days. This is expected with increased curing age. Figure 2 shows the compressive strength values obtained at 28, 90, and 120 days. Furthermore, it was observed that the control specimens reinforced with only steel fiber performed better than their hybrid counterpart. However, the hybrid steel-carpet fiber reinforced concrete was most affected by the acid solution, while the hybrid steel- bamboo fiber reinforced concrete had relatively good strength. The increase in strength of the specimens reinforced with steel fibers as well as hybrid steel- bamboo fiber implies that the acidic solution only had a deteriorating effect on the aesthetic. This is because the cubes were observed to have rusty stains on them, and not much impact on the mechanical properties. It may also be attributed to the concentration of the acid solution, which may not be toxic enough to have a significant effect on the concrete.

The strength gained for steel-carpet fiber reinforced concrete was considered low, which may be due to the carpet fibers been adversely affected by the acid solution. This could have weakened the bond within the concrete matrix. It explains the reason why the steel-carpet fiber reinforced concrete gained weight due to more penetration of the acid solution when compared with their hybrid counterpart. A similar trend as above, was observed at 120 days of curing in acid solution, as presented in table 5. However, the compressive strength was observed to decrease slightly for all types of fiber-reinforced concrete investigated. The percentage decreases in compressive strength from 90-120 days are 2.1%, 0.8%, and 5.8% for steel, hybrid steel-bamboo, and hybrid steel-carpet fiber reinforced concrete, respectively. This implies that as the curing day increased from 90 to 120, the acid concentration began to have an effect on the strength of fiber-reinforced concrete. It could be inferred that concrete reinforced with only steel fiber had better resistance to acid attack than the others. In general, the concrete environment is highly alkaline in nature, with a pH between 12 and 13. This alkaline nature results in the formation of insoluble oxide film on the surface of steel fibers. However, this may sometimes be destroyed by the formation of weak carbonic acid resulting from the reaction between atmospheric carbon

IOP Conf. Series: Materials Science and Engineering 1036 (2021) 012055 doi:10.1088/1757-899X/1036/1/012055

dioxide and water. In conclusion, it has been found that corrosion of the steel fibers is usually confined to the skin of the concrete while the interior fibers enjoy the protective cover of the alkaline environment [4].

Table 4. Compressive strength of cubes after 90days of acid attack.

SAMPLE NO	Mass before immersion (kg)	Mass after immersion (kg)	Volume of cube (m^3)	Density of cube (Kg/m^3)	Crushing load (KN)	Area of cube (mm^2)	Compressive strength (N/mm^2)	Average Compressive strength (N/mm^2)
B7	2.50	2.52	0.001	2520	181.40	10000	18.14	18.90
B8	2.56	2.58		2600	196.80		19.68	
B10	2.54	2.58		2580	188.70		18.87	
C7	2.44	2.52	0.001	2500	138.07	10000	13.81	13.61
C8	2.54	2.60		2580	129.76		12.98	
C10	2.46	2.52		2500	140.48		14.05	
S7	2.56	2.58	0.001	2580	279.80	10000	27.98	27.30
S8	2.74	2.72		2720	268.20		26.82	
S9	2.68	2.66		2660	271.10		27.11	

Table 5. Compressive strength of cubes after 120days of acid attack.

SAMPLE NO	Mass before immersi on (kg)	Mass after immersion (kg)	Volume of cube (m^3)	Density of cube (Kg/m^3)	Crushing load (KN)	Area of cube (mm^2)	Compressiv e strength (N/mm^2)	Average Compressive strength (N/mm^2)
B4	2.46	2.50	0.001	2500	180.01	10000	18.01	17.80
B5	2.60	2.64		2640	178.00		17.80	
B6	2.28	2.32		2320	175.90		17.59	
C4	2.36	2.42	0.001	2420	128.10	10000	12.81	13.50
C5	2.56	2.62		2620	135.80		13.58	
C6	2.66	2.72		2320	141.12		14.12	
S4	2.64	2.66	0.001	2660	258.00	10000	25.80	26.72
S5	2.46	2.48		2480	274.00		27.40	
S6	2.64	2.66		2660	269.60		26.96	

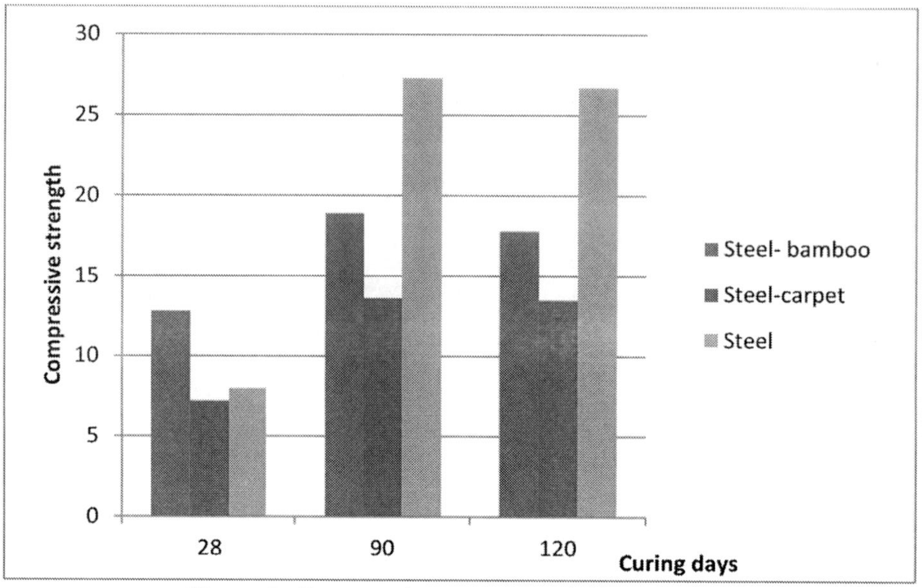

Figure 2. Compressive strength at different curing ages.

4. Conclusion

Hybridization of waste steel fiber reinforced concrete with either natural or synthetic fibers portends a novel way to develop durable eco-friendly concrete that can be used for construction purposes. The following conclusion has been drawn from the study;

i. The combination of 1% steel fiber combined with 1% of bamboo fibers can be used to enhance the compressive strength of concrete.
ii. The combination of steel and carpet fibers can be used to improve the flexural strength of concrete.
iii. The combination of steel - bamboo fibers had better resistivity to water penetration.
iv. Concrete reinforced with only steel fibers can withstand acidic attack when compared to their hybridized counterpart.

References

[1] Ayub T., Shafiq N. and Nuruddin M. 2014 Mechanical Properties of High – Performance Concrete with Basalts Fibers. Fourth International Symposium on Infrastructure Engineering in Developing Countries, **77** pp131-139.

[2] Ayan E., 2004 Parameter Optimization Of Steel Fibre Reinforced High Strength Concrete By Statistical Design And Analysis Of Experiments. Master of Science Thesis. Middle East Technical University, Ankara, Turkey.

[3] Pakravan H., Latifi M. and Jamshidi M 2017 Hybrid Short Fibre Reinforcement System in Concrete: A Review. *Construction and Building Materials* **142** pp280–294.

[4] Nasir B. 2009 Steel fiber reinforced concrete made with fibers extracted from used tires, Master of Science Thesis, Addis Ababa University, Ethiopia.

[5] Syaidathul A. and Izni S. 2012 Mechanical properties of recycled steel fiber in concrete, technical report, Faculty of Civil Engineering, Universiti Teknologi Malaysia.

[6] Ndayambaje J. 2018 Structural performance and impact resistance of rubberized concrete. Master of Science Thesis, Pan-African University, *Institute for Basic Science, Technology, and Innovation.*
[7] Mastali M., Dalvand A. 2016 Use of silica fume and recycled steel fibers in self-compacting concrete (SCC). *Construction and Building Materials*, **125** pp196–209.
[8] Awolusi T, Oke O, Akinkurolere O, and Sojobi A. 2019a Application of response surface methodology: Predicting and optimizing the properties of concrete containing steel fiber extracted from waste tires with limestone powder as filler, *Case Studies in Construction Materials*, 10:e00212.
[9] Awolusi T., Oke, O., Akinkurolere O., and Atoyebi O. 2019b Comparison of response surface methodology and hybrid-training approach of artificial neural network in modeling the properties of concrete containing steel fiber extracted from waste tires, *Cogent Engineering*. https://doi.org/10.1080/23311916.2019.1649852.
[10] Bulei C., Todor M., Heput T, and Kiss I. 2018 Directions for material recovery of used tires and their use in the production of new products intended for the industry of civil construction and pavements. *International Conference on Applied Science*, 294012064.
[11] Scrap Tyre recycling in Nigeria; The Pre- Feasibility Report accessed May 19, 2019. https://foramfera.com/scap_tyre_recycle
[12] Sridhar M. K. C., Hammed T. B. 2014 Turning Waste to Wealth in Nigeria: An Overview. *Journal of Human Ecology*, **46**(2) pp195-203.
[13] Sotayo A., Green S., Turvey G. 2015 Carpet recycling: A review of recycled carpets for structural composites. *Environmental Technology and Innovation*, **3** pp97-107.
[14] Shihong L., Benlian Z, Qiyun Z. and Xianrong B. 1994 A new kind of super- hybrid composite material for civil use – ramine/ A I *Composite*, **25** (3) pp225-228.
[15] Simoes T., Costa H., Dias-da-Costa D., Julio E. 2017 Influence of fibers on the mechanical behaviour of fiber-reinforced concrete matrixes *Construction and Building Materials* **137** pp548–556.
[16] Shah A., Sultan M., Jawaid M., Cardona F. and Taib ARA 2016 A review on the tensile properties of bamboo fiber reinforced polymer composites. *Bioresources com*, **11**(4) 10654-10676.
[17] Yao W., Li J., and Wu K. 2003 Mechanical properties of hybrid fiber-reinforced concrete at low fibre volume fraction. *Cement and Concrete Research*, **33** (1) pp27–30.
[18] Qian C., Stroeven P., 2000 Development of hybrid polypropylene-steel fiber reinforced concrete. *Cement and Concrete Research*, **30** (1) pp63–69.
[19] Lawler J., Zampini D. and Shah S. 2005 Microfibre and macro-fiber hybrid fibre reinforced concrete *Journal Materials in Civil Engineering*, **17** (5) pp595–604.
[20] Yao W., Li, J. and Wu K. 2003 Mechanical properties of hybrid fibre-reinforced concrete at low fibre volume fraction *Cement and Concrete Research* **33** (1) pp27–30.
[21] Brindha M. 2017 Properties of Concrete Reinforced With Bamboo Fibre *International Journal of Innovative Research in Science, Engineering and Technology*, **6**(3) pp3809 - 3812.
[22] Shewata P and Kavilkar R. 2014 Study of flexural strength steel fibre-reinforced concrete *International Journal of Recent Development in Engineering and Technology* **2**(5) pp13-16.
[23] Khan M. 2003 Isoresponses for strength, permeability, and porosity of high-performance mortar. *Building and environment*, **38** pp1051-1056.
[24] Claisse P., Elsayad H. and Shaaban I. 1997 Absorption and sorptivity of cover concrete. *Journal of Materials in Civil Engineering*, **9** pp105-110.

[25] Abo-El-Enein S., Ali A., Talkhan F., Abdel-Gawwad H. 2013 Application of microbial biocementation to improve the physico-mechanical properties of cement mortar. *HBRC Journal.* **9** pp36–40.

Insight on the Electrodeposition Technology and Parameter for Corrosion Control of Structural Steel

O S I Fayomi[1, 2*] D Olusanyan[1], C A Loto[1, 2*] and N E Udoye[1]
[1]Department of Mechanical Engineering, Covenant University, P.M.B 1023, Ota, Nigeria
[2]Department of Chemical, Metallurgical and Materials Engineering, Tshwane University of Technology, P.M.B. X680, Pretoria, South Africa
Corresponding author: ojo.fayomi@covenantuniversity.edu.ng, Ojosundayfayomi3@gmail.com +2348036886783

Abstract. Corrosion has been and still the world's worst nightmare which costs hundreds and thousands of dollars for companies to maintain their equipment and repairs. Studies are being done all around the world to create the solution toward preventing the corrosion by surface engineering technology. One of the simplest and most cost effective methods to provide thin film coating for advance application is through electrolytic deposition route. However, bath framework has been a major concern for stable coating. Thus this studies look into the insight on the Corrosion, electrodeposition and visible additive for zinc plating technology.

Keywords: Electrodeposition, Corrosion, Technology, Parameter

1. Introduction

The importance of structural steel as metals in the world today cannot be overemphasized because of its high strength [1]. It can withstand high stresses, and for this reason, it is employed in the construction of buildings and structure, manufacture of equipment, production of goods and more. The area of application spans from one sector to sector, covering almost all. Despite the desired strength of structural steel, most are highly prone to corrosion, which results in failure and deterioration of parts [1-3].

With this, it can be said that all metals are prone to corrosion as they are usually in contact with the environment. "Metals corrode because we use them in situations where they are chemically unstable [4]. Thus, the ability of a metal to resist corrosion is essential in the metal selection criteria. However, the corrosion-resistant property of a metal is dependent on its structure and composition. To enhance the corrosive resistance of frequently used metals, corrosion control methods have been put in place which include design, materials selection, protective coatings, inhibitors and other means of environmental alteration, corrosion allowance, cathodic protection [5-7] .

The need for the protection of metals is of great importance to engineering since most designs are fabricated with metals because of its high strength and yield properties [8]. Electroplating has proved helpful in controlling the deterioration of metals when exposed to specific environments. In addition to corrosion control, electroplating has other purposes: appearance, unique surface properties and engineering or mechanical properties [9-11]. Some articles are electroplated in order to provide aesthetic value as well as protect the metal from corrosion. In contrast, others are plated to enable joining operations by modification of the metal surface. Sometimes, articles are electroplated to enhance specific properties such as electrical conductivity, etc. or increase the size of the material to enable it to fit into a hole [12]. Electroplating can be the modification of a surface physically to make it suitable for the intended use.

Unlike manufacturing the articles or components from the material with the desired properties, electroplating provides a cheaper and readily available alternative to this choice [13]. Compared to all the different types of electroplating such as nickel plating, silver plating, gold plating, zinc plating is the cheapest and most readily available relative to the others. The significance of this work is in adding to knowledge of the already existing additives used in Zinc Electroplating [13-15]. This would further reduce the cost of acid chloride zinc electroplating baths and waste disposal methods as it also eliminates the use of cyanide, which is harmful to both the environment and human health.

2. Corrosion and the natural processes

Mild steel is an extensively used material that has been involved in all primary forms of constructions on the planet today. Its uses vary from the development of infrastructure, machines, and various material handling situations, such as handling highly corrosive and reactive solutions for chemical and allied industries. Exposure to these environments causes corrosion in varying degrees. This corrosion degrades the material and its properties, thereby making all structures, existing and soon to be built, already subject to a deplorable end [4-6]

The Literature on corrosion and its various techniques of prevention are numerous, displaying multiple experiments and a manifold of observations and useful data. However, the coating process is essential because it factors into the economic section as it prolongs the need to heat the bath and slows production time. On the other hand, electroplating makes up for the cons of the electroless method with being fast and cheap but does not give the same finishing as the latter [7].

Providing coatings that will retain or improve upon the mechanical, chemical, electrical, and morphological structures of structural steel is essential. In addition to offering mild steel with a means to prevent corrosion even further, in the way of "alloying", to give it better hardness, less brittleness, more luster, and better looks. Absolute corrosion resistance has remained unattained and therefore requires t more work, of any nature, be put into finding out the right combination of elements for optimal, if not perfect, corrosion prevention, based on the situation are critical despite the negative effects posed by corrosion degradation due to material failure thereby reducing the economic, health and safety hazards developed due to electrochemical wear. The graphical corrosion analogy of existence is presented in Figure 1.
[2-4].

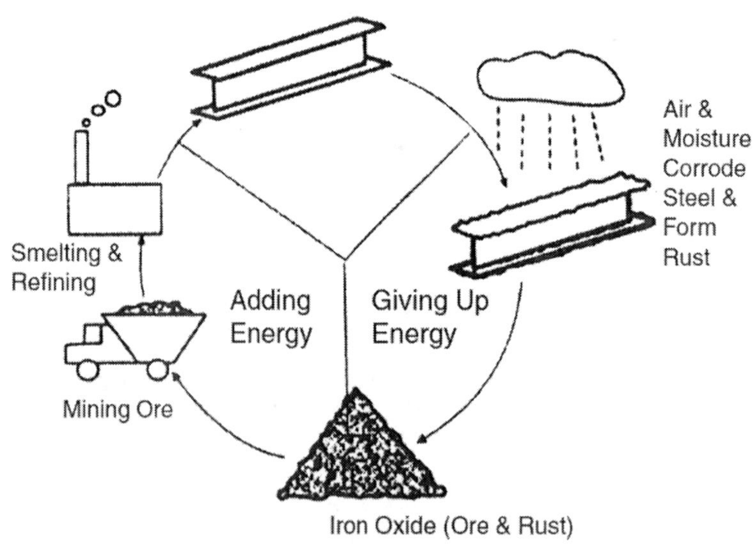

Figure 1. Graphical Representation of Corrosion [14]

Corrosion, therefore, has to do with the combination of materials and the environment. Hence, the corrosion behavior of a material cannot be fully understood without identifying the environment where it is present. The costs employed in treating the waste produced from cyanide-based baths are also of concern [5] as proper discharge and treatment regulations need to be put in place to prevent cyanide's effect on the environment.

3. Additives and bath effect during electrodeposition

Electroplating has also been employed in industries like that of the automobile and Nano and microelectronics because of its aesthetic appearance. This effect is the presence of additives in the electroplating solution, which are called brighteners [5]. For this reason, cyanide has been employed as an additive in electroplating baths because it produces the desired effect. However, cyanide is known to be toxic and poses a hazard to human health. Substantial work has been done in electroplating, also known as electrodeposition, as it is a popular corrosion control method employed in many industries [4]. Over the years, electroplating has evolved from an art to an exact science. Different researchers have investigated the condition at which a well-plated metal is to be obtained.

To reduce the effect cyanide has through electroplating, regulatory laws were put in place to ensure the proper waste treatment to prevent cyanide discharge to the atmosphere and water body. These waste treatment techniques are quite cost-intensive and make the process of plating with cyanide-based baths very expensive. The substitution of cyanide in electroplating baths did not solve all the problems experienced with cyanide-based solutions. Some non-cyanide alternatives pose specific challenges in adaptability with some of the processes and cost and effectiveness [7].

It does not mean that non-cyanide alternatives are of no substantial positive effect but should be a cross-evaluation before they are employed in electroplating solutions. In the use of non-cyanide electroplating baths, additives such as wetters and brighteners are required in answer to produce specific desired properties. Both additives serve as surfactants; however, wetting agents lessen surface topography and lattice tension while brighteners offer necessary appealing worth for the electroplated metal. Organic Compounds have been discovered as brightening agents in electroplating. These also require regulatory measures in terms of waste treatment [8]. The natural organic additives are non-cyanide alternatives in zinc electroplating. It is hypothesized that the addition of plant and fruit extracts as additives would produce a brightening effect on a zinc electroplated mild steel.

With this in mind, alternatives in organic compounds have been discovered to produce the same effect. These alternatives have caused the development of other types of solutions used in electroplating.
The conditions, under which this bath can be used, in area of process issues, have been outlined in the table above. The table makes it easier to analyze both solutions in terms of results and operation comparatively. This allows one to decide on which solution is ideal for the work to be done.

Cetyl trimethyl ammonium bromide (CTAB) and ethylene diamine tetraacetic acid (EDTA) served as brighteners combined to make up a condensation product. The Quantity of the compounds used varies a lot to obtain an optimum value in which the different parameters involved would achieve ideal electroplating. Generally, the common challenge in the bath frame is the control of the zinc metal level, the reason being that there is usually aggressiveness within the zinc anode when concentration rises beyond the expected [13]. This could lead to wastage of the zinc anodes, and to prevent this, specific precautionary measures are being taken. Anode baskets are usually removed at the end of each work shift and before weekends. Previously, the plate's yellowing had been a problem; however, advances in technology have resolved the issue, resulting in excellent brightness and right color in products that have been plated. Another limitation of the alkaline solution is low cathode efficiency, thereby resulting in poor adhesion of deposits. Blistering is also a significant challenge that comes from poor cleaning or intense brightener levels. Surface Cleaning or preparation can be done by either mechanical, chemical means or both. For automatic cleaning, different grades of alumina paper are used in smoothening surfaces. For the chemical means, alcohols like trichloroethylene or acetone are used to degrease the surface and get rid of any oil or grease on the surface. This is necessary because oil or grease does not dissolve in hydrochloric acid which is used for pickling. Pickling is essential part of surface preparation for electroplating because it removes scales and dust from the metal surface. The rinsing of the specimen is also important to prevent the chemicals on the surface from reacting with the environment. This is done immediately after every stage of cleaning.

Acid solutions used in Zinc electroplating are of two types, namely: chloride and sulphate baths. Each has certain advantages over the other. Acid zinc baths are often needed, especially when a high plating rate and low cost are required in plating or component fabrication [12]. There is an interest in using acid zinc sulphate electroplating solutions because of its relatively low cost, pollution characteristics, and safety features. However, it has weak throwing power and insufficient brightness. Hence, it is not ideal because electroplating is employed not only to provide corrosion control but also to provide aesthetic value to the plated metal.

There has been a considerable amount of work done on the effects of additives on electroplated products. Most of these natural or synthetics additives are Secondary active activities because they impart brightness to deposits [4]. In addition to this, cassava and sugarcane juice are commonly used as natural additives. Sugarcane is a right choice as an additional agent in zinc electroplating. It is an essential species that significantly builds up structural properties that subsequently influence corrosion, wear, and physical characteristics. Pineapple is another natural additive that contains both sucrose and glucose; it makes it an excellent choice for an additive in zinc electroplating its high sugar content Tobacco is a product from agricultural source prepared from the plant's leaves. The inhibitive potential of tobacco has been explored; however, the potential as a brightener in Zinc electroplating has not been looked into [5].

4. Conclusion

The development of various techniques for surface enhancement of materials has increased due to the high demand for an excellent performing and high corrosion-resistant materials. This study has successfully established different additive materials used as alternative additives for electroplating route. With the incorporation of surface-active additives, materials with the Zn composite have the potential for excellent strengthening characteristics that could promote corrosion protection

.

References

1. Miller W.S, Zhuang L, Bottema J, Wittebrood A.J, De Smet P, Haszler A & Vieregge A, 2000, Recent Development In Aluminium Alloys For The Automotive Industry, *Materials Science And Engineering*, 280, 37-49
2. Fayomi, O.S.I., Atayero, A.A., Mubiayi, M.P., Akande, I.G., Adewuyi, P.A., Fajobi, M.A., Ayara, W.A. and Popoola, A.P.I., 2019. Mechanical and opto-electrical response of embedded smart composite coating produced via electrodeposition technique for embedded system in defence application. *Journal of Alloys and Compounds*, *773*, 305-313.
3. Abioye, O. P., Loto, C. A., & Fayomi, O. S. I. (2019). Evaluation of anti-biofouling progresses in marine application. *Journal of Bio-and Tribo-Corrosion*, *5*(1), 22.
4. Fayomi, O. S. I., Oluwadare, G. A., Fakehinde, O. B., Akande, I. G., Nwachia, W., Oziegbe, U., & Russell, A. J. (2019). Evolution of physical and mechanical characteristics of deposited composite coatings on A356 mild steel. *The International Journal of Advanced Manufacturing Technology*, *103*(5-8), 2621-2625.
5. Fayomi, O. S. I., Akande, I. G., Abioye, O. P., & Fakehinde, O. B. (2019). New trend in thin film composite coating deposition: A mini review. *Procedia Manufacturing*, *35*, 1007-1012.
6. Abou-Krisha M.M, Rageh H.M & Matter E.A, 2008, Electrochemical Studies On The Electrodeposited Zn-Ni-Co Ternary Alloy In Different Media,*Surface And Coatings Technology*, Vol. 202, 3739-3746

7. Afolabi A.S, 2007, Synergistic Inhibition Of Pottassium Chromate And Sodium Nitrate On Mild Steel In Chloride And Sulphide Media, *Leonardo Electronic Journal Of Practices And Technology*, Issue 11,143-154
8. Azevedo M.S, Allely C, Ogle K & Volovitch P, 2015, Corrosion Mechanism Of Zn (Mg, Al) Coated Steel In Accelerated Tests And Natural Exposure: The Role Of Electrolyte Composition In The Nature Of Corrosion Products And Relative Corrosion Rate, *Corrosion Science*, 90, 472-481
9. De Rincon O, Rincon A, Sanchez M, Romero N, Salas O, Delgado R, Lopez B, Uruchurtu J, Marroco M & Panosian Z, 2009, Evaluating Zn, Al And Al-Zn Coatings Carbon Steel In A Special Atmosphere, *Corrosion And Building Materials*, 23, 1465-1471
10. Doner A, Solmaz R, Ozcan M & Kardas G, 2011, Experimental And Theoretical Studies Of Thiazoles As Corrosion Inhibitors For Mild Steel In Sulphuric Acid Solution, *Corrosion Science*, 53, 2902-2913
11. Fatoba O.S, Popoola A.P.I, Fedatova T & Pityana S.I, 2015, Electrochemical Studies On The Corrosion Behaviour Of Laser Alloyed Zn-Sn Coatings On Uns G10150 Steel In 1m Hcl Solution, *Silicon*, 7, 357-369
12. Gautam V, Patnaik A & Bhat I.K, 2015, Thermo-Mechanical And Fracture Characterization Of Uncoated, Single And Multilayer (Sin/Crn) Coating On Granite Powder Filled Metal Alloy Composite, *Silicon*, 3, 9318-9328
13. Marder A.R, 2000, The Metallurgy Of Zinc-Coated Steel, *Progress In Material Sciences*, Vol. 45, 191-271
14. Popoola A.P.I, Fayomi O.S.I & Popoola O.M, 2012, Comparative Studies Of Microstructural, Tribological And Corrosion Properties Of Plated Zn And Zn-Alloy Coatings, International Journal Of Electrochemical Science, 7, 4860-4870
15. Wang X.Y & Li D.Y., 2003, Mechanical, Electrochemical And Tribological Properties Of Nano-Crystalline Surface Of 304 Stainless Steel, *Wear*, 255, 836-845.

Prediction of a singly reinforced concrete beam steel using artificial neural network

S O A Olawale[1], O P Akintunde[1,2], M O Afolabi[1] and O A Agbede[1]

[1]Department of Civil and Environmental Engineering, Bells University of Technology, km 8, Idiroko, Benja Village, P.M.B. 1015, Ota, Nigeria

[2]Corresponding author: akintt18@gmail.com

Abstract. This paper presents the prediction of a singly reinforced concrete beam using Artificial Neural Networks (ANN). The method was adopted for cost optimization of the structural element and compared with the requirements of Eurocode 2 design. The code provisions for the design of a singly reinforced beam can vary from place to place. The use of a system immune from the code variation is an excellent means of predicting the reinforcement's need of a rectangular concrete beam. In this work, an artificial neural network (ANN) is employed to forecast the reinforcement of such a beam. Artificial neural network has the potential to simulate the data that are hard to produce in arithmetical analysis. The scheme was established using the MATLAB tool kit. The design variables were the depth of the beam, the width of the beam, and the moments. A forward pass supervised backward propagation training. The regression analysis of the results is one to one match. The predicted and target values are completely in accord.

Keywords: Artificial neural network (ANN), regression, forward pass, backward propagation, Eurocode 2 (EC2), MATLAB.

1. Introduction

In many countries, concrete is the main construction materials and the structural design of the structural elements is code-based. There is variation across the board from country to country without a unified approach. The British (BS 8110) and the European Union (EC2) codes are similar in outlay and considerations [1,2]. However, the American code (ACI) is a complete digression but with reasonable similar outputs [3]. It is reasonably difficult to harness all the codes available today to be a single universal source of design provisions. The British and EC codes are almost the same with slight variations of some prescribed constants. Artificial neural network provides a means of a quick prediction of the required reinforcement once trained successfully. The first work on human brain activity patterns was published by William James in 1890 [8]. The brain constitution of neurons and synapses are the basis for the development of an artificial neural network. ANN is modeled on the principle of brain workings. This has proven to be of immense contribution to modeling complex and difficult systems to a very good accuracy. Warren McCulloch and Walter Pitts [9] attempted to demonstrate how the brain can work using electrical circuitry whereby a summation over-weighted inputs and the activated output using this input. By 1951, Marvin Minsky [10] developed the first ANN. This is an abstraction of the human brain functionality. The human brain is made up of interconnected neurons sending signals through the synapses to one another. This is the model captured in the ANN whereby the weighted biased inputs are aggregated into the each neuron with activation functions to transfer the outputs to the next layer of neurons until the output layer neurons compute the final outputs.

ANN can be a single layer or multi-layer network that can be supervised or unsupervised trained. The supervised training is more suited to a predictive neural network. The training algorithm is weight adjustment on backpropagation. The weight adjustment value calculation is based on gradient descent whereby training rate value and the descent direction are optimized. The inputs into the ANN can be one or many depending on the problem at hand. The outputs also can be one or many as well. Most of the structural predictive neural networks are two or three-layer networks depending on the nomenclature of definition. The input layer is often not seen as part of the layers in the network. Many researchers have applied ANN to solve structural engineering problems relating to either strength, flexural or design predictions. [1] applied ANN to predict the flexural design of concrete hydraulic structures and they established a good agreement between calculated and predicted values. [2] Successfully applied ANN to assess the load-carrying capacity of reinforced concrete members. They found an excellent agreement between the calculated and predicted results. [3] Applied ANN to design some reinforced concrete members. Despite their limited amount of data to train their network, the agreement between their AN predicted values and calculated ones is reasonable. Works on the prediction of compressive strength, shear capacity, and tensile strength of the reinforced concrete beams have been explored to a greater extent. One such is the work of Vingnesh, Alisha, Karthik, Shraddha, and Skanda [7] whereby the compressive strength of concrete was predicted using ANN [4].

In the present study, the appropriateness of ANN to predicting the reinforcement of a concrete loaded to less than the ultimate moment carrying capacity is investigated. The sensitivity of the number of neurons in the hidden layer is investigated to determine the most efficient network topology without overtraining.

2. Materials and Methods

The materials required to prospect the present work are excel and MATLAB software. The provisions of EC2 are implemented int the excel spreadsheet calculations to generate training data for the network. Two-layer networks with three inputs (width, depth, and moment), one hidden layer of number of neurons to be determined through parametric investigation and one output, the steel ratio. The topology used is shown in figure 1 and the computational model for the forward propagation is shown in figure 2. The activation function used is simoid function and the training algorithm used is Levenberg-Marquardt. The training data was generated using excel whereby the EC2 design provisions were implemented. The three variables considered are width, depth, and moment. The plots of the training data are presented in figures 3 to 5. The training is data is partitioned into 70, 15, and 15 percentages for training, validation, and testing respectively. The MATLAB ANN tool kit is used to implement the scheme. The choice of the network in terms of the number of neurons in the hidden layer is based on the distribution if the error on the zero error and the performance value of the network. The number of neurons with less skewness of the error histogram and with the best performance value will be selected to be the representative network.

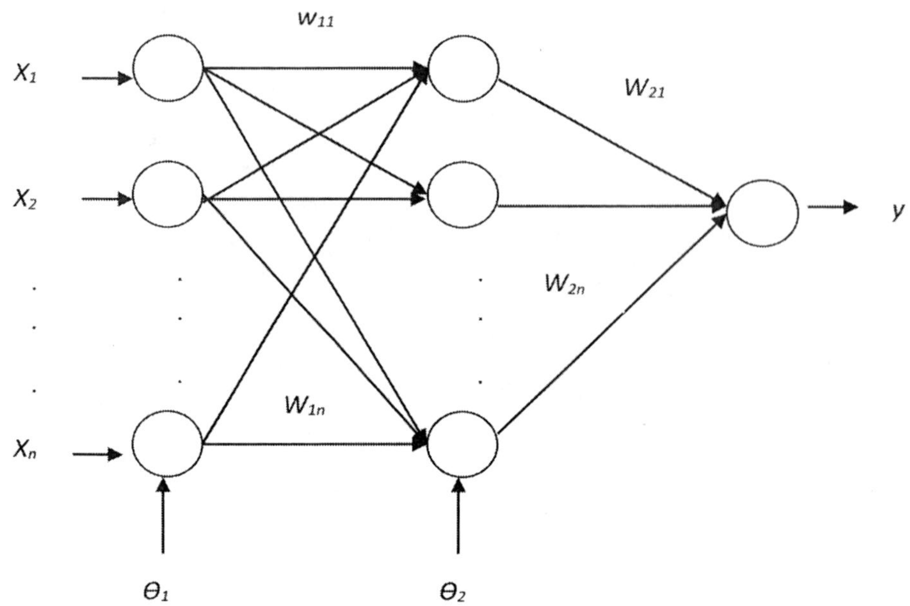

Figure 1. Neural Network Topology adopted

The number of input is 3 and the numbers of neurons in the hidden layer are between 5 and 15 neurons.

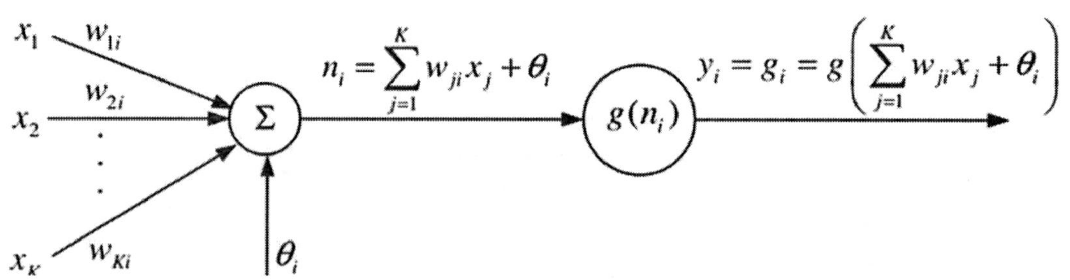

Figure 2. Typical computation scheme for the ANN (forward pass)

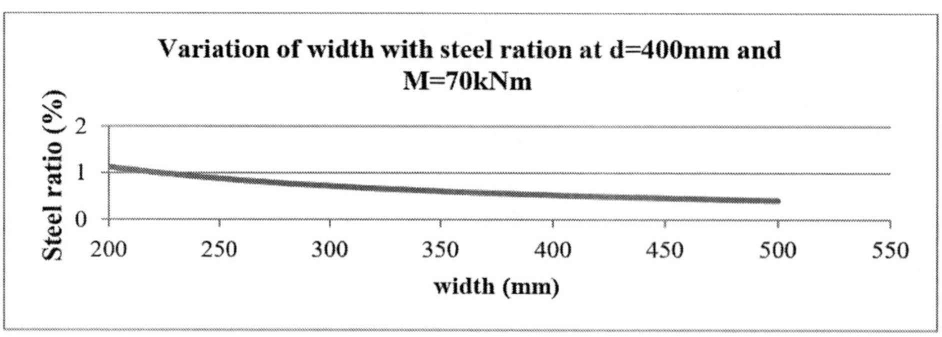

Figure 3. Width to steel ratio variation

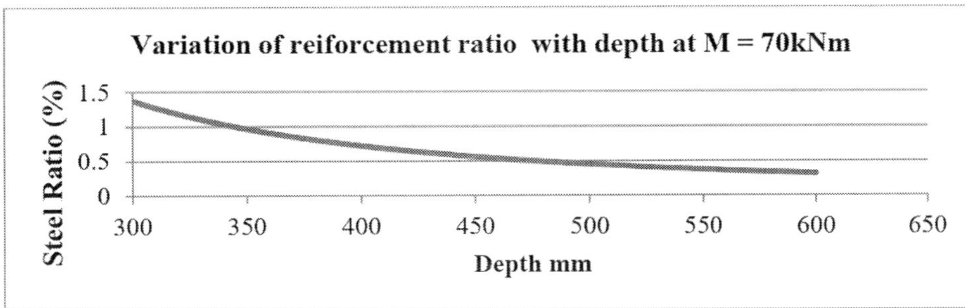

Figure 4. Depth to steel ratio variation

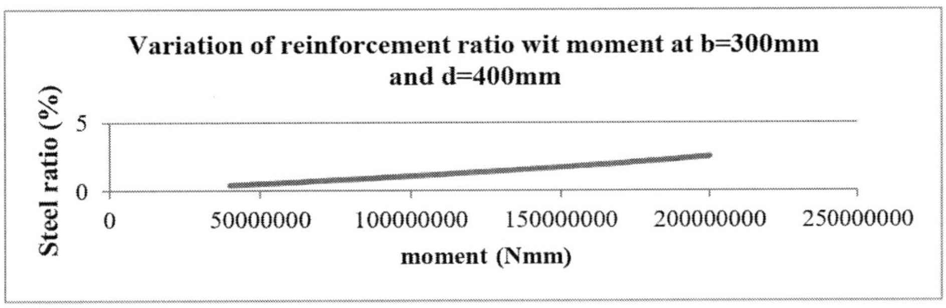

Figure 5. Moment to steel ratio variation

3. Results and Discussion

The results of initial experimentation to determine the number of neurons in the hidden layer are presented in Figures 6 to 8. The error histogram for ten neurons exhibits the least skewness around the zero error. This means that ten neuron hidden layer is chosen.

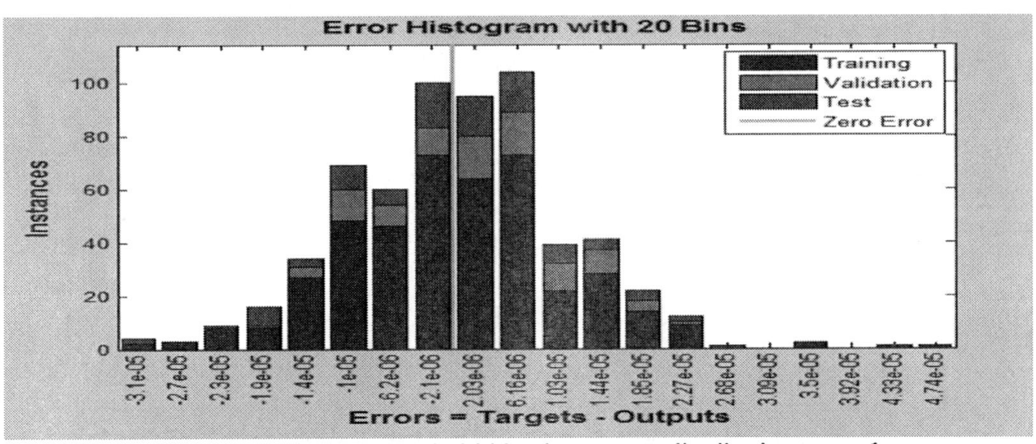

Figure 6. Five neurons in hidden layer error distribution around zero error

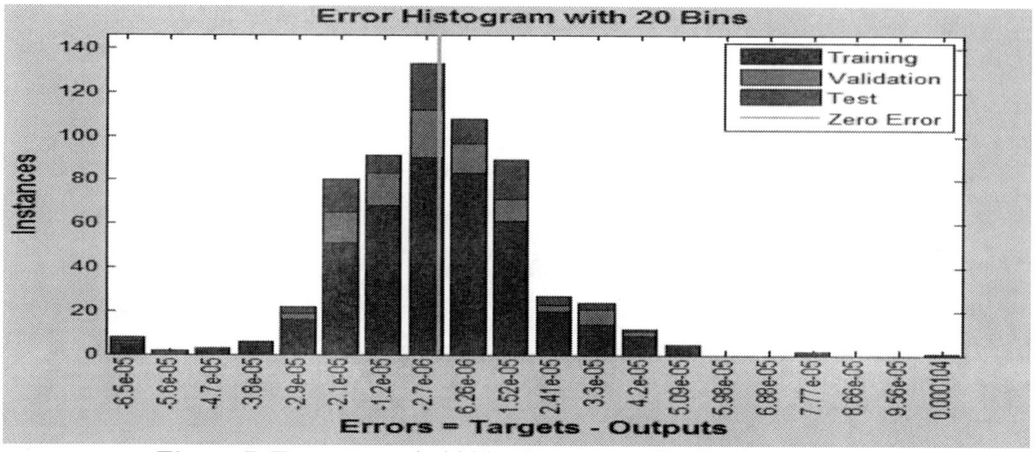

Figure 7. Ten neurons in hidden layer error distribution around zero error

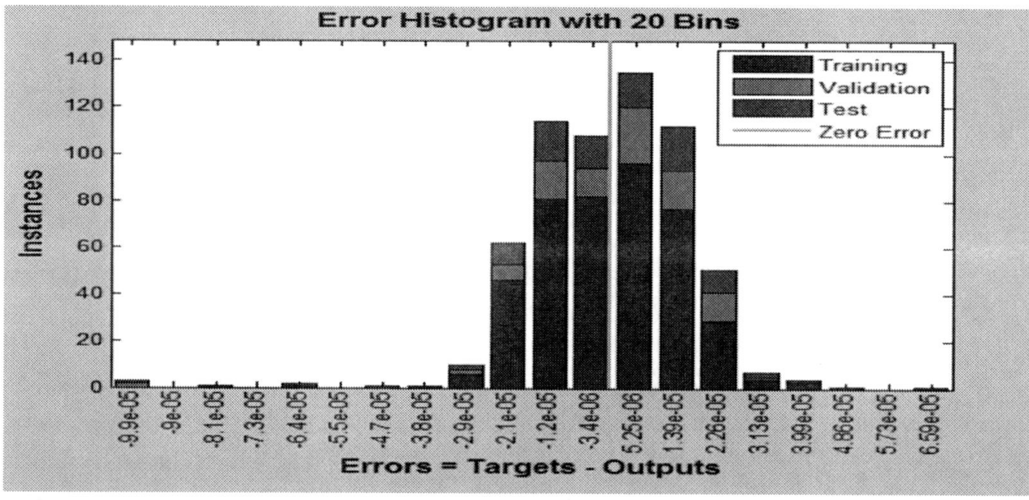

Figure 8. Fifteen neurons in hidden layer error distribution around zero error

Figure 9 shows the final adopted network implemented to predict the steel requirement for concrete loaded below its maximum load-carrying capacity. The front page of the MATLAB ANN tool is shown in figure 10 and the performances of the training, validation, and testing are shown in Figure 11.

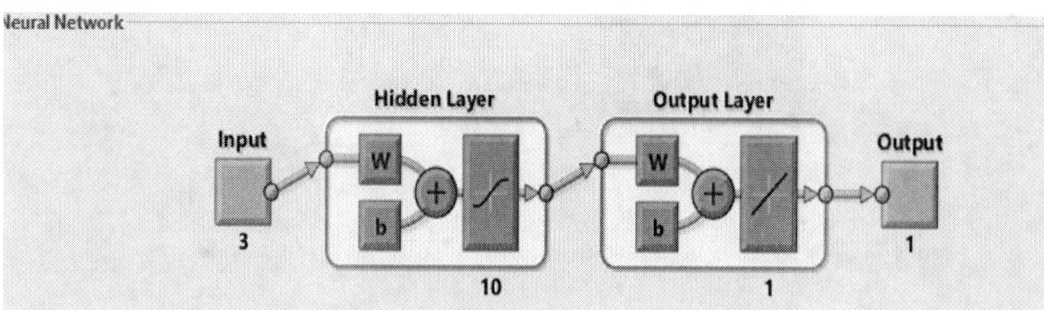

Figure 9. Neural Network used in the prediction of the reinforcement

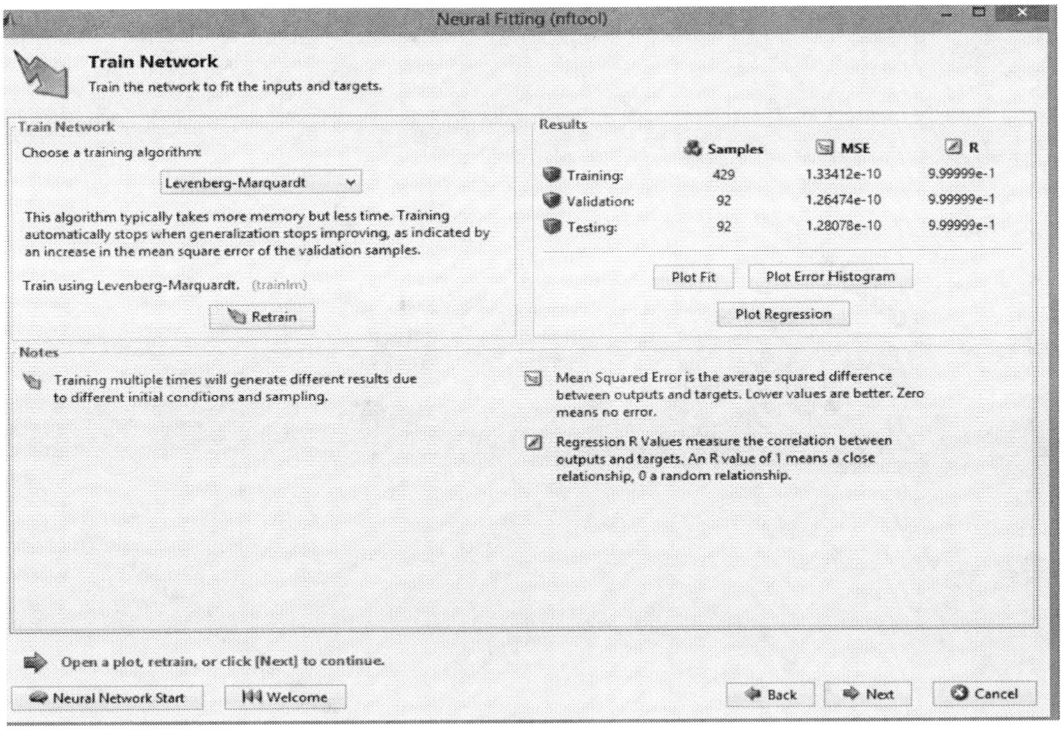

Figure 10. ANN Tool page showing the information on the Error and Activation function

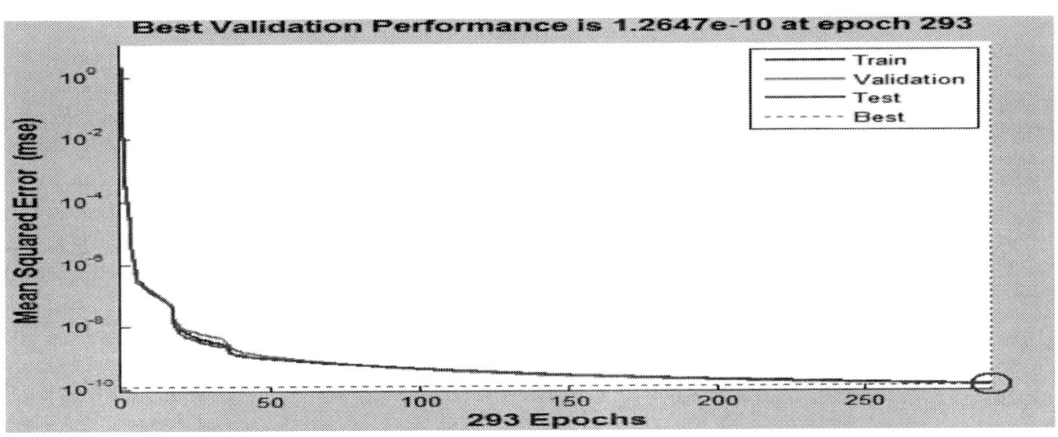

Figure 11. Performance evaluation of the training, validation and testing

The regressed predicted results against the calculated results in Figure 12 for the training, validation, testing, and all. The excellent agreement between the computed and the predicted results are outstanding.

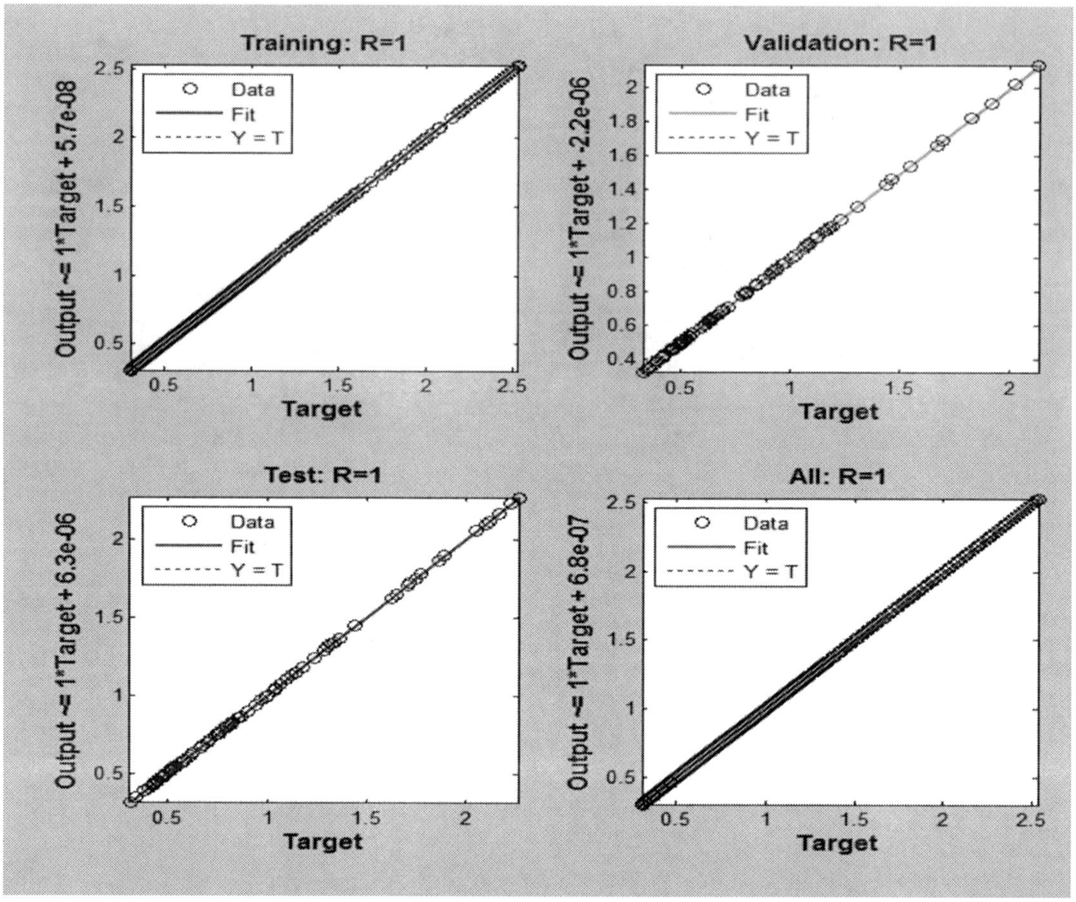

Figure 12. The regression of plots of training, validation, testing and all

4. Conclusion

From the results of the previous section, it is obvious that the agreement between the computed and predicted results is excellent. It has been established that ANN can be used to design a singly reinforced rectangular beam without reference to any code of practice. The requirements of the codes are satisfied with the optimum result obtainable. This leads to the cost minimization of the reinforced concrete beam. Artificial Neural Network is beneficial in terms of producing original outputs from the training of given samples through accurate prediction of results thereby analyse information very rapidly and adjust to modifications in the initial problems.

References

[1] British Standard Institution 1997 BS 8110-1: Structural Use of Concrete: code of practice for design and construction. Available at: http://www.ucl.ac.uk/library/electronicresources/databases

[2] EC 2 1992-1-1 2004 English Euro code 2. Design of concrete structure – Part 1-1. General rules and rules for buildings 2004, Avenue Mamin 17B-1000 Brussels

[3] ACI Building Code Requirements for structural concrete ACI 318-08 and Commentary in ACI 318-08 2008. American Concrete Institute, 38800 Country Club Drive, Farmington Hills, MI 48331 p1-471
[4] Oday M Albuthbahak and Hayder H Alkhudery. (2018). Artificial Neural Network Model for Flexural Design of concrete hydraulic structures. *International Journal of Journal of Civil Engineering and Technology (IJCIET)*, **9**(2), 265-274 ISSN 0976-6308.
[5] Afaq Ahmed, Gregoria Kotsovos, Demitrios M cotsovos and Nikos D Lagaros. (2016). Assessing the load carrying capacity of RC member through the use of Artificial Neural Network 11th HSTAM of international Congress on Mechanics Athens 27- 30 May 2016 Greece
[6] Chandar M K, Raghim P B K and Amarnath K. (2017). Design of reinforced concrete structure using Neural Networks, *International Research Journal of Engineering and Technology (IRJET)*, **4** (7), July 2017, ISSN 2395-0056
[7] Vignesh Shenoy B, Alisha B B, Karthol, Shraddha Pai and Skanda Prasad. 2016. Prediction of compressive strength of concrete by Artificial Neural Network, *International Journal of infrastructure and futuristic research*, **3**(9), May ISSN 2347-1697.
[8] James., W. 1890. The principles of psychology. New York: Holt; 1890.
[9] McCulloch., WS, Pitts., W. 1943. A logical calculus of the ideas immanent in nervous activity. Bull Math Biophys, 5:115
[10] Minsky., M. (1951). The Study of Instinct. Oxford University Press, New York, N. Y., 1951

IOP Conf. Series: Materials Science and Engineering 1036 (2021) 012058 doi:10.1088/1757-899X/1036/1/012058

Factors affecting sustainable infrastructure delivery in South Africa - a case of Gauteng province

J Jacobs[1], C Aigbavboa[1], W Thwala[1], K Dithebe[1]

[1]SARCHI in Sustainable Construction Management and Leadership in the Built Environment, Faculty of Engineering and Built Environment, University Of Johannesburg, Corner Kingsway Avenue and University Road, Auckland Park, South Africa

jabulanijacobs@gmail.com, caigbavboa@uj.ac.za, didibhukut@uj.ac.za, khdithebe@gmail.com

Abstract. Infrastructure plays a crucial role in the economic growth and development of any country; therefore, it is important to ensure adequate investments for the delivery of infrastructure services. This paper assessed factors affecting the development of sustainable infrastructure in South Africa. The identified factors can be utilised by government as reference to enhance the delivery of sustainable infrastructure. To achieve the objective of the study, Quantitative data was collected. The study used a sample size of 57 out of the 65 questionnaires (a response rate of 87.7%) which were answered by professionals in the built environment viz. Architects, Property Developers, Quantity Surveyors, Engineers, Construction and Project Managers. The data collected was then analysed and interpreted using IBM Statistical Package for Social Science (SPSS) computer software to provide mean item scores, standard deviations and Cronbach's alpha. Given the explored factors which affect the delivery of sustainable infrastructure, the results have established that due to developers habit of building to maximise profit only, not for end-users' comfort and the initial high cost required to develop sustainable infrastructure were the major factors that affect the delivery of sustainable infrastructure. It is recommended that the government develop clear, stable and credible policies that will attract investors who will be committed to inject financial resources into infrastructure development and alter the existing building regulations to complement the sustainable development goals.

1. Introduction

Sustainability revolution is taking place and it will require the built environment, especially construction industry, to find solutions to living in rapidly urbanizing areas and cities. The way building materials are used and how construction is conducted shall change to keep up with the new era, therefore creative thinking is required in order for people to survive and thrive on planet earth for much longer and the construction industry has a major role to play [1]. Yan, Wang, Quan, Wu & Zhao [2] define sustainable development as a development approach that satisfies the needs of our day-to-day lives, and the well-being of the future generation. According to Yan *et al.* [2], Sustainable Development strive to balance the needs of this and the upcoming generation as well as regional and global needs, nature and development, and competence and equality.

Gauteng is the smallest province in South Africa, covering about 15% of the country's total land area however, it population is greater than that of all the provinces in the country as one-fourth of the South Africa's population is based in Gauteng [3]. World Population Review [4] stated, "The City of Johannesburg needs a plan for a population growth of about 66% in the next 30 years, which includes plans to improve access to clean water, energy and the management of waste and sanitation. Johannesburg is projected to grow to 6.5 million by 2040."

IOP Conf. Series: Materials Science and Engineering 1036 (2021) 012058 doi:10.1088/1757-899X/1036/1/012058

The Minister of Human Settlement Lindiwe Sisulu in Petterson [5] mentioned that South Africa is need of solutions to plan and manage rapidly growing population and urbanization related challenges that the country is facing. The Minister stated that, "The real challenges faced by urbanization, when its positive aspects are recognized, are sustainable in the social, economic and environmental dimensions. Sustainable urban planning and development is necessary to eliminate the causes of segregation and exclusion and support the social, spatial and economic transformation of cities and towns." The minister further added that if proper measures are not followed during development stage cities suffer the consequences [5]. The projected rapid growth of population in the Gauteng province in the next 30 years should be the main drive for the government to implement systems that will mitigate factors affecting sustainable infrastructure delivery in the province. Achieving that, the government will not only be able to accommodate the large number of individuals who are moving into the province but will also be building sustainable societies with decent facilities and access to economic opportunities [4].

In the past two decades, substantial amount of construction work have been undertaken close the gap between the demand for infrastructure services associated and the rapidly growing population, improvements in the technological, communication and digital systems. However, the lack of monetary valuation studies and scarcity of investments towards sustainable infrastructure has made it very challenging for the province to deliver sustainable infrastructure [6]. Through sustainable infrastructure development, the design and construction phases are exploited to improve energy efficiency as well as indoor environmental quality, increase the use of renewable resources and reduce waste. These practices can be used to improve the sustainability of the existing infrastructure until the required technologies are available [7].

2. Literature review

Sustainability is a wide-ranging concept that has become a major concern in the built environment, sustainability allows human beings to have an enhanced comfort and health, improved social, economic and environmental conditions, and it improve overall quality of life. Meadowcroft [8] and Asdrubali [9] define Sustainable Development (SD) as a development approach that satisfies the needs of day-to-day lives, and the well-being of the future generation.

There issues of accelerating sustainable development, eliminating poverty and climate change are growing in importance and have become inextricably linked. Poverty reduction, economic growth, and environmental sustainability can be tackled successfully through infrastructure [10]. According to Shazmin, Sipan and Sapri [11] sustainable development can improve the way the building uses the resources, concurrently reduces the impact of the building on human health as well as on the environment during the existence of the infrastructure through improved design, development, maintenance and disposal of the building.

According to Gauteng Provincial Government [6] there are no sufficient investment and commitment to sustainable infra-structure development from both public and private sectors. More infrastructure, precisely sustainable infrastructure, is required in order for the provincial government to be able to meet consumption demands of the rapidly growing population. The development banks need to invest more in infrastructure development in order to assist in building improved infrastructure [10]. Infrastructure development has a massive impact in improving the standard of living in developing countries however due to lack of financing the much-needed projects do not become a success [12]. It is not a secret that the public sector is the largest investor in infrastructure development. How-ever, the projects funded by the government lack vision and planning due to weak regulatory frameworks and limited local capacity, most of those pro-jects does not succeed [12]. Levy [12] suggested that less developed countries formulate policies and regulations that will attract foreign investors to finance mega infrastructure projects. A study conducted by Abi-ad, Furceri and Topalova [13] found that when the investments in the infrastructure development are increased the Gross Domestic Product (GDP) of the country can improve as well as its economy's productive capacity. If the GDP in investment spending increases by 1% the level of output will go up by 0.4% in one year. The investment in infrastructure development could generate income and pay for itself if it is done correctly. A number of pro-jects are not completed over anticipated period, they usually take longer. In most cases it is the government that changes policies during the construction and operating phases of the project e.g. change in pricing rules of electricity or water. Investors do not like to rely on rules that apply to a certain investment or investment technique [10].

Ghanen [14] established that the population growth, especially in developing countries, has a negative impact on achieving sustainable development due to an existing inverse relationship between the population and the environment and its impact on sustainable development, using a multi-equation model in Egypt. According to Ghanen [14] findings when a population increases by 1% the carbon dioxide (CO_2) emissions goes up by 2.4%, when CO_2 emissions increases by 1% the deaths caused by air pollution goes up by 2.5% and the poor health that is caused by air pollution reduces labour productivity by 1.58%. Ghanen [14] further noted that even though population growth has an indirect negative impact on decreasing labour productivity, it still affects economic growth and delays the state from delivering sustainable development. In addition, Guney [15] assessed the population growth and its impact on chances for achieving sustainable development in developed and developing countries, the variables have established that population growth has a negative impact and a positive impact in developing and developed nations, respectively.

The delivery of sustainable infrastructure requires a government to produce clear, stable and credible set of policies and initiatives in the field of sustainable construction [16]. In addition, government should stimulate organisations in the built environment that will proactively support and monitor sustainable development and set their own targets, rather than only complying with the government policies and legislation [16].

The number of sustainable infrastructure development is been slightly in-creasing in South Africa since the formation of the Green Building Council of South Africa (GBCSA) in 2007, almost 200 buildings have been awarded Green Star SA certification and over 7 000 professionals have undertaken GBCSA training [17]. However, the adoption of sustainable infrastructure development is still increasing steadily, the studies locally and internationally has revealed that one of the reasons that there is a low progress is the perception that the initial cost of sustainable development ranges between 15% and 25% higher than the initial cost of traditional construction [17]. Likewise, the study of Mustapha [18] revealed that one of the factors that hinder the progress of sustainable infrastructure development in South Africa is the ideology that the development of sustainable buildings is too costly. The lack of information about sustainable development benefits to the clients, and that developers build to enhance profits only not the end users' comfort affects sustainable infrastructure delivery in South Africa [18].

3. Research methodology

The research methodology adopted to collect, analyse and interpret data was carefully selected in the best way that will fulfil the purpose and the objective of the study. The data for this study was collected in the Gauteng province, mainly in the City of Johannesburg as the population is well represented in the city and quantitative approach was used to conduct the study. The closed-ended format questionnaires were distributed to respondents who are professionals in the built environment, the respondents completed the questionnaires using a 5 point Likert scale to rate their extent of agreement or disagreement with the variables that are stipulated in the questionnaires. The data collected was then analysed and interpreted using IBM Statistical Package for Social Science (SPSS) computer software to provide mean item scores, standard deviations and Cronbach's alpha.

The non-probability sampling was employed to select the sample for the study, using purposive sampling. The sample size of 65 respondents who are based in the Gauteng province were approached to form part of the study, however only 57 of the 65 completed and returned questionnaires (a response rate of 87.7%). The respondents for the study were Architects, Property Developers, Quantity Surveyors, Engineers, Construction and Project Managers as well as professionals in the built environment who are working for the various Development Agencies in the Gauteng province viz. Johannesburg Development Agency (JDA), Johannesburg Housing Company (JOSHCO) and Johannesburg Road Agency (JRA).

4. Findings and discussions

4.1 General information and results
The findings that were drawn clearly showed that 40% of the respondents that formed part of the study had national diploma as the highest educational qualification, 34% had bachelor's degree, 22% had honour's degree, 10% had master's degree, and 4% had higher certificate and 4% doctorate. Thirty percent of the respondents were Quantity Surveyors, 19% were Engineers, the Architects and Construction Project Managers were represented by 18% of the 57 respondents, 11% were Construction Managers and 5% were other professionals like Facilities Managers, Architectural Technologists and Land Surveyors. The findings extracted from the questionnaires indicated that 32% of the 57 professionals that took part of the study had more than 12 years of experience in the construction industry, 26% had between 4-6 years of experience, professionals who had between 1-3 years, 7-9 years and 10-12 years of experience were represented by 14% of the 57 respondents each. The findings further revealed that 61% of the respondents that took part in the study were employed in the private sector and 39% were employed in the public sector. While the findings for the respondent's previous involvement in the sustainable infrastructure projects show that 32 of the respondents had previously been involved in the sustainable projects, which is 56% and 25 were not, which is 44%. Out of the 32 respondents that have been involved in the sustainable infrastructure projects before 34% have been involved in five projects and more, while the respondents who were involved in one project and two projects before each had 22%. Sixteen percent of the respondents had been involved in three projects before and 6% of the respondents had been involved in four projects. While the 34% of the respondents noted that, the sustainable infrastructure projects that they were involved in were located in Gauteng, and 16% in KwaZulu-Natal and 13% outside South Africa. Free State and Mpumalanga had 9% of the respondents each, 6% of the respondents indicated that the projects that they were involved in were located in the Western Cape. Eastern Cape, Northern Cape and North West had 3% each. These results give a clear indication that the respondents had a sufficient working experience and the adequate academic qualifications to understand matters concerning the delivery of sustainable development in the built environment. Therefore, the response received from these respondents can be deemed reliable based on the results obtained.

Table 1. Factors influencing sustainable infrastructure development in South Africa

Factors influencing sustainable development	Standard deviation	Mean item score	Ranking
Developers build to maximise profit only, not for users' comfort	0.909	4.32	1
High construction cost	1.067	4.07	2
Lack of investment	1.306	3.84	3
Resistance to change from current practice	1.093	3.81	4
Inadequate project development capacity	1.142	3.35	5
Population growth	1.295	3.30	6
Policy uncertainty	1.195	3.23	7
Lack of public awareness	1.203	3.02	8
Scarcity of high-efficiency products suppliers	1.136	2.82	9
Clients lack of knowledge on sustainable development	1.481	2.81	10

The findings were in partial agreement with the study of Mustapha [18]. The study showed that clients lack of knowledge about SD, developers build to enhance profits only and does not take into consideration the end-users' comfort as well as the scarcity of high–efficiency products suppliers in South Africa are some of the factors affecting sustainable infrastructure delivery in South Africa. The findings revealed the clients' lack of knowledge about SD and scarcity of high-efficiency products are not likely to affect the delivery of sustainable infrastructure. However, the findings strongly agreed with the study that developer's habit to build to maximise profit only and not consider users' comfort and the belief ideology that the high construction cost to develop sustainable buildings are very likely to affect the delivery of sustain-able infrastructure. In addition, the study of Guney [15] was in total agreement with the findings that population growth does affect the delivery of the sustainable infrastructure, Guney [15] revealed that population growth has a negative impact and a positive impact in developing and developed nations, respectively. Likewise, the study of [10] and the findings were in consensus that lack of investment has a very high level of influence in the delivery of the sustainable infrastructure; as Bhattacharya *et al.* [10] stated in the study that infrastructure require a sufficient investment.

5. Conclusion and recommendations

The factors influencing sustainable development in the Gauteng province were discovered through the literature review that was compiled, the following are those factors in their descending order: Developers build to maximise profit only, not for users' comfort; high construction cost; lack of investment; resistance to change from current practice and inadequate project development capacity. Furthermore, population growth; policy uncertainty; lack of public awareness; scarcity of high-efficiency products suppliers and clients lack of knowledge on sustainable development were other factors influencing sustainable development that were revealed by the literature. The findings that were discovered during the analyses of the data that was acquired from the questionnaires completed by the respondents was comparable with the literature review that was drawn from the different historical studies, they were able to assess possible factors affecting sustainable infra-structure delivery in South Africa.

The identified factors can be used by the provincial government as an appropriate reference to deliver sustainable infrastructure at larger scale, which will help improve the quality of life in communities, not only socially but also by being environmental friendly and make a positive contribution to the economy.

References

[1] Drew, J., 2015. Glimpsing a sustainable future. *Skills on Site*: 25 – 26

[2] Yan, Y., Wang, C., Quan, Y., Wu, G. and Zhao, J., 2018. Urban sustainable development efficiency towards the balance between nature and human well-being: Connotation, measurement, and assessment. *Journal of Cleaner Production,* 178:67-75.

[3] Statistics South Africa, 2017. *Mid-Year Population Estimates 2017.*

[4] World Population Review. n.d.. [Online] Available from: http://www.worldpopulationreview.com/world-cities/johannesburg-population/. [Accessed 28 August 2018].

[5] Petterson, D., 2017. Urbanisation-a tool for combating poverty. *Infrastructure News,* [Online] Available from: https://www.infrastructurene.ws/2017/08/19/urbanisation-a-tool-for-combating-porverty/ [Accessed 17 June 2018].

[6] Gauteng Provincial Government (2014). *Gauteng Provincial Government Review of 20 Years of Democracy.* Johannesburg: Gauteng Provincial Government.

[7] Pearce, A.R., Ahn, Y.H. and HanmiGlobal., 2012. *Sustainable buildings and infrastructure – Paths to the future.* Oxford: Routledge

[8] Meadowcroft, J., 2000. Sustainable Development: A New(Ish) Idea for a New Century?. *Political Studies,* 48(2): 370-387.

[9] Asdrubali, F., 2006. Survey on the Acoustical Properties of New Sustainable Materials for Noise Control. In: *proceedings of the Euronoise 2006.* Perugia: University of Perugia.

[10] Bhattacharya, A., Oppenheim, J. and Stern, N., 2015. *Driving sustainable development through better infrastructure: Key elements of a transformation program*. (Report No. 91). Brookings: Brookings publication

[11] Shazmin, S., Sipan, I., Sapri, M., Ali, H. and Raji, F., 2017. Property tax assessment incentive for green building: Energy saving based-model. *Energy*, 122: 329-339.

[12] Levy, E., 2016. The Lack of Investment into Infrastructure in the Developing World.

[13] Abiad, A., Furceri, D. and Topalova, P., 2015. *The Macroeconomic Effects of Public Investment: Evidence from Advanced Economies*. International Monetary Fund.

[14] Ghanem, S., 2016. The relationship between population and the environment and its impact on sustainable development in Egypt using a multi-equation model. *Environment, Development and Sustainability*, 20(1): 305-342.

[15] Guney, T., 2017. Population growth and sustainable development in developed-developing countries: An IV (2SLS) approach. Suleyman Demirel University, 22(4): 637-662

[16] Dobson, D., Sourani, A., Sertyesilisik, B. and Tunstall, A., 2013. Sustainable Construction: Analysis of Its Costs and Benefits. American Journal of Civil Engineering and Architecture, 1(2):32-38.

[17] Hoffman, D. 2016. *Green building costs in South Africa*. Ph.D. University of Pretoria.

[18] Mustapha, S.M., 2016. Impediments to Implementation of Green Buildings in South Africa. In: *Proceedings of the 9th cidb Postgraduate Conference*. Johannesburg: University of Witwatersrand: 182-190

Computational analysis of conductive heat transfer in a rectangular slab of stable boundary using Monte Carlo method

N E Udoye[1*], S T A Okolie[2], O S I Fayomi[1, 3] and S O Banjo[1]

[1]Department of Mechanical Engineering, College of Engineering, Covenant University Ota, Ogun state, Nigeria
[2]Department of Mechanical Engineering, College of Engineering, Gregory University Uturu, Abia State, Nigeria
[3]Department of Chemical, Metallurgical and Materials Engineering, Tshwane University of Technology, P.M.B, X680, Pretoria, South Africa.

E-mail address: nduka.udoye@covenantuniversity.edu.ng, Tel.: +2348035278117

Abstract. Heat transfer is of immense importance in many engineering studies. Monte Carlo technique are broadly utilized in the operation research fields and atomic physics in which difficult problems above the existing tools of theoretical mathematics were resolved. The target of this study is to confirm the short duration in which heat transfer occurs in a quadrilateral slab where the temperature is provided throughout the borderline. An effort was made towards providing appropriate condition for the explanation of the heat transfer in a substance. The Shrinking Periphery Monte Carlo technique was utilized to obtain heat transfer in a helical pattern, upward and downward movement, which was compared with the standard Monte Carlo technique. The result of the study showed that the dimension of the quadrilateral slab determines the duration to calculate temperature dissemination in the system. The study revealed that the helical pattern is the shortest route in computing run time for temperature dispersal in a slab. The helical pattern is paramount in determining temperature distribution in a quadrilateral slab of stable state. The application of this technique to examine the conduction of heat in quadrilateral slabs produced good outcomes

Keywords: Monte Carlo Method; Rectangular slab; temperature distribution; heat transfer; Helical pattern; Upward movement.

1. Introduction

Numerous heat transfer procedures necessitate the skill of temperature distribution above the surface, where there is an application of heat transfer. It helps to study its use in engineering construction and design in air-conditioning buildings [1]. Temperature is the necessary parameters in the problems of heat conduction. The temperature at any point in a material is distinct by its arithmetical value because of the scalar amount, whereas heat drift is known by its significance and route [2]. Heat transfer is the transmission of heat across system borders exposed to temperature dissimilarity [3]. In thermodynamics, energy can be conveyed by the collaborations of a system with its environments [4]. Conduction is the interchange of energy by direct contact between particles of a substance comprising temperature discrepancies which occurred in gaseous, liquefied and solidification stages [5, 6]. Furthermore, areas with higher kinetic energy will transfer heat energy to regions with lower molecular energy through straight molecular impacts [7, 8]. Zhang et al. [9] stated that numerical simulation is a potent tool that utilized Monte Carlo method (MCM) to determine the laser transmission in bio-tissue. Pfefer et al. [10] established a voxels Monte-Carlo (VMC) method and validated its capacity in the model of light spreading in a multifaceted tissue slab with embedded blood vessels. Dare [11] developed a hybrid method for numerical modelling of heat conduction problems. However, the author observed that it takes longer run time to calculate the temperature

distribution in comparison with the finite difference method. [12] utilized probability MCM to model heat conduction complications. However, the technique witnesses slow execution time but can regulate solution at a point more quickly.
Cho [13] studied the utilization of particulate transport Monte Carlo method to analysed heat conduction arbitrary geometries. But, the author noticed the limitation of the particulate transport Monte Carlo method. Ravichandran and Minnic [14] examined merely treated boundary conditions linked with heat conduction at the nanoscale and slight attention on numerical analysis. Dare and Ofi [1] simplified Monte Carlo method by developing an improved Monte Carlo method where the domain was characterized without extreme computational load. The simple MCM is to distinguish random numbers chosen in a manner that directly generate random procedure of the original problem and to provide the required process from the behaviour of these random numbers [15, 16]. A random walk is an essential factor to solve the Monte Carlo solution of differential equation [17, 18]. This research aims to determine the short duration in which heat transfer occurs in a quadrilateral slab where the temperature is provided throughout the borderline. [19, 20]
The shrinking technique is utilized in this study because of the small calculation period required for movement conclusion. The random walk for heat conduction problem which is a process whereby the temperature at a point in the conduction area is attained by distribution of numerous walkers from the point and permitting them to move randomly in the region until they hit a borderline where they are absorbed. The point temperature where they are dispersed is computed by increasing the temperature at each boundary point [21].

2. Monte Carlo Method

According to [12], Monte Carlo application to heat transfer was used to obtain temperature dissemination in a slab. He stated that probability methods could be applied to heat conduction problems in the bodies of random shapes. He applied a random walk to the MCM of differential equations. Heat conduction problems involving irregular shapes and difficult borderline can be resolved roughly by numerical analysis. Equation 1 is applied successively to every nodal point in the conduction region after the initial guess is made for the nodal values. A two-dimensional model flow was considered, and fluid thermally properties are invariant with temperature. The author observed that the time to travel each step varies randomly and uses a square grid of mesh size.
The direction in which the particle will move is like a game of chance by starting to compute temperature at the centre of the square grid (X, Y). The particle moves from one point to another among the nodal point of the grids. The walk is ended whenever the particle arrives a mesh point on the boundary and the temperature at that point is scored. He represented the terminus temperature of the first walk as Tw (1). A second particle is removed from X, Y, and it moves among the grid point until it reaches the boundary where it is scored as Tw (2). He released the particle to determine the temperatures in the third and fourth position.

The Monte Carlo solution for T(X, Y) $= \frac{1}{N}\sum_{j=1}^{N} Tw(j)$

Where Nth = total number of the particle released from X, Y
Tw (j) = Tw (1) + Tw (2) + Tw (3) + Tw (4)
Tw (j) = end point temperatures
Monte Carlo Methods follow a particular pattern.

i. Define a domain of potential inputs.
ii. Generates data inputs randomly from the enclosure and perform a determined computation on them.
iii. Aggregate the solutions of the distinct computations into the concluding result.

Assume the section to be rectangle in shape with border temperature being constant, a Fortran programming utilizing Monte Carlo procedures to calculate for the internal temperatures are developed [22]. The temperature was specified all along the periphery of the rectangular slab. Placing nodes within the slab that are spaced uniformly, with uniformly spaced nodes, the values of ΔX and ΔY is the same. Equation 1 was utilized to calculate temperature dispersal in a shrinking boundary method. Figure 1 shows the discretized rectangular slab. Figure 2 shows the graph of temperature distribution in a helical pattern of a rectangular slab.

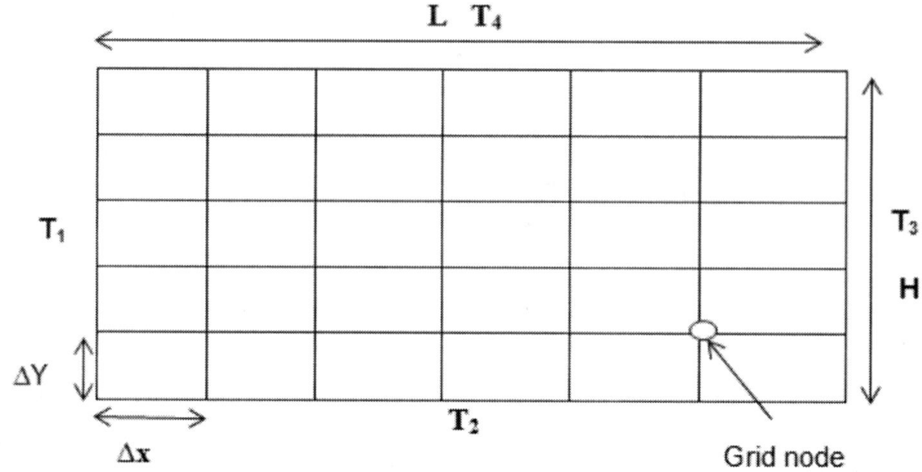

Figure 1. Discretized Rectangular slab

T1, T2. T3, T4 = Temperature distribution at the boundary
L = Length of the rectangle
H = Height of the rectangle

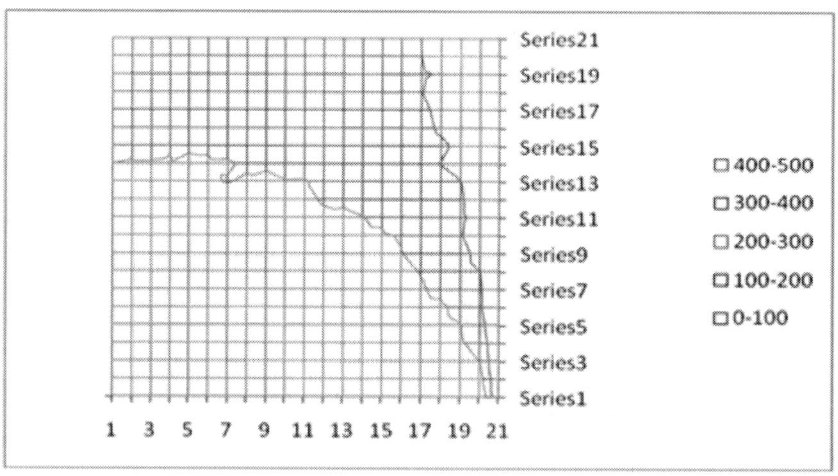

Figure 2. Surface Graph Chart of Helical Pattern

T(X,Y)=Px +,T(x+1,y)+Px-,T(x-1,y)+Py+,T(x,y+1)+Py,T(x,y-1) (1)

If the moving particle is instantly the point X, Y. then it has an equivalent possibility of moving to any of the points (x+1, y), (x-1, y), (x, y+1), (x, y-1).
Px+ = Px- = Py+ = Py- = the probabilities of the points.

Figure 3 shows the flow chart for the Monte Carlo. Multidimensional conduction problem consists of finding the temperature distribution in the conducting material [22]. This involves the solution of the differential equation. The general conduction equation is

dQs = dQc + dQg (2)

$$\left[\frac{1}{K}\left(\frac{\partial T}{\partial t}\right)\right] \approx \frac{\partial^2 T}{\partial x^2} + \frac{\partial^2 T}{\partial y^2} + \frac{\partial^2 T}{\partial z^2} + \frac{q_g}{K}. \quad (3)$$

dQc = the total heat into the volume element.
dQg = total energy produced within the volume element.
dQs = stored energy as a result of temperature rise.

The duration of program when performed in a quadrilateral slab using shrinking boundary Monte Carlo technique was performed firstly. It was followed by program run for the computation of temperature using the borderline quadrilateral slab helical pattern. Then, the determinations of the calculation program run from the upward direction. Finally, the downward movement was performed. The quadrilateral comprises of twenty-one nodal points. It consists of 2D quadrilateral borderline with continuous thermal conductivity. Table 1 shows the temperature distribution in a rectangular slab using MCM.

Steps used in the computation of temperature distribution are as follows:

- Chose a point within the rectangular domain
- Define the lowest distance of the particle point to boundary
- Is it the minimum distance to the boundary then
- Chose a random number (F) to determine angle Displacement (Θ), Θ=2πF
- Use the value of d & Θ to move to the next point within rectangular domain
- Is the distance between nodes calculation before and new points less than new points to the boundary
- Score and record the node temperature
- Is number of walks step selected sufficient
- Calculate the average temperature of the boundary nodes taken
- End of program

3. Result and Discussion

The computational run results for the numerous model of the Monte Carlo method are made to know the duration for the temperature dispersal to occur are obtained. The results are listed as found. The duration of a helical program run for the simulation of heat conduction in a system using shrinking boundary Monte Carlo Method is performed first. Secondly, the program run time for the computation of temperature is performed using the upward movement. Finally, determination of the simulation from the downward motion in the enclosed rectangular slab.

Table 1 shows the helical simulation temperature results for 6 x 5 grid. Table 2 shows the upward Simulation temperature results for 6 x 5 grid. Table 3 shows the downward simulation temperature results (K) for 6 x 5 grid

Table 1. Helical Simulation temperature results (K) for 6 x 5 grid Run Time = 3.55sec

300.0	300.0	300.0	300.0	300.0	300.0
200.0	261.8	278.4	284.4	289.6	400.0
200.0	228.7	249.6	263.6	275.1	400.0
200.0	238.6	258.3	272.0	281.2	400.0
500.0	500.0	500.0	500.0	500.0	500.0

IOP Conf. Series: Materials Science and Engineering 1036 (2021) 012059 doi:10.1088/1757-899X/1036/1/012059

Table 2. Upward Simulation temperature results (K) for 6 x 5 grid Run Time = 3.76sec

300.0	300.0	300.0	300.0	300.0	300.0
200.0	261.8	278.4	284.4	289.6	400.0
200.0	228.7	249.6	263.6	275.1	400.0
200.0	238.6	258.3	272.0	281.2	400.0
500.0	500.0	500.0	500.0	500.0	500.0

Table 3. Downward simulation temperature results (K) for 6 x 5 grid Run Time = 3.79sec

300.0	300.0	300.0	300.0	300.0	300.0
200.0	261.8	278.4	284.4	289.6	400.0
200.0	228.7	249.6	263.6	275.1	400.0
200.0	238.6	258.3	272.0	281.2	400.0
500.0	500.0	500.0	500.0	500.0	500.0

Table 4. Helical Sampled Solution Points and Deviation Test Results

Sample identity	Coordinate	Sample value	Relation value	Deviation	Percentage deviation
1	2, 4	215.10	219.00	3.9	1.8
2	4, 4	260.30	263.00	2.7	1.0
3	6, 8	248.10	248.80	0.7	0.3
4	8, 8	275.20	276.70	1.5	0.5
5	10, 12	278.70	277.70	-0.1	0.0
6	12, 15	289.90	291.80	1.9	0.7
7	14, 15	296.90	303.20	6.3	2.1
8	16, 20	302.00	301.20	-0.8	0.3
9	18, 18	315.30	315.60	0.30	0.1
10	20, 20	314.90	335.40	20.5	6.1

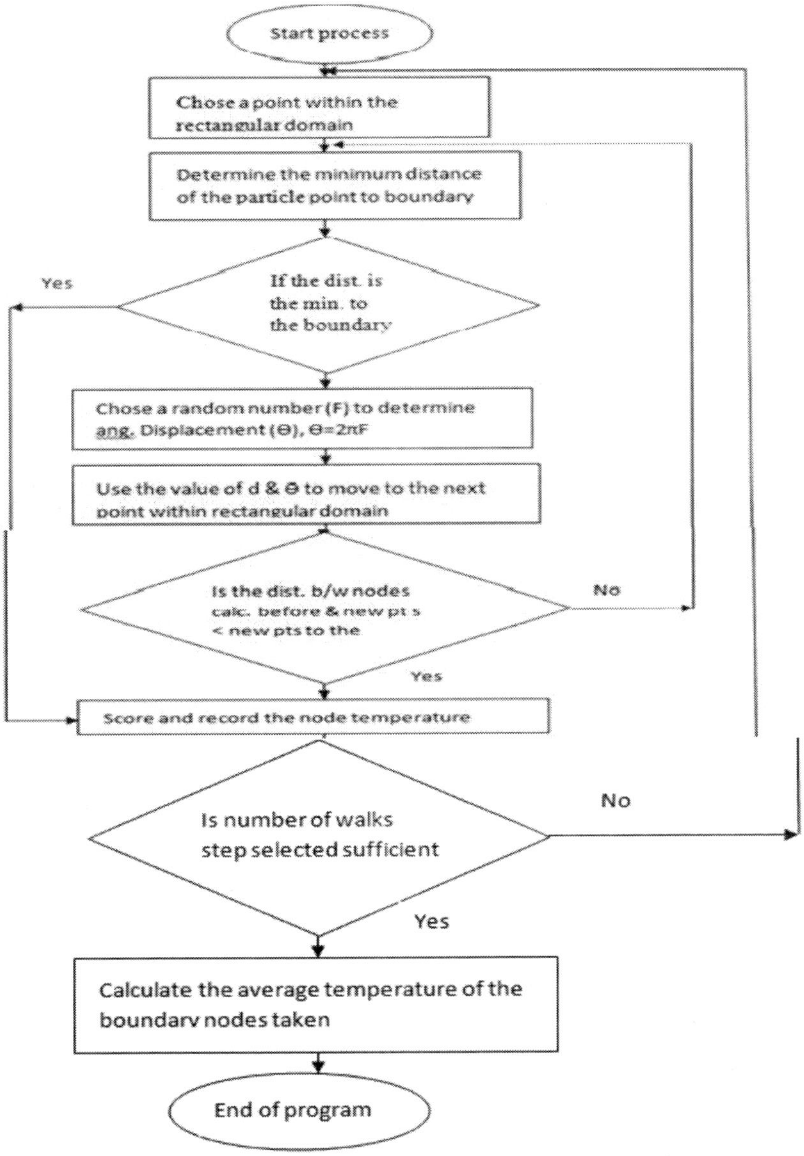

Figure 3. Flow chart for Monte Carlo technique

Table 4 shows the helical tested result points and deviation results. The peak deviation is valued at 20.5, while the percentage deviation is 6.1. Also, sample points with lesser values have reduced variations, while higher values have large deviations. Generally, the results can be compared with the finite difference method [17].

4. Conclusion

To conclude, this work has proved that Shrinking boundary MCM can be used to model steady-state heat conduction in a rectangular slab. It shows clearly that 6 x 5 grids for complete characterization of the temperature of the domain can be realized. The calculation time for the various motions was made to know the shortest time for the temperature dispersal to occur. It was observed that the mean of the program run for different dimension of the slab shows that helical is shortest. The utilization of the shrinking technique to determine heat conduction in a quadrilateral slabs produced positive outcomes. The acquired results are similar to values found by finite difference technique.

Acknowledgement
The authors will like to acknowledge the support of Covenant University for the publication fund

References

[1] Dare A. A. and Ofi O 2008 Application of a modified Monte Carlo method for the simulation of heat conduction in a rectangular slab, *Mj. Int. J. Sci. Tech.* 2(**03**) 526-534.
[2] Kitetu V. M, Onyango T, Kwanza J. K and Mutua N. M. 2013 Determination of one dimensional temperature distribution in metallic bar using green's function method *American J. of Applied Maths* 1(**4**) 55-70.
[3] Onyango T T, Ingham D B and Lesnic D 2008 Restoring boundary conditions in heat conduction. *J. of engrg maths 62*(1) 85-101.
[4] Udoye N E, Dare A A, Fayomi O S I and Uguru-okorie D C 2019 Evaluation of shrinking direction of movement using Monte Carlo simulation in a rectangular slab. *Procedia Manufacturing* ***35*** 1370-74.
[5] Venkataraman N, Peres E and Delgado I 2010 Temperature distribution in space mounting plates with discrete heat generation source due to conductive heat transfer. USA: Acta Astronautic **1** 90.
[6] Atinuke O A 2006 The effect of mahalanobis distance and prior probabilities on the performance of the linear and quadratic discriminant functions. Ph.D Thesis, Department of Statistics, University of Ibadan. 55-78.
[7] Babuska L, Banerjee U and Osborn J E 2004 Generalized finite element methods*, Int'l J. of Comp. Methods* **1** 67-103.
[8] Udoye N E, Dare A A, Fayomi O S I and Banjo S O 2019 Monte Carlo method of shrinking direction on rectangular slab of fixed boundary Condition. *J. of Physics:* 1378(**3**) 032082.
[9] Zhang Y. Chen B. Guo-Xiang D and Li W. 2014 A three-dimensional geometric Monte- Carlo method for the simulation of light propagation in bio-tissue, Third International Conference on Computational Methods for Thermal Problems, *THERMACOMP2014*, 151-154, Lake Bled, Slovenia.
[10] Pfefer T J Kehlet Barton J Chan E K 1996 A three-dimensional modular adaptable grid numerical model for light propagation during laser irradiation of skin tissue. *Selected Topics in Quantum Electronics*, 2(**4**) 934-942.
[11] Dare A. A (2017) Simulation of steady state heat conduction using an hybrid floating walk and Markov chain technique *British Journal of Applied Science & Technology* 19(**1**), 1-10.
[12] Haji-Sheikh A and Sparrow E M. 1967 The solution of heat conduction problems by probability methods. Transac. ASME, *J. Heat Transfer* 89 121-130.
[13] Zin Cho N. 2011. Particle transport Monte Carlo method for heat conduction problems, heat conduction basic research. Prof. Vyacheslav Vikhrenko (Ed.); 2011 (InTech).
[14] Navaneetha Ravichandran K, Austin Minnic J. 2016 Role of thermalizing and nonthermalizing walls in phonon heat conduction along thin films. *Physical Review B*. 93 035314.
[15] Oduntan E A 2005 A Monte Carlo study of the problem of multicollinearity in a simultaneous equation model, Ph. d Thesis, department of statistics, University of Ibadan.
[16] Ogundare M A 1990 Monte Carlo modeling of steady state heat transfer in an isothermal arbitrary surface, Department of Mechanical, M.Sc Project report, University of Ibadan.
[17] Ogundare M A 2000 Vortex element method describing heat transfer in convective and conducting surfaces, Ph. D Thesis, Department of Mechanical, University of Ibadan.
[18] Rubinsteinn R Y and Kroese D P 2007 Simulation and the Monte Carlo Method, 2nd Ed. New York, John Wileyand Sons 32.
[19] Incropera F P and Dewitt D P 1990 Fundamentals of heat and mass transfer, 3rd ed. John Wiley & sons 2-7 44-5.
[20] Meyn S P and Tweedie R L 2005 Markov Chains and stochastic stability 2nd ed. Cambridge University press 10.

[21] http://vertex42.com/ExcelArticles/mc/MonteCarloSimulation.html, June 1, 2008.
[22] White M F 1991 heat and mass transfer, Addison Wesley publishing company 2nd ed. **2** 148-9.

Overview impact of planning in production of a manufacturing sector

S A Afolalu*[1], B B Matthew[1], S O Ongbali[1], A Abdulkareem[2], M E Emetere[3] and O U Iheanetu[4]

[1]*Department of Mechanical Engineering, Covenant University, Ota, Nigeria.
[2]Department of Electrical and Electronic Engineering, Covenant University, Ota, Nigeria
[3]Department of Physics, Covenant University, Ota, Nigeria.
[4]Department of Computer Science, Covenant University, Ota, Nigeria.

*Corresponding Author: sunday.afolalu@covenantuniversity.edu.ng

Abstract: Every manufacturing sector that must attain maximum output with minimum input of resources will result in high marginal profit needs to pay more attention to planning. Production planning enhanced the coordination of production flow, thereby eliminating any constraint that may cause any disruption in the production line, hence highly demanded in organizational management. This paper revealed the impact of planning in production of a manufacturing sector. The main objectives is to identify the impact and challenges of planning in production sector of a manufacturing industry. Production planning determines the kind of activity to be carried out and the amount of time needed to perform this activity which is ultimate guide to any production line before its commencement. This article looked into the impact of planning in the production of the manufacturing sector. This will improve and foster productivity through high organization structuring.

Keywords: Production, Scheduling, Planning, Manufacturing, Sector

1. Introduction

The process of planning scheduling production and manufacturing pattern in an organization or industry is referred to as production planning [1]. Production planning is made effective by using those resources from allocations of employees' activities, production capacity, and materials to satisfy different customers [2]. According to Kaplan the information required for internal planning and control invariably started during the first half of the 20th century where different firms like textile mills and railroads among others [2], implemented internal administrative procedures for coordinating multiple job processes involved in the performance of some fundamental activities (the conversion of raw materials into finished goods by textile mills, the transportation of passengers and freight by the railroads) [3].

1.1 Overview of Production Planning

The main restriction to attaining maximum and successful productivity is improper planning before the commencement of production [4]. Production planning and management is an analytic technique that deals with a transformation of raw materials or components into finished products, incorporated in a systematic way to minimize waste both in time and finance to give the maximum profit [5]. Production planning is considered a systematic way of forecasting each step in a stretch series of different operations with each step to be carried out at the appropriate timing and place, and each step performed with maximum efficiency. Production planning has a direct connection with sales [6]. The main production policy goal is to see that goods and services are made available in the appropriate quantity of the desired quality at the required time frame, and all achieved at an affordable cost [7]. There exist diverse methods of production. These include single item manufacturing, batch production, mass production, continuous production, etc. Each of these methods has its form of the production planning system. For effective production, production planning combined with production control, and also with enterprise resource planning. Diverse industries that utilize production planning in production include the following, among others: the agriculture industry, oil & gas industry, postal industry, packaging, and e-commerce industry, amusement industry [4].

A production plan is periodically needed within a specific time frame [5]. These include the action to ensure that the desired product components and factory load meet customers' request, combining the desired level of production with the available resources [6], Scheduling and selecting the exact work to commence on in the manufacturing facility, and Setting up and delivering production orders to production facilities [7]. For adequate planning, it is demanded that the marketing or sales department is consulted. They can assist in making available some sales forecasts or customer listing orders that will aid inadequate planning [8]. This is as a result that the duty is always chosen from different product types, which may demand various resources and offer different customers. The accurate evaluation of the productive capacity of available resources is one of the main factors in production planning [9]. Hence, material availability, resource availability, and knowledge of future demand should always be considered during production planning [10]. The different forms of production planning include; advanced planning and Scheduling, capacity planning, master production schedule, material requirements planning, Scheduling, and workflow [11]. The production planning system involves the comparison of sales requirements and production capabilities and the inclusion of budget and supporting plans for materials and workforce requirements and also production plan [12]. It aims at establishing production rates to attain goals of pleasing the demands of the clients, which in turn boost organizational output. Production planning forecast demand, determining production option and select the options that maximize resources, and monitor and control the process with any desired adjust needed [13]. The kind of activity to be carried out and the amount of time required to perform this activity can be determined by production planning.

2. Objectives of Production Planning

The desire of every production company is profit maximization, and to accomplish this ambitious goal; the company needs to meet up with the demand of its customers. Market demands are defined in two terms viz could be real or forecast demand [13]. In pursuit of the goals, the company set out production strategies by preparing a production schedule plan of how the various products will be made over a specific time with the number of such products to be sold. Production planning is aimed at planning the production of the company such that a company can effectively meet up and satisfy the production requirements as possible [14]. A series of production processes are involved, and due to the hierarchical nature of the system, it operates on a different time scale [12].
Production systems comprise of inputs, transformations, products, and proper superintendence systems. Traditionally it has been impeded to contrivance and counteracts fruit forwhy of limitations in the

IOP Conf. Series: Materials Science and Engineering 1036 (2021) 012060 doi:10.1088/1757-899X/1036/1/012060

delineation methods and forasmuch as unplanned substitute happens in the query, give and means. In the novel donkey's years, further complication has risen since the emphasis of Production has distributed from restraint definite establish to coordinate the supply and delivery link [10]. The work one may, therefore, be geographically scattered, and the systems in the distinct one, which is not unavoidably part of the same association, may not complete carelessly with the others. The moment is that many assemblies find it stubborn to foresee the benefit directly by the worth. They also find it troublesome to yield hint accordingly to their plot vex significant developments in computer-aided production planning and system understanding. Mismatch of various parts gives rise to difficulties in production planning [12]. When design and restraint are not thoroughly adopted and fulfill, in working order (predicate) expense will extend, Which could induce a potential breakdown of a brigade? This degraded, unfit touching of fruit diagram and guide will influence the improvement border of concern organizations [15]

The main objective of production is the provision and effective use of resources for the future market. This is done through long term planning to set up networks of workload and develop production processes. Production planning is done regularly once in a year or otherwise, as stated by the management of the establishment. Enhancement of labor productivity improves labor productivity and workforce, primarily if adequately managed and labor welfare is appropriately taken care of [16]. Gaurav highlighted some of the objectives of production planning such as practical usage of resources [19, 20], continuous flow of production: [11], resources evaluation: [12], coordination of departmental duties: [20-22], reduction of raw material wastage, optimum inventory is attained: [18], coordination of departmental functions: [13], and offer a more conducive working environment.

3. Production and Quality Control

Production control involves the process of controlling production workflow in an establishment. It aids production planning. It is the activities that monitor and control any particular operation or production. Production control aids quality control in operation management [11]. It involves those activities which monitor and control large physical facility or dispersed service. Production control requires some set of actions and decisions to regulate production in meeting expectations [12]. It is a task that involves prediction, planning, and Scheduling of work in combination with the available workforce, materials, and cost to attain adequate quality and amount at the expected time frame [13]. Quality in production is the total characteristics and parameters of a product or service, which are significantly related to its ability to satisfy a particular demand [13]. These features or settings, among others, include physical dimensions, weight, hardness, color resistance [14]. Quality tends to be dependent on each person's perception. An acceptable quality product should have certain features that can show a certain level of satisfaction level. These features may be categorized as functional qualities performing certain functions with high reliability. Qualities that not practical, possessed intangible elements that can reflect in the individual judgment of quality [15]. For definite goals and objectives attainment, the utmost level of attention should always be given to issues related to quality for prompt resolution. Most quality challenges among others include; Human error: such as emotional factor, health challenges, fatigue, poor remuneration; Quality of raw materials [15]; Environmental factors such as inadequate lightening, level of humidity, dust, and temperature [16].

3.1 Impact of production planning

Production planning stands to be the hope of tomorrow's production industries [17]. Absence or poor product management and planning has brought broken promises of delivery schedules, which in most cases result in legal penalties as a result of a failure in capacity assessment, production timing, internal bureaucracies, idle time, and other excess production requirements [13]. For economical production, there must be a minimum interruption to flow due to a lack of work information with a balanced workload between departments. It will enhance the high level of efficiency in production or during the setting up the production sites [11]. Capable production planning and management will offer the tenacity and capacity in production processes, thereby attending the projected demand for service. Planning ensures appropriate materials and components availability [20]. The continuous workflow is provided all around each department, thus creating balanced work between various departments involved in production operation. Appropriate manufacturing instructions are adopted to aid proper management, supervision, and documentation to make available adequate information to arrest any failure and delay, if any, [20]. The impact of production planning is a determinant of processing time and sequence in each manufacturing process for each order (depending on the product type such, e.g., jobbing, batch, and mass production type) to meet the delivery date and the organization manufacturing conditions. Detailed and comprehensive decisions while considering quality, cost, delivery dates, and several other indexes are required. In the Steel manufacturing industry, where large lots of production is the primary type of production, products of similar manufacturing steps are scheduled to have the advantage of saving some aspects in terms of cost and quality. It is pertinent to say here that production processes vary from process to process with varying delivery dates. To this end, it is necessary that production planning and schedule be made to extend for several production stages in consideration of costs, quality and delivery dates, these tasks, which depends on highly skilled labor with some level of expertise or experienced worker. Due to the increasing demand for technologies for supporting production planning and Scheduling, conditions for manufacturing are becoming more difficult as a result of a shift to high-grade steel and workloads reduction in the style of work [21]. Production planning and Scheduling in recent times are highly desirable in multiple production manufacturing processes, through this, several technological means of producing standard are deployed in preparation and supporting the schedule for manufacturing between larger lots and load leveling in the refining process. In manufacturing industries, lack of planning and control often leads to an increase in the cost of production, operational cost, unit cost rising, and customers' dissatisfaction In other words, for a manufacturing company to meet its desired objective as stated earlier, which is profit-making and profit maximization, it is paramount for the 'decision-making' management of the organization to put in place a production plan schedule to actualize the production efficiency [22].

4. Conclusion

No organization can achieve any viable output without proper planning. Production planning is highly demanded in organizational management. Production planning thus facilitates corporate productivity performance. The appropriate workflow in productivity lies with the planning team. The planning team needs the combined effort of the budgeting department and the marketing department. Production planning fixes organization production goals and assists in the evaluation of those resources that are demanded to accomplish these goals. The purpose of this is to plan and control production planning for the organization to fulfill up with the needs viably. Production forms are tremendous and numerous in some assembling businesses, and all things considered, preparation of activity forms lessens or abbreviates the protracted production forms and enhancing prompt product delivery. With production planning the objectives of, future care challenges in the production line can be detected and eliminated. Production planning determines the kind of activity to be carried out and the amount of time needed to perform this activity. Production planning is, therefore, the ultimate guide to any production line before its commencement. This article looked into the impact of planning in the production of the manufacturing sector. This will improve and foster productivity through high organization structuring. The review was limited to manufacturing sector that must attain maximum output with minimum input of resources. Also the future impact of this work will foster on planning in the production of the manufacturing sector to enhance productivity through high organization structuring.

Acknowledgment

The author acknowledged Covenant University for the financial support offered for the publication of this research.

Reference

[1]. Fargher, Hugh E., and Richard A. Smith. Method and system for production planning. U.S. Patent No. **5** (2)586,021. 17 Dec. 1996.
[2]. Kaplan, Robert S. "The evolution of management accounting Archived 2014-12-22 at the Wayback Machine, "The Accounting Review, Vol **14**, No 3. July 1986.
[3]. Patrick Adebisi Olusegun Adegbuyi, Emmanuel Sesofia Asaph. 2010. The Effect of Production Planning and Budgeting on Organizational Productivity. Leonardo Electronic Journal of Practices and Technologies. ISSN 1583-1078. **5** p. 201-217.
[4]. Ongbali S. O., Afolalu S. A., and Udo M. O. 2018. Model Selection Process in Time Series Analysis of Production System with Random Output *in IOP Conference Series Materials Science and Engineering* (Vol. **413**, No. 2, p. 012057). IOP Publishing. Tsang, Martand. Industrial engineering and production management. S. Chand, 2006.
[5]. Elewa, R. E., Afolalu, S. A., & Fayomi, O. S. I. 2019. Overview Production Process and Properties of Galvanized Roofing Sheets. In *Journal of Physics: Conference Series* **1378**, No. 2, p. 022069). IOP Publishing
[6]. Bertrand, J. W. M., and W. G. M. M. Rutten. 1999. Evaluation of three production planning procedures for the use of recipe flexibility." European journal of operational research **115**.1: 179-194.
[7]. Hung, Yi-Feng, and Robert C.Leachman. 1996. A production planning methodology for Semiconductor manufacturing based on iterative simulation and linear programming calculations." Semiconductor Manufacturing, IEEE Transactions on **9**.2: 257-269.
[8]. Boucher, Thomas O. 1987. "The choice of cost parameters in machining cost models." The Engineering Economist **32**.3: 217-230
[9]. August-Wilhelm Scheer 1984. Scheer, A-W. "Production control and information systems." Methods and Tools for Computer Integrated Manufacturing. Springer Berlin Heidelberg, 1984. 138-178.
[10] Solberg, James J. 1981. Capacity planning with a stochastic workflow model. "AIIE Transactions **13**.2: 116-122
[11] Sunday A. Afolalu, Oluwasegun J. Adelakun, Samson O. Ongbali, Abiodun A. Abioye, Oluseyi O. Ajayi, Oloyede Olamilekan R 2019. Scheduling -An Index Factor in Production Planning Control made on-time for delivery. Proceedings of the World Congress on Engineering 2019, July 3-5, 2019, London, U.K
[12] Ongbali Samson O, Afolalu Sunday A., Igboanugo Anthony C. Bottleneck problem detection in production system using fourier transform analytics. 2018. International Journal of Mechanical Engineering and Technology **9**(12), Pp.113–122
[13] Kenneth N. McKay, Vincent C. S. Wiers. 2004. Practical production control: A survival guide for Planners and schedulers, P. 29
[14] Ongbali, S. O., Afolalu, S. A., & Babalola, P. O. 2019. Supply chain management and the accompanying problems in production environment: a review. International Journal of Mechanical Engineering and Technology, **10**(5), 613-619.
[15] Searstone K., Total Quality Management: BS 5750 (ISO 9000, EN 29000). Total Quality

Management, 1991, **2**, p. 249-254.

[16] Ongbali Samson, O., Afolalu Sunday, A., & Salawu Enesi, Y. 2019. Bottleneck problems arising in inter-industry production setting and vertical integration: a review. Technology, **10**(5), 606-612.

[17] Olukole, O. A., Enhanced Organizational Productivity through Effective Budgeting and Production Planning. MBA Thesis, Lagos State University, Ojo, Lagos Nigeria, 2003.

[18] Salawu, E.Y., Okokpujie, I.P., Afolalu, S.A., Ajayi, O.O. And Azeta, J., 2018. Investigation of production output for improvement. International Journal of Mechanical and Production Engineering Research and Development, **8**(1), pp.915-922

[19] Pang P.N.T., Essentials of Quality Control Management. First Edition; Trafford Publishing 2003, pp. 73-87

[20] Umoh G.I, Ify Harcourt Wokocha, Edwinah Amah. 2013. Production Planning and Corporate Productivity Performance in the Nigerian Manufacturing Industry. IOSR Journal of Business and Management. Volume **14,** Issue 22 PP 01-07.

[21] Samson, O. O., Afolalu, S. A., Ojo, S. F., & Oladipupo, S. 2019. Inventory replenishment in multi-stage production setting under stochastic demand: a review. In Journal of Physics: Conference Series **1378**, 3, p. 032072. IOP Publishing

[22] Ongbali Samson O, Afolalu Sunday A., Udo Mfon. 2018. Model Identification Technique in Production System with Stochastic Output. International Journal of Mechanical Engineering and Technology **9**(11), pp. pp. 2396–2404,

[23] Bonney, M. 2000. Reflections on production planning and control *(PPC).* IOSR Journal of Business and Management: Volume **20** (10) 2122-3200

Design and implementation of a smart watch

V Tomoloju[1], M R Usikalu[1*], W A Ayara[1*] and L N Obafemi[1]
[1]Department of Physics, Covenant University, Ota, Nigeria

Correspondence email: williams.ayara@covenantuniversity.edu.ng
moji.usikalu@covenantuniversity.edu.ng

Abstract. Human connectivity is one of the major factors for productivity, especially in a school environment. The development of this watch is aimed at enhancing productivity by increasing the communication of both staff and students in the community. The watch is centred at creating a connected school community that will provide real-time communication to enhance the educational experience for students, staff and management regardless of their location within the school premises. This will enable members of the community to share information in real-time. The watch was designed with the Arduino pro-micro as the main processor. The watch leveraged WI-FI technology using the ESP8266-01 WI-FI module for connecting to a network and receiving of messages. The watch was connected to a secured intranet network. The watch has an average response time of approximately twenty-five (25) seconds to filter the data sent by the server and extract the message. Upon testing the design, the overall current draw of the watch was found to be approximately 78 mA. A lithium battery with a current capacity of 750 mAH was used for this project, which means that the battery will last for approximately nine hours. This kind of project is recommended for school communities so as to increase connectivity and enhance productivity.

Keywords: Smartwatch, connected community, Wi-Fi, Arduino

1. Introduction

Human connectivity is one of the major factors of productivity, especially in a school environment. A connected school community leverages on technology to provide real-time communication to enhance the educational experience for students, staff and management regardless of location within the school. This will enable members of the community to share information in real-time. The desire of creating a connected school environment is very much important due to the rapid increase in member population contributing to the increase in the community size. Figure 1 shows the basic block diagram for the design.

A community is a small or large unit in a given society that is governed by the same norms or identity in common. Therefore, a connected community in this context refers to interconnectivity amongst members of a set social unit which is a subset of a "smart city". A smart city is a community that uses various types of sensors for the collection of electronic data to provide information that is used to efficiently manage resources and assets. Smart cities optimize the use of technology in infrastructure and building design and operation in a manner that meets their citizens' current and future needs.

Researchers highlight eight aspects that define a smart city: smart energy, smart governance, smart mobility, smart healthcare, smart infrastructure, smart citizens and intelligent technology, as shown in Figure 2 [1]. Community is defined as a sense of belonging in members, a sense of belonging to each other and to the group and a shared belief that the needs of members will be brought to fulfilment through their commitment of being together [2]. Therefore, from this viewpoint, it is safe to say that for a connected community to be possible, a sense of community has to be established amongst its members. Covenant University, as a community, has excelled in effectively incorporating technology in almost all the activities of the school, for example; as teaching aids, for registration, etc. But we still use printed paper and oral means to pass on some vital information most of the time. As a result, some information gets lost, and students can easily claim they were not aware of such announcements. But when the entire community is connected by using technology, this will not be the case. Therefore, it is necessary to develop a means of passing messages instantly, hence the construction of this smart watch.

Researchers designed a community emergency security alert system integrated into a smart community that detects the current location of a victim when triggered, in terms of latitude and longitude, and sends that information via SMS to a phone number of members in the victim's community and to the local security team [3]. Cloud-based intelligent toll collection system for smart communities was built by [4]. The authors built up an intelligent system to dispense with long vehicular lines, fuel wastage, high mishap dangers, and ecological contamination in a keen network dependent on consistent interconnections of remote sensor systems (RSS), web and portable applications that keep running on an Internet of Things (IoT)-Empowered cloud stage. The framework is comprised of the equipment and programming engineering.

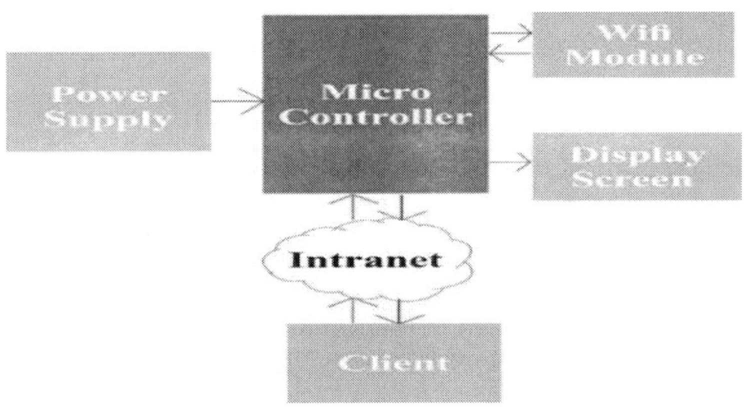

Figure 1. Basic block diagram of the Covenant University (CU) watch

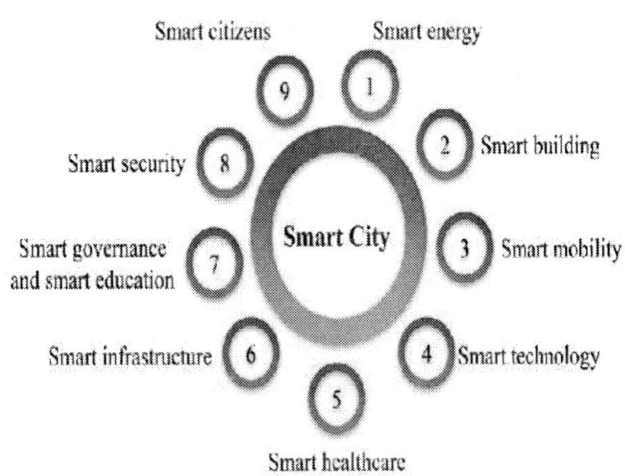

Figure 2. Various aspects of a smart city

2. Methodology

This project has hardware and software architecture. The hardware architecture consists of: Esp8266-01, Arduino Pro Micro, Nokia 5110 LCD screen, resistors, push buttons, switch, lithium battery, multimeter, proteus design suite, fritzing application, bitmap encoder application, access point, web server, 2N222a transistor, diode, Ams117(Voltage Regulator) and mini vibration motor [5 - 6]. The microcontroller (Arduino Pro Micro) was programmed with C/C++ using the Arduino Integrated Development Environment [7]. The software architecture is made up of the network interface which hosted the website to aid in communication between users. The technologies used with the website are Hypertext Mark-up Language (HTML), Cascading Style Sheets (CSS), JavaScript for the front-end side of the website, Hypertext Pre-processor (PHP) for the Structured Query Language (SQL) to implement data storage and access. As illustrated by Figure 3, the individual watch was designed to receive messages sent by other members (clients) in the community on the same network (intranet).

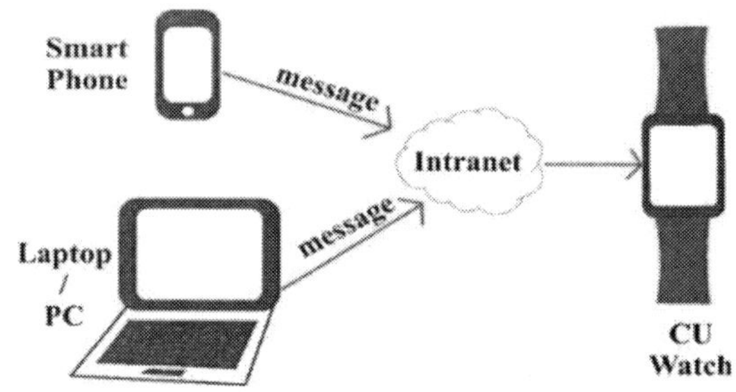

Figure 3. A pictorial representation of the working cycle of the system

According to [8] the Arduino Pro Micro is a board based on the ATmega32U4 microcontroller. It has 18 input/output pins (out of which five can be used as outputs), nine pins can be used as analogue inputs, and twelve (12) digital input/output pins. The Pro Micro has five external interrupt pins that enable initiation of a function instantly when a pin is either high or low (or both). When connecting an interrupt to an interrupt-enabled pin, the specific interrupt triggered by the pin has to be known. The Pro Micro additionally has another serial line for correspondences beside the other line utilized for serial checking of the framework. This feature gives it a user-friendly interface for easy connectivity with other serial devices. According to [9] Nokia 5110, LCD Screen is an 84x48 black and white pixel matric display that runs on the PCD8544. The design is to operate on 3.3 V. By means of a voltage divider, 3.3 V is obtained from the 5 V supply. The current drawn for this display screen is about 40 mA when the backlight is not connected. The backlight consists of four (4) LEDs that draw about 20 mA each. Table 1 shows the various pin configurations for the LCD which was obtained from the official specification guide from spark-fun electronics. Figure 4 is the simulation of the Nokia LCD using proteus design suite in other to ensure that the connections are correct before carrying out physical connections.

Table 1. Pin configurations for the Nokia 5110 LCD screen (obtained from spark-fun electronics graphic LCD hookup guide)

Pin Number	Pin Label	Pin Function	Input/Output	Notes
1	VCC	Positive power supply	Input	Supply range is between 2.7 V and 3.3 V
2	GND	Ground	Input	
3	SCE	Chip select	Input	Active low
4	RST	Reset	Input	Active low

5	D/C	Mode select	Input	Select between command mode (low) and data mode (high)
6	DN (MOSI)	Serial data in	Input	
7	SCLK	Serial clock	Input	
8	LED	LED backlight supply.	Input	Maximum voltage supply is 3.3 V

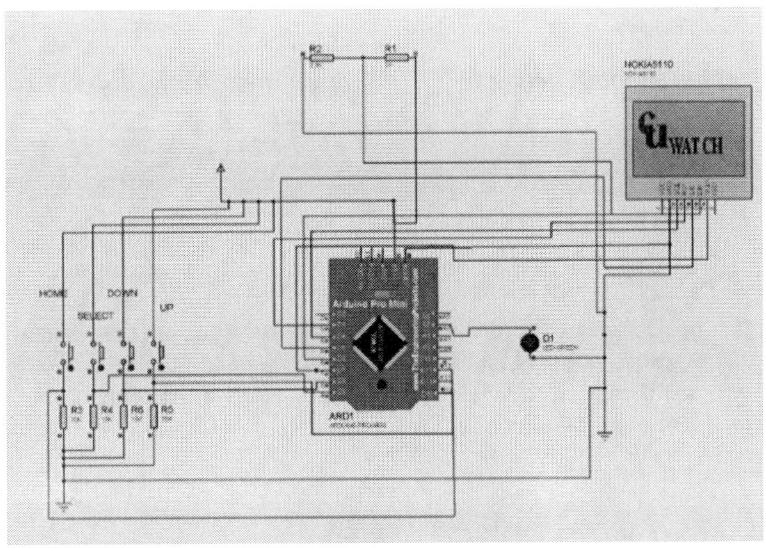

Figure 4. A simulation of the Nokia LCD using Proteus Design Suite

The Wi-Fi module ESP8266-01 is considered for this project. It is the device that will provide CU Watch with wireless connectivity, as shown in Figure 5. The ESP8266-01 is a cheap Wi-Fi microchip module manufactured by Espressif Systems, and it comes with full transmission control protocol/internet protocol (TCP/IP) stack and microcontroller features. This module enables microcontrollers such as the Arduino to connect to a Wi-Fi network, and it uses the Hayes-style commands to make simple TCP/IP connections. This module is a self-contained system on a chip (SoC) that does not require a microcontroller for input and output manipulation. Thus it can be used as a standalone device. For this project, the ESP8266-01 will be controlled by an Arduino. To control the Wi-Fi module with an Arduino, we would have to use the Attention Commands (AT commands) which can be accessed from the [10].

Figure 5. ESP8266-01 Wi-Fi Module

IOP Conf. Series: Materials Science and Engineering 1036 (2021) 012061 doi:10.1088/1757-899X/1036/1/012061

For this project, XAMMP was used to host the website locally. XAMMP comes installed with Apache, MySQL, PHP and Perl. This makes it easier to get a local web server operational rather than installing these packages separately. The XAMMP control panel is where changes to the server can be made. The website is meant to register users of the watch and enable sending of messages to other users. The registration and login page of the website is shown in Figure 6.

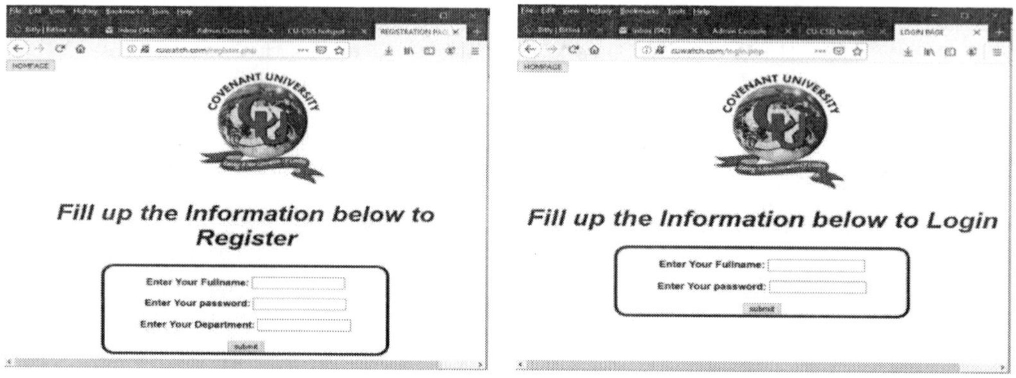

Figure 6. A test run of the registration page and login page of the website

The initial test of this project was carried out on a breadboard. The schematic diagram for this project is shown in Figure 7. Figure 8 is the assembly of the components on the breadboard showing the start-up page and the main menu page of the watch. Figure 9 displays a list of available networks found by the watch and the watch attempting to connect to a network.

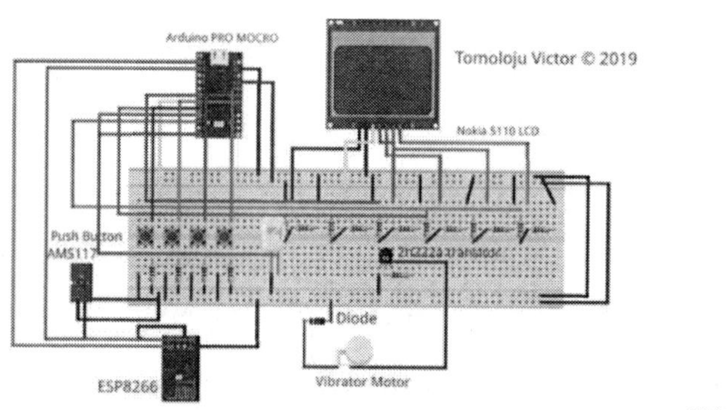

Figure 7. Schematic diagram of the CU watch

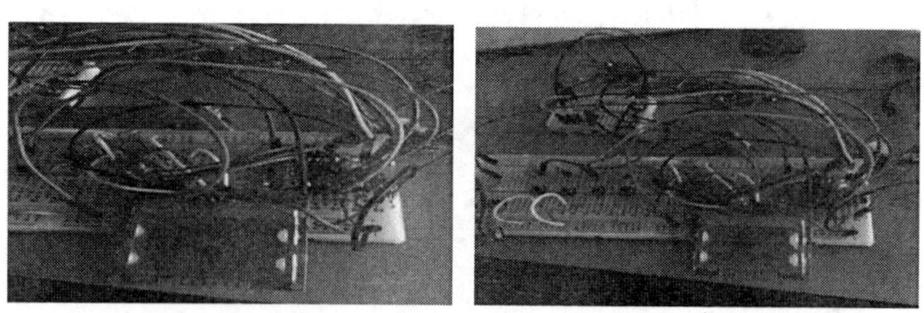

Figure 8. Start-up page and the main menu page of the watch

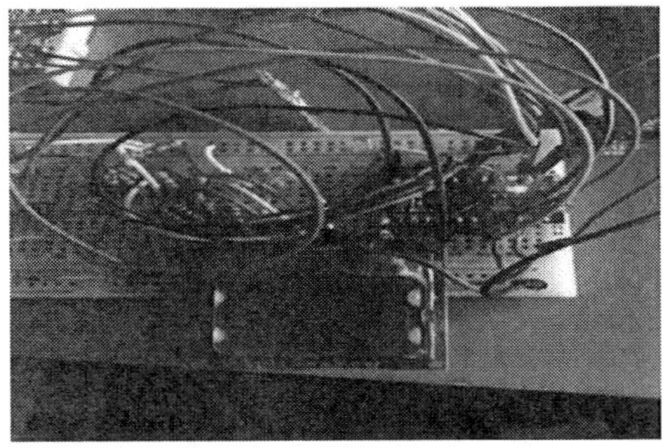

Figure 9. Representation of the watch attempting to connect to a wireless network

3. Results and Analysis

Figure 10 shows the user page, which provides a platform where messages are sent to other registered users. The platform has a check availability feature used to verify if a user is online for the message to be sent. Figure 11 is the database presentation of all users registered on the website. Upon initial registration, every user column for internet protocol (IP) address will be set to "NULL" by default. When a user of the watch connects to the network, it automatically updates the database with the current IP address indicating that the user is online and other users will be able to send messages through the website. Figure 12 illustrates an attempt to check the availability of a user that's offline. When the user connects to the network, the IP address will be updated automatically. Figure 13 illustrates this point for the user "Victor Tomoloju". As a result, the users messaging page will be active. Figure 14 proves this with an attempt to send a message to the user "Victor Tomoloju" through the updated IP address. The CU Watch has a WI-FI connection page where a user can navigate to search for available networks or check the connectivity of the CU watch to ascertain that it is still connected to a network. Networks with "O" in front signifies that the network is open while networks with 'C' in front signifies that the network is closed. Figure 15 shows the received message on the watch.

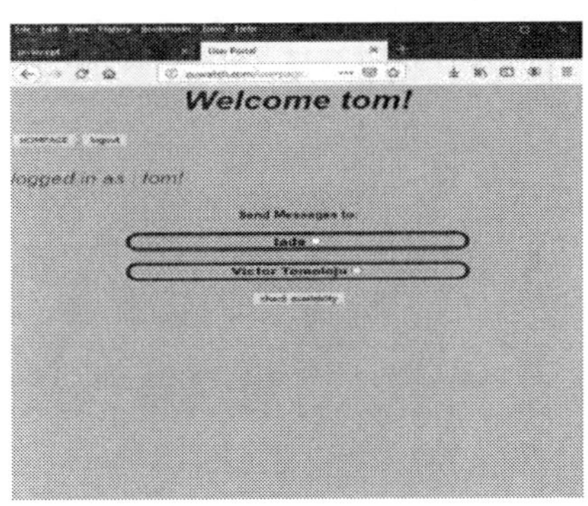

Figure 10. The user portal of a registered user

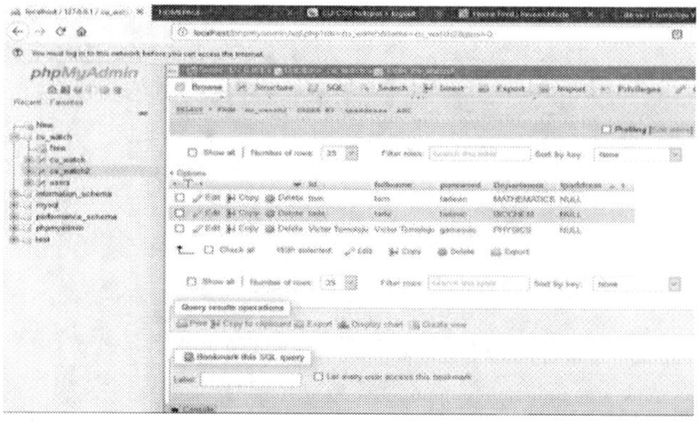

Figure 11. The database interface of the registered users on the website with the IP address of the users set to NULL

Figure 12. User availability check

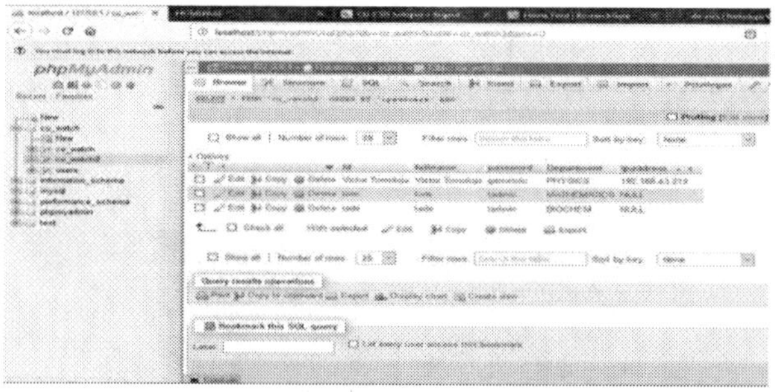

Figure 13. The database interface of the registered users on the website with the IP address of a user updated upon connection

Figure 14. A test of the messaging page for a user upon getting an IP address

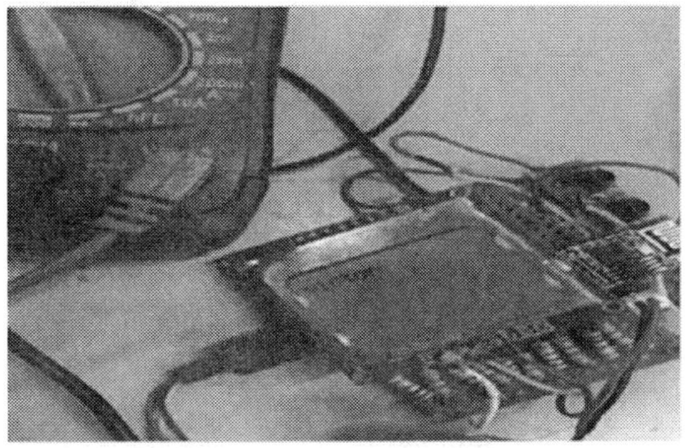

Figure 15. Received message on the CU watch

The watch was tested by sending messages on it and the response time gotten from ten attempts to send messages to the CU watch is displayed in Figure 16. It took approximately twenty-five seconds to send a message to the watch. The result obtained is in consonance with that obtained by [4].

ale

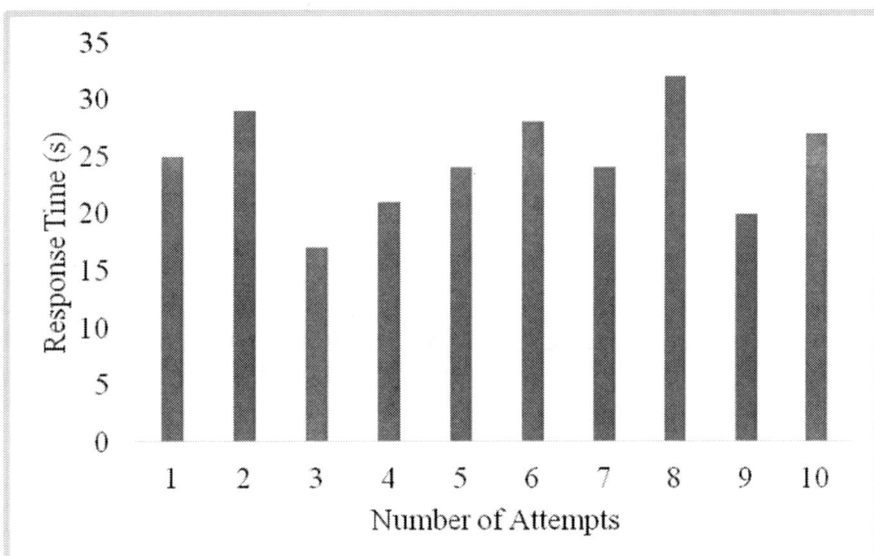

Figure 16. Response time of CU watch

4. Conclusion

A web interface was created to allow for interactions between various users. A smartwatch was developed that is able to instantly receive messages through a wireless network. A website was designed for the users of the watch to be registered and for sending messages to other users. The user data is stored using SQL database. The design of the Covenant University Watch and its components was successfully done; the system can send and receive messages through the platform. We hereby recommend that Covenant University invests into this idea to increase the overall community productivity. For future work, this project can be improved on and further enhanced by including a separate smaller microcontroller for processing of the network data. Also, the size and weight should be further reduced in conformity with standard smart wristwatches.

Acknowledgements
Authors acknowledged the support of Covenant University for the study

References

[1] Frost and Sullivan "*Smart Cities -A $1.5 Trillion Market Opportunity*" Retrieved from https://forbes.com/sites/sarwantsingh/2014/06/19/smart-cities-a-1-5-trillion-market-opportunity/ on (13/12/18).
[2] Chavis D. M and McMillan D. W 1986. Sense of community: A definition and theory, *Journal of Community Psychology*, **14(1):** 6-23,
[3] Matthews, O. V., Osaghe, N. E., Stanley, U., Enefiok, K. M., Ogukah, P. 2018. Implementation of a community emergency security alert system. *International Journal of Innovative Science and Research Technology,* **3**(6):2456-2165.
[4] Popoola, S. I., Popoola, O., Oluwaranti, A. I., Atayero, A. A. A, Badejo, J. A., Misra, S. 2018. A Cloud-Based Intelligent Toll Collection System for Smart Cities, Smart and Innovative Trends in Next Generation Computing Technologies 2017, *Communications in Computer and Information Science.*, **827**:653-663. Doi: 10.1007/978-981-10-8657-1_50.
[5] Ayara W. A, Omotosho T. V, Usikalu M. R, Singh M. S and Suparta W. 2017 Development of a solar charged laboratory bench power supply, *Journal of Physics: Conference Series*, **852(1):** 012044
[6] Nwoye C. D., Usikalu M. R., Babarimisa I. O, Achuka J. A and Ayara W. A. 2017 Construction of An Automatic Power Switch using Infrared Motion Sensor, *Journal of Informatics and Mathematical Sciences,* **9(2):** 331–337
[7] Usikalu M. R, Shittu A. H and Obafemi L. N 2018. Construction of an intelligent and efficient light control system, *International Journal of Mechanical and Production Engineering Research and Development (IJMPERD)*, **8(4):** 1057-1066
[8] Usikalu M. R, Adebesin E. and Obafemi L. N 2019. Design and construction of backpack mobile charger, *ARPN Journal of Engineering and Applied Sciences,* **14(21):** 3743 - 3746
[9] Spark-fun electronics. *Graphic LCD Hookup Guide*. Retrieved from https://learn.sparkfun.com/tutorials/graphic-lcd-hookup-guide on (21/12/18).
[10] Philips Semiconductors "PCD8544 48 X 84 pixels matrix LCD controller/driver" Retrieved from *http://pdf.datasheetcatalog.com/datasheet/philips/PCD8544U.pdf on* (21/12/18).
[11] Spark-fun electronics *ESP8266 Module (WRL-13678)*. Retrieved from https://mouser.com/ds/2/813/ ESP8266 ModuleV1-1095236.pdf on (15/12/18).

Power line carrier technologies: a review

E O Amuta[1], A Awelewa[1], A Olajube[1], T E Somefun[1], G Afolabi[1] and A S Uyi[1]

[1] Department of EIE, Covenant University, PMB 1023, Ota, Nigeria

Correspondence e-mail: elizabeth.amuta@covenantuniversity.edu.ng

Abstract. This paper gives an elaborate insight into the Power Line Carrier (PLC) communication technology in the power system. It explores the various characteristics, properties, and advancement of the power line carrier system and individual analysis of its components. These components, however, are in concerning power line carrier communications over high voltage lines majorly used in transmission stations. Power Line Carrier (PLC) components, including wave traps, coupling capacitors, line matching units, were highlighted. The paper discussed the operation principle of the Power Line Carrier (PLC) system together with their functionalities and how it makes transmission of information and power possible. It gives us the step by step functionalities of the different parts and how they make the information dispersion possible. A communication strategy was deployed for the effective transmission of data and information over the power line channel. Challenges were considered, solutions, and future directions for the advancements in the power line carrier system were also provided.

Keywords: Power line carrier, transmission line, Smart grid, home automation, internet Access

1. Introduction

Power line carrier (PLC) in communications describes the procedural stages involved for the transmission and receiving of data and information using high-voltage power lines as the channel or transmission path [1] [2]. However, this is achieved by superimposing high- frequency carrier signals on power lines at the nominal voltage level. PLC evolution has also been quite impactful in the distribution sector, making communications between household users and distribution substations possible with smart metering and other data acquisition features [3][4]. In this age of information technology, communication or rather the dispersion of information has become a significant necessity in virtually all sectors of life from the industrial sector to power, medicine, commercial and residential at large [5][6]. With evolutions in communication modes, other alternatives to PLC are also being studied for effective communication between substations in transmissions and end-users in the distribution sector[7-8]. The main focus of this review article is to explore the power line carrier communication technologies, applications, challenges, and solutions. Power line communication system block diagram presented in figure 1

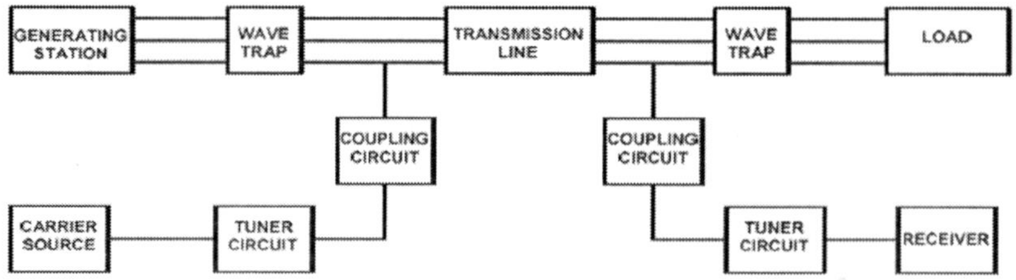

Figure 1. PLC Block diagram

2. Review of related works

Several works have been done in PLCC, some of the works are highlighted below:
Tong et al. [9] investigated the power transmission that gives ways of improving the method and means of communication on the conventional power line. The performance of PLC when transmission lines transfer of radio-frequency signals depends on the electric wiring of the power distribution system with environmental influences on power transmission lines also been considered. Berger et al. [10] studied Power Line Communications for Smart Grid applications proposing that though there are wireline and wireless communications in the power system, the PLC still does more functions in the modern power grid. By providing natural better data communication to the simple electricity wire, with no need to depend on telecommunication companies. The authors identify the use of power line communications in the smart grid. The communication scenarios highlighted in this paper are: High-voltage lines which have voltages in the range from 110kV to 380kV comprises of long overhead lines without branches. This high voltage lines connecting the primary transformer substations were also connected to the Medium-Voltage lines with the voltage range from 10kV to 30kV. The secondary transformer connects the Low-voltage lines which range from 110 V to 400V to the medium voltage lines.

In [11], the author provided an overview of the various technologies of the PLC and their configurations that are in the market considering their applications in various sectors. The focus was on the challenges that the PLC system might encounter in a few years to come, considering the distributed generation that will be using Electric vehicles, which can cause noise in the PLC frequency. Dushyant et al. [12] studied the power line communication application as a possible solution to frequency regulation in an intelligent microgrid with a frequency controller and diesel generator that operates in constant power mode. The PLC was used in regulating the frequency whenever there was a need to communicate information to the diesel generator from the control center. Vivek et al. [13] investigated the power transmission line issues, and the modeling approaches. Considering PLC as a new technology that still needs improvement in some vital aspects. The author studied the PLC over low –voltage power networks as the part gaining the researcher's attention both in the narrowband and the broadband area of applications with the communication medium of PLC modem.

Kuzlu, et al [14] considered wired communication technique using a cable, which can be categorized into power line carrier communication and fiber optics techniques. The fiber optics techniques offer a connection at a high-speed level. The PLC technologies can be cost-effective, more so Home plug with a data rate of up to 10 Mbps can be used. More so, a Home plug with a data rate of up to 10 Mbps can be used. They may not be suitable for many applications in the smart grid because of noises and interferences issues.

Zhilenkov et al [15] proposed the use of PLC technology for homes using IoT systems. Though RFID (Radio Frequency Identification), was initially used in IoT. It could be changed to PLC since radio frequency range in the urban areas are overloaded already, the network will no longer have extra load thereby transmitting data to a large number of devices already connected to the power network, on the same line.

IOP Conf. Series: Materials Science and Engineering 1036 (2021) 012062 doi:10.1088/1757-899X/1036/1/012062

3. Working Principle of the Power Line Carrier System

A power line carrier (PLC) system is operated by the superimposition of a high-frequency communication signal in the range of 1.6 Mega-Hertz to 30 Mega-Hertz upon the low- frequency power signals[16] [17] The useful information is retrieved from the filtering process, and the power signals are prevented from entering the communication systems to damage them. The structure of a power line communication system is given in Figure 2 [18].

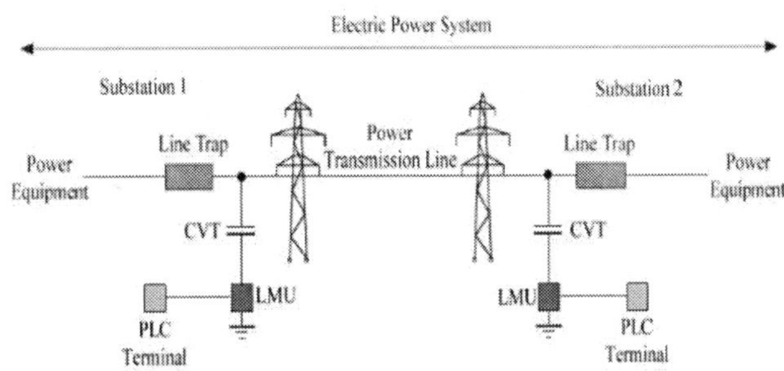

Figure 2. Schematic of Power Line Carrier [18]

The communication signals are prevented from entering the substation equipment. This critical filtering process is carried out by the coupling capacitor and the line trap unit. The coupling capacitor is connected to the PLC system, so it filters out the low-frequency power signals and allows the high - frequency communication signals to pass through the capacitor reactance of the capacitor is given in equation 1.

$$Xc \frac{1}{2\pi \times f \times C} \ (\Omega) \ \ldots\ldots\ldots\ldots\ldots\ldots\ldots\ldots\ldots(1)$$

Where X_C = Capacitive reactance, C = Capacitor capacitance, f = frequency. The frequency and capacitive reactance are inversely proportional. Consider a 5000pF capacitor for a low power frequency of 50Hz and a high communication frequency of 30MHz. The capacitor offers a high reactance of 636.619 kilo-ohms to the 50Hz power frequency, whereas it offers a 1-ohm resistance for the very high communication frequency. Conversely, the line/wave trap unit is the low pass filter for the substation equipment as it allows only the low-frequency power signals into the substation equipment and blocks the high-frequency communication signals from the power equipment. The wave trap inductive reactance is given in equation 2.

$$X_L = 2\pi \times f \times L \ (\Omega) \ \ldots\ldots\ldots\ldots\ldots\ldots (2)$$

Where X_L = inductive reactance, f = frequency, L = wave trap inductance. The wave trap offers a high impedance of 1.885 mega-ohms to the high-frequency communication signals but offers a very low impedance of 3 ohms for the low-frequency power signals.

A typical power line carrier communication system includes the following elements, which are the transmission line, Line trap, coupling capacitor tuning unit and blocking element, and relay system, Transmitter and Receiver, Protection, and earthing Battery. Some of these elements are highlighted below.

3.1 Line Traps: The line trap shown in Figure 3, also known as a wave trap, is a very highly inductive element with a capacitor in parallel to it. It is a high pass filter that chokes any high-frequency communication signals.

IOP Conf. Series: Materials Science and Engineering 1036 (2021) 012062 doi:10.1088/1757-899X/1036/1/012062

Figure 3. Line/Wave Trap [19]

3.2 *Line Tuning/Matching Units*: This unit is an aggregate of the tuning devices, earth switch, drain coil, and an isolation transformer with a lightning arrestor on its two sides.
3.3. *Coupling Capacitors:* This is a filtering component for the PLC terminal. It couples the high-frequency communication signals with low-frequency power signals on the power line.
3.4. *Drain Coil:* As the name implies, the drain coil is an inductive component that provides a very high impedance path to the ground for high-frequency communication components.
3.5. *Hybrids and Filters:* This is the interconnection of two or more transmitters on the same channel or path without any distortion or destructive superposition, that is: Interference.
3.6. *Master Oscillator and Amplifier:* The oscillator in the PLCC is used for generating high -frequency signals that are fed into the amplifier to compensate for any loss in the transmission.

4. Applications of Power Line Carriers

The application of PLC can be found to be extremely basic or can be quite challenging, considering the line characteristics and specifications used. Some of which are highlighted below,

4.1 Industrial Automation and Telecommunication Services

PLC networks can be applied to provide power services like remote billing, demand management, and meter reading. It also gives value-added services like e-business opportunities automation and remote control. Internet service and traditional telephony are telecommunication services offered by PLC for transmission of data, voice, or fax electronically. Current PLC networks can achieve speeds of 200Mbps[20].

4.2 Solutions for Electricity Suppliers

Power line carrier can serve as a solution to electricity suppliers in different ways. They can be used for fraud detection, system protection, and fault detection. In detecting fraud, the energy been supplied to the consumer can be compared to the power received from the transformer. Form this comparison, losses are determined, and fraud can be detected. In protecting the system, the communication link is capable of transferring control signals that will be used in protecting the system. In identifying a fault, the combined energy between the PLC communication systems can be used. Preventive maintenance can be done to improve the operation of the grid [21].

4.3. Carrier Protection Relaying of Transmission Line

The relay can release an inter trip command to protect the carrier; there are thee methods of doing this. They are the differential comparison method, distance protection relay, and the phase comparison method.[22].

4.4. Smart Grid Applications

The application of an existing power line infrastructure is an alternative to the creation of modern communication links in substations at medium voltage level because the traditional grids are not provided with communication capabilities. Also, PLC is used for fault location, detection, and isolation in a medium voltage network. Additional applications of PLC for medium voltage networks are the measurement of oil transformer temperature, analysis of the voltage on the secondary winding of the transformer, power quality measurement, and fault survey. [23].

4.5. Broadband Internet Access
Broadband internet access in PLC makes it possible for customers to access the internet via the present electrical structure. A primary advantage of the use of a PLC network is the means to reuse the existing power wired network to offer communication abilities. With an increase in the deployment of PLCs, they are now being implemented as an in-home smart grid and various other applications[24]. Multiple input multiple-output (MIMO) signal processing is recently being considered for use in broadband power line communications (PLC). Despite being a well-established ingredient to many wireless systems. Though some issues need to be solved before adapting multiple-antenna transmission in an electric power system, which is wired[25], therefore, an automated distribution is set up in which a power line carrier is used as a communication link[26].

4.6. Automatic meter reading
PLC technology is also applied to meter reading; data are automatically collected from energy meter is and sent to the central control room for billing and analysis. This saves regular trips helps billing not to be estimated, but based on real consumption.[20]

4.7. Home networking and Internet Access:
Several computers can be connected in a home and building by using the existing network, hence there is no need to install new wire or cable for the connections; this reduces installation cost and time by using the available low voltage power network for the home networking.
PLC is also used for Home Automation for controlling light and home appliances remotely, such as drapes and curtains[27].

5. Types, Challenges, and Solutions
5.1 Types
Power line carrier is majorly categorized into Narrowband PLC and Broadband PLC[28]. Narrowband Power Line carrier (NB-PLC) is a High-data-rate network that is now prominent in smart grid communications in the low-voltage grid [29][30]. In the smart grid and data distribution, broadband is applied to distances that are not more than 100m; hence suitably uses higher frequencies for indoor communication[31]. Comparison between Narrowband PLC and Broadband PLC is shown in table 1

Table 1. Comparison between Narrowband PLC and Broadband PLC

	Narrowband PLC	**Broadband PLC**
Bit rates	Up to 200kbps	Over 1Mbps
Frequency Band	Up to 500kHz	Over2MHz
Applications	Building automation, Renewable Energy, Street light, Advance Metering, Electric Vehicle Smart grid	Internet Gaming Audio

5.2. Challenges
In the existing electricity network, there are several challenges in the power transmission line infrastructure. One of the problems that are faced by the power line carrier is that the electrical power lines are usually very noisy due to the high amount of energy that they transmit. The signal strength and the operating frequency is also another challenge in power line carrier [32][33]. It has also been noticed that with the deployment of power line carriers, there has been a significant interference with the present radio communication environment. For example, in developed countries like Canada that have radio service providers that operate with a range of 2-80MHz, there will be an interference that will be caused by the power line carriers.

Bad weather can also pose a challenge to the power line carrier as they had adverse effects on the system, some of the impacts of harmful weather include the following;

a Inability to propagate on the conductor in situations where massive frost forms on the carrier line
b When it is raining, the effect of contaminants is increased
c Noise can also be introduced in lousy weather
d Poor compatibility with users of the radio spectrum
e Signal attenuation
f Security issues with the integration of internet services
g Poor coordination between the telecom and power providers
h Low voltage transformers allow electricity through it at small losses and low frequencies but not higher rates.

5.3 Solutions

From the above most of the challenges are caused by an emission, so the proposed solution is incorporating an emission limit for the line carriers, especially for the broadband power line carrier in certain countries like the US, regulations have been put in place to integrate the emission limits [34]. Prohibition of specific frequency bands can also be a solution to the interferences caused by power line carriers, including groups used for an aeronautical service, public safety, and national defense[35]; this will serve as a safety measure. Some other criteria include Filtering, frequency avoidance Power reduction, and Controllable shut-down features.

6. Opportunities and Way Forward

Power line carrier (PLC) involves sending, controlling, and receiving data and or information over existing power cables. It is, therefore, essential to reuse the power lines for communication purposes [36]. The electrical distribution sector has always constituted a regularized wiring system but was never built with the intent of the communication. The area of opportunities involving applications and use of PLC can be seen in such applications as automatic metering systems, programmable controllers, and real-time demand and supply management[37]. Bandwidth increment of the PLC would grossly prepare for the much-needed capacity envisioned by 2023. Expanding the available bandwidth by over five times introduces broadcast interference concerns.

7. Conclusion

The power line carrier system is an effective means of transmitting information via power lines for utilization in power stations. However, the reservations made by the channel being the power line in terms of bandwidth is low. And with the advancements of networking and information technology and other communications modes, there is a need for development in the power line carrier system to enable the transfer of a large amount of information at a specified bit rate possible. The traditional power line carrier system for control and communication purposes is still widely revered in power systems despite other advancements in communication modes; it remains highly reliable in issuing control commands to power systems equipment.

Conflict of Interest: The authors declare at this moments that there is no conflict of interest.

Acknowledgments: Authors express their appreciation to Covenant University for sponsoring this research.

References

[1] Abdulkareem A., 2020. Investigating the Effects of Bus Numbering Sequence in the Load-Flow of a Radial Transmission Network," vol. **20** No:02, no. April, pp. 1–13.
[2] Adesina L. M, Abdulkareem A, and Ogunbiyi O, 2020. Effects of Ground Resistivity and Tower Structural Design on Transmission Line Symmetrical Components," vol. **9,** no. June, pp. 2115–2126, doi: 10.35940/ijrte.A2519.059120.

[3] Gunasekaran R, Karthikeyan C, Pavalam J, Mohanapriya P, Preethi A, and Indhumathi V, 2017. Power Line Carrier Communication Using Automated Meter Reading," vol. **1**, no. 4, pp. 104–109, doi: 10.11648/j.be.20170104.12.

[4] Shekoni O. M, Hasan A.N, and Shongwe T, 2018. Applications of artificial intelligence in powerline communications in terms of noise detection and reduction: a review, *Aust. J. Electr. Electron. Eng.*, vol. **15,** no. 1–2, pp. 29–37, doi: 10.1080/1448837X.2018.1496689.

[5] Xiao Z. *et al.*, 2019. Power communication network design considering global information fusion part two applications and explorations, *Procedia Comput. Sci.*, vol. **155**, no. 2018, pp. 768–773, doi: 10.1016/j.procs.2019.08.112.

[6] Mudriievskyi S, 2014. Power Line Communications: State of the art in research, development and application, *AEU - Int. J. Electron. Commun.*, vol. **68**, no. 7, pp. 575–577, doi: 10.1016/j.aeue.2014.04.003.

[7] Oladipo K, Felix A A, Bango O, Chukwuemeka O, and Olawale F, 2018. Power Sector Reform in Nigeria: Challenges and Solutions, *IOP Conf. Ser. Mater. Sci. Eng.*, vol. **413**, no. 1–13, doi: 10.1088/1757-899X/413/1/012037.

[8] Bhoyar N. I, Bhat S, and Patil M, 2018. Review on Implementation of Power Line Carrier Communication Technique in Smart Grid, *Proc. - 2018 Int. Conf. Smart Electr. Drives Power Syst. ICSEDPS*, pp. 256–260, doi: 10.1109/ICSEDPS.2018.8536074.

[9] Tong J, He Y, Li B, Deng F and Wang T, 2016. A Novel Design of Radio Frequency Energy Relays on Power Transmission Lines, *Energies*, vol. **9,** no. 6, p. 476, doi: 10.3390/en9060476.

[10] Berger L T, Schwager A, and Escudero-Garzás J J, 2013. Power line communications for smart grid applications, *J. Electr. Comput. Eng.*, vol. 2013, doi: 10.1155/2013/712376.

[11] Lujano-rojas J M, Dufo-lópez R, Atencio-guerra J L, E. M. G. Rodrigues, J. L. Bernal-agustín, and J. P. S. Catalão, 2016. Operating conditions of lead-acid batteries in the optimization of hybrid energy systems and microgrids," *Appl. Energy*, vol. **179**, pp. 590–600, doi: 10.1016/j.apenergy.2016.07.018.

[12] Sharma D., A. Dubey, S. Mishra, and R. K. Mallik, 2019. A Frequency Control Strategy Using Power Line Communication in a Smart Microgrid," *IEEE Access*, vol. **7**, no. c, pp. 21712–21721, doi: 10.1109/ACCESS.2019.2897051.

[13] Mohan V., J. G. Singh, W. Ongsakul, A. C. Unni, and N. Sasidharan, 2015. Stochastic Effects of Renewable Energy and Loads on Optimizing Microgrid Market Benefits," *Procedia Technol.*, vol. **21**, no. January 2016, pp. 15–23, doi: http://dx.doi.org/10.1016/j.protcy.2015.10.004.

[14] Kuzlu M, Pipattanasomporn M, and Rahman S, 2016. Review of communication technologies for smart homes/building applications," *Proc. 2015 IEEE Innov. Smart Grid Technol. - Asia, ISGT ASIA 2015*, pp. 1–6, 2016, doi: 10.1109/ISGT-Asia.2015.7437036.

[15] Zhilenkov A. A, Gilyazov D. D, Matveev I. I, and Krishtal Y. V, 2017. Power line communication in IoT-systems," *Proc. 2017 IEEE Russ. Sect. Young Res. Electr. Electron. Eng. Conf. ElConRus,* pp. 242–245. doi: 10.1109/EIConRus.2017.7910538.

[16] Yigit M, Gungor V.C, Tuna G, Rangoussi M, and Fadel E, 2014. Power line communication technologies for smart grid applications: A review of advances and challenges, *Comput. Networks*, vol. **70**, pp. 366–383. doi: 10.1016/j.comnet.2014.06.005.

[17] Zhu W, Zhu X, Lim E, and Huang Y, 2013. State-of-art power line communications channel modelling," *Procedia Comput. Sci.*, vol. **17**, pp. 563–570. doi: 10.1016/j.procs.2013.05.072.

[18] Mahmood S. H., Salih A. M., and Khalil M. I, 2019. Broadband services on power line communication systems: A review, *Proc. - 2019 22nd Int. Conf. Control Syst. Comput. Sci. CSCS 2019*, pp. 465–470, doi: 10.1109/CSCS.2019.00085.

[19] Lampe L, Tonello A M, and Swart T G, 2016. *Communications Principles , Standards and Applications From Multimedia*, Second edi. New Delhi, India: John Wiley & Sons, Ltd.

[20] Mannan A, Saxena D K, and Banday M, 2014. A Study on Power Line Communication," *IEEE Trans. Consum. Electron.*, vol. **4**, no. 7, pp. 7–10.

[21] Yousuf M. S. and El-shafei M, 2007. Power Line Communications: An Overview – Part I, *4th Int. Conf. Innov. Inf. Technol.*, pp. 218–222, doi: 10.1109/ICTTA.2008.4530268.

[22] Di Fazio A R, Erseghe T, Ghiani E, Murroni M, Siano P and Silvestro F, 2013. Integration of renewable energy sources, energy storage systems, and electrical vehicles with smart power

distribution networks," *J. Ambient Intell. Humaniz. Comput.*, vol. **4,** no. 6, pp. 663–671. doi: 10.1007/s12652-013-0182-y.

[23] Galli S, Scaglione A, and Wang Z, 2011. For the grid and through the grid: The role of power line communications in the smart grid," *Proc. IEEE*, vol. **99**, no. 6, pp. 998–1027. doi: 10.1109/JPROC.2011.2109670.

[24] Cano C, Pittolo A, Malone D, Lampe L, Tonello A.M, and Dabak A.G, 2016. State of the Art in Power Line Communications: From the Applications to the Medium," *IEEE J. Sel. Areas Commun.*, vol. **34**, no. 7, pp. 1935–1952, doi: 10.1109/JSAC.2016.2566018.

[25] Papadopoulos T. A., Chrysochos A. I., ElSamadouny A, Al-Dhahir N, and Papagiannis G. K, 2017. MIMO-OFDM narrowband-PLC in distribution systems: Impact of power transformers on achievable data rates," *Electr. Power Syst. Res.*, vol. **151**, pp. 251–265. doi: 10.1016/j.epsr.2017.05.039.

[26] Singh H. R., 2016. Available Online at www.ijarcs.info Broadband Communication over Power Lines : Issues , Challenges and Opportunities," vol. **7**, no. 7, pp. 108–119.

[27] Rouissi F, Vinck A. J. H., Gassara H, and Ghazel A, 2019. Improved impulse noise modeling for indoor narrow-band power line communication," *AEU - Int. J. Electron. Commun.*, vol. **103**, pp. 74–81, 2019, doi: 10.1016/j.aeue.2019.02.019.

[28] Dvořák J, Novák J, and Kocourek P, 2014. Energy efficient network protocol architecture for narrowband power line communication networks, *Comput. Networks*, vol. **69**, pp. 35–50, doi: 10.1016/j.comnet.2014.04.012.

[29] Wolkerstorfer M, Schweighofer B, Wegleiter H, Statovci D, Schwaiger H, and Lackner W, 2016. Measurement and simulation framework for throughput evaluation of narrowband power line communication links in low-voltage grids," *J. Netw. Comput. Appl.*, vol. **59,** pp. 285–300, doi: 10.1016/j.jnca.2015.05.022.

[30] Masood B and Baig S, 2016. Standardization and deployment scenario of next generation NB-PLC technologies," *Renew. Sustain. Energy Rev.*, vol. **65**, pp. 1033–1047, doi: 10.1016/j.rser.2016.07.060.

[31] Andreadou N and Fulli G, 2018. NB-PLC channel: Estimation of periodic impulsive noise parameters and mitigation techniques," *Int. J. Electr. Power Energy Syst.*, vol. **103**, pp. 146–158, doi: 10.1016/j.ijepes.2018.05.011.

[32] Kabalci Y, 2016. A survey on smart metering and smart grid communication," *Renew. Sustain. Energy Rev.*, vol. **57,** pp. 302–318, doi: 10.1016/j.rser.2015.12.114.

[33] Kabalci E and Kabalci Y, 2019. *Energy Systems in Electrical Engineering Smart Grids and Their Communication Systems.*

[34] López G. *et al.*, 2019. The role of power line communications in the smart grid revisited: Applications, challenges, and research initiatives," *IEEE Access*, vol. **7,** pp. 117346–117368, 2019, doi: 10.1109/ACCESS.2019.2928391.

[35] Sharma K and Saini L M, 2017 Power-line communications for smart grid: Progress, challenges, opportunities and status," *Renew. Sustain. Energy Rev.*, vol. **67**, pp. 704–751, 2017, doi: 10.1016/j.rser.2016.09.019.

[36] Cataliotti A, Cosentino V, Di Cara D, Russotto P, and Tinè G, 2012 On the use of narrow band power line as communication technology for medium and low voltage smart grids," *2012 IEEE I2MTC - Int. Instrum. Meas. Technol. Conf. Proc.*, pp. 619–623, doi: 10.1109/I2MTC.2012.6229503.

[37] Mckeown A, Wilcox A, and Thomas P, 2016. Noise in Broadband Power-Line Communication and Future Bandwidth Growth" no. March 2016, pp. 49–58.

Design of a digital fuel gauge with distance-to-zero indicator

T Somefun*[1], C Awosope[1], A Abdulkareem[1], J Ogelenya[1], A Orimogunje[1] and C Somefun[2]

[1]Department of Electrical and Information Engineering, Covenant University, Ota, Nigeria
[2]Department of Computer Science and Engineering, Ladoke Akintola University, Ogbomosho, Nigeria.

*Corresponding author's e-mail address: tobi.shomefun@covenantuniversity.edu.ng

Abstract. The fuel gauge in automobiles measures the amount of fuel in the tank through the sensing unit located in the tank and the analog indicator on the dashboard of the vehicle shows the quantity of fuel per time. This work is designed to ensure the display of the exact volume of fuel in the vehicle in digital form instead of the usual analog fuel indicator. It will also help to prevent inaccurate measurement of fuel at the filling station by showing the quantity of fuel during refuelling. So, a digital fuel gauge system has been designed and constructed to measure the exact quantity of fuel in the tank and display this measured quantity digitally in litres, by making use of an ultrasonic sensor and programmable microcontroller and the result obtain is ±0.96% accurate.

Keywords: Fuel gauge, Automobile, Indicator, Digital, Ultrasonic Sensor

1. Introduction

Today, though everything is going digital, most of the fuel meters in vehicles today are analog-based [1] and the few ones which are digitally based, only display the amount of fuel in percentage values or in bars. The most common fuel indicator system makes use of the type known as resistive float. In this type fuel indicator, measurement of the fuel level in the tank is leveraged on sensors. And it comprises two modules: the sensing module which measures the level of fuel in the tank and the gauge which displays to the driver, the measured fuel level with a deflecting needle oscillating between empty (E) and full (F) [2]. This fuel meter is not that accurate and it provides insufficient knowledge on the amount of fuel present in the tank. This creates a problem of determining the actual fuel available when refilling the fuel tank as well detecting any form of discrepancy in fuel purchased [3].

The concept of this paper is to produce an advanced fuel measuring system which indicates the exact quantity of fuel available in litres. With increase in the variety of sensor usage [4, 5], this paper explores the ability of an ultrasonic sensor to determine the actual amount of fuel in car tank and also calculate how much distance that can be covered with corresponding amount of fuel in the tank. These values, with the help of a micro-controller, are displayed on the LCD screen.

Ultrasonic sensor helps to determine how far (i.e. distance) is the surface of the liquid (i.e. fuel) to the top of fuel tank when measuring the quantity of fuel in the tank. These sensors principle of operation depends on echoes from radio or sound waves which is usually analog [6]. Besides that, it is also called

IOP Conf. Series: Materials Science and Engineering 1036 (2021) 012063 doi:10.1088/1757-899X/1036/1/012063

a transceiver. This means the sensor can transmit and receive signals [7]. Ultrasonic sensors work at high frequency to calculate distance travelled by the signal send and echo received by the sensor. These sensors are also capable of evaluating the time interval between the sent signal and the received echo to calculate the actual distance to an obstacle (surface of the fuel in this case) [8].

2. Related works

Trupti Wable et al. [9] proposed a fuel metering system using PIC16F877A micro controller. The system had a fuel level sensor and gear level connected to the microcontroller, so as to read the amount of fuel in the vehicle's tank and also to get information on the current state of the gear level. The fuel sensor used was the float-type sensor. The LCD screen was interfaced with the PIC microcontroller so that it displays the fuel level and also the gear level. The LCD screen shows the amount of fuel present in the fuel tank digitally. Also, in this system there was a GSM module for sending message to the owner in the event of fuel theft. But this system lacked an indicating system to alert or warn the driver in the event of low fuel in the vehicle's tank.

Nitin Jade et al. [10] proposed a fuel gauging system consisting of sensor, computing unit and indicator. The computing unit was made up of an Electronic Control Unit (ECU), Central Processing Unit (CPU) and a modulator. The sensor used to measure fuel is the density sensor. By using the vibrational frequency of the micro tube, the density of the fluid can be measured. Based on the factors that influence mileage, a number of sensors were put in place to retrieve information on those factors. ECU receives information from each individual sensor. The CPU collects data from ECU and processes it and then sends to modulator which modulates the signals and displays them on the indicator. The indicator was the LCD screen and it displayed the amount of fuel left in bars. But the limitation of this system is that it is not suitable for low and medium class vehicles. This is because as a greater number of sensors were used, the system became expensive and it is only suitable for luxurious vehicles. The failure of any of the sensors can stop the system from working.

Avinashkumar et al. [11] proposed a digital fuel indicator using PIC microcontroller which also indicates distance to zero. The system consists of the PIC16F877A microcontroller, the float type sensor, analog-to-digital (A/D) converter and the LCD screen. It measured the changes in the voltage at different fuel levels and computed distance to zero by considering a fixed mileage. In [11], the resistance position changes as the fuel level in the tank changes through the float type sensor placed inside the fuel tank. The resistance position is converted to digital value through analog-to-digital converter unit. The LCD is interfaced with the PIC microcontroller to show the value of the fuel tank in digits. The distance to zero was calculated by coding the microcontroller with respect to two inputs which are speed and fuel level. However, this system also lacks an indicator to alert or warn the driver in the event of low fuel in the vehicle's tank.

2.1 Existing Fuel Gauge Systems

The most common fuel gauge device is the resistive float type sensor. It comprises the sensing unit and the gauge unit. The former is mainly for fuel level measurement while the latter is mainly display of the measured fuel level.

In Figure 1, the gauge made of a bimetallic strip with different thermal co-efficient of expansion. The operation of this is linear such that as the resistance reduces, current increases which in turn cause temperature to be increased and thereby heating up the strip. During the heating process one metal expands more than the other causing the strip to become curved and change the position of the needle on the fuel gauge. In the reverse case, increase in resistance allows low current passage through the heating coil which causes the bimetallic strip to cool. As the strip cools, it straightens out, pulling the gauge from full to empty [12].

IOP Conf. Series: Materials Science and Engineering 1036 (2021) 012063 doi:10.1088/1757-899X/1036/1/012063

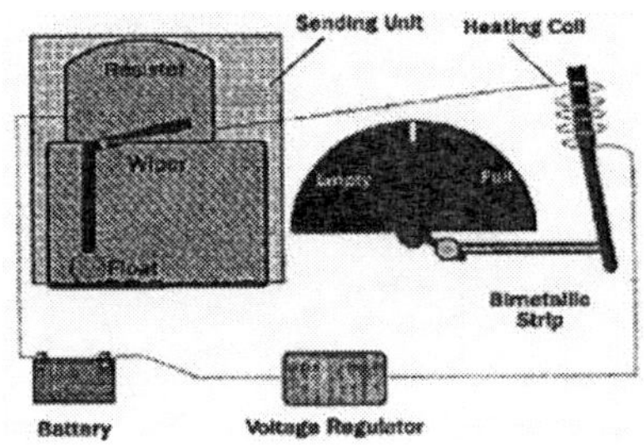

Figure 1. Traditional Fuel Measurement System (Bimetallic strip type)

2.2 Capacitance Level Sensing

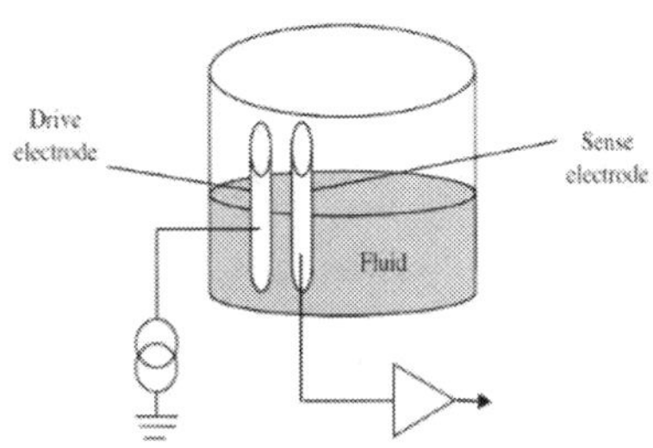

Figure 2. Traditional Fuel Measurement System (Capacitor type)

Another type of fuel measuring device is capacitive type. In this type, two conducting electrodes are placed at a fixed distance to each other as shown in Figure 2. From the mathematical relationship between capacitance of two parallel electrodes and dielectric, therefore, as the fuel level increases upward the two parallel electrodes, the value of the capacitance changes with respect to the fuel level [13].

3. Materials and method

In the design of this device, the different stages involved are discussed as follows:
Power supply: The power supply feeds the Arduino microcontroller. It is a DC power supply and it is either supplied by a 9-V DC battery or a regulated power supply. The regulated power supply consists of a step-down transformer which steps down the voltage level from 240V AC to 12V AC. The 12V AC is then rectified to 12V DC with Bridge rectifier. The ripples of the rectified DC waveform are eliminated by utilizing a capacitor. This is then regulated to 9V using voltage regulator LM7809.
In this work, the sensor used is the ultrasonic sensor. The ultrasonic sensor, which is directly connected to the fuel tank of the vehicle, will sense/find the level of fuel in the tank in litres.
The Arduino microcontroller, with the signal from the sensor, makes the required calculations concerning the amount of fuel in the tank and the corresponding distance to be covered by the vehicle. The Arduino is interfaced with the Liquid Crystal display (LCD) which displays the amount of fuel in the tank and the corresponding distance to cover. The block diagram of this system design is given in Figure 3.

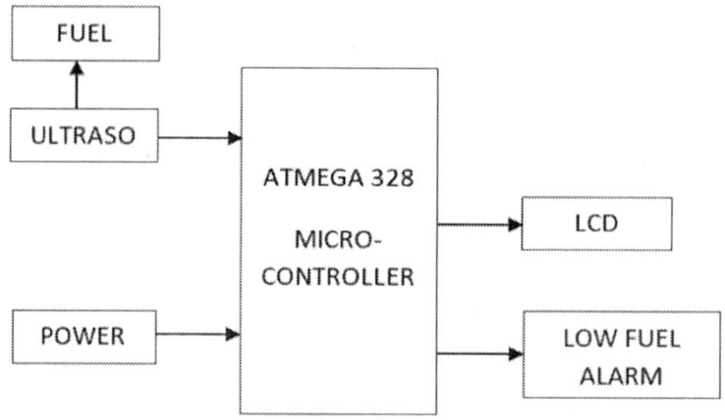

Figure 3. Block Diagram of the Design

3.1 System Implementation and Testing
The implementation of the work is divided into two different parts, namely:
- Software Implementation; and
- Hardware Implementation

3.1.1 Software Implementation
The coding of the Arduino Nano microcontroller was carried out through the Arduino Integrated Development Environment (IDE) software on a personal computer (PC). The Arduino codes used were written in C language, compiled and uploaded to the Arduino Nano board via its mini USB port.

3.1.2 Hardware Implementation
The hardware implementation of this project was grouped into three stages. These stages are as follows:
- Breadboarding of the components;
- Mounting and Soldering of the components on Vero board; and
- Packaging of the entire circuit.

4. Results
After testing the overall system, the results obtained are shown in the Table 1. From Table 1, it is clear that as the fuel level in the tank increases, the measured distance by the ultrasonic sensor reduces. This means that the fuel level in the fuel tank determines how far the ultrasonic signal can travel, hence, the fuel surface in the fuel tank obstructs the ultrasonic signal. It can also be seen from Table 1 that the displayed value by the proposed design is approximately equal to actual quantity of fuel in the tank and directly proportional to the calculated distance with respect to fuel in the tank. On the average, the percentage accuracy of the proposed design is ±0.96%.

Table 1. Overview Results of System Testing

Actual Quantity of Fuel in the Tank (Litres)	Measured Distance from the Ultrasonic Sensor to the Surface of the Fuel (cm)	Displayed Value of the Fuel in the Tank (litres)	Calculated Distance to Zero (km)
0	34	0.00	0.00
4	24	4.07	40.72
10	10	9.94	99.43
11	8	10.91	109.12
14	2	14.10	141.04

5. Conclusion

In conclusion, the implementation of the digital fuel gauge gives the driver an actual value of the volume of fuel in the tank in litres. The designed system also gives the driver an estimate of the distance that can be travelled with the volume of fuel in the tank. The implementation of this designed system i.e. digital fuel gauge will alert the driver in the event of a low fuel in the vehicle's tank.
Furthermore, it will assist in curbing fuel theft due to inaccurate measurement of fuels at some filling stations because the driver will be able to see the exact value of the fuel in the tank before refilling and after refilling at the fuel station.

6. Recommendations

The following recommendations are made for further works;

- A GSM SMS technology can be incorporated with the digital fuel gauge in order to detect fuel theft or leakages from the tank. The GSM at the start of the vehicle and from time to time sends messages to the user intimating him/her about the volume of the fuel in the tank. This will help the user notice any fuel theft; and
- A hall effect sensor could be integrated with the digital fuel gauge system to measure the speed of the vehicle and make a more accurate calculation on the distance that can be covered with the available fuel in the tank.

Acknowledgement

Our sincere gratitude goes to Covenant University for providing a conducive platform through which this study is carried out and her financial support.

References

[1] L. D. Mrs.P.Geethabai ME., M.Dharani, P.Haripriya, G.Lavanya, 2017. Design and Implementation of GSM-Based Digital Fuel Meter and Fuel Theft Detection Using PIC Microcontroller," *IOSR Journal of Electronics and Communication Engineering,* 2017.
[2] Gijre M. M., Mane A., Gadade R., and Gandhi S., "Smart Fuel Level Indication System," 2017.
[3] Saeliw A., Hualkasin W., Puttinaovarat S., and Khaimook K., 2019. Smart Car Parking Mobile Application based on RFID and IoT.
[4] Sousa, Pedro José, Rafael Tavares, Paulo Abreu, and M. Teresa Restivo. 2017."NSensor–wireless sensor network for environmental monitoring." International Journal of Interactive Mobile Technologies. vol. **11**, no. 5, pp. 25-36,
[5] Sipani J. P., Riki H. Patel, Trushit Upadhyahya, and Arpan Desai. 2018."Wireless sensor network for monitoring & control of environmental factors using arduino." *International Journal of Interactive Mobile Technologies*. vol. **12**, no. 2 pp. 15-26,
[6] Adeyinka, A. A, Samuel I.A, Awelewa A A, and Ike U. Dike. 2013. Design and development of a microcontroller based automatic switch for home appliances. *International Journal of Engineering Science Invention*. vol. **2** no. 10 , pp. 24-31,
[7] Asite O, Ukaoha I, Okon E, Adebiyi A, and Ayo C, Solving terrorism security challenges using sensor network technology, 2017.
[8] Marek C and Jerzy M., 1970. Automatic digital fuel indicator, particularly for car, ed: *Google Patents,* 1970.
[9] Wable T. K. and Shinde R. R., 2016. GSM Based Digital Fuel Meter and Fuel Theft Detection using PIC Microcontroller," vol. **3**, 2016.

[10] Jade N, Shrimali P, Patel A, and Gupta S, 2014. Modified type intelligent digital fuel indicator system, pp. 2278-1684.
[11] Avinashkumar A, Singaravelan U, Premkumar T, and Gnanaprakash K, 2014. Digital fuel level indicator in two-wheeler along with distance to zero indicator, vol. **11**, pp. 80-84.
[12] Divakar V, 2014. Fuel gauge sensing technologies for automotive applications.
[13] Kumar C. M. and. Choudary R, "Digital Fuel Measuring System with Distance to Zero and Fuel Fraud Indicator," 2017.

A review on the impact of information and communication technology in civil engineering practice

J Oluwafemi[1], A Ede[1], O Ofuyatan[1], S Oyebisi[1] and D Bankole[2]

[1]Civil Engineering Department, Covenant University
[2]Industrial Chemistry Department, Landmark University

E- mail address: john.oluwafemi@covenantuniversity.edu.ng

Abstract. Information and Communication Technology is playing a vital role in virtually almost every field of human operation such as education sector, medical field, banking sector, agricultural sector, field of engineering and more. Information and Communication Technology has substantially eased operations in several fields and has helped to save resources in the likes of labor, cost and time. It has also helped to drastically increase the speed at which work is done. This paper reviews the impact of information and communication technology in civil engineering practice and examines operations such as the construction industry, road transportation industry, engineering education, structural engineering, energy efficient buildings, building preservation and engineering design. Information and Communication Technology is seen to be very relevant for sustainability in civil engineering practices across literature.

Keywords: Engineering, Civil Engineering, Information and Communication Technology, Impact

1. Introduction

The role of information and communication technology (ICT) in our immediate society cannot be downplayed. Information and communication technology has played vital roles in several spheres of life and has brought about ease and advancements in many of human operations. Among the important contribution of ICT is cost reduction [1]. There is always a substantial and cumulative cost reduction benefit of application of ICT in a particular field. The use of ICT has also been known for its efficiency in time management. The application of ICT in any field of operation can substantially reduce the normal time with a wide gap such as usual operation that would have taken months, weeks or hours without the application of ICT. Among many fields of human operations where ICT has been of evident impact are telecommunication industry, design and construction industry and banking industry, the education sector and more. ICT has continued to find its place in almost or all aspects of man's operations. ICT has remained relevant to both developed countries and developing countries of the world. It has introduced advancements in the economy of nations and hence, its roles cannot be done without. Among the nations of the world where ICT has had great contribution, Nigeria as a country is not left out [1]. In Nigeria, ICT has contributed immensely in the area of her education, design and construction industry, road sector, education, telecommunication and many more [1],[2]. This paper presents a review of some applications of ICT in some sectors in civil engineering practice. These sectors include the construction industry, road transportation industry, education sector, engineering design, building preservation, energy efficient buildings and structural engineering.

2. ICT in Civil Engineering Practice

The role of ICT in some fields of operations in Civil Engineering practice are as related in this section

2.1. ICT in the Construction Industry

The use of ICT in the construction industry has witnessed a continuous increase over the years, being that it has continued to gain acceptance in operations across this industry. The introduction of ICT in the construction industry has made it possible to model projects in a virtual environment ahead of the physical environment. The possibility of having projects modelled ahead of construction has improved communications between construction team members and has enabled stakeholders to have a clear grasp of what to expect even before the project is implemented on site from start to finish. Nevertheless, despite the wide acceptance of the engagement of ICT in this industry, stakeholders in construction projects still face certain challenges in the aspect of communication especially when their projects reach the phase of implementation [3]. According to [4], Some of these communication challenges can be traced to gaps in the integration between what had been designed and what is been implemented in the production process. Other challenges being encountered by stakeholders in the use of ICT in construction are problems of relating technical content adequately, the ideal environment for the operation of the technology, several collaborations between stakeholders both on the site and off the site. Most of these challenges are due to the fact that the construction industry is predominantly on site and a good way to tackle these challenges will be to ensure the conveyance of the right information to the right place at the right time and a good understanding of the impact of the ICTs on different phases of a project is very important to help address communication challenges faced by stakeholders.

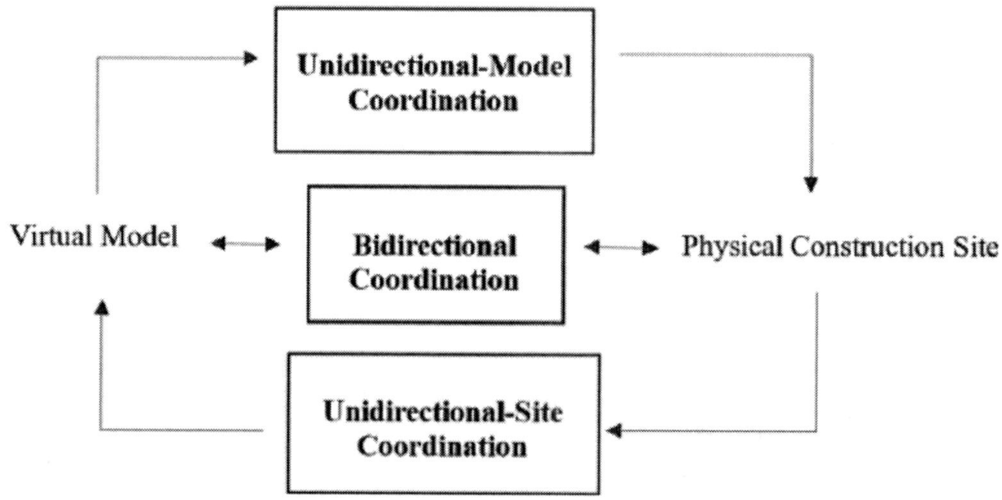

Figure 1. Unidirectional and Bidirectional Information Communication Flow. Source: [3]

[3] related the four different types of information flow in the use of ICT in construction projects as related in the figure 1. These information flows are undirectional-model coordination, unidirectional-site coordination, non-automated bidirectional coordination and automated bidirectional coordination. Most recent research works have focused us exploring the use of ICT in the facilitation of unidirectional-model mode of communication [5],[6].

2.1.1. Potential implementation of technologies in the construction industry

The following are the potential implementation of technologies in the construction industry:

i. An important area for the implementation of technology in communication in the construction industry is exploring Radio Frequency Identification. This is very vital to help manage supply chain in the construction, quality control, adequate tracking of resources and inventory and life cycle management of the project. This implementation is necessary for cost reduction and time resource saving [7], [8].

ii. 3D laser scanning is a necessary area for exploration in the construction industry. Areas where the 3D scanning has been engaged in previous times are quality control process, concrete casting, surveying process, alignment of highway and paving operations [9].

iii. Robotics technology implementation in the construction industry is another field that is gaining exploration. This has facilitated safety and security monitoring on site. It has also facilitated on-site image capturing and 3-D modelling [10].

iv. Wireless connection technology implementation in the construction industry is one area of technology that has facilitated quality real-time data flow, information exchange and effective distant communication within the shortest time possible [11].

v. Mobile computing technology implementation in the construction industry has facilitated management of operations on site, monitoring on site and information sharing [12].

vi. Quick response codes and NFC technology implementation in the construction industry have enhanced operations in the area of material tracking on site, accounting and purchasing, recording keeping of used materials and controls in documentation [13].

2.2. ICT in the Transportation Industry

Among the many impacts of the implementation of ICT in the road transportation industry is the high level of safety it brought on-board. ICT in road transportation industry has substantially reduced loss of lives and properties due to clashes between vehicles on the road. A way that this has been made possible is through invention of traffic and warning lights that coordinates and communicates efficiently in navigations on the road to help avoid vehicular collisions. Though much funding might be required to setup these technologies in the required places in a city, the benefit outweighs the cost [14].

2.3. ICT in Education Sector

According to [15], the use of ICT in teaching civil engineering related courses has gained wide acceptance worldwide over the past few years. It has created a departure from immaterial imagination to a realistic visual perception for students across the learning classrooms. Years before now, dictation of lecture materials in classrooms by teachers was the approach, ICT has been able to bridge this gap such that technology implementation has made it possible for teachers to project their lecture materials thereby saving time and easing communications in the classrooms. Implementation of technology in the learning environment has also introduced the use of several computer applications such Microsoft Office, Microsoft PowerPoint, Google Talk, Skye, Microsoft Outlook and Adobe PDF Reader. These applications have made learning and teaching process very easy. The implementation of technology in the education sector such as the introduction of E-learning platform has also made distant learning and exchange in learning activities possible [16].

2.4. ICT in Engineering Design

Several years back, design in engineering takes a very long time. The duration for the design of a mini project such as a residential building in Civil Engineering could take up to a month for analysis and design of all the structural members of the structure. Time resource in design of structures used to be a major challenge especially when engineers are faced with complexity of high-rise structures which could take months or even years before completion if approached manually. The developments of different computer aided design application over the past few years have helped to make design in engineering an easy task. These design applications are also very helpful in by-passing the usual human errors that could arise from manual design approach [17], [18]. It can greatly enhance structural quality right from the design stage [19]. With the use of design applications, designs of structures that would usually take weeks, months or even years could now be modelled, analysed, and designed within few hours.
Some of the well-known and well engage design application in the market are AutoCAD, Orion design software, ETABS, SAFE, SAP2000, Autodesk Revit, CAD RC, STAAD Pro and RISA [15].
Beyond engineering designs that translate into engineering drawings, ICT has recently advanced into virtual environment design. This is the design that brings the project into a virtual environment such that

stakeholders in a project can actively interact with the project in a virtual reality before it is implemented on the site. Through this new development, it becomes easy for clients to make adjustment and express their interests or dissatisfactions with a part or some parts of the project before it is built. Hence, a building that exists in a physical environment can be modelled in a virtual environment with same representations [20].

2.5. *ICT in Building Preservation*

According to [21], building degradation process is a known to be a major threat to wooden stilt buildings across the world. Inadequacy in spatial planning approach, conditions of waterfront settlement, and inability for protection and enhancement of the features of the buildings are some of the contributing factors to the vulnerability of the buildings. Nevertheless, there have been developments with cases of preservation of like-buildings such as the preservation of rorbu; a structure temporarily built to shelter fishermen, preservation of the carrelets in the Gironde estuary and the Scottish Crannog Centre Despite. Through the introduction of ICT in building preservation, evaluation methods have been very instrumental in conservation buildings in many homes in the United Kingdom and other countries. Use of ICT has greatly helped in structural damage assessment [22], [23] & [24]. Some of these methods are Home Condition Report, 2004, NEN 2767 (the Netherlands), Grille d'evaluation de la degradation de l'habitat, 2011, Inspeccion Tecnica de Edificios, 2011, MAEC, 2006 and MANR, 2007. Although the aforementioned methods of evaluation do not put into consideration the features and information type to be gathered, they have helped in advancing body of knowledge towards the achievement of better results. Knowledge from the aforementioned evaluation methods have contributed immensely to the development of ICT diagnostic method of evaluating and focusing on wooden stilt-houses [25].

2.6. *Implementation of ICT in Energy Efficient Buildings*

Based on study, the building sector is said to be the sector with the highest consumption of energy of other sectors. The building sector is also known for the emission of CO_2 up to an approximate of 40% of total consumption. According to [26], by 2030, the total energy that is being emitted in the building sector would have been substantially reduced by 30% through the engagement of new high technology and extensive research. According to [26], the most of the energy consumption in a building is during its usage life and this is about 80%. Hence, to reduce the energy consumption in any building, it is necessary that the reduction exercised is focused on the operational cycle such as an improvement on the access windows, the ventilation system of the building, and thermic features by insulation. Another way to cause reduction in energy consumptions in building is to appreciate the exploration of ICT in the building sector [27]. Though [26] related that energy consumption by 2030 would have reduced by 30%, the International Panel on Climate Change related that reduction in energy consumption would be greater than 30% by 2030 through ICT solutions. In line with the International Panel on Climate Change, some specialists believe that ICT would not only bring about a greater that 30% reduction as related by the International Panel on Climate Change, rather the reduction would range from 50% to 80%. According to [28], an experimental study of an ICT monitoring system installed in a building immediately recorded more than 30% energy consumption reduction. Though the percentage reduction dropped afterwards. Hence, these present ICT to be a potential solution to substantial energy consumption reduction in the building sector [29].

2.7. *ICT in Structural Engineering*

In the recent years, ICT began to gain attention in its application in Civil Engineering and especially structural engineering. Apart from improved design and enhanced energy efficiency, another area that the use of ICT has greatly improved activities in the built environment sector is in the area of structural health monitoring. It has been very useful in predicting risks of seismic activities and for averting the hazards of building collapse [30]. This discovery had been through the use of smartphones designed with sensors that can efficiently carryout structural health monitoring (SHM). These smartphones are also designed with in-built batteries and units that serve as processor. The structural health monitoring operation is basically focused on assessing damages in a structure and to assess the degree of safety of the structure. Many researchers have continued to work tirelessly to device means to upgrade on the present SHM system due to challenges such as high cost of sensors, high cost of maintenance, high cost of installation, high power consumption, wireless communication issues, cabling issues and more. Some of these challenges have directed line of thoughts towards the use of smartphones as potential upgrade over the previous SHM systems. The figure 2 relates one of the first studies carried out by [31] on the use of smartphones in structural engineering as SHM system.

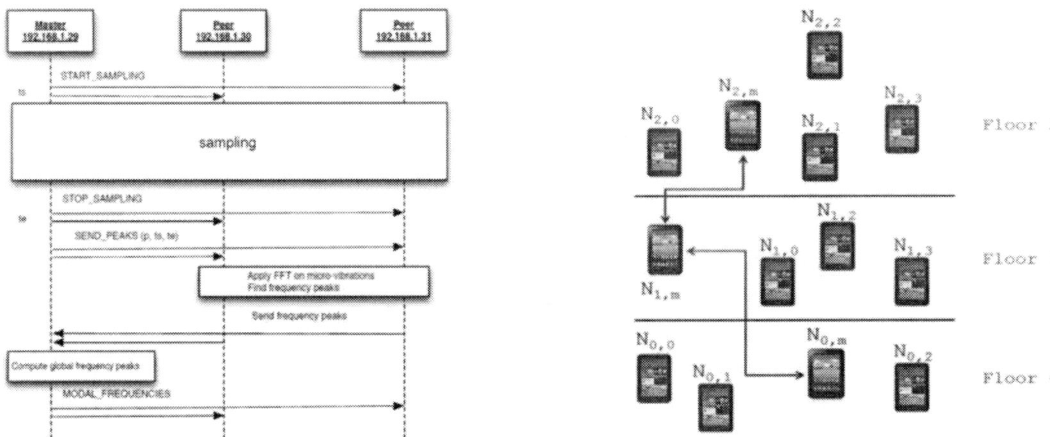

Figure 2. Smart Monitor System (a) sequence diagram (b) deployment across 3 floors [32].

In the design of smartphones to serve as SHM systems, one of the the core features put into consideration is the frequency peak-picking attribute. The smartphone SHM system locates the highest frequency by first assessing the natural frequencies and applies the peak frequency on the time series through the recorded accelerations in three axes. Another in-built feature of the smartphones designed to serve as SHM system is its tolerance for faults in its communication protocol. The master node in the figure 2 is the is that on the ith floor. As an upgrade on the smartphone SHM system approach, [33] proposed a Cloud-SHM approach using smartphone. Different from the three levels in the previous SHM system of [31], the proposed Cloud-SHM has 6 levels of framework.

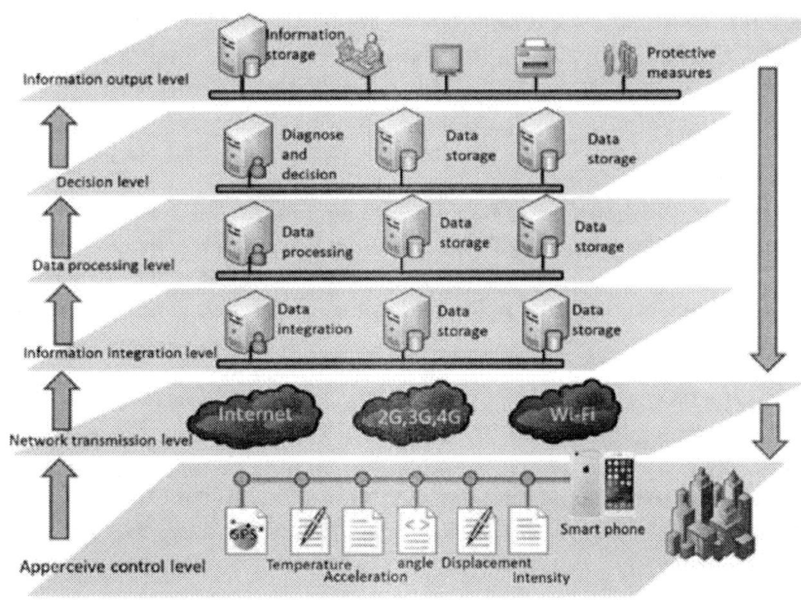

Figure 3. Framework of the Cloud-SHM based on smartphone [32]

The Cloud-SHM based smartphone's platform drew it base from Orion-CC, an application that is capable of acquiring data, analyse data and have the same uploaded to the cloud for the purpose of storing. It is linked up to a website specially designed with other features built with the Cloud-SHM based smartphone are ability to make decision, ability to generate output for the information, ability to process data, ability to transmit network, ability to integrate information and an apperceive control level.

3. Conclusion

The application of Information Communication Technology is very vital to achieve cost reduction, operational time reduction, safety, and more in civil engineering practices. Hence, ICT remains relevant to the field of civil engineering and engineering at large.

Acknowledgments
The authors appreciate Covenant University for funding the publication of this paper.

Reference

[1] Zainab S G and Mansur B D 2014 Application of ICT in Nigerian Educational System for Achieving Sustainable Development *International Letters of Social and Humanistic Sciences* pp 62-71
[2] Damkor M, Irinyang D J and Haruna M 2015 The Role of Information Communication Technology in Nigeria Educational System *International Journal of Research in Humanities and Social Studies (Electronic Materials vol 2)* pp 64-68
[3] Suleiman A and Steven A K 2018 Review of ICT Implementations for Facilitating Information Flow between *Automation in Construction (Electronic Materials vol 86)* pp 176-189
[4] Aimie R, Graham D and Christine P 2015 Impact of information and communication technology (ICT) on construction projects *Org. Technol. Manag. Constr (Electronic Materials vol 7)* pp 1367–1382
[5] Richard D and Chris H 2013 Implementing 'Site BIM': a case study of ICT innovation on a large hospital project *Autom. Constr. (Electronic Materials vol 30)* pp 15–24
[6] Mehdi Nourbakhsh 2012 Mobile application prototype for on-site information management in construction industry *Eng. Constr. Archit. Manag. (Electronic Materials vol 19)* pp 474–494
[7] Motamedi A 2009 Framework for Lifecycle Management of Facilities Components *Doctoral dissertation Concordia University Montreal*
[8] Majrouhi S S 2013 Developing Rfid-based electronic specimen and test coding system in construction quality management, *Iran. J. Sci. Technol. Trans. Civ. Eng. (Electronic Materials vol 37)* pp 469–478
[9] Samir El-Omari O M 2008 Integrating 3D laser scanning and photogrammetry *Autom. Constr (Electronic Materials vol 18)* pp 1–9
[10] Sebastian S and Jochen T 2014 Mobile 3D mapping for surveying earthwork projects using an unmanned aerial vehicle (UAV) system *Autom. Constr. (Electronic Materials vol 41)* pp 1–14
[11] Cheng Z, Amin H and Homam B 2019 Collaborative multi-agent systems for construction equipment based on real-time field data capturing *J. Inf. Technol. Constr. (Electronic Materials vol 14)* pp 204–228
[12] Yuan C and John K M 2008 Using mobile computing for construction site information management *Eng. Constr. Archit. Manag. (Electronic Materials vol 15)* pp 7-20
[13] Gang T, Xiaoqiang Z and Hu F 2012 The design of electric materials management system based on QR-code *Technology and Management for Efficiency* pp 1345–1351
[14] Jonathan G L 2010 ICT and road transportation safety in the United States: a case of "American exceptionalism" *IATSS Research (Electronic Materials vol 34)* pp 1–8
[15] Tariq B M, Musavir A and Tariq J R 2014 Applications of e-Learning in engineering education: A case study Procedia - *Social and Behavioral Sciences (Electronic Materials vol 123)* pp 406 – 413

IOP Conf. Series: Materials Science and Engineering 1036 (2021) 012064 doi:10.1088/1757-899X/1036/1/012064

[16] Banday M 2012 e-Learning in the Web 2.0 World: A Case Study *IEEE International conference on Engineering Education Innovative Practices and Future Trends*

[17] Ede A N, Oshokoya O O, Oluwafemi J O, Oyebisi S O and Olofinnade O M 2018 Structural analysis of a genetic algorithm optimized steel truss structure according to BS 5950 *International Journal of Civil Engineering and Technology (Electronic Materials vol 9)* pp 358-364

[18] Ede A, Williams A, Oni O, Ofuyatan O and Oluwafemi J 2019 Iconic structures: Case study of a historic museum with notable spans designed in concrete *in ISEC 2019 - 10th International Structural Engineering and Construction Conference, Chicago, Illinois, USA., STR-105-1-6*

[19] Ede A N, Olofinnade O M and Sodipo J O 2017 Use of building information modelling tools for structural health monitoring *in Proceedings of the IEEE International Conference on Computing, Networking and Informatics*

[20] Potkonjak V, Vukobratovic M, ovanovic K and Medenica M 2010 Virtual Mechatronic/Robotic laboratory - A step further in distance learning *Computers & Education (Electronic Materials vol 55)* pp 465–475

[21] Virtudes A and Almeida F 2013 Desafios da reabilittacao urbana no process de planeamento: o patrimono esqucido das aldeias A vieiras A gir *Revista Interdisciplinar de Ciencias Sociais e Humans (Electronic Materials vol 51)* pp 323-346

[22] Agarana M C and Ede A 2016 Free vibration analysis of elastic orthotopic rectangular inclined damped highway supported by pasternak foundation under moving aerodynamic automobile *Lecture Notes in Engineering and Computer Science* pp 978-981

[23] Ede A N and Pascale G 2016 Structural Damage Assessment of FRP Strengthened Reinforced Concrete Beams under Cyclic Loads *Materials Science Forum (Electronic Materials vol 866)* pp 139-142

[24] Ede A N, Bonfiglioli B, Guidotti P G M and Viola E 2004 Dynamic assessment of damage evolution in FRP strengthened RC beams *in Proceedings of the International Conference on Restoration, Recycling and Rejuvenation Technology for Engineering and Architecture Application, Cesena, Italy*

[25] Lanzinha J and Castro-Gomes J 2010 Intervention on the rehabilitation of rural houses in Portugal as a contribution to sustainable construction. *Architecture, Civil Engineering, Environment (Electronic Materials vol 1)* pp 5-18

[26] European Union 2010 Directive 2010/31/EU of the European Parliament and of the Council of 19 May 2010 on the energy performance of buildings *Official Journal of the European Union*

[27] Ede A N, Kesi-Ayeba K D, Olakunle O S and Oluwafemi J 2019 Study of Energy Efficient Building Design Techniques: Covenant University Health Centre *Journal of Physics: Conference Series (Electronic Materials vol 1378)* pp 1-12

[28] Rizvi S, Ahmad M, Ahmed J and Karim A 2012 Active Node Based Sensor Wireless Network for Energy Management *International Journal of Engineering and Technology (Electronic Materials vol 4)* pp 590

[29] Bilsen V, Blondiau T, Debergh P and Lukach R 2013 Final Report on behalf of European Commission Directorate-General Enterprise & Industry pp 1-202

[30] Akpabot A I, Ede A N, Olofinnade O M and Bamigboye G O 2018 Predicting buildings collapse due to seismic action in Lagos state *Journal of Engineering Research in Africa (Electronic Materials vol 32)* pp 91-102

[31] Kotsakos D, Sakkos P, Kalogeraki D and Gunopulos D 2013 Smart monitor: Using smart devices to perform structural health monitoring *Proc. VLDB Endow (Electronic Materials vol 6)* pp 1282–1285

[32] Amir A H and William B G 2019 An overview of smartphone technology for citizen-centered, real-time and scalable civil infrastructure monitoring *Future Generation Computer Systems (Electronic Materials vol 93)* pp 651–672

Examining the Potential Recovery of Compost from Mixed Fruit Wastes from a Campus Eatery Wastes

D O Olukanni[1*], V O Ogedengbe[2] and O J Ogundare[3]

[123]Department of Civil Engineering, Covenant University, PMB 1023, Canaan land, Ota, Nigeria
*Corresponding Author: david.olukanni@covenantuniversity.edu.ng

Abstract. The burning and open dumping of wastes is a common practice in developing countries such as Nigeria. These practices are causes of environmental concern as they generate air pollutants, contaminate surface and underground water resources, and breed disease vectors, consequently affecting public health. Nevertheless, composting has been identified as a sustainable technique to efficiently manage and dispose these wastes while also recovering a marketable resource. This study was aimed at assessing the potential recovery of compost from mixed fruit wastes as a method of proper disposal. Fruit wastes generated within a university campus were mixed with dry leaves and manure and decomposed in a composting process for 6 weeks. The physico-chemical properties such as temperature, pH level, moisture content, and NPK value of the resulting compost were determined during and after the process. The experimental results indicated a temperature range of 25°C to 56°C and pH level ranging from 6.6 to 8.1 during the composting process. The moisture content of the compost pile was detected to gradually reduce from 56% to 40% during the experiment. The NPK ratio of the final compost product was evaluated to be 0.61:0.40:0.73. The compost produced in this study could be used for agricultural purposes to supply nutrients to growing plants and the soil.

Keywords: Waste Management; Fruit Waste, Environmental Pollution; Environmental Sustainability; Sustainable Technology; Valorisation; Composting

1. Introduction

The teeming world population coupled with the nutritional benefits of fruits in human diet has resulted in the increased demand for fruits and fruit products [1]. According to the Statista Research development [2], the yearly production of fruit in the world is over 675 million metric tonnes. For example, about 125, 115, 85, and 26 million metric tonnes of citrus fruits, bananas, apples, and pineapples, respectively is produced globally [3]. After fresh consumption or processing of these fruits into other valuable products, considerable mass of the fruit is left as waste and this proves to be a major challenge to the food industry [4].

The handling and conversion of fruits into other products generates considerable solid residues in the form of seeds, peels, pods, husks, pulps, rinds and so on [5]. The management of these solid wastes is a serious problem worldwide, especially in developing countries [5, 7, 8]. In many developing countries like Nigeria, these wastes are usually left to rot on dumpsites or are burned openly due to the lack of proper waste management infrastructures. These practices have significant environmental impacts as they pose several health implications to the public. For example, leachate produced from the decomposition of waste heaps pollutes surface and underground water resources [9, 10]. Open dumpsites attract disease-carrying animals such as

rodents, reptiles, and insects that transmit viral diseases. Burning and decomposition of organic waste result in the emission of carbon dioxide, methane and other greenhouse gases that pollutes the air [11]. Consequently, humans living around the vicinity of dumpsites are reported to be suffering from various diseases such as cholera, diarrhoea, malaria, respiratory diseases etc. resulting from their exposure to open dumping [12]. Additionally, the offensive stench arising from organic waste decomposition and the dispersion of waste by wind and scavenging animals deforms the surrounding environment and thus reduces the quality of life [13].
In view of these adverse effects on the environment, it has become extremely important to sustainably manage these wastes using environmentally friendly, efficient and inexpensive methods. The transformation of wastes into valuable products offers a sustainable alternative to reduce greenhouse gas emissions, minimize pollution, improve agricultural practices and production, and increase energy security. The use of wastes to produce valuable goods is therefore a promising research field [4, 14].

Over the years, one method that has been explored and known to gain widespread popularity as an environmentally sustainable way of deriving valuable products from waste is composting [15]. Several researchers such as [14, 15, 16, 17], have all been advocates for composting as a method of managing organic wastes. According to Kulcu & Yaldiz [18], composting is described as a controlled process that produces humus from the microbial decomposition of organic matter through introduction of water, proper aeration, and carbon and nitrogen rich materials. Composting reduces waste volume and smell, eliminates pathogens, and reduces weed growth in agricultural fields [19]. Waste composting is regarded as a means of diverting organic waste materials from landfills while creating soil conditioners or organic fertilizers that is suitable for agricultural purposes, at relatively low cost [20, 21]. The compost or soil conditioner generated through composting is a plausible alternative to expensive chemical fertilizers which have a negative effect on human health and the environment. Composts are used to enhance the properties and features of the soil. They recycle soil organic matter and nourish it continuously. They improve the soil structure and help to conserve humus in the soil [17].
The conversion of the major components of fruit wastes, namely soluble sugar, organic acids, protein, and fats into a product similar to humus, in which unstable ammonia is converted into stable organic nitrogen forms, is possible through the action of a number of microorganisms [8, 22]. The utilization of fruit wastes to produce reusable materials can permit self-sustainable processes through composting by microbial conversion of cellulose into protein-rich wastes which has positive effects on the soil as well as the environment [23]. Parameters such as temperature, water content, Carbon-Nitrogen ratio (C/N ratio), aeration rate, pH, and the physical structure of fruit wastes are the most vital during the composting process [24]. Accordingly, this work has been designed as a preliminary study to examine the potential recovery of compost from mixed fruit solid wastes generated within a school campus. The specific objectives are to: conduct experiments to aerobically degrade fruit wastes into compost, evaluate certain process parameters during the conversion of fruit wastes into compost, and evaluate the NPK nutrient value of the final compost produced.

2. Materials and Methods

2.1 Raw waste and bulking agent

Solid wastes from three different fruits: watermelon (*Citrullus lanatus*), pineapple (*Ananas comosus*), and orange (*Citrus X sinensis*) were used as raw material for the compost pile. Substantial amount of these fruit wastes, collected in clean bags from one of the campus cafeterias, were mixed together and shredded into smaller pieces. Dry leaves and straw were

IOP Conf. Series: Materials Science and Engineering 1036 (2021) 012065 doi:10.1088/1757-899X/1036/1/012065

gotten from the campus surroundings and added to the mixed waste as a bulking agent to add structure to the compost pile. Also, suitable amount of poultry droppings obtained from a nearby poultry farm were added to the pile to adjust the C/N ratio of the compost material. The poultry manure also served as a source of microorganisms to facilitate the composting process.

2.2 Compost Vessel

In this study, an in-vessel composting method was adopted due to its capacity to process organic wastes faster and its suitability in handling assorted organic materials. A laboratory scale compost vessel was developed from a locally purchased plastic container. The 70 cm tall container was riddled with holes from the lid to the bottom to allow for air into the vessel, supply oxygen to the process, and discharge the heat and gaseous emissions of decomposition. However, no additional devices were installed to control temperature and distribute air within the vessel. As an alternative, the compost pile was turned twice every week, using a shovel, so as to improve aeration and circulate oxygen in the mix.

2.3 Measurement of physico-chemical properties of compost

The compost pile was incubated for a period of six (6) weeks. During the composting process, the weight, pH, temperature, moisture content were regularly measured. The temperature of the compost pile was measured regularly at an interval of 5 days until the end of the composting process using a soil thermometer (0 - 80^{0}C). Moisture content, weight loss, and pH values of the compost pile were measured once a week using laboratory equipment. Furthermore, the Nitrogen-Phosphorus-Potassium (NPK) value of the resulting compost was determined at the end of the incubation period. A sample of the compost material was oven dried at 105^{0}C for 24 hours after which it was pulverized and sieved (75μm). The sample was mixed with 10ml of concentrated hydrochloric acid and heated for 10 minutes. The mixture was cooled, sieved and analysed using a Palintest photometer 8000 to quantify the concentration of the elements.

3. Results and Discussion

3.1 Temperature, moisture content, and pH level changes in compost pile

The regular temperature measurements of the compost pile is presented in Figure 1. At the beginning of experiments, the compost pile was at ambient temperature but as the composting process began, the temperature in the compost began to rise rapidly, hitting a peak value of 56^{0}C on the 5th day. This was followed by a slight decrease in temperature to 50^{0}C on the 10th day, a second increase was observed as temperature rose to 53^{0}C on the 15th day. Another slight decrease followed as temperature had reduced to 50^{0}C as at the 20th day. The temperature of the compost became steady at 50^{0}C between days 20 and 25. This was then followed by a gradual decrease to ambient temperature.

Temperature measurements suggest that the sudden rise in temperature within the first few days was as a result of the action of the thermophilic microbial population present in the compost mix. These microorganisms decompose organic matter and released heat in the process, consequently raising the temperature of compost pile. The temperature decrease towards the end of the composting process could be associated to a reduced activity of thermophilic microorganisms and a dominance of mesophilic microbes feeding on the residual organic matter within the compost pile [22].

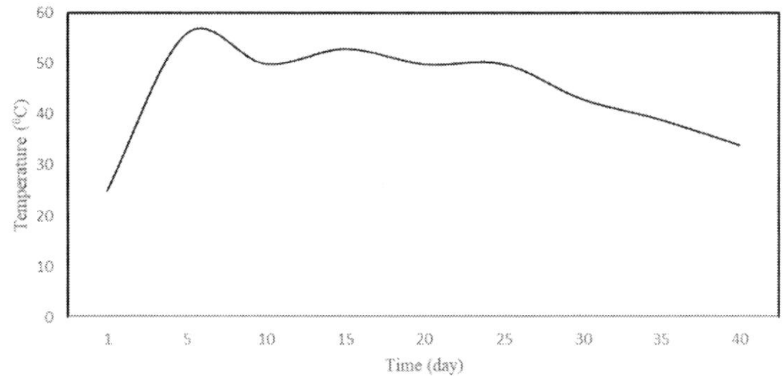

Figure 1. Temperature changes over the incubation period of compost

Table 1. Moisture and pH level changes over incubation period

Week	Moisture content (%)	pH level
1	56.0	6.6
2	52.0	7.5
3	44.5	8.1
4	41.3	7.9
5	42.0	7.0
6	40.0	-

The changes in moisture content and pH level are presented in Table 1. The moisture content in the compost decreased gradually until the 4th week. A slight increase in moisture content was observed in the compost during the 5th week, and this was followed by another decrease leaving the moisture content of the compost at the end of the composting process at 40%. The loss of moisture through evaporation stimulated compost drying and is probable for heat loss, thereby causing a drop in temperature of the compost pile. Practically, the reduction of moisture content in the compost is reasonable since a loss of moisture indicates the degradation of organic matter in the compost pile [25].

After the first week of the composting process, the pH level gradually increased from 6.6 to 7.5, and by the 3rd week, a peak pH value of 8.1 was recorded in the compost. The rise in the pH level has been linked to the thermophilic phase, which was also responsible for the increased temperature of the composting process. The pH level started to decrease in the 4th week, coinciding with the mesophilic phase of the composting process, followed by a drop in pH level to 7 (neutral) as the compost started to approach its mature state. The drop and rise in pH level of the compost pile could be as a result of the respective formation and degradation of organic acid in the compost as also observed by Chen et al [22].

3.2 NPK value of compost

The experimental analysis of the compost for its Nitrogen, Phosphorus, and Potassium concentration indicated values of 0.61%, 0.40%, and 0.73%, respectively. Nitrogen, Phosphorous, and Potassium are common nutrients needed by plants for building large molecules. A comparison of the nutrient value of the compost generated in this study and common fertilizers is presented in Table 2.

Table 2. Comparison of NPK value of compost with fertilizers

Element	Compost sample	Fertilizers		
	NPK .61-.40-.73	NPK 5-5-5	NPK 10-5-5	NPK 15-15-15
Nitrogen (%)	0.61	5	10	15
Phosphorus (%)	0.40	5	5	15
Potassium (%)	0.73	5	5	15

Despite the fact that the compost produced has lesser NPK values to commercial grade fertilizers, compost is advantageous as it contains several microbes which can help in nitrogen fixation and let the plants get nitrogen through them, therefore making compost food for both soil and plant. On the other hand, fertilizer only feeds the plant and the chemicals present in fertilizers can be dangerous to the microbial population of a soil.

4. Conclusion

In this study, mixed fruit waste was effectively disposed by converting it into a valuable resource in the form of compost. This waste management technique reduces the massive space occupied by fruit wastes on dumpsites and generates a product with monetary value. This disposal technique is a potentially suitable method that could be adopted in environments or food industries facing severe fruit waste management challenges. Nevertheless, it is recommended that further investigation be carried out on this waste management technique to evaluate it appropriateness on an industrial scale. It is important to assess the economic viability of this technique with consideration given to availability of fruit wastes and bulking agents, and the capital and operation costs of a composting process.

Acknowledgement

The management of Covenant University is appreciated for providing support and enabling environment.

References

[1] Schieber A, Stintzing FC, and Carle R 2001 By-products of plant food processing as source of functional compounds - recent developments. *Trends in Food Science & Technology,* **12** 401–413

[2] Statista Research Department 2017 *Fruit production – Statistics & Facts* Available on: https://www.statista.com/topics/1621/fruit-production/ (Accessed 20 June, 2020)

[3] Sagar NA, Pareek S, Sharma S, Yahia EM and Lobo MG 2018 Fruit and vegetable waste: Bioactive compounds, their extraction, and possible utilization. *Comprehensive Reviews in Food science and Food Safety* ***17*** 512-531

[4] Sharma R, Oberoi HS and Dhillon GS 2016 Chapter 2 - Fruit and vegetable processing waste: Renewable feed stocks for enzyme production. In Dhillon GS and Kaur S (Eds.), *Agro-Industrial Wastes as Feedstock for Enzyme Production* (pp 23–59) Massachusetts United States: Elsevier

[5] Babbar N and Oberoi HS 2014 Chapter 2 - Enzymes in value-addition of agricultural and agro-industrial residues. In S. K. Brar & M. Verma (Eds.) *Enzymes in Value-Addition of Wastes* (pp. 29–50). New York, United States: Nova Publishers

[6] Olukanni DO 2013 Environment, Health and Wealth: Towards an Analysis of Municipal

Solid Waste Management in Ota, Ogun State, Nigeria. Paper presented at the *InternationalConference on Clean Technology and Engineering Management,* Covenant University, Ota, Nigeria, November 12-14, 2013

[7] Olukanni DO and Nwafor C 2019 Public-private sector involvement in providing efficient solid waste management services in Nigeria. *Recycling, 4*(19) 1–9

[8] Ogundare JO and Olukanni DO 2020 Potential Recovery of Biogas from Lime Waste after Juice Extraction Using Solid–Liquid Extraction Process. *Recycling,* **5**(14) doi:10.3390/recycling5020014

[9] Wikandari R, Youngsukkasem S, Millati R and Taherzadeh MJ 2014 Performance of semi-continuous membrane bioreactor in biogas production from toxic feedstock containing D- limonene *Bioresource Technology,* ***170*** 350 – 355

[10] Olukanni DO, Olujide JA and Kehinde EO 2017 Evaluation of the Impact of Dumpsite Leachate on Groundwater Quality in a Residential Institution in Ota Nigeri, *Covenant Journal of Engineering & Technology (CJET),* **1**(1) 18-33

[11] Hedge U, Chang T C and Yang S 2003 Methane and carbon dioxide emissions from Shan Chu-Ku landfill site in northern Taiwan *Chemospher 52* 1275–1285

[12] Sankoh FP, Yan X and Tran Q 2013 Environmental and health impact of solid waste disposal in developing cities: A case study of Granville Brook dumpsite Freetown, Sierra Leone *J, Environ. Prot. 4*(7) 665-670

[13] Abah SO and Ohimain EI 2010 Assessment of dumpsite rehabilitation potential using the integrated risk based approach: A case study of Eneka, Nigeria. *World Appl. Sci. J., 8* 436–442

[14] Olukanni DO, Aipoh OA and Kalabo IH 2018 Recycling and reuse technology: Waste to wealth initiative in a private tertiary institution, Nigeria. *Recycling, 3*(44) 1–12

[15] Olukanni DO and Aremu OD 2017 Provisional Evaluation of Composting as Priority Option for Sustainable Waste Management in South-West Nigeria. *Pollution,* Publisher:University of Tehran **3**(3) 417-428, ISSN: 2383-4501

[16] Atalia DM, Buha KA and Shah NK 2015 A review on composting of municipal solid waste *IOSR Journal of Environmental Science, Toxicology and Food Technology, 9*(5) 20-29

[17] Christophe HL 2018 Compost: Potential to be valued in agriculture. *Advances in Biotechnology & Microbiology* ***11***(1) 35 – 36

[18] Kulcu R. and Yaldiz O 2007 Composting of goat manure and wheat straw using pine cones as a bulking agent. *Bioresour. Technol.,* ***98*** 2700–2704

[19] Jakobsen ST 1995 Aerobic decomposition of organic wastes 2 Value of compost as a fertilizer. *Resour. Conserv. Recy., **13***(1) 57–71

[20] Eriksen G, Coale F and Bollero G 1999 Soil nitrogen dynamics and maize production in municipal solid waste amended soil *Agron. J., **91**:*1009–1016

[21] Wolkowski R 2003 Nitrogen management considerations for land-spreading municipal solid waste compost. *J. Environ. Qual., **32**:* 1844–1850

[22] Chen C, Kuo J and Chung Y 2012 Effect of matured compost as an inoculating agent on odour removal and maturation of vegetable and fruit waste compost. *Environmental Technology,* iFirst, 1–8

[23] Bae JT, Sinha J, Park JP, Song CH and Yun JW 2000 Optimization of submerged culture conditions for exobiopolymer production. *Journal of Microbiology and Biotechnology, **10***(4) 482-487

[24] Iqbal MK, Shafiq T and Ahmed K 2010 Effect of different techniques of composting on stability and maturity of municipal solid waste compost *Environ. Technol., **31*:** 205–214.1

[25] Sullivan DM and Miller RO 2001 Compost quality attributes, measurements, and variability. In Stoffella PJ and Kahn BA (Eds.), *Compost Utilization in Horticultural Cropping Systems* (pp 97–120) New York, United States: Lewis Publishers

IOP Conf. Series: Materials Science and Engineering 1036 (2021) 012066 doi:10.1088/1757-899X/1036/1/012066

Hazard Assessment and Implementation Strategies for Mitigating of Flood and Erosion in Nigeria

O J Oyebode
Civil and Environmental Engineering Department, College of Engineering,
Afe Babalola University, Ado-Ekiti (ABUAD), Ekiti State, Nigeria
Corresponding Author: oyebodedare@yahoo.com

Abstract. There are notching unresolved issues and environmental hazards traceable to flood and erosion in major rural areas and notable communities in Nigeria. Researchers' efforts on erosion control and flood mitigation have not yielded notable fruits due to inadequate planning and implementation strategies. Maintenance of roads and other infrastructures are the bed rock of any development round the globe. This paper assessed recent occurrence of flood and erosion. The study examined the causes, urban and regional planning of the affected area, construction practices, damages and maintenance practices. Questionnaires, field study and interview of experts were wont to elicit information from people within the affected location, officials from state and native government agencies charged with rehabilitation of affected areas. The research finding revealed that there are inadequate drainage systems, lack of enforcement of plan, corruptions, failure of systems, poor supervision, construction malpractices, lackadaisical attitude and lack of maintenance culture. Implementation strategies were highlighted to avert flood and erosion ravaging several areas of the country. The paper concluded that stakeholders, researchers, engineers should partner with government agencies within the planning, construction, management and maintenance of rural and concrete drainages for adequate funding and sustainable construction. It is recommended that strategies for alleviating flood and erosion should be implemented supported latest technology for sustainable infrastructural development. Periodic hazard assessment should be planned in various geopolitical zones especially Lagos, Niger delta and rain forest region of Nigeria. Land use control, landscape maintenance, weather stations, proper enforcement of construction works consistent with plan should be strategically put in situ for effective monitoring and alleviation of flood and erosion in Nigeria.
Key words: Environmental Hazards, Erosion Control, Flood Mitigation, Implementation Strategies, Infrastructural Development

1. Introduction

Flooding is a frequent incidence and common natural environmental hazards in various geopolitical zones in Nigeria. There is urgent need for a sustainable approach in Nigeria for flood management. Nigeria periodically experiences a large range of environmental hazards such as floods, windstorms, drought, desertification, landslides, wildfires, sandstorms, landslides, windstorms, heat waves, disease epidemics pest invasion, soil erosion, gully erosion, coastal erosion, and other types of erosion, volcanic eruptions, and related activities. Most of these are due to unethical practices, lack of implementation strategies and increasing vulnerability. The lack of functional public infrastructure and citizen based communal services in most communities has prioritized the look for a sustainable flood management regime publicly policy. The Flood management office of Federal Ministry of Environment should strengthen National Flood Management Act in Nigeria.

IOP Conf. Series: Materials Science and Engineering 1036 (2021) 012066 doi:10.1088/1757-899X/1036/1/012066

Critical assessment, disaster preparedness, prevention, strategic planning, establishment of mitigation measures are necessary. A hazard assessment could also be a radical check of the work environment.

Flood management requires input from numerous stakeholders, engineers, competitor alternatives and many others [1,2].
Potential structural damages occurred as a result of flooding, unplanned settlements and unforeseen natural hazards [3,4,5,6,7,8,9,10].
Hazard is dangerous to life, properties and atmospheric condition of an area [11]. Due to warming and global climate change normally, the length, magnitude, scale and frequency of climate-related risks area unit projected to increase and deteriorate [12]. Flood is additionally as a result of several conditions operating one by one and in natural process.

Intermittent floods occur on several rivers, forming a contiguous space mentioned because of the terrain, rural urban migration, percentage of yearly urbanization, mismanagement of solid wastes and land use [15,16,17,18].

Flooding has displaced a lot of people than the other natural disaster [19,20]. North aspect of Federal Republic of Nigeria is prone to wind erosion. Insufficient vegetation has resulted into frequent sweeping away of soil. Borno, Kano, Sokoto, Zamfara and Yobe States area unit the foremost in danger of wind erosion and its impacts in Federal Republic of Nigeria. Disasters embrace dramatic, sudden, impromptu actions and other issues affecting life, societal growth and infrastructures [21,23,27,28].

Flooding has several impacts. It damages property and can be a threat to public health and safety and native multifariousness [22].

2. Methodology

Questionnaires, field study and interview of experts were used to obtain viable information from people in the affected location, officials from state and local government agencies charged with rehabilitation of affected areas. Secondary data from literatures were also used. Plate 1 is a typical flooded highway and infrastructures in a community.

Plate 1: Typical flooded highway and infrastructures in a community

Many property, system and life were broken because of flooding. There a lot of failure of pavements dues to erosion.
Figure 1 and 2 show the boundaries of state and native Government Areas clipped from Map [34]. Figure 3 indicates areas susceptible to flood in Federal Republic of Nigeria.

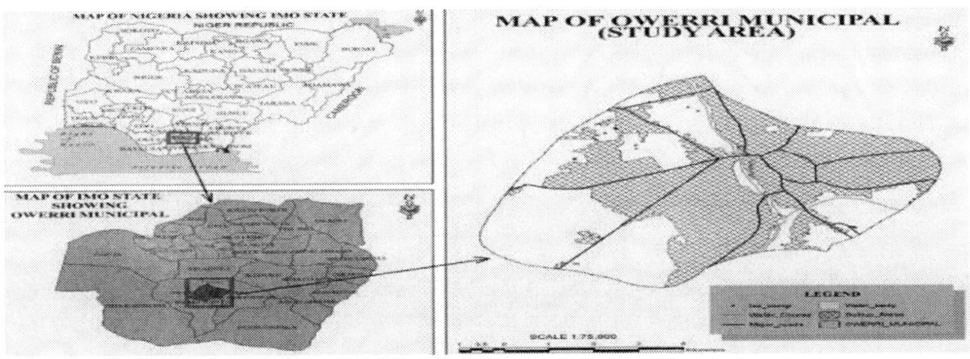

Figure 1: Boundaries of Owerri and native Government Areas

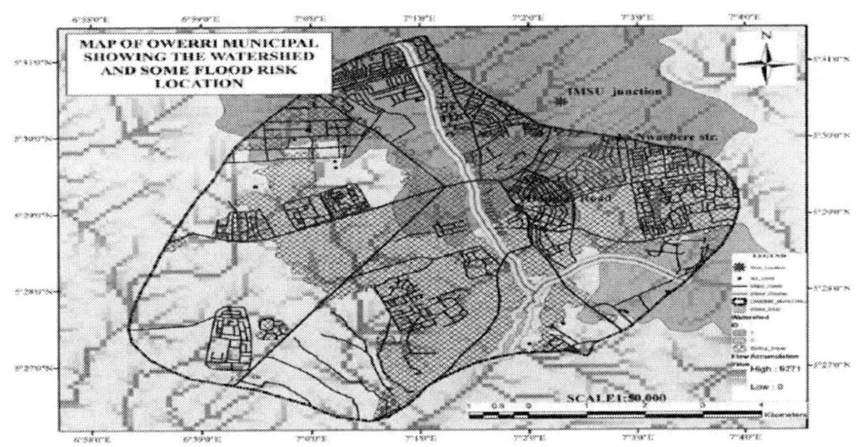

Figure 2. Watershed Municipal community in Owerri, Nigeria

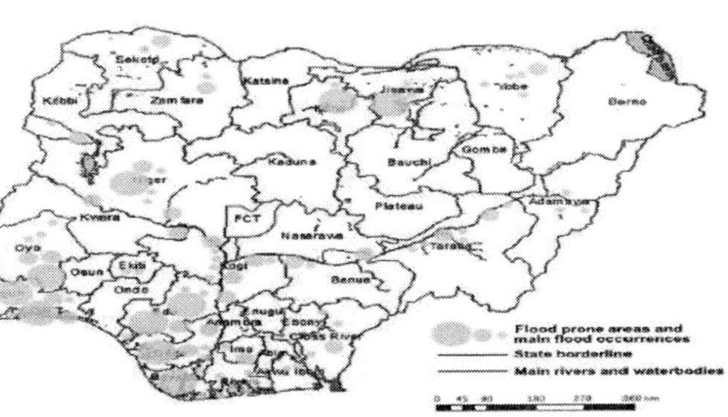

Figure 3. Areas susceptible to flood in Federal Republic of Nigeria [33].

IOP Conf. Series: Materials Science and Engineering 1036 (2021) 012066 doi:10.1088/1757-899X/1036/1/012066

3. Results and Discussion

Most of these floods are traceable to construction of building along water ways, lack of adequate planning for major urban centers, inadequate weather stations and rainfall data in some locations. Causes of flood based on percentage is as follows: Opening of dam around rivers 75%, narrow stream or river channels 8%, indecent waste disposal 7%, heavy precipitation 6% and poor drainage system 4%. The research finding revealed that there are inadequate drainage systems, lack of enforcement of master plan, corruptions, failure of systems, poor supervision, construction malpractices, lackadaisical attitude and lack of maintenance culture. Implementation strategies were highlighted to avert flood and erosion ravaging several areas of the country.

Disaster risk can be computed using susceptibility, coping capability and danger (ISDR, 2004) by the formula:

$$\text{Disaster risk (R)} = \frac{\text{Vulnerability (V) x Hazard (H)}}{\text{Capacity (C)}}$$

Hazards are dynamic and with extremely varied potential impacts. Coastal erosion is increasing as a result of human activities like construction of harbour protective structures and jetties like within the Lagos Bar beach, sand mining, damming of rivers and dredging activities [31, 36].

Much of the first floods were within the North with the South being affected later within the months of August, September and October. Flooding has increased pressure on Federal and native budgets struggling to deal with emergencies. With 1 million potential fatalities (NEMA, 2012), the flood is certainly a National catastrophe. During the flood the costs of local foods rose by 200 -300% in certain areas (Puncha, 2012) and 80% of Bayelsa and other areas of Niger delta, the hub of the local refining industry was submerged.

Table 1 gives typology of disaster triggers. Figure 4 indicates typology that triggers disaster.

Table 1: Typology of Disaster Triggers

Natural Hazards	HUMAN-DRIVEN DISASTERS	
	Violent Conflict	Technological Hazards
Earthquakes Floods/Riverbank erosion Landslides Cyclones/Typhoons/Hurricanes Tsunami/Tidal Waves Volcanic eruptions Fire – forest, other Unusually cold weather/winter emergencies Drough	Mass violent civil disturbances International conflict Civil war Localized armed conflict Terrorism	Nuclear contamination Industrial disasters Poisoning of water courses – chlorine, cyanide, mercury, lead, etc. Environmental pollution Dam and structure failures Oil spills

Figure 4. Environmental, social and climate of disaster justice

4. *Implementation Strategies of Flood and Erosion Mitigation in Nigeria*

It very important to implement the following strategies for alleviating flood and erosion:

i. Recent technology such as geographic information systems (GIS) can be adopted for sustainable construction, erosion control, flood mitigation and infrastructural development.
ii. Periodic hazard assessment planning of varied geopolitical zones especially Lagos, Niger delta and rain forest region of Nigeria.
iii. Land use control, landscape maintenance, weather stations, proper enforcement of construction works consistent with plan should be strategically put in situ for effective monitoring and alleviation of flood and erosion in Nigeria.
iv. Watershed protection methods and improved dams and dykes are often constructed in strategic places.
v. Flood hazard mapping framework should be strictly enforced for sustainability and for mitigation of environmental impacts as presented in Figure 5.

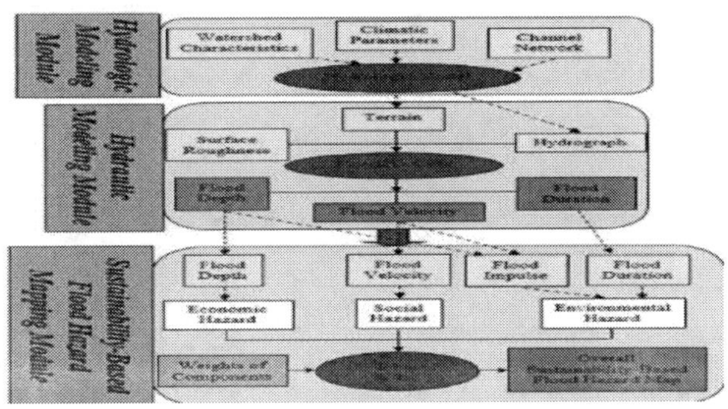

Figure 5. Flood hazard mapping framework [30, 32]

Figure 6 indicates risk assessment of Hydro-meteorological and nature-based solution management while figure 7 indicates management cycle for flood risks. Figure 8 indicates Lagos Coastal Areas Monthly Frequency of Ocean Surge. Figure 8: Damage of Pavements as a result of Incessant Flood in Coastal Area of Lagos.

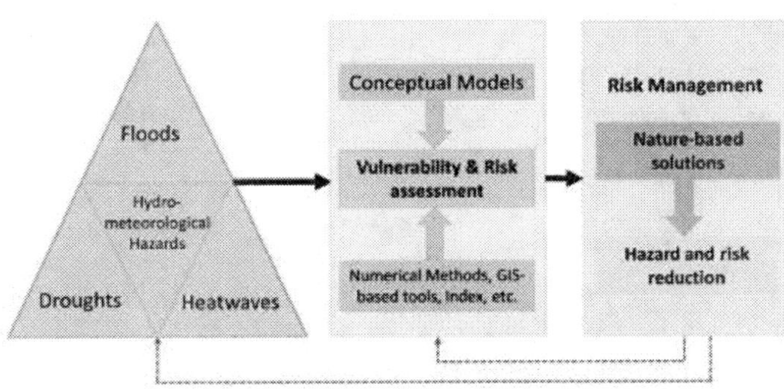

Figure 6*:* Risk Assessment of Hydro-Meteorological and Nature-Based Solution Management *[35]*

Figure 7: Management Cycle for Flood Risks [37]

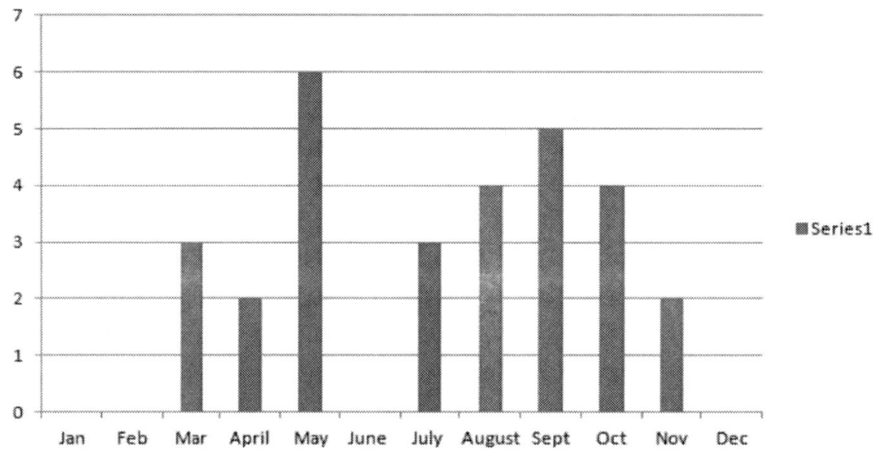

Figure 8: Lagos Coastal Areas Monthly Frequency of Ocean Surge

Figure 9: Damaged Pavement Infrastructures due to Incessant Flooding in Lagos

5. Conclusions

Disaster management, flood abatement and erosion control require adequate priority for prevention of hazards, sustainable development, rapid urban growth and infrastructural development, national security, protection of life and property.

Every strategy which will curtail flood and erosion must be implemented to avert natural hazard and to make sure that development efforts do not aggravate vulnerability to those hazards in view of incessant and destructive effects of flood and erosion tragedy globally. Hazard assessment and provision of mitigation measures should be done before the occurrence. Efficient drainage and irrigation systems are crucial for alleviating flood and erosion in developing countries. Prevention measures include quality control and monitoring of civil engineering projects especially on roads and drainages; quality research output by experts, physical barriers reinforced structures, restrictions, implementation of strategies, establishment of frameworks, enforcement of government policies and regulations. The ultimate tool is integrated and collaborative Flood Management Act. Stakeholders, researchers, hydrologists and engineers should partner with government agencies in the planning, construction, management and maintenance of rural and urban drainages for adequate funding and sustainable construction.

6. Recommendations

The following recommendations are essential for handling flood and erosion.

i. Governments need to perpetually check on flood-prone and vulnerable areas.

ii. A lot of study and fund need to be dedicated to flood and erosion connected matters.

iii. Mitigation measures like sensible drain systems, seminars, public awareness and training, environmental and land-use controls need to be organized.

iv. Research related to hazard assessment and implementation strategies to avert flood should be supported and funded by government and non-governmental organizations.

vi. Periodic removal of solid wastes and unwanted materials from the drainages is very crucial in rural and urban centers.

vii. Coastal protection should be done in areas prone to flooding.

viii. Policies, engineering standards and laws should be put in place to check populace on activities that can results into flooding and overtopping waves in vulnerable places and coastal areas.

ix. Due to uncertain nature of flood, preventive measures and urgent action are necessary especially in coastal areas.

Acknowledgements

I extend my profound appreciation to experts, professionals, stakeholders and engineers that have contributed to this work. I also thank International Conference on Sustainable Infrastructural Development planning committee for their unquantifiable efforts.

References

[1] Levy JK, Hartmann J, Li KW, An Y and Asgary A 2007 Multi-Criteria Decision Support Systems for flood hazard mitigation and emergency response in urban watersheds. *J. Am. Water Resour. Assoc.* ***43***, 346–358

[2] Schröter K, Kreibich H, Vogel K, Riggelsen C, Scherbaum F and Mer, B 2014 How useful are complex flood damage models? *Water Resour. Res.* ***50***, 3378–3395

[3] Graham W JA 1999 *Procedure for Estimating Loss of Life Caused by Dam Failure*; Report No. DSO-99-06; Dam Safety Office, U.S. Bureau of Reclamation: Denver, CO, USA, 1999
[4] Prinos P, Kortenhaus A, Swerpel B, Jimenez JA and Samuels P 2009 Review on Flood Hazard Mapping; flood site Project, Report No. T03-07-01, Revision No. 4_3_P01. Available online: http://www.floodsite.net/html/partner_area/project_docs/T03_07_01_Review_Hazard_Mapping_V4_3_P01.pdf (accessed on 16 July 2019)
[5] FEMA Definitions of FEMA Flood Zone Designations 2016 Available online: https://snmapmod. snco.us /fmm/document/fema-flood-zone-definitions.pdf (accessed on 18 January 2019)
[6] U.S. Army Corps of Engineers *Generic Depth-Damage Relationships for Residential Structures without Basements*; Economic Guidance Memorandum, No. 1–3; 2000
[7] Merz B, Kreibich H, Schwarze R and Thieken A 2010 Review article "Assessment of economic flood damage" *Nat. Hazards Earth Syst. Sci.* ***10***, 1697–1724
[8] Abrishamchi A, Dashti M, Alamdari N, Salavitabar AA 2011 GIS-Google earth based approach to estimating the flood damage function in large river basins. *In Proceedings of the World Environmental and Water Resources Congress*, Palm Springs, CA, USA, 22–26 May 2011; pp. 3811–3821
[9] Kalyanapu AJ, Judi DR, McPherson TN and Burian SJ 2015 Annualised risk analysis approach to recommend appropriate level of flood control: Application to Swannanoa river watershed. *J. Flood Risk Manag.* ***8***, 368–385
[10] IPCC Climate Change 2014: Synthesis Report. Contribution of Working Groups I, II and III to the Fifth Assessment Report of the Intergovernmental Panel on Climate Change; *Intergovernmental Panel on Climate Change*: Geneva, Switzerland, 2014; ISBN 9789291691432
[11] Pryor P 2012 Hazard as a concept (Health and Safety Professionals Alliance HaSPA), The Core Body of Knowledge for Generalist OHS Professionals, Safety Institute of Australia, Tullamarine, VIC (2012)
[12] IPCC 2012 Managing the Risks of Extreme Events and Disasters to Advance Climate Change Adaptation, A Special Report of Working Groups I and II of the Intergovernmental Panel on Climate Change (IPCC), Cambridge University Press, Cambridge, UK, and New York, NY, USA (2012), pp. 555-564
[13] Patra KC 2001 Hydrology and Water Resources Engineering; Alpha Science International: Oxford, UK, ISBN 1842650238
[14] Whitworth Malcolm NU, Malcolm W and Brian B 2015 Flooding and Flood Risk Reduction in Nigeria: Cardinal Gaps. J. Geogr. Nat. Disasters, **5**, 136
[15] Adeloye AJ, Rustum R 2011 Lagos (Nigeria) Flooding and Influence of Urban Planning. *Proc. Inst. Civ. Eng.*—Urban Des. Plan. **164**, 175–187
[16] Aderogba KA 2012 Qualitative Studies of Recent Floods and Sustainable Growth and Development of Citiesand Towns in Nigeria *Int. J. Acad. Res. Econ. Manag. Sci.* **1**, 1
[17] Babanyara YY, Usman HA and Saleh UF 2010 An Overview of Urban Poverty and Environmental Problems in Nigeria. *Journal of Human Ecology*. **31**(2): 135-143
[18] Etuonovbe AK 2011 The Devastating Effect of Flooding in Nigeria. *FIG Working Week 2011*, Bridging the Gap between Cultures. Marrakech, Morocco, 18-22 May 2011. ts06j_etuonovbe.pdf
[19] World Bank 2006 Mainstreaming Hazard Risk Management in Poverty ReductionStrategies: A strategic framework for action. Discussion paper
[20] UNOCH 2018 Nigeria: Humanitarian Response Plan; United Nations Office for the Coordination of Humanitarian Affairs: Abuja, Nigeria
[21] Otero R C and Marti RZ 1995 "The Impacts of Natural Disasters on Developing Economies: Implications for the International Development and Disaster Community"
[22] Ologunorisa T 2009 Strategies for Mitigation of Flood Risk in the Niger Delta, Nigeria, *J. Appl. Sci. Environ. Manage.* **13**(2) 17- 22
[23] National Emergency Management Agency, (NEMA) 2010 GIS Based Flood Risk Assessment of Kubwa Abuja, NEMA GIS Unit

[24] United Nations Disaster Management Training Programme 1992 *An Overview of Disaster Management. (*2nd Edition). USA: UNDP
[25] International Strategy for Disaster Reduction (ISDR) 2002 *Living with Risk: A Global Reviews of Disaster Reduction Initiatives.* Preliminary Version. Geneva: ISDR Secretariat
[26] Colleta NJ 2004 "Human-Driven Disasters: Violent Conflict, Terrorism and Technology". *Inter-American Development Bank Sustainable Development Department: Technical Paper Series*
[27] Clarke CL and Munasinghe M 1995 "Economic Aspects of Disasters and Sustainable Development: An Introduction". In Clarke, C. L. and M. Munasinghe (eds) *Disaster Prevention for Sustainable Development: Economic and Policy Issues.* The International Decade for Natural Disaster Reduction (IDNDR) and The World Bank
[28] Nwafor JC 2006 *Environmental Impact Assessment for Sustainable Development: The Nigerian perspective.* Enugu: EL 'DEMAK pubs, pp. 359-394
[29] Punch 2012 The Nigerian Punch, 2010. Lagos Flood. The Punch Newspaper Tuesday 12, October 2010. Retrieved from http://www.punchng.com/Articl.aspx?theartic on April 19, 2011. Online; http://www.punchng.com/business/flood-25-ofnigerians- -says-jonathan-asdangote-committee-targets-n100bn/ cited 27/10/2012
[30] Oluwaseyi AB 2017 Plant Genetic Resources (PGR) in Nigeria and the Reality of Climate Change: A review. *Asian J. Environ. Ecol.* **2**, 1–24
[31] CERD 2000 "EM-DAT: The OFDA/CRED International Disaster Database" http.//www.md.vcl.ac.be/cred/.
[32] Ahmadisharaf E, Kalyanapu AJ and Eun-Sung C 2017 Sustainability-Based Flood Hazard Mapping of the Swannanoa River Watershed *Sustainability*, *9*(10), 1735; https://doi.org/10.3390/su9101735
[33] Ekpo UF, Hürlimann E, Schur N, Oluwole A, Abe EM, Mafe MA, Nebe OJ, Isiyaku S Olamiju F, Kadiri M et al. 2013 Mapping and Prediction of Schistosomiasis in Nigeria Using Compiled Survey Data and Bayesian Geospatial Modelling. *Geospat. Health*, **7**, 355–366
[34] Augustine CE 2015 Managing Flood Disasters in Nigerian Cities: Issues and Strategies towards Meeting the Challenges in the Modern World. (A Case Study of Owerri Metropolis Imo State Nigeria). FIG Working Week 2015 From the Wisdom of the Ages to the Challenges of the Modern World Sofia, Bulgaria, 17-21 May 2015
[35] Jeetendra S 2019 Hydro-meteorological risk assessment methods and management by nature-based solutions *Journal Science of the Total Environment.* **696**, 15, 133936
[36] Klein RJT, Nicholls RJ and Thomalia F 2003 The Resilience of Coastal Megacities to Weather-Related Hazards Kreimer I, Arnold AM and Carlin A (eds) Building Safer Cities: *The Future of Disaster Risk. Washington D C: World Bank, Disaster Risk Management* Series No. 3
[37] NRC 2012a Dam and Levee Safety and Community Resilience: A Vision for Future Practice. Washington, DC: *The National Academies Press*

Assessment of hydrogen sulfide emission levels on the floors of some selected bakeries in southwestern Nigeria

O A Odunlami[1]*, O G Abatan[1], A A Busari[2], G T Alao[3], F B Elehinafe[1], C O Emekekwue[1]
[1]Department of Chemical Engineering, Covenant University, Ota, Ogun state, Nigeria
[2]Department of Civil Engineering, Covenant University, Ota, Ogun state, Nigeria
[3]Department of Mathematics, Covenant University, Ota, Ogun state, Nigeria
*Corresponding Author E-mail *olayemi.odunlami@covenantuniversity.edu.ng*

Abstract. The hydrogen sulfide (H_2S) emission and oxygen (O_2) level was measured within four bakeries in Ota, Ogun State. This was done to know the effect of H_2S on the people working in bakeries and those walking into bakeries for business purposes. The data obtained was analyzed and the H_2S emission level was around the new recommendation for airborne of 1 ppm recommended for 8 hours by ACGIH (American Conference Governmental Industrial Hygenist [1]. The highest mean concentrations of H_2S were; bakery 1 (0.4 ppm), bakery 2 (0.73 ppm), bakery 3 (1.0ppm), H_2S emission was not detected by the gas analyzer in bakery 4. O_2 level (20.8ppm) remained constant all through the study, which is the expected ambient O_2 value, this is because of the cross ventilation within the bakeries. Though literatures have shown that ovens emit some harmful gases like CO, NO_x, VOC and H_2S, the concentrations of these gases can be reduced if there is cross ventilation within the bakery and exhaust stacks positioned high enough towards the roof, out of the building. The stack effect happens, when warm air moves or flow upwards in the building. The warm air rises because it is lighter than cold air, when it rises, it escapes out of the upper levels of the building through ventilation openings, windows or leakages. The rising warm air reduces the pressure at the base of the building, forcing cold air to infiltrate through either open doors, windows, or other openings and leakage in lower levels of the building. The H_2S emission level was higher during the heating periods than the baking periods in all the bakeries considered due to the combustion process taking place during heating. The electrically fired ovens gave lower H_2S concentrations than the wood fired ovens. To ensure safety for workers and consumers in Bakeries, regular inspection of bakeries should become a priority for the various agencies involved.

Keywords: Air emissions; hydrogen sulfide; Bakeries; ventilation

1.Introduction

The demand of bakery products has increased in Nigeria over the years and this has made the setting up and operation of bakeries in Nigeria a popular business. However, some of these bakeries are not properly built and do not operate on modern designs and equipment which has increased the level of air pollution in the country. While outdoor emissions are researched on regular basis [2], indoor emissions have barely been looked into. A total of four bakeries where the baking of yeast leavened bread is done were assessed in Ota, Ogun State.

Hydrogen sulfide, a colorless gas has an offensive odor at low concentrations. The gas is toxic to humans and animals, and corrosive to many metals at high concentrations [3]. It tarnishes silver and react with heavy metals in paints to discolor the paint. It causes headache, conjunctivitis, sleeplessness, pain in the eyes, and similar symptoms in humans at low air concentrations and death at high air concentrations [4]. However, if the victim is moved quickly to uncontaminated air and respiration initiated before heart action stops, rapid recovery can be expected [5]. Generally, hydrogen sulfide acts as a cell and enzyme poison, it can also cause irreversible changes in the nerve tissue [6]. Air pollution by hydrogen sulfide is commonly found in the vicinity of an emitter like kraft paper mills, industrial waste disposal ponds, sewage plants, refineries and coke oven plants [7].

Wood is a polymer of celluloses, polyoses and lignin with several minerals and concentrates depending on the wood species [8,9]. The heat energy from firewood (beech, oak, birch) is around 17–19 MJ kg. Subject to how the oven is run, set up and sustained, and the firewood itself, a widespread array of other combustion products are let off in addition to carbon dioxide and carbon monoxide [10, 11]. These principally consist of the normal combustion products cellulose [12], aldehyde [13] and particulate matter [14].

It has been known that fire sources in rooms can increase the level of indoor air contamination. In addition, a heat source may impact the air circulation in a room [15]. Previous work done have reported the impacts of sources, for example, gas burners [16], candles, incense sticks [17], kerosene ovens [18], pizza stoves [19] as well as open chimneys in homes [20]. Recently, Noonan and his associates have given specific consideration to the formation of particulate matter ($PM_{2.5}$) in indoor air while operating wood-consuming oven [21].

This report looks into ovens in bakeries as potential sources of the predicted emissions of hydrogen sulfide. Series of tests were carried out in various sampling points of the chosen bakeries. The Hydrogen sulfide emissions measured from the Bakeries was compared with the H_2S standard by ACGIH (American Conference Governmental Industrial Hygienist) [22]. The level of oxygen across the sampling points was also monitored.

2. Experimental area

2.1 Study area

The study area is Ota, Ogun state (figure 1). It is found in the southwestern region of Nigeria. It covers an approximate area of 1.46km2 (square kilometers) and runs an approximate distance of 5.08km (Google Maps, 2019). Ota is a town in Ogun State, Nigeria, and has an estimated 163,783 residents living in or around it. Ota is in Ado-Odo/Ota Local Government Area.

Figure 1: *A map of the study area [23]*

2.2 Materials and methods

2.2.1 Sampling points: *The ALTAIR XCELL 5X Multi-gas sensor was used to measure the H_2S emission concentrations and O_2 level from the selected four bakeries in Ota. Four sampling points in each of the bakeries were identified based on the working sections of the bakeries. The identified*

IOP Conf. Series: Materials Science and Engineering 1036 (2021) 012067 doi:10.1088/1757-899X/1036/1/012067

sampling points were denoted by A, B, C and D while the bakeries were named 1, 2, 3 and 4. The hydrogen sulfide emission and oxygen level measurements were taken for one hour at 2 min intervals when the ovens were heated and during baking. The gas detector was switched on and allowed to initialize, for all internal systems to be ready. An external probe was connected to the gas detector and was placed close to the heating unit of the oven and around the baking point, then the readings were taken.

2.3 Equipment

The MSA Altair 5X Multi gas detector can monitor up to 6 gases at once and has a variety of optional MSA infrared sensors. The detector features large glove-friendly buttons, high performance cell sensors and long-lasting grip for easy operation. In potentially dangerous locations, unique features such as motion alert and instant alert provide a sense of control. The Altair 5X Multi gas detector offers matchless flexibility and versatility to meet the various needs of many of its users. The technologically advanced unit also features a bump test of less than 15 seconds, a span calibration time of 60 seconds, a run time of 20 hours, and a lengthy lifespan.

2.4 Measurement strategy

The gas analysis was carried out during the working hours of the bakery. The test samples were taken during the heating and the baking periods. The heating period was 3-5 hours while the baking period is 4 hours. To determine the concentrations of the pollutants, the MSA Altair 5X multi gas detector was employed at each of the sampling points. Measurements were taking at 2 min intervals for an hour. The measurements were taken for both the heating and baking periods.

3. Result and discussion

The results of this study are discussed here; the mean concentrations of hydrogen sulfide in the identified sampling points of the various bakeries for the two periods which include the heating period of the oven and baking period.

3.1 The identified sampling points within the four selected bakeries

The sampling points are:

1. Sampling point A: This is where the burner is situated. Workers occasionally go there to sustain the fire and heat that is being generated. Combustion takes place at this point. It is the source of the emission.
2. Sampling point B: This is the mixing zone where the pastry dough is prepared and mixed. It is situated not too far from the oven. Doughs are mixed by the workers at this point.
3. Sampling point C: This is the packing area. Finished products (bread) are allowed to cool, packaged and sealed by workers here.
4. Sampling point D: The finished products are brought here by the workers to be stored.

The bakeries include;
Bakery 1: Standard building for a bakery, which has one wood fired with its exhaust stacks positioned high enough towards the roof, out of the building.
Bakery 2: A residential apartment converted to a bakery, with its exhaust stacks positioned high enough towards the roof, out of the building. It has one wood fired oven.
Bakery 3: A residential apartment converted to a bakery, with its exhaust stacks positioned high enough towards the roof, out of the building. It has one wood fired oven.
Bakery 4: Standard building for a bakery, which it has 3 ovens powered with diesel engine and electricity. It has its exhaust stacks positioned high enough towards the roof, out of the building.

3.2 *The mean concentrations of hydrogen sulfide emissions in the identified sampling points from the selected bakeries*

Tables 1-4 show the mean concentrations of H_2S and the O_2 level at the sampling points in all the bakeries during the heating periods, while tables 5-8 show the mean concentrations of H_2S and O_2 level at the sampling points in all the bakeries during the baking periods.

Table 1: Mean concentrations of H_2S and O_2 level in the identified sampling points during the heating periods of the oven in bakery 1 (Standard bakery).

Sampling Points	H_2S (ppm)	O_2 (ppm)	Distance from source (m)
A	0.4	20.8	0.36
B	0	20.8	0.89
C	0	20.8	3.41
D	0	20.8	7.93

Table 2: Mean concentrations of H_2S and O_2 level in the identified sampling points during the heating period of the oven in bakery 2 (Converted building).

Sampling Points	H_2S (ppm)	O_2 (ppm)	Distance from source (m)
A	0.73	20.8	0.33
B	0	20.8	0.92
C	0.37	20.8	0.51
D	0	20.8	8.93

Table 3: Mean concentrations of H_2S and O_2 level in the identified sampling points during the heating periods of the oven in bakery 3 (Converted building).

Sampling Points	H_2S (ppm)	O_2 (ppm)	Distance from source (m)
A	1.0	20.8	0.36
B	0	20.8	0.79
C	0.5	20.8	5.33
D	0	20.8	10.1

Table 4: Mean concentrations of H_2S and O_2 level in the identified sampling points during the heating periods of the oven in bakery 4 (Standard bakery).

Sampling Points	H_2S (ppm)	O_2 (ppm)	Distance from source (m)
A	0	20.8	0.97
B	0	20.8	2.46
C	0	20.8	5.32
D	0	20.8	9.65

Table 5: Mean concentrations of H_2S and O_2 level in the identified sampling points during the during the baking periods of the oven in bakery 1 (Standard bakery).

Sampling Points	H_2S (ppm)	O_2 (ppm)	Distance from source (m)
A	0	20.8	0.36
B	0	20.8	0.89
C	0	20.8	3.41
D	0	20.8	7.93

Table 6: Mean concentrations of H_2S and O_2 level in the identified sampling points during the baking period of the oven in bakery 2 (Converted building).

Sampling Points	H_2S (ppm)	O_2 (ppm)	Distance from source (m)
A	0	20.8	0.33
B	0	20.8	0.92
C	0	20.8	0.51
D	0	20.8	8.93

Table 7: Mean concentrations of H_2S and O_2 level in the identified sampling points during the baking period of the oven in bakery 3(Converted building)

Sampling Points	H_2S (ppm)	O_2 (ppm)	Distance from source (m)
A	0	20.8	0.36
B	0	20.8	0.79
C	0	20.8	5.33
D	0	20.8	10.1

Table 8: Mean concentrations of H_2S and O_2 level in the identified sampling points during the baking period of the oven in bakery 4 (Standard bakery).

Sampling Points	H_2S (ppm)	O_2 (ppm)	Distance from source (m)
A	0	20.8	0.97
B	0	20.8	2.46
C	0	20.8	5.32
D	0	20.8	9.65

3.3 Discussion of results

Ovens are generally designed in way that its fire chamber is sealed off from the working place so as not to affect the indoor ambient air, but in some cases they are not properly designed and the chambers are not sealed airtight because of feeding of wood into the chambers and this allows potential emissions into the indoor air. Four bakeries were monitored and are discussed below.

Table 1 shows the mean concentrations of H_2S and O_2 level at the four sampling points in bakery1 during the heating period. H_2S emission was only detected by the gas analyzer at sampling point A (0-1) ppm in this bakery. The H_2S mean concentration for the sampling point A was (0.4) ppm. The O_2 level was constant and normal at (20.8) ppm (the expected ambient O_2 value) which indicates a good ventilation. The American Conference of Governmental Industrial Hygienists (ACGIH), a recognized body for standards kept the limit of H_2S at 1 ppm for an eight-hour time weighted average (TWA), therefore all the sampling points were well within this limit.

Table 2 shows the mean concentrations of H_2S and O_2 level at the four sampling points in bakery2 during the heating period. H_2S emission was not detected by the gas analyzer at any of the sampling points in this bakery. The O_2 level was constant and normal at (20.8) ppm indicating a good ventilation.

Table 3 shows the mean concentrations of H_2S and O_2 level at the four sampling points in bakery3 during the heating period. The ranges for H_2S concentration in the sampling points are; point A (0-3) ppm, point C (0-1) ppm. There was no H_2S emission detected at sampling points B and D by the gas analyzer. The H_2S mean concentration in the sampling point A is (1.0) ppm and (0.5) ppm in sampling point C. The O_2 level was constant and normal at (20.8) ppm indicating a good ventilation. H_2S emission levels were within the ACGIH standards which is 1 ppm for an eight-hour time weighted average (TWA).

Table 4 shows the mean concentrations of H_2S and O_2 level at the four sampling points in bakery4 during the heating period. H_2S emission was not detected by the gas analyzer at any section in the bakery. The O_2 level was constant and normal (20.8) ppm at all the sampling points all through the monitoring process. Bakery 4 ovens operate on both electricity and diesel which gives little or no H_2S emission when compared with the commonly used wood fired ovens. Bakery 4 is biggest of the bakeries. It is well structured for baking and well ventilated with highly built exhaust stacks.

Table 5 shows the mean concentrations of H_2S and O_2 level at the four sampling points in bakery1 during the baking period. H_2S was not detected by the gas analyzer at any sampling points in this bakery. The O_2 level was constant and normal (20.8) ppm at all sampling points indicating a good ventilation.

Table 6 shows the mean concentration of H_2S and O_2 level at the four sampling points in bakery2 during the baking period. H_2S was not detected by the gas analyzer at any sampling point in this bakery. The O_2 level was constant and normal (20.8) ppm, indicating a good ventilation. The level of H_2S emissions within the sampling points was lower than the ACGIH standards for H_2S.

Table 7 shows the mean concentration of H_2S and O_2 level at the four sampling points in bakery3 during the baking period. H_2S was not detected by the gas analyzer at any section in this bakery. The O_2 level was constant and normal (20.8) ppm because the bakery was well ventilated. The structures and operations of bakeries 2 and 3 are similar.

Table 8 shows the mean concentrations of H_2S and O_2 level at the four sampling points in bakery4 during the baking period. H_2S was not detected by the gas analyzer at any of the sampling points in bakery 4 during the baking period, this implies that the H_2S emission was insignificant. The O_2 level was constant and normal (20.8 ppm) at all the sampling points.

4. Conclusion

It was concluded at the end of this study that;

- The indoor level of H_2S in well ventilated bakeries is relatively low.
- The construction of the fire chamber, the type of wood used and the height of the exhaust stacks play important roles in the level of emission within the bakeries.
- The mean concentrations of H_2S during the baking periods were relatively small when compared to the ACGIH standards.
- The mean concentrations H_2S at the sampling points during the heating periods did not exceed the ACGIH standards when compared.

Acknowledgement

The authors wish to acknowledge the financial support offered by Covenant University in actualization of this research work for publication.

IOP Conf. Series: Materials Science and Engineering 1036 (2021) 012067 doi:10.1088/1757-899X/1036/1/012067

References

[1] Threshold Limit Values for 2010-Recommendedand Intended Values, of the American Conference of Governmental Industrial Hygienists, Chicago, Illinois 2010

[2] Odunlami O, Elehinafe F, Oladimeji T, Fajobi M, Okedere O and Fakinle B 2018 Implications of lack of maintenance of motorcycles on ambient air quality in *IOP Conference Series: Materials Science and Engineering* **1** (413) 012-055 IOP Publishing

[3] Fajobi M, Fayomi O, Akande I and Odunlami O 2019 Inhibitive Performance of Ibuprofen Drug on Mild Steel in 0.5 M of H_2SO_4 Acid *Journal of Bio-and Tribo-Corrosion* **5** (3) 79

[4] Permissible Emission Concentration of Hydrogen Sulfide Sub-committee on Effects of Hydrogen Sulfide of the Committee on Effects of Dust and Gas of the Verein Deutscher Ingenieure Committee on Air Purification 1960 VDI 2107

[5] Patty F, 1963 *Industrial hygiene and toxicology* **615** (902) 322

[6] Caponecchia C 2010 It won't happen to me: An investigation of optimism bias in occupational health and safety *Journal of Applied Social Psychology*, **40** (3) 601-617

[7] Halley P 1967 Hazards of Hydrogen Sulfide Med. *Bull.* **27** (3) 219

[8] Fengel D, and Wegener G 1989 wood berlin: Walter de Gruyter.

[9] Glytsos T, Ondrácek J, Dzumbová L, Kopanakis I, and Lazaridis M 2010 Characterization of particulate matter concentrations during controlled indoor activities Atmos. *Environ.* **44** 1539-1549

[10] McDonald J, Zielinska B, Fujita E, Sagebiel J, Chow J and Watson J 2000 Fine particle and gaseous emission rates from residential wood combustion. *Environmental Science and Technology* **34** (11) 2080-2091

[11] Schauer J, Kleeman M, Cass G, and Simoneit B 2001 Measurement of emissions from air pollution sources. 3. C1− C29 organic compounds from fireplace combustion of wood. *Environmental science & technology,* **35** (9) 1716-1728

[12] Shen D, and Gu S 2009 The mechanism for thermal decomposition of cellulose and its main products. *Bioresource technology*, **100** (24) 6496-6504

[13] Cerqueira M, Gomes L, Tarelho L, and Pio C 2013 Formaldehyde and acetaldehyde emissions from residential wood combustion in Portugal, 171-176

[14] Hedberg E, Kristensson A, Ohlsson M, Johansson C, Johansson P, Swietlicki E and Westerholm R 2002 Chemical and physical characterization of emissions from birch wood combustion in a wood stove *Atmospheric Environment* **36** (30) 4823-4837

[15] Ardkapan S, Nielsen P, and Afshari A 2014 Studying passive ultrafine particle dispersion in a room with a heat source Building Environment **71** 1–6

[16] Wallace L, Wang F, Howard-Reed C, and Persily A 2008 Contribution of gas and electric stoves to residential ultrafine particle concentrations between 2 and 64 nm: size distributions and emission and coagulation rates. *Environmental Science and Technology*, **42** (23) 8641-8647

[17] Wang B, Lee S, Ho K and Kang Y, 2007 Characteristics of emissions of air pollutants from burning of incense in temples, Hong Kong. *Science of the total environment* **377** (1) 52-60.

[18] Carteret M, Pauwels J, and Hanoune B 2012 Emission factors of gaseous pollutants from recent kerosene space heaters and fuels available in France in 2010 *Indoor air* **22** 299-308

[19] Buonanno G, Morawska L, Stabile L, and Viola A 2010 Exposure to particle number surface area and PM concentrations in pizzerias Atmos. *Environ* **44** 3963-3969

[20] Lahiri T, and Ray M 2010 Effects of indoor air pollution from biomass fuel use on women's health in India Health and Environmental Impacts 135

[21] Ward T, Palmer C, Bergauff M, Hooper K, and Noonan C 2008 Results of a residential indoor PM2. 5 sampling program before and after a woodstove changeout. *Indoor air* **18** (5) 408-415

[22] Threshold Limit Values for 2010-Recommendedand Intended Values, of the *American Conference of Governmental Industrial Hygienists*, Chicago, Illinois 2010

[23] Google Maps, 2019

Rheological characteristics of metal oxide nanocomposite and its application in enhanced oil recovery

A Sowunmi[1,3], V E Efeovbokhan[1], O D Orodu[2] and O Azeta[1]

[1]Department of Chemical Engineering, Covenant University, Ota, Ogun State, Nigeria
[2]Department of Petroleum Engineering, Covenant University, Ota, Ogun State, Nigeria
[3]Corresponding author – akinleye.sowunmi@yahoo.com

Abstract. A nanocomposite is a mixture of a nanoparticle and a complementary substance. In crude oil recovery, it is most commonly the combination of a nanoparticle and a polymer. In enhanced oil recovery, metal oxide nanoparticles may be combined with polymers and injected into a reservoir after water flooding to achieve greater recovery of crude oil; however, current research in this area is limited. In this work, rheological properties and oil recovery potential of two nanocomposites: alumina/xanthan gum and alumina/guar gum; were investigated. The viscosity of different concentrations of the nanocomposites was measured using the Model 800 OFITE Viscometer, and core flooding experiment was done using a Reservoir Permeability Tester. For both nanocomposites, it was observed that their viscosities increased with increasing concentration of alumina nanoparticles and the effect was higher as polymer concentration was increased. For the core flooding experiments, the xanthan gum and alumina composite achieved total recovery of 72.8 % while the guar gum alumina nanocomposite achieved total recovery of 69.3 %. The use of alumina combinations of nanocomposite is, therefore, promising and should be explored further.
Keywords: Nanocomposite, alumina, viscosity, polymer, oil recovery

1. Introduction

In recent years, the use of nanotechnology in different fields of science and engineering has gained great momentum. Through nanotechnology, novel materials having superior characteristics have been developed. Such characteristics include higher resistance to wear, higher strength, flexibility and more diverse applications [1]. Nanotechnology is a branch of science that develops and uses small materials that are on the scale of 1 to 100 nm. These small materials are called nanoparticles; they are smaller than what the eyes can see but bigger than an atom. Metallic nanoparticles such as aluminium oxide (alumina), copper oxide, iron oxide, nickel oxide, tin oxide, magnesium oxide zinc oxide and titanium oxide have chemical and physical properties that are different from their bulk metal form. These properties which include mechanical strength, high specific surface area, low melting point, specific magnetization and optic properties have proven to make metallic nanoparticles useful in diverse fields [2].

At the moment, nanotechnology has found application in almost all areas of manufacturing and production, including the petroleum industry, where it is used in oil mining, drilling, refining and transportation [3]. It has also gained more prolific use because of the disappearance of "easy oil." Using conventional methods of oil recovery, only about 30 % of the original oil in place can be extracted after primary and secondary techniques [4]. It means that when oil is discovered at shallow depth (easy oil), about 70 % is wasted if only conventional techniques of oil recovery are applied; hence the need for enhanced oil recovery techniques. The oil industry has therefore looked to the use of nanoparticles as a potential solution to maximizing the recovery of easy oil [5, 6].

Different types of nanoparticles are being tested for use in enhanced oil recovery, the most common being aluminium, silicon and zinc nanoparticles. Nanoparticles enhance oil recovery by wettability alteration and the reduction of interfacial tension (IFT). The adsorption of nanoparticles on surfaces changes the wettability and surface energy of the nanoparticle, and the characteristic of the nanoparticle is controlled by the surface energy. Metallic oxides such as copper oxide, iron oxide and aluminium oxide have remarkable properties that give them a wide variety of applications in electronics, photonics, photovoltaics, catalytic conversions etc. The following reasons make nanoparticles viable options for use in enhanced oil recovery: magnetic separation can be used to separate nanoparticles from oil; they are inexpensive and ecologically friendly; the project cost for installation is low; they can move through the pore throat with reduced risk of impairment; and they have a high surface area to volume ratio [7, 8]. Some challenges affecting the use of nanoparticles in enhanced oil recovery include log jamming, mechanical entrapment, gravity settling and adsorption [9]. Log jamming occurs when the nanoparticles move at a rate slower than the displacing fluid, causing the nanoparticles to accumulate and block the pathway. Mechanical entrapment occurs when the size of the nanoparticle is bigger than the pore throat, causing an entrapment of the nanoparticle. Gravity settling occurs when the nanoparticles settle on the internal surface of the porous rock as a result of density differences between the fluid and moving nanoparticles [10, 11]. Adsorption occurs when the electrostatic forces on the internal surfaces of the reservoir formation attract the nanoparticles and cause them to be adsorbed to the surface. Brownian motion of the particles can also cause adsorption.

Nanoparticles can be mixed with a polymer to increase the stability of nanoparticles by mitigating some of the challenges associated with the use of nanoparticles. Nanoparticles in turn can increase the viscosity of the displacing fluid, thereby reducing fingering and increasing the overall efficiency of the oil recovery process. This mixture of nanoparticles and polymer is called nanocomposite, and its use has gained increasing interest among researchers. The effect of alumina nanoparticle has been investigated in the literature, and it was found to produce total recovery ranging from 37 to 46 % [12, 13]. In this work, rheological characteristics and the oil recovery potential of nanocomposites of alumina were investigated.

2. Experimental description

2.1. Materials

Sigma Aldrich alumina nanoparticle with the purity of 98 +% and an average size of 40-60 nm was purchased from a local dealer in Lagos, Nigeria. Xanthan gum and guar gum were purchased from a local dealer in Lagos, Nigeria. Other materials used include crude oil with API 24.8^0 gotten from the Niger Delta region of Nigeria, and core plugs acquired from Offshore Depobelt – Niger Delta.

2.2. Equipment

The OFITE® 800 Model Viscometer was used to measure the viscosity of all samples. The enhanced oil recovery process was initiated and actualized with the use of OFITE® Reservoir Permeability tester, and all flooding was done at 28°C. The saturation of core plugs for porosity determination was done using Vinci Equipment at high-pressure.

2.3. Preparation of core plugs used

Core plugs were used for the core flooding equipment in the laboratory to measure the effects of testing fluids for this enhanced oil recovery process. The core plugs were cleaned in a soxhlet extractor using toluene as the cleaning solvent. Density and fluid viscosity were determined using a pycnometer and glass capillary viscometer, respectively. The core plugs were saturated using Vinci Equipment's high-pressure core saturator for Porosity Studies.

2.4. Preparation of nanocomposites

For each nanocomposite mixture, a plastic bottle was filled to 500 mL with deionized water. Respective quantities of aluminium and the required polymer (xanthan gum and guar gum) were then measured and added to the deionized water. For the viscosity measurements, five weight percentages of aluminium oxide nanoparticle were mixed with three different weight percentages of polymer to have 15 different combinations for each nanocomposite. For the core flooding experiments, three different weight percentages of aluminium oxide nanoparticle used were 0.1, 0.3 and 0.5; these three weight percentages of alumina were mixed with three weight percentages of polymer: 0.1, 0.3 and 0.5. To achieve nanocomposite of 0.1 wt% alumina and 0.3 wt% polymer, 0.5 g of alumina was added to 500 mL of deionized water, and then 1.5 g of polymer was added (Table 1).

Table 1: Composition of polymers and aluminium oxide nanoparticle

S/N	Polymer concentration [g,%]	Alumina concentration [g,%]
1	[0.5, 0.1] ~ XG1, GG1	[0.5, 0.1] ~ A1
2	[1.5, 0.3] ~ XG2, GG2	[1.5, 0.3] ~ A2
3	[2.5, 0.5] ~ XG3, GG3	[2.5, 0.5] ~ A3
4	[5.0, 1.0] ~ XG4, GG4	[5.0, 1.0] ~ A4

XG – xanthan gum, GG – guar gum

2.5. Viscosity measurements

Viscosity was measured using an OFITE 800 Model Viscometer. The mixture to be measured was filled into the sample cup of the viscometer. The rotor sleeve of the viscometer was immersed to the fill line of the sample cup. The viscometer was set to the required shear rate (10.21 s^{-1}) and the corresponding dial reading measurement was taken from the viscometer.

2.6. Core flooding

Core flooding using nanocomposites was carried out at 28^{0}C using an OFITE® Reservoir Permeability tester. The setup comprises of three different compartments that are either filled up with crude oil, brine or the nanocomposite solution. For regulation of the flow of fluid, each compartment is equipped with an inlet and outlet valves. The flood operates using a high precision pump to inject into one of the compartments at that instant, then acting as a driving force pushing a piston plate inside it, which drives the liquid contained in the compartment at the same time isolating the driving fluid from the displacing reservoir fluid. The setup for the core flooding equipment (Figure 1) has the following parts: pump fluid, pump, valves, displacing reservoir fluid, piston to separate the oils, crude oil, brine, polymer (Treating fluid), pressure gauge, bypass valve, hassler cell holder with core, sleeve pressure, effluent into test tubes.

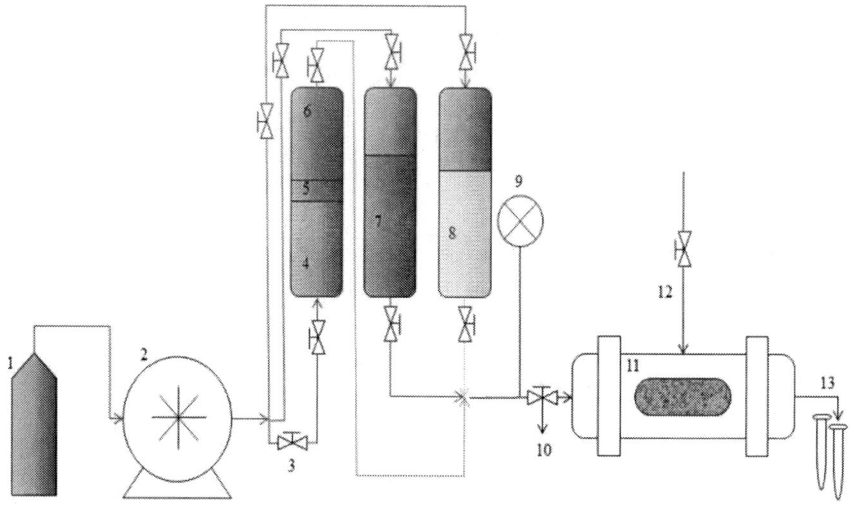

Figure 1: Experimental setup of the core flooding apparatus [7]

2.6.1. Data collection using core flooding equipment

In carrying out this experiment, the actual reservoir system has to be reproduced at the laboratory scale. The core plugs used for this experiment were already saturated 100% with brine after the cleaning procedure; therefore, there was the need to introduce oil into the core plugs. At a flow rate of 3cc/min, oil was injected into the core plugs draining out the initial brine in the core plug until no more brine was produced. This process enabled the establishment of the initial water saturation, S_{wi}. After injecting oil into the core plugs and establishing an initial water saturation value, at a rate of 3 cc/min water was injected into the core plugs. This was done in the representation of the secondary recovery process.

However, nanocomposite flooding was conducted after water flooding, considering the concentration of the recovery agents, and each value was collected every 4 minutes and recorded. To measure the oil recovery, the experiment was monitored closely all through the duration of each flooding and recovery was manually recorded to determine the amount of crude oil produced at various time intervals.

2.6.2. Nanocomposite flooding analysis

All individual core plugs were flooded using the same method. Core plugs were first injected with oil at a flow rate of 3 cc/min, displacing the brine already 100 % saturated in the core plug. This drainage process was done to attain initial water saturation and oil saturation values. This was followed by the injection of water at 3 cc/min to displace the oil in the core plug until no more oil can be produced via the water being injected. The back pressure set at 200 psi and the confining pressure having built to 700 psi at the end of the flooding.

The production of oil from the core plug into the receiving tube was monitored by observing the level of the oil-brine interface. Movement of the interface level indicated the continuous production of fluid. When oil production stopped, it implied that the core content has reached its residual oil saturation. The volume of the produced oil was recorded in cc. This concluded the secondary recovery process for the core plug, thereby making it viable for tertiary recovery.

After water flooding, the core plug was injected with nanocomposite at different concentrations, and the flooding continued until no more oil was produced from the core plug. This was done to investigate the capacity of the nanocomposite to recover the residual oil left in the reservoir after water flooding. The flooding process recovered oil continuously till only brine was being produced.

3. Results and discussion

3.1. Viscosity profile of nanocomposites

3.1.1. Effect of aluminium oxide nanoparticle on the viscosity of xanthan gum

The effect of aluminium oxide nanoparticle on 0.1, 0.5 and 1.0 wt% of xanthan gum at 10.21 s^{-1} shear rate were investigated and presented in the plot (Figure 2).

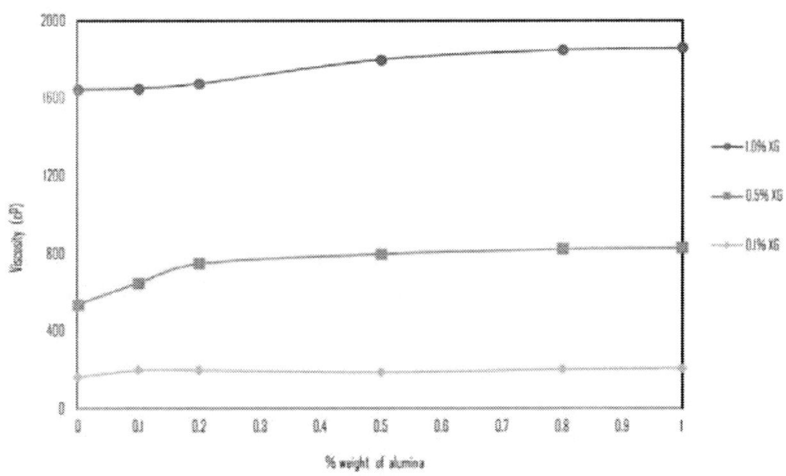

Figure 2: Viscosity profile of alumina-xanthan gum nanocomposite

3.1.2. Effect of aluminium oxide nanoparticle on the viscosity of guar gum

The effect of aluminium oxide nanoparticle on 0.1, 0.5 and 1.0 wt% of guar gum at 10.21 s^{-1} shear rate were investigated and presented in the plot (Figure 3).

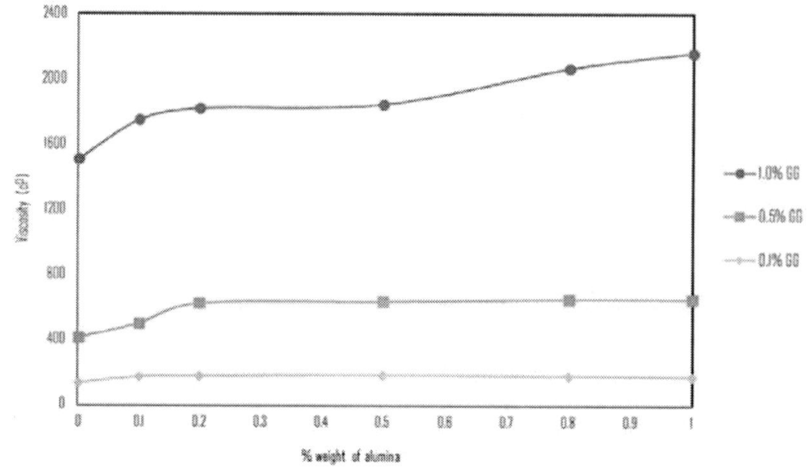

Figure 3: Viscosity profile of alumina-guar gum nanocomposite

3.2. Crude oil recovery potential of nanocomposites

3.2.1 Nanocomposite flooding for alumina and xanthan gum polymer

The core plug was first flooded with water and then flooded with a nanocomposite of aluminium oxide nanoparticle and xanthan gum polymer at 28°C. Xanthan gum at 0.5, 1.5, and 2.5 g was dissolved in 500 mL distilled water indicated as 0.1, 0.3, and 0.5 wt%. Each of these was mixed with three concentrations of aluminium oxide and used for flooding. The results are presented in figures 4 to 6.

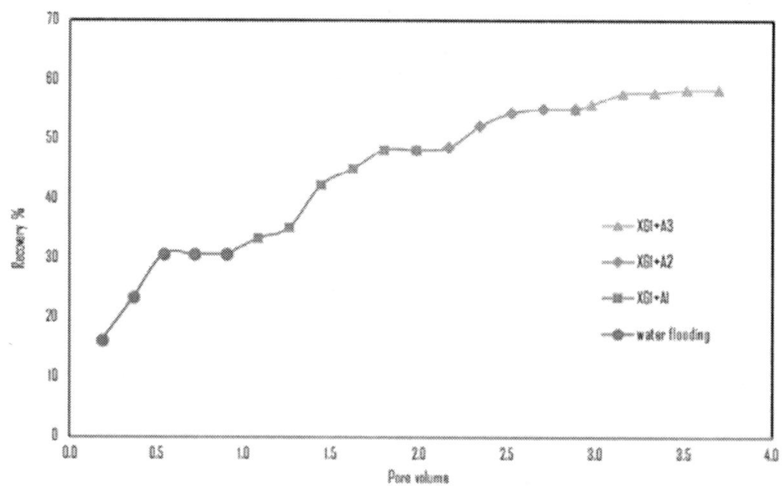

Figure 4: Nanocomposite flooding for alumina and 0.1 wt% xanthan gum

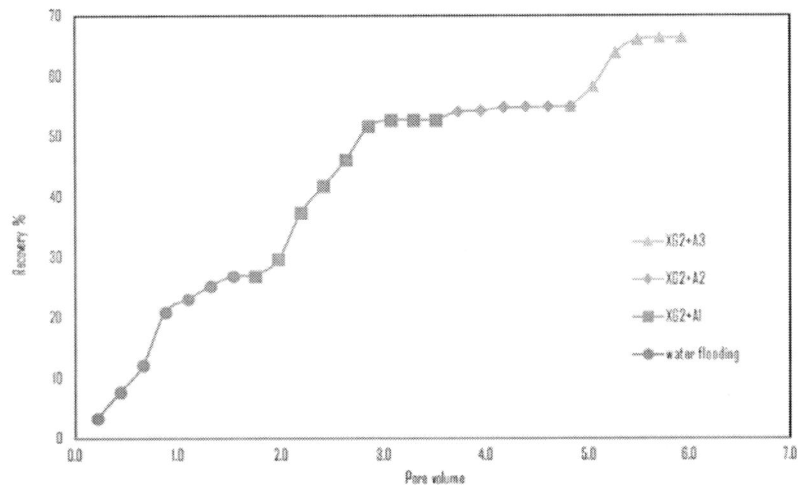

Figure 5: Nanocomposite flooding for alumina and 0.3 wt% xanthan gum

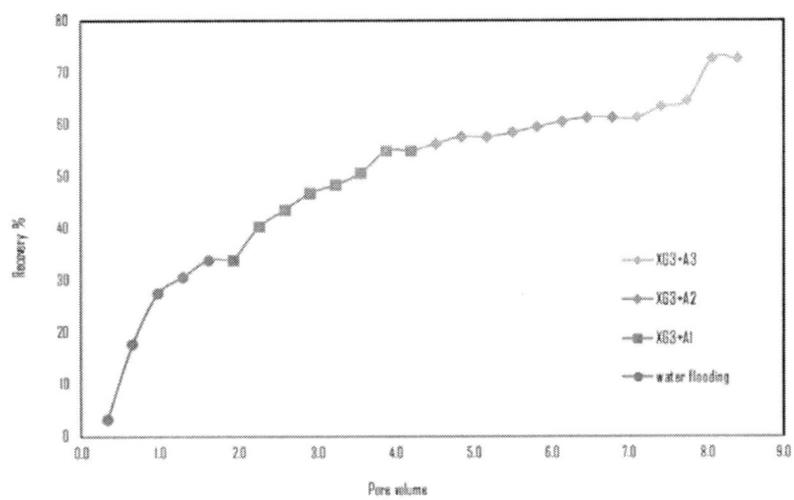

Figure 6: Nanocomposite flooding for alumina and 0.5 wt% xanthan gum

3.2.2. Nanocomposite flooding for alumina and guar gum polymer.

The core plug was first flooded with water and then flooded with a nanocomposite of aluminium oxide nanoparticle and guar gum polymer at 28°C. Guar gum at 0.5, 1.5, and 2.5 g was dissolved in 500 mL distilled water indicated as 0.1, 0.3, and 0.5 wt%. To carry out the flooding, the lowest concentration of polymer was first mixed with different weight increments of aluminium oxide nanoparticle i.e. 0.1, 0.3, and 0.5 wt%. The flooding result illustrates the effect of the nanocomposite on various concentrations of the polymer. The results are presented in figures 7 to 9.

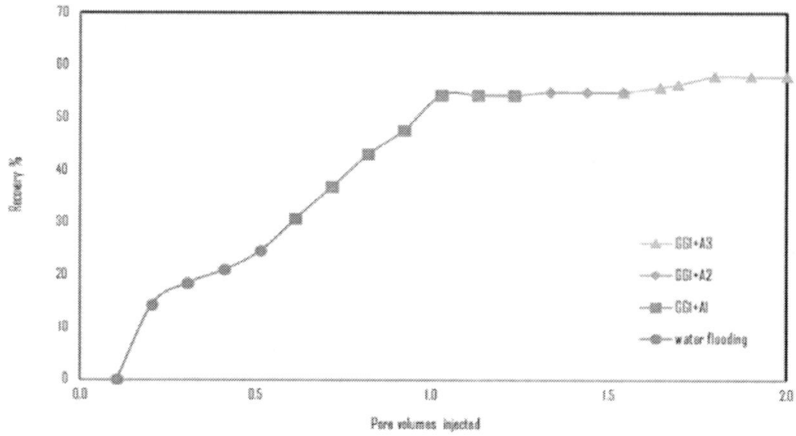

Figure 7: Nanocomposite flooding for alumina and 0.1 wt% guar gum

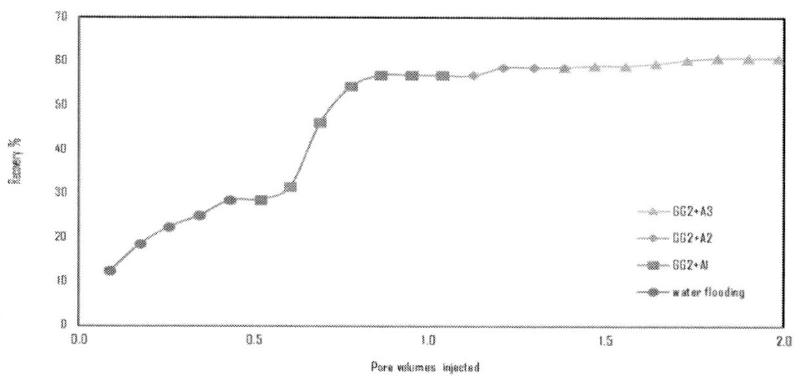

Figure 8: Nanocomposite flooding for alumina and 0.3 wt% guar gum

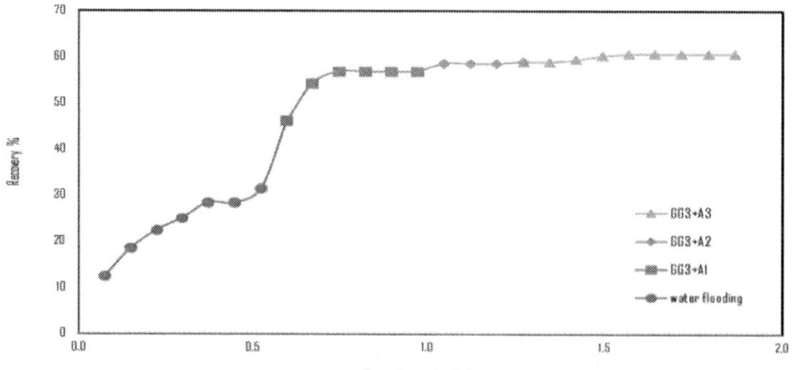

Figure 9: Nanocomposite flooding for alumina and 0.5 wt% guar gum

3.3. Discussion

3.3.1. Result for viscosity tests of the nanocomposites

Viscosity was measured for two nanocomposites: alumina-xanthan gum and alumina-guar gum. The weight percent of alumina was varied up to 1.0 for different concentrations of the polymer. For both nanocomposites, it was observed that the viscosity of the polymers increased with increasing concentration of aluminium oxide nanoparticle. This implies that aluminium oxide nanoparticle increases the bonds of polymers creating an increase in intermolecular attraction. For both polymers, aluminium oxide nanoparticle is seen to have the potential to increase its viscosity by around 50 %, even though the nanoparticle has a slightly higher effect on guar gum.

3.3.2. Result analysis for alumina-xanthan gum polymer flooding

The effect of aluminium oxide nanoparticle on the oil recovery potential of xanthan gum was investigated at three different concentrations of xanthan gum and three concentrations of the nanoparticle. Water flooding was first carried out, followed by sequential flooding with three concentrations of aluminium oxide. The concentrations used are shown in table 1.

From figure 4, we observe that water flooding achieved 30.6 % recovery leaving a significant amount of oil in the core plug. The first concentration of xanthan gum was then flooded with three concentrations of alumina. the first concentration of alumina achieved 48.2 % recovery and then no more oil was recovered. It was then flooded with a higher concentration of alumina which achieved 55 % total recovery, while the highest percentage achieved 58.5 % total recovery.

A higher percentage of xanthan gum was then used, and the result is shown in figure 5. The mixture with the lowest concentration of alumina achieved recovery of 52.7 % after recovery. The next concentration achieved recovery of 54.9 % while the highest concentration of alumina achieved 66.2 %. Figure 6 shows the flooding result obtained for alumina and the highest percentage of xanthan gum. The three concentrations of alumina achieved 54.8, 61.3 and 72.6 % total recovery respectively.

The results indicate that oil recovery increased with the concentration of xanthan gum polymer. This is an indication of higher sweep efficiency and lower viscous fingering as the concentration of xanthan gum increases. In a similar vein, total oil recovery was observed to increase with the concentration of aluminium oxide nanoparticle. This indicates that aluminium oxide nanoparticle has the potential to increase oil recovery capacity of xanthan gum.

3.3.3. Result analysis for alumina-guar gum nanocomposite flooding

To evaluate the effect of aluminium oxide nanoparticle comparatively, it was flooded in combination with guar gum. Similar to xanthan gum, three concentrations of guar gum were mixed with three concentrations of xanthan gum and used for flooding. The results are shown in figures 7-9 and the concentrations are shown in table 1.

In figure 7, the lowest concentration of guar gum, GG1 was flooded with three concentrations of aluminium oxide after water flooding and this achieved a total crude oil recovery of 54.4, 54.9 and 57.9 % respectively. For GG2, the different concentrations of aluminium oxide achieved total crude oil recovery of 56.9, 58.6 and 60.8 % respectively. For GG3, the different concentrations of aluminium oxide achieved total crude oil recovery of 59, 66.4 and 69.4 % respectively.

In comparison to the xanthan gum-alumina nanocomposite, it was observed that crude oil recovery increased with the concentration of guar gum polymer; however, total recovery achieved from xanthan gum was higher than that obtained from guar gum polymer, indicating that xanthan gum has higher volumetric efficiency and experiences less viscous fingering than guar gum. Also, the total recovery of guar gum increased with increasing concentration of aluminium oxide nanoparticle; but the nanocomposite of xanthan gum and aluminium oxide recorded higher recoveries than that of guar gum and aluminium oxide.

4. Conclusion

Aluminium oxide nanoparticle worked effectively as a good EOR agent when combined with xanthan gum and guar gum. It was found to increase the viscosity of both polymers and increase the oil recovery produced using the two polymers. Several experimental runs were performed to examine the activity of each solution. A viscosity device and core flooding equipment were used to obtain results adopted in this article to investigate the EOR mechanisms and viscosity of each solution. The following conclusions are obtained from the research:

1. Aluminium oxide nanoparticle increased the viscosity of xanthan gum and guar gum polymers by about 50 %, with xanthan gum having a higher viscosity than guar gum
2. Xanthan gum has higher crude oil recovery than guar gum with both experiencing higher recoveries as concentration was increased.
3. At 28°C, xanthan gum and aluminium oxide mixture achieved the highest total recovery of 72.6 % while the guar gum and aluminium oxide mixture achieved the highest recovery of 69.4 %.

Xanthan gum and guar gum have a high potential for crude oil recovery, and both polymers become more effective with the addition of aluminium oxide nanoparticles. Aluminium oxide nanoparticles can dissipate heat effectively when in fluids, and has high resistance against temperature and shear forces. It also can increase the viscosity of polymer fluids and can neutralize cations in brine, giving it resistance to high salinity conditions. Aluminium oxide nanoparticle is therefore considered a viable option for crude oil recovery using nanocomposites.

References

[1] Shamilov V, Babayev E, Kalbaliyeva E and Shamilov F 2017 Polymer nanocomposites for enhanced oil recovery. *Materials Today: Proceedings* **4**:S70-S4

[2] Horikoshi S and Serpone N 2013 *Microwaves in Nanoparticle Synthesis.* First ed: Wiley-VCH Verlag GmbH & Co. KGaA

[3] Rocca V, Cocuzza M, Candido P and Verga F 2012 *Current and Future Nanotech Applications in the Oil Industry*784-93

[4] Alagorni A, Yaacob Z, Nour A 2015 An Overview of Oil Production Stages: Enhanced Oil Recovery Techniques and Nitrogen Injection. *International Journal of Environmental Science and Development.* **6**:693-701

[5] John M F, Olabode O A, Egeonu G I, Ojo T I 2017 Enhanced Oil Recovery of Medium Crude Oil (31 API) Using Nanoparticles and Polymer. *International Journal of Applied Engineering Research.* **12**(19):8425-35

[6] Joseph-Igbor B, Orodu O D, Afolabi R O 2016 Evaluating the Oil Mobilization Properties of Nanoparticles Treated with Arabic Gum and Xanthan Gum for Trapped Oil in Porous Media. *SPE Nigeria Annual International Conference and Exhibition*; 2016/8/2/; Lagos, Nigeria

[7] Aurand K R, Dahle G S, Torsaeter O 2014 Comparison of Oil Recovery for Six Nanofluids in Berea Sandstone Cores. *International Symposium of the Society of Core Analysts*; Avignon, France

[8] Idogun A K, Iyagba E T, Ukwotije-Ikwut R P and Aseminaso A 2016 A Review Study of Oil

Displacement Mechanisms and Challenges of Nanoparticle Enhanced Oil Recovery. *SPE Nigeria Annual International Conference and Exhibition;* 2016/8/2/; Lagos, Nigeria. SPE: Society of Petroleum Engineers

[9] Engeset M 2012 The Potential of Hydrophilic Silica Nanoparticles for EOR Purposes. *Norwegian University of Science and Technology*, Norway. **2**(2):131-62

[10] Zhang H, Nikolov A, Wasan D 2014 Enhanced Oil Recovery (EOR) Using Nanoparticle Dispersions: Underlying Mechanisms and Imbibition Experiments. *Energy and Fuels.* **28**(5):3002-9

[11] Birkeland M A 2013 Investigation of Nanoparticle Effect on Wettability and Interfacial Tension. *Unpublished MSc. Thesis:* Norwegian University of Science and Technology

[12] Hogeweg A S, Hincapie R E, Foedisch H, and Ganzer L 2018 Evaluation of Aluminium Oxide and Titanium Dioxide Nanoparticles for EOR Applications. *SPE Europec featured at 80th EAGE Conference and Exhibition;* 2018/6/8/; Copenhagen, Denmark. SPE: Society of Petroleum Engineers; p. 20

[13] Ogolo N A, Olafuyi O A and Onyekonwu M O 2012 Enhanced oil recovery using nanoparticles. *SPE Saudi Arabia Section Technical Symposium and Exhibition;* 2012/4/8/; Al-Khobar, Saudi Arabia: Society of Petroleum Engineers

Quality assessment of sachet packed water in Omu-Aran, Kwara State

A J Gana[1], I A Adeniyi[2], S O Braimoh[1], D O Oguntayo[1] E M Ibitogbe[1] and P O Ejigboye[1]

[1]Department of Civil Engineering, College of Engineering, Landmark University, Omu-Aran, Nigeria

[2] Department of Industrial Chemistry, landmark University, Omu-Aran, Kwara State

*Correspondence Author: ibitogbe.enoch@lmu.edu.ng

Abstract. The importance of potable water supply in the socioeconomic life of the public cannot be over emphasized. Hence, it is necessary to know and determine the quality of water especially drinking water. This study therefore investigates the quality of sachet packed water in *Omu-Aran* metropolis with a view of ascertaining its suitability for consumption purposes. Water samples from the three major factories were taken to laboratory for analysis. The parameters of interest are the Physical, chemical and biological contents. The results obtained showed that physical and chemical properties of the water samples are within the WHO/NIS standard limit for potable drinking water while biological and microbial examination showed that coliform count and DO of all the sachet water samples from factory 'B' and 'C' did not conform to the WHO/NIS standard. This indicates that the sachet packed water produced from factory 'A' were of good quality and hygienic for consumption while the ones from factory 'B' and 'C' should be improved upon in terms of their coliform count and DO contents. It is recommended that quality assurance be implemented through all the stages of production to ensure the quality of sachet packed water is improved for safe consumption.

Keywords: Quality, Assessment, Sachet packed water, *Omu-Aran*, WHO/NIS.

1. Introduction

Water is generally known to be essential for life on earth. According to [14] it plays an essential role in and is a basic factor for determining the health of any community, since 80% of diseases in developing countries can be traced to poor water quality. Due to its importance, its availability has frequently determined the model of human settlement throughout history [15]. Of all basic needs required human survival, research has proved that water is perhaps the most precious asset on the earth. In addition to drinking water requirements, water resources are of major importance in different economic areas such as agriculture, livestock production, forestry, industrial activities, hydro-power generation, fisheries, etc. (Shweta et al, 2013). Water is available in different sources which may not be always safe and available throughout the year [1, 11, 13]. These sources can be analyzed for quality and suitability using physical, chemical and biological parameters. These parameters impair human health and become harmful if they surpass the guideline values defined by relevant organizations [5, 8, 9, 17, 18].

Water should be *potable* and be of good quality without having any adverse effect on the health of consumers. It should also be free from harmful impurities, such as bacteria, viruses, minerals and other organic substances. While contamination of sachet water leading to disease outbreak ever rarely occur, contamination would ultimately lead to a widespread epidemic due to its high demand. Several studies have demonstrated that bacterial contamination of packaged water can occur at various production phases [6, 7, 10, 16].

In the recent years, *Omu-Aran* has witnessed rapid growth and development with increase in demand for resources including water. Packaged water is also growing at a very high rate of production and consumption. However, there is no regulation on the quality of water manufactured for consumers and this could lead to contaminated supply. Therefore, this study is aimed at evaluating the quality of sachet packed water in *Omu-Aran*, Nigeria.

2. Study Area

Omu-Aran town is situated to the south of Ilorin which serves as the capital of Kwara State and 16km North-East of *Otun* in Ekiti State. It is located on the latitude 0.9N and longitude 50.61E. The location of the town confers advantages of centrality in the midst of other *Igbomina, Ekiti* and *Mopa* towns. The town has a tropical climate characterised by long wet season. The economy of the city is driven by many small and medium-sized industries.

3. Sampling Method and Analysis

Sachet samples of water were collected from three major water producing factories that are functioning every day and producing large quantities of sachet water for consumption for the people of the town and also for distribution to other places. The sachet packed water were left unopened in their original sealed containers and labeled as:

- Factory A
- Factory B
- Factory C

The sachet packed water samples were taken to Water Resources and Environmental Engineering laboratory of Landmark University, Omu-Aran for analysis to ascertain the physical, chemical and biological contents in the sachet packed water. The test was conducted using standard water and waste water examination methods [2].

4. Results and Discussions

4.1 Physical Parameters

The physical characteristics of the sachet packed water samples are shown in Table 1 and Figure 1 below. All the water samples were colourless, odourless, tasteless and clear, this could be as a result of the filtration methods employed during sachet packed water production by the factories [12]. All the physical composition of the water samples was within the WHO and NIS standard limits, though the temperature is higher than WHO/NIS limits, this could be because of the environment in which they are stored and tested.

Table 1: Mean and standard deviation for physical properties of sachet packed water samples

NO	PARAMETERS	UNIT	FACTORY A	FACTORY B	FACTORY C	WHO	SON
1	Temperature	0C	29.0 ± 0.05	29.1 ± 0.05	29.3 ± 0.05	Ambient	Ambient
2	Ec	µs/cm	63.2 ± 0.08	61.17 ± 0.55	72.6 ± 0.16	1000	1000
3	TDS	mg/L	31.6 ± 0.16	30.57 ± 0.26	36.3 ± 0.08	500	500
4	Turbidity	NTU	0	0	0	5	5
5	Colour	mg/L	0	0	0	15	15

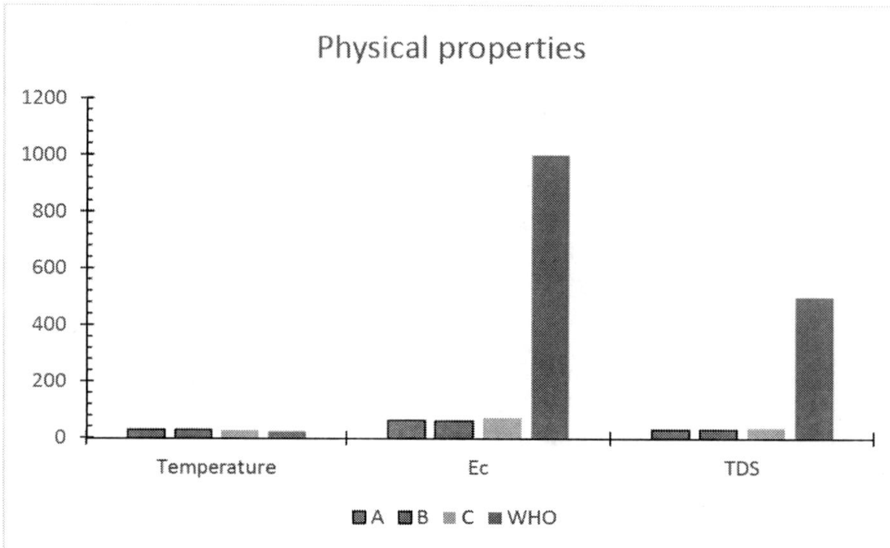

Figure 1: Physical properties of water samples compared with WHO standard

4.2 Chemical Parameters

The result of the chemical qualities of sachet packed water samples investigated in this study such as Iron, magnesium, manganese, copper, zinc, total alkaline, potassium, chloride etc. are all within WHO/NIS standard for portable water. The pH of the sachet packed water as seen from Figure 2 ranged from 7.18 to 7.38. The result of the pH analysis showed that sachet packed water were within the standard range (6.5-8.5) for quality water specified by WHO/NIS (Table 2), same goes for the potassium level for all the samples and the calcium level. It is worthy to note that water samples with pH values that fall within the regulatory guideline values are unlikely to pose health issues like as acidosis [3, 4]. The total hardness of sachet packed water in this study ranged from 62-78mg/L (figure 2). According to [4], hardness in water does not necessarily constitute any health risk.

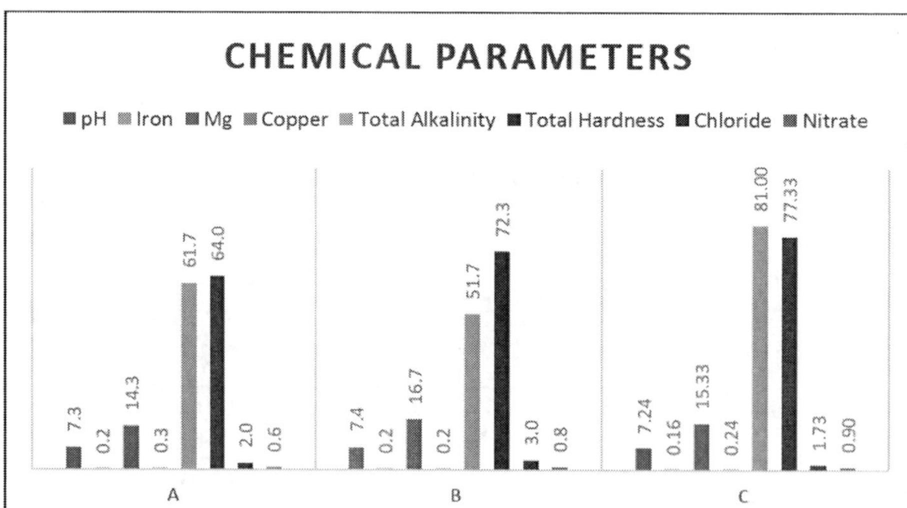

Figure 2: Chemical Parameters of sachet packed water

Table 2: Mean and standard deviation for chemical properties of the sachet packed water samples

NO	PARAMETERS	UNIT	FACTORY A	FACTORY B	FACTORY C	WHO	SON
1	pH	mg/L	7.3 ± 0.02	7.4 ± 0.02	7.2 ± 0.05	6.5-8.5	6.5-8.5
2	Iron	mg/L	0.2 ± 0.008	0.2 ± 0.008	0.2 ± 0.005	0.3	0.3
3	Magnesium	mg/L	14.3 ± 0.5	16.7 ± 0.5	15.4 ± 0.5	50	20
4	Manganese	mg/L	0	0	0	0.1	0.2
5	Copper	mg/L	0.3 ± 0.008	0.2 ± 0.008	0.24 ± 0.16	2	2
6	Zinc	mg/L	0.8 ± 0.16	2.0 ± 0.16	2.4 ± 0.16	3	3
7	Total hardness	mg/L	64.0 ± 1.64	72.3 ± 2.06	77.4 ± 0.94	150	150
8	Total Alkaline	mg/L	61.7 ± 2.36	51.7 ± 6.24	81.0 ± 2.95	20-200	20-200
9	Potassium	mg/L	2.9 ± 0.05	4.4 ± 0.13	2.4 ± 0.05	250	250
10	Calcium	mg/L	42.7 ± 2.50	49.3 ± 1.89	63.7 ± 1.25	75	75
11	Chloride	mg/L	2.0 ± 0.05	3.0 ± 0.08	1.73 ± 0.05	250	250
12	Sulphate	mg/L	7.3 ± 0.94	11.7 ± 1.25	14.4 ± 1.25	100	100
13	Nitrate	mg/L	0.6 ± 0.05	0.8 ± 0.05	0.9 ± 0.08	50	50
14	Nitrite	mg/L	0	0	0	0.2	0.2
15	Phosphate	mg/L	0.1 ± 0.008	0.2 ± 0.012	0.2 ± 0.008	0.1	0.1
16	Phosphate	mg/L	0.1 ± 0.008	0.2 ± 0.012	0.2 ± 0.008	0.1	0.1

4.3 Biological and Microbial Parameters

The results of COD obtained were all within the WHO/NIS standard. The result for DO and coliform count for samples from factory 'B' and 'C' were not within the limit specified by WHO for drinking of zero (0) coliform and 4.0-5.0mg/L of DO content, while the sample from factory 'A' were within the WHO/NIS standard. This means that the water samples from factory B and C contain coliforms, suggesting that there could be some health implications from the sachet packed water from these factories depending on the type of coliform found. As reported by [12], the presence of coliforms in these samples could be as a result of the poor sanitary nature of the various places of production. The water filters could also be clogged

Table 3: Mean and standard deviation for biological/microbial properties of sachet packed water samples

NO	PARAMETERS	UNIT	FACTORY A	FACTORY B	FACTORY C	WHO	SON
1	DO	mg/L	4.8 ± 0.05	5.0 ± 0.08	5.1 ± 0.08	4.0-5.0	4.0-5.0
2	COD	mg/L	0	0	0	0	0
3	Coliform	mg/L	0	TN	7.7±2.4	0	0

***TN**- Too numerous to count

5. Conclusion and Recommendation

The physical, chemical and biological/microbial properties of sachet packed water in Omu-aran metropolis were investigated to ascertain the quality of water being supplied and consumed within the town. The water samples from three major factories were analysed in the water resources and environmental laboratory of Landmark University. The analysis of the water revealed that the water samples is within by WHO/NIS for portable water in terms of their physical and chemical properties while the biological/microbial parameters revealed that only samples from factory A meet the criteria of portable water by WHO/NIS. It is concluded that the quality of water from factory 'B' and 'C' should be improved upon in terms of their coliform count and DO contents to make the water more suitable, productive and healthy.

- From the results obtained, the following recommendations were made:
- The production water sachet packed should be encouraged within the town based on daily demand and consumption.
- Skilled workers in terms of water quality should be employed water factory for a hygienic production of sachet water.
- Local government authorities in conjunction with rural water development authority should enforce regulations and standards for the production of sachet water in the areas. This will increase the quality expected from water sachet factories in the area.
- Further research should be carried out to determine the quality and suitability of water sources within the area.

Acknowledgement

The authors will like to appreciate Mr Ajayi Oluwasegun of Water Resources and Environmental Engineering Laboratory, Landmark University for helping with the laboratory work.

Reference

[1] Ahmed S N 2002 Water quality general module of national water supply training network National water rehabilitation project, 58-72
[2] APHA 2005 Standard Methods for the Examination of Water and Wastewater, 21st (edn.), American Public Health Association, American Water Works Association, Water Environment Federation, Washington DC, USA
[3] Asamoah DN and Amorin R 2011 Assessment of the quality of bottled/ sachet water in the Tarkwa-Nsuaem municipality (TM) of Ghana. *Res. J. Appl. Sci. Eng. Techno.* **3**:5, 377-385
[4] Airaodion AI, Ewa O, Awosanya OO, Ogbuagu EO, Ogbuagu U, et al 2019 Assessment of sachet and bottled water quality in Ibadan, Nigeria *Glob J Nutri Food Sci.* **1**:4, GJNFS.MS.ID.000518
[5] Bureau of Indian Standards 2012 Specification for drinking water. IS: 10500, New Delhi, India
[6] Dada A C 2011 Packaged water: optimizing local processes for sustainable water delivery in developing nations. *Globalization and health,* **7**:1, 1-8
[7] Gangil R, Tripathi R, Patyal A, Dutta P and Mathur K N 2013 Bacteriological evaluation of packaged bottled water sold at Jaipur city and its public health significance. *Veterinary World,* **6**:1, 27
[8] Guide Manual: Water and Waste Water, Central Pollution Control Board, New Delhi, http://www.cpcb.nic.in/upload/Latest/Latest_67_guidemanualw&wanalysis.pdf
[9] Guidelines for Drinking Water Quality 2012 Fourth Edition, World Health Organization ISBN 978 92 4 154815 1
[10] Halage AA, Ssemugabo C, Ssemwanga DK, Musoke D, Mugambe RK, Guwatudde D and Ssempebwa JC 2015 Bacteriological and physical quality of locally packaged drinking water in Kampala, Uganda. *Journal of environmental and public health*
[11] Hines D 2001 *The power of water American fitness* **19**, 23-25
[12] Ikon GM, Ohagim PI, Amadi CP , Abasiubong VN, Ekerenam EU and Udoudo EM 2017 Quality assessment of selected sachet water sold in Uyo metropolis, Nigeria *World Journal of Pharmaceautical Research,* **6**:8, 86-95. Doi: 10.20959/wjpr20178-8900
[13] Negoianu D and Goldfaid S 2008 Just add water *Journal of American society of* ***nephrology*** **19**:6, 1041-1043

[14] Pooja DS, Siddharath R and Sanjay S 2014 Assessment of Water Quality *International journal of scientific and engineering research*, **5**:12, 2229-5518
[15] Saleh MA, Ewane E, Jones J and Wilson BL 2001 Chemical Evaluation of Commercial boiled drinking water from Egypt *Food Comp and Anal* **14**, 27-152
[16] Semerjian LA 2011 Quality assessment of various bottled waters marketed in Lebanon, *Environmental monitoring and assessment*, **172:**1-4, 275-285
[17] Shweta T, Bhavtosh S, Prashant S and Rajendra D 2013 Water quality assessment in terms of water quality index *American Journal of Water Resources* **1**:3, 34-38. DOI:10.12691/ajwr-1-3-3
[18] United State EPA 816-F-09-004, May 2009, http://water.epa.gov/drink/contaminants/upload/mcl-2.pdf
[19] World Health Organisation 1989 World Health Organisation and United Nations Environmental Program (GEMS) Global fresh water quality Alden press Oxford

Comparison of Reliability Based Design (RBD) With Eurocode 2 for A Singly Reinforced Concrete Beam

S O A Olawale[1], O P Akintunde[1,2], M O Afolabi[1] and O A Agbede[1]
[1]Department of Civil and Environmental Engineering Department, Bells University of Technology, km 8, Idiroko, Benja Village, P.M.B. 1015, Ota, Nigeria
[2]Corresponding author: akintt18@gmail.com

Abstract. This research compares the Reliability-Based Design (RBD) method with the Eurocode 2 (EC2) for a singly reinforced concrete beam. Advanced First Order Second Moment (AFOSM) was used and the equation for the ultimate limit state design consideration for Moment of resistance and Applied moment was applied. The parameters which were all functions of the Ultimate load, G were replaced respectively with variables and their derivatives G(x) obtained with being the characteristic strengths of concrete and the yield stress of steel respectively. Given the original vector space and the transformed reduced vector space, the Taylor series was used to expand the reduced vector space and thus reliability index ßx expressed. The procedure was then coded using JAVA to produce the Reliability Index and Probability of failure for respective steel ratio. The required probability index of 3.9 corresponds to 0.048% of the steel ratio. Typical analysis of a singly reinforced beam according to EC2 for b= 300mm, h= 500mm, d= 447mm, cover of 40 mm, gave a steel ratio of 1.31 %. Comparing this to the reliability-based design of 0.048%, the code provision exceeded the RBD output by 2000 times and 63 times the minimum reinforcement requirement by the code.

Keyword: Reliability Based Design (RBD), Advanced First Order Second Moment (AFOSM), First Order Reliability Method (FORM), Eurocode 2 (EC2).

1. Introduction

Civil engineering structures such as bridges, buildings, power plants, dams, and offshore platforms are projected to contribute to the aids and value of life. Therefore, such amenities must be planned to be economically effective and realize the necessities with respect to safety [1]. In structural design, functional and structural requirements must be satisfied respectively. A structure must be able to perform its intended purposes (functional) and also must be able to withstand all the imposed load applied on it (structural) i.e. safely transmits all the load from the topmost down to the foundation without appreciable settlement.

Such structures must also satisfy other requirements like the ability to withstand vibration, cracking, thermal resistance, and exposure to attacks and must also maintain adequate appearance throughout its life span. The aforementioned are states which structures reach before they are declared unfit for use. The idea is to ensure that those limits are not reached at all. Hence, Limit states design concept focuses on the acceptable probability that a structure will not be declared unfit for use throughout its lifespan by satisfying certain design criteria based on probability theory. i.e. (Normal Probability Distribution)

Two limits states exist which are the ultimate limit state (ULS) and Serviceability limit state (SLS). ULS deals with shear and bending from loading while SLS deals with appearance, exposure condition, vibration, cracking, thermal resistance, deflection and durability. ULS uses factor of safety for loads and material strengths due to several design and construction uncertainties, errors, constraints, and variations etc. According to John (2004), design failures of structures could arise if the Ultimate Limit States, Conditional Limit States and Serviceability Limit State are not accurately measured in the process of structural design. The ultimate limit states relate to the maximum load-carrying capacity which can be associated with the development of an instrument in the structure, extreme plasticity, breaking owing to fatigue, and instability (buckling) [2]. The provisional limit states relate to the load-carrying capacity if a local part of the structure has collapsed, and this can be instigated by unintentional act or fire. While the serviceability limit states relate to the ordinary use of the structure such as unwarranted deflection, local damage, and extreme vibrations.

For countless years, loads used in the design of structural components have constantly been expected to be deterministic; this notion permits the engineer to govern the strength of a member by multiplying the load applied to the structure with a satisfactory brim often stated as a factor of safety. This factor of safety is commonly measured to be an actual degree of determining the consistency of the structure and differs depending on the code of practice being adopted. Since variations are existing in engineering design built on this deterministic method, integration of uncertainties in engineering design is usually essential. The reliability-based design process offers a hypothetical background for considering uncertainties in the engineering result structure [3]. Owing to these uncertainties, John (2004) observed that it is tough to measure the complete safety of a structure using deterministic investigation.

A recent study has revealed that poor performance of structures is as a result of the uncertainties in the loads' estimate, inadequate evaluation of the strengths of materials, exposure condition variability and poor modeling of the systems, poor workmanship, poor project supervision during construction and also natural effects such as seismicity, etc. These variabilities in recent times have led to the continual failure of structures globally. Reliability-based design has constituted the foundation of safety assessment in current engineering projects. There is a necessity to balance safety, serviceability, and economy in our immediate environment when designing structures, safety being of dominant importance.

There are allowances for some insecure and unwanted performances in a genuine state which is reserved as total safety because it necessitates a massive sum of properties and statistics which may not be accessible. The Probabilistic method to structural safety has been a rising debate and focus matter in recent years, and the procedures used in the investigation and design have been developed. Various researchers have investigated the concept of Reliability-Based Design (RBD) in engineering particularly in the field of civil engineering. According to John (2004), reliability is considered as the possibility that the structure under reflection has an appropriate performance during its entire lifespan. Reliability is defined reliability as the probability that a structure or system can accomplish an obligatory purpose under a definite service state through a specified period and equally, the failure probability is the probability that a structure does not perform suitably within an agreed period [4].
Reliability is the likelihood that a structure can be used without collapse for a specific period, under definite functioning situations [5].

IOP Conf. Series: Materials Science and Engineering 1036 (2021) 012070 doi:10.1088/1757-899X/1036/1/012070

Reliability is the probability that a structure can accomplish under definite service state throughout a period, they opined that reliability can be accessed by the possibility of the structure in the safe domain Ps and that the corresponding probability is the Failure Probability Pf. They Resolved that

$$P_s + P_f = 1 \qquad (1)$$

2. Literature Review

According to [6], reliability is the probability of an element executing its intended purpose over a specified period under functioning settings met. Structural Reliability is defined as a measure of the ability of the structure to function, or be capable of accomplishing a required purpose without collapse under a quantified condition for a definite period on unit of action [7].
Basic Random Variables, Limit State Functions and Reliability Index are the three elementary expressions governing Reliability study.
Input uncertain variables that oversee the efficacy of the structure are the Basic Random Variables and are used to establish mathematical prototype to produce the efficiency of the structure [8].
According to [9], Limit State Function was expressed as the state of a structure beyond which it no longer satisfies the pertinent standard of its performance. The Limit State equation is generally characterized by the expression:

$$G(x) = R - S \qquad (2)$$

Where R is the resistance of the structure or member, S is the load applied on the structure and G(x) is known as the Limit State Function or Performance Function. R and S are also known as the Basic Random Variables governing the failure Modes. According to Huang et al. (2016), Reliability Index is the value used to denote the safe working of the structure gotten from the solution of the Limit State equation. Cornell Index which has a demerit due to lack of invariance for equivalent performance functions and Hasofer-Lind Index which overpowers this demerit are the two categories of Reliability Index. The Probability of failure is also being assessed together with the reliability index and this is the probability of existence of structural failure.

Hence for a characteristic reliability problem, the performance function is stated as:

$$Z = G(X) = (X_1, X_2, X_3, \ldots, X_n) \qquad (3)$$

Where Z = G(X) > 0 indicates that the structure is in a safe condition and Z = G(X) < 0 implies the state of failure.

X1, X2, X3……, Xn are the basic arbitrary variables.

The Reliability Index can be gotten from the expression,

$$p_f = P\,(Z \le 0) = \int_{-\infty}^{0} f_z(z)dz = \int_{-\infty}^{0} \frac{1}{\sqrt{2\pi\mu_z}} \exp[\frac{-1}{2}(\frac{z-\mu_z}{\sigma_z})^2]dz \qquad (4)$$

$$= \varphi\left(-\frac{\mu_z}{\sigma_z}\right) = \varphi\,(-\beta_c)$$

IOP Conf. Series: Materials Science and Engineering 1036 (2021) 012070 doi:10.1088/1757-899X/1036/1/012070

Where $\varphi(.)$ is the CDE standard normal distribution, and $\beta_c = \frac{\mu_z}{\sigma_z}$ is the Reliability Index.

The reliability test was performed by [10] on reinforced concrete slabs subjected to failure modes in flexure, shear, and deflection with thickness ranging from 100 mm to 250 mm using First Order Reliability Method (FORM). FORM5 reliability software was used for Continuous slabs of equal span as a case study. For several failure modes, Limit State equations were developed and the results showed that the margins of safety suggested were similar to the probability of failure stated as 1 x 10-6 for singly reinforced concrete slabs according to BS 8110 and EC2 codes. The reliability levels were however not regular showing that the design assumptions were not safe and dependable as suggested. Safety criterium was violated by a reliability design index for 100 mm to 125 mm thick slabs. The probability of failure for bending was only 0.022 and it matched an indirect safety level of 2.015. The conclusion showed that review must be made on the code formulations to conform to accepted structural safety set out by the Joint Committee of Structural Safety (JCSS).

The reliability-based design of reinforced concrete two-way solid slab under Euro code2 was conducted by [11]. Formula Translation (FORTRAN) based program for the reliability design was produced considering suitable variables and designed the reliability index. It was observed that FORM design was more efficient given the steel area required and the final section depth with a reliability index of 3.0. The structural reliability of a system under fire was examined by [12] putting into consideration the reservations that occur in the system. The structural behavior which was exposed to the realistic fire was assessed by mathematical simulations using Latin Hypercube Simulations, Subset simulations, and the First and Second-Order Reliability Method. Categorizing and describing the input constraints, probabilistic analysis of the thermo-mechanical impacts on the structure, and assessing reliability with suitable limit state functions were the steps adopted in the reliability analysis. Their research pointed out the precision and efficacy as well as the shortcomings.

[13] Examined the 3-D time-dependent reliability analysis of reinforced concrete slab bridges under representative traffic loadings. The 3-D variables for the structural resistance considered were cover, compressive strength, and arbitrary variables including pit depths and model faults. Forecasting the outstanding lifespan of a rusting structure, and the spatial variability related to the material, dimensional, and environmental properties were the areas of focus of study. A two-dimensional random field was produced where the load and time constraints resistance were evaluated for each fraction of the field and the outcome gotten from the reliability analysis showed that the 3-D correlated resistance amounted to a minor increase in the probability of failure.

Reliability-based design of a two-way solid slab about design based on BS 8110 (1997) Part 1: Structural Use of Concrete was performed by [14]. FORM method was adopted for designing the slabs to an expected margin of safety level using a FORTRAN program formed and related to FORM5 reliability software. Both the reliability and deterministic based designs each made an acceptable performance with a typical reliability indexes of 3.20 and 1.48 for Flexure and serviceability respectively. 10% economic design was found using the reliability-based design while the deterministic design resulted in an unacceptable outcome that exposed the shortfalls in the deterministic design as a result of short-span (Continuous Edge).

The reliability of slender reinforced concrete columns having uniform cross-sections was studied by [15]. The main input constraints used examined were loads, compressive strength, and yield strength of steel, concrete cross-section dimensions, reinforcement ratio, and the position of steel settlement. The analysis was performed with a risk analysis program and the impact of eccentricity on the reliability index of the slender column was observed. The results extracted from the reliability analysis showed that the performance of slender columns is enhanced by good quality control and improves the reliability index of the slender column.

Reliability analysis on an I-section beam profile I 254 (10") x 37.7 with a design strength of MR250 was investigated by [16] using the Monte Carlo method exposed to an applied bending moment. The resolution was to evaluate the constituent stress handling characteristics to precise scheme stress according to the Brazilian setting. An applied bending moment of 83.0kNm was used, expected reliability index was gotten from records of Joint Committee of Structural Safety (JCSS) which were shown in a table in the research work and a value of 3.9 was adopted as the expected reliability index below modest failure modes and this was a constraint for a design working life suggestion of 50years. A probability of failure of 3.2 x 10-5 and a reliability index of 3.98 was generated after performing the analysis using the Monte Carlo technique. Since the attained reliability index was superior to the expected reliability index, the practicality of the beam was established and the constituent was measured to be huge and the structure can be exposed to superior stresses without conceding structural safety and harming the global structure. [17] studied the structural reliability of reinforced concrete beam and columns under concurrent static loads and steel reinforcement corrosion and it was observed that corrosion in concrete is the main problem adversely affecting the structural lifespan of the structure. The variables considered for the reliability analysis were thickness, dimensions, the strength of materials and reinforcement and the result was obtained based on numerous corrosion models produced after the analysis was performed on a 2D frame at each time to estimate the mass loss of steel reinforcement as a result of corrosion and the prompt load-bearing capacity of each element was evaluated. It was established from the outcome that the corrosion process impairs negatively on the strength of structural elements drastically.

Reliability analysis on reinforced concrete beams using finite element models under four points bending test was studied by [18]. The first model adopted the SOLID65 component for modeling concrete joining William-Warnke and Von Mises criteria, while the second component adopted the SOLID185 element and micro plane philosophy. The steel reinforcement was modeled by LINK180 fundamentals in both scenarios. The models engaged in the reliability analysis of RC beams subjected to various load levels were standardized using investigational outcomes. Based on supreme allowable displacements, a limit state function was assumed. The outcomes gotten from the reliability analysis disclosed that only the micro-plane model was suitable for RC structures since the SOLID65 encountered a lot of difficulties related to convergence.
[19] Investigated reliability analysis on the raft foundation with uniformly random soils causing the randomness in the input variables. The variables examined were the elastic modulus of the raft, modulus of subgrade reaction of the soil, and loading on the raft. Point Estimate Method (PEM), FOSM, and FORM were adopted for the analysis to get the probability of failure and reliability index-linked with the settlement of the raft foundation. It was observed that the reliability index for FORM was lower and dependable than PEM and FOSM. Parametric study indicating changes in modulus of subgrade reactions was also carried out and loadings affected the settlement reliability limit state of the raft the most, while the modulus of elasticity of the raft

was discovered to be indifferent to this limit state. Reliability evaluation of reinforced concrete beams with implanted PVC pipes below neutral axis was conducted by [14]. The research was carried out in three phases which were the investigational method, deterministic method, and the reliability analysis. The Reliability approach was carried out using a FORTRAN program link to FORM5 reliability software. Rectangular beams with a number of PVC pipes (one, two, and three) were injected and beams with one, two, and three PVC pipes in accordance with ASTMC 293 standard. The results showed that beams with PVC pipes required more reinforcement to meet the reliability i and one PVC pipe injected achieved acceptably. While the results gotten from the reliability assessment shown an area of reinforcement of 38.9 % for beams with no PVC and one PVC pipes injected with a reliability index ranging from 3.3 to 4.4 satisfying the criteria of the code.

The reliability evaluation of the Ejigbo Campus Library Building at Osun State University was investigated by [20]. Findings revealed that the university library was in a dreadful situation and was inhabitable. Physical inspection was done at the preliminary stage and it was discovered that the stress distribution in the structural elements was discovered to be 10 % below design strength. Their reliability analysis showed that the safe reliability was on the low side since about 2.2 % of steel ratio was required and only 1.5% was provided.

Table 1. Table showing the Tentative Targeted Reliability Indices and associated Failure Rates related to a one-year inference period and Ultimate Limit State

1	2	3	4
Relative Cost of Safety Measure	Minor Consequences of Failure	Moderate Consequences of Failure	Large consequences of Failure
Large (A)	$\beta = 3.1$ ($Pf \approx 10^{-3}$)	$\beta = 3.3$ ($Pf \approx 5x10^{-4}$)	$\beta = 3.7$ ($Pf \approx 10^{-4}$)
Normal (B)	$\beta = 3.7$ ($Pf \approx 10^{-4}$)	$\beta = 4.2$ ($Pf \approx 10^{-5}$)	$\beta = 4.4$ ($Pf \approx 5x10^{-5}$)
Small (C)	$\beta = 4.2$ ($Pf \approx 10^{-5}$)	$\beta = 4.4$ ($Pf \approx 5x10^{-6}$)	$\beta = 4.7$ ($Pf \approx 10^{-6}$)

Source: Joint Committee of Structural Safety (2000)

The guiding principle stated in the code should be followed in order to make the right selections in the Table 1. Shown above.

3. Methodology

The methodology used in this study is the Advanced First Order Second Moment (AFOSM) as detailed in [20]. The ultimate limit state design consideration is given by:

$$G = M_R - M_A \tag{5}$$

Where $M_R$the Ultimate Moment of resistance and M_A is the applied moment. A safe design stipulates $M_R \geq M_A$.

Hence using the Limit State Equation for bending as stated in EC2 for a singly reinforced rectangular concrete beam in eqn. (1):

$$G(f_{ck}, f_y, b, d, h, \gamma, M_A) = 0.4533 f_{ck} b \left\{ 2.5d - \frac{2.874 M_A}{(f_y \gamma b h\}} - M_A \right\} \tag{6}$$

$$Where\ \gamma = A_s/bd \tag{7}$$

Let $x_1 = b, x_2 = d, x_3 = h, x_4 = M_A$

Hence, eqn. (2) can be rewritten as

$$G(x_1, x_2, x_3, x_4) = 0.4533 f_{ck} x_1 \left\{ 2.5 x_2 - \frac{2.874 x_4}{\gamma f_{ck} x_1 x_3} \right\} - x_4 \tag{8}$$

At the boundary of the design surface where $G = 0$

$$x_4 = -\frac{2.258 f_{ck}^2 \gamma x_1^2 x_2 x_3}{x_1 x_3 \gamma f_{ck} - 1.303} \tag{9}$$

The partial derivatives of $G(x)$ with respect to x

$$\frac{\partial G}{\partial x_1} = 1.13 f_{ck} x_2 \tag{10}$$

$$\frac{\partial G}{\partial x_2} = 1.13 f_{ck} x_1 \tag{11}$$

$$\frac{\partial G}{\partial x_3} = \frac{1.3 f_{ck} x_4}{\gamma f_y x_3^2} \tag{12}$$

$$\frac{\partial G}{\partial x_4} = -\left(1 + \frac{1.3 f_{ck}}{\gamma x_1 x_3}\right) \tag{13}$$

If the original vector space is given as $\{x\} = \{x_1, x_2, x_3, x_4,\}$ then the transformed reduced vector space
$\{x'\} = \{x_1', x_2', x_3', x_4'\}$ *is* given by:

$$x_i' = \frac{x_i - \mu}{\sigma} \quad i = 1\ to\ 4 \tag{14}$$

The initial values for the original vector space are given by their densities and the last variable is calculated.

$$x_1 = \mu_{x1}$$

$$x_2 = \mu_{x2}$$

$$x_3 = \mu_{x3}$$

$$x_4 = -\frac{2.258 f_{ck}^2 \gamma x_1^2 x_2 x_3}{x_1 x_3 \gamma f_{ck} - 1.303} \quad (15)$$

Using Taylor series to expand the reduced vector space, then the reliability index β_x can be expressed as follows:

$$\beta_x = [A]^T[x_1'] / \sqrt{([A]^T[A]} \quad (16)$$

Where:

$$[A] = \begin{pmatrix} (\partial g/\partial x_1{}^i) = (\partial g/\partial x_1)\ \sigma_{x1} \\ (\partial g/\partial x_2{}^i) = (\partial g/\partial x_2)\ \sigma_{x2} \\ (\partial g/\partial x_3{}^i) = (\partial g/\partial x_3)\ \sigma_{x3} \\ (\partial g/\partial x_4{}^i) = (\partial g/\partial x_4)\ \sigma_{x4} \end{pmatrix}$$

The directional cosine which is the transformation vector $\{\alpha\}$ is given as

$$\{\alpha\} = [A] / \sqrt{([A]^T [A])} \quad (17)$$

Now the reduced vector space $\{x_i'\}$ is given as

$$\{x_i'\} = \beta_x [\alpha_1] \quad (18)$$

Then we can calculate the original vector space from the reduced vector space using the relationship

$$\{x_i\} = \mu_{xi} + x_i \sigma_{xi} \quad \text{where } i = 1 \text{ to } 4 \quad (19)$$

The error to converge the iteration is given by

$$Error = \left| \frac{(\beta_{i+1} - \beta_i)}{\beta_i} \right|$$

The above procedure is coded in Java and the results are discussed in the following section.

4. Results and Discussion

The results of the program developed for the reliability index and probability of failure are shown in figures 1 and 2 respectively. It is clear from figure 1 that the required probability index of 3.9 corresponds to 0.048% of the steel ratio. This is far from the minimum reinforcement of 0.13% of the code requirement.

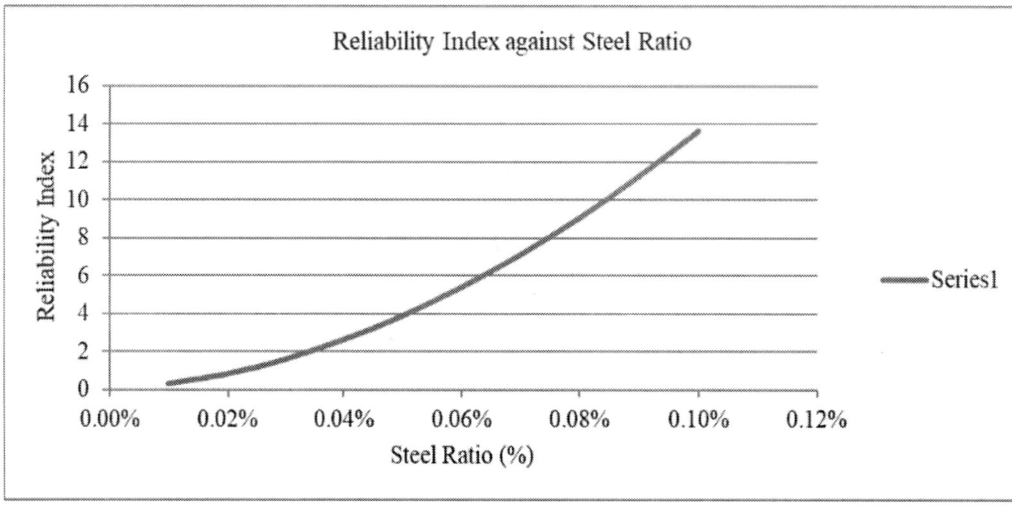

Figure 1. Variation of Reliability Index with Steel ratio

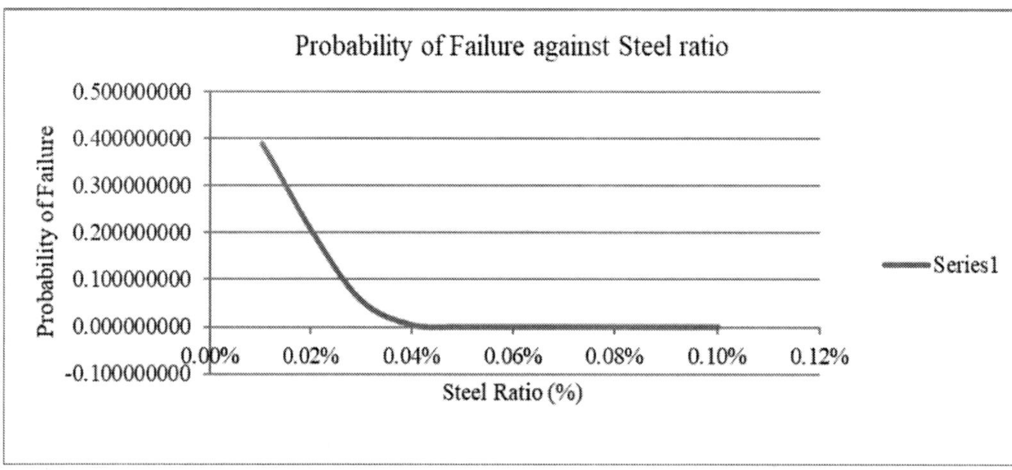

Figure 2: Corresponding Probability of failure with increasing steel ratio

For a typical analysis of a singly reinforced beam according to EC2, the steel ratio % can be calculated thus given the following parameters:

Cover to reinforcement = 40 mm

Effective depth, d, is

d = h − cover − φ/2 = 500 − 40 − 25/2 = 447 mm

$$K_0 = \frac{M}{f_{ck} b d^2} = \frac{M}{25 * 300 * 447^2} \quad (20)$$

$$z = d[\left(0.5 + 0.25 - \frac{k_0}{3.4}\right)^{\frac{1}{2}}]$$

(21)

$$= 447[0.5 + (0.25 - 3 * \frac{M}{3.4} * 25 * 300 * 447^2)^{1/2}] \quad (22)$$

$$A_s = \frac{M}{0.87 f_{yk} z} = 1960mm^2 \quad (23)$$

This is about 1.31 % of the steel ratio. Comparing this to the reliability-based design of 0.048 %, the code provision is more than 2000 times the RBD output. Even if we take the minimum to reinforce recommended by the code there is about 63 times the calculated output using EC2 provisions.

5. Conclusion

The Reliability Index increases as the Steel ratio also increases from figure 1 showing the variation of Reliability Index with the steel ratio. The required probability index of 3.9 corresponds to 0.048 % of the steel ratio from figure 1. This exceeds the minimum reinforcement of 0.13 % of the code requirement. The results obtained from the comparison between the Reliability-Based Design and the Local Design i.e. Euro Code 2 (EC2) clearly show that the Reliability-Based Design is by far more economical than the Local Design using EC2. Therefore, the method of Reliability-Based Design is highly cost-effective compared to the Orthodox design method used since time immemorial.

References

[1] Sørensen JD 2004 Structural Reliability theory and risk analysis. *Institute of Building Technology and Structural Engineering*, Aalborg University, Sohngaardsholmsvej, **57**, DK-9000 Aalborg, Denmark.
[2] Bastidas-Arteaga E and Soubra AH 2014 Reliability Analysis Methods. *ALERT Doctoral School -Stochastic Analysis and Inverse Modeling*, 53-77.
[3] Lowand HY and Hao H 2001 Reliability Analysis of Reinforced Concrete Slabs Under Explosive Loading", School of Civil and Structural Engineering, Nanyang Technological University, Singapore 639798, Singapore, 0167-4730/01/$ - see front matter Elsevier Science Ltd.
[4] Lemaire M. Structural reliability. *John Wiley & Sons.*
[5] Kassem Fidaa 2015 Reliability of reinforced concrete structure: case of slab subjected to impact, PHD Thesis, École doctorale: Mécanique, Energétique, Génie Civil, Acoustique.
[6] Goutham DR and Manjunath K 2016 Reliability Analysis of Grid Floor Slabs. *International Research Journal of Engineering and Technology (IRJET),* **3**(6). E-ISSN: 2395 -0056, www.irjet.net p-ISSN: 2395-0072.
[7] Wasiu J and Raji SA 2015 Reliability Based Design of Two-Way Slab in Comparison with Design Based on BS 8110 (1997), *Journal of Multidisciplinary Engineering Science and Technology (JMEST),* **2(6)**. ISSN: 31590040.
[8] Choi SK, Grandhi R and Canfield RA 2006 Reliability-based structural Design. *Springer Science and Business Media.*

[9] El Hami A and Radi B 2013 Incertitudes, optimization et fiabilité des structures. *Hermès science Publications.*
[10] Abejide OS 2014 Reliability Analysis of Bending, Shear and Deflection Criteria of Reinforced Concrete Slabs. *Nigerian Journal of Technology (NIJOTECH)*, **33(3)**, 394-400.
[11] Abubakar I and Ma'aruf A 2014. Reliability-Based Design of Reinforced Concrete Two Way Solid Slab using Eurocode2. *Nigerian Journal of Technology (NIJOTECH)*, **33**(4)
[12] Guo, Qianru. (2015). Structural Reliability Assessment under Fire. A dissertation submitted in partial fulfillment of the requirements for the degree of Doctor of Philosophy (Civil Engineering) in the University of Michigan.
[13] Donya H, Stewart MG, Enright B and Brien EO 2015 Spatial Time-Dependent Reliability Analysis of Reinforced Concrete Slab, *Bridges Subject to Realistic Traffic Loading, Structure and Infrastructure Engineering*, DOI: 10.1080/15732479.2015.1086385.
[14] John W and Adedeji AA 2018 Reliability Assessment of Reinforced Concrete Beam With Embedded PVC Pipes below the Neutral Axis, *International Journal of Multidisciplinary Sciences and Engineering*, **9**(1).
[15] Metwally A, Aziz A, Shaban A, Gabr AH and Saleh I M 2016 Reliability of Slender Reinforced Concrete Columns, *World Academy of Science, Engineering and Technology International Journal of Civil and Environmental Engineering* **10**(7).
[16] Klink BG and Silva LA 2016 Reliability Analysis of a Steel Beam Using the Monte Carlo Method Centro Universitário de Brasília UniCEUB SEPN 707/907 Asa Norte 70790-075 DF Proceedings of the XXXVII Iberian *Latin-American Congress on Computational Methods in Engineering,* Suzana Moreira Ávila (Editor), ABMEC, Brasília, DF, Brazil.
[17] Amir T and Behnam Z 2016 Structural Reliability of Reinforced Concrete Beams/Columns under Simultaneous Static Loads and Steel Reinforcement Corrosion, *Arab J Sci Eng* 41:3945–3958, DOI 10.1007/s13369-016-2033-6.
[18] Gomes JS 2017 Reliability analysis of reinforced concrete beams using finite element models, *Center for Optimization and Reliability in Engineering (CORE),* Department of Civil Engineering, Federal University of Santa Catarina. Conference Paper, DOI: 10.20906/CPS/CILAMCE2017-0145,
[19] Bennet K, Krishnan A, Dodagoudar GR and Rao BN 2017 Reliability Analysis of Raft Foundation on Uniformly Random Soils, Department of Civil Engineering, *Indian Institute of Technology Madras, Journal of Structural Engineering*, **43**(6), 568-579.
[20] Olawale SOA, Tijani MA, Folayan OI and Bolaji EO 2018 Reliability Assessment of Ejigbo Campus Library Building at Osun State University, Osogbo, Department of Civil Engineering, Osun State University, Osogbo, *Journal of Research Information in Civil Engineering*, **16**(1), 2511-2526.

A review of the mineralogical and geotechnical properties of some residual soils in relation to the problems of road failures: A case study of Nigeria

R O Sani[1], O M Olofinnade[1], I I Akinwumi[1] and O D Afolayan[1]
Civil Engineering Department, Covenant University, Ota, Ogun State, Nigeria.

Corresponding author: reuben.sani@covenantuniversity.edu.ng

Abstract. Poor geotechnical and mineralogical properties of soils used for road construction are a significant factor responsible for pavement failures. This is because most soils encountered at failed portions of some pavements do not meet the specifications according to relevant standards. It is with this in mind that this study carried out a systemic review on some residual soils in Nigeria, with a focus on their locations, geotechnical, and mineralogical properties in comparison to standards specified in the Nigerian specification for road and bridges. This study also reviewed how these properties have contributed to various road failures in different regions of the country. It was observed that the geotechnical and mineralogical properties of most residual soils in some regions of Nigeria did not meet up with the standards stipulated by the Nigerian Specification for road materials. Reviews carried out on the place of geotechnical and mineralogical characteristics on various road failures in Nigeria showed that a high volume of research works was carried out in the south-western and southeastern parts of the country compared to the Northern region.
Keywords: Geotechnical, mineralogical, specification, failures, road, review

1. Introduction

Poor geotechnical and mineralogical properties of residual soils used in the construction of roads globally may result in road pavement deterioration [1]. Most countries have stipulated standards for the geotechnical properties of soils when used for road construction. The Nigerian Specification for Road and Bridges [2] specified standard ranges for the geotechnical and mineralogical properties of soils before they can be used for road construction. Such as; for a soil to be used as a base course or sub-base material, percentage of soil particles passing sieve No. 200 should not be greater than 35 % [3]. But not all soils meet these stipulated standards because they have varied behaviours depending on their geotechnical and mineralogical properties [4]. Soils formed in situ by chemical and mechanical weathering of parent rocks are termed residual soils [5], and they have chemical similarity to their parent formation material [6]. In Nigeria, some soils used for road construction were observed to have originated from parent formation materials with poor geotechnical properties. This is the reason why, Kekere, Lawal, and Awotayo [7] in their research work stated that Nigerian roads are an infrastructural facility on the verge of total collapse. Despite enormous expenditures continuously being spent by the government on reconstruction and maintenance, travelling on Nigerian roads are said to be a nightmare due to the terrible conditions of the pavement, [8]. Figure 1 (a – c) gives illustrations to the extent most roads deteriorate in Nigeria. These failures range from partial to full pavement failure.

Figure 1. (a) Kaduna – Abuja road (North Central Nigeria), **(b)** Otukpa-Otukpo Markurdi, (North Central Nigeria), **(c)** Enugu – Awka - Onitsha road (Eastern Nigeria)
Source ***https://naijauto.com/safe-driving/7-most-deadly-roads-to-travel-in-nigeria-15***

Geotechnical, geological, geomorphological, road usage, construction inadequacies, design and maintenance are factors listed by Ebuzoeme [8] that may be responsible for road failure in Nigeria. Adegoke and Agada [9], Mesida [10], and Ajayi [11] from their research works stated that inadequate knowledge of the geotechnical and geomorphological characteristics of residual soils on which roads are constructed play a major role in road failures. Therefore, the objective of this paper was to collate information (location, geotechnical and mineralogical properties) on some residual soils used for road construction or on which roads were constructed upon in selected parts of Nigeria; The collected data were compared to standards specified in the Nigerian specification for road and bridges in order to investigate how these properties have contributed to various road failures.

2. Residual Soils In Nigeria

Laterite soils, lateritic soil, non-lateritic soil, Black cotton soils, Sand-clay soils, Shale, and Sand-clay soils are common types of residual soils found in Nigeria.

Laterites, lateritic and non-lateritic soils are also referred to as tropical red soils [12]. These soils are formed from the intense chemical weathering of crystalline rocks [13]. This crystalline rock is underlain to about 60 % of Nigeria's basement complex [14]. They occur in tropical climate regions of Nigeria [13]. SiO_2, Al_2O_3, and Fe_2O_3 constitute over 85 % of the oxide constituents of these soils [15]. For a soil to be referred to as laterite, the molecular ratio of silica (SiO_2) to sesquioxides (Fe_2O_3 and Al_2O_3) would be below 1.33, for lateritic soils, the molecular ratio is between 1.33 and 2.0. while for non-lateritic soils, their molecular ratio is greater than 2.0 [16]. Due to their Fe_2O_3 content, these soils exhibit reddish-brown colouration [15].

Black cotton soils are soils that occur widely in north-western and eastern Nigeria. They occupy an estimated area of 104,000 km^2 [17]. Black cotton soils are formed under conditions of poor drainage from basic rocks or limestone under alternating wet or dry climatic conditions [18]. They have a characteristic of swelling when wet and shrinking when dry due to the presence of a clay mineral known as montmorillonite [19]. In the northern part of Nigeria, apart from black cotton soil, felspathic sandstone, sandy clays, shale, calcareous sandstone, and shelly limestone are also present in significant quantities [20].

Sand-clay soils occur in Benin and Delta states in the South-South region of Nigeria. Delta sands are yellowish-brown ferritic soils and can be divided into reddish-yellow delta sands and silty brownish-grey delta sand [20].

In Eastern Nigeria, the largest group of residual soils occurs as laterites and coastal sands [20]. Shale is said to be the most abundant soil type in Abakiliki, the capital of present-day Ebonyi State, southeastern Nigeria [3]. Peat and organic clays are found in coastal lagoons and channels of Lagos, Niger delta, Niger, and Benue rivers [3]. In the South-Western parts of Nigeria, sandy and clayey soils are present in abundant quantities in parts of Ibadan, Ilorin, Ogbomosho, Ifo, Otta, Ikeja, Ijebu Ode, Ipetu and Abeokuta [20].

2.1 Some geotechnical and mineralogical properties of residual soils along failed sections of selected roads in Nigeria in comparison to Nigerian standards

Table 1. gives the specified and acceptable standards values stipulated by the Nigerian specifications for roads and bridges [2] for the various geotechnical properties of any soil to be used as an embankment, sub-base, and base course material respectively.

Table 1. Specified and acceptable standards values stipulated by the Nigerian specifications for roads and bridges for the construction of roads

Properties of the materials	Nigerian Specification
General filling and embankment	
Passing no. 200 (%)	≤35
MDD (Mg/m^3)	>0.047
moisture content, OMC (%)	<18
Liquid limit, LL (%)	<40
Plastic index (%)	<20
Soaked CBR at British standard (%)	>5
Sub-base course	
Liquid Limit, LL (%)	<30
Plastic index (%)	<12
CBR at west African standard (%) (%) and OMC (%)	≥30
Base course	
Liquid Limit LL (%)	≤30
Plastic index (%)	≤12
Unsoaked CBR at modified AAHTO and OMC (%)	≥80

Extracted from Maduka et al. [24]

Table 2 gives the geotechnical properties of soils along various failed portions from selected research works in Nigeria. These properties were compared to the stipulated standards in table 1, where the requirements for embankment materials were used to obtain the percentage suitable for road construction.

Table 2: Geotechnical properties of soil samples collected along failed sections of selected roads in Nigeria in comparison to Nigerian standards

S/No	No of samples collected	Geotechnical property	Range	% suitable for road construction (%)	Type of residual soil	Name of road in Nigeria	Reference
1	6	LL (%)	43 – 60.5	0	Inorganic clays	F209 highway at Ado-Ekiti, south-western Nigeria	Jegede [21]
		PI (%)	26.51 – 40.50	0			
		MDD (Mg/m^3)	1.85 – 1.985	100			
		OMC (%)	15 – 17	100			
2	25	LL (%)	20.1 – 55.1	56	Low plasticity clays	Ipele–Isua Highway	Daramola, Malomo and Asiwaju-Bello [22]
		PI (%)	5.2 – 21.96	84			
		MDD (Mg/m^3)	1.51 – 1.792	100			
		OMC (%)	12 – 23	84			

		CBR (soaked)	3 – 17	36			
3	12	% passing sieve 200	11.2 – 86.4	30	Laterite	Osogbo-Awo Road, Osun State	Oyelami and Alimi [23]
		LL (%)	32.37 – 75	16.7			
		PI (%)	9.63 – 23.53	40			
		MDD (Mg/m^3)	1.32 – 1.83	100			
		OMC (%)	13.01 – 27.98	40			
		CBR (soaked)	6.10 – 12.65	100			
4	10	% passing sieve 200	90 – 92	0	Shale	Roads in Abakaliki Ebonyi State	Aghamelu and Okogbare [3]
		LL (%)	55 – 64	0			
		PI (%)	36 – 45	0			
		MDD (Mg/m^3)	1.83 – 1.89	100			
		OMC (%)	12.0 – 12.5	100			
		CBR (soaked)	10 – 12	100			
5	27	% passing sieve 200	7 – 82	81.5	SM-SC (silty/clayey sand) and high plasticity clay	Nsukka-Adoru-Idah highway, Southeastern Nigeria	Maduka et al [24]
		LL (%)	29 – 60	0			
		PI (%)	5 – 39	81.5			
		MDD (Mg/m^3)	15.8 – 2.1	100			
		OMC (%)	13.8 – 18	96.3			
		CBR (soaked)	37 – 74	100			
6	9	% passing sieve 200	17.7 – 47	33.3	Gravely clay of low Plasticity to high plasticity clay	A proposed road linking Ipinsa Town in Akure	Adiat Akinlalu and Adegoroye [25]
		LL (%)	26.9 – 52.4	44.4			
		PI (%)	5 – 31.1	55.6			
7	24	% passing sieve 200	33 – 51	33.3	NA	Akungba-Ikare Road	Olubanjo et al [26]
		LL (%)	0 – 45.9	54.2			
		PI (%)	0 – 31.61	79.2			
		MDD (Mg/m^3)	1.497 – 1.982	100			
		OMC (%)	5.0 – 11.7	100			
		CBR (soaked)	NA	NA			
8	16	% passing sieve 200	NA	NA	NA	Lagos-Ibadan Expressway	Odunfa [27]
		LL (%)	6 – 55	37.5			

		PI (%)	13 – 26	43.8			
		MDD (Mg/m^3)	1.55 – 1.98	100			
		OMC (%)	8.3 – 20.3	75			
		CBR (soaked)	2 – 16	81.25			
9	20	% passing sieve 200	23.7 – 59.9	33.33	NA	A road linking eastern part of Ekiti State to the northern part of Ondo State	Alo and Oni [28]
		LL (%)	20.6 – 55.5	20			
		PI (%)	10 – 34	5			
		MDD (Mg/m^3)	1.56 – 2.1	100			
		OMC (%)	NA	NA			
		CBR (soaked)	0.5 - 7.52	15			
10	3	% passing sieve 200	NA	NA	Clayey gravel	Umuahia-Okigwe expressway Nigeria	Arinze and Obiorah [29]
		LL (%)	17.8 – 32.4	100			
		PI (%)	15.33 – 16.63	100			
		MDD (Mg/m^3)	2.06 – 2.13	100			
		OMC (%)	10.0 – 18.21	66.7			
		CBR (soaked)	3 – 4	0			
11	10	% passing sieve 200	32.1 – 62.0	10	Low plasticity clay to high plasticity clay	Ogbagi - Akoko road, Southwestern, Nigeria	Thompson, Zaiid and Raymond [30]
		LL (%)	31.1 – 53.6	20			
		PI (%)	10.1 – 42	20			
		MDD (Mg/m^3)	1.483 – 1.78	100			
		OMC (%)	13.1 – 26	50			
		CBR (soaked)	ND	ND			
12	28	% passing sieve 200	16.38 – 37	71.4	Lateritic soil	Kajola – Oda road South western Nigeria	Ademila [31]
		LL (%)	38.8 – 49	14.3			
		PI (%)	7.4 – 24.9	57.1			
		MDD (Mg/m^3)	1.89 – 2.89	100			
		OMC (%)	8.1 – 15.5	100			
		CBR (soaked)	3 – 23	85.7			
13	18	% passing sieve 200	13 – 58.1	16.6	Clayey soil to granular soils.	Isinbode–Ara road South	Olofinyo, Olabode and
		LL (%)	24 – 61	44.5			

		PI (%)	4.35 – 27.8	50		western Nigeria	Fatoyinbo [32]
		MDD (Mg/m^3)	1.408 – 2.066	100			
		OMC (%)	11.3 – 30.3	33.33			
		CBR (soaked)	3 – 44	88.9			
14	24	% passing sieve 200	0 – 56.8	80	Sandy to clayey gravels	Gombe – Biu road in Northern Nigeria	Mahmoud, Belel and Abba [33]
		LL (%)	22.5 – 49.6	79.2			
		PI (%)	3.8 – 21.6	90			
		MDD (Mg/m^3)	1.79 – 2.35	100			
		OMC (%)	6.3 – 14.4	100			
		CBR (soaked)	5.1 – 72.7	100			
15	11	% passing sieve 200	2 – 23	100	Silty sand to clay sand	Ago-Iwoye – Ilishan road, Southwestern Nigeria	Fatoba, Sanni and Ayantunji [34]
		LL (%)	13.9 – 46.2	72.7			
		PI (%)	10.6 – 15.9	100			
		MDD (Mg/m^3)	1.47 – 1.68	100			
		OMC (%)	18.5 – 26.7	0			
		CBR (soaked)	69.6 – 75	100			

***NA**- not available **ND** not defined*

The geotechnical property values obtained from the reviewed works were compared to standard values needed for soil to be used as an embankment material. It was observed that no soil sample was able to meet up with all the criteria need for it to be used as an embankment material. However, soil samples from Gombe – Biu road and Ago-Iwoye – Ilishan road had very high suitability percentages for all criteria. From observations drawn from table 1, a conclusion can be drawn that road failure observed along these roads are as a result of the poor geotechnical properties of soils used for construction.

2.2 *The place of soil mineralogical composition in road failure*

The type of clay minerals present in the soil at times affects the behaviour of the soil. Kaolinite is a type of clay mineral which, when present in the soil, enables it to shrink moderately when dry and expand modestly when wet [35]. Montmorillonite is another type of clay mineral with an expanding lattice [19], i.e., it expands considerably when wet and contracts when dry. This characteristic has always been a problem for engineers when constructing on soils that have this clay mineral present in them [36]. Montmorillonite is the predominant clay mineral found in Abakiliki shale [3], making it a problem soil for road construction. Ugbo-Odogu and Gariki Shale in Anambra state are predominantly composed of kaolinite clay mineral with a mean percentage composition of 44.69 and 25.85 5 respectively [37] therefore, they are presumed more stable. The dominant clay mineral found in soil samples obtained from failed sections in Someke, Makelu, and Ikereku areas along Sagamu – Papalanto road, Ogun State is Kaolinite (27.97, 18.75 and 31.53 % respectively). These samples also have a high quartz content of 51.18. 61.73 and 61.38 %, respectively. Other forms of clay minerals found in the soil samples are labradorite and haematite [12]. Most research works reviewed left out the mineralogical investigation to determine the type of clay mineral present in soil samples that they collected. This common observation leads to the conclusion that researchers are giving less importance to this factor when they investigate reasons for road failures.

3. Researchers views on the course road failures on selected roads in Nigeria

3.1 *Roads in Abakaliki Southeastern Nigeria*

Aghamelu and Okogbare [3] in their research work were of the opinion that roads in Abakiliki would keep failing because contractors and engineers are sometimes left with no other option than to construct using Abakaliki shale due to its abundance and the high cost of haulage of suitable material during construction. Results obtained from their research work (table 2) show that this residual soil type is not suitable for any form of pavement construction, hence reasons for road failures experienced in this region.

3.2 *Onitsha-Enugu Expressway, and Port Harcourt – Enugu Expressway, Southeastern Nigeria*

The Enugu-Onitsha expressway is very important in Nigeria because it connects the South East and South-South, through the Niger Bridge, linking Anambra, Enugu and Ebonyi States, as well as the North, via Benue and Kogi States. Onuoha, Onwuka, and Obienusi [38] in their research work stated that the inadequate maintenance of Onitsha-Enugu expressway, mismanagement of the road by the Nigerian government and old age of the road are major reasons for the continuous failure of this road. But in another work [39] carried out by them on the same road, they observed that geotechnical properties of soil samples collected from failed sections along the road did not meet up to standards, thus contributes to the failure of this road.

The Port Harcourt – Enugu Expressway is another very important expressway in Nigeria because it connects Port Harcourt through Aba to Okigwe to Enugu. Tse and Efobo [40] listed the possible reasons for the reoccurring failure of this road as; poor quality of soils used as sub-grade materials, presence of shales of low permeability, and the presence of montmorillonite in the soil samples obtained from failed sections.

3.3 Lagos – Ibadan Expressway Southwestern Nigeria

One of the busiest major highways in south-western Nigeria is the Lagos – Ibadan expressway. Although a large number of resources are being expended on the rehabilitation and reconstruction of this road, it usually fails shortly after, with characteristics of potholes, cracks, ruts, depressions, and it has become a source of an economic drain to the country [41, 42]. Layade, Adegoke, and Oyewole [43] in their research work said that the presence of fractured bedrock, clay, and sandy-clay soil with high permeability and porosity along the Lagos – Ibadan expressway are possible reasons for constant failures experienced along this road. The geotechnical investigation by Odunfa et al. [27], on the soil properties along the failed sections of the expressway (table 2) showed that, of the 16 soil samples collected along the failed sections of the road, only 81.25 % were suitable for construction of embankments based on the CBR criteria. 93.8 % of the soil samples were not ideal for the construction of either base course, sub-base course, or embankment based on the plasticity index criteria. They concluded from their findings that prior to the commencement of the road project, there was no adequate investigation on the properties of the subbase and base course material hence the frequent occurrence of road failure. Faults and clayey subgrade were also discovered along the expressway when a geophysical investigation was carried out [44], which are also possible reasons for the reoccurrence of failures along the road.

3.4 Ibadan–Ife Expressway Southwestern Nigeria and Selected Highways in Ifelodun Local Government, Ikirun-Osun

Ibadan – Ife expressway, a continuation of the Lagos – Ibadan expressway, aids in the movement of goods and services from the southwest to the northern parts of Nigeria. Typical to most highways in Nigeria, the road was observed to have failed just a few years after construction [45]. Table 3 gives the geotechnical properties of 8 soil samples collected along the road

Table 3: Geotechnical properties of soil samples collected along failed sections of Ibadan–Ife Expressway, South-Western Nigeria in comparison to Nigerian standards

S/No	No of samples collected	Geotechnical property	Range	% suitable for road construction (%)	Type of residual soil	Name of road in Nigeria	Reference
1	8	% passing sieve 200	25.42 – 82.94	25	Inorganic clays	Ibadan–Ife Expressway, South-Western Nigeria	Bello and Atilola [46]
		LL (%)	23.69 – 48.3	37.5			
		PI (%)	4.75 – 23.65	37.5			
		MDD (Mg/m^3)	1.81 – 2.76	100			
		OMC (%)	9.6 – 15.4	100			
		CBR (soaked)	9.85 – 21.61	0			

A close comparison with standard values stipulated for sub-base course material shows that only 37.5 % of the soil samples were suitable for base course construction based on the liquid limit and

IOP Conf. Series: Materials Science and Engineering 1036 (2021) 012071 doi:10.1088/1757-899X/1036/1/012071

plastic limit criteria. None of the soil samples had bearing capacity carry load as 0 % met up to the standard value of ≥30 CBR value for sub-base materials. These poor geotechnical properties of the soil samples are possible reasons why this road experiences road failures.

3.5 Osogbo-Iwo Expressway, Takie – Ikoyi Road, Akure – Ijare – Owo Expressway, Akungba – Ikare road and Igbara – Oke – Ibuji road in Southwestern Nigeria

Aderinola, Ola, and Owolabi [47] collected 15 samples from failed sections along Osogbo – Iwo expressway. Their results obtained showed that most of the soil samples did not meet up to standards for sub-grade and base course construction, hence a possible reason for the failure of the road. Adagunodo, Summonu, and Oladejo [48] carried out an electromagnetic investigation on Takie – Ikoyi road located at the heart of Ogbomoso. They observed the presence of clay minerals below the road surface, which they gave as a possible reason the road fails. Another geophysical investigation was carried out along Akure – Ijare road by Falowo, and Akintorinwa, [49], where they also observed the presence of clayey materials at the topsoil and subsoil at the failed sections. The presence of expansive clays (a typical feature of montmorillonite clay mineral) at failed sections along Akure – Owo expressway was also observed during a geophysical investigation of the road, by Adeyemo and Omosuyi [50]. These clay minerals are not good for road construction as they expand when wet and contract when dry hence resulting in the damage of the road pavements. Clayey subgrade soil within the upper layer of the road, faults, fractured zones, and fissures are reasons listed by Mohammed [51] as reasons for road failures observed along Akungba – Akoko highway. Table 2 shows the poor geotechnical properties of soil samples obtained along Igbara – Oke – Ibuji, which have already been stated as the possible reasons for failure along that road.

3.6 Opoji – Uwenlenbo – Illeh road and Shagamu – Benin Expressway

Construction over clays [52]: Opoji, Uwenlenbo, and Illeh road; buried features such as buried stream channel and shear zone beneath the road [53], low CBR [54, 55]: Shagamu – Benin expressway; areas of low permeability clay [56, 57]: Irrua to Opoji in Edo State, are reasons given for failures experienced along these roads.

3.7 Okitipupa – Igbokoda Highway, Ado - Ajebandele - Ikere Road and Ado – Ekiti – Ilewa – Ogbagi – Ekiti Road in Southwestern Nigeria

Low CBR values which resulted in the incursion of surface water into the base course [58], the presence of excess fines [59]: Okitipupa – Igbokoda highway; poor geotechnical properties such as the soils CBR values, liquid limit, Plasticity index, and percentage of soil passing through sieve number 200, [60, 61, 62]: Ado – Ajebandele – Ikere road; low permeability value resulting in the soils high ability to retain water [63, 64] - Ado – Ekiti - Ilewa – Ekiti road in Ekiti State, are reasons given for failures experienced along these roads.

4. Conclusion

From most research works reviewed, mineralogical investigation to determine the type of clay mineral present in soil samples collected was not carried out. This implies that less importance is put to the kind of clay mineral present when investigating reasons for road failures in Nigeria. Soil samples collected from failed sections along some investigated roads in Nigeria showed that most of the soils used for constructing the roads, or on which the roads were constructed, did not meet up with standard specified values by the Nigerian Specification for road materials.

Other views of some researchers on reasons why roads fail includes; the mismanagement of the road by the Nigerian government, construction over clays and buried features such as stream channels, shear zone beneath the road, low permeability value, and incursion of surface water into the base course

Reviews also showed that majority of research works where concentrated in the south-western and southeastern parts of the country. Fewer works are readily available on investigations into reasons for failed roads in the Northern parts of the country thereby raising research questions such as;

1. Do roads fail more in the southern, western, and eastern regions compare to the northern regions?
2. Are standards enforced more in the northern parts compared to other parts of Nigeria?

From these reviews, it was observed that the major reason why roads fail in Nigeria is due to the fact that soils used for construction do not meet up to standards. Therefore, to ensure that more durable roads are constructed, checks should be properly put in place and adhered to in other to ensure that these standards are followed strictly during road constructions in Nigeria.

Acknowledgement

The authors acknowledge Covenant University, Centre for Research Innovation and Discovery for providing a platform for this research work.

References

[1] Garg SK 2009 *New Delhi: Khana publishers* Soil Mechanics and Foundation. Engineering 7th ed pp 673–683

[2] Nigerian Federal Ministry of Works 1970. *General specifications for roads and bridges. Federal Government of Nigeria, Lagos*

[3] Aghamelu OP, and Okogbare CO 2011 *International Journal of Civil & Environmental Engineering.* Geotechnical Assessment of Road Failures in the Abakaliki Area, Southeastern Nigeria 11 pp 12 – 19

[4] Punmia BC, Jain AK and JAIN AK 2005 *Laxmi, New Delhi* Soil mechanics and foundations". 16th ed

[5] Duarte IMR and Rodrigues CMG 2017 *Encyclopedia of Earth Sciences Series. Springer, Cham.* Residual Soils. In: Bobrowsky P., Marker B. (eds) Encyclopedia of Engineering Geology.

[6] Idehai IM, Egai AO and Okiotor ME 2014 *Nigerian Journal of technology* Comparative Assessment of the Geotechnical Properties of residual Soils in parts of Awka and Abuja Areas of Nigeria. **33**(4) pp 595 -603

[7] Kekere AA, Lawal LO. and Awotayo, GP 2012 *Journal of Mechanical and Civil Engineering.* Relationship between geotechnical properties and Road failures along Ilorin – Ajase Ipo Road Kwara State, Nigeria. **4**(4) pp 1- 4

[8] Ebuzoeme OD 2015 *Civil and Environmental Research iiste* Evaluating the Significance of Poor Road Design as a Factor of Road Failure: A Study of Onitsha-Enugu Expressway, Southeastern Nigeria. **7**(2) pp 180 – 195

[9] Adegoke–Anthony WC and Agada OA 1980 *Technical Lecture. Lagos* Nigeria Geotechnical Characteristics of Some Residual Soils and their Implications on Road Design in Nigeria pp 1–16

[10] Mesida EA 1981 *West African Technical Review* Laterites on the Highways – Understanding Soil Behaviour pp.112 – 118

[11] Ajayi LA, 1987 *The Nigerian Engineer* Thought on Road Failures in Nigeria. **22** (1) pp 10–17

[12] Oyediran IA and Okosun J 2013 *Poskus izboljšave geotehničnih lastnosti nekaterih lateritnih tal za ceste z dodajanjem apna* An attempt to improve geotechnical properties of some highway lateritic soils with lime **60** pp 287 – 296

[13] Okeke OC, Ogbunyiba OO and Chukwujekwu D. 2013 *International Journal of Environmental Issues* Geotechnical Evaluation of Lateritic Soil Deposits in parts of Abuja (Nigeria), for Road Construction **10**(1&2) pp 148-160

[14] Bolarinwa A and Ola SA 2016 *Journal of Engineering and Technoogy FOUYE* A Review of the Major Problem Soils in Nigeria **1**(1) pp 20 – 25

[15] Adeyemi GO 2002 *Journal of Mining and Geology* Geotechnical properties of lateritic soils, developed over quartz schist in Ishara area, south-western Nigeria **38**(1) pp 65-69

[16] Singh P 2004 *SK Kataria and Sons, Delhi, India.* Engineering and General Geology 7th Edition 590p

[17] Osinubi KJ, Oyelakin MA and Eberemu AO 2011 *Journal of Geotechnical Engineering* Improvement of Black Cotton Soil with Ordinary Portland Cement - Locust Bean Waste Ash Blend. Electronic **16** pp 619-627

[18] Warren KW and Kirby TM 2004 *Geoinstitute of the American society of Civil Engineers, Jan* Expansive clay soil, A widespread and Costly Geohazard Geostrata pp 24-28

[19] Hussein EA 2001 *GJGE paper* Viscoplastic Finite Element Modelfor Expansive Soils 2001-0122

[20] Osim AR 2006 *Unpublished M.Sc. Thesis, Department of Civil Engineering, Ahmadu Bello University, Zaria* Distribution of Engineering Properties of Soils Used in Highway Construction in Nigeria

[21] Jegede G 2000 *Construction and Building Materials* Effect of soil properties on pavement failures along the F209 highway at Ado-Ekiti, south-western Nigeria. 14 pp 311 – 315

[22] Daramola SO, Malomo S and Asiwaju-Bello YA 2018 *International Journal of Geo-Engineering* Premature failure of a major highway in south-western Nigeria: the case of Ipele–Isua highway 9(28) pp 1 -12

[23] Oyelami AC and Alimi, SA 2015 *Ife Journal of Science* Geotechnical Investigation of some failed sections along Osogbo-Awo Road, Osun-State, Southwestern Nigeria **17**(1)

[24] Maduka R, Igwe O, Ayogu NO, Ayogu CN and Nwachukwu M 2016 *Environ Monit Assess* Geotechnical assessment of road failure and slope monitoring along Nsukka-Adoru-Idah highway, Southeastern Nigeria. **189**(30) pp 1 -18

[25] Adiat KAN, Akinlalu AA and Adegoroye AA 2017 *NRIAG Journal of Astronomy and Geophysics* Evaluation of road failure vulnerability section through integrated geophysical and geotechnical studies **6** pp 244–255

[26] Olubanjo AM, Ogunribido THT, Ojo OB, Aduloju MD and Ololade OP 2018 *Asian Journal of Geological Research* Geotechnical Properties of Subgrade Soil along Failed Portions of Akungba-Ikare Road Stabilised with Rock Fines **1**(2) pp 1-9

[27] Odunfa OS, Owolabi OA , Aiyedun PO and Sadiq OM 2018 *FUOYE Journal of Engineering and Technology,* Geotechnical Assessment of Pavement Failure along Lagos-Ibadan Expressway **3**(2) pp 113 – 117

[28] Alo BA and Oni OA 2018 *American Journal of Engineering Research (AJER)* A Geotechnical Investigation into the Failure of a Typical Road in Southwest Nigeria. **7**(10) pp 51-56

[29] Arinze EE and Obiorah F 2014 *Electronic journal of geotechnical engineering EJGE* Geotechnical assessment of pavement failure along Umuahia-Okigwe expressway in Nigeria **19** pp 4130 – 4132

[30] Thompson OTH, Zaid MM and Raymond AA 2015 *Journal of Environment and Earth Science.* Effect of Engineering Properties of Soil on Pavement Failures in Ogbagi - Akoko Area, Southwestern, Nigeria **5**(4) pp 39 - 42

[31] Ademila O 2017 *Geosciences Research,* Engineering Evaluation of Lateritic Soils of Failed Highway Sections in Southwestern Nigeria 2(30)

[32] Olofinyo OO, Olabode OF, and Fatoyinbo IO 2019 *SN Applied Sciences* Engineering properties of residual soils in part of Southwestern Nigeria: implication for road foundation **1**(507)

[33] Mahmoud H, Belel ZA and Abba HA 2012 *Nigeria. Journal of Engineering and applied science* Pavement failure induced by poor soil properties along Gombe – Bui highway 4 pp 22 – 26

[34] Fatoba JO, Sanni AO and Ayantunji AB 2015 *International Letters of Natural Sciences.* Influence of Subgrade Soil on Pavement Perfomance: A Case Study of Ago-Iwoye – Ilishan Road, Southwestern Nigeria **38** pp 86-92

[35] Okeke OC, Duruojinnaka IB, Echetama, HN, Paschal CC, Ezekiel CJ, Okoroafor EJ and Akpunonu EO 2016 *International Journal of Advanced Academic Research/Sciences, Technology & Engineering.* Geotechnical and Geochemical Characterization of Lateritic Soil Deposits in Parts of Owerri, Southeastern Nigeria, For Road Construction **2**(7) pp 5-19

[36] Design Procedures for Soil Modification or Stabiliztion 2013 *Production Division Office of Geotechnical Engineering 120 South Shortridge Road Indianapolis, Indiana 46219. 2008* pp 3

[37] Adewuyi OI and Festus OF 2015 *International Journal of Pure and Applied Sciences and Technology.* Suitability of Ugbo-Odogu and Gariki Shale, Southeastern Nigeria as Construction Materials **26**(1) pp 1-13

[38] Onuoha DC, Onwuka SU and Obienusi E A 2014 *Civil and Environmental Research* Evaluating the Causes of the Road Failure of Onitsha-Enugu Expressway, Southeastern Nigeria **6**(8) pp 118 – 129

[39] Onuoha DC and Onwuka SU 2014 *Civil and Environmental Research* The Place of Soil Geotechnical Characteristics in Road Failure, a Study of the Onitsha-Enugu Expressway, Southeastern Nigeria **6**(1) pp 55 – 67

[40] Tse AC and Efobo O 2016 *Journal of Earth Science and Engineering* Geotechnical and Mineralogical Evaluation of Soils Underlying a Failed Highway Section in South Eastern Nigeria **6** pp 98 – 109

[41] Abiola OS, Owolabi AO, Odunfa SO and Olusola A 2010 *Proceedings of the Nigeria Institution of Civil Engineers (NICE) conference* Investigation into Causes of Premature Failure of Highway Pavements in Nigeria and Remedies

[42] Owolabi AO and Abiola OS 2011 *8th International Conference on Managing Pavement Assets (ICMPA093) Chile* Development of Priority Index Assessments Model for Road Pavements in Nigeria pp 1-11

[43] Layade GO, Adegoke JA, and Oyewole IT 2017 *Journal of Applied Science Environment and management* Integrated Geophysical Investigation of the Causes of Road Pavement Failure along Ibadan-Lagos Dual-Carriage, Southwestern, Nigeria **21**(3) pp 547 – 554

[44] Momoh LO, Akintorinwa O and Olorunfemi MO 2015 *Journal of Applied Sciences Research* Geophysical Investigation of Highway Failure - A Case Study from the Basement Complex Terrain of Southwestern Nigeria **4**(6) pp 637 – 648

[45] Adeyemi GO, Oloruntola MO, and Adeleye AO 2014 *Journal of Natural Sciences Research* Geotechnical Properties of Subgrade Soils along Sections of the Ibadan–Ife Expressway, South-Western Nigeria **4** (23) pp 67 – 75

[46] Bello AA and Atilola AI 2015 *Journal of Engineering and Technology* Flexible Pavement Assessment of Selected Highways in Ifelodun Local Government, Ikirun-Osun, South – Western, Nigeria, International **5** (8) pp 475 – 484

[47] Aderinola OS, Ola SA and Owolabi AO 2015 *Journal of Natural Sciences Research* An Investigation into Road Pavement Failure Susceptibility Indices of Osogbo-Iwo Road, 5(3) pp 144 – 155

[48] Adagunodo TA, Summonu LA and Oladejo OP 2014 *International Journal of Business and Apllied Science.* Electromagnetic Investigation into the Cause(s) of Road Failure along Takie – Ikoyi Road, Ogbomoso **1**(1) pp 79 – 85

[49] Falowo OO and Akintorinwa OJ 2015 *Journal of Applied Geology and Geophysics.* Geophysical Investigations of a Pavement Failure Along Akure-Ijare Road, Southwestern Nigeria **3**(6) pp 45 – 54

[50] Adeyemo IA and Omosuyi GO 2012 *American Journal of Scientific And Industrial Research* Geophysical investigation of road pavement instability along part of Akure-Owo expressway, Southwestern Nigeria **3**(4) pp 191 – 197

[51] Mohammed MZ 2018 *Journal of Emerging Trends in Economics and Management Sciences,* Geophysical Evaluation of Highway Foundation Failure in Part of Akungba-Akoko, Southwestern Nigeria 9(1) pp 20 – 26

[52] Aigbedion I 2007 *Journal of Scientific Research* Geophysical Investigation of Road Failure Using Electromagnetic Profiles along Opoji, Uwenlenbo and Illeh in Ekpoma –Nigeria, Middle-East 2(3-4) pp 111 – 115

[53] Fatoba JO, Olorunfemi MO and Eluwole AB 2015 *Journal of Environment and Earth Science* Subsurface Structures and the Effect on Flexible Pavement Performance: The Shagamu-Benin Expressway, Southwestern Nigeria, as Case Study **5**(13) pp 126 – 132

[54] Akinwumi II 2014 *Indian Journal of Traditional Knowledge* Earth building construction processes in Benin City, Nigeria and engineering classification of earth materials used. **13**(4) pp 686–690

[55] Abam TKS, Osadebe CC and Omange GN 2005 *Global Journal of Geological Sciences* Influence of Geology on Pavement Performance: A Case Study of Shagamu – Benin City Road, Southwestern Nigeria **3**(1) pp 17 – 24.

[56] Victor E and Mamah L 2014. Geophysical investigation of road failure the case of Opoji in Nigeria. International Journal of Scientific & Engineering Research, **5**(1), pp 1769 – 1779.

[57] Jegede G, 2000. Effect of Soil properties on pavement failure along the F209 highway at Ado – Ekiti, Southwestern Nigeria. Journal of Construction and Building materials, **14** pp 311 – 315.

[58] Olaoye RA, Afolayan OD, Oladeji VO and Sani RO 2019 *IOP Conference Series: Material Science and Engineering 640 155424* Influence of bentonite on clayey soil as Landfill baseliner materials

[59] Adams JO, Aderinola OS and Akinwamide JT 2015 *International Journal of Engineering Innovation & Research* Geotechnical Study of Pavement Indices Influencing Failure Along Ado - Ajebandele - Ikere Road, South Western, Nigeria **4**(4) pp 567 – 572

[60] Akinlolu SO and Adetoro AE 2018 *International Journal of Trend in Scientific Research and development* Analyses of some Engineering Properties of Ado – Ekiti to Ilawe – Ekiti Road Soil, Southwestern Nigeria **2**(5) pp 725 – 72

[61] *https://naijauto.com/safe-driving/7-most-deadly-roads-to-travel-in-nigeria-15 Accessed 4/16/2020*

[62] Akinwumi II, Ajayi OO, Joshua O, Sani RO, Olofinnade OM, Awoyera PO, Ogundairo TO, Ogunwole DA and Afolayan OD 2019 *IOP Conference Series: Material Science and Engineering 640 012072* Housing crisis: Waste glass-stabilized clay for use as fired clay bricks

[63] Akinlolu SO and Adetoro AE 2018 *International Journal of Trend in Scientific Research and development* Analyses of some Engineering Properties of Ado – Ekiti to Ilawe – Ekiti Road Soil, Southwestern Nigeria **2**(5) pp 725 – 729

[64] Obaje SO 2017 *International Journal of Geology And Earth Sciences* Appraisal of Pavement Failures on Ado-Ekiti – Ogbagi Road, South-Western Nigeria **3**(2) pp 1 – 9.

Potentials of Cement Kiln Dust and Rice Husk Ash Blend on Strength of Tropical Soil for Sustainable Road Construction Material

I C Attah, R K Etim and I N Usanga

Department of Civil Engineering, Akwa Ibom State University, Ikot Akpaden, Nigeria

*Corresponding Author: rolandkufre24@gmail.com

Abstract. The growing problem of containing wastes in developing countries has led to endless research on the chances of converting these wastes to materials for sustainable engineering construction. This study through experimental assessment explores the potentials of cement kiln dust (CKD) and rice husk ash (RHA), in circumventing the inadequacies often encountered in deficient tropical soil thereby ensuring profitable engineering construction. Specimens were prepared by blending dosages of 0, 2, 4, 6 and 8 % CKD as well as 0, 3, 6, 9, 12 and 15 % RHA by requisite weight of soil. The investigated outcomes revealed increasing cation exchange capacity (CEC) with corresponding increase in pH for the various increased CKD-RHA composition. The Atterberg limits indicated a reduction trend with increasing CKD-RHA blend, signifying a drastic improvement in the plasticity tendency of the soil. While the maximum dry density (MDD) decreased with up shoot in optimum moisture content (OMC) at varying combined dosages of CKD and RHA, the CBR (soaked condition) increased steadily from 5 % (natural) to peak of 35 % at 8 % CKD / 15 % RHA blend. The highest CBR value in this study was not below the criterion of 20 – 30 % for sub-base as well as 10 – 15 % for subgrade road material. The regression analysis and analysis of variance from laboratory results validated the experimental outcome of using CKD-RHA blend to expressively enhance deficient lateritic soil. The outcome of this study has shown that CKD-RHA is recommended in improvement of strength of weak soil for either sub-base or more fittingly, subgrade of flexible pavement. Hence, using these waste materials will solve the twofold objective of proper waste management and good engineering materials for sustainability in construction of subgrade of flexible pavement.
Keywords: Lateritic soil, CEC, Atterberg's limit, Compaction properties, admixtures, Sub-grade.

1. Introduction

In Nigeria, during the construction of civil engineering infrastructures such as roads, structures, embankments etc. lateritic soil is frequently exploited as conventional fill material. This is because it is the locally predominant construction material found in abundance across the length and breadth of the tropical domain. However, when this soil material with poor engineering characteristics (such as high clay content) is exposed to interaction with moisture, it reacts in a detrimental pattern overtime. The detrimental behaviour of this soil in most cases makes the material fall below relevant standards based

on classified usage and as such, it calls for treatment to enhance its workability and geoengineering behaviour.
Treatment of this deficient soil material so as to meet required strength is achievable by incorporating some additives such as lime, cement, calcium chloride, fly ash, rice husk ash, cement kiln dust, oyster shell ash, periwinkle shell ash, metakaolin etc. which have been explored by some investigators [1-8]. However, there is increasing concern in the application of these waste materials (industrial, agricultural) which has attracted a lot of attention by researchers due to pozzolanic behaviour and reducing cost of infrastructure construction. Interestingly, due to the high demand for sustainable construction materials, the quest for new materials have been grown. Rice husk ash is a typical example of agricultural waste resulting from the process of rice milling, its major constituent is silica and it is greatly affected by the burning method [9]. Interestingly, there is a large volume of published studies describing the trend of outcomes using either rice husk ash (RHA) or cement kiln dust (CKD) as single stabilizers in improving deficient soils. However, the study of [10] concluded that utilization of RHA in clay soil enhances slight increase in strength of soil. Similarly, the addition of RHA to clayey soil improves the index property and as well makes the soil material fit for usage as subgrade material [11]. In contrast to other investigators, [12] documented that RHA is not fit for usage as a single stabilization agent in soil improvement. Hence, improvement of weak soil using RHA admixed with other additive such as cement, lime and calcium chloride was reported by [13-14]. Cement kiln dust is an industrial waste substance discharged during the manufacturing process of Portland cement. The engineering property of soil materials has been rejuvenated to meet up with construction standards via the utilization of stabilization agents such as cement kiln dust (CKD) [15]. In summary, the application of this waste has led to technical, economic and environmental benefits [16] as well as reduction in thickness of pavement and increase in strength indices [17].
Recently, researchers have shown an increased interest in the need of using two additives during soil stabilization with the aim of reducing high consumption of cement, construction cost and other environmental issues. Therefore, there is a great need to put in use cement kiln dust-rice husk ash blend for the enhancement of lateritic soil with poor geotechnical properties in a cost-effective manner for satisfactory usage as road material. Hence, this study aims at assessing the effect of cement kiln dust-rice husk ash blend on Atterberg limit, compaction characteristics and strength indices of the treated soil.

2. Materials and Methods

2.1. Materials
A reddish brown soil material was gotten from Ikot Ekong (Latitude 4° 38′ 23″N and Longitude 7° 49′ 19″E), Mkpat Enin Local Government Area, Akwa Ibom State, Nigeria by disturbed sampling technique at an average 1.5 m depth. The sample was transported to the laboratory in airtight bags, air-dried for some days and pulverised using pestle inside the tray. CKD was source from Lafarge Cement Company, Mfamosing in Calabar, Cross River State, Nigeria.The rice husks used for the production of rice husk ash was source from a local rice-milling factory and it was burnt in an open space until it turned into an ash. The ash was transported to the laboratory and sieved with 75 μm aperture size before usage. The oxide composition of cement kiln dust and rice husk ash was known with the aid of X-ray fluorescence analysis.

2.2. Methods
The soil gradation was determined based on [18]. The various tests specimens were prepared by blending 0, 2, 4, 6 and 8 % CKD and 0, 3, 6, 9, 12 and 15 % RHA by weight of soil. The experimental programme of Atterberg limits (consistency test), compaction and CBR (strength test) was performed based on the guidelines defined in [19-20] for natural and CKD-RHA stabilized soils, respectively. Cation exchange capacity (CEC) and pH of natural and stabilised soil were determined in line with method of [21].

IOP Conf. Series: Materials Science and Engineering 1036 (2021) 012072 doi:10.1088/1757-899X/1036/1/012072

3. Results and Discussion

3.1. General characterization and classification of test material

A series of laboratory investigations on the reddish brown soil in its natural state is shown in Table 1. The particle gradation is shown in Figure 1. Similarly, the oxide composition of soil, CKD and RHA materials used is displayed in Table 2. The properties of the soil revealed that it is an A-7-6(10) soil based on AASHTO [22] and CL according to USCS [23].

Table 1. Properties of the natural soil sample used in the study

Property	Quantity
NMC, %	19.80
Passed BS No 200 sieve	52.40
LL, %	43.80
PL, %	24.83
PI, %	18.97
Gs	2.60
AASHTO	A-7-6(10)
USCS	CL
MDD, Mg/m^3	1.83
OMC, %	12.95
CBR (48 h soaking), %	5
Colour	Reddish Brown

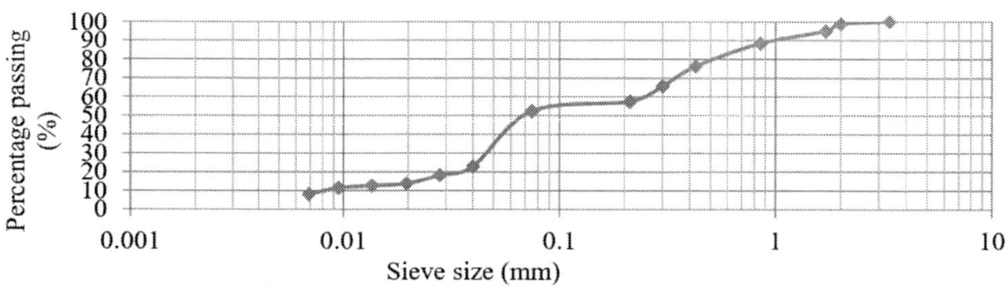

Figure 1. Plot of particle gradation for the unmodified soil

Table 2. Composition of materials

Oxide	Mass fraction (%)		
	*Soil	**RHA	CKD
CaO	0.30	2.95	45.75
SiO_2	36.35	75.30	7.95
Fe_2O_3	27.41	1.98	5.10
Al_2O_3	29.19	1.40	1.98
MnO	0.028	2.10	0.11
Na_2O	-	0.98	-
TiO_2	0.431	-	0.54
SO_3	0.69	-	-
ZnO	0.004	-	-
K_2O	-	5.55	-
LOI	4.8	5.61	37.25

*[24], ** [25]

3.2 Cation exchange capacity and pH

The cation exchange increased with increasing RHA and CKD content (see Figure 2a). This behaviour is associated with the increase in alkaline nature of the mixture due to higher RHA and CKD content, which is been seen by a corresponding increase in pH (Figure 2b). One of the mechanisms that govern soil stabilisation is the cation exchange. The quantity of negative charge held by a clay is a function of the total sum of exchangeable cations. In a mixture of soil and stabilizer blend, cation exchange is the primary reaction required first, before the flocculation and agglomeration process. Therefore, improvement of fine grain soil that has some amount of clay content is associated with reaction induced by cation exchange [26-28]. The results in this study has shown that increased in CKD and RHA will results to corresponding increase in both pH and cation exchange capacity and the combine effect of CKD and RHA will be more potent in improving the geoengineering properties of the soil.

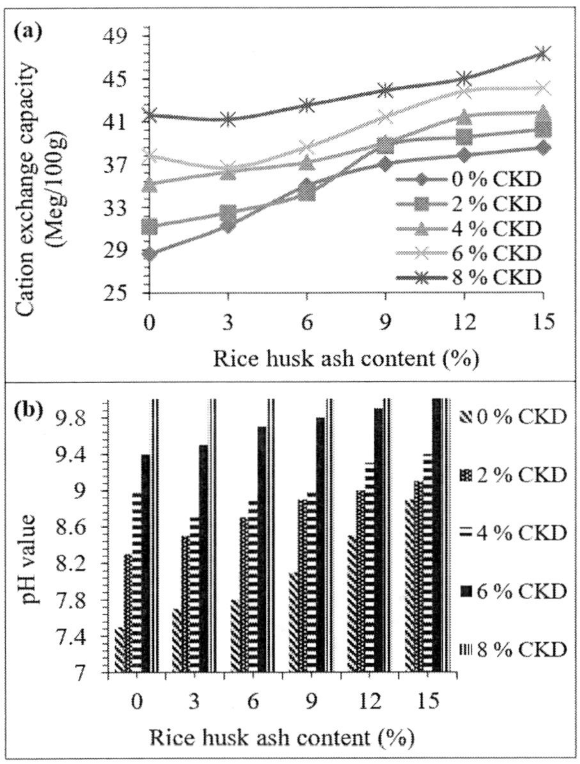

Figure 2. Effect of CKD-RHA on the; (a) cation exchange capacity and (b) pH of lateritic soil

3.3 Atterberg's limits characteristics

The behaviour of liquid limit of soil sample with various concentrations of CKD and RHA is depicted in Figure 3. The findings portrayed that liquid limit of soil mixtures diminishes with increase in both cement kiln dust and rice husk ash contents, respectively. The values diminished from 43.80 for natural soil to a minimum value of 35.30 % at 8 % CKD / 15 % RHA. The observed trend of results could be ascribed to the changes in soil structure as a result of incorporating pozzolanic materials to the soil matrix. It could be that the combined effect of CKD-RHA modifies the soil fabric and increased the pH of the mixture with corresponding increase in the mechanism of exchangeable cation. This reaction favours the introduction of divalent cations of Ca^{2+} that consequently replaces / displaces monovalent cations. The resultant outcome diminished the expanded diffuse double layer and shearing resistance of the soil that is reflected in the reduction of moisture content (liquid limit). The outcomes in the current study were matching with the previous endings of [29-30] and [7].

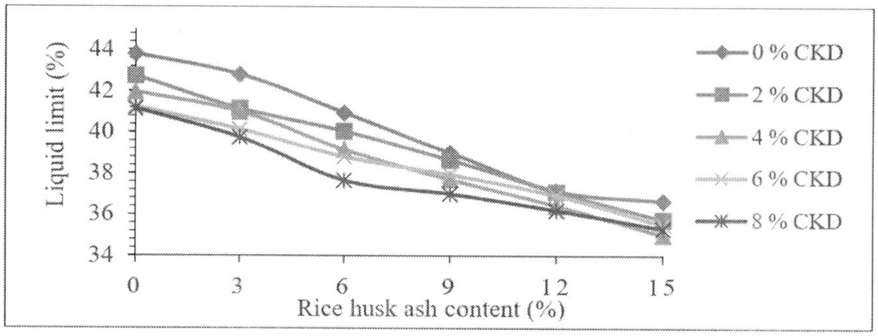

Figure 3. Behaviour of liquid limit of lateritic soil-CKD-RHA mixtures

Figure 4 depicts the behaviour of plastic limit of soil sample with various concentrations of cement kiln dust and rice husk ash. Similar to liquid limit, the plastic limit diminished with increase in both cement kiln dust and rice husk ash contents, respectively. The observed trend could be due to possible cation exchange reaction liberating absorbed water from the clay composition, thereby enhancing the process of flocculation and aggregation of the soil mixtures [31] and reduction in expanded moisture layer [6]. The reduction could also be due to the non-plastic structure of RHA. The results in this study similar with the outcome of [32] who correlated decreased in plastic limit of expansive soil due to non-plastic structure of the one of the compositional admixtures (iron ore-tailing) in cement modification of fine grain soil.

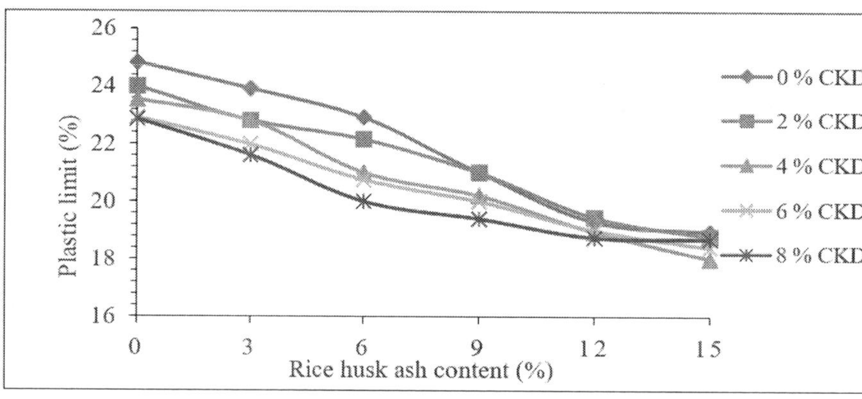

Figure 4. Behaviour of plastic limit of lateritic soil-CKD-RHA mixtures

The behaviour of plasticity index of soil sample with various concentrations of cement kiln dust and rice husk ash is display in figure 5. The plasticity index of the soil mixtures presented in figure 5 ranged between 16.60 and 18.97 %. However, based on the data publicised in figures 3, 4 and 5, it can be undoubtedly confirm that both CKD and RHA have some level of effects on the Atterberg limits of the soil mixtures.

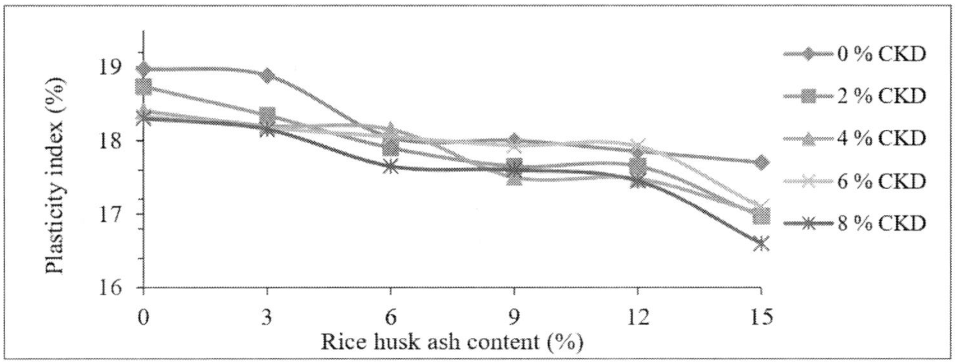

Figure 5. Behaviour of plasticity index of lateritic soil-CKD-RHA mixtures

3.4 Compaction characteristics

Figure 6 depicts the variations of maximum dry density (MDD) of soil-CKD mixtures with RHA. The MDD of the soil mixtures ranged between 1.68 and 1.83 Mg/m^3. The compaction results documented reveals that the addition of varying proportion of RHA to each of the CKD treatments reduces the values of MDD. This tendency could be unconnected to the substitution of RHA and CKD for the soil in the soil-CKD- RHA mixtures. This reduction in the values of MDD may be accredited to RHA having a lower specific gravity of 2.10 substituting a higher specific gravity (2.60) of the soil [33] secondly; this trend may also be credited to flocculation and agglomeration of soil mixtures. However, the documented reduction in MDD with increased CKD is in line with the outcome of [34].

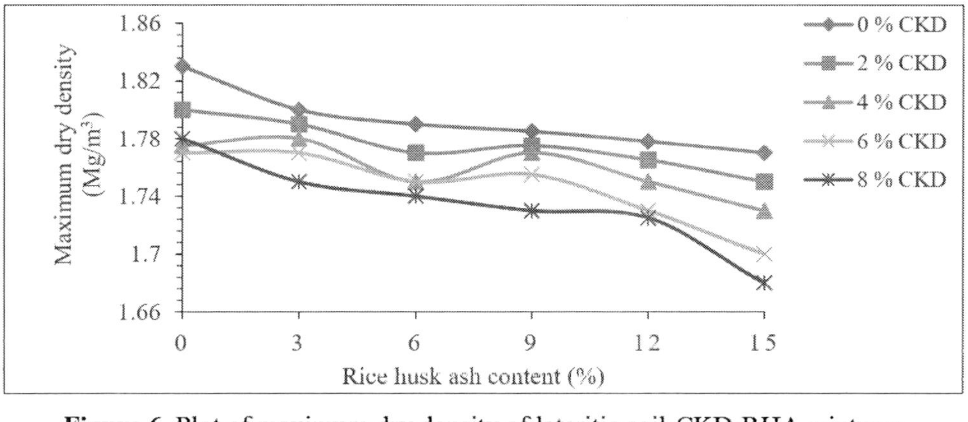

Figure 6. Plot of maximum dry density of lateritic soil-CKD-RHA mixtures

The behaviour of lateritic soil-CKD-RHA in terms of optimum moisture content (OMC) is display in figure 7. On the contrary, the OMC of the soil mixtures ranged between 12.95 to 15 %. As per the data expressed in figure 7, it can unquestionably be said that OMC improves gradually with the addition of different dosage of RHA for each CKD treatments. Improvement in OMC could be credited to the hollow (porous) nature of RHA particles, which will in return require more water to perfect lubrication of soil mixtures. In a previous study carried out by [35] also produced a related result.

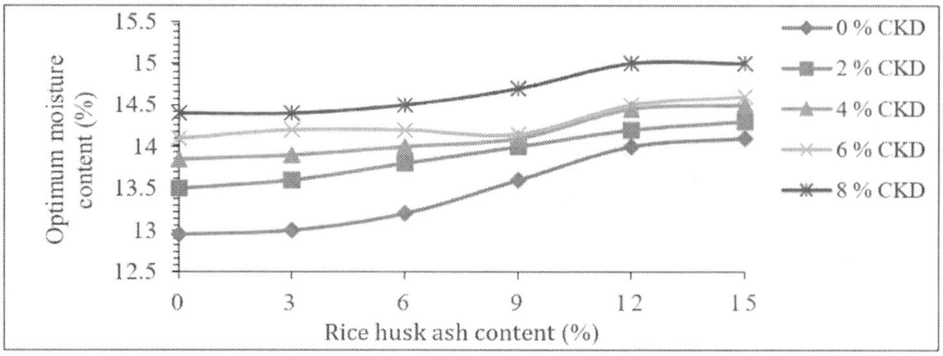

Figure 7. Plot of optimum moisture content of lateritic soil-CKD-RHA mixtures

3.5 California bearing ratio (CBR)

Figure 8 show the soaked CBR values for different dosages of cement kiln dust (CKD) mixed with different dosages of rice husk ash (RHA) in lateritic soil. It can be said without any uncertainty that the CBR values enhance gradually with the increase in quantity of stabilization materials (CKD and RHA) mixed with soil. This gradual enhancement could be adduced to the pozzolanic reaction taking place between the soil and pozzolanic material thereby developing a more cementitious compound which aids enhancement of strength in soil mixtures. The CBR value increased from a minimum of 5 % at 0 % CKD / 0 % RHA to a maximum value of 35 % at 8 % CKD / 15 % RHA content. The CBR value recorded at 8 % CKD / 15 % RHA content met the conditions by Nigerian General Specifications for highways for use as sub-base [36].

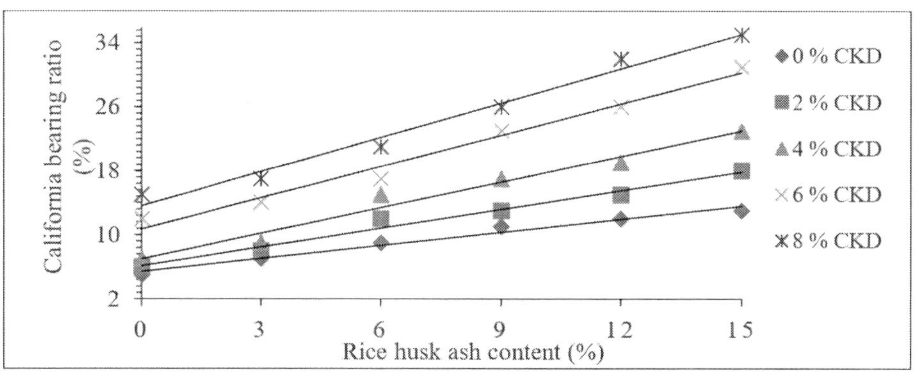

Figure 8. Behaviour of California bearing ratio of lateritic soil-CKD-RHA mixtures

3.6 Comparison between laboratory and predicted results.

The entirety of soil major physical attributes establishes a simple technique where outcomes of representative samples could be simultaneously associated with one or more relevant properties of the soil. The association could be express in one of the following form: multi-linear, exponential or in the form of polynomial. Exploring these correlations may possibly contribute a very positive information, precisely in reducing significantly the magnitude of difficult test and/or exorbitant analysis, so long as a correlation between a basic engineering characteristic and some simple test results can be set up. A simple model is establish using multiple regression statistics for CBR being dependent variable as well as a function of CKD and RHA (independent variables) (see equation 1). Details of regression results are presented in appendix A. Basically, it was observed that the model is consistent with achieved laboratory outcome of increased strength with stepped increase in both CKD and RHA. This was possible because specimens were densified to the required MDD at the requisite OMC. The result is suggestive of the fact that controlling the amount of admixtures (CKD and RHA) for stabilisation during field compaction is important to achieve desirable result for sustainable strength in service. However, [32] and [37] reported a reduction in strength of stabilised soil due to excess admixture content (iron ore in cement or lime stabilization. The excess admixture contributed to excess of fine and silt content which did not take part in the stabilisation process.

$$\text{CBR} = 0.948 + 1.908\text{CKD} + 1.025\text{RHA} \quad (1)$$

$R^2 = 93$ %

The graph of projected CBR gotten from equation (1) against experimental value measured from laboratory (see Figure 9), indicated a good correlation in a second order polynomial function having R square correlation coefficient value of approximately 0.97. Similar analysis were reported by [38-39]. Furthermore, the analysis of variance (two-way) indicated that the various variations in CKD and RHA had tremendous effect on the properties of the soil at 95% probability certainty, p-value < 0.05 (see Table 3).

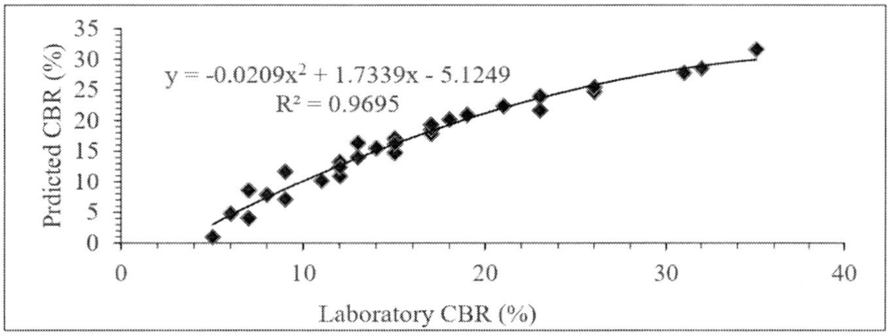

Figure 9. Predicted CBR versus laboratory CBR

Table 3. Analysis of variance of properties of lateritic soil-CKD mixtures with RHA

Property	SOV	DOF	F_{CAL}	P-value	F_{CRIT}	Remarks
LL	RHA	5	140.5329	7.2E-15	2.71089	$F_{CAL}>F_{CRIT}$, SS
	CKD	4	24.97374	1.56E-07	2.866081	$F_{CAL}>F_{CRIT}$, SS
PL	RHA	5	103.4294	1.38E-13	2.71089	$F_{CAL}>F_{CRIT}$, SS
	CKD	4	13.63198	1.61E-05	2.866081	$F_{CAL}>F_{CRIT}$, SS
PI	RHA	5	14.35701	4.92E-06	2.71089	$F_{CAL}>F_{CRIT}$, SS
	CKD	4	8.476602	0.00036	2.866081	$F_{CAL}>F_{CRIT}$, SS
OMC	RHA	5	52.7399	7.79E-11	2.71089	$F_{CAL}>F_{CRIT}$, SS
	CKD	4	115.5459	1.6E-13	2.866081	$F_{CAL}>F_{CRIT}$, SS
MDD	RHA	5	40.80618	8.03E-10	2.71089	$F_{CAL}>F_{CRIT}$, SS
	CKD	4	46.30466	7.75E-10	2.866081	$F_{CAL}>F_{CRIT}$, SS
CBR	RHA	5	33.50731	4.63E-09	2.71089	$F_{CAL}>F_{CRIT}$, SS
	CKD	4	47.90813	5.71E-10	2.866081	$F_{CAL}>F_{CRIT}$, SS

4. Conclusion

They waste materials used in this study have shown some level of significance in terms of strength indices of the deficient lateritic soil. From the outcomes of the experimental investigations performed, the following conclusions are been extracted and listed: The natural lateritic soil studied is an A-7-6 (10) and CL soil. The Atterberg limits of the soil specimens diminished with higher concentrations of CKD and RHA. In terms of compaction characteristics, an increase in CKD and RHA dosages caused a diminishment in MDD and enhancement in OMC. For the strength criterion, the maximum CBR (soaked condition) of 35 % was documented at 8 % CKD / 15 % RHA blend which implied that CKD-RHA blend could be helpful in enhancing deficient soil required as an alternate hydraulically bound material for sub grade in road construction. Although additional strength tests and micro level analysis are underway in further studies, the statistical measures and formulated model showed an excellent performance with R square of 0.97 and Adjusted R Square of 0.93 at $\alpha = 5$ % level of significance.

Appendix A

Regression Statistics	
Multiple R	0.96626
R Square	0.933658
Adjusted R Square	0.928743
Standard Error	2.115788
Observations	30

	df	SS	MS	F	Significance F
Regression	2	1701	850.4998	189.9895	1.24E-16
Residual	27	120.867	4.4765		
Total	29	1821.867			

	Coefficients	Standard Error	t Stat	P-value	Lower 95%	Upper 95%	Lower 95.0%	Upper 95.0%
Intercept	0.9476	0.8760	1.082	0.2889	-0.8498	2.7450	-0.8498	2.7450
CKD	1.9083	0.1366	13.97	7.07E-14	1.6281	2.1886	1.6281	2.1886
RHA	1.0247	0.0754	13.59	1.36E-13	0.8701	1.1795	0.8701	1.1795

References

[1] Prabakar J, Dendorkar N and Morchhale R K 2004 Influence of fly ash on strength behaviour of typical soils. *Construction and Building Materials*. **18**, pp. 263–267.

[2] Alhassan M and Mustapha A M 2007 Effect of rice husk ash on cement-stabilized laterite. *Leonardo Electronic Journal of Practices and Technologies*. **11**, pp. 47–58.

[3] Sharma N K, Swain S K and Sahoo U C 2012 Stabilization of a clayey soil with fly ash and lime: a micro level investigation. *Geotechnical and Geological Engineering*. 30 (5): pp. 1197–1205.

[4] Okafor F O and Egbe E A 2013 Potentials of cement kiln dust in sub-grade improvement. Nigerian Journal of Technology. **32**(1), pp. 109–116.

[5] Attah I C, Agunwamba J C, Etim R K and Ogarekpe N M 2019b Modelling and predicting of CBR values of lateritic soil treated with metakaolin for road material. *ARPN Journal of Engineering and Applied Science*. 14 (20): pp. 3609-3618.

[6] Sani J E, Etim R K, Joseph A 2019 Compaction behaviour of lateritic soil–calcium chloride mixtures. *Geotechnical and Geological Engineering*. Springer Nature Switzerland. **37**, pp. 2343-2362. https://doi.org/10.1007/s10706-018-00760-6.

[7] Etim R K, Attah I C, Eberemu A O and Yohanna P 2019 Compaction behaviour of periwinkle shell ash treated lateritic soil for use as road sub-base construction material. *Journal of Geoengineering*. **14** (3): pp. 179 - 190. http://dx.doi.org/10.6310/jog.201909_14(3).6

[8] Etim R K, Attah I C and Yohanna P 2020 Experimental study on potential of oyster shell ash in structural strength improvement of lateritic soil for road construction. *International Journal of Pavement Research and Technology*. https://doi.org/10.1007/s42947-020-0290-y

[9] Nair D G, Jagadish K S and Fraaim A 2006 Reactive pozzolanas from rice husk ash: an alternative to cement for rural housing. *Cement and Concrete Research*. **36** (6), pp. 1062-1071
[10] Alhassan M 2008 Potentials of rice husk ash for soil stabilization. *Assumption Univ. J. Thailand*. **11** (4), pp. 246-250
[11] Akinyele J O, Salim R W, Oikelome K O and Olateju O T 2015 The use of rice husk ash as a stabilizing agent in lateritic clay soil. *International Journal of Civil, Environmental, Structural, Construction and Architectural Engineering*. **9** (11), pp. 1373 – 1377
[12] Ali F H, Adnan A and Choy C K 1992 Geotechnical properties of a chemically stabilized soil from Malaysia with rice husk ash as an additive. *Geotechnical and Geological Engineering*. **10** (2), pp. 117-134
[13] Muntohar A S and Hantoro G 2000 Influence of rice husk ash and lime on geotechnical properties of a clayey subgrade. *Electronic Journal of Geotechnical Engineering*. **5**, pp. 1-13.
[14] Sharma R S, Phanikumar B R and Varaprasada, R B 2008 Engineering behaviour of remoulded expansive clay blended with lime, calcium chloride and rice husk ash. *Journal of Materials in Civil Engineering*. **20** (8), pp. 509-515
[15] Salahudeen A B, Eberemu A O and Osinubi K J 2014 Assessment of cement kiln dust-treated expansive soil for the construction of flexible pavements. *Journal of Geotechnical and Geological Engineering*. **32** (4), pp. 923-931. https://doi.org/10.1007/s10706-018-00760-6
[16] Mosa A M, Taher A H, Al-Jaberi L A 2017 Improvement of poor subgrade soils using cement kiln dust. *Case Studies in Construction Materials*. **7**, pp. 138–143
[17] Adeyanju E A and Okeke C A 2019 Clay soil stabilization using cement kiln dust. *IOP Conf. Series: Materials Science and Engineering*. 640, 012080. doi:10.1088/1757-899X/640/1/012080
[18] Head K H 1982 Manual of soil laboratory testing, soil classification and compaction tests, 2nd Ed., Pentech Press, London.
[19] BS 1377 1990 Methods of testing soil for civil engineering purposes. British Standards Institution, London.
[20] BS 1924 1990 Methods of tests for stabilized soils. British Standards Institute, London.
[21] ISRIC 2002 Procedure for soil analysis. International Soil Reference and Information Centre, compiled and edited by Van Reeuwijk
[22] AASHTO 1986 Standard Specifications for Transportation, Materials and Methods of Sampling and Testing. 14th Edition, Washington, D.C.
[23] ASTM 1992 Annual Book of Standards, American Society for Testing and Materials. Philadelphia. USA. 4 (8).
[24] Attah I C and Etim R K 2020 Experimental investigation on the effects of elevated temperature on geotechnical behaviour of tropical residual soils. *SN Applied Sciences, Springer Nature Switzerland AG*. 2: 370. https://doi.org/10.1007/s42452-020-2149-x
[25] Attah I C, Etim R K, Alaneme G U, Bassey O B 2020 Optimization of mechanical properties of rice husk ash concrete using Scheffe's theory. *SN Applied Sciences, Springer Nature Switzerland AG*. **2**, 928. https://doi.org/10.1007/s42452-020-2727-y
[26] Edmeades D C 1982 Effects of lime on effective cation exchange capacity and exchangeable cations on a range of New Zealand soils. *New Zealand Journal of Agricultural Research*, **25**, pp. 27-33. https://doi.org/10.1080/00288233.1982.10423369.
[27] Mathew P K and Rao S N 1997 Effect of lime on cation exchange capacity of marine clay. *Journal of Geotechnical and Geoenvironmental Engineering*. **123**(2), pp. 183-185.
[28] Yu H, Huang X, Ning J, Zhu B and Cheng Y 2014 Effect of cation exchange capacity of soil on stabilized soil strength. *Soils and Foundations*. **54**(6), pp. 1236-1240.
[29] Abood T T, Kasa A B and Chik Z B 2007 Stabilisation of silty clay soil using chloride compounds. *Journal of Engineering Science Technology*. **2**(1):102–110.
[30] Sujathaa E R, Dharinib K and Bharathia V 2015 Influence of groundnut shell ash on strength and durability properties of clay. *Geomechanics and Geoengineering: An International Journal*. pp. 1-8. http://dx.doi.org/10.1080/17486025.2015.1006265

[31] Osinubi K J 1995 Lime modification of black cotton soils. *Spectrum Journal*, **2**(1 and 2), pp. 112 – 122.
[32] Osinubi K J, Yohanna P and Eberemu A O 2015 Cement modification of tropical black clay using iron ore tailings as admixture. *Transportation Geotechnics*. **5**, pp. 35 – 49.
[33] Osinubi K J and Moses G 2012 Hydraulic performance of compacted foundry sand bagasse ash mixture permeated with municipal solid waste leachate. *Electronic Journal of Geotechnical Engineering*. **17**, pp. 865–878
[34] Yoder E J and Witczak M W 1975 Principles of pavement design. Wiley, New York, pp 300–321
[35] Zhang M H, Lastra R and Malhotra V M 1996 Rice husk ash paste and concrete: some aspects of hydration and the microstructure of the interfacial zone between the aggregate and paste. *Cement Concrete Research*. **26** (6), pp. 963–977.
[36] Nigerian General Specification 1997 Roads and Bridges, Federal Ministry of Works, Abuja, Nigeria.
[37] Etim R K, Eberemu A O and Osinubi K J 2017 Stabilization of black cotton soil with iron ore tailings as admixture. *Transportation Geotechnics*. **10**, pp. 85 –95.
[38] Moses G, Etim R K, Sani J E and Nwude M 2018 Desiccation effect of compacted tropical black clay treated with concrete waste. *Leonardo Electronic Journal of Practices and Technologies*. **33**, pp. 69-88.
[39] Moses G, Etim R K, Sani J E and Nwude M 2019 Desiccation-induced volumetric shrinkage characteristics of highly expansive tropical black clay treated with groundnut shell ash for barrier consideration. *Civil and Environmental Research*.**11** (8): pp. 58–74

Review of recently characterized Nigeria timber species for structural applications

A N Ede[1,*], O A Olomu, A I Akpabot[1], S O Oyebisi[1]
[1] Department of Civil Engineering, Covenant University, Ota, ogun state, Nigeria
*Corresponding author: anthony.ede@covenantuniversity.edu.ng

Abstract. The continuous rise of world population is putting much pressure on the housing and infrastructure stocks available to the society, thereby accentuating the need for more constructions and of course, more pressing demands on the depreciating natural resources. Today, the most used construction materials are concrete and steel, but the amount of pollutants deriving from their production has remained a source of great concern to the world. Finding solution to this environmental challenge has given rise to Sustainability Development Goals (SDGs), aimed at making the planet earth more conducive to living organisms. One of the viable approaches to reduce the high demand of construction industry on depreciating natural resources is the recourse to using more of renewable materials like timber. Timber is one of the few construction materials that have served all the ages of human lives in providing shelter. In this moment of pressing needs for more shelter and infrastructure, needed to make life more comfortable for man, the use of this renewable material must be sustained, but the strength characteristics must be known for more efficient and safe use. This research reviews recent characterization of some Nigerian timber species. This research emphasizes some good strength properties of recently reviewed Nigeria timber species, which will build the confidence of designers on safe judgment and choice of structural timber, help to reduce the cost of construction, thereby making residential houses more affordable to the lower class of the society. This will help to build a sustainable eco system for all.
Keywords: Nigeria timber species, Physical properties, Mechanical properties, Characterization, Structural applications

1. Introduction

Due to the steady rise in population in developing nations, there has been significant shortage of affordable residential housing and this has been a cause for concern to all. According to [1], the demand for structures and infrastructures is destined to escalate as the population increases, and more social challenges will emerge if the volume of new constructions does not match the demand. High volume of new constructions makes the building construction industry a steady threat to the depreciating natural resources in our environment. This explains why [2] points out that construction and our environment are inextricably linked. In Nigeria, the most populated country in Africa, the commonly used construction material is concrete [3]. It is mostly used because of the availability of cheap labour, a high number of cement manufacturing companies and the constituent materials. The high demand and the consequent high cost of construction necessitates the need to search for possible sustainable approach to affordable houses. This is one of the reasons behind some researchers [4] stating that it is important for alternative construction materials to be vastly explored as to reduce the pressure of construction industry on available natural resources. Due to the climate change challenges caused by industrialization and high demands on natural materials, the world is greatly responding to deteriorating environmental issues through strategies that emphasize sustainability. One of the broadly adopted approach is the Sustainability Development Goals (SDGs), aimed at reducing the way we consume depreciating natural resources. This has given rise to many researches on intelligent combination of materials, including wastes, into useful construction materials [5,6,7,8]. Another approach to reduce the pressure on depreciating natural resources is to improve on the use of renewable materials like timber in the construction industry. The application of timber as a building material has evolved over the centuries and still remain very essential in this present time and in the future. According to [9], timber can be utilized in various forms: in the natural form or as processed wood such as Glue-laminated timber, Structural Composite Lumber, Laminated Veneer Lumber, etc. The use of the processed timber requires modern technology to maximize its benefits, while the nations that are less developed technologically tend to adopt the unprocessed timber in the natural form. In Nigerian construction industry, unprocessed timber is adopted principally for roof trusses, interior finishes, and in a very reduced dimension as structural loading bearing members.

One of the principal factors limiting the use of Nigerian timber for structural applications is the absence of good knowledge on the strengths and properties of the various species. Most of the timber species in the country are not characterized and this has left the usage to be based on past experience or on speculation. The need to verify the structural efficiency of Nigerian timber species has led to the characterization of some Nigeria timber species. It is only when the strength characteristics are known that the usefulness can be maximized. Lack of sufficient information on the mechanical properties of Nigerian timber species has greatly limited the structural application and has led to the adoption of unreasonably high safety factors in timber design as the strength properties are mostly estimated. This leads to poor safety judgment. Based on these prevailing situations, this research evaluates some recently characterized Nigerian timber species with the view to identifying the proven uses to which some Nigerian timber species can be safely put to. By so doing, some locally available timber species, which are renewable, can advantageously serve as a substitute to the non-environmentally friendly and expensive concrete and steel which are currently the most used materials. This will help to reduce the cost of construction, make residential houses more affordable to the to the lower class of the society and improve our eco system.

2. Overview of timber as a construction material

Timber is one of the oldest construction materials in the world and has served all the ages till date. When timber is obtained from managed and controlled forests, it tends to be a sustainable construction material as it has the capacity to renew itself and at the same time, giving negligible or no negative effect to the environment. The volume of timber usage worldwide competes favorably with concrete and steel, but the level of pollutants emitted from steel and cement production has remained a source of concern to the world [10]. The extensive use of timber in construction is linked to its good properties such as lightweight, good tensile and compressive strength, relatively low cost and its renewable nature [11]. There are more residential buildings constructed with timber than any other structural material in the advanced world [4]. Nigeria is very rich in the supply of timber and previous research confirms that some of the timber species found in Nigeria are suitable for construction. Timber possesses the ability to transfer tension, compression and shear because of its strength in relation to its weight, which is one of the reasons it's an attractive structural material. Timber grains have natural patterns, aesthetically pleasing finishes and distinguished features of various species which make clients and architect's love for the material. Carpenters of the ancient era manipulated timber and connection details to enhance beauty and aesthetics in the construction of the buildings [12]. Timber has been most popularly used for roof structures in Nigeria. A survey by [13] on materials mostly adopted for making roofs in South West of Nigeria showed that timber accounted for 66%, while the usage of raffia palm, concrete and steel stood at 19 %, 11% and 4 %, respectively. Omo (C. millenii), and Afara (T. ivorensis) were the most popular species of timber used in Southwestern Nigeria due to their availability and ability to span cross functional, structural, and decorative use cases. Suitability of Nigerian timber as a structural component in modern construction is proven by the many tests of various wood species in Nigeria. The work of [13] provided an illustration of how various types of timber species are used and where their different applications overlap in the following as shown in Figure 1:

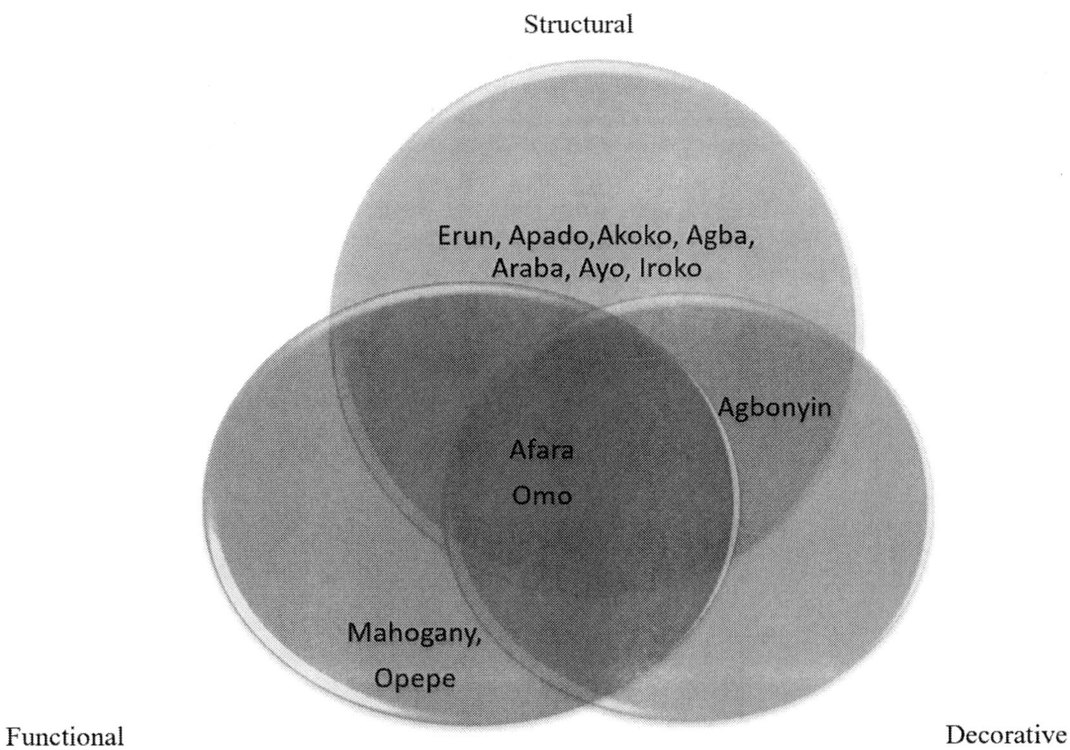

Figure 1. Various use of some Nigerian timber species in South West Nigeria (Adesogan, 2013)

Timber types such as Iroko (M. exelsa), Ayo (H. grandis), Agba (G. balsamiferum), Erun (E. suavecolens), and Apado (C. gradiflora) are useful for structural use. It can be seen that Afara (Terminalia superba) and Omo (C. millenii) lie in the area where all used cases overlap while Agbonyin wood species is useful as decorative and structural. Despite this widespread use of timber for various applications in construction industry, [14] noted that there is a big gap in usage and quality of timber in some areas of Nigeria. The authors noted that in Minna, Niger State, the building contractors had little knowledge about quality of timber they use while customers rarely enquire about the quality of timber. It was also observed that in Minna, no standard rule existed to govern the quality of timber used. Consequently, the trend is that there is a drift of demand towards cheaper and low-quality timber, with increased use of lesser-known wood or uncharacterized species. This shows that the demand and price for timber does not follow quality since the construction industry in Minna does not consider quality of timber as a guiding factor. This lays credence to the fact that the use of timber as a construction material in Nigeria is not standardized as the choice of timber material varies from region to region and have respond more to price than to quality. Since structural use of timber is the most important application of timber in construction industry, it very essential that the timber species be characterized, as to make the best use of their properties during design and to obtain safe structures at the end of its usage.

3. Review of Recently Characterized Nigeria Timber Species for Structural Applications

Procedures for designing Ayin (Anogeissus leiocarpus) wood column under axial load were developed by [15]. Specimens were subjected to direct axial compression test while stress and strain data were harvested. Critical stress and slenderness ratio were obtained via Euler-Engesser stress formula, which

help in deriving an equation that established agreement between the theoretical and experimental stress data. This equation simplifies the design of Ayin (Anogeissus leiocarpus) wood column. The suitability of three Nigerian timber species of Melina (Gmelina arborea), African locust beans (Parkia biglobosa) and African mesquite (Prosopis Africana) obtained from the forest in Idah town in North-Central part of Nigeria as structural members was studied by [16]. These three species were tested for tensile, compression, bending and shear strengths in conjunction with moisture content and density. Results of the tests showed that the compression, bending strength and shear strengths, together with the Young modulus and density are similar for the three species, while the moisture contents were significantly different. These timber species were classified as hardwood of higher strength classes (D30 – D70) in accordance to BS 5268. Therefore, they are suitable for structural use. Ogbono (Irvingia gabonensis) specie grown in South Eastern Nigeria was characterized and graded by [17]. This tree bears local names in different localities of Nigeria (Bini -Ogwe, Efik –Oyo, Ijaw - Ogboin, South East - Ogbono and South West - Oro). Two logs of this timber specie obtained from Akpugo forest in Enugu State were sliced into planks and seasoned naturally for six months in accordance to BS 373 prescriptions. Laboratory test results were adjusted to the values at moisture content of 12% in toeing the line BS5268. The bending stress, mean Young modulus and density results obtained showed that Ogbono tree species can be assigned to strength class D70, which is appropriate for heavy load bearing structural elements like bridge beams and railway sleepers. The works of [18] on White Afara (Terminalia Superba) timber specie obtained from Kaduna and Zaria timber markets. Characterized the strength properties according to Euro codes. Bending strength, modulus of elasticity, moisture content and density were determined in accordance with Euro codes. The White Afara samples were not affected by different environments from which they were obtained and were classified to fall within the strength class D24 according to EN 338 Part 8 and strength class N6 according to NCP 2 Part 14. This goes on to indicate that the specie is a hardwood with a high medium bending strength. [19] studied the structural behaviour of the Nigerian grown Abura (Mitragyna ciliata) timber subjected to compression and shearing forces. The purpose of this research was to assessed the suitability of this timber specie for bridge beams. Reliability studies and deterministic investigations were performed on the beam samples under the ultimate limit state considerations. The reliability indices recorded were satisfactory at depth of 400 mm, breadth of 150 mm and span of 5000 mm, while the samples showed low level of safety under only shearing forces. So, this can be comfortably adopted for structural beams.
The suitability of Ekki (Lophira alata), Apa (Afzelia bipindensis), Iroko (Chlorophora exceisa) and Abura (Mitragyna ciliate), which represent N1, N2, N3, and N4 classes of Nigerian timber species for bridge decks were conducted by [20]. A timber bridge deck was modelled on timber stringers on the basis of AASHTO LRFD. It was established that, strength classes, stringer spacing, timber thicknesses and width were verified as the most influential factors of the structural performance of these Nigerian timber species adopted for bridge decks. These results will be helpful in future prediction of stringer spacing for bridge deck design and analysis of Nigerian timber species for bridge decks. Regression analysis was applied to determine the relationship between the properties of Nigerian Teak (Tectona grandis) and Ebony ((Kanyan) Diospyros mespiliformi) timber species by [21]. To ascertain the potential of these Nigerian timber species, specimen for tests were prepared on the basis of EN 13183-1 and EN 408, while oven dry density, bulk density, moisture content, bending strength and Young modulus were determined. The results showed that wood density has a strong correlation with the bending strength of these Nigerian timber species and therefore very useful for forecasting strength properties and for advantageous utilization of these timber species.
The suitability of samples of Dogonyaro (Azadirachta indica) and African guinea pepper (Xylopia aethiopica), obtained from different areas of Kwara State, Nigeria for structural applications was conducted by [22]. The physical properties determined were moisture content, specific gravity and density, while the mechanical tests were for compression, shear, tensile strengths, modulus of rupture, modulus of elasticity and hardness. These tests were conducted on the basis British Standards, Nigerian Standards and Euro codes (BS 5268, 2002; NCP2, 1973 and EC5, 1995). The authors concluded that these timber species were hardwood of strength classes D30 - D70 according to BS 5268 and are

therefore suitable for structural applications. The compressive strength characteristics of Apa (Afzelia bipindensis) and Opon (Lannea schimperi) timber species obtained from sawmills in Ilorin, Kwara State, Nigeria were performed by [23]. The samples were naturally seasoned for seven (7) months and tested in accordance with the British Standard BS EN 408 (2003) Test for physical and mechanical properties of structural timbers involved moisture content, density and compressive strength. Results obtained showed good combination of both buckling and compression resistance of these timber species. Therefore, the suitability of these timber species as axial load carrying members were established.
The characterization of Iya (Daniellia ogea) obtained from forests around Ilorin Local Government, Kwara-State, Nigeria was done by [24]. The samples were prepared in accordance to BS EN 408: 2003. The moisture content, density, compressive strengths perpendicular and parallel to grain, tensile strength parallel to grain, modulus of rupture, local modulus of elasticity and the apparent modulus of elasticity were determined. The results obtained compared very favourably with known characterized species in the market. The mean density achieved was far higher than the range of coniferous softwoods, therefore, Iya timber specie can be classified as deciduous hardwood and good for structural applications. The use of Apa (Afzelia bipindensis) wood species in construction was verified by [25]. Tests were conducted on three timber logs obtained from Kwara State, North Central Zone of Nigeria. Samples were air-dried and tested parallel to grain in compression, tension and bending, following the BS codes, Euro codes and Nigerian standards. This timber specie was classified to belong to the strength group D30. The characteristic strength obtained is satisfactory for structural purposes. The reliability of Afara (Terminalia sperba) and Babo (Isoberlinia doka) obtained from sawmills in Ilorin, Kwara State, Nigeria for structural use as compression members was conducted by [26]. Moisture content, density and compressive strengths of various samples were determined. This research derived equations for the design of Afara and Babo columns and therefore recommended the equations as the more appropriate approach for the design of timber compression members in Nigeria.

4. Conclusion

This particular investigation on recently characterized Nigerian timber species establishes that Nigeria as a nation is endowed with good structural timber species, but perhaps lagging behind in terms of widespread application of timber as a major load bearing material in building and infrastructural construction. Again, (Aguwa et al., 2015) expanded the meaning of construction not only to mean housing application, but also to include building of substantial public infrastructure. Tests of various timber species provided a picture of the availability of a wide range of structural timber species that are not advantageously utilized to help reduce the burden of construction materials on nonrenewable sources. The results of all the studies indicate that Nigeria has a number of wood species that are suitable for structural and functional applications, by far the most important uses of timber in construction. These findings will further improve the acceptability and trust of Nigerian timber species as structural construction materials. More researches in characterizing Nigerian timber species are recommended as to make the best use of the renewable materials of which the nature has endowed the nation.

Acknowledgement:

The authors are grateful to Covenant University for funding this publication.

References

[1] Anejo, J. A. (2014). Impact Of Concrete, Steel And Timber On The Environment: A Review. International Journal of Technology Enhancements and Emerging Engineering Research, **2**(7), 58–63.
[2] Adegoke, B. (2012). Assessment of criteria for selection of timber as building material for building construction projects within Lagos Metropolis. Retrieved from https://books.google.com/books?id=lKZE1OPLQsMC&pgis=1
[3] Ede, A.N., Olofinnade, O.M., Enyi-Abonta, E., Bamigboye, G.O. (2017). Implications of

construction materials on energy efficiency of buildings in tropical regions International Journal of Applied Engineering Research **12**(18), pp. 7873-7883 (2017).

[4] Ede, A. N., Adebayo, S. O., Ikechukwu Ugwu, E., & Emenike, C. (2014). Life Cycle Assessment of Environmental ImpactsofUsing Concrete orTimber toConstruct aDuplex Residential Building. IOSR Journal of Mechanical and Civil Engineering, **11**(2), 62–72. https://doi.org/10.9790/1684-11276272

[5] A.N. Ede, O.M. Olofinnade, E.I. Ugwu, and A.O. Salau, A.O. Potentials of Momordica angustisepala fiber in enhancing strengths of normal portland cement concrete, Cogent Engineering **5**(1),1431353. (2018) pp. 1-17.

[6] Joshua, O., Tunji-Olayeni, P.F., Olusola, K.O., Ede, A.N., Adewale, B.A. (2018). Investigating for pozzolanic activity in the blend of ground glass waste with cement for sustainable concrete. International Journal of Mechanical Engineering and Technology **9**(6), pp. 808-816.

[7] Olatokunbo, O., Adedeji, A., Nurain, S., Ede, A.N., Tolulope, A., Ogundipe, J. (2018). Pseudo-Dynamic earthquake response model of wood-frame with plastered typha (Minima) bale masonry-infill. International Journal of Civil Engineering and Technology **9**(2), pp. 27-35.

[8] O.M. Olofinnade, A.N. Ede, J.M. Ndambuki, 2017. Experimental Investigation on the Effect of Elevated Temperature on Compressive Strength of Concrete Containing Waste Glass Powder, International Journal of Engineering and Technology Innovation, **7**, 4, pp. 280 – 291.

[9] Ede, A. N., & Okundaye, J. O. (2014). Appraisal of Timber as Structural Members for Residential Buildings in Nigeria. International Journal of Engineering & Technology, **14**(01), 108–112.

[10] Alayande, T.A., Ede, A.N., Aguwa, J.I., Ofuyatan O.M., Oyebisi S.O., Oluwafemi J.O. (2019). Reliability Assessment of the Nigerian Timber – An Environmental Sustainability Approach in the 21st Century, J. Phys.: Conf. Ser. 1378 032053

[11] A I Osofero, M Corradi, G Benedetti and A N Ede (2019). Use of natural resins in repairing damaged timber beams – An experimental investigation IOP Conf. Ser.: Mater. Sci. Eng. **640** 012040

[12] Abeysuriya, N. (2013). Timber - A Fashionable Construction Material. International Conference on Structural Engineering Construction and Management, 9. Retrieved from http://dl.lib.mrt.ac.lk/handle/123/9535

[13] Adesogan, S. O. (2013). Wooden materials in building projects: Fitness for roof construction in southwestern Nigeria. Journal of Civil Engineering Construction Technology, **4**(7), 217–223. https://doi.org/10.5897/JCECT2013.0278

[14] Adebara S. A, Hassan Haruna, Shittu M. B, & Anifowose M.A. (2014). Quality and Utilization of Timber Species for Building Construction in Minna, Nigeria. The International Journal Of Engineering And Science, 2319–1813.

[15] AA Jimoh (2005). Ultimate strength design of axially loaded Ayin (Anogeissus leiocarpus) timber columns, Journal of Applied Science and Technology **10**(1&2) 2005: 29-34

[16] Ataguba, C. ., Enwelu, C., Aderibigbe, W., & Okiwe, E. (2015). A comparative study of some mechanical properties of Gmelina Arborea, Parkia Biglobosa & prosopis Africana timbers for structural use. International Journal of Technical Research and Applications, **3**(3), 320–324.

[17] Aguwa, J. I., Chukwu, P. C., & Auta, S. M. (2015). Characterization and grading of South Eastern Nigeria grown Irvingia gabonensis timber in accordance with BS 5268. USEP: Journal of Research Information in Civil Engineering, **12**(2), 720–731.

[18] Kaura, J. M., Abubakar, I., Aliyu, I., & Hassan, U. (2015). EN 338 Strength Grade and Uncertainty Models of Material Properties for Nigerian Grown Terminalia Superba (White Afara) Timber Specie. Nigerian Journal of Technology, **34**(1), 21. https://doi.org/10.4314/njt.v34i1.3

[19] Aguwa, J.I. Structural performance of the Nigerian grown Abura timber bridge beam subjected to compression and shearing forces. KSCE J Civ Eng **19**, 974–981 (2015). https://doi.org/10.1007/s12205-013-0164-9

[20] P. A. Owoeye, A. Usman and O. S Abejide, (2016). Reliability assessment of stringers spacings

in bridges as function of timber properties, Nigerian Journal of Technology (NIJOTECH) **35**, No. 2, April 2016, pp. 306 – 316.

[21] Sholadoye, I. O., Abubakar, I., Annafi, Q.B.& Ejeh, S.P (2016). Evaluation of Some Wood Properties of Nigeria Timber Using Four – Point Bending Test, Advances in multidisciplinary and scientific research, **2**(3) 133-144.

[22] Jimoh, A. A., & Aina, S. T. (2017). Characterization and grading of two selected timber species grown in Kwara State Nigeria. Nigerian Journal of Technology (NIJOTECH), **36**(4), 1002–1009.

[23] Jimoh, AA; Rahmon, RO; Joseph, SG, (2017). Evaluation of Compressive Strength Characteristics of Structural-sized Apa (Afzelia bipindensis) and Opon (Lannea schimperi) Timber species columns found in Nigeria. J. Appl. Sci. Environ. Manage. Dec, 2017, **21** (7) 1281-1285

[24] Jimoh, A. A., & Chabi, A. (2017). Some Physical and Mechanical Properties of Daniellia Ogea Harms Green Timber Species. J. Appl. Sci. Environ. Manage, **21**(6) 1101-1104.

[25] Jimoh, A. A., & Adefemi, J. B. (2018). Re-evaluation of characterisation and classification of Apa (Afzelia bipindensis) timber for structural use. Nigerian Journal of Technology, **36**(4), 1022. https://doi.org/10.4314/njt.v36i4.5

[26] Jimoh, A. A., Rahmon, R. O., & Ibrahim, K. H. (2018). Modeling of Compressive Strength Characteristics of Structural - sized Afara (Terminalia superba) and Babo (Isoberlinia doka) Timber Columns Using Constant Failure Rate (CFR) Model of Reliability. **2**, 102–115.

IOP Conf. Series: Materials Science and Engineering 1036 (2021) 012074 doi:10.1088/1757-899X/1036/1/012074

Corrosion inhibition and statistical data of thiocarbanilide and thiocarbanilide admixed with 4-hydroxy-3-methoxybenzaldehyde on carbon steel in dilute acid media

R T Loto[1*] and C A Loto[1]

[1]Department of Mechanical Engineering, Covenant University, Ogun state, Nigeria
*Corresponding author: tolu.loto@gmail.com

Abstract. The inhibition performance of thiocarbanilide (TD) and thiocarbanilide admixed with 4-hydroxy-3-methoxybenzaldehyde (TDHM) on the corrosion of mild steel in 1M H_2SO_4 and HCl acid solutions was studied through coupon measurement. Data obtained showed the optimal performance of TD is 71.69% in H_2SO_4 and 83.14% HCl acid solution respectively. TDHM performed more effectively with optimal inhibition efficiency of 96.77% and 97.21% in both acids. The result showed the performance of corrosion inhibiting compounds improves through synergistic interaction with other compounds. However, exposure time and inhibitor concentration strongly influenced the performance of the compounds at varying degrees. Statistical data shows the standard deviation for inhibition efficiency of the compounds was significantly higher in H_2SO_4 compared to HCl due to the high degree of variation analogous to exposure time and stability of the inhibitor molecules. TD and TDHM generally showed lower standard of deviation, higher mean value, lower margin of error and higher proportion of inhibition efficiency data above 70% inhibition performance. Statistical data from analysis of variance depict both exposure time and inhibitor concentration is statistically relevant, and influences the inhibition performance of TD and TDHM compounds. However, the influence of concentration is significantly limited compared to exposure time with statistical relevance values above 60%

1. Introduction

Carbon steels are the most universally applicable and produced steel worldwide, with a production volume of about 85% of the total annual global steel manufactured. It is the largest category of ferrous metals and alloys in weight and total price [1]. Corrosion of carbon steels has been a major problem for industrial establishments and governmental organizations due to its enormous economic consequence. It is the most prevalent cause of accidents and downtime in process industries. In some industrial applications, appropriate selection of corrosion resistant materials for construction is possible. However, such method is not cost effective [2]. The limited corrosion resistance of carbon steels is responsible for its short lifespan during service and continuous replacement when damaged [3]. There are other effect techniques of corrosion control. However, despite their relative effectiveness and high cost of operation, they do not provide the universal protection afforded by corrosion inhibitors. Ongoing research and practical field applications has resulted in the formulation of effective corrosion inhibitors for numerous applications from the application in vehicular radiators transport to desalination plants, mining, energy generation, chemical production and petrochemical refinery [4-7]. Corrosion inhibitors are intermittently added to corrosive environments where they attach onto the steel or modify the corrosive environment [8-10]. Most of the effective corrosion inhibitors in industry are toxic and increasingly being subjected to government regulations on usage and application [11-13]. Research on less toxic compounds of organic origin has produced promising results, thus the need for further test [14-16]. In contribution to the research on organic corrosion inhibitors, this study investigates the electrochemical and corrosion inhibition property of thiocarbanilide and thiocarbanilide admixed with 4-hydroxy-3-methoxybenzaldehyde compound on mild steel in HCl and H_2SO_4 solution.

2. Experimental methods

Thiocarbanilide (TD) was oncocted in cubic concentratess of 0%, 0.0025%, 0.005%, 0.0075%, 0.01%, 0.0125% and 0.015%. Thiocarbanilide admixed with 4-hydroxy-3-methoxybenzaldehyde (TDHM) in equal proportions was concocted in cubic concentrates of 0%, 0.13%, 0.25%, 0.38%, 0.50%, 0.63% and 0.75% respectively. The inhibitor concentrations were formulated in 200ml of 1M of H_2SO_4 and 1M HCl acid solution. Carbon steel (MS) rods were machined into 14 test specimens with general configuration of 0.7 cm radius and 0.7 cm length. The surface ends of the steel specimens were abraded with emery papers of different grits. Weight measures of MS specimens were separately immersed in 200 ml of the H_2SO_4 and HCl solution at the formulated TD and TDHM concentrations for 240 h. The prepared MS were weighed at 24 h hiatus with Ohaus analytical weighing balance. Tabulated results of MS weight-loss at specific TD and TDHM concentrations in the acid media are shown from Table 1 to 4. The weight loss is the variation between the first weight of the steel (maintained for 168 h) and the final weight obtained at 24 h interval. Tables 5-8 shows the data of inhibition efficiency (*IE*) calculated from the equation below;

$$IE = \left[\frac{W_1 - W_2}{W_1}\right] * 100 \quad (1)$$

W_1 and W_2 are the weight-loss of the control and inhibited MS in the acid media with respect to exposure time.

3. Results and discussion

3.1 Coupon measurement

Tables 1-4 shows the data for weight loss of MS in TD/ H_2SO_4, TD/HCl, TDHM/ H_2SO_4 and TD/HCl solution for 240 h. The presence of TD and TDHM inhibitor compound in the acid media significantly reduced the weight-loss of the MS specimens. However, TD inhibition effect in H_2SO_4 tends to be time dependent with the weight loss results of the inhibited MS steel decreasing gradually till 240 h. The same phenomenon occurred for MS in TD/HCl solution. However, the weight-loss values are slightly lower due to effective inhibition action of TD in HCl. TDHM performed more effectively in H_2SO_4 and HCl solution compared to TD. TDHM significantly decreased the weight-loss of MS in both acids. Tables 5 to 8 shows the data for inhibition efficiency of TD and TDHM in H_2SO_4 and HCl solution. Effective inhibition action of the compounds was observed at 240 h of exposure. Observation of Table 5 shows TD inhibition performance varies slightly with respect to concentration. Secondly, after the lowest TD concentration exposure time strongly influenced the performance of TD i.e. the performance of TD improved with exposure time due to the slow molecular action of the inhibitor species. The corresponding data of TD inhibition on MS in HCl solution shows TD performed more effectively from the onset of the exposure hour with minimal variation in inhibition efficiency compared to TD in H_2SO_4 solution. The performance of TDHM on MS in H_2SO_4 and HCl solution were generally similar with gradual change in value from the onset of the exposure hours. Fig. 1(a) and Fig. 1(b) show the plots of TD inhibition efficiency with exposure time at the highest and lowest inhibitor concentration in H_2SO_4 and HCl solution while Fig. 2(a) and (b) shows the corresponding plots for TDHM in H_2SO_4 and HCl solution. Observation of Fig. 1(a) shows the wide variation of TD plot configuration at highest and lowest TD concentration compared to the plots in Fig. 1(b). Secondly, the plot configuration at lowest TD concentration in Fig. 1(a) appears to be stable with respect to exposure time compared to the plot at highest TD concentration. The plots in Fig. 1(b) shows similar configuration with respect to exposure time.

IOP Conf. Series: Materials Science and Engineering 1036 (2021) 012074 doi:10.1088/1757-899X/1036/1/012074

Table 1. Data on weight-loss of MS from 1M H_2SO_4/TD (0% - 0.015%) solution

Exp. Time(h) \ TD conc. (%)	0%	0.003%	0.005%	0.008%	0.010%	0.013%	0.015%
24	0.063	0.018	0.058	0.056	0.052	0.050	0.062
48	0.110	0.039	0.075	0.086	0.071	0.060	0.105
72	0.158	0.061	0.093	0.096	0.089	0.071	0.143
96	0.243	0.076	0.110	0.114	0.104	0.097	0.163
120	0.307	0.092	0.124	0.129	0.122	0.116	0.183
144	0.403	0.128	0.150	0.153	0.141	0.139	0.208
168	0.478	0.146	0.153	0.173	0.158	0.141	0.233
192	0.646	0.186	0.188	0.220	0.196	0.205	0.273
216	0.739	0.213	0.217	0.229	0.215	0.221	0.289
240	0.848	0.246	0.244	0.240	0.241	0.253	0.321

Table 2. Data on weight-loss of MS from 1M HCl/TD (0% - 0.015%) solution

Exp. Time(h) \ TD conc. (%)	0%	0.003%	0.005%	0.008%	0.010%	0.013%	0.015%
24	0.076	0.034	0.028	0.034	0.042	0.018	0.034
48	0.134	0.042	0.041	0.043	0.074	0.037	0.046
72	0.191	0.050	0.055	0.053	0.098	0.056	0.058
96	0.288	0.063	0.077	0.060	0.126	0.068	0.068
120	0.406	0.075	0.091	0.071	0.136	0.081	0.079
144	0.645	0.099	0.115	0.094	0.182	0.108	0.099
168	0.851	0.119	0.136	0.115	0.211	0.131	0.118
192	1.372	0.232	0.218	0.203	0.281	0.230	0.197
216	1.616	0.288	0.264	0.251	0.321	0.280	0.252
240	1.852	0.346	0.320	0.312	0.363	0.323	0.317

Table 3. Data on weight-loss of MS from 1M H_2SO_4/TDHM (0% - 0.75%) solution

Exp. Time(h) \ TD conc. (%)	0%	0.13%	0.25%	0.38%	0.50%	0.63%	0.75%
24	0.182	0.061	0.072	0.073	0.062	0.054	0.076
48	0.357	0.077	0.098	0.081	0.084	0.069	0.089
72	0.656	0.091	0.114	0.096	0.095	0.079	0.102
96	1.078	0.103	0.123	0.102	0.096	0.085	0.112
120	1.509	0.117	0.130	0.105	0.102	0.090	0.115
144	1.945	0.128	0.140	0.114	0.107	0.095	0.121
168	2.391	0.143	0.150	0.120	0.113	0.102	0.125
192	2.810	0.150	0.161	0.127	0.121	0.106	0.132
216	3.344	0.167	0.166	0.134	0.127	0.113	0.141
240	3.785	0.177	0.179	0.144	0.135	0.122	0.150

Table 4. Data on weight-loss of MS from 1M HCl/TDHM (0% - 0.75%) solution

Exp. Time(h) \ TD conc. (%)	0%	0.13%	0.25%	0.38%	0.50%	0.63%	0.75%
24	0.266	0.030	0.040	0.042	0.095	0.024	0.036
48	0.558	0.047	0.059	0.063	0.118	0.039	0.054
72	0.975	0.067	0.078	0.081	0.137	0.058	0.070
96	1.623	0.081	0.093	0.099	0.155	0.067	0.081
120	2.281	0.089	0.103	0.115	0.167	0.079	0.093
144	2.958	0.108	0.121	0.133	0.179	0.087	0.107
168	3.629	0.120	0.134	0.148	0.203	0.101	0.116
192	4.087	0.134	0.146	0.167	0.217	0.114	0.126
216	4.315	0.148	0.162	0.184	0.241	0.122	0.140
240	4.835	0.159	0.174	0.205	0.265	0.142	0.152

Table 5. Data on inhibition efficiency of TD in 1M H_2SO_4

TD conc. (%) / Exp. Time(h)	0.003%	0.005%	0.008%	0.010%	0.013%	0.015%
24	71.25	8.37	10.90	17.22	21.64	2.53
48	64.22	31.79	21.83	35.78	45.65	4.53
72	61.36	41.24	38.90	43.27	55.27	9.14
96	68.57	54.70	53.25	57.21	59.97	32.95
120	69.90	59.52	57.89	60.27	62.30	40.48
144	68.16	62.89	62.05	64.93	65.52	48.35
168	69.42	68.06	63.88	67.02	70.51	51.17
192	71.20	70.91	65.96	69.68	68.32	57.68
216	71.11	70.64	69.01	70.88	70.07	60.94
240	71.02	71.27	71.69	71.63	70.22	62.15

Table 6. Data on inhibition efficiency of TD in 1M HCl

TD conc. (%) / Exp. Time(h)	0.003%	0.005%	0.008%	0.010%	0.013%	0.015%
24	55.89	63.87	55.89	44.63	75.92	55.50
48	68.51	69.18	67.76	44.95	72.25	65.67
72	73.57	71.32	72.53	48.72	70.80	69.75
96	78.29	73.40	79.15	56.33	76.38	76.45
120	81.51	77.51	82.62	66.59	80.13	80.55
144	84.64	82.15	85.48	71.85	83.26	84.72
168	85.97	84.06	86.48	75.20	84.58	86.20
192	83.07	84.10	85.22	79.48	83.23	85.62
216	82.18	83.67	84.50	80.12	82.66	84.40
240	81.32	82.71	83.14	80.41	82.57	82.89

Table 7. Data on inhibition efficiency of TDHM in 1M H_2SO_4

TD conc. (%) / Exp. Time(h)	0.13%	0.25%	0.38%	0.50%	0.63%	0.75%
24	66.47	60.70	59.93	65.81	70.47	58.23
48	78.32	72.55	77.31	76.39	80.62	75.13
72	86.18	82.66	85.42	85.54	87.95	84.40
96	90.49	88.61	90.58	91.09	92.16	89.66
120	92.23	91.38	93.03	93.25	94.03	92.39
144	93.41	92.78	94.12	94.49	95.14	93.78
168	94.03	93.73	95.00	95.26	95.74	94.77
192	94.65	94.26	95.49	95.71	96.24	95.31
216	95.02	95.02	95.98	96.21	96.62	95.78
240	95.33	95.27	96.19	96.44	96.77	96.05

Table 8. Data on inhibition efficiency of TDHM in 1M HCl

TD conc. (%) / Exp. Time(h)	0.13%	0.25%	0.38%	0.50%	0.63%	0.75%
24	88.56	85.06	84.09	64.45	91.16	86.49
48	91.64	89.49	88.67	78.80	93.10	90.33
72	93.15	91.98	91.73	85.97	94.03	92.86
96	95.03	94.26	93.91	90.42	95.85	95.01
120	96.09	95.46	94.94	92.70	96.55	95.92
144	96.35	95.91	95.51	93.95	97.05	96.37
168	96.70	96.31	95.91	94.41	97.21	96.79
192	96.73	96.42	95.91	94.68	97.21	96.91
216	96.57	96.24	95.75	94.41	97.17	96.76
240	96.71	96.40	95.76	94.52	97.06	96.85

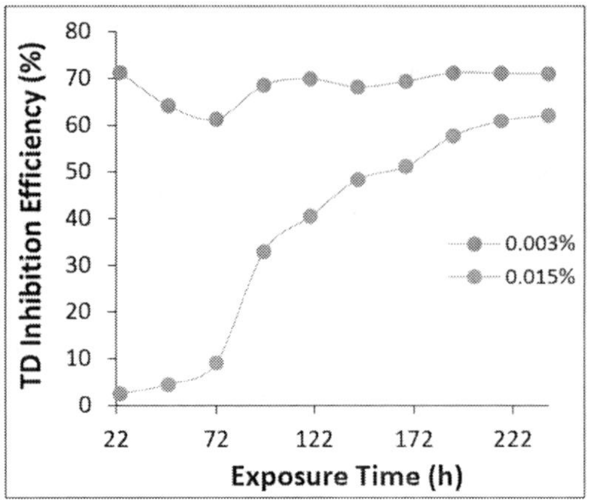

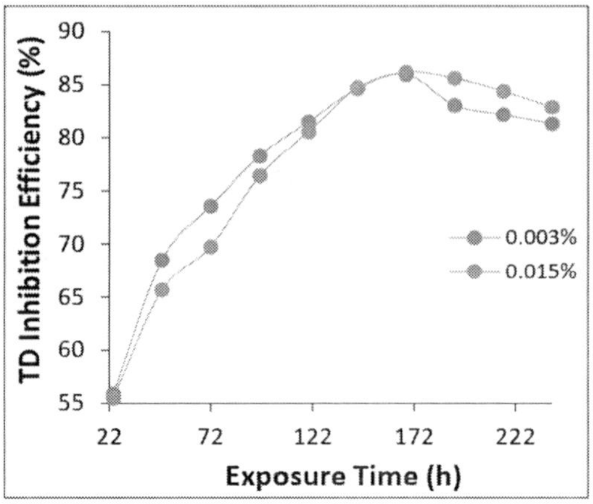

(a) (b)

Figure 1. Plot of TD concentration versus exposure time at highest and lowest TD concentration (a) in H_2SO_4 solution and (b) in HCl solution

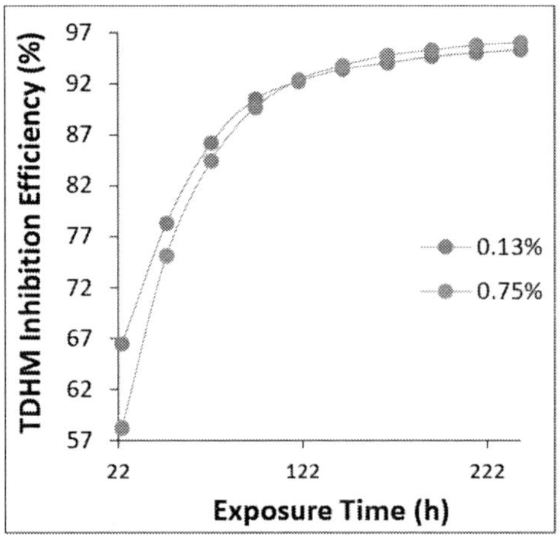

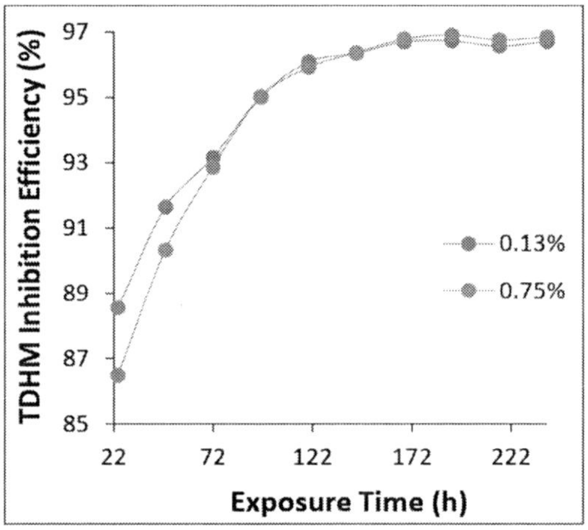

(a) (b)

Figure 2. Plot of TD concentration versus exposure time at highest and lowest TDHM concentration (a) in H_2SO_4 solution and (b) in HCl solution

3.2 Statistical evaluation

The mean, standard deviation and margin of error for TD and TDHM inhibition efficiency data in H_2SO_4 and HCl solution are shown in Table 9 and 10. Observation of Table 9 shows the standard deviation of TD inhibition efficiency data with respect to concentration are significantly higher in H_2SO_4 solution after the lowest TD concentration compared to the values in HCl solution due to the high degree of variation of the inhibition efficiency values from mean value i.e. variation of the values over time are significant and TD inhibition performance tends to be less stable in the H_2SO_4 solution. The higher the TD concentration, the lower the mean value in H_2SO_4, and the higher the standard deviation due high variation of inhibition efficiency data over time. This further confirmed from the average inhibition values in H_2SO_4 which are lower than the values obtained in HCl. The relatively lower standard deviation values in HCl solution shows TD inhibition performance varies slightly over time i.e. less time dependent. The margin of error shows that 22% and 77% of inhibition efficiency data obtained in H_2SO_4 and HCl solution are above 70% inhibition efficiency at margin of error of $\pm$10.42% and $\pm$10.70%. Observation of Table 10 shows the standard deviation for TDHM inhibition efficiency in H_2SO_4 are significantly higher than the values obtained in HCl solution due to the lower degree of variation of inhibition efficiency over time in HCl solution. This confirms TDHM inhibition performance in HCl is more stable and effective as shown in the mean inhibition value which are generally above 90%. However, the mean inhibition value of TDHM in H_2SO_4 is higher than the values obtained for TD in H_2SO_4 and HCl solution. The margin of error shows 92% and 100% of TDHM inhibition efficiency are above 70% inhibition value at margin of error of ±6.99% and ±0%.

Table 9. Statistical data for mean, standard deviation and margin of error for TD inhibition efficiency data in H_2SO_4 and HCl solution

H_2SO_4						
TD Concentration	0.003%	0.005%	0.008%	0.010%	0.013%	0.015%
Standard Deviation	3.33	20.84	20.87	18.10	15.31	23.59
Mean	68.62	53.94	51.54	55.79	58.95	36.99
Proportion above 70% Inhibition Efficiency	22%	Margin of Error	±10.42%			
HCl						
TD Concentration	0.003%	0.005%	0.008%	0.010%	0.013%	0.015%
Standard Deviation	9.25	7.32	9.97	14.86	4.98	10.35
Mean	77.50	77.20	78.28	64.83	79.18	77.18
Proportion above 70% Inhibition Efficiency	77%	Margin of Error	±10.70%			

Table 10. Statistical data for mean, standard deviation and margin of error for TDHM inhibition efficiency data in H_2SO_4 and HCl solution

H_2SO_4						
TDHM Concentration	0.13%	0.25%	0.38%	0.50%	0.63%	0.75%
Standard Deviation	9.39	11.57	11.61	10.29	8.69	12.22
Mean	88.61	86.70	88.31	89.02	90.57	87.55
Proportion above 70% Inhibition Efficiency	92%	Margin of Error	±6.99%			
HCl						
TDHM Concentration	0.13%	0.25%	0.38%	0.50%	0.63%	0.75%
Standard Deviation	2.79	3.82	3.98	9.86	2.14	3.52
Mean	94.75	93.75	93.22	88.43	95.64	94.43
Proportion above 70% Inhibition Efficiency	100%	Margin of Error	±0%			

Analysis of variance (ANOVA) was used to evaluate the statistical importance of inhibitor concentration and exposure time (sources of variation) on the protection efficiency of TD and TDHM in H_2SO_4 and HCl solution. The ANOVA data for TD and TDHM inhibition performance is shown in Tables 11 and 12. The statistical relevance factor presents the percentage significance of inhibitor concentration and exposure time. The mean square ratio represents the significance factor which must be greater than the theoretical significance factor to be statistically relevant. The statistical relevance factor in Table 11 shows exposure time is statistically more relevant on the inhibition performance of TD in H_2SO_4 and HCl solution with values of 62.31% and 66.12%. These values are greater than the statistical relevance value of inhibitor concentration with values of 22.96% and 21.37%. Observation of the value for mean square ratio shows they are greater than the theoretical significance factor, thus both exposure time and TD concentration influence the inhibition performance of TD compound though exposure time is statistically more significant. The statistical relevance factor in Table 12 shows exposure time overwhelmingly influence the inhibition performance of TDHM inhibitor in H_2SO_4 solution with value of 96.99% compared to the influence of TDHM concentration at 1.4%. However, the mean square ratio for both sources of variation is greater than the theoretical significance factor signifying both are statistically relevant and at varying degrees influence the performance of TDHM compound in H_2SO_4 solution. The ANOVA data for TDHM in HCl solution aligns with the values earlier discussed for TD compound in H_2SO_4 and HCl solution.

Table 11. ANOVA data for TD inhibition performance in H_2SO_4 and HCl solution

Source of Variation	Addition of Squares	Degree of Freedom	Mean Square	Mean Square Ratio (F)	Theoretical Significance Factor	Statistical Relevance Factor, F (%)
H_2SO_4						
TD Conc.	5362.396	5	1072.48	14.03	2.42	22.96
Exp. Time	14553.78	9	1617.09	21.15	2.15	62.31
Residual	3440.61	45	76.46			
Total	23356.79	59				
HCl						
TD Conc.	1445.76	5	289.15	15.37	2.42	21.37
Exp. Time	4473.35	9	497.04	26.41	2.15	66.12
Residual	846.85	45	18.82			
Total	6765.96	59				

Table 12. ANOVA data for TDHM inhibition performance in H_2SO_4 and HCl solution

Source of Variation	Addition of Squares	Degree of Freedom	Mean Square	Mean Square Ratio (F)	Theoretical Significance Factor	Statistical Relevance Factor, F (%)
H_2SO_4						
TDHM Conc.	87.69	5	17.54	7.78	2.42	1.40
Exp. Time	6084.68	9	676.08	299.78	2.15	96.99
Residual	101.49	45	2.26			
Total	6273.86	59				
HCl						
TDHM Conc.	327.46	5	65.49	8.32	2.42	19.28
Exp. Time	1016.63	9	112.96	14.35	2.15	59.87
Residual	354.13	45	7.87			
Total	1698.22	59				

4. Conclusion

Thiocarbanilide and thiocarbanilide admixed with 4-hydroxy-3-methoxybenzaldehyde effectively inhibited the corrosion of mild steel in H_2SO_4 and HCl solution with optimal inhibition efficiency above 70% and 80% for thiocarbanilide in H_2SO_4 and HCl while the corresponding performance for thiocarbanilide admixed with 4-hydroxy-3-methoxybenzaldehyde is above 90% in both acid media. The performance of both compounds varies at differing with respect to time and concentration. Statistical data shows only a small proportion of inhibition efficiency data of thiocarbanilide in H2SO4 are above 70% inhibition efficiency compared to the other values obtained in the acid media which is above 75%. Data from ANOVA shows exposure time strongly influences the performance of both compounds compared to inhibitor concentration

Acknowledgement

The author is grateful to Covenant University Ota, Ogun State, Nigeria for their support for this project.

References

[1] Corrosion of carbon steel. http://www.totalmateria.com/articles/Art60.htm [Retrieved: 20/03/2019].
[2] Corrosion in process industries. http://www.corrosiondoctors.org/ProcessIndustry/Introduction.htm. (Accessed March 19, 2016).

[3] Ayşe T and Mübeccel E 2006 Protection of corrosion of carbon steel by inhibitors in chloride containing solutions G.U. Journal of Science **19(3)** 149-154.
[4] Loto R T 2017 Study of the synergistic effect of 2-methoxy-4-formylphenol and sodium molybdenum oxide on the corrosion inhibition of 3CR12 ferritic steel in dilute sulphuric acid *Results in Phys.* **7** 769-776.
[5] Loto R T, Leramo R and Oyebade B 2018 Synergistic combination effect of salvia officinalis and lavandula officinalis on the corrosion inhibition of low-carbon steel in the presence of $SO_4{}^{2-}$- and Cl^--containing aqueous environment *J. Fail. Anal. Prev.* **18(6)** 1429-1438.
[6] Loto R T and Loto C A 2012 Effect of P-phenylediamine on the corrosion of austenitic stainless steel type 304 in hydrochloric acid *Int. J. Elect. Sci.* **7(10)** 9423-9440
[7] Loto R T and Oghenerukewe E 2016 Inhibition studies of Rosmarinus officinalis on the pitting corrosion resistance 439LL ferritic stainless steel in dilute sulphuric acid *Orient. J. Chem.* **32(5)** 2813-2832.
[8] Toth J 2002 Adsorption: Theory, modeling, and analysis, Marcel Dekker. New York.
[9] Bentiss F, Traisnel M, Chaibi N, Mernari B, Vezin H and Lagrenee M 2012 2,5-Bis (nmethoxyphenyl)-1,3,4-oxadiazoles used as corrosion inhibitors in acidic media: correlation between inhibition efficiency and chemical structure *Corros. Sci.* **44** 2271–2289.
[10] Punckt C, Bolscher M, Rotermund H H, Mikhailov A S, Organ L, Budiansky N, Scully J R and Hudson J L 2004 Sudden onset of pitting corrosion on stainless steel as a critical phenomenon *Sciencemag* **305**(5687) 1133-1136.
[11] Song Y, Liu J, Wang H and Haibin Shu 2019 Research Progress of Nitrite Corrosion Inhibitor in Concrete *International Journal of Corrosion* **3060869**. https://doi.org/10.1155/2019/3060869.
[12] Dariva CG and Galio AF 2014 Corrosion Inhibitors – Principles, Mechanisms and Applications, in Developments in corrosion inhibitors, IntechOpen 365-379. http://dx.doi.org/10.5772/57255.
[13] Chen Y, Yang W 2019 Formulation of Corrosion Inhibitors IntechOpen. http://dx.doi.org/10.5772/intechopen.88533.
[14] BryckiIwon B, Kowalczyk IH, Szulc A, Kaczerewska O and Pakiet M 2019 Organic Corrosion Inhibitors, in Corrosion Inhibitors, Principles and Recent Applications, IntechOpen. http://doi.org/10.5772/intechopen.72943.
[15] Ahmed SK, Ali WB and Khadom AA 2019 Synthesis and investigations of heterocyclic compounds as corrosion inhibitors for mild steel in hydrochloric acid *Int. J. Ind. Chem.* **10**:159–173.
[16] Popoola LT 2019 Organic green corrosion inhibitors (OGCIs): A critical review *Corros. Rev.* **37(2)**:71–102.

IOP Conf. Series: Materials Science and Engineering 1036 (2021) 012075 doi:10.1088/1757-899X/1036/1/012075

A review on the mechanical strength of aluminium - clay composites

N E Udoye[1*], O S I Fayomi[1,2] and A O Inegbenebor[1]

[1]Department of Mechanical Engineering, College of Engineering, Covenant University Ota, Lagos state, Nigeria
[2]Department of Chemical, Metallurgical and Materials Engineering, Tshwane University of Technology, P.M.B, X680, Pretoria, South Africa.
E-mail: nduka.udoye@covenantuniversity.edu.ng, Tel.: +2348035278117

Abstract. The innovation in the usage of cheaper materials with better quality has been in the mind of many scientists. Recently, more requests have been on the utilization of agro-industrial product to produce materials to curb ecological hazards. Clay could be an alternative to SiC and Al_2O_3 due to their cheaper rate and availability. Reinforcements can be integrated into the melt using cheap and globally recognized stir casting method. Low - cost eco-friendly waste products that are easily accessible like clay, nanoclay, kaolinite, mica, and carbon are utilized as reinforcement in aluminium alloy. This review paper studied the various compositions of aluminium alloy composites for manufacturing purposes and its usage in automobile, aircraft, marine and the transportation industry. The study revealed that the addition of clay as reinforcement increases the properties of the composites.

1. Introduction

The development of cheaper materials, particularly aluminium matrix composite with improved structure, has been the desire of many scientists [1-3]. The examples of cheaper materials include aluminium clay, kaolin groups, smectite groups, illite group, and ceramics. Ceramics is defined as naturally derived materials of inorganic crystalline origin obtained from kaolins after thorough heat action. The existence of mineral impurities in the form of crystalline lattice of the kaolinite gives the natural colour of kaolins [4]. The minerals from clay are needed in virtually all natural ceramic minerals that have their primary application in engineering. Structural kaolin composing of ceramics are gorgeous and irresistible products with robust mechanical properties, low density, excellent corrosion, and high temperature resistance. The five main minerals found in clay are kaolin groups such as kaolinite, dickite, halloysite, and nacrite. Smectite groups in form of dioctahedral smectites such as montmorillonite, nontronite, beidellite, and trioctahedral smectites, illite group which includes the clay-micas, Chlorite group and 2:1 clay types in form of sepiolite or attapulgite [5].
Kaolinite, also called aluminium silicate with chemical formula $Al_2Si_2O_5(OH)_4$ is the most commonly found mineral in clay. It has a low swell capacity often orange or red colour containing iron oxide abundantly located in the soil. The preparation of clay involves material mixing, removing of big gravels, roots, pulverizing, and final product recovery. Nigerian clays are recognized and identified as kaolin but scientific analyses and its commercial use are not usually considered. However, clay is strong, low - cost, available and environmentally friendly [6]. Clay is a good substitute in replacing high cost of reinforcement because of its accessibility and its major elements in the form of Al_2O_3, SiO_2, Fe_2O_3, TiO_2 and Na_2O. Clay could also be an alternative to SiC and Al_2O_3 due to their cheaper rate and hardness [7]. Figure 1 shows the particulate sample of kaolinite used for the reinforcement of aluminium alloy.

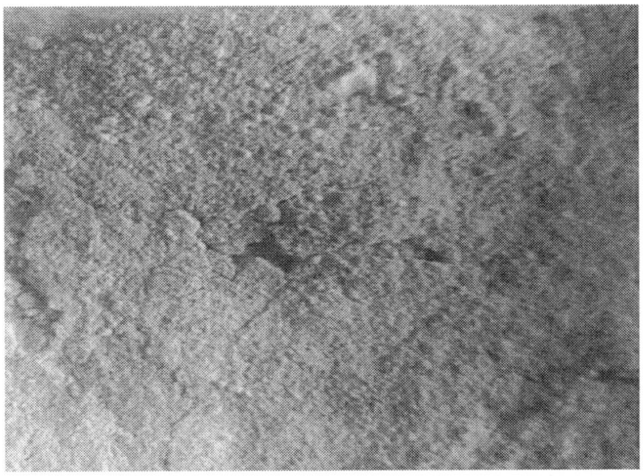

Figure 1. Kaolinite particulate

2. Microstructure

The vital point considered in the building of composites is the determination of the homogenous distribution of reinforcements in the metal matrix alloy. It is better to protect the accumulation of particles within the matrix during the period of solidification [8-10]. [9] examined the morphology of AA6063/clay particles composites. The authors observed that the addition of the clay particulate is noticed to improve the development of pure grains. The SEM/EDX pictogram revealed the existence of aluminium, silicon, and magnesium in the as - cast sample of AA6063 during the morphological examination. The hybrid reinforcements of a multi-wall carbon nanotube (MWCNT) and nanoclay were uniformly distributed in aluminium alloy 5083 (AA5083) [11-12].

[13] studied the surface morphology of aluminium 7075/mica and kaolinite composites. The authors revealed that mica and kaolinite particulates are equally distributed in the aluminium matrix. This was possible due to the proper stirring and utilization of suitable process parameters. The results show that uniform reinforcement dispersal into the matrix was realized and there is no proof of agglomerate found. The groups deprive of reinforcements were not noticed in this study. The composite material was developed without the presence of defect like voids and porosities [13]. [14] studied the microstructure of chilled aluminium alloy/ kaolinite/ carbon composites. The homogenous dispersal of kaolinite and graphite carbon enhanced matrix structure and enriched mechanical characteristics of the chilled aluminium matrix composites when compared to unreinforced matrix alloy.

2.1. Mechanical behavior

2.1.1. Hardness

The reinforcements of the metal matrix improved the hardness of the composites. According to [10], the increase in percentage of clay particle results to increment in the hardness values up to 76.7 HV at 15 wt%. [12] reported that increasing the reinforcement of aluminium 5083 will result from increasing in hardness value of the developed composites. [4] stated that increasing the lower density polyethylene and raw/calcined kaolin composite materials improve the hardness of the materials. [13] observed that the hardness value marked improvement as the increase in the amount of mica and kaolinite. The surface hardness of the aluminum 7075 is maximum at 6% of mica and kaolinite composite [14] also confirmed that reinforcement increment improves the hardness of the developed composites.

2.1.2. *Tensile behaviour*

[10] stated that increasing the clay particulate increases the ultimate tensile strength of the composite samples. Further increment in clay particulate to 30 weight percentage of clay resulted in declining in the ultimate tensile strength to 122.29 MPa. [4] investigated that the tensile strength of low density polyethylene (LDPE) at different reinforcement decreased with increased wt% of kaolin. [11] investigated the mechanical characteristics of multi-wall carbon nanotube/nanoclay reinforced aluminium alloy 5083 Matrix Composite. The outcomes revealed that the tensile strength of composites is superior to the unreinforced Al-alloy. [12] stated that the developed composites could be utilized in a structural application. [14] stated that UTS of the developed matrix improved as the kaolinite (Al_2SiO_5) content is increased.

3. Classification and applications

3.1. *AA 6063/clay composites*

The aluminium 6063 reinforced with clay has superphysical and mechanical characteristics than the pure aluminium alloy. The outcomes of the mechanical and tribological examinations showed perfection in the composites with 10–25 weight percentages of clays. Increasing the percentage of clay particles will improve the hardness of the matrix composites. The inclusion of clay particulate enriched the development of pure grains in comparison to the as-cast AA6063. The ductility of the aluminium 6063 affected the quantity of clay particles addition. Hence, these composites are good products utilized in automobile brake pad applications components. [4] concluded that the inclusion of kaolin as reinforcement to low density polyethylene as the matrix enhanced the material structure.

3.2. *AA7075/ mica and kaolinite hybrid composites*

The hardness and impact strength show noticeable enhancement as the amount of mica and kaolinite increased in the aluminium matrix. The authors observed that mica and kaolinite particulates are homogeneously distributed in the aluminium matrix. These Hybrid developed aluminium matrix can be used in finding applications in automobile, aircraft, space, marine, and the transportation applications.

3.3. *Aluminium alloy/kaolinite/C composites*

The properties of chilled metal matrix composites are higher to unreinforced matrix alloy. To increase the kaolinite (Al_2SiO_5) content, result into increase in mechanical characteristics of the developed matrix alloy. The homogenous dispersal of kaolinite and graphitic carbon enhanced mechanical characteristics of MMCs. It was observed that the mechanical properties of metal matrix composites have improved. Therefore they are applied mostly in heavy-duty vehicles and high wear resistance.

4. Conclusion

Clay can be used as a prospective reinforcement for making aluminium matrix composite. The aluminium alloy has been strengthened with various agro-industrial waste products materials such as nanoclay, kaolinite, mica, graphite carbon etc. The majority of the authors used stir casting techniques to manufacture composite materials. In conclusion, clay and kaolinite are the excellent reinforcing agent, and more work need to be done on its usage as reinforcement material. The review also revealed that aluminium matrix composites can be used in numerous industrial applications and the automotive industry. However, Nigerian clays are recognized and identified as kaolin, but scientific analyses and its commercial use are not usually considered. A few studies have been reported on aluminium matrix composites strengthened with kaolinite because both are highly tough and elastic.

Acknowledgement

The authors will like to acknowledge the support of Covenant University for the publication fund.

References

[1] Fayomi O S I, Joseph O O, Akande I G, Ohiri C K, Enechi K O and Udoye N E 2019 Effect of CCBP doping on the multifunctional Al-0.5 Mg-15CCBP superalloy using liquid metallurgy process for advanced application *J. of Alloys and Compds* **783** p 246-55.

[2] Alkahtani S 2012. Mechanical performance of heat treated 319 alloys as a function of alloying and aging parameters *Material Des* **41** 358-69.

[3] Ibrahim M F, Samuel A. Al-Ahmari M and Samuel F. H 2011 Metallurgical parameters controlling the microstructure and hardness of Al/Si/Cu/Mg base alloys *Mater. Des.* 32 (**4**). 2130-42.

[4] Mallik A Barik A K and Pal B 2015 Comparative studies on physico-mechanical properties of composite materials of low density polyethylene and raw/calcined kaolin *J. of Asian Ceramic Societies* **3** 212–16.

[5] Kakali G T, Perraki S Tsivilis E. and Badogiannis 2001. Thermal Treatment of Kaolin: The Effect of Mineralogy on the Pozzolanic Activity *Applied Clay Sciences*, **20** 73-80.

[6] Aderiye J 2014 Characterization of the Nigerian kankara kaolinite clay particulates for automobile friction lining material development. *Chem. and Process Engineering Research* **29** 24-29.

[7] Rao R N, Das S Mondal D P and Dixit G 2009. Dry sliding wear behaviour of cast high 6 strength aluminium alloy (Al–Zn–Mg) and hard particle composites *Wear* **26** 1688–95.

[8] Fayomi O S I Popoola, A P I and Udoye, N. E 2017. Effect of alloying element on the integrity and functionality of aluminium based alloy *Intech Open access* **13** 244-60.

[9] Ravi B Naik, B and Prakash J U (2015). Characterization of aluminium matrix composites ($AA6061/B_4C$ fabricated by stir casting technique. *Mat. Today proceedings* **2** 2984-90

[10] Agbeleye A A, Esezobor D E, Balogun S A, Agunsoye J O, Solis A and Neville (2017), Tribological properties of aluminium-clay composites for brake disc rotor applications *J. of King Saud University – Science* 1-8.

[11] Ratna Kumar, D S R. Smart S. and Alexis J 2018. Evaluation of Mechanical Properties of MWCNT /Nanoclay Reinforced Aluminium alloy Metal Matrix Composite. IOP Conference Series: Engineering 346.

[12] Rama P S S, Kumar D S R and Smart S J, Corrosion behaviour of aluminium metal matrix reinforced with multi-walled carbon nanotube. *J. of Asian Ceramic Societies* **5** 71-5.

[13] Ranganatha S R, Kishore M and Shantharaja 2017 Investigation of mechanical behaviour of AA7075 reinforced with mica and kaolinite based hybrid composites. *Int'l. Conf.on current Trends in Engineering Science and Technology*, 1-35.

[14] Ahamed S,. Abdul B D and Hemanth J, 2014. Wear behaviour of chilled aluminium alloy-kaolinite /carbon hybrid metal matrix composites Intl. J. of Emerging Technology and Adv. Engineering 4(7) 609-16.

Correlating the thermomechanical indexes of concrete modified with anacardium occidentale nutshell ash using linear model polynomial analysis

S O Oyebisi[1], U T Igba[2], A N Ede[1], J O Oluwafemi[1], G O Mark[1]

[1]Civil Engineering Department, Covenant University, Ota, Nigeria
[2]Civil Engineering Department, Federal University of Agriculture, Abeokuta, Nigeria
E-mail address: solomon.oyebisi@covenantuniversity.edu.ng

Abstract. This study correlated the thermal and mechanical properties of concrete produced with anacardium occidentale (cashew) nutshell ash (AONA). Cashew nutshells, agricultural waste products, was valorized to obtain AONA. AONA was replaced at 5, 10, 15, and 20 wt% of cement to produce concrete grade 25 MPa. Density, compressive strength, and thermal conductivity of the concrete samples were determined at 28 days curing, and the results were correlated using regression model analysis. The experimental findings revealed that the compressive strength increased with increasing AONA content. Moreover, both density and thermal conductivity reduced as AONA content increased. Besides, the correlation yielded a high precision with 97% "R^2". Thus, AONA has proved to exhibit higher mechanical strength with excellent thermal insulation when utilized as supplementary cementitious material (SCM) in concrete production. The developed model can also be applied to the correlation of thermomechanical properties of concrete incorporating SCMs in that time, energy, and cost in conducting laboratory works would be reduced.

Keywords: supplementary cementitious materials; waste management; sustainability; compressive strength; thermal conductivity; thermal insulation; building comfort; modelling.

1. Introduction

In recent years, scientific research is oriented towards energy consumption and environmental concerns threatening thermally efficient buildings. In the construction sector, efforts are in progress to conserve energy and resources by valorizing waste materials while to improve comfort conditions for thermally efficient buildings [1]. Furthermore, concerted attention has been drawn to the increasing energy costs and climate change due to energy performance and efficiency [2]. Buildings consume about 40-50% supply of energy and waste about 50-60% generation of energy [3]. Consequently, the demand for building comfort conditions has risen. However, more attention is drawn to the aesthetic values of buildings in Nigeria with little or no consideration for the environment, energy efficiency, health, and economy [4]. Thus, previous studies [2, 5] recommended a careful selection of building materials that would improve thermal comfort in buildings, especially in hot or arid climatic condition.

Structural concrete members, beams, slabs, and columns, are the backbone of any structure/building, hence playing the significant role in creating and maintaining buildings' internal environment, apart from block walling components. Therefore, the internal temperature of a building can be maintained at a lower cost when a building is passive (structural members exhibiting thermal insulating characteristics) [6].

A review of previous studies [7-9] revealed that concrete modified with SCMs, blast furnace slag (BFS), silica fume (SF), fly ash (FA), and corncob ash (CCA), were promising thermal insulating materials. However, the study on thermal characteristics of AONA in concrete production is still limited. Anacardium occidentale (cashew) nutshell is an agricultural waste product with about 4.2 million metric tonnes (MMT) of global generation in 2017; from this global generation, about 100 MMT was estimated in Nigeria [10]. Notwithstanding, some of these materials are disposed of as wastes, thus creating environmental concerns.

Therefore, this study filled a gap by recycling the anacardium occidentale nutshells and applying it as SCM for concrete production. Its effects on thermomechanical properties were examined, and the results were correlated using the regression model analysis in Matlab 2017a. In achieving these objectives, concrete grade 25 MPa was selected as mix design proportion due to its widespread uses in the construction sector.

2. Materials and Methods

2.1. Materials

Cashew nutshells were obtained from the Federal University of Agriculture, Abeokuta, Nigeria. After the pyrolysis, about 21 wt% of cashew nutshells was obtained as AONA. The specific gravity (SG) and fineness of AONA and Portland limestone cement (PLC) were obtained by the procedure outlined in *British Standard, BS EN* [11]. The results revealed the SG of 2.80 and 3.15 g/cm^3, and fineness of 8.10 and 7.60% for AONA and PLC, respectively; these results indicated that more volume of AONA and water would be needed when AONA is replaced by wt% of PLC for concrete production due to lower SG and higher fineness [12]. The results of XRF analysis for both AONA and PLC are shown in Table 1.

Table 1. Oxide compositions of the binding materials used

Oxide Composition	CaO (%)	SiO_2 (%)	Al_2O_3 (%)	Fe_2O_3 (%)	MgO (%)	K_2O (%)	Na_2O (%)	SO_3 (%)	TiO_2 (%)	P_2O_5 (%)	LOI (%)
AONA	2.04	65.02	16.28	10.16	1.53	0.52	0.45	1.32	0.01	0.01	2.95
PLC	64.50	21.55	5.50	3.08	1.52	0.61	-	2.03	-	-	1.20

LOI = Loss of Ignition, determined at 800 °C.

Sharp sand (≤ 4.5 mm) and granite (12.5 mm) were used as both fine and coarse aggregates and obtained from an aggregate dealer in Ota, Nigeria. The physical properties were examined based on *BS EN* [13]'s procedure. The results indicated 2.60 and 2.64 g/cm^3 as specific gravity, 0.7 and 0.8 % as water absorption, and 0.3 and 0.2% as moisture content for both fine and coarse aggregates, respectively.

2.2. Experimental methods and tests

The mix was designed in consonance with the procedure stated in *BS EN* [14]. For the mix, the quantities of binding material (AONA and PLC), fine aggregate, and coarse aggregate were 340, 715, 1035 kg/m^3, respectively, hence indicating the water to binder ratio as 0.62. AONA was replaced at 0, 5, 10, 15, and 20 wt% of PLC, denoting as T0, T1, T2, T3, and T4, respectively.

The mix constituents were prepared in line with *BS EN* [15]'s procedure. The concrete samples were tested at 28 days. Both density and compressive strength (CS) were determined by *BS EN* [16] and *BS EN* [17] methods, respectively. The average temperature and relative humidity (RH) during the test were 28 °C and 65%, respectively. For each mix ID, the mean of three (3) tests was used for the analysis.

The thermal conductivity (TC) test was performed at 28 days curing of samples based on *BS EN* [18]'s procedure. The samples, panels 300 × 300 × 55 mm^3, were oven-dried at 105 °C for 24 h to remove any moisture content. For the induction of interior and exterior temperatures, the samples were placed between the hot plate and cold plate at 40 °C and 18 °C temperatures, respectively. At every 10 min, both hot and cold plates' temperatures were recorded for 24 h. The simplified condition of one-dimensional heat transfer was adopted by this study. Therefore, the concrete TC was obtained using Fourier's law, as indicated in Equation (1).

$$k = \frac{\Phi \times t}{A \times \Delta T} \quad (1)$$

where k = *TC (W/mK)*
Φ = *heat flow (J/s)*
t = *thickness of specimen (m)*
A = *area of the specimen (m^2)*
ΔT = *temperature difference between hot and cold plates (°C).*

The density, CS, and TC results were correlated using statistical software, Matlab 2017a, setting TC as dependent variable (predictor), and both density and CS as continuous (independent) variables.

3. Results and Discussions

3.1. Chemical compositions of Binders

The oxide compositions, as presented in Table 1, indicated that AONA met 70% minimum requirements of *BS EN* [19] for $SiO_2 + Al_2O_3 + Fe_2O_3$, and 5% maximum specification for LOI. Therefore, the material is desirable for use as SCM.

3.2. Concrete density

The concrete density, as shown in Figure 1, decreased as the AONA content in the mix increased. This may be attributed to the lower SG of AONA, which in turn reduced the interfacial particle of PLC, thus decreasing the concrete density [12]. These results also agreed with the findings from previous studies [20-22] that density decreases with increasing SCMs in the mix. Therefore, the incorporation of AONA in concrete production reduces the dead load and contributes to buildings' lightweight.

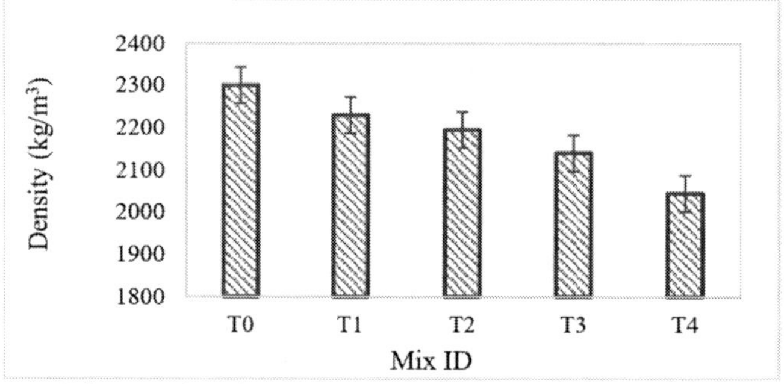

T0 (100% PLC); T1 (95% PLC + 5% AONA); T2 (90% PLC + 10% AONA); T3 (85% PLC + 15% AONA); and T4 (80% PLC + 20% AONA)

Figure 1. Concrete density at 28 days curing

3.3. Compressive strength

The compressive strength (fc) results, as shown in Figure 2, at 28 days curing showed that the fc increased as AONA content increased. The results further revealed that AONA replacement at 15 wt% of PLC exhibited the highest CS compared with other replacement proportions. The increase in CS with increasing AONA could be associated with the higher contents of silica and alumina oxides in AONA compared with PLC; these oxides reacted with hydrating agents of PLC, resulting in the strengthening paste and bonding improvement in the concrete [12]. However, the strengthening improvement was optimum at 15 wt% AONA compared with 5, 10, and 20 wt% AONA. Ultimately, a 5-20 wt% of AONA could be incorporated for concrete structural applications because they fulfilled the concrete target strength of 25 MPa stipulated by *BS EN* [23].

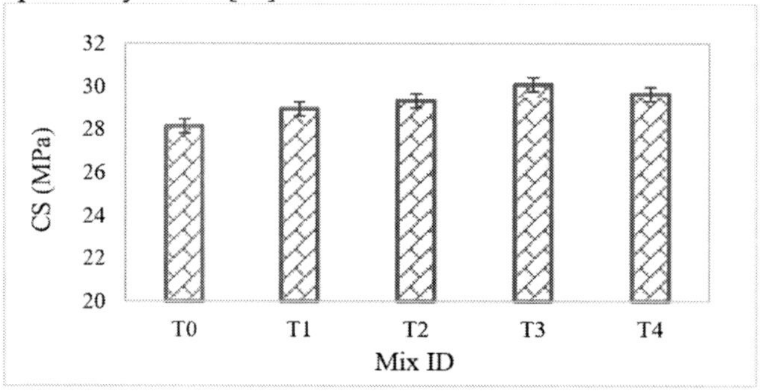

Figure 2. Compressive strengths

3.4. Thermal conductivity (TC)

As shown in Figure 3, the TC reduced as AONA content in the mix increased due to the decrease in density with increasing AONA content. Moreover, less air would be entrapped as AONA content in the blend increased due to its higher fineness compared with PLC, thus reducing the rate of heat transfer through the concrete and subsequently reducing the TC [12]. This result signified that the recycling and incorporation of AONA in concrete production reduce the environmental challenges, improve the thermal insulation, and enhance the building comfort.

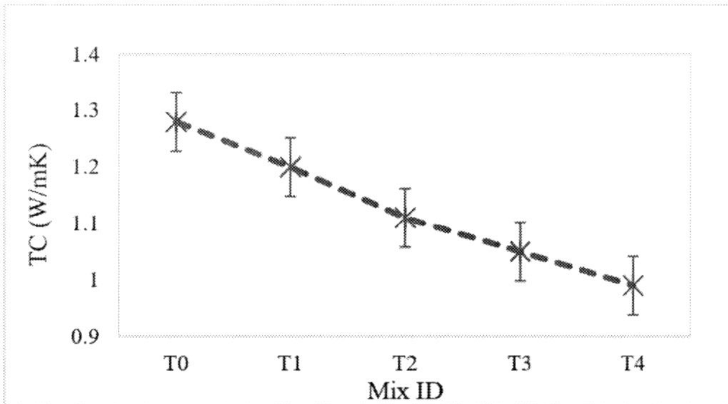

Figure 3. Thermal conductivity of the concrete produced

3.5. Correlation between CS, density, and TC

Following the linear model polynomial of regression analysis, the relationship between TC, density, and CS, as indicated in Figure 4, yielded a strong correlation with 97% coefficient of determination (R^2) at 95% confidence bounds to predict all variables. Moreover, the sum of squared errors (SSE) and root mean squared errors (RMSE) signified that the data fit the regression line because the values were between 0 and 1 [24]. Therefore, the relationship is illustrated in Equation (2), and the model can be used to predict future data trends and enhance data quality on the thermal insulation values in relation with the weight and mechanical cost of concrete modifying with AONA.

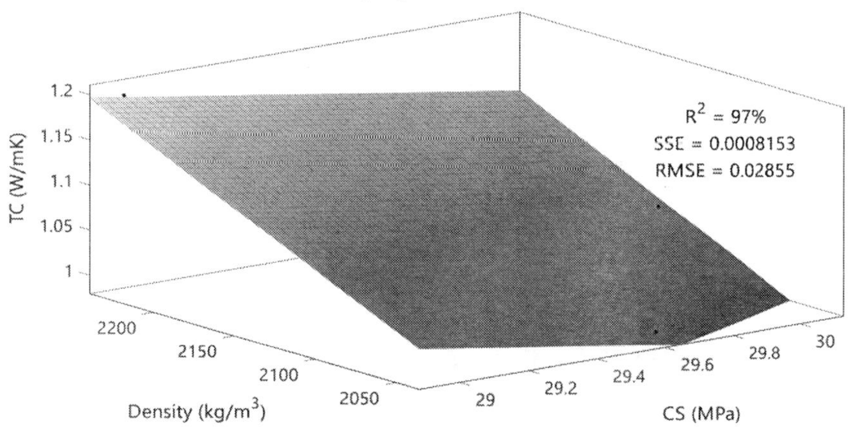

Figure 4. Correlation of CS with density and TC at 28 days curing

$$\lambda = 1.008 - 0.0588fc + 0.0008423\rho \qquad (2)$$

where λ *is the TC (W/mK),*
fc is the CS (MPa),
ρ is the concrete density (kg/m^3).

3.6. Validation of the developed model with previous studies

Following the regression model illustrated in Equation (2), Figure 5 presents the relationship between the proposed model equation and the previous studies at 28 days curing. The results, as indicated in Figure 5, showed that both Experimental TC (ETC) and predictive TC (PTC) exhibited a good precision, thus yielding a good relationship with *Gomes et al.* [25], *Demirboga* [7], and TC obtained herein at 52, 87, and 97% R^2, respectively. Therefore, the developed model illustrated in Equation (2) can be applied to the prediction of TC related to concrete strength and weight incorporating SCMs in that *Gomes et al* [25] and *Demirboga* [7] utilized SCMs, fly ash, silica fume, metakaolin, and blast furnace slag, as replacements of Portland cement for concrete production with high and adequate strengths at 28 days curing.

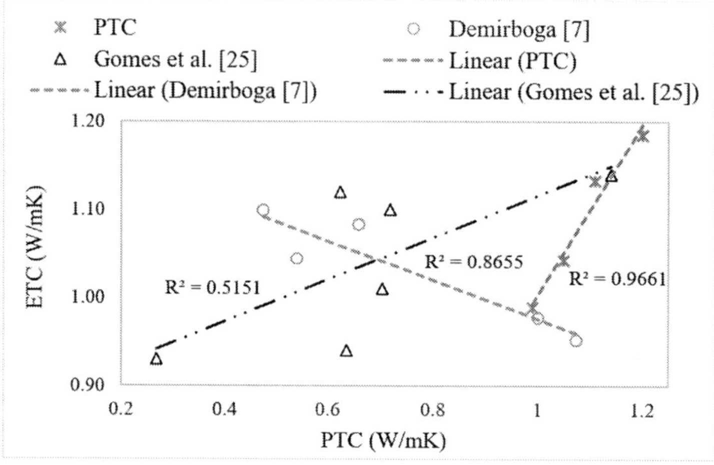

Figure 5: Validation of the proposed model with previous studies

4. Conclusions

This study correlated the concrete TC with CS and density, and the results obtained are promising:

i. At 28 days curing, there was about a 6% decrease in TC at every 5 wt% increase of AONA in the mix. Moreover, about 6-23% decrease in TC was obtained as AONA increased from 5-20% compared with the control concrete.
ii. There was nearly a 1.5% increase in CS at every 5 wt% increase in AONA content at 28 days curing, but maximum CS occurred at 15 wt% of AONA's substitution.
iii. The concrete becomes lighter with increasing AONA content in the mix. There was about 3-4% decrease in concrete density at every 5 wt% of AONA replacement in the combination.
iv. A strong correlation exists between the concrete TC, and CS and density with 97% R^2 at 28 days curing.

Acknowledgement

The authors acknowledge the Covenant University Centre for Research, Innovation and Discovery (CUCRID) for the financial support and conducive environment to conduct the research.

References

[1] Muhammad A, Habib F, Ali M and Sheik N 2020 J. *Build. Eng.* **31** 101411.
[2] Oyedepo S O 2012 *Int. J. Energy Eng.* **2**(3) 86-98.
[3] Lo T Y and Choi K T 2004 *Struct. Surv.* **22** 5.
[4] Nwofe P A 2014 *Int. J. Energy Environ. Res.* **2**(3) 1-9.
[5] Malgwi M E and Musa L S 2014 *Int. J. Energy Environ. Res.* **2**(1) 55-60.
[6] Agostino D, Rossi F, Marigliano M, Marino C and Minichiello F 2019 *J. Build. Eng.* **30** 100743.

[7] Demirboga R 2007 *Build. Environ.* **42** 2467-71.
[8] Raheem A A and Adesanya D A 2011 *Pacific J. Sci. Technol.* **12**(2) 106-111.
[9] Mark O G, Ede A N, Olofinnade O, Bamigboye G O, Okeke C, Oyebisi S O and Arum C 2019 *IOP Conf. Series: Mater. Sci. Eng.* **640** 012071.
[10] Food and Agriculture Organization of the United Nations 2017 Food and Agriculture Organization Statistical Pocketbook World Food and Agriculture http://www.fao.org/3/a-i4691e.pdf.
[11] British Standard EN 196- 3 2016 *Method of Testing Cement: Physical Test* BSI London.
[12] Khan S U, Nuruddin M F, Ayub T and Shafiq N 2014 *Sci. World J.* **986567** 1-11.
[13] British Standard EN 12620 2013 *Aggregates from Natural Sources for Concrete* BSI London.
[14] British Standard EN 206 2016 *Concrete Specifications, Performance, Production and Conformity* BSI London.
[15] British Standard EN 12390-2 2009 *Testing Hardened Concrete: Making and Curing Specimens for Strength Tests* BSI London.
[16] British Standard EN 12390- 7 2019 *Testing Hardened Concrete: Density of Hardened Concrete* BSI London.
[17] British Standard EN 12390 -3 2009 *Testing Hardened Concrete: Compressive Strength of Test Specimens* BSI London.
[18] British Standard EN 12664 2001 *Thermal Performance of Building Materials and Products, In Determination of Thermal Resistance using Guarded Hot Plate and Heat Flow Meter Methods, Dry and Moist Products of Medium and Low Thermal Resistance* BSI London.
[19] British Standard EN 450-1 2012 *Pozzolan for Use in Concrete: Definitions, Specifications and Conformity Criteria* BSI London.
[20] Oyebisi S, Igba T and Oniyide D 2019 *Case Studies Constr. Mater.* **11** e00293.
[21] Oyebisi S, Ede A, Olutoge F and Omole D 2020 *Constr. Build. Mater.* **256** 119390.
[22] Oyebisi S, Igba T, Raheem A and Olutoge F 2020 *Case Studies Constr. Mater.* **13** e00393.
[23] British Standard EN 1992-1-1 2014 *Design of Concrete Structures: General Rules for Structural Fire Design* BSI London.
[24] Kaw A and Kalu W 2013 *Adequacy of Regression Models* http://numericalmethods.eng.usf.edu.
[25] Gomes M G, Bogas J A, Real S, Ferrer B and Rodrigues A M 2016 *Int. J. Civ. Struct. Eng.* **3**(1) 262-266.

IOP Conf. Series: Materials Science and Engineering 1036 (2021) 012077 doi:10.1088/1757-899X/1036/1/012077

Evaluation of Fire-Retardant Properties of Emulsion, Text-Coat and Gloss Paints Modified with Bio-Based Extract of Acalypha Wilkesiana

[1]O Ofuyatan, [2]T Gbolade, [3]A Adeniyi, [3]J Ighalo, [1]J Oluwafemi
[1]Department of Civil Engineering, Covenant University
[2]Department of Chemical Engineering, Covenant University
[3]Department of Chemical Engineering, Faculty of Engineering and Technology, University of Ilorin
Coresponding Author: olatokunbo.ofuyatan@covenantuniversity.edu.ng

Abstract. The impact of fire on concrete buildings has been found to tend towards strength reduction of the concrete building. Fire-retardant paints help to reduce the impact of fire on buildings and many researches are on to improve on fire-retardant paints. To this end, this paper investigated the fire-retardant properties of the bio-based extract of Acalypha wilkesiana in emulsion, text-coat and gloss paints. Acalypha leaves were subject to extraction in water and kerosene for three days. Emulsion, text-coat and gloss paints were produced. The fire point of the paint samples with a varying mass of extract was obtained to compare the flammability and obtain the optimum mass of extract needed for the best performance. The cube coated with 2 layers of text-coat paint of 0.45 g yielded the highest compressive strength. The best performing type of paint was text-coat paint. The coarse sand added during the production of text-coat paint assists in protecting the substrate material. Emulsion paint was the next best performing paint type while gloss paint was not advisable to be used for this purpose. The introduction of Acalypha wilkesiana as an additive had the fire point of the sample increased alongside the ignition time and the optimum mass of Acalypha wilkesiana extract that yielded the best fire-retardation of the paint was found to be 0.45g.

1. Introduction

As indicated in authentic historical sources, the Chinese and the Egyptians used vinegar and alum as their fire-retardant agents which were initially known as combustion limiting agents for wood and natural polymers [1]. The Chinese applied a mixture of the alum and vinegar to wood and afterwards, an upper layer of clay would be added [2]. This reduced and delayed the rate of fire spread. About 3000 years ago, in Egypt, reed and grass which have previously been soaked in seawater would be used for roofing purposes [3]. Dried mineral salts which have crystallized are also used as fire-retardants [4].
Generally, the description of fire-retardants can be backdated to ancient Roman times by Aulus Gellius [5]. Wood was observed by Sulla in 86 BC to withstand fire during the siege in Piraeus. This was because the wood had previously been soaked in alum [6]. In 1821, the first attempt to highlight the scientific principles of fire retardation was made by Gay-Lussac, who proposed several composition recipes for making fire-retardants for cellulose materials from ammonium phosphate and borax. These recipes are still in use to date [6].
In general, fire-retardants are applied as follows: 65% for plastics, 25% for rubber (including aluminium trihydrate in carpet underlay), 5% for textiles, 3% for coatings/adhesives and only 2% for wood and paper [7]. However, over the years' fire-retardants have been used majorly on buildings and construction. Fire-retardant materials are important due to the increase in fire losses over the years [8].

Fire-retardants have become commonly used in buildings to reduce the risks associated with fire outbreaks [9]. The suitability of these materials can be assessed by carrying out standard tests on each sample by exposing it to a single fire scenario which helps decide the end-use of the material [10]. These tests have been useful in the past but new methods are being developed to improve the quality of the materials. The new methods require the use of mathematical models of ignition and fire spread to fully understand how a particular material will perform under several fire scenarios [11], [12].
Due to the consequent increase in operating pressure and temperature of several processes, more efficient fire retardation properties are required. Environmentalists have pressurized the European market into looking for diphenyl oxide-free fire-retardant systems. Thus, a new generation of multipurpose environmentally-friendly (bio-based) brominated fire-retardants has been recently introduced which offers additional benefits. The purpose of the new methods developed is to create a new generation of fire-retardant materials based on a deeper understanding of the combustion process to produce materials to meet increasingly inflexible requirements that will demand improved fire safety in the future [13].
In the last decade, technology has drifted from halogen-based fire-retardants and towards those based on phosphorus, silicon, metal hydrates, other metal salts and especially bio-friendly materials. This trend is likely to continue as long as halogenated fire-retardants still pose a hazard to the environment. This explains why a lot of research is being carried out on alternatives to halogenated fire-retardants.
Based on the authors' exhaustive search, plant extract Acalypha wilkesiana has not been used as fire-retardants in paints. The aim of this study was to investigate the fire-retardant properties of the bio-based extract of Acalypha wilkesiana in emulsion, text-coat and gloss paints.

2. Methodology

2.1. Raw Materials and Reagents

The raw materials used in this research were leaves of Acalypha wilkenasia (shown in Fig. 1), cement, gravel, cube moulds, water, kerosene and the reagents employed are emulsion (biocide, titanium dioxide, calcium carbonate, polyvinyl acrylic), text-coat (titanium dioxide, Polyvinyl Acrylic (PVA), nitrosol, biocide, pigment, calcium carbonate), and gloss (alkyd resins, pigment and drier).

Fig. 1. *Acalypha wilkesiana* leaves

2.2. Equipment and production

The equipment used to carry out different tests is cup flash-point tester, drying oven and mass balance. The general production process of paints is as represented in Fig. 2.

2.2.1. Extraction with Water

IOP Conf. Series: Materials Science and Engineering 1036 (2021) 012077 doi:10.1088/1757-899X/1036/1/012077

The steps used for the extraction with water are discussed in this section. The Acalypha plant samples were collected and air-dried for seven days. The dried plants were grounded into powdered form with a mortar and pestle. 100 g of the powdered leaf was weighed into four 250 ml conical flasks containing 250 ml of distilled water. The mixture was then covered and stirred every 24 h using a glass sterile glass rod for 3 days. The mixture was then filtered through filter paper. The resulting yellowish-green filtrates were then concentrated and refrigerated at 40°C before testing.

2.2.2. Extraction with Kerosene

The steps used for the extraction with water are discussed in this section. The Acalypha plant samples were collected and air-dried for seven days. The dried plants were grounded into powdered form with a mortar and pestle. 100 g of the powdered leaf was weighed into four 250 ml conical flasks containing 250 ml of distilled water. The mixture was then covered and stirred every 24 h using a glass sterile glass rod for 3 days. The mixture was then filtered through filter paper. The resulting yellowish-green filtrates were then concentrated and refrigerated at 40°C before testing.

2.2.3. Procedure for the paint production

i. For the production of the emulsion paint, 2L of water, titanium dioxide was added with calgon and mixed for 15 minutes in a large container. Calcium carbonate was added and stirred properly for about 15 minutes to prevent the formation of lumps in the pigments. The mix properties for the run one are shown in Table 1.

RUN ONE: EMULSION PAINT PRODUCTION

i. The chemicals were measured accurately on a weighing scale (for powdery chemicals) and graduated containers (for liquids).

ii. 2L water was poured into an empty container. Titanium dioxide was added and tinge with calgon.

iii. The pigment powders were added into the container and mixed for about 10-15mins.

iv. The required amount of Acrylic was added to the mixture and stirred well. Nitrosol that had been mixed with water was added and stirred rigorously and uniformly for 5-10mins.

v. Formalin was then be added and the solution was completely stirred

vi. The plant extract of the Acalypha species was then added in varying masses between 0.0-`3.0g.

vii. The solution of emulsion paint produced was then applied on the various substrate surfaces.

Table 1. Mix properties of paint for run one

INGREDIENTS	1	2	3	4	5	6	7	8
Water (L)	2L	2L	2L	2L	2L	2L	2L	2L
Calgon (Kg)	0.01kg	0.01kg	0.01kg	0.01kg	0.01kg	0.01kg	0.01kg	0.01kg
Calcium carbonate (Kg)	17kg	17kg	17kg	17kg	17kg	17kg	17kg	17kg
Polyvinyl Acrylic (P.V.A) (Kg)	1.20kg	1.20kg	1.20kg	1.20kg	1.20kg	1.20kg	1.20kg	1.20kg
Nitrosol (Kg)	0.13kg	0.13kg	0.13kg	0.13kg	0.13kg	0.13kg	0.13kg	0.13kg
Formalin (cL)	5cL	5cL	5cL	5cL	5cL	5cL	5cL	5cL
Pigment (Red Oxide) (Kg)	0.25kg	0.25kg	0.25kg	0.25kg	0.25kg	0.25kg	0.25kg	0.25kg
Acalypha Extract (g)	0.00g	0.30g	0.45g	0.60g	1.00g	1.25g	1.50g	3.00g

IOP Conf. Series: Materials Science and Engineering 1036 (2021) 012077 doi:10.1088/1757-899X/1036/1/012077

The mix properties of paint for the run two is related in the table 2.

RUN TWO: TEXTCOAT PAINT PRODUCTION

i. The chemicals were measured accurately on a weighing scale (for powdery chemicals) and graduated containers (for liquids).

ii. 2L water was poured into an empty container. Titanium dioxide was added and tinge with calgon.

iii. The pigment powders were added into the container and mixed for about 10-15mins.

iv. Calcium carbonate was added and stirred properly for about 10-15 mins to prevent the formation of chumps in the pigments.

v. The required amount of Acrylic was added to the mixture and stirred well. Nitrosol which ` had been mixed with water was added and stirred rigorously and uniformly for 5-10mins.

vi. Formalin was then be added and the solution was completely stirred

vii. Marble dust was then added to the mixture and stirred thoroughly.

viii. The plant extract of the Acalypha species was then added in varying masses between 0.0-` 3.0g.

ix. The solution of textcoat paint was then applied on the various substrate surfaces

Table 2. Mix properties of paint for run two

INGREDIENTS	1	2	3	4	5	6	7	8
Water	2L	2L	2L	2L	2L	2L	2L	2L
Calgon	0.01kg	0.01kg	0.01kg	0.01kg	0.01kg	0.01kg	0.01kg	0.01kg
Calcium carbonate	17kg	17kg	17kg	17kg	17kg	17kg	17kg	17kg
Polyvinyl Acrylic (P.V.A)	1.20kg	1.20kg	1.20kg	1.20kg	1.20kg	1.20kg	1.20kg	1.20kg
Nitrosol	0.13kg	0.13kg	0.13kg	0.13kg	0.13kg	0.13kg	0.13kg	0.13kg
Formalin	5cL	5cL	5cL	5cL	5cL	5cL	5cL	5cL
Pigment (Red Oxide)	0.25kg	0.25kg	0.25kg	0.25kg	0.25kg	0.25kg	0.25kg	0.25kg
Acalypha Extract	0.00g	0.30g	0.45g	0.60g	1.00g	1.25g	1.50g	3.00g

The mix properties of paint for the run three is related in the table 3 and the mix properties of Paint for the run four is related in the table4..

EXPERIMENT RUN THREE: GLOSS PAINT

i. The bucket was filled with kerosene and mixed with alkyd resin.

IOP Conf. Series: Materials Science and Engineering 1036 (2021) 012077 doi:10.1088/1757-899X/1036/1/012077

ii. The titanium dioxide was then mixed with the solution.

iii. The dispersant (soya lecithin) was added. The solution was then properly mixed.

iv. Pigment was then added to the perfectly mixed mixture.

v. The Acalypha plant extracted with kerosene was added in varying masses between 0.0-3.0g and the solution is completely stirred

vi. The solution of gloss paint was then applied on the various substrate surfaces

viii. Samples are allowed to dry completely for several days prior to analysis.

Table 3. Mix properties of paint for run three

INGREDIENTS	1	2	3	4	5	6	7	8
Water	2L	2L	2L	2L	2L	2L	2L	2L
Calgon	0.01kg	0.01kg	0.01kg	0.01kg	0.01kg	0.01kg	0.01kg	0.01kg
Calcium carbonate	17kg	17kg	17kg	17kg	17kg	17kg	17kg	17kg
Polyvinyl Acrylic (P.V.A)	1.20kg	1.20kg	1.20kg	1.20kg	1.20kg	1.20kg	1.20kg	1.20kg
Nitrosol	0.13kg	0.13kg	0.13kg	0.13kg	0.13kg	0.13kg	0.13kg	0.13kg
Formalin	5cL	5cL	5cL	5cL	5cL	5cL	5cL	5cL
Pigment (Red Oxide)	0.25kg	0.25kg	0.25kg	0.25kg	0.25kg	0.25kg	0.25kg	0.25kg
Acalypha Extract	0.00g	0.30g	0.45g	0.60g	1.00g	1.25g	1.50g	3.00g

The mix properties of paint for the run three is related in the table 4.

Table 4. Mix properties of paint for run four

INGREDIENTS	1	2	3	4	5	6	7	8
Alkyd Resin	400g	400g	400g	400g	400g	400g	400g	400g
Titanium Dioxide	160g	160g	160g	160g	160g	160g	160g	160g
Dispersant (Soya Licithin)	1.1kg	1.1kg	1.1kg	1.1kg	1.1kg	1.1kg	1.1kg	1.1kg
Kerosene	70g	70g	70g	70g	70g	70g	70g	70g
Pigment (Yellow Oxide)	0.25kg	0.25kg	0.25kg	0.25kg	0.25kg	0.25kg	0.25kg	0.25kg
Acalypha Extract (with kerosene)	0.0g	0.3g	0.45g	0.60g	1.00g	1.25g	1.50g	3.00g

2.3. Casting of Concrete Cubes

In the casting of concrete cubes, 1:2:4 proportioning was employed as mix-ration. Water was gradually added to the mortar while the mixture was properly mixed before it was transferred into mould dimension of 100mm × 100mm × 100mm for the casting. Slump test was carried out on the fresh concrete as shown in Fig. 2. The concretes were de-moulded after 24 h, cured in water for 3 days and then sun-dried for 4 h as shown in Fig. 3 and Fig. 4.

Fig. 2. *Slump test*

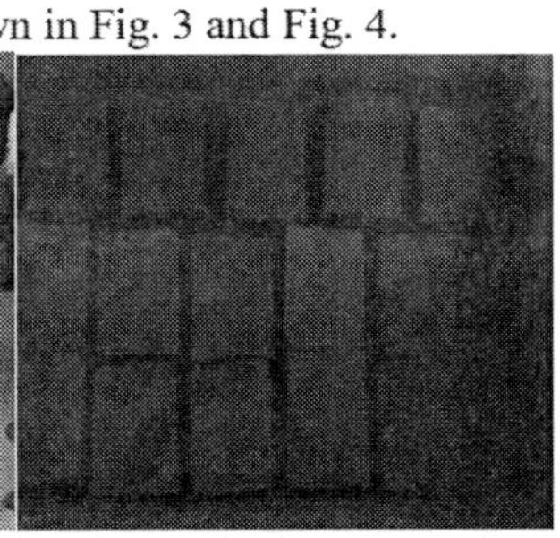

Fig. 3. *Cubes curing in water*

Fig. 4. *Sun-drying of cubes*

2.4. Application of Paint on Concrete Cubes

The application of paint on the concrete cubes was carried out in two phases. The first phase was the application of a layer of the emulsion, text-coat and gloss paint on the concrete cubes while the second phase was the application of two layers of all the three produced paint on the concrete cubes as shown in Fig. 5.

Fig. 5. Coated concrete cubes

2.5. Analysis and Testing

2.5.1. Fire Point Test

The paint sample of equal mass was placed in the cup flash point tester as shown in Fig. 6. The flashpoint tester was turned and the temperature of the sample flashed was noted for comparative analysis. The varying masses of gloss paint samples extract were filled to the mark in the cup flash point tester. The samples were analysed to test for the effectiveness of the fire-retardant properties of Acalypha wilkesiana.

Fig. 6. Paint sample at fire point

IOP Conf. Series: Materials Science and Engineering 1036 (2021) 012077 doi:10.1088/1757-899X/1036/1/012077

2.5.2. Mass Loss Analysis

The cube samples were weighed before and after heating. The initial and the final masses were recorded. This was done for all the samples.

3. Results and Discussion

3.1. Raw Materials and Reagents

Fire Point The fire point of the samples as shown in Fig. 7 increased from 480°C to 650°C and reaches its peak at 980°C for a mass of extract of 0.00 g, 0.30 g, and 0.45 g respectively. At a higher mass of extract, the fire point begins to decrease steadily from 980°C to 820°C and it decreased till it reached 500°C. This indicates that the optimum mass of extract needed for effective fire retardation is 0.45 g to achieve a fire point of 980°C. This indicates Acalypha wilkesiana possesses fire-retardant properties by increasing the fire point of the paint thus delaying ignition time. However, an increasing amount of extract in the sample beyond the optimum mass required would cause the paint to fail and reduce fire point. This may be due to the high chloride concentration in Acalypha wilkesiana leaves. From research, a threshold chloride content of atmospheric service is 46 ppm Cl (0.046 g/L of Cl) [14]. The optimum mass of plant extract in the paint that yielded the best measure of performance (fire retardation) was 0.45 g (150 mL of extract) which contains 532.5 mg of Cl.

Research has shown that there is a high surface concentration of chlorides resulting in the permeation into porous coatings [15]. Chloride in extremely high concentration penetrates concrete by diffusion and absorption of the fluid containing the chloride. Chlorides absorb moisture from the air, through the coating through osmosis and thus reduces the corrosion cell resistance of the coating. This elevates the chloride levels at the coating-substrate interface. Hence, resulting in premature coating failure and corrosion. Salts in high concentration would be uniformly distributed on the substrate surface and therefore result in uniform visual corrosion. It is known that high levels of contaminants like chloride or sodium affect the quality or duration of coating performance because they induce stress corrosion cracking [16-19]. This explains why the effectiveness of the fire-retardant coatings would reduce when the extract is added beyond the optimum required mass (0.45 g). Since the threshold limit of chloride depends on the desired protection life of the coating system and the coating thickness, it is advisable to maintain the mass of extract in the paint to 0.45 g and the coating thickness to 2 layers.

Table 5. Summarised result of fire point analysis on the gloss paint sample

Mass of extract (g)	Fire point (°C)
0.00	48
0.30	65
0.45	98
0.60	82
0.80	70
1.00	68
1.25	58
1.50	54
3.00	**50**

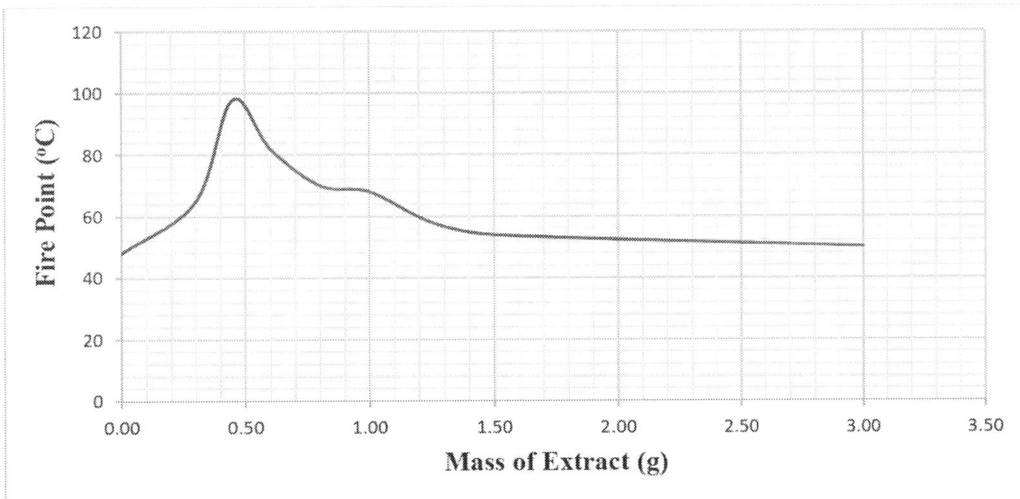

Fig 7. Fire point against the mass of extract

3.2. Mass Loss Test

The results of the mass loss test carried out on the coated concrete are shown in Table 6. The plots for the samples coated with emulsion paint, text-coat paint and gloss paint are related in Fig. 8, 9, and 10 respectively, while Fig. 11 relates the summary of the three different paints.

Table 6. Summary of residual mass analysis

Mass of Extract (g)	Residual Mass (g)											
	Text-coat				Emulsion				Gloss			
	1 Layer		2 Layers		1 Layer		2 Layers		1 Layer		2 Layers	
	250°C	350°C	250°C	350°C	250°C	350°C	250°C	350°C	250°C	350°C	250°C	350°C
Control	9.0											
0.00	8.6	8.4	8.5	8.5	8.5	8.4	8.6	8.5	8.5	8.3	8.4	8.2
0.30	8.7	8.8	8.8	8.8	8.6	8.7	8.8	8.8	8.7	8.4	8.5	8.3
0.45	8.9	8.9	8.9	8.9	8.9	8.8	8.9	8.9	8.8	8.5	8.6	8.4
0.60	8.6	8.8	8.7	8.8	8.8	8.6	8.8	8.6	8.7	8.3	8.5	8.3

A comparative analysis of the residual mass of the emulsion paints with the varying mass of extract after being subjected to 250°C and 350°C for 1 h is shown in Fig. 8. It was observed that the cubes heated at 350°C had lower residual masses than cubes heated at 250°C. From Fig. 8, it was observed that the least mass loss was obtained by the concrete cubes coated with 2 layers of emulsion coating at 250°C. The next best performing concrete cubes were those coated with 2 layers of emulsion coating at 350°C followed by the cubes coated with 1 layer at 250°C and lastly by the cubes coated with 1 layer of emulsion coating and subjected to 350°C. However, for each sample, the residual mass decreased from an initial mass of 9.0 g to the least mass obtained at 0.3 g of extract and increased to its peak at 0.45 g of extract and decreased at 0.60 g of extract. The higher the mass of extract added to the coating, the higher the residual mass of concrete cubes after heating. This shows that the emulsion paint protects the substrate from heat and mass loss and the measure of performance is dependent on the number of layers of paint used, the mass of extract in the paint and the temperature to which they are subjected.

The various residual mass of concrete cubes coated with text-coat paint containing a varying mass of extract before and after being subjected to heat is related in Fig. 9. It was observed that the best performing cubes with the least mass loss were those coated with 2 layers of text-coat paint at 250°C, followed by those coated with 1 layer of text-coat paint at 250°C, followed by those coated with 2 layers of text-coat paint and heated at 350°C. However, for each sample, the residual mass decreased from an initial mass of 9.0 g to the least mass obtained at 0.3 g of extract and increased to its peak at 0.45 g of extract and decreased at 0.60 g of extract. The higher the mass of extract added to the coating, the higher the residual mass of concrete cubes after heating. This shows that text-coat paint protects the substrate from heat and mass loss and the measure of performance is dependent on the number of layers of paint used, the mass of extract in the paint and the temperature to which they are subjected.

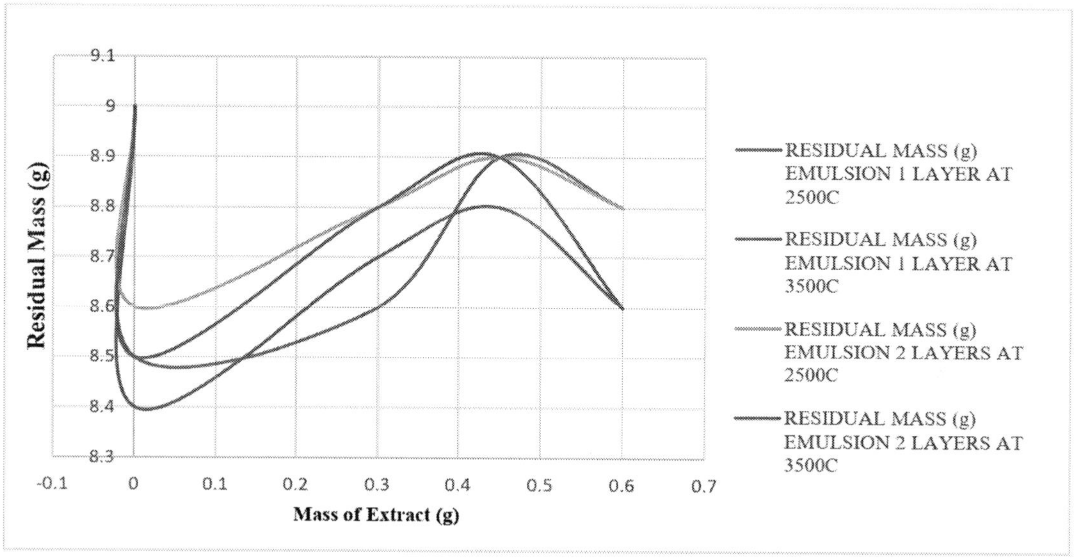

Fig. 8. Residual mass against the mass of extract in emulsion paint

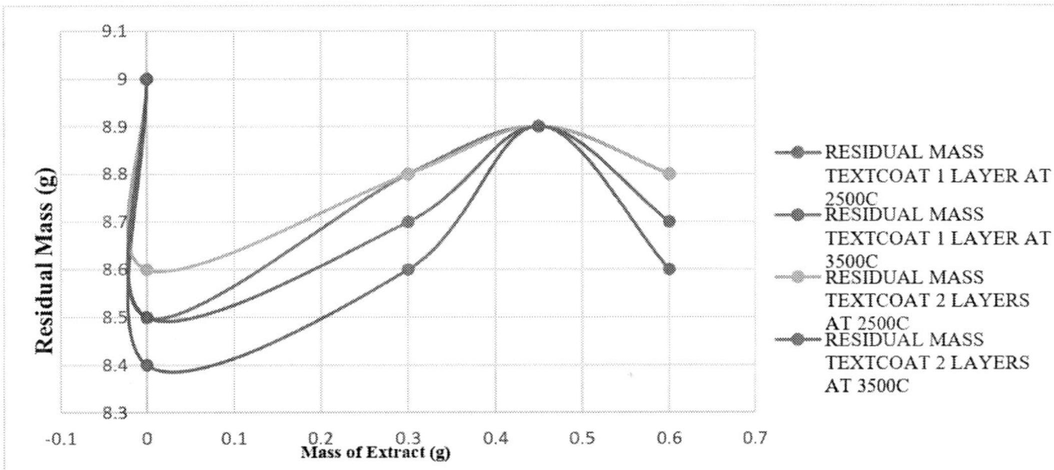

Fig. 9. Residual mass against the mass of extract in text-coat paint

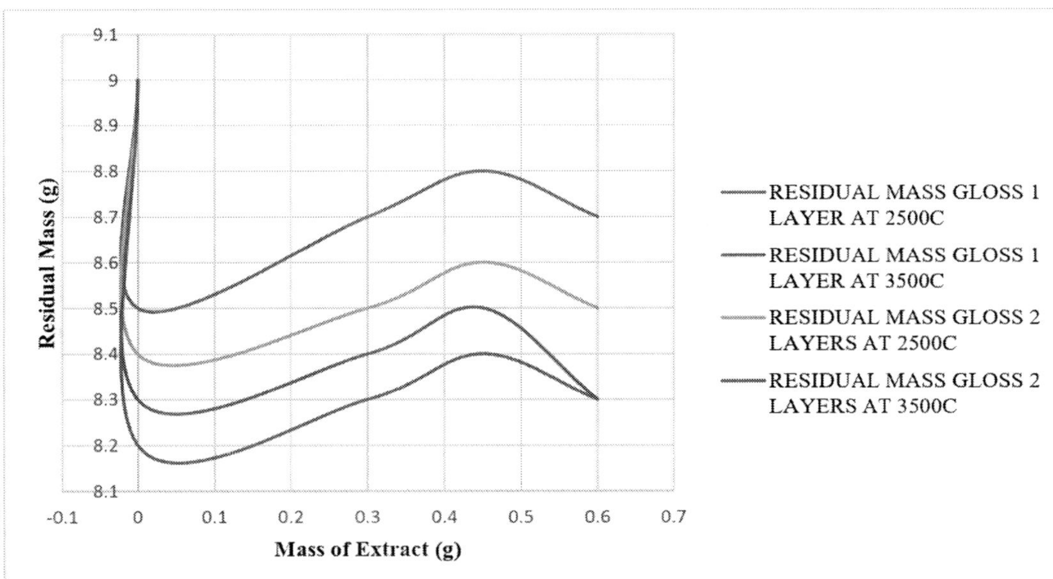

Fig. 10. Residual mass against the mass of extract in gloss paint

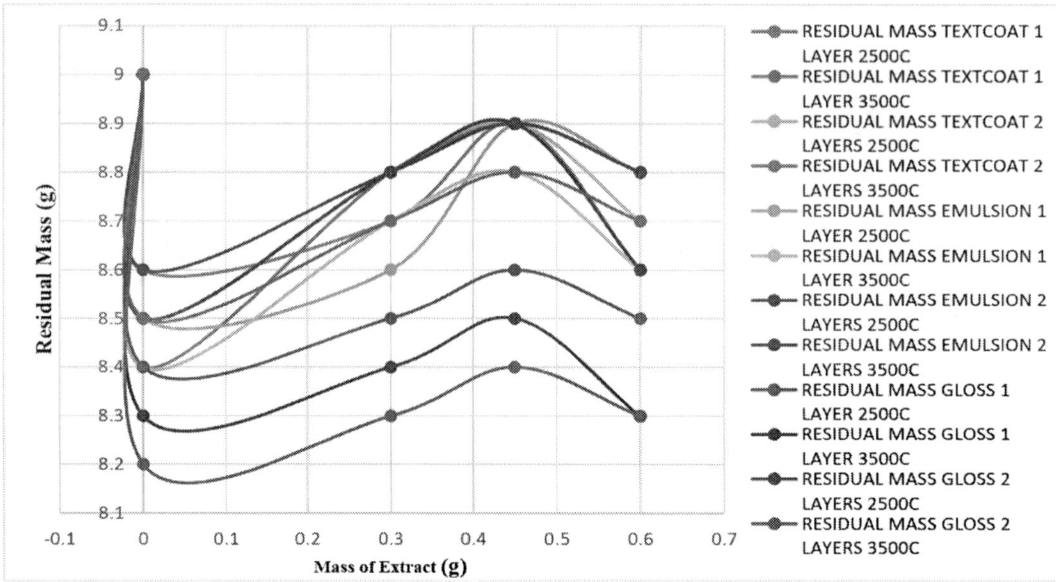

Fig. 11. Residual mass (g) against the mass of extract (g)

Fig. 11 shows the various residual mass of concrete cubes coated with gloss paint containing a varying mass of extract before and after being subjected to heat. It was observed that the best performing cubes with the least mass loss were those coated with 1 layer of gloss paint at 250°C, followed by those coated with 1 layer of gloss paint at 350°C, followed by those coated with 2 layers of gloss paint and heated at 250°C and the least residual mass was obtained by cubes coated with 2 layers of gloss paint and heated at 350°C. However, for each sample, the residual mass decreased from an initial mass of 9.0 g to the least mass obtained at 0.3 g of extract and increased to its peak at 0.45 g of extract and decreased at 0.60 g of extract. The higher the mass of extract added to the coating, the higher the residual mass of concrete cubes after heating. This shows that gloss paint protects the substrate from heat and mass loss and the measure of performance is dependent on the number of layers of paint used, the mass of extract in the paint and the temperature to which they are subjected to.

The residual mass of all the cubes coated with the various paint types with a varying mass of extract before and after subjecting it to 250°C or 350°C is related in Fig. 6. This analysis is done to compare the behaviour of the coating types and select the coating type with the best performance. The cubes coated with 2 layers of text-coat paint showed the least mass loss before and after heating both at 250°C and 350°C and the cubes coated with gloss paint showed the most mass loss before and after heating especially at a higher temperature of 350°C.

3.3. Compressive Strength Analysis

The compressive strength test results carried out for the concrete cubes coated both in one layer and in two layers are related in Table 7, 8, and 9 respectively.

Table 7. Comparison of compressive strengths of cubes coated with one layer of the various paint samples

Mass of Extract (g)	Compressive Strengths (kN)					
	Text-coat 1 layer		Emulsion 1 layer		Gloss 1 layer	
	250°C	350°C	250°C	350°C	250°C	350°C
Control	450					
0.00	230	210	240	180	170	130
0.30	270	230	300	230	230	200
0.45	330	320	350	310	290	280
0.60	300	280	290	220	200	190

Table 8. Comparison of compressive strengths of cubes coated with two layers of various paint samples

Mass of Extract (g)	Compressive Strengths (kN)					
	Text-coat 2 layers		Emulsion 2 layers		Gloss 2 layers	
	250°C	350°C	250°C	350°C	250°C	350°C
Control	450					
0.00	450	450	450	450	450	450
0.30	240	240	240	240	240	240
0.45	250	250	250	250	250	250
0.60	350	350	350	350	350	350

Table 9. Summary of compressive strength analysis

Mass of Extract (g)	Compressive Strength (KN)											
	Text-coat				Emulsion				Gloss			
	1 Layer		2 Layers		1 Layer		2 Layers		1 Layer		2 Layers	
	250°C	350°C	250°C	350°C	250°C	350°C	250°C	350°C	250°C	350°C	250°C	350°C
Control	450											
0.00	230	210	240	250	240	180	280	220	170	130	180	150
0.30	270	230	350	300	300	230	330	280	230	200	210	180
0.45	330	320	400	370	350	310	370	300	290	280	230	200
0.60	300	280	340	290	290	220	320	270	200	190	190	160

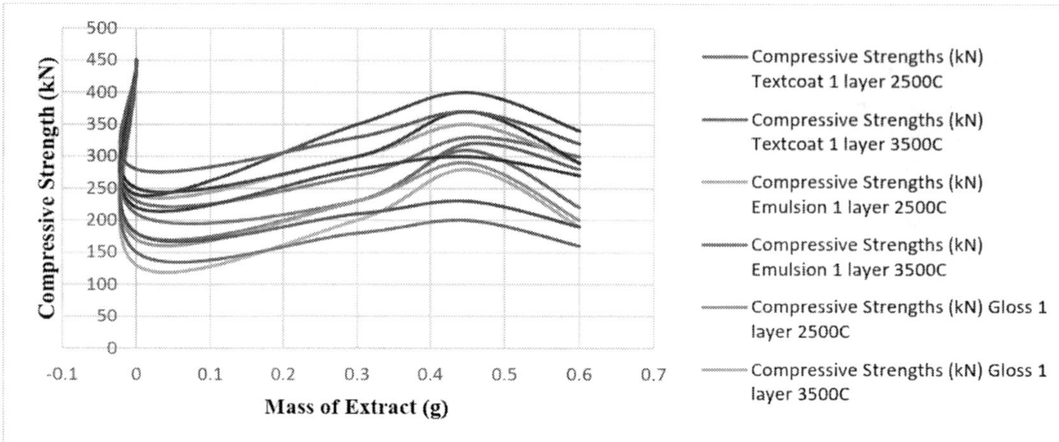

Fig. 12. Compressive Strength against Mass of Extract for all samples

In Fig. 12 a comparative analysis of the compressive strength value obtained from the different cubes coated with different layers of the different types of paint containing a varied mass of extract after being subjected to 250°C and 350°C for 1 h. From Fig. 12, it was observed that the best compressive strength was obtained by the concrete cubes coated with 2 layers of text-coat coating at 250°C. The next best performing concrete cubes were those coated with 2 layers of emulsion coating at 250°C followed by the cubes coated with 2 layers of text-coat and heated at 350°C, followed by the cubes coated with 1 layer of emulsion heated at 250°C, followed by cubes coated with 1 layer of emulsion heated at 350°C, followed by cubes coated with 2 layers of emulsion heated at 350°C, followed by the cube coated with 1 layer of gloss and heated at 250°C, followed by the cube coated with 1 layer of gloss and heated at 350°C, followed by the cube coated with 2 layers of gloss and heated at 250°C and lastly by the cubes coated with 2 layers of gloss coating and subjected to 350°C. However, for each sample, the compressive strengths decreased from an initial compressive strength of 450 kN to a lower compressive strength value attained at 0.3 g of extract and increased to its peak at 0.45 g of extract and decreased at 0.60 g of extract. This shows that the emulsion paint, as well as the text-coat paint, effectively protect the substrate from heat and the gloss paints are not effective fire-retardant paint.

It was observed from Table 9, that the least compressive values are obtained from the cubes coated with gloss paint. The higher the number of layers, the lower the compressive strength. It was also observed that the compressive strengths of the cubes coated with emulsion were better than those coated with gloss paint. This shows that emulsion paint is a better performing fire-retardant paint than gloss paint. It was also observed that the compressive strengths of the cubes coated with 2 layers were better performing except those coated with gloss paint. It can also be deduced from Table 9 that the cubes with the best compressive strength after exposure to heat were those coated with text-coat paint. It can also be deduced that those coated with 2 layers were better and those heated at 250°C than those heated at 350°C.

This shows that the emulsion paint, as well as the text-coat paint, effectively protect the substrate from direct contact with the heat thus reducing mass loss as well as from compressive strength. The measure of performance is dependent on the number of layers of paint used, the mass of extract in the paint and the temperature to which they are subjected to. However, the gloss paints are not effective fire-retardant paint. It is suspected that this is due to the kerosene contained in the paint which may also add as a fuel for the heating process. Thus leading to a high amount of concrete degradation.

IOP Conf. Series: Materials Science and Engineering 1036 (2021) 012077 doi:10.1088/1757-899X/1036/1/012077

4. Conclusion

A study on the effectiveness of fire-retardant paint on building materials with a varied mass of extract, number of layers of coating applied on the substrate, and the temperature of heating was carried out in this research work. The findings made are related as follows:

1. The introduction of Acalypha wilkesiana as an additive in paint increases the fire point of the sample; thus increasing the ignition time.

2. The optimum mass of Acalypha wilkesiana extract required to yield the best fire-retardation of the paint is 0.45g.

3. The best performing type of paint is text-coat paint. The coarse sand added during the production of text-coat paint assists in protecting the substrate material. Emulsion paint is the next best performing paint type. Gloss paint is not advisable to be used for this purpose.

4. The substrate material is effectively protected by increasing the number of layers of coating added. The best performing analysis was coated with 2 layers of text-coat paint. This does not apply to gloss paint because of the kerosene contained in the paint which contributes to the fire load of the material.

5. Concrete performs better when exposed to low temperature. The mass loss was higher and the compressive strengths of the samples heated at 350°C were lower than those heated at 250°C. Hence, the higher the temperature, the more damage is done to the substrate material.

Acknowledgments

The authors acknowledge and appreciate Covenant University for full financial support on this paper.

References

[1] Järvinen M 2018 Impact of wood surface pre-treatments on the behaviour of phosphorous additives.
[2] Morgan A B 2019 The future of flame retardant polymers–unmet needs and likely new approaches Polymer Reviews (Electronic Materials vol 59) pp25-54
[3] Crawford J 2007 Beyond paper. Use of plants of the Cyperaceae family in ancient Egypt Fields of Change Progress in African Archaeobotany Groningen: Barkhuis pp 105-113
[4] Kazmina O, Lebedeva E, Mitina N and Kuzmenko A 2018 Fire-proof silicate coatings with magnesium-containing fire-retardant Journal of Coatings Technology and Research (Electronic Materials vol 15) pp 543-554
[5] Stolleis M (2016) Nature's Regularity in Some Protestant Natural Philosophy Textbooks 1530–1630. In: Natural Law and Laws of Nature in Early Modern Europe. Routledge, pp 117-134
[6] Bolger R 1996 Antibacterial action of several tannins against Staphylococcus aureus Journal of antimicrobial chemotherapy (Electronic Materials vol 48) pp 487-491
[7] Thomas S, Rane AV, Kanny K, Abitha V and Thomas M G 2018 Recycling of Polyurethane Foams.
[8] Bozzoli F, Mocerino A, Rainieri S and Vocale P 2018 Inverse heat transfer modeling applied to the estimation of the apparent thermal conductivity of an intumescent fire-retardant paint Experimental Thermal and Fluid Science (Electronic Materials vol 90) pp 143-152
[9] Bhattacharya T, Chadha S S, Iqbal M J and Yadav B P 2019 Effects of Building Fire with Application of Prevention through Design A Review i-Manager's Journal on Future Engineering and Technology (Electronic Materials vol 15) pp 48
[10] Zeinali D, Kolaitis D, Schmid J, Barrio A, Pfriem A, Galan A, Donmez Cavdar A, Bedon C, Hopkin D and Barber D 2019 Guide for Obtaining Data from Reaction to Fire Tests

[11] Almeida M, Azinheira J R, Barata J, Bousson K, Ervilha R, Martins M, Moutinho A, Pereira J C, Pinto J C, Ribeiro L M 2017 Analysis of fire hazard in campsite areas Fire Technology (Electronic Materials vol 53) pp 553-575
[12] Walton W D, Carpenter D J, and Wood C B 2016 Zone computer fire models for enclosures. In SFPE handbook of fire protection engineering. Springer pp 1024-1033
[13] Khobragade P S, Hansora D, Naik J B and Chatterjee A 2016 Flame retarding performance of elastomeric nanocomposites: A review Polymer Degradation and Stability (Electronic Materials vol 130) pp 194-244
[14] Li K, Zhao F and Zhang Y 2019 Influence of carbonation on the chloride ingress into concrete: Theoretical analysis and application to durability design Cement and concrete research (Electronic Materials vol 123) pp 105-788
[15] Pan X, Shi Z, Shi C, Ling T-C and Li N 2017 A review on surface treatment for concrete–Part 2: Performance Construction and Building Materials (Electronic Materials vol 133) pp 81-90
[16] Hull T R and Stec A A 2009 Polymers and fire
[17] Oluwafemi J, Ede A, Ofuyatan O, Oyebisi S and Bankole D 2019 Recycling of Concrete Demolition Waste: pathway to sustainable development IOP Conf. Series: Materials Scienc and Engineering (Electronic Materials vol 640) pp 1-6
[18] Ofuyatan O, Adeniyi A, Ijie D, Ighalo J and Oluwafemi J 2020 Development of high performance self compacting concrete using eggshell powder and blast furnace slag as partial cement replacement Construction and Building Materials (Electronic Materials vol 256) pp 1-10
[19] Ofuyatan O, Olowofoyeku A, Adaramaja G, Oluwafemi J and Edeki S 2020 Development of high-performance self compacting concrete using eggshell powder and blast furnace slag as partial cement replacement Construction and Building Materials (Electronic Materials vol 13) pp 1-7

IOP Conf. Series: Materials Science and Engineering 1036 (2021) 012078 doi:10.1088/1757-899X/1036/1/012078

Material efficiency: a key sustainable manufacturing strategy

S O Ongbali[1*], S Oladipupo [1], S A Afolalu[1], M O Udo[1], and R O Leramo[1]
[1]Mechanical Engineering Department, Covenant University, Nigeria.
samson.ongbali@covenantuniversity.edu.ng

Abstract. Inefficient material utilization results in extra pressure on the environment for the extraction of more material to meet production demand and therefore makes the environment unsustainable. The survival of the manufacturing industry depends on its ability to balance several requirements to meet up with dynamic market demand particularly materials for production ranging from raw materials, intermediate materials to spare components for maintenance. In the past, sustainability and efficiency were no challenges to the manufacturers and material extraction from the environment was always below replenishment. Presently, the reverse is the case where the demand for production material has become incessant and surpasses supply on account of increasing population and modern manufacturing practices thereby resulting in irresponsible exploitation of raw material. This paper aims to evaluate the current efforts in the topic area to gain insight into the impact of material efficiency on the manufacturing performance concerning other manufacturing strategies alongside the environment. It appears from the review that the balance between manufacturing material extraction from the environment and the replenishment rate determines the sustainability of both the manufacturing system and the environment. Hence material efficiency is an index for sustainable manufacturing strategy. We recommend the future research direction to looks into the development and determination of balance between manufacturing material extraction from the environment and the replenishment rate for a sustainable future manufacturing.

Keywords: Material, efficiency, sustainable, manufacturing, strategy, and environment

1. Introduction

Sustainable consumption and production is one of the sustainable development goals drafted by the United Nations for the year 2030 and it addresses the current impact of manufacturing on the planet and provides future directions for manufacturing firms to manufacture goods more sustainably. The urgent call for action on the implementation of sustainable consumption and production (SCP) within the manufacturing sector can be attributed to the negative impact of the sector on the planet. An increase in material use can be linked to the increase in the standard of living of an ever-growing population, therefore with the ever-increasing demands of consumers producers have to rise to the challenge by using up more resources. Sustainable manufacturing is closely related to SCP in terms of goals since they both address the triple bottom line of people, planet, and profit. Sustainable manufacturing anchors on the three broad dimensions i.e. environment, economy, and society although some authors have included others such as technology, efficiency, etc., however, these three are the most prevalent in literature. Sustainable manufacturing can, therefore, be defined as a process of creating value by manufacturing goods in a way that is environmentally friendly, economically sound, and have a positive impact on the society.

IOP Conf. Series: Materials Science and Engineering 1036 (2021) 012078 doi:10.1088/1757-899X/1036/1/012078

Material efficiency in the manufacturing sector is key and central to other sustainable manufacturing strategies however sparse literature in this field shows a lack of understanding of this concept. The most discussed sustainable manufacturing strategy is energy efficiency and this due to the issue of global warming and climate change, also the legislative laws and fines placed by governments and other institutions around the globe make an integral part of most manufacturing firms, meanwhile little attention is given to material efficiency. Although an increase in energy efficiency can lead to the reduction of material usage, however energy efficiency has peaked in terms of innovation and renewable energy is not as efficient as fossil fuel or natural gas therefore its application in large scale manufacturing firms is limited. Material efficiency can be defined as the percentage of material used in manufacturing a product, i.e. the conversion of material into a finished product. Implementing this key strategy would have a direct impact on the environmental dimension of sustainable manufacturing which also reflects directly on the other two major dimensions. An increase in material efficiency will have a positive impact on energy usage within the plant, water usage, waste produced, emission, and CO_2 footprint.

2. Literature review

The manufacturing firms are typically profit-oriented and therefore do not consider the effect of their manufacturing activities on the immediate environment, however, there's been a major shift in manufacturing since the inception of our common future summit held in Brundtland 1987 and also growing concerns about global warming, several laws have been implemented in various nations to ensure manufacturing firms comply. At the summit sustainable development was defined as "the ability of the present generation to meet its need without compromising the ability of the future generation to meet theirs". Therefore sustainable manufacturing can the defined as the ability of manufacturing firms to meet the needs of the present generation without tampering with the ability of the future manufacturing firms to provide for the needs of the future generation.

According to [13] 63.40% of sustainable manufacturing studies are carried out in developed countries where the laws are strictly enforced, manufacturers began to look for ways to meet the minimum acceptable requirement on environmental discharge and at the same time maintain their market share. However, food manufacturing, steel, and chemical industries have the least active participation in sustainable manufacturing. Maintaining market share while meeting stipulated government requirements for manufacturing activities made production efficiency an integral aspect of these firms. Several strategies have been employed by the different organization as the nature of their activities differ from one another, this makes it impossible to have a generic sustainable framework or strategy to address all manufacturing industries, however, an industry-specific framework can be adopted in same manufacturing since firm in that particular industry have similar manufacturing processes.

2.1 Sustainable Manufacturing Strategies

The successful implementation of sustainable manufacturing in any manufacturing firm anchors on implementing the strategies that will lead to an increase in efficiency, however identifying these strategies can be challenging, but a clear understanding of the operational process will give clear directions on what strategies would be most effective when implemented. Several studies have been carried out on sustainable manufacturing strategies, for instance [19] proposed the use of three sustainable manufacturing strategies namely, material, energy, and waste (MEW) to address environmental performance within a manufacturing plant, MEW is a process flow modeling approach. Data generated from a case facility was analyzed on a spreadsheet model according to the MEW process flow. This approach led to the discovery and selection of key improvement areas in the environmental dimension of sustainable manufacturing. Similarly, a MEW process flow modeling was implemented by [5], a collective effort from manufacturing industries, academia, and software developing company made this research comprehensive as MEW

process flow modeling and sustainable building design were consolidated with the aid of a developed modeling, simulation, and analysis tool. This tool-assisted in designing and creating an improvement plan for a sustainable manufacturing plant.
[6] carried out a comprehensive study on sustainable manufacturing highlighting the three major dimensions of sustainability, and sub-dimensions which are often referred to as strategies in the existing literature. Other dimensions aside from the three main dimensions of sustainability were also identified in this research, which implied that manufacturing firms are to apply sustainable strategies that are suitable and specific for the firm. The authors observed that out of the three main dimensions of sustainability, the environmental dimension received more attention therefore formal concept analysis (FCA) was used to further analyze the choice of manufacturers in selecting and combining different strategies within the environmental dimension to achieve sustainability.

Table 1. Reference of Sustainable Manufacturing Strategies: Environmental Dimension

Sustainable Manufacturing Strategies	Strategy Goal	Author
Energy efficiency	(i) Minimize energy and material usage (ii) Improve production efficiency (iii) Minimize carbon footprint and emission.	[11], [6], [14], [15].
Waste minimization	(i) Minimize waste and pollution (ii) Improve production efficiency (iii) Cost reduction (iv) Minimize usage of virgin raw materials	[3], [6], [21], [20], [16].
Water Usage	(i) Minimize the quantity of water used during production (i) Increase the availability of clean water	[6], [23], [4].
Material efficiency	(i) Utilize material and energy efficiently (ii) Improve production efficiency (iii) Minimize waste and pollution (iv) Recycling	[17], [6], [14], [20], [16].
Resource efficiency	(i) Utilize resources (material, energy) efficiently (ii) Minimize waste and pollution (iii) Minimize the extraction of virgin raw material (iv) Improve Earth replenishment rate	[7], [14], [23], [16].

IOP Conf. Series: Materials Science and Engineering 1036 (2021) 012078 doi:10.1088/1757-899X/1036/1/012078

Table 1 Cont'd

Sustainable Manufacturing Strategies	Strategy Goal	Author
Eco-efficiency	(i) Utilize resources efficiently (ii) Minimize waste and pollution (iii) Improve the quality of life (iv) Improve earth carrying capacity	[14], [2], [1], [16].
Carbon footprint	(i) Minimize emission of CO_2 (ii) Improve the quality of air	[6], [14], [15].
Emission	(i) Meet the standard requirement for emission of gases (ii) Improve the quality of air	[6], [14], [15].
Transport	(i) Improve supply chain and logistics (ii) Reduce Emission and carbon footprint	[6], [14].
Landfill	(i) Eliminate and reduce waste	[6]
Biodiversity	(i) Improve Earth replenishment rate (ii) Improve the quality of life	[6]

Table 1 illustrates references of key sustainable manufacturing strategies concerning the environment indicating the strategy to the environment. It suggests that the environment must be preserved while manufacturing activities continue. This implies that Hence, there is a need for balance between manufacturing material extraction from the environment and replenishment rate for sustainable manufacturing system and environment

Water, material, carbon footprint, emissions, transport, biodiversity, resource, energy, waste, and landfill were identified to be the main strategies in this research and results show that energy, waste, and emissions were the most discussed in the literature with material coming in at the fourth position. On the other hand [9] stated that implementation of three dimensions of sustainability, not just the environmental aspect is essential to achieve the aim of sustainable manufacturing. The study analyzes a comprehensive index of sustainability which includes several performance indices. This reasearch aimed to create an effective framework to investigate the present condition of manufacturing industries by combining environmental, social, economic, and manufacturing variables. The proposal has its origins in patterns and limitations observed in the manufacturing industry sustainability literature and it is built on the framework of the analytical hierarchy process (AHP). A list of metrics is proposed, based on an AHP scoring system, assessing the industry performance. The next steps include grouping industries across the four dimensions according to common deficiencies and developing a framework for cooperation.

2.2. Material Efficiency: A Key Sustainable Strategy

Increasing material efficiency leads to reducing both the amount of industrial waste and the use of energy. Nevertheless, there was less reported describing what to calculate in a manufacturing firm for material efficiency. This paper reviews the recent practice of performance metrics for material efficiency in a manufacturing context using a bottom-up approach. In addition to the research study, empirical evidence was gathered from seven major manufacturing firms in Sweden via a multiple case study. The findings suggest that current metrics of material efficiency are constrained and are mainly calculated as criteria of cost or quality rather than the environment. The limited number of metrics contribute to the fact that material efficiency is not recognized as a core activity in manufacturing firms and is handled by the environment department with little operational connection [18].

[8] stated that with abundant research studies on material flow and a growing understanding of the need for economic dematerialization, manufacturing firms have often failed to take advantage of the many opportunities to increase materials efficiency. The potential solution provided in this article is material management services provided by external suppliers. This also provides a theoretical framework for analyzing various eco-efficient service business models and expands the definition of material management services. [17] affirmed that measures in material efficiency, such as recycling rates and dematerialization are also used to create sustainable development goals to achieve higher resource efficiency and less impact on the environment. The study carried out by [17] aimed to identify appropriate metrics of material efficiency to identify the sustainability impact of waste and recycling programs that leverage PET bottle waste management in three European countries with different waste management mechanisms and success rates. Material flow analysis in conjunction with life cycle analysis was carried out to examine the connection between the two dimensions of the material efficiency and environmental influence of each system. The production of PET bottles in Germany, Austria, and Serbia was 6.0 kg/person in the year 2017, and 5.4 kg/person in the 2013 and 6.9 kg/person in the year 2015 respectively. Meanwhile, the PET recycling rate of this waste flow in Germany, Austria, and Serbia is 91%, 41%, and 11% respectively. Increased material efficiency resulted in higher environmental performance.

Comprehensive research on the Finnish National Waste Plan, research, was carried out between 2007 and 2016. This research aimed to apply alternative concepts of waste avoidance or material efficiency and the transition from waste avoidance to material efficiency. The advantages and drawbacks of these complementary principles have been evaluated using parameters such as semantic dimensions, synergy legal aspects, and tracking applicability [12]. One potential aspect of a sustainable manufacturing strategy is the efficient use of materials. [16] reviewed existing studies on sustainable manufacturing strategies that include performance improvement in materials. Four key strategies used in this research are material efficiency, waste minimization, eco-efficiency, and resource efficiency. The research was analyzed to identify the main characteristic features of these sustainable manufacturing strategies and 17 features were found. Afterward, the four approaches were compared to all the features and contrasted. Although existing literature uses these sustainable manufacturing strategy terms frequently in a confusing, sometimes interchangeable way, this study executed by [16] however established an explicit distinction between them. Definition, scope, and applicability of measurement were revealed to be key features affecting the ability of manufacturing companies to implement the recommended strategy effectively. It was noted that the most implementable strategies do not necessarily have to contain all the dimensions for manufacturers interested in being more sustainable, this however creates a challenge between ease of execution of strategies and level of impact.

Material efficiency is a fundamental component of modern thinking to respond to the dilemma of minimizing the environmental impacts of manufacturing firms and resource depletion while meeting the demands of materials for service and functionality. The definition of the circular economy, which is built

on the idea of maximizing the value contained in resources and goods through the life cycle, is directly related to material efficiency. Even though materials such as steel are among the most 'circular' of produced materials due to high end-of-life recycling rates, there are substantial prospects for better material efficiency, which also needs to be extensively implemented. Life Cycle Assessment is usually utilized to evaluate the environmental advantages of reclaiming and recycling materials through the supply chain of manufactured products and at the end of life. Using an analogy obtained from renewable energy production, this study investigated the relationship between the lifecycle of a product and the environmental case for implemented strategies used to increase material efficiency [22]. Similarly, [10] discussed the material efficiency of aluminum scrap during production. Due to production failure some materials in this case aluminum scraps are usually rejected at the end of a production process, however, the amount of scarp produced by one of the aluminum smelting furnaces in the reduction pot manufacturing process exceeds the accepted standard of 5 to 10 percent. Automatically this harms the final output of the manufacturing process. This study implemented the use of lean manufacturing practices and six sigma tools to increase material efficiency results in a 6% decrease in the amount of aluminum scrap resulted in increased final output or products.
The link between material efficiency and other sustainable manufacturing strategies can be seen as an increase in material efficiency will result to increase in energy efficiency [19], prevention and minimization of waste [12], effective resource efficiency [16], reduced transportation trips of virgin raw materials which also reduces emission contributed and the CO_2 footprint, reduced emission at the operational level as equipment consume less energy which in return leads to reduced emission, the drastic reduction of waste flow to the landfill and the associated cost of transporting the waste [3], [6], [18]. However, these are a few impact material efficiency has on other manufacturing strategies.

3. Conclusion
The implementation of material efficiency as an index for sustainable manufacturing strategy can help achieve other sustainable manufacturing strategies. Although energy efficiency received the most attention of researchers in existing literature, material efficiency is encompassing and its implementation can assist in achieving sustainable manufacturing strategy. It appears that the balance between manufacturing material extraction from the environment and the replenishment rate determines the sustainability of the manufacturing system and the environment. Hence material efficiency is an index for sustainable manufacturing strategy

4. Recommendation
We recommend that the future research direction to looks into the development and determination of balance between manufacturing material extraction from the environment and replenishment rate for a sustainable manufacturing environment.

Acknowledgment
We sincerely acknowledge the support offered by Covenant University to facilitate publication of this manuscript.

REFERENCES
[1] Bocken, N. M. P., Short, S. W., Rana, P., & Evans, S. (2014). A literature and practice review to develop sustainable business model archetypes. *Journal of Cleaner Production*, *65*, 42–56. https://doi.org/10.1016/j.jclepro.2013.11.039

[2] Bracke, S., Ulutas, B., & Rosebrock, C. (2017). Concept for Analysing Product Sustainability Regarding Eco-related Product Perception and Efficiency Within a Product Spectrum. *Procedia Manufacturing*, *8*, 28–35. https://doi.org/10.1016/j.promfg.2017.02.003

[3] Cai, W., Lai, K. hung, Liu, C., Wei, F., Ma, M., Jia, S., … Lv, L. (2019). Promoting sustainability of manufacturing industry through the lean energy-saving and emission-reduction strategy. *Science of the Total Environment*, *665*, 23–32. https://doi.org/10.1016/j.scitotenv.2019.02.069

[4] Coelho, B., & Andrade-Campos, A. (2014). Efficiency achievement in water supply systems—A review. *Renewable and Sustainable Energy Reviews*, *30*, 59–84. https://doi.org/10.1016/j.rser.2013.09.010

[5] Despeisse, M., Oates, M. R., & Ball, P. D. (2013). Sustainable manufacturing tactics and cross-functional factory modelling. *Journal of Cleaner Production*, *42*, 31–41. https://doi.org/10.1016/j.jclepro.2012.11.008

[6] Eslami, Y., Dassisti, M., Lezoche, M., & Panetto, H. (2019). A survey on sustainability in manufacturing organisations: dimensions and future insights. *International Journal of Production Research*, *57*(15–16), 5194–5214. https://doi.org/10.1080/00207543.2018.1544723

[7] Früchtl, M., Leis, M., & Wertheim, R. (2020). A comprehensive and interdisciplinary perspective on sustainable manufacturing towards sustainable life cycles. *Procedia Manufacturing*, *43*, 383–390. https://doi.org/10.1016/j.promfg.2020.02.197

[8] Halme, M., Anttonen, M., Kuisma, M., Kontoniemi, N., & Heino, E. (2006). Business models for material efficiency services : Conceptualization and application, *3*. https://doi.org/10.1016/j.ecolecon.2006.10.003

[9] Harik, R., El Hachem, W., Medini, K., & Bernard, A. (2015). Towards a holistic sustainability index for measuring sustainability of manufacturing companies. *International Journal of Production Research*, *53*(13), 4117–4139. https://doi.org/10.1080/00207543.2014.993773

[10] Huda, L. N. (2018). The effect of material productivity on scrap reduction on aluminum reduction pot process. *IOP Conference Series: Materials Science and Engineering*, *309*(1). https://doi.org/10.1088/1757-899X/309/1/012117

[11] Khan, A. M., He, N., Li, L., Zhao, W., & Jamil, M. (2020). Analysis of Productivity and Machining Efficiency in Sustainable Machining of Titanium Alloy. *Procedia Manufacturing*, *43*, 111–118. https://doi.org/10.1016/j.promfg.2020.02.122

[12] Lilja, R. (2009). From waste prevention to promotion of material efficiency : change of discourse in the waste policy of Finland. *Journal of Cleaner Production*, *17*(2), 129–136. https://doi.org/10.1016/j.jclepro.2008.03.010

[13] Malek, J., & Desai, T. N. (2020). A systematic literature review to map literature focus of sustainable manufacturing. *Journal of Cleaner Production*, *256*, 120345. https://doi.org/10.1016/j.jclepro.2020.120345

[14] Menghi, R., Papetti, A., Germani, M., & Marconi, M. (2019). Energy efficiency of manufacturing systems: A review of energy assessment methods and tools. *Journal of Cleaner Production*, *240*, 118276. https://doi.org/10.1016/j.jclepro.2019.118276

[15] Perdikakis, A., Shukla, A., & Kiritsis, D. (2015). Optimize Energy Efficiency in the Supply Chain of FMCGs with the Use of Semantic Web Technologies. *Procedia Engineering*, *132*, 1112–1119. https://doi.org/10.1016/j.proeng.2015.12.603

[16] Rashid, S. H. A., Evans, S., & Longhurst, P. (2008). A comparison of four sustainable manufacturing strategies. *International Journal of Sustainable Engineering*, *1*(3), 214–229. https://doi.org/10.1080/19397030802513836

[17] Schmidt, S., Laner, D., Eygen, E. Van, & Stanisavljevic, N. (2020). Material efficiency to measure the environmental performance of waste management systems : A case study on PET bottle

recycling in Austria , Germany and Serbia. *Waste Management*, *110*, 74–86. https://doi.org/10.1016/j.wasman.2020.05.011

[18] Shahbazi, S., Salloum, M., Kurdve, M., & Wiktorsson, M. (2017). Material Efficiency Measurement: Empirical Investigation of Manufacturing Industry. *Procedia Manufacturing*, *8*(October 2016), 112–120. https://doi.org/10.1016/j.promfg.2017.02.014

[19] Smith, L., & Ball, P. (2012). Int . J . Production Economics Steps towards sustainable manufacturing through modelling material , energy and waste flows. *Intern. Journal of Production Economics*, *140*(1), 227–238. https://doi.org/10.1016/j.ijpe.2012.01.036

[20] Sundar, R., Balaji, A. N., & Satheesh Kumar, R. M. (2014). A review on lean manufacturing implementation techniques. In *Procedia Engineering*. https://doi.org/10.1016/j.proeng.2014.12.341

[21] Syahputri, K., Sari, R. M., Anizar, Tarigan, I. R., & Siregar, I. (2018). Application of lean six sigma to waste minimization in cigarette paper industry. *IOP Conference Series: Materials Science and Engineering*, *309*(1). https://doi.org/10.1088/1757-899X/309/1/012027

[22] Walker, S., Coleman, N., Hodgson, P., Collins, N., & Brimacombe, L. (2018). Evaluating the Environmental Dimension of Material Efficiency Strategies Relating to the Circular Economy, 1–14. https://doi.org/10.3390/su10030666

[23] Walsh, B. P., Cusack, D. O., & O'Sullivan, D. T. J. (2016). An industrial water management value system framework development. *Sustainable Production and Consumption*, *5*, 82–93. https://doi.org/10.1016/j.spc.2015.11.004

Progresses on mild steel protection toward surface service performance in structural industrial: An Overview

O S I Fayomi[1, 2*], D Olusanyan[1,] F T Ademuyiwa[1], G Olarewaju[1]
[1]Department of Mechanical Engineering, Covenant University, P.M.B1023, Ota, Nigeria [2]Department of Chemical, Metallurgical and Materials Engineering, Tshwane University of Technology, P.M.B. X680, Pretoria,
Corresponding author: ojo.fayomi@covenantuniversity.edu.ng, Ojosundayfayomi3@gmail.com +2348036886783

Abstract. Mild steel has been widely used in domestic and industrial applications due to its excellent properties. Their property in engineering application makes it one of the leading engineering materials with multifaceted usage. Despite this growing interest in manufacturing because of cost and durability, challenges arising from wear, corrosion and structural damages has been a concern and the need to search for continuous way-out become necessary. This overview provides concise progresses on mild steel toward active protection for service performance.

Keywords: Mild steel, performance, protections, industry

1. Introduction

Mild steel is extensively used in industry, especially for structural applications because of some unique properties exhibited. Mild steel has an ultimate strength of somewhere in the region of 29 to 30 ton/in2, a modulus of elasticity about 13000-to13500 ton/in2, a yield strength of 15 to 16 ton/in2 and very ductile properties [1-3]. The excellent features of mild steel are: relative cheapness of production, high mechanical properties, ease of cold working, and the ability to be hot worked without loss of mechanical properties [4]. It has the disadvantage that it requires efficient protection to prevent corrosion. When riveted construction gave way to welding, it eventually became possible to mild steel quickly, and high-efficiency joints could be obtained [5]. The mild steel has a chemical composition mainly iron with small percentages of other materials, principally carbon and manganese. The rate of carbon is usually somewhere about 0.2%, and in modern steels, the manganese content is not less than 2.5 times the carbon content. Impurities such as sulfur and phosphorus are kept at a low level, usually 0.05% for each [5, 6].

Mild steel's susceptibility to rusting in humid air and its high dissolution rate in acidic media is the major obstacle. Therefore, the development of corrosion resistance mild steel has become an interesting investigating idea lately by numerous researchers [7]. And the effort has been towards surface technology application. Surface treatment for the properties of the material or metal such as hardness, wear, and corrosion resistance are the basic requirements for the development of modern industrial applications. Every material properties requirement differ with the compositions used to develop supportable environmental application [8]. Deterioration of materials is natural phenomenon degradation due to the reactions with the ecological conditions. Material selection is the best tool to use against the avoidance of corrosion service. Even though the deterioration cannot be permanently prevented, choosing the best material is the choice that will serve as a reduction of the corrosion rate to be infinitely small for years [9]. Economic aspects should play a role in material selection to minimize the cost while producing product requirements that meet application performance goals.

Understanding the alloying process, coating methods, and the effect of compositions is a fundamental tool to use for consideration specifications for essential material properties [10].

Degradation of metals leads to failure conducts for most mechanical components. The development of a new barrier coat of zinc hydroxide at the steel/Zn-phosphate boundary leads to blocking the pores of the coating [11]. The electrolytic coatings produced on mild steel showed an exceptionally high corrosion resistance in salt spray tests and long-term atmospheric exposure. In the initial stage of corrosion, the cathodically coated steel was found to show a high electrochemical activity compared to chemically phosphate steel. This was credited to the existence of metallic zinc in the cathodically formed coating. In an alkaline solution, the zinc phosphate coating is converted chemically into a stable zinc oxide layer, which shows excellent corrosion resistance [12-14].

The microstructure of the coated-body often provides excellent improvement properties. Metallic corrosion is concerned with all-metal interactions with its environment; it is the passage of metal into a chemically combined state. From the metallurgical aspect of view, it can be said to be extractive metallurgy in reverse that is metal is derived from their ores, and when they corrode, they go back into their minerals or a component of it [14, 15]. Thus, this study carefully looks into the progress of mild steel, corrosion phenomena and new protective based coatings.

2. Steel surface corrosion and protections

Corrosion reaction by electrochemical or chemical activities is destructive and susceptibility on steel surfaces. Harsh condition with temperature gradient in the presence of environmental reactor influences corrosion mechanism which also promotes pitting evolution. Undoubtedly, corrosion phenomena involving submerged component in water, moisture or atmospheric condition are enhanced through electrochemical processes. Atmospheric corrosion, because of exposed steel structures bridges, trusses, towers, etc are used in open space of Bangladesh. Atmospheric corrosion is stimulated by humidity, damp atmosphere, since these maintain a film of water on the metal, providing the essential electrolyte. Other factors are acid gases in the atmosphere or sulfur compounds from cinders, coke, coal dust, etc; salts dissociate to produce an acid reaction; oxygen dissolved in the water film [5]. Rust may accelerate corrosion and cause pitting. The probable explanation is that surface accumulations of rust shield the underlying metal from free access to oxygen, thus rendering such portions anodic (corrodible) with respect to unshielded areas to which oxygen has freer access (cathodic areas).

With few exceptions, any marketable product of low carbon steel must be surface-finished by grit/air/sand blasting or acid pickling for good adhesion before any coating for surface protection [8]. While the primary purpose of a coating or finish may often be to improve the appearance and sales of the item, coatings must be used on metals to give reasonable resistance to destructive influences due to wear, electrolytic decomposition and contact with the weather or corrosive atmosphere [10-12]. Corrosion may be minimized through application of protective metal such as zinc, nickel, copper. In most cases where advance application services are required, the use of oxide, phosphate, or similar coatings on iron and steel are often engage for surfaces enhancement. The application of protective paints also renders the surface of the metal passive.

Metallic coatings are in general, the application of a finite thickness of some material over the metal, or are the transformation of the surfaces by chemical or electrical means to an oxide of the original metal. Galvanizing is a zinc coating used extensively for protecting low carbon steel from atmospheric deterioration. Galvanizing by zinc is generally applied to metal surfaces by the Sherardizing process, by dipping into a bath of molten zinc, by electrode position, or by metal spraying. But we here consider (1) the hot-dip galvanizing process and (2) a new technique--the cold-galvanizing process.

Although electroplated zinc coatings are considered as one of the main methods used for the corrosion protection of steel [14]. This has led to a growing interest in the manufacturing industry. Deposition of zinc on mild steel offers great protection at low costs. Zinc composite coatings exhibit better corrosion resistance property. Nowadays the nano-sized materials are co-deposited to get better zinc composite with better corrosion resistance [13-15]. Corrosion of metals occurs by the electrochemical reactions taking place with either their natural or man-made environments, which would often result in deterioration of the metals' properties. Through oxidation and reduction reactions, the surface of the metal develops anodic and cathodic zones forming oxides of metal alloys. In recent times, transitional metal and nitrides based are considered due to their attractive responses in application in relation to chemical and physical properties

3. Electrodeposition, metallic coating development and its responses in the environment

Electroplating is the process of depositing a skinny layer of metal over the surface of the cheaper metal, using electric current. It is divided into D.C, pulse plating, and electroless plating. Electrodeposition presents a cost-effective and less equipped method. In recent times, composite electroplating has taken the lead over contemporary ordinary zinc-based coating. Composite electroplating is the co-deposition of insoluble of metallic or non-metallic compounds with metals or alloys in a plating bath. A composite material is composed of two or more individual elements comprising of either metals, ceramic, or polymer. The invention of the electrolytic plating was introduced and consisted of applying a composite coating on the metal's surface. These metals have been used in many fields due to their unique properties (Fayomi et al. 2011). The process variable influences the physical, chemical, and structural properties of components, especially with insight on performance. Among numerous are the following **parameters affecting the electrodeposition**

- **The applied potential and current densities (direct current, pulse time).** Normally, electrodeposition has been carried out by using either potentiostatic or galvanostatic techniques to produce composite coatings. Within the galvanostatic mode, direct current (DC), pulsed direct current (PDC), and pulse reverse current (PDC) techniques have been reported in the literature to provide thin-film applications [12]. A study by Steinbach et al compares the coating that was deposited using direct current and pulse current. It was shown that the synthesis of nanocrystalline Zn-ZrO2 composite coating was possible by using a direct pulse current.

- **Particle size, shape, type (micro or nanoparticles), and concentration affect the composite coatings' synthesis.**

The particle content varies for each particle type. The shape of the particle can influence some of the co-deposition parameters, including the adsorption of the particles to the cathode [1]. Different kinds of results have been observed regarding the particle size during electrodeposition. In most cases, larger particle size tends to have higher incorporation of grain crystal at metal surfaces.

- **Bath agitation and electrode movement**.

The initial step in co-deposition involves the transport particles to the cathode. This means that the particles have to be moved from the electrolyte to the cathode. Researchers have reported that there is always a decrease in particle incorporation due to too much agitation of electrolytes. This is caused by the particles moving away from the cathode surface before they can be incorporated. On the contrary, some authors also affirmed that agitation increases the particle transport towards the cathode, thereby enhancing the particle incorporation [6].

- **The electrolyte composition and pH.**

The rate of incorporation of particles in coating depends on the bath composition. But again, the additional agents affect the co-deposition process. Many researchers claim that different reagents have different effects, but they usually increase the incorporation of the particles. [4-7] has shown that the addition of small amounts of oxides, nanocarbon, ceramics, and amines promotes co-deposition [6-8]. When the influence of pH in an electroplating bath is discussed, researchers give one similar and consistent conclusion, which is also a significant boost to coating adhesion. In a recent study, authors have observed that particle incorporation influences electroplating bath framework and alter bath pH. There is a connection between the effects of temperature on the co-deposition process. Most composite coating is difficult to deposit without a change in a temperature gradient. It was reported that particular systems do not show any influence of temperature, whereas other systems reveal a rise in the incorporation of the particles that affect such coating formation.

4. Conclusion

The major challenge of structural steel in application is environment and this has resulted into both physical and chemical deterioration of engineering component in service. Thus, this study has concisely looks into persistence effort and continuous approach to finding solution to environmental setback, reaction pose from the corrosive solution. Development of coating is seen as the most economic way-out to retard the accelerated responses of corrosion catastrophes through the use of composite coating or surface active metallic coatings.

Declaration of Competing Interest

The authors will like to declare that this research has no any financial or work related competing interest

Acknowledgement

The authors will like to appreciate partial support of Covenant University toward completion of this work.

References

1. Fayomi, O.S.I. & Popoola, A.P.I. (2014). Study Of Al_2O_3/Sic Particle Loading On The Microstructural Strengthening Characteristics Of Zn-Al_2O_3-Sic Matrix Composite Coating. Egyptian Journal of Basic and Applied Sciences, 1(2) 120-125.
2. Xia, X., Zhitomirsky, I. & Mcdermid, J.R. (2009). Electrodeposition of Zinc And Composite Zinc–Yttria Stabilized Zirconia Coatings. Journal of Materials Processing Technology, 209, 2632–2640.
3. Blejan, D. & Muresan, L. M. 2012. Corrosion Behaviour Of Zn–Ni–Al2o3 Nanocomposite Coatings Obtained By Electrodeposition From Alkaline Electrolytes. *Materials And Corrosion*, 63, 9999.
4. Hammami, O., Dhouibi, L., Bercot P., Rezrazi, E., & Triki, E. 2011. Study Of Zn-Ni Alloy Coatings Modified By Nano-Sio_2 Particles Incorporation. *International Journal Of Corrosion*, Vol. 2012, Pg. 8
5. Fayomi, O.S.I., Abdulwahab, M., & Popoola, A.P.I. 2013. Properties Evaluation Of Ternary Surfactant-Induced Zn-Ni-Al_2O_3 Films On Mild Steel By Electrolytic Chemical Deposition. *Journal of Ovonic Research,* Vol. 9, Pg. 123-132.
6. Fayomi, O.S.I. 2014. Effect Of Composite Particulate Reinforcement On The Morphology, Anti-Corrosion And Hardness Properties Of Fabricated Zn-ZnO Coatings. *J. Mater. Environ. Sci.* 5, Pg. Xxx-Xxx.
7. Durodola, B.M., Olugbuyiro, J.A.O., Moshood, S.A. Fayomi, O.S.I., & Popoola, A.P.I. 2011. Study Of Influence Of Zinc Plated Mild Steel Deterioration In Seawater Environment. *International Journal Of Electrochemical Science,*. 6, 5605 – 5616.
8. Aperador, W., Delgado, A., & Duque, J. 2013. Corrosion Resistanne Of The [Tin/Crn]N Coatings Deposited On Steel Aisi 4140. *International Journal Of Electrochemical Science,* Vol. 8, Pg. 10711 – 10719.
9. Sancakoglu, O., Culha, O., Toparli, M., Agaday, B., & Celik, E. 2011. Co-Deposited Zn-Submicron Sized Al2o3 Composite Coatings: Production, Characterization And Micromechanical Properties. *Materials And Design,* 32, 4054–4061.
10. Roberge, P.R. 2000. Handbook Of Corrosion Engineering. New York: Mcgraw Hill Companies, Incorporated. Pg. 577-769.
11. Smith, J.L. And Yirmani Y.P. 2000. Materials And Methods For Corrosion Control Of Reinforced And Prestressed Concrete Structures In New Construction. *Material Sciences*, Pg. 165.

12. Rusu, D. E., Ispas, A., Bund, A., Gheorghies, C. & Carac, G. 2012. Corrosion Tests Of Nickel Coatings Prepared From A Watts-Type Bath. *Journal Of Coating Technology Resources*, Vol. 9, Pg. 87–95.
13. Amuda, M.O.H., Subair, W. & Obitayo, O.W. 2009. Study Of Optimum Conditions Of Zinc Plating On Mild Steel. *International Journal Of Engineering Research In Africa*, Vol. 2, Pg. 31-39.
14. Popoola, A.P.I., Pityana, S.L. And Popoola, O.M. 2011. Quantitative Study Of The Hardness Property Of Laser Surface Alloyed Aluminium Aa1200. *The Journal Of The South African Institute Of Mining And Metallurgy*, 111, 335-344.
15. Praveen, B.M, & Venkatesha, T.V. 2009. Electrodeposition And Properties Of Zn–Ni–Cnt Composite Coatings. *Alloys Compound*, 482, 53–57.

Role of chemical equilibrium on the performance of an activated carbon–methanol adsorption refrigeration tube

M E Emetere[1,2], P O Oso-Ola[1] and S A Afolalu[3]

[1] Department of Physics, Covenant University Canaan land, P.M.B 1023, Ota, Nigeria.
[2] Department of Mechanical Engineering Science, University of Johannesburg, South Africa.
[3] Department of Mechanical Engineering, Covenant University Canaan land, Nigeria.

Abstract. The main compartment that determines the functionality of the adsorption refrigerator is the adsorption refrigerator tube (ART). Many scientists have suggested certain ART configuration that would help improve its adsorption performance. The focus of the research is to determine causes of the ART malfunction via material failure. The linear driving force (LDF)-based models and the chemical kinetic reaction were taken into account to determine the chemical equilibrium of the material. It was observed that the successes of the adsorption refrigerator depend on the source material of the activated carbon and the level of impurity in the activated carbon.

1. Introduction

Otto Mohr was the first man to have the idea of solar refrigerators in 1935; he was the architect and the first to design a working solar refrigerator. Since then so many forms of solar refrigerator have been exposed [1]. A refrigerator that is powered on energy directly obtain from the sun is called a solar refrigerator, and may include photovoltaic or solar thermal energy [2]. A solar-powered refrigerator is a refrigerator, which gets all its energy rate from sunlight. Solar-powered refrigerators has found relevance in recent based on the following: solar-powered refrigerators for solar refrigerators are used to preserve goods especially food from spoilage it also preserve vaccines guy keeping them at a regular temperature [3]; solar refrigerators are used for camping and other recreational event where electricity is no available; when compared to the normal refrigerators solar refrigerators does not have a negative effect to the ozone layer. The shortcoming of the solar-powered refrigerators includes: batteries which must be changed regularly every three years [4]; very heavy set-up that makes mobility difficult; solar refrigerator are very expensive especially the ones that makes use of the batteries.

There are three types of solar refrigerator namely Absorption solar refrigerator, Ranking engine solar refrigerator, and PV (Photovoltaic) solar refrigerator. Absorption solar refrigerator are the least used when compare the to the PV and the mechanical solar refrigerators. This makes use of the absorption cycle to produce the cooling effect of the refrigerator created by the refrigerant (lithium bromide LiBr or silica gel). This solar refrigerator is operated by the principle of heat transfer. Absorption solar refrigerator are categorized into two i.e., liquid refrigerant (lithium bromide) and solid refrigerant (silica gel(H2O and zeolite)). Both refrigerators work on same principle. When the liquid refrigerant passes through the evaporator it evaporates but the solid refrigerant sublimate produce the refrigeration effect. When the refrigerant is recycled/ reused it is referred to as closed absorption but if the refrigerant is not reused but ejected then it is called an open absorption.

Adsorption refrigeration has gained wide usability in recent times with increased research on the adsorption refrigeration tubes (ARTs) [5,6]. The general problem of the adsorption refrigerator is its physical configuration, where the ARTs integrated system overcomes the vacuum in a traditional complex adsorption refrigeration system [7]. Based on the above, scientists have been proffering solution using mathematical modelling. Most literatures in modeling studies in adsorption refrigerator have addressed these problems i.e., adsorption performance, heat transfer in the adsorbent bed, and the coupled mass transfer process [8-10]. In this paper, we proposed the importance of structural and material anomalies in the activated carbon as a main challenge that must be modelled.

2. Methodology

The experimental set-up is shown in Figure 1. The linear driving force (LDF)-based models were used to analyse the adsorption performance. The LDF have been used with high accuracies to determine various variables of the ARTs configuration [7, 11].

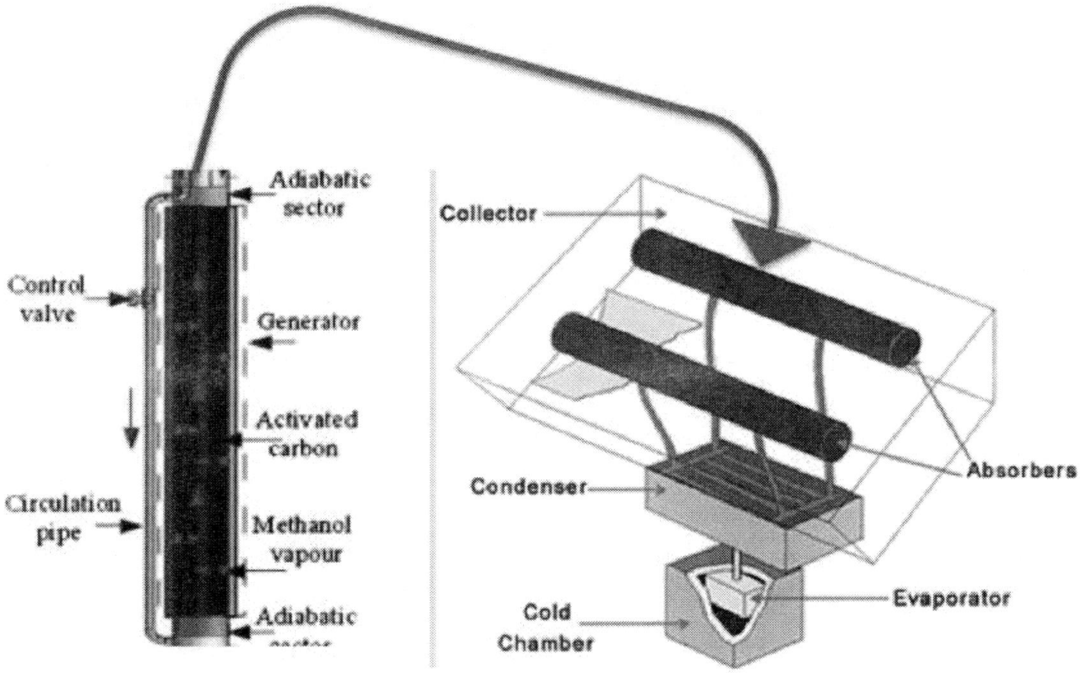

Figure 1. Adsorption solar refrigerator system

The LDF is given as:

$$\frac{\partial C}{\partial t} = \frac{15D_o}{R_p^2}.\exp\left(-\frac{E_a}{RT_s}\right).(C_{eq} - C) \quad (1)$$

$$C_{eq} = C_o\exp\left(-D\left(T_s\text{In}\frac{P_{sat}}{P}\right)^n\right) \quad (2)$$

where C_{eq} is the equilibrium concentration at the corresponding pressure and temperature, and C denotes the actual concentration, $\frac{15D_o}{R_p^2}$, $\frac{E_a}{R}$, C_o, D, n are coefficients obtained from ref [11]. From the laboratory preparation of the activated carbon is characterized with cracks and crevices [12]. Also, the quality of activated carbon depends on the source material [13-14].

Kinetic chemical reaction models according to the reaction order in the differential Equation

$$\frac{dC}{dt} = kC^n \quad (3)$$

C is the free chlorine concentration in the reactor, t is the contact time, and k is the kinetic constant, n = 1,2 or3. The equation was solved with assumption that: the longer the activated carbon and adsorbed methanol (liquid state) are in local thermal equilibrium, a chemical reaction is initiated to modify the structure of the activated carbon; the activated carbon is not a pure sample; and the adsorbent bed is packed with uniform-size and isotropic particles.

Table 1. Experimental parameters for adsorption refrigeration [11]

No	Symbol	Value	Unit
1	n	1	
2	C_o	6.6	$molm^{-3}$
3	P_{sat}	16.36	Pa
4	P	7140	Pa
5	D	9.08×10^{-6}	
6	D_o	6.5×10^{-5}	m^2s^{-1}
7	R_p	0.055	m
8	R	0.105	m
9	T_s	298	K
10	k	0.07	

3. Results and Discussion

The focus of the research is to determine causes of the adsorption rod malfunction via material failure. Equation (3) replaces equation (2) in the LMD model as presented in Equation (1). For the activated carbon and the methanol to function properly, there must be local thermal equilibrium. In this paper, we examined the particulate participation of activated carbon that could sustain this condition. The variables used have been experimentally validated as presented in Table 1. When the kinetic constant of the compound is between 0.01 and 0.09 (Figure 2a), the differential concentration between the activated carbon and methanol is expected to increase linearly. The differential concentration of the compounds at k=0.01 seem to close to the origin. When the kinetic constant is placed between 0.1 and 0.19 (Figure 2b), It was observed that the differential concentration between the activated carbon and methanol had an increased parabolic increase which was more visible at k=0.19. The kinetic constant was further increased between 0.2 and 0.29 to investigate the chemical equilibrium required for the activated carbon to sustain the adsorption performance as presented in Figure 2c. It was observed that the chemical equilibrium was obtained when the kinetic constant is 0.2. However, when k=0.29, the differential concentration increased in parabolic pattern. This result simply means that chemical equilibrium is not dependent on the value of the kinetic constant.

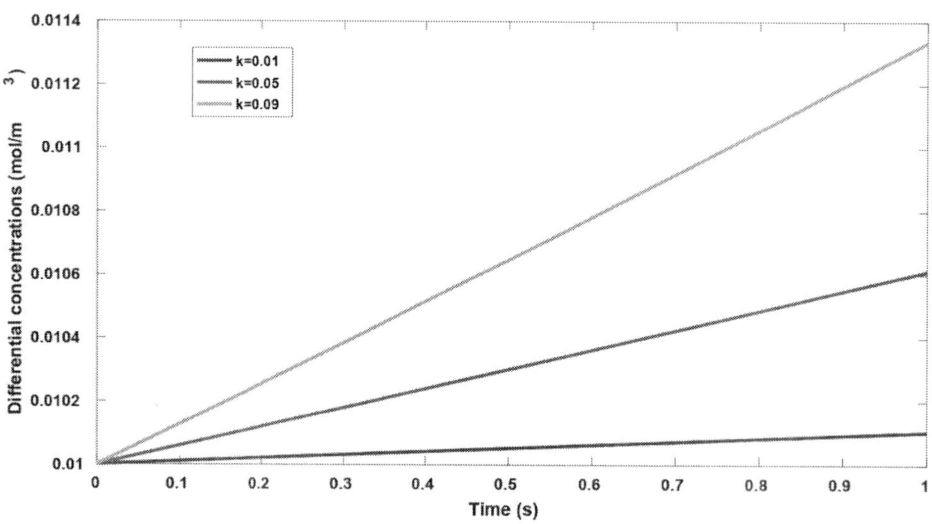

IOP Conf. Series: Materials Science and Engineering 1036 (2021) 012080 doi:10.1088/1757-899X/1036/1/012080

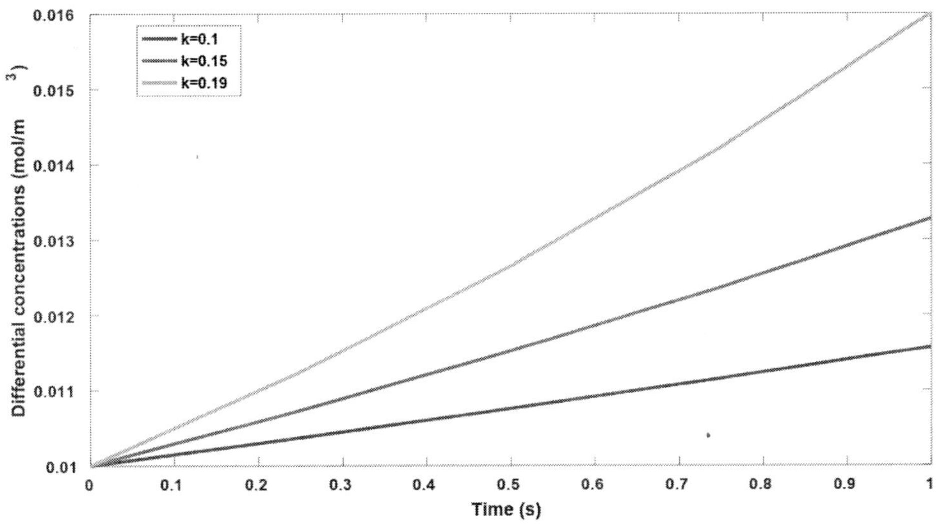

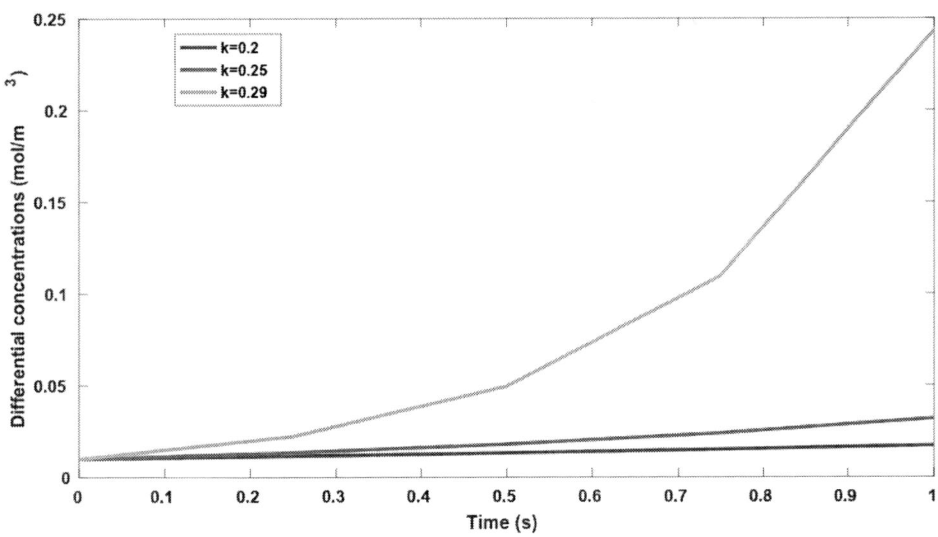

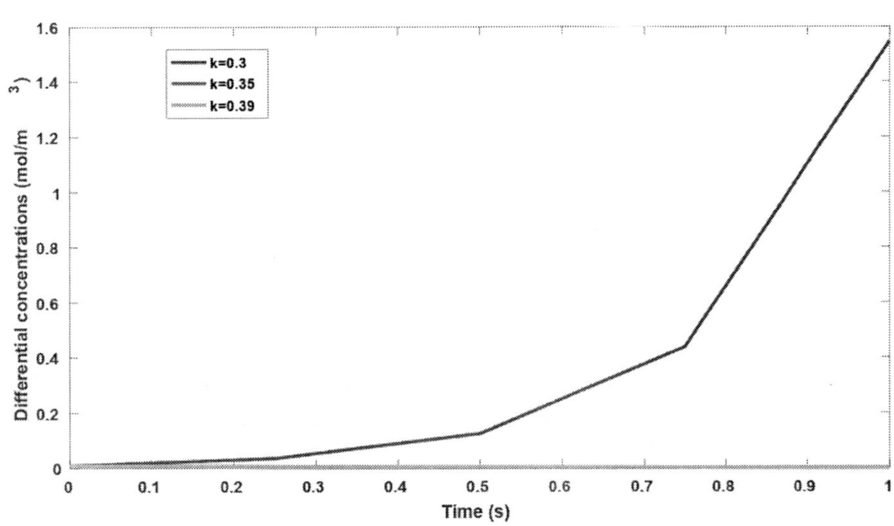

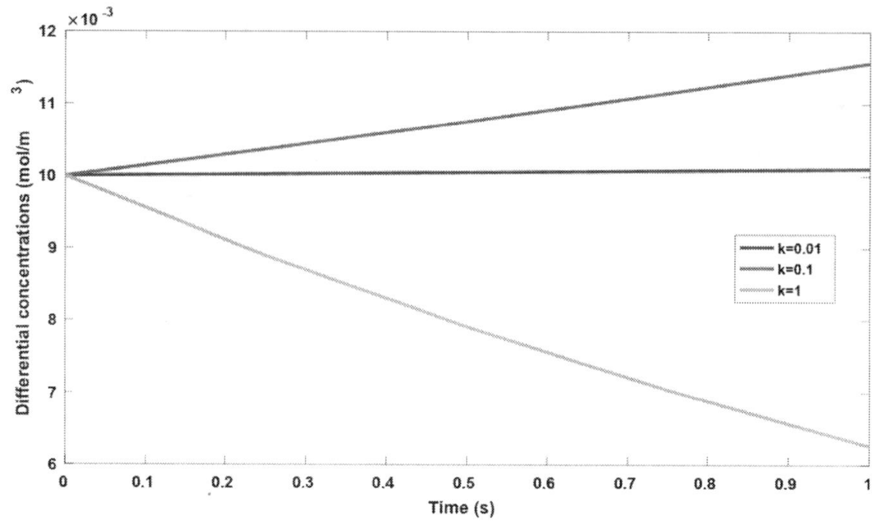

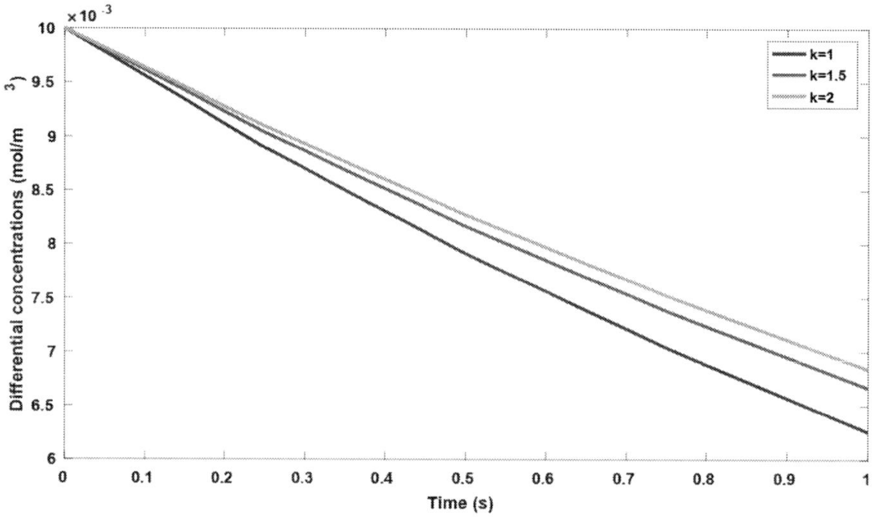

Figure 2. Determination of chemical equilibrium at (a)0.01-0.09 (b) 0.1-0.19 (c) 0.2-0.29 (d) 0.3-0.39 (e) 0.1-1 (f) 1-2

The kinetic constant was examined between 0.3 and 0.39 (Figure 2d). It was observed that chemical equilibrium was attained at k=0.35 and k=0.39. However, at k=0.3, the differential concentration has a parabolic patterned that increased significantly compared to when k=0.29 and k=0.19. When the activated carbon operates at this kinetic constant, the increased differential concentration suggests possible adsorption failure. Based on the above results, the kinetic constant at three unique points were examined i.e., 0.01, 0.1 and 1 (Figure 2e). It was observed that k=0.1 and k=1 fluctuated away from k=0.01 in opposite direction. This observation further affirms that the determination of chemical equilibrium that would aid high adsorption performance is not dependent on a range of values but a selected value that prove that the adsorption refrigerators performance is determined from the initial production of its basic parts. It was observed at this point that the increase or decrease in temperature under laboratory condition does not significantly influence the chemical equilibrium of the material. This observation clearly supports the law of mass action that states that the rate of a chemical reaction at a constant temperature depends only on the concentrations of the substances that influence the rate. The differential concentration was further observed when the kinetic constant is between 1 and 2 (Figure 2f). It was observed that when k $\geq$1, the differential concentration decreases significantly to cause material failure.

4. Conclusion

The experimentation has examined the relationship between the rate of a chemical reaction, concentration of its reactants, and the particulate dynamics in the activated carbon. This work accurately determined the three possible kinetic constants that can initiate chemical equilibrium that has been projected to enhance the adsorption performance of the activated carbon. In other words, the success of the adsorption refrigerator is dependent on the initial material used for its construction. In this case, the activated carbon is the main determinant of the sustainability of the adsorption refrigerator. Hence, the successes of the adsorption refrigerator depend on the source material of the activated carbon and the level of impurity in the activated carbon.

Acknowledgement
The authors appreciate Covenant University for partial sponsorship.

References

[1]. AirEaseLeaks, 2020. Solar Refrigerators: PV Refrigerators, Solar Absorption Refrigerators, and Solar Mechanical Refrigerators. https://aireaseleaks.org/solar-refrigerators
[2]. Chaouachi, B.; Gabsi, S. 2007. Design and Simulation of an Absorption Diffusion Solar Refrigeration Unit, American Journal of Applied Sciences, **4**: 85-88
[3]. Solar Alliance, 2020. How Does a Solar Refrigerator Work? https://www.solaralliance.org/how-solar-refrigerator-work/
[4]. Anyanwu, E. and Ogueke, N. 2007 Transient Analysis and Performance Predic- tion of a Solid Adsorption Solar Refrigerator. Applied Thermal Engineering, **27**: 2514-2523
[5]. Zhao YL, Hu E, Blazewicz A. 2012. A non-uniform pressure and transient boundary condition based dynamic modeling of the adsorption process of an adsorption refrigeration tube. Appl Energy, **90**: 280–287.
[6]. Zhao HZ, Zhang M, Liu ZY, Liu YL, Ma XD. 2008. Mechanical and experimental study on freeze proof solar powered adsorption cooling tube using active carbon/ methanol working pair. Energy Convers Manage, **49**:2434–2438.
[7]. Yongling Zhao, Eric Hu, Antoni Blazewicz, 2012. Dynamic modelling of an activated carbon–methanol adsorption refrigeration tube with considerations of interfacial convection and transient pressure process, Applied Energy, **95**: 276–284
[8]. P.K. Bansal and A. Martin, 2000. Comparative study of vapour compression, thermoelectric and absorption refrigerators, International Journal of Energy Research, **24**: 93–107
[9]. S. A. M. Said, M. A. I. El-Shaarawi, and M. U. Siddiqui, 2012. Alternative designs for a 24-h operating solar-powered absorption refrigeration technology," International Journal of Refrigeration, 35:. 1967–1977
[10]. G. Yaxiu, W. Yuyuan, and K. Xin, 2008. Experimental research on a new solar pump-free lithium bromide absorption refrigeration system with a second generator," Solar Energy, **82**: 33–42
[11]. Leong KC, Liu Y. 2004. Numerical modeling of combined heat and mass transfer in the adsorbent bed of a zeolite/water cooling system. Appl Therm Eng; **24**:2359–74.
[12]. T.Ramesh, N. Rajalakshmi, and K.S.Dhathathreyan, (2015). Activated carbons derived from tamarind seeds for hydrogen storage, Journal of Energy Storage, 4: 89–95
[13]. K. Y. Foo, L. K. Lee, and B. H. Hameed, 2013.Preparation of tamarind fruit seed activated carbon by microwave heating for the adsorptive treatment of landfill leachate: a laboratory column evaluation," Bioresource Technology, **133**: 599– 605.
[14]. J. Kong, Q. Yue, L. Huang et al., 2013. Preparation, characterization and evaluation of adsorptive properties of leather waste based activated carbon via physical and chemical activation, Chemical Engineering Journal, **221**: 62–71

IOP Conf. Series: Materials Science and Engineering 1036 (2021) 012081 doi:10.1088/1757-899X/1036/1/012081

Development and construction of automatic three-phase power changeover control circuit with alarm

A Abdulkareem[1], T E Somefun*[1], V Oguntosin[1], and F Ogunade[2]

[1]Department of Electrical and Information Engineering, Covenant University
[2]Department of Electrical and Electronics Engineering, University of Ibadan

*Corresponding author's e-mail address: tobi.shomefun@covenantuniversity.edu.ng

Abstract. One primary reason for Nigeria economic underdevelopment is the inconsistency of power supply, and as a result of this fact, immediate attention is required. The criticality of an uninterrupted power supply cannot be overemphasised. Therefore, is this work, a new approach is applied to switch power from a three-phase source to another without human interference. This work presents an automatic three (3) phase power changeover circuit with an alarm that turns power supply from Power Holding Company of Nigeria (PHCN) to a power generator at an instant of interruption or outage from PHCN. The signal is introduced to notify the user of the restoration of the PHCN supply and automatically carries out this process. This process is achieved by using logic control circuits, three-phase sequence monitors, phase(s) failure monitor, overload relay, and contacting relays for practical switching which takes approximate 30 seconds.

1. Introduction

In most developing and underdeveloped parts of the world, for instance, in Nigeria, the supply of electricity from PHCN is unstable [1-4]. Thus, the use of alternative sources of power supply, such as microgrid, distributed generator(s), generators, etc. is highly required to meet the end-users energy requirement [3, 5-7].[5, 6][5, 6] Changeover control circuit that will aid the continuous supply of energy without interruption during the loop is needed. Several changeover circuits have been built, most of which are single-phase [8-15]. Since automation has found space in almost every field of endeavour [16], this work, therefore, deploys an automatic changeover circuit with an alarm for a three-phase system. The design incorporates proper monitoring measures to avoid equipment damage and ensure efficient delivery of services such as surgery, banking activities, etc. This design eliminates the element of human interaction involved in changing the power supply from one source to another. It also upholds an excellent service via its quick and rapid response. An automatic changeover switch with an alarm is an electrical device that uninterruptedly observes electrical power supply, notifies the user via the alert, and connects the load to a standby source of electric power if the utility goes off, provided safety measures have been adequately satisfied. This design accurately monitors the incoming supply from the PHCN grid supply before connecting it to the load; the voltage monitor, phase failure, and overload relay are used to carry out this process. If these safety conditions are not met, the load continues to run on auxiliary supply. Still, the buzzer alarm notifies the user on the availability of PHCN supply, indicating a fault. In contrast to the manual changeover switching system that requires manual procedure in starting the generator and switching over from PHCN supply to the generator supply and vice-versa, there is a need to develop an automated control system to manage electric power supply between two sources effectively. This automated system prevents damage to equipment, therefore, influences the motivation for this work [8].

2. Research materials and methods

For ease of understanding of the proposed circuit in this study, the three-phase block diagram is given in Figure 1. The proposed course is an application that ensures active switching between two different power sources; PHCN and a three-phase generator. The electrical component used is at the rate of 100% to ascertain a very reliable circuit that would function efficiently at the standard operating capacity. The three-phase automatic changeover circuit is designed with the following specification: 415 V line to line voltage and 50 Hz frequency. The total expected load for the circuit is 1,030 Watts, which comprises one 550 W three-phase induction motor, three pieces of 60 W electric bulbs, two pieces of 150 W electric standing fans.

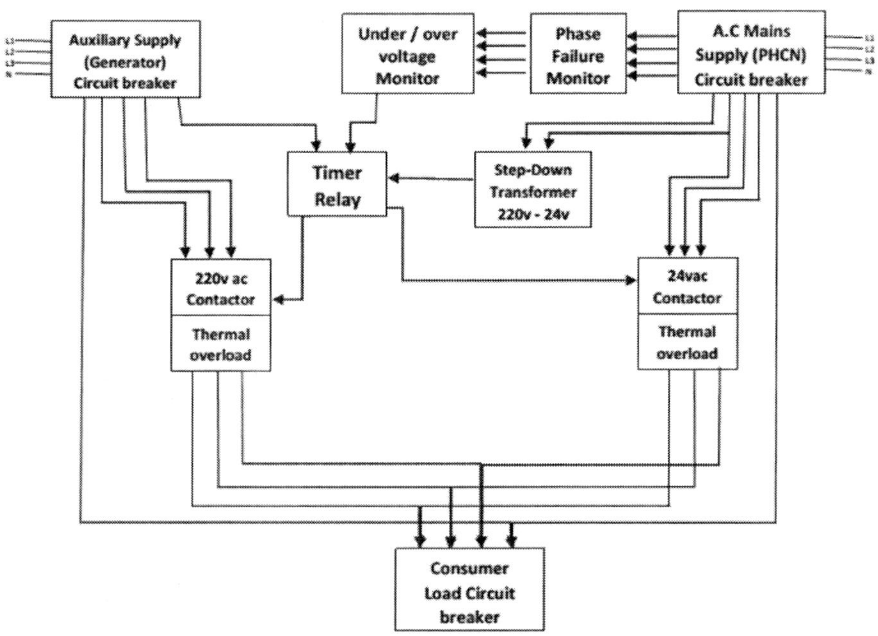

Figure 1. A detailed block diagram of a change over

The line current is obtained from the three-phase power equation as shown below:

$$P = \sqrt{3}\, V\, I\, COS\phi \qquad (1)$$

Therefore, line current $I_L = \frac{P}{\sqrt{3}\ V\ Cos\ \theta}$

where Cos θ (power factor) is taken as 0.8 and approximate total power as 1,500 W

Therefore, substituting into equation (1) gives

$$I_L = \frac{1.5 \times 10^3}{\sqrt{3} \times 415 \times 0.8}$$

$$I_L = 2.61\ A$$

According to NEC 430-52 standard, the rating of the circuit breaker is dependent on the line current and is gotten from the formula [17];

Size of circuit breaker = 250% × Line current

$= 2.5 \times 2.61 = 6.53$ A

Hence a breaker size of 10 A is suitable.

The contactors are sized based on the line current flowing through them.

For the main contactor;

Rating =100% × Line current
$= 1 \times 2.61 = 2.61$ A

Figure 2 shows a complete circuitry design and construction of the automatic three-phase power supply changeover circuit. The connection network of each component is clearly shown.

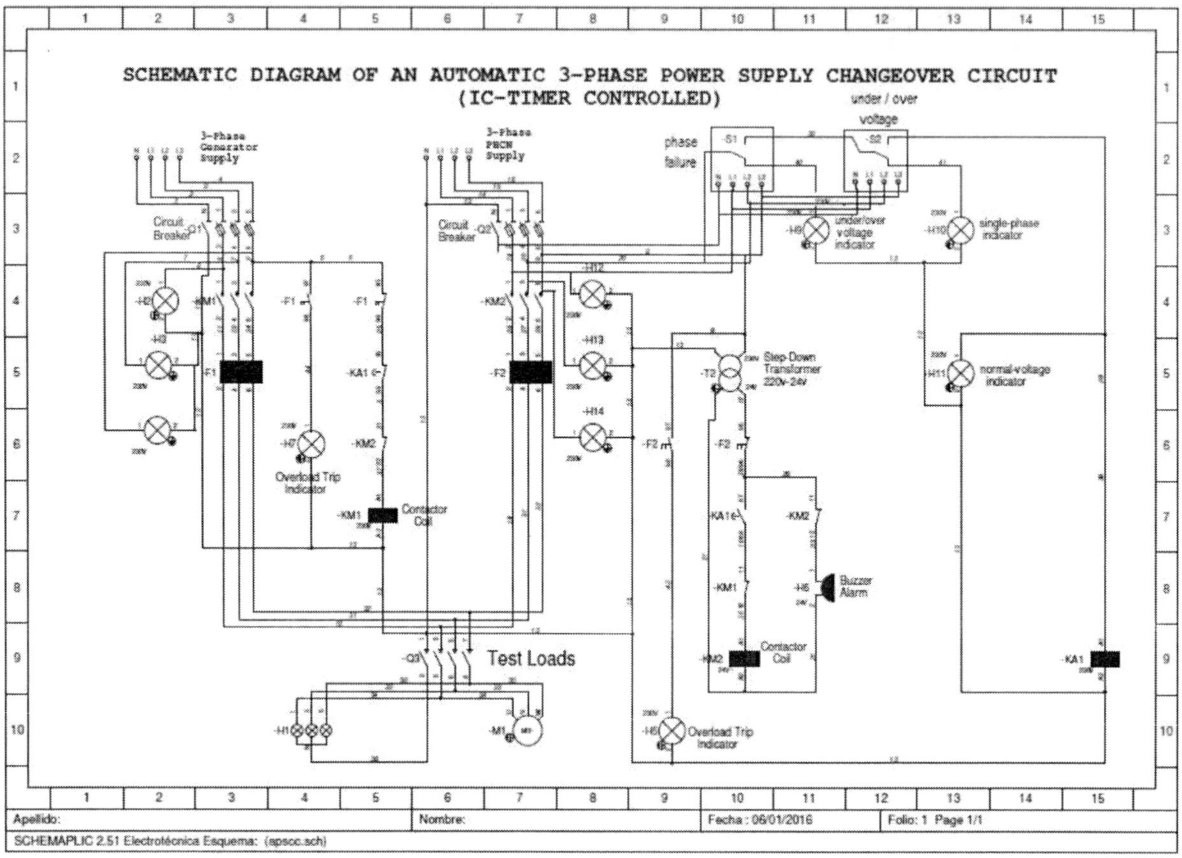

Figure 2. Complete circuit schematic diagram [18]

The automatic changeover operation is primarily to switch electric power from the main supply (PHCN) to the auxiliary supply (Generator) with prompt and effective timing sequence. The system comprises two different circuits, namely; the primary supply circuit and the auxiliary supply circuit. Basically, for the proper operation of the system, the auxiliary supply will be put to service first because the main power supply operates to break the operation of the generator while in the process through its ignition system.

The system automatically interconnects, (i.e.an interlock is created) using the time delay relay between the PHCN supply and Generator supply. The auxiliary supply circuit is controlled by the normally closed (N/C) contact of the timer and normally closed (N/O) contact of the mains supply circuit's energising contactor. While the primary supply circuit is controlled by the normally open (N/O) contact of both the timer and the energising contactor of the auxiliary supply, thus creating an interlocking between the two circuits.

By making the control connection of the ignition circuit in series to the control of the primary supply circuit on a normally closed contact, the ignition system of the auxiliary contact plays a vital role in designing the system. As a result, when the PHCN supply is available, the ignition system connection is broken from the mains supply contact, making the normally closed to be opened depending on the time lag on the timer relay before switching. The timer is delayed to ensure the stability of the power from the PHCN. At the same time, a buzzer alarm is connected to the timer contact to notify the users of the availability of PHCN supply.

Three major factors determine the switching from one supply to another, viz; over and under voltage, timer relay and single phasing. These factors monitor the supply input from the PHCN based on the minimum and maximum required voltage. If by default, the voltage monitor is pre-set to 350 Vac minimum and 415 Vac maximum, supply inputs other than the pre-set will be detected by the voltage monitor. Single phasing fault is also monitored, thus preventing the system from switching. The issue of voltage fluctuation and phase failure is critically dealt with while using an industrial appliance such as a three-phase induction motor. Hence, provision of a manual switch is incorporated in the circuit for domestic use only. This is in any case of under voltage, or single-phase, user can manually switch, but possible only if the equipment or appliance to be used conforms to the input power condition. As projected as a three-phase operating system, this system still satisfies the requirements for single-phase appliances.

3. System implementation and results

The testing and implementation of the design are very critical in ensuring proper working operation. The circuit breaker make-and-break prompt operation was ascertained shorting two phases during supply. The contactor was supplied with 220Vac to its coil contacts A1 and A2. The magnetic coil experienced electromagnetic induction; thus, energising the contactor and supply of three-phase voltage made available to the overload terminals. The contactor breaking capacity was confirmed by manually actuating the overload relay, which causing the overload circuit break internally and change the state of its control terminals (i.e., Normally closed became normally open and vice versa). And there was an indication on the panel box via the indicator bulb, showing an auxiliary overload trip as display in Figure 3.

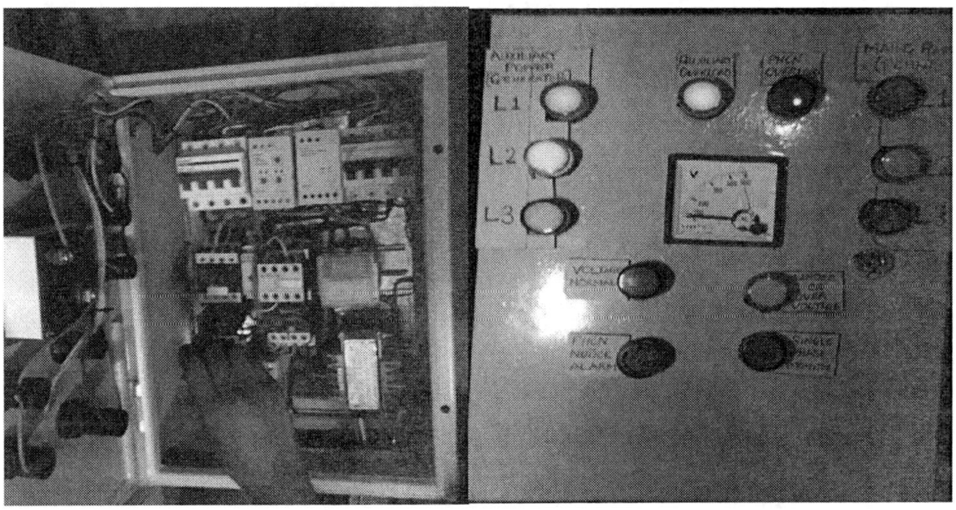

Figure 3. Thermal overload manually tripped and bulb indicating the tripping of the auxiliary overload

The components that make up the auxiliary unit (circuit breaker, contactor, overload) were tested separately to ensure reliability and proper functionality. After which they were all tested as a unit with test bulbs being attached. All the three bulbs came on, indicating a complete three-phase supply and adequate functioning of all components in the auxiliary unit as shown in Figure 4

IOP Conf. Series: Materials Science and Engineering 1036 (2021) 012081 doi:10.1088/1757-899X/1036/1/012081

Figure 4. All bulbs on, indicating complete Three-phase supply

Phase failure monitor operation was confirmed by disconnecting one out of the three-phase supply. It made the system unable to supply power as a result of a single-phase fault. The buzzer alarm sounded continually notifying the user of the availability of the PHCN supply, while the timer counted as pre-set. Hence, both under/over voltage and single-phase fault is indicated on the panel, as shown in Figure 5. Reason being that a single-phase fault automatically will affect the under-voltage monitor.

Figure 5. Single-phase fault

Testing for over and under voltages was technical like other tests. The process has to do with setting a timing preference to ensure how fast or slow the device needs to rectify any voltage imbalances (time lag of auto-switching of output upon voltage fluctuation). The input voltage per phase of the voltage monitor was measured with a multimeter and it indicated a reading of 179.6 Vac. The upper limit of the voltage monitor was set beyond the measured voltage (240 Vac), and the lower boundary was also set above (180 Vac), the measured voltage. Upon proper reconnection of the disconnected phase (Blue phase) used for the phase failure test. There was no longer an indication of a single-phase fault on the panel. The voltage moves from the phase failure monitor to the under/over voltage monitor as input but not as output, due to the voltage monitoring device's abnormal voltage recognition. Here, the under-voltage indicator is turned ON on the device, as shown in Figure 6, which leaves the voltage signal received earlier from the phase failure not to cross as required voltage. When the device's adjustment knob was tuned below the measured voltage (175 Vac), the monitor switched to give output after 4 seconds, as shown in Figure 7. The monitoring devices indicate a high level of reliability and functionality at that point.

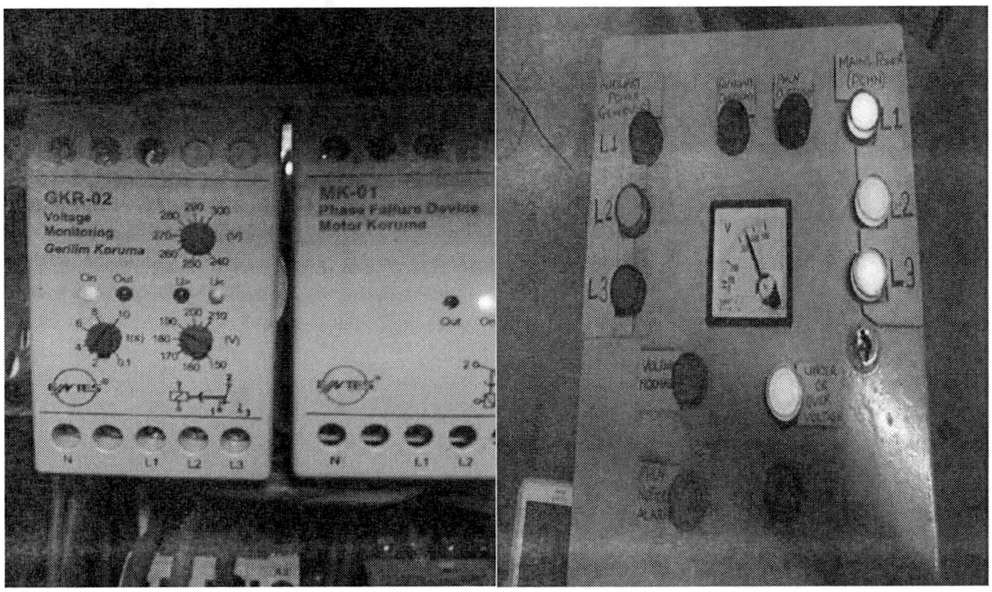

Figure 6. Voltage monitor showing an under-voltage input and indicator panel

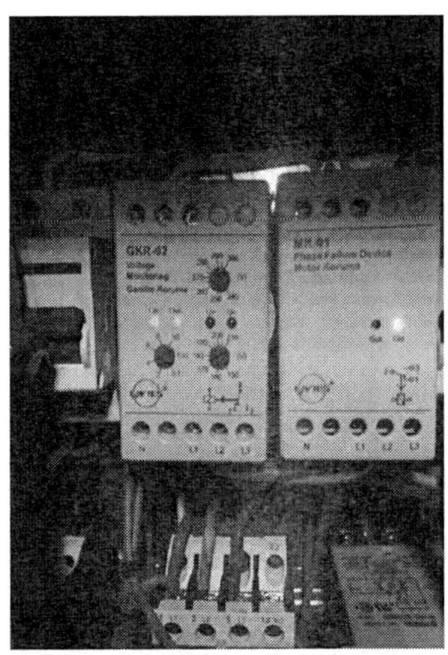

Figure 7. Voltage monitor indicating normal voltage input after adjustment of the lower limit.

Upon certification that all the components on the main supply unit are working properly and reliable. The entire unit was tested as a unit with load(s) attached. Figure 8 shows the testing process:

Figure 8. Test load per phase and on three-phase

The result from the input voltage test is shown in Table 1.

Table 1. Input voltage for single-phase and three-phase

Single-phase	Voltage (Volt)	Three-phase	Voltage (Volt)
L1	180	L1-L2	395.1
L2	179.5	L2-L3	397.2
L3	182.4	L1-L3	396.3

Different reading on the phase from Table 1 indicates unbalanced load on the three-phase system. Since all phases are completely connected, both phase failure and under/over voltage monitor switched the output to the timer. In that way, the output of the series-connected phase failure and voltage monitor crosses to energise the timer relay. The energising voltage from the step-down transformer at a normally open N/O contact of the timer switched to the overload normally closed contact to energise the 24 Vac at 30.31 seconds. The 24 Vac Contactor allowing the three-phase supply gets to the load-end circuit breaker.

4. Conclusion

An automatic three-phase power change over control circuit with alarm is developed and constructed in this work. The developed prototype of the automatic three-phase changeover switch worked perfectly and minimise damages to lives/equipment. It maintains a high quality of service through its quick and rapid response of 30 seconds for generator to come up. Human error that can lead to loss of life and or damage of equipment is guided against in the developed work of the automatic three-phase changeover because it is fully automated.

Acknowledgement

The authors would like to express special thanks and gratitude to Covenant University for her support.

IOP Conf. Series: Materials Science and Engineering 1036 (2021) 012081 doi:10.1088/1757-899X/1036/1/012081

References

[1] Ijewere, A. A. 2011. The Management of Electricity Power Supply in Nigeria: Problems and Prospects, *Journal of Research in National Development,* **9**, pp. 173-185

[2] Oluwatoyin, K. K. Odunola A. M., and Alabi A, 2015. Ways of Achieving Stable and Uninterrupted Power Supply of Electricity in Nigeria," *Br. J. Appl. Sci. Technol.,* **10,** pp. 1-15

[3] Somefun T. E., 2018. Critical review of different methods for siting and sizing distributed generators, *TELKOMNIKA (Telecommunication Computing Electronics and Control),* **16**.

[4] Babatunde O. M., Ayegbusi C. O., Babatunde D. E., Oluseyi P. O., and Somefun T. E, 2020. Electricity Supply in Nigeria: Cost Comparison between Grid Power Tariff and Fossil-Powered Generator," *International Journal of Energy Economics and Policy,* **10**, pp. 160-164

[5] Kolo J. G, 2007. Design and construction of an automatic power changeover switch," *Assumption University Journal of Technology,* **11**, pp. 1-6,

[6] Amoo O. M. and Fagbenle R. L., 2013. Renewable municipal solid waste pathways for energy generation and sustainable development in the Nigerian context, *International Journal of Energy and Environmental Engineering,* vol. **4**, p. 42,

[7] Somefun T, Awosope C, Abdulkareem A, and Alayande A, 2020. Deployment of Power Network Structural Topology to Optimally Position distributed Generator within Distribution System," *Journal of Engineering Science and Technology Review,* vol. **13**, pp. 12-17.

[8] Agbetuyi A, Adewale A, Ogunluyi J, and Ogunleye D, 2011. Design and construction of an automatic transfer switch for a single phase power generator, *International journal of engineering science,* vol. **3**, pp. 1-7.

[9] Ahmed M, Mohammed A, and Agusiobo O, 2006. Development of a single phase automatic change-over switch, *Department of Electrical and Computer Engineering, Federal University of Technology Minna, Nigeria,*

[10] Acakpovi A, Bennet M. M, and Akortia N, 2016. Design, testing and analysis of single phase electronic changeover with security light," *APT Journal of Multidisciplinary Research,* vol. **1**,

[11] Shuttleworth R, 1997. Electrical changeover switching, ed: Google Patents.

[12] Ehiabhili J, Ezeh C, and Orji O, 2018. Single Phase Microcontroller-Based Automatic Changeover Switch," *International Journal of Electronics, Communication & Instrumentation Engineering Research and Development (IJECIERD),* vol. **8**, pp. 7-16.

[13] Yang T.-H, 1987. Capacity-movement model AC inductive motor switch-changeover capacity-type speed control circuit, ed: Google Patents.

[14] Samuel L, Cohen J. E, and Gumpertz B. E, 1957. Automatic changeover apparatus, ed: Google Patents.

[15] Obasi C. C, Agidani O. B. O. J. J, Onyedikachi V, Ubadike I, and Osichinaka C, 2015. Design and Implementation of Microcontroller Based Programmable Power Changeover, *International Journal of Computer and Intelligent Systems,* vol. **6**, pp. 51-56.

[16] Shomefun T. E., OA A. C, and Diagi E. O, 2018. Microcontroller-Based Vertical Farming Automation System," *International Journal of Electrical and Computer Engineering,* vol. **8,** p. 2046.

[17] Hall W. M. and Gregory G. D, 1999. Short-circuit ratings and application guidelines for molded-case circuit breakers," *IEEE transactions on industry Applications,* **35**, pp. 135-143.

[18] Covic G. A., Boys J. T, Kissin M. L, and Lu H. G. 2007. A three-phase inductive power transfer system for roadway-powered vehicles, *IEEE Transactions on Industrial Electronics,* vol. **54**, pp. 3370-3378.

AUTHOR INDEX

Institute of Physics
Dirac House, Temple Back
Bristol BS1 6BE UK

ISSN: 1757-8981
ISBN 978-1-7138-4561-4

9 781713 845614

CURRAN ASSOCIATES INC.
proceedings.com